S0-ADW-398

ANNALS OF THE NEW YORK ACADEMY OF SCIENCES

Volume 508

EDITORIAL STAFF
Executive Editor
BILL BOLAND
Managing Editor
JUSTINE CULLINAN
Associate Editor
M. K. BRENNAN

The New York Academy of Sciences
2 East 63rd Street
New York, New York 10021

PHYSIOLOGICAL NMR SPECTROSCOPY:
FROM ISOLATED CELLS TO MAN

ANNALS OF THE NEW YORK ACADEMY OF SCIENCES
Volume 508

PHYSIOLOGICAL NMR SPECTROSCOPY: FROM ISOLATED CELLS TO MAN

Edited by Sheila M. Cohen

The New York Academy of Sciences
New York, New York
1987

Cover figure courtesy of J. den Hollander and P. Luyten.

Library of Congress Cataloging-in-Publication Data

Physiological NMR spectroscopy: from isolated cells to man/edited by Sheila M. Cohen.
p. cm.—(Annals of the New York Academy of Sciences; v. 508)
Based on a conference held by the New York Academy of Sciences in New York City on Sept. 24–26, 1986.
Includes bibliographies and index.
ISBN 0-89766-412-4. ISBN 0-89766-411-6 (pbk.)
1. Nuclear magnetic resonance spectroscopy—Congresses.
2. Metabolism—Regulation—Research—Technique—Congresses.
3. Physiology—Research—Technique—Congresses. I. Cohen, Sheila M. II. New York Academy of Sciences. III. Series.
[DNLM: 1. Cells—metabolism—congresses. 2. Nuclear Magnetic Resonance—congresses. 3. Spectrum Analysis—methods—congresses.
W1 AN626YL v. 508/QD 96.N8 P578 1986]
Q11.N5 vol. 508
[QP519.9.N83]
500 s—dc19
(574.87'028]
DNLM/DLC
for Library of Congress

SP
Printed in the United States of America
ISBN 0-89766-412-4 (cloth)
ISBN 0-89766-411-6 (paper)

ANNALS OF THE NEW YORK ACADEMY OF SCIENCES

Volume 508
November 30, 1987

PHYSIOLOGICAL NMR SPECTROSCOPY:

FROM ISOLATED CELLS TO MAN[a]

Editor and Conference Organizer
SHEILA M. COHEN

CONTENTS

[a]This volume is the result of a conference entitled Physiological NMR Spectroscopy: From Isolated Cells to Man, which was held by the New York Academy of Sciences in New York City on September 24–26, 1986.

POSTER PAPERS

Financial assistance was received from:

- AMERICAN CYANAMID
- ABBOTT LABORATORIES
- BRUKER MEDICAL INSTRUMENTS
- BURROUGHS WELLCOME CO.
- CIBA-GEIGY
- GENERAL ELECTRIC COMPANY/MAGNETIC RESONANCE BUSINESS
- GENERAL ELECTRIC COMPANY/NMR INSTRUMENTS
- HOFFMAN-LA ROCHE INC.
- STUART PHARMACEUTICALS/DIVISION OF ICI AMERICAS, INC.
- INTERNATIONAL BUSINESS MACHINES
- LILLY RESEARCH LABORATORIES
- MCNEIL PHARMACEUTICAL
- MERCK SHARP & DOHME/ISOTOPES (MSD ISOTOPES)
- MERCK SHARP & DOHME RESEARCH LABORATORIES
- MONSANTO COMPANY
- NATIONAL CANCER INSTITUTE
- NORWICH EATON PHARMACEUTICALS, INC.
- THE PROCTER & GAMBLE COMPANY
- SANDOZ CORPORATION
- SIEMENS MEDICAL SYSTEMS, INC.
- THE UPJOHN COMPANY
- VARIAN/FREEMONT OPERATIONS

Introduction

SHEILA M. COHEN

Merck Sharpe & Dohme Research Laboratories
Rahway, New Jersey 07065

The recent, rapid maturation of *in vivo* NMR spectroscopy as an approach to basic problems of concern to biochemists, physiologists, and clinician-investigators prompted the organization of The New York Academy of Sciences conference, Physiological NMR Spectroscopy: From Isolated Cells to Man. The conference focused on the role of physiological NMR spectroscopy in promoting our understanding of biochemical metabolic regulation and bioenergetics in intact cells, perfused organs, experimental animal models, and man. The conference provided a forum for an in-depth examination of *in vivo* NMR approaches to the study of metabolism in brain, liver, kidney, heart, and skeletal muscle; of ion transport in cells and tissue; of metabolism in tumors and cancer cells; and of metabolism in microorganisms. To complement the conference's session on clinical applications of NMR spectroscopy and to indicate future directions, the methods recently introduced for the spatial localization of resonances in man and in experimental models were presented by the instrument specialists who developed them.

The increased availability of NMR spectrometers appropriate for physiological NMR studies marks a general expansion of the field. The conference was thus especially timely in bringing together the pioneers of this frontier area of research to exchange the accumulated sum of their experience and to share the philosophy of their innovative approaches to problems that *in vivo* NMR methods are uniquely able to address. As the papers in this volume will suggest, physiological NMR spectroscopy provides a window into biochemical metabolic regulation that is unlike any other. It seems possible that by the correlation of investigational and clinical NMR studies, a number of fundamental biochemical questions will be better defined and understood.

NMR: Physical Principles and Current Status as a Biomedical Technique

EDWIN D. BECKER AND CHERIE L. FISK

National Institutes of Health
Bethesda, Maryland 20892

The last few years have seen an explosive growth in the application of nuclear magnetic resonance (NMR) to the study of living systems, ranging from cells in tissue culture, through perfused organs to intact animals and human beings. Increasingly, NMR data provide detailed insights into metabolic processes. The use of NMR imaging already rivals or surpasses other imaging modalities, such as X-ray computed tomography, in its capacity to define anatomical features in humans and to serve as a leading diagnostic tool. Some biologists and medical practitioners view NMR as a new technique because it has only recently made an inpact in their fields. But NMR in bulk materials was discovered in 1946,[1-3] and in the four decades since then it has moved gradually from a phenomenon of interest to physicists to an essential tool in chemistry, biochemistry, and now physiology and medicine.

This article provides a summary of the fundamentals of NMR, with emphasis on those aspects of importance for *in vivo* NMR studies and descriptions of some of the techniques currently in use. We do not provide a detailed account of NMR theory here, since such presentations are available in a number of textbooks.[4-9]

BASIC PRINCIPLES OF NMR

Although some aspects of NMR require a quantum mechanical formulation, many features of the technique can be adequately and conveniently discussed by means of a classical picture, such as that given in FIGURE 1. We see that the macroscopic magnetization, **M**, is made up of an ensemble of individual nuclear magnetic moments (μ_i), each precessing about the magnetic field $\mathbf{B}_0$, which is oriented along the z axis. At equilibrium, the phase of each precessing nucleus is random and thus **M** is oriented along the z axis. The whole purpose of an NMR experiment is to tip the magnetization away from the z axis and to generate a component in the xy plane in which all the nuclei are in phase. This occurs when a radio frequency (RF) field $\mathbf{B}_1$ is introduced at the proper (resonance) frequency. When the RF field is removed, the component is left to precess in the xy plane and it generates a signal in the radio frequency pickup coil placed in this plane. The component will not precess indefinitely, but decays with time. The magnetization returns to its equilibrium position in which **M** lies along the z axis with no component in the xy plane. This return to equilibrium is a result of relaxation processes characterized by two time constants: T_1, the spin-lattice or longitudinal relaxation time, for return along the z axis; and T_2, the spin-spin or transverse relaxation time, which results from the dephasing of all the nuclei in the xy plane. T_1 relaxation is a critical parameter in an NMR experiment. Only when **M** has returned to its equilibrium position can the experiment be repeated with maximum sensitivity. This becomes crucial in experiments on insensitive nuclei such as ^{13}C, which are often repeated many times to improve the signal-to-noise ratio. T_2 relaxation is important in

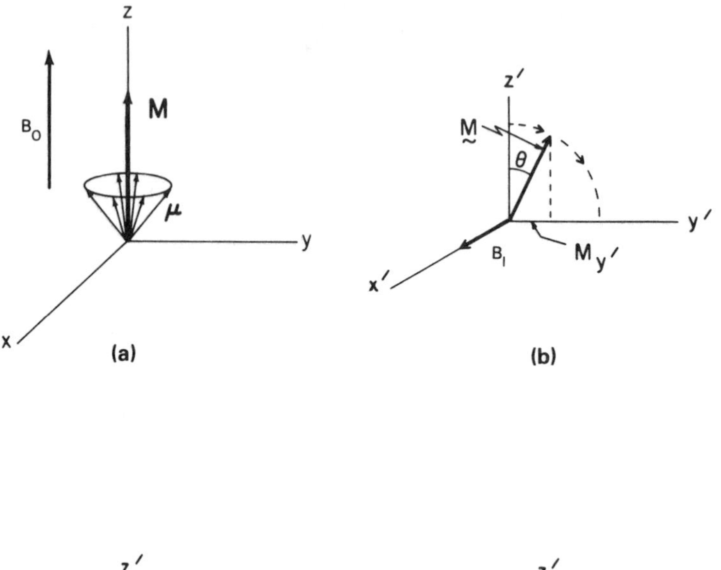

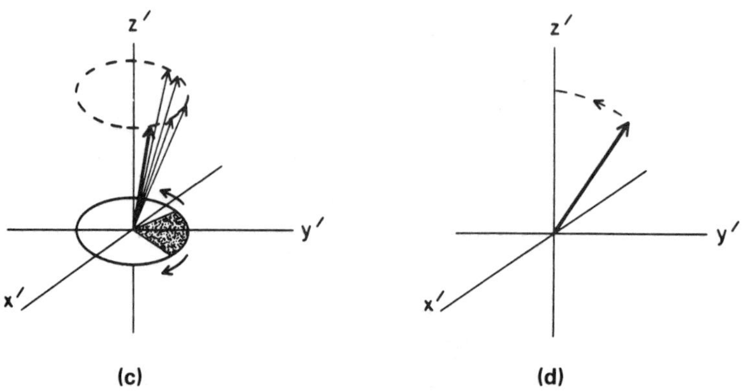

FIGURE 1. (a) Precession of an ensemble of magnetic moments of spin quantum number $I = \frac{1}{2}$ (e.g. ^{1}H, ^{13}C, ^{31}P), and the resultant macroscopic magnetization **M**. Only the excess spins in the low energy state (parallel to $\mathbf{B}_o$) are shown. (b) **M** tipped into the *xy* plane by interaction with the radio-frequency field $\mathbf{B}_1$. (c) Spin-spin relaxation: dephasing of the components of magnetization in the *xy* plane. (d) Spin-lattice relaxation: restoration of the *z* component of magnetization. For simplification, processes (b)–(d) are shown in a reference frame rotating at the Larmor frequency.

determining the lineshapes of NMR signals. Both types of relaxation are of importance in the study of molecular structure and dynamics.

The fundamental relationship of NMR is the well known Larmor equation

$$\nu_o = \frac{\gamma}{2\pi} B_o$$

showing that the precession frequency (ν_o) is directly proportional to the applied magnetic field, with the constant of proportionality γ, the magnetogyric ratio, characteristic of the particular nucleus being studied. NMR frequencies fall in the radio frequency (RF) range of about 1 to 600 MHz. Although many nuclides are susceptible to NMR study, only a few are commonly used in *in vivo* studies, as indicated in TABLE 1. ^{1}H is frequently studied because of its high intrinsic sensitivity, nearly 100% abundance, and relatively narrow lines, but it is so ubiquitous that ^{1}H NMR spectra *in vivo* are frequently complex superpositions of the spectra of many individual compounds (including H_2O solvent). ^{19}F provides excellent sensitivity when it can be introduced in appropriate compounds in sufficiently high concentration. The sensitivity of ^{31}P is an order of magnitude lower, but the presence of phosphorus in many compounds of biochemical and metabolic importance has led to a large emphasis on the use of this isotope. ^{13}C and ^{15}N can be very useful when introduced in high enrichment in spite of their low inherent sensitivities. Isotopes with electric quadrupole moments, such as ^{23}Na, usually give rise to somewhat broad lines, but the sensitivity of linewidth to environmental perturbations can be turned to advantage.

FACTORS AFFECTING NMR SPECTRA

If the Larmor equation described completely the origin of NMR frequencies, the technique would be of relatively little value for biochemical and physiological studies. The factors that make NMR of interest to us are primarily: (1) chemical shift, (2) scalar (spin-spin) coupling, (3) magnetic dipole-dipole interactions, and (4) electric quadrupole interactions (for nuclides with quadrupole moments). All four of these interactions affect the appearance of NMR spectra. In the solid phase, each can be observed in suitable circumstances and is found to be anisotropic, i.e., the values depend upon orientation of molecules relative to the applied magnetic field. In fluid media, which are largely those currently studied in physiological applications, the

TABLE 1. Magnetic Nuclei Often Utilized for Biological Studies[a]

Nucleus	NMR Frequency[b] (MHz)	Spin (I)[c]	Natural Abundance (%)	Relative Sensitivity at Natural Abundance[d,e]
^{1}H	100.00	1/2	99.98	100.0
^{19}F	94.08	1/2	100.00	83.3
^{23}Na	26.45	3/2	100.00	9.2
^{31}P	40.48	1/2	100.00	6.6
^{39}K	4.67	3/2	93.10	0.05
^{13}C	25.14	1/2	1.11	0.02
^{17}O	13.56	5/2	0.04	0.001
^{15}N	10.13	1/2	0.37	0.0004
^{2}H	15.35	1	0.02	0.0002

[a] For additional information see Appendix B in Becker.[4]

[b] The frequency given corresponds to a magnetic field of 2.35 Tesla (23.5 kG).

[c] The nuclear spin is given in units of $h/2\pi$ where h is Planck's constant. Nuclei with $I > 1/2$ are "quadrupolar" and have NMR spectral and relaxation characteristics that are distinctly different from nonquadrupolar nuclei.

[d] This value corresponds to the relative sensitivity for equal numbers of nuclei (at a given field strength) multiplied by the natural abundance.

[e] The relative sensitivity of nuclei such as ^{13}C and ^{15}N may be substantially increased by the use of heteronuclear double-resonance techniques.

orientation dependence of dipolar and quadrupolar interactions averages out, but these interactions still affect relaxation.

Chemical shifts arise from the effects of different and distinctive electronic configurations (shieldings) on the nuclei in a molecule. Shielding (σ) affects the Larmor equation causing a frequency shift:

$$\nu_o = \frac{\gamma}{2\pi} B_o(1 - \sigma).$$

Thus, a given nuclide in different chemical environments normally shows different NMR frequencies—an NMR spectrum. Often the chemically shifted NMR lines are split into multiplets by the second factor mentioned above, the scalar coupling. The magnetic moments of all nuclei orient in at least two directions with respect to the magnetic field $\mathbf{B}_o$, and it is the sensitivity of one nucleus to the spin state of a neighbor that gives rise to scalar coupling. Scalar interactions are mediated by the electrons in chemical bonds, and thus are often called "indirect" spin-spin coupling.

There is another kind of interaction between two nuclear magnetic moments that occurs directly, not by means of the bonding electrons, and is termed the direct dipole-dipole interaction. It is simply a classical interaction between two nuclei through space, which depends upon the strength of their magnetic (dipole) moments and upon the orientation they make relative to each other and to the magnetic field (FIG. 2). This interaction leads to broad lines in solids, but when molecules tumble rapidly, as they do in liquids, the dipole interactions average, and thus lines in liquids are sharp. In a similar manner, those nuclei that have electric quadrupole moments undergo interactions between the quadrupole moment and a surrounding electric field gradient formed by the electrons in the molecule. Again, the result is an extremely broad line in the solid, but a marked narrowing if the molecules tumble rapidly in fluid media.

RELAXATION PROCESSES

Nuclear relaxation, including both T_1 and T_2 processes, depends on random molecular motions, which are described by a correlation time τ_c, i.e., the average time a molecule takes to tumble through one radian or to translate roughly a molecular diameter. In liquids τ_c is short; the effect of this rapid motion is to average to zero each of the anisotropic interactions described earlier—chemical shift anisotropy, dipole-dipole interactions, and quadrupole interactions—and sharp lines are observed. However, even though these interactions have been averaged over a sufficiently long time period, they have not disappeared. Although the spectrum displays only the average effect, the interactions are in fact present; and as the molecule tumbles, all can cause local fields that fluctuate (with components at the Larmor frequency) and cause the nuclei to relax. These interactions—dipolar, quadrupolar, and chemical shift anisotropy—are the most common mechanisms of relaxation. Paramagnetic substances, including trace metal contaminants or oxygen, have very large magnetic dipole moments and thus can be dominating factors in the relaxation of any system where they are present.

For nuclei that have quadrupole moments (e.g., ^{2}H, ^{14}N, ^{23}Na), quadrupolar relaxation almost always dominates. It depends on the magnitudes of the quadrupole moment, the surrounding electric field gradient, and the correlation time. Relaxation studies can thus be used to probe molecular motions and the symmetry of the electronic environment around the nucleus. Nuclei with $I \geq 3/2$, such as ^{23}Na, have two or more relaxation components, and hence display line shapes that may be complex when molecules tumble relatively slowly or move in anisotropic environments. In certain instances, part of the NMR signal is so greatly broadened that it is not observed. Examples of "invisible" sodium are reported in various living systems.

For nuclei without quadrupole moments, such as ^1H, ^{13}C, and ^{31}P, dipolar interactions usually provide the most efficient relaxation mechanism. The effectiveness

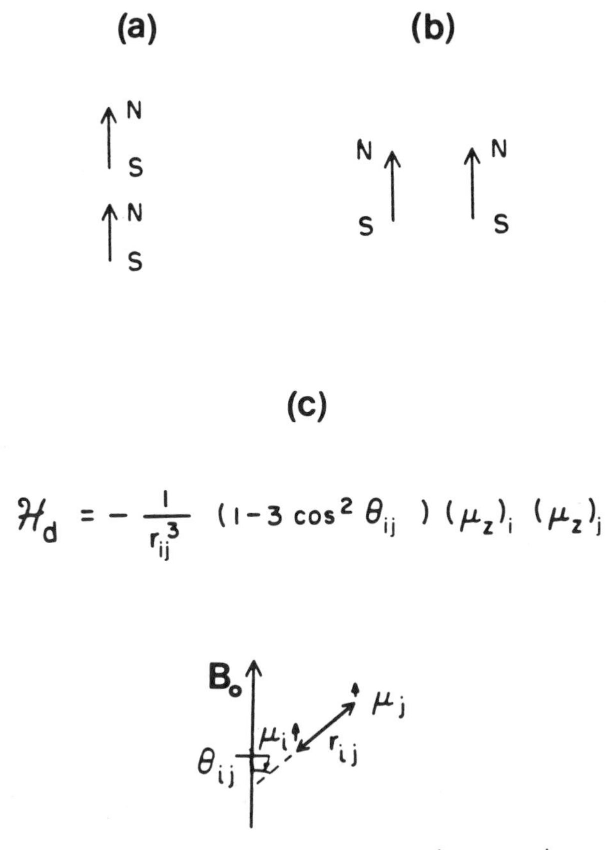

FIGURE 2. Illustration of the interaction between two nuclear magnetic moments. (a) An orientation of two spin-½ nuclei (μ_i and μ_j) resulting in an attractive force between north and south poles, and (b) an orientation resulting in a repulsive force. (c) The energy of interaction for the two nuclear magnetic moments (oriented at an angle θ_{ij} in the magnetic field $\mathbf{B}_o$) as described by the dipolar Hamiltonian operation, H_d.

of this interaction for relaxation depends on the square of the static dipolar interaction; thus

$$1/T_1^d \propto \gamma_I^2 \gamma_S^2 / r_{IS}^6.$$

Because of the quadratic dependence on the magnetogyric ratios, those nuclei with large magnetic moments are most effective in relaxation. Likewise, the inverse sixth

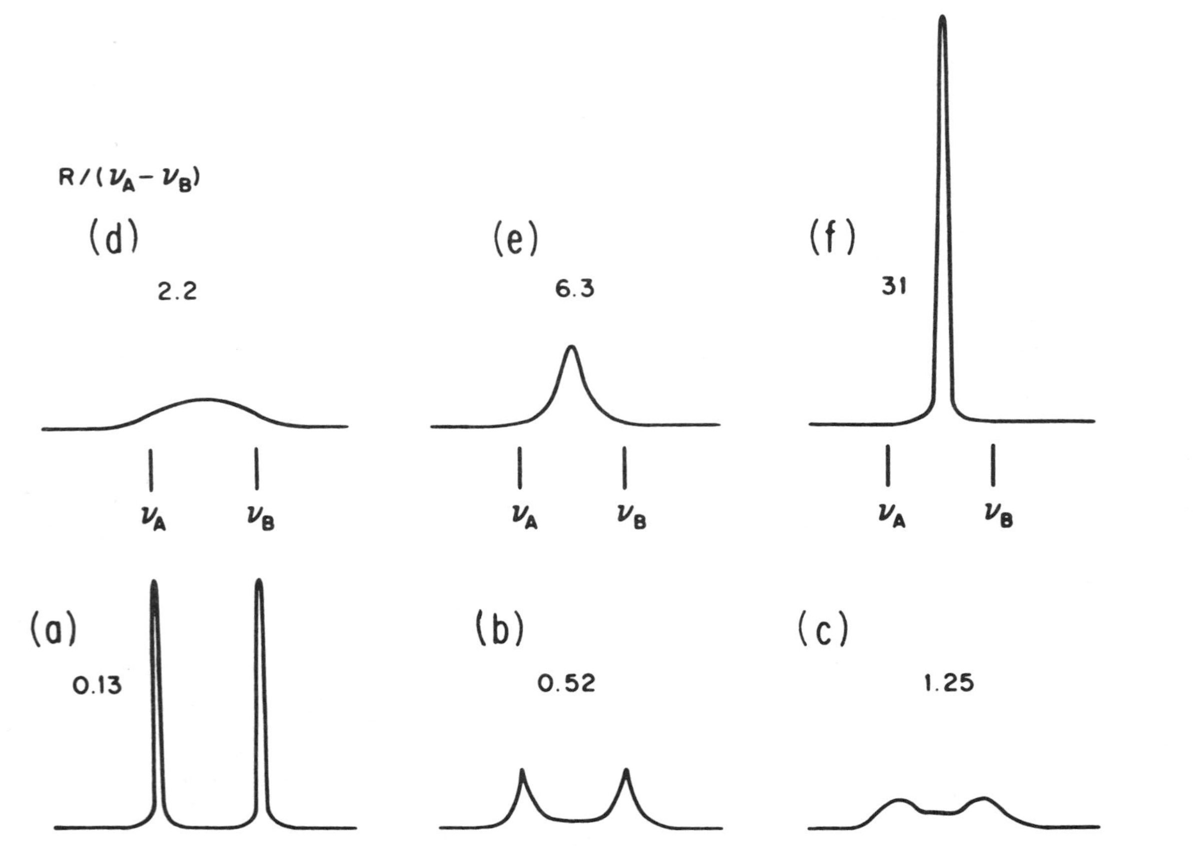

FIGURE 3. Calculated line shapes for the NMR signal of a nucleus exchanging between two equally populated states. As the exchange rate (R) increases relative to the frequency difference between the two states ($\nu_A - \nu_B$), the spectrum changes from two sharp resonances (a) to a single average resonance (f). Intermediate exchange rates result in broadened lines (b–e).

power of the distance between the nuclei (r_{IS}) implies that nuclei relatively close to each other contribute most significantly to relaxation. Usually nuclei that relax each other are within the same molecule, but they also can be in separate molecules that come close together. Dipolar relaxation can be separated from other mechanisms by measurement of the nuclear Overhauser effect (NOE).

The anisotropy of the chemical shielding can also lead to relaxation as molecules tumble. It can be shown that this relaxation mechanism depends on B_o^2.[4-9] In biological NMR studies it becomes particularly significant for ^{31}P measurements at high magnetic field strength.

Another important concept in understanding NMR spectra in biological systems is chemical exchange. If a nucleus can exist in two or more distinct magnetic environments, with different resonance frequencies and/or different relaxation times, separate NMR signals will be observed if the nucleus exchanges only slowly among the environments. However, if there is sufficiently rapid exchange (usually much faster than the difference in resonance frequencies), then only the average frequency (or the average relaxation time) is observed. For exchange at intermediate rates, broad lines usually result (FIG. 3).

TABLE 2. Signal/Noise (S/N) for Various Spectrometers

Year	Spectrometer Frequency (MHz)	S/N[a]
1961	60	6
1965	100	30
1969	220	80
1978	200	300
1978	360	800
1985	500	3600

[a]Strongest peak in the methylene proton signal of ethylbenzene, 1% by volume. Single scan or single pulse.

IMPROVEMENTS IN NMR TECHNIQUES

The tremendous increase during the last few years in the use of NMR to study biological systems clearly has resulted from technological developments that fall into three categories: (1) improvements in magnet design leading to higher magnetic fields and larger sample volumes; (2) improvement in instrument sensitivity; and (3) development of pulse Fourier transform methods. Over the last 20 or 25 years the commonly available magnetic fields and resonance frequencies have increased by about a factor of 10, with a corresponding increased separation of spectral lines to permit the extraction of information that was masked previously. Moreover, during the same period, we have experienced an improvement of a factor of about 600 in sensitivity (TABLE 2). The gain in sensitivity is partly due to the factor of 10 increase in Boltzmann distribution from the higher magnetic field, but it is also due to better probe design and higher quality electronic components. In addition to the sensitivity gain of almost three orders of magnitude illustrated in TABLE 2, NMR has benefited from the replacement of the older continuous wave (cw) techniques with pulse Fourier transform (FT) methods. FT methods provide at least another order of magnitude gain in sensitivity through far more efficient gathering of data. During the last 15 years, pulse FT methods have become routine and provide great versatility in NMR experiments.

NMR studies in living systems began in force in the mid-1970s with investigations of metabolic processes in cells kept alive in nutrient media, then moved rapidly to studies of excised small animal tissue that could be perfused with nutrients in an NMR sample tube. Studies of whole animals and of humans became possible only with the advent of sufficiently large magnets with the requisite field strength and homogeneity. Currently whole body size magnets for human studies (bore diameter about 1 m) are limited to about 2 tesla (2 T) in field strength, while those for animals (up to 40 cm bore) are available at 4.7 T (200 MHz NMR frequency for ^{1}H).

The ability to place a large, living subject inside a magnet is insufficient to obtain meaningful NMR data. Interest usually focuses on a particular tissue or organ, or more likely, on a small part of an organ. Thus, a means of localizing the signal is needed. One straightforward technique is to use a surface coil, i.e., a radio frequency coil designed to fit over the part of the surface of the body that is of interest. The surface coil may be used as a detector coil, with the RF pulses applied uniformly by another much larger coil that surrounds the entire subject, or it may be both transmitter and receiver. The shape of the coil and the characteristics of the pulse sequences used permit studies up to a few centimeters below the surface, and continual advances are being made to increase the precision in localizing NMR signals using these coils. In addition to surface applications, coils can be implanted surgically at appropriate places in experimental animals or introduced in catheterization procedures.

A more general method for localizing NMR signals depends upon the principles of NMR imaging, in which linear magnetic field gradients are applied in specific directions across the sample. With suitable FT techniques, a two-dimensional image can be created that faithfully represents the spatial distribution of the particular nucleus being studied. Typically, ^{1}H images are obtained because of the high concentration of ^{1}H in water and fat. Although applications of NMR imaging are not included in this paper, later articles will describe the ways in which combinations of gradients in the static magnetic field, gradients in the RF field, and cleverly designed pulse sequences can be used in combination to localize NMR signals to very small volumes within large, living subjects. From these small volumes located deep within the body, NMR spectral information can be obtained that bears directly on important physiological processes.

One final development in NMR technology that will play an increasingly important role in *in vivo* studies is the use of sophisticated multiple pulse experiments, including two dimensional (2-D) and multiple quantum spectroscopy. 2-D NMR can be used to correlate certain spectral features, providing an important tool for the interpretation of complex spectra and the study of species that are undergoing exchange or chemical reaction. Multiple quantum techniques can be used for spectral editing, i.e., elimination from the spectrum of certain unwanted resonances, such as those from overlapping peaks or even solvent water. In addition, modern methods permit a transfer of polarization (intensity) from one nucleus to another, so that insensitive nuclei, such as ^{13}C, can be studied with almost the same sensitivity as ^{1}H. These methods, which are now well established in chemical investigations, are only beginning to be used with living systems, but their use is destined to increase.

REFERENCES

1. BLOCH, F. 1946. Nuclear induction. Phys. Rev. **70:** 460
2. BLOCH, F., W. W. HANSEN & M. PACKARD. 1946. The nuclear induction experiment. Phys. Rev. **70:** 474.

3. Purcell, E. M., H. C. Torrey & R. V. Pound. 1946. Resonance absorption by nuclear magnetic moments in a solid. Phys. Rev. **69:** 37.
4. Becker, E. D. 1980. High Resolution N.M.R. 2nd edit. Academic Press. New York.
5. Farrar, T. C. & E. D. Becker. 1971. Pulse and Fourier Transform NMR. Academic Press. New York.
6. Gadian, D. G. 1982. Nuclear magnetic resonance and its applications to living systems. Oxford University Press. New York.
7. Harris, R. K. 1983. Nuclear Magnetic Resonance Spectroscopy. A Physicochemical View. Pitman. London.
8. Slichter, C. P. 1978. Principles of Magnetic Resonance. 2nd edit. Springer-Verlag. Berlin.
9. Abragam, A. 1961. The Principles of Nuclear Magnetism. Clarendon Press. Oxford.

Contributions of ^{13}C and ^{1}H NMR to Physiological Control

ROBERT G. SHULMAN

Department of Molecular Biophysics & Biochemistry
Yale University
New Haven, Connecticut 06511

There will be only a few presentations on ^{13}C in this volume. Most will be on phosphorus and its great applicability, and a certain number on protons. ^{13}C has all the odds against it when you consider it from an NMR viewpoint. It has a natural abundance of only 1.1% and even an equal number of nuclei give a very weak ^{13}C NMR signal compared to the proton. An equal number of ^{13}C and proton nuclei, with the same relaxation properties and in a non-conducting solution, will give a ^{13}C signal about 60 times weaker, so that multiplying weak intensities with low abundance, things are pretty bad. However, compared to phosphorus, ^{13}C is only a factor of about four lower and we do pick up some nuclear Overhauser effects at times so that equal numbers of ^{13}C and ^{31}P have comparable intensity. David Hoult has pointed out that large conducting samples like tissue are the source of noise so that the signal-to-noise disadvantage of ^{13}C compared to ^{1}H is, under these conditions, only a factor of 12 instead of 60, but that is still a big difference.

What are the advantages of ^{13}C compared to ^{31}P NMR and compared to ^{14}C, which is, of course, a common radiolabel? There are two different kinds of experiments discussed in this brief survey of the ^{13}C situation. The first measures the naturally abundant ^{13}C nuclei. Because of its low sensitivity you need a very high concentration of a compound to see ^{13}C in natural abundance. Two compounds that can be observed in mammalian systems in natural abundance are glycogen and fatty acids. The ^{13}C resonances of glycogen, to our considerable surprise, are 100% visible.[1] We don't know why, since the glycogen molecule is very big with molecular weights up to 10^8, and yet quite a number of very definite experiments hydrolyzing the glycogen, one way or another, show that the glucose peaks that appear are equal in intensity to the glycogen peaks that were there.

Because of the low natural abundance one can, and very often does, resort to the second kind of experiment, i.e. the use of labeled precursors whence one can follow the label as it goes through different pathways. Following pathways, particularly in the liver, brain, and in microorganisms, has been a very informative use of ^{13}C. The kinds of information that you can get from that method will be illustrated in Sheila Cohen's talk in which she will discuss in more detail the chemical information that you get.[2,3] The information that is available in an eleven-minute accumulation of ^{13}C spectra of a perfused liver is very impressive and we certainly would like to have that information available *in vivo*. It is the purpose of the rest of this talk to accept this kind of informative spectrum as a standard and to discuss the obstacles in obtaining these spectra *in vivo*.

The first problem to discuss is resolution because while there is excellent resolution at 8.4 Tesla, we must work with whatever resolution is possible at 2 to 4 Tesla, the fields available for human studies. Second, I will discuss the problem of decoupling because when you see a ^{13}C spectrum, particularly one as clear as these high field spectra, the protons have been decoupled to simplify the spectrum, and there are concerns about heat deposition *in vivo* from decoupling. Third, I will discuss the problem of localization—showing that it is necessary and then indicating ways in which it is being solved. Fourth, I will discuss the cost, since people always ask how much it costs to do an experiment using ^{13}C-labeled substrates in a human. Then actually as part of the question of cost is the fifth question of the signal-to-noise ratio of the carbon spectra which I posed earlier. The carbon resonance is intrinsically weak, what can be done to increase the signal-to-noise ratio?

Starting with the resolution, I think you can see in Sheila Cohen's paper that there is resolution to burn at 8.4 T. Even glutamine and glutamate carbons could be separated, particularly at carbon 4, which is next to the carbon 5 amide or the carboxyl group, which is the only difference between the two compounds.

M. Stromski, Jeff Alger, and Fernando Arias Mendoza took a ^{13}C spectrum in our lab *in vivo* of the rat liver at 1.9 Tesla with labeled alanine infused into the living animal.[4] That spectrum is compared to the spectrum of the same liver measured at 360 MHz *in vitro* in a recent publication. The resolution at 1.9 T was four times worse, as you would expect because the linewidths are independent of the field, but the resolution even at 1.9 T was really very good. In it we could resolve the different carbons of glutamate and aspartate, we could watch the glucose being formed, and could see the different peaks of glucose. Hence resolution is really not too serious a problem and of course we are looking forward in the next few years to even higher fields for human studies.

Glycogen ^{13}C NMR spectra will be discussed by Gene Barrett, when he describes studies they are doing of heart glycogen and its repletion using ^{13}C-labeled glucose. It has been possible to observe the natural abundance glycogen peak in the rabbit, at 1.9 T and this has been published by Alger *et al.*[5] In the ^{13}C spectrum of glycogen in the perfused liver, carbon 1 (C-1) of glycogen is a few parts per million downfield from the glucose C-1 peaks, whereas all the other glycogen peaks coincide with the corresponding glucose peaks. Hence the available resolution gives essentially a glycogen C-1 peak, an area where C-2 through 5 fall but are not individually resolved and then a peak at the position of C-6 (FIG. 1). There is interference in the last two regions in the spectrum from glycerol peaks. The central CHOH of glycerol comes in the same region as the CHOH's of carbons 2 through 5 of glycogen and the terminal CH_2OH of glycerol is very close in chemical identity and therefore chemical shift to the CH_2OH of C-6 of the glucose moieties of glycogen. Hence there is some overlap in these regions with glycerol but here there's no overlap at the C-1 glycogen peak and its intensity is a very good measure of the amount of glycogen present. There is a beautiful spectrum of hepatic glycogen in humans taken by Tom Jue on our new 1-meter diameter, 2.1 T spectrometer.[6] We have not yet decoupled that spectrum for reasons that I will discuss. One of the features of glycogen comes from the fact that carbohydrate physiology in humans is not as well understood as I thought it was when I first went into this field. I thought they knew everything and they don't. If you consider the simple process of glycogen repletion in a fasted animal and if you start with glucose labeled at carbon 1, there are two major pathways by which that glucose can form glycogen. The first is it goes right into the liver where it becomes glucose 6-phosphate, glucose 1-phosphate, and then is deposited as glycogen. In this path the carbon label that starts at carbon 1 of glucose will end up at carbon 1 of glycogen and you'll get a single glycogen peak that you can monitor. Also, we must consider flux through the Cori cycle in which glucose goes into muscle where it becomes a three-carbon compound like lactate or alanine, which then is carried to the liver where it goes through the gluconogenic pathway to synthesize glucose with a complicated pattern of labeling. In that pattern of labeling C-1–labeled glucose will form the C-3–labeled alanine, which will in turn label carbons 1, 2, 5, and 6 of the glucose moieties of glycogen. Consequently we have essentially two

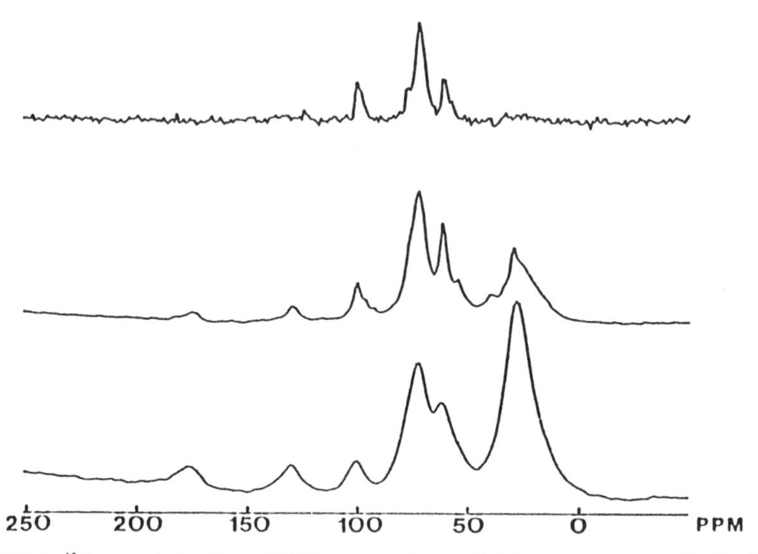

FIGURE 1. ^{13}C natural abundance NMR spectra taken at 1.9 T. Lowest spectrum: Rabbit liver *in vivo* using a surface coil. 6,000 accumulations with a pulse repeat time of 0.1 sec. Total accumulation time of 10 min; 60 Hz line broadening filter used in processing. Middle spectrum: Excised liver from the rabbit used for the spectrum below. Observed with a five-turn, 3 cm diameter solenoid with a perpendicular Helmholtz coil decoupler: 66° pulse repeated every 0.1 sec; spectral width, 10 kHz; 1,024 time-domain data points; 6,000 accumulations, total accumulation time of 10 min; 30 Hz line-broadening filter used in processing. Upper spectrum: Glycogen extracted from the liver of the rabbit used for the spectra below. Accumulation parameters for the excised liver; 10 Hz line-broadening filter used in processing; Assignments: 170–180 ppm, carboxyl carbons of fatty acids; 125–135 ppm, unsaturated carbon of fatty acids; 95–105 ppm, glycogen C_1 carbon; 65–80 ppm, glycogen $C_{2\text{-}5}$ and glycerol C_2; 60–65 ppm, glycogen C_6, and glycerol $C_{1,3}$; 55.5 ppm, methyl group of choline; 40.0 ppm, methylene groups of choline and ethanolamine; 10–40 ppm, saturated carbons of fatty acids.

vectors. We have one vector, which is the glycogen labeling observed when glucose goes directly to glycogen and labels C-1. The other vector comes from glucose through the Cori cycle from where it labels C-1, 2, 5, and 6. By looking at the labeling in the glycogen and breaking these three peaks up into two vectors, which is possible, we can measure the relative fluxes through both pathways. This has considerable advantages since it can be done *in vivo* continuously. Another advantage (not illustrated) is you can get from this the amount of unlabeled glycogen by other NMR methods.[7]

The second problem, decoupling, will be discussed very briefly because this has not yet been solved, but it certainly is clear how it can be solved. Decoupling is needed because when you look at ^{13}C, nearby protons are J-coupled and they split up the spectrum so that it too is overlapping and difficult to interpret. The problem with decoupling is that it may sometimes require too much power to decouple the protons broadly. The FDA guidelines, which are suggested guidelines not rules, are that there should be no more than 8 watts per kg of energy deposited in any local volume and 0.4 watts per kg averaged over the whole body. I don't want to mention names because I don't know if he wants me to, but someone in this audience did use power levels of this sort in a foreign country to decouple someone's arm (whose name will also not be mentioned). In the ^{13}C spectrum they got very beautiful decoupled peaks and the decouplee is walking around happy and well—he assures me. In fact since he used no more than the suggested power level I can say that it was Walter Aue, which was sufficient, using an MLEV sequence, to decouple the human arm. In the event that more decoupling power is needed the physiologists point out to us that in vigorous exercise, which can be prolonged for a long period of time, the energy dissipation is of the order of 40 watts per kilogram over much of the body. This is two orders of magnitude above the present guidelines (of 0.4 watts/Kg over the whole body) so that there is every reason to believe that after serious studies are done (and they are being done in a number of laboratories) one can use more power than the present guidelines, which are very conservative.

The perfused liver spectra are from a rat liver in a 20 millimeter diameter NMR tube. How do we deal with liver and other organs in the body? Of course we have to localize and that will be, as I say, the subject of many talks in this volume. The localization in the liver has to go in rather deep. We have seen from Paul Bottomley[8] and from Peter Styles and George Radda,[9] phosphorus spectra of organs within the body, i.e. liver and heart, well resolved and well identified (also from Jan den Hollander[10] lately). Localization was achievable using two different kinds of methods that I'll mention for your information, but their details and beauty will be described by the inventors.

First are methods that take advantage of the B_1 gradient. They use surface coils or concentric surface coils without any gradient in the static B_0 field. David Hoult and Robin Bendall have been very active in developing those methods.

Then there are the methods that depend upon gradients in the static magnetic field such as Aue's volume selective excitation, ISIS from Ordidge, SPARS from den Hollander, and those all will be discussed. Finally there is the DRESS sequence of Bottomley, which uses a combination of both, a surface coil with B_0 field gradients.

We come to the question of the cost of ^{13}C experiments in the human body. The optimum infusion rate of glucose is one gram per kilogram/hour. If you infuse more than this the body excretes the glucose you put in, if you infuse less you end up with a lower specific activity of the glucose in the blood. The cheapest price that we've had is about \$140/gram for ^{13}C-labeled glucose, so an experiment on a 70 kilogram person would cost \$10,000. That's too much. What can we do to lower that? Let us look a little bit ahead: The basic cost of the ^{13}C in glucose is about one-twentieth of the cost of the glucose in which it is incorporated (TABLE 1). Large batch processing should certainly bring that price down several fold.

Finally, the second improvement would be to increase the sensitivity so that less labeled material is needed. Methods have been developed that we have used *in vivo*

TABLE 1. Cost of Human Experiment[a]

Method	Possible Reductions Theoretical Factor	Realistic Factor
Reduced synthesis costs	10–1,000 [20 for 1 – ^{13}C Glu]	2×
Cost of raw material	Laser methods (?)	1.5×
Proton observe ^{13}C decouple	20–60	5×
Larger sample; better coils	[Should compensate for loss of filling factor]	1×
	Product	15×
	Possible Cost/Exp	\$600

[a]Cost of running experiment. NMR images cost ~\$600. Spectroscopic studies should cost the same, or more. Therefore, ^{13}C label can become comparable with cost of running equipment.
70 kg Human needs 70 g [1 – ^{13}C]glucose and 70 × \$140/g = \$10,000.

where you observe protons and decouple carbon.[11] In these you have the sensitivity of the proton next to the carbon and use the proton to monitor what the carbon is doing. There is about a factor of ten available there, and as indicated in the TABLE there are other small factors that one can hope to obtain so that it is fairly realistic to expect a 10- to 100-fold reduction in the cost of the experiment. If possible this would bring the price down to several hundred dollars per experiment. Since it costs about a thousand dollars to take a scan of a patient with an NMR imaging spectrometer, if we can reduce the cost of the ^{13}C to less than a thousand dollars it would not be a substantial increase in the cost of the experiment.

I think that pretty much describes the expectations of the ^{13}C NMR for human studies. The proton is very promising in the future and you will hear that at different points in this volume.

REFERENCES

1. SILLERUD, L. O. & R. G. SHULMAN. 1983. Structure and metabolism of mammalian liver glycogen monitored by ^{13}C NMR. Biochemistry **22:** 1087–1094.
2. COHEN, S. M. 1987. Ann. N.Y. Acad. Sci. (This volume).
3. COHEN, S. M. & R. G. SHULMAN. 1982. ^{13}C NMR studies of gluconeogenesis in hepatocytes from euthyroid rats and in perfused mouse liver: *In situ* measurements of pyruvate kinase flux and pentose cycle activity. *In* Noninvasive Probes of Tissue Metabolism. J. Cohen, Ed.: 119–147. John Wiley & Sons. New York.
4. STROMSKI, M. E., F. ARIAS-MENDOZA, J. R. ALGER & R. G. SHULMAN. 1986. Hepatic gluconeogenesis from alanine: ^{13}C NMR methodology for *in vivo* studies. Magn. Res Med. **3:** 24–32.
5. ALGER, J. R., K. L. BEHAR, D. L. ROTHMAN & R. G. SHULMAN. 1984. J. Magn. Res. **65:** 334–337.
6. JUE, T., J. LOHMAN, R. ORDIDGE & R. G. SHULMAN. 1987. Abstract Soc. Mag. Res. Med. (August, 1986). (In press.)
7. SHULMAN, G. I. *et al.* 1985. Mechanism of liver glycogen repletion *in vivo* by NMR spectroscopy. J. Clin. Invest. **76:** 1229–1236.
8. BOTTOMLEY, D. A., T. H. FOSTER & R. D. DARON. 1984. Depth resolved surface-coil spectroscopy (DRESS) for *in vivo* ^{1}H, ^{31}P and ^{13}C NMR. J. Magn. Res. **59:** 338–312.
9. STYLES, P. & G. K. RADDA. 1987. Ann. N.Y. Acad. Sci. (This volume).
10. LUYTEN, P. R. & J. A. DEN HOLLANDER. 1986. Observation of metabolites in the human brain by MR spectroscopy. Radiology **161:** 795–798.
11. ROTHMAN, D. L., K. L. BEHAR, H. P. HETHERINGTON, J. A. DEN HOLLANDER, M. R. BENDALL, O. A. C. PETROFF & R. G. SHULMAN. 1985. ^{1}H observe and ^{13}C decouple spectroscopic measurements of lactate and glutamate in the rat brain *in vivo*. Proc. Natl. Acad. Sci. USA **82:** 1633–1637.

DISCUSSION OF THE PAPER

M. W. WEINER: My question concerns the safety of decoupling for carbon experiments. For human research, one is not necessarily bound by the FDA guidelines, the manufacturers are, but we're only bound by common sense and by our institutional research boards. What do you think is a safe figure that clinical investigators could present to their review boards saying that this is a safe figure in terms of watts?

R. G. SHULMAN: I could give you the number, but I'd be quoting Eleanor Adair. She said 4 watts per kilogram. What we're planning to do in collaboration with Dr.

Adair is "instrument" animals and deposit that much energy at our frequencies in animals—up to at least 4 watts per kilogram and actually to go higher than that and then to settle at a human safe level in our own laboratory that we can present to our review board.

R. S. BALABAN: This may be true for all the control, healthy people, but what about people with compromised peripheral circulation, or compromised circulation of a particular organ where this power is positive?

SHULMAN: This is the problem. This is why people are concerned about the local heating and the guidelines that require lower levels for local heating because what will happen to the overall flow, the systemic flow? In general though, the thing that gives me considerable sense of safety is that the temperature sensors are on the surface in any method used, there's more heat deposited on the surface from surface coils and therefore by the time someone is being heated to the point of discomfort and says it's hot on the surface, and then you get them out. But we haven't done that yet.

MOONEY: I want to comment on the regional aspect and dissipation of heating due to decoupling. Blood flow is of course very interesting and very important, but you compare it to physiology with exercise—with vigorous exercise—and that's perhaps a bit misleading because then you have a very large blood flow. I think guidelines are also proposed on the basis of parts of the body where there is no blood flow and you heat these parts of the body as well. That's the thing that you have to take care of.

SHULMAN: Yes, that is the difference between the guidelines for systemic heating and for localized heating, to be sure that there's blood flow to the local region.

DR. EGAN: In some instances, can you alleviate the coupling problem by putting deuterium on the carbon to reduce the magnitude and maybe even have self coupling (i.e., the coupling would in effect be removed) if the deuterium T_1s are sufficiently short?

BALABAN: Dr. Alger would like to answer that.

J. R. ALGER: I have talked to a few people who have used mass-Spec for analysis of deuterium. They told me that introduction of that much deuterium, say stoichiometric amounts, and deuterium-labeled carbon would be inappropriate for human studies because of the deuterium isotope effects. There would simply be too much deuterium.

D. G. GADIAN: I would like to ask about the spectral resolution at 2 tesla, relative to 8 tesla. I thought at least for some of the lines that the limitations would have been field inhomogeneity in parts per million. So that if you go down in field the inhomogeneity in parts per million stays the same and therefore the spectral resolution should not be too much worse at low field.

SHULMAN: Yes, that's a good point. When we compared the spectra, there was a small contribution from the inhomogeneity which was a little worse at high field than at low field. When I compared the line widths in the *in vivo* liver with the surface coil placed on the exposed liver in open abdomen preparation (that's at 1.9 tesla), with that same liver taken out at 8.4 tesla, instead of the resolution being 4.5 times worse it was about 3 to 3.5 times worse because lines were a little sharper at the lower field. But it wasn't linear with the field; it didn't compensate enough for the field differences. However, better shimming in the large low field magnets may be possible, as you suggest.

NMR Studies of Methanogens: What Good Is a Cyclic Pyrophosphate?[a]

MARY F. ROBERTS, JEREMY N. S. EVANS,
CYNTHIA J. TOLMAN, AND DANIEL P. RALEIGH

Chemistry Department
Francis Bitter National Magnet Laboratory
Massachusetts Institute of Technology
Cambridge, Massachusetts 02139

INTRODUCTION

Methanogens are fastidious anaerobes whose metabolism centers around the reduction of CO_2 to CH_4.[1,2] H_2 provides reducing equivalents and the overall reduction appears to be a membrane-related process. The electron transport generated by this process ("methanogenesis") drives ATP synthesis in these archaebacteria. Many aspects of cellular metabolism are unclear in these organisms. *Methanobacterium thermoautotrophicum* strain ΔH is the most commonly studied methanogen. It is a thermophile (62°C optimum growth temperature) that uses CO_2 as the sole source of carbon for growth as well as methane production. More reduced C_1 compounds or more complex carbon species are not sufficient to support cell growth. Neither the Calvin cycle[3] nor the full reductive tricarboxylic acid cycle[4] operates in carbon fixation in *M. thermoautotrophicum*. Instead, a different scheme, the "activated acetic acid" pathway, has been proposed for fixation of C_1 units into C_2 and higher order carbon compounds.[5] Based on the work of Wood and co-workers with *Clostridium thermoaceticum*,[6] this scheme has one CO_2 molecule sequentially reduced to a CH_3 unit while another CO_2 is converted to a bound CO by the enzyme CO dehydrogenase.[7] These two different C_1 units are then condensed into an acetyl moiety (acetyl CoA), which is subsequently converted to pyruvate, then phosphoenolpyruvate, and on to more complex molecules.

NMR spectroscopy is a useful tool for investigating biological pathways. Both ^{31}P and ^{13}C NMR spectroscopy have been used to study cellular metabolism in a wide range of microorganisms.[8–11] Information can be obtained on development of pH gradients across cell membranes, glycolysis, ATP levels, carbon assimilation, etc. Our early ^{31}P NMR studies of *M. thermoautotrophicum* ΔH[12] led us to isolate and identify a novel cyclized pyrophosphate, 2,3-cyclopyrophosphoglycerate (CPP), whose intracellular concentration under growth conditions in excess phosphate is exceptionally high (20–100 mM). Because CPP has no chromophore, it escaped detection for many years. In the ^{31}P NMR spectrum, it is the major observable phosphorus species. CPP does not function as a high energy phosphate for substrate level phosphorylation of ADP.[13] Since methanogens produce polyphosphate, CPP is also unlikely to be a polyphosphate substitute.[13,14] CPP is found at high levels only in a subset of all methanogens (methanobacteria and methanobrevibacter), which have a rigid pseudo-

[a]Supported by the Basic Research Department of the Gas Research Institute (Contract 5083-260-0867). The NMR spectra were obtained at the Francis Bitter National Magnet Laboratory, M.I.T., and in the Biochemistry Department at Tufts University Medical School.

murein cell wall.[13,15] The carbohydrate content of the cell wall is high and requires large carbon fluxes during cell growth. The same class of methanogens also has high intracellular K^+ (1–2 M).[16] These factors led us to suspect that CPP was somehow involved in carbon assimilation in methanogens.

^{13}C NMR spectroscopy (both solution and solid-state) is an ideal technique for investigating carbon assimilation in *M. thermoautotrophicum*.[17] By growing methanogens on ^{13}C–labeled compounds and analyzing the labeling pattern of molecules we can monitor specific ^{13}C uptake, label distribution, and turnover. For these studies three basic experimental NMR approaches have been taken: (1) preparation of ethanol extracts of cells (to isolate small molecules from macromolecular components) at fixed labeling time points for very high resolution spectra, (2) isolation of cell debris after a labeling experiment for analysis of ^{13}C distribution by solid-state NMR techniques, and (3) direct observation of suspensions of intact cells at 60°C or 4°C for a more continuous time course of label incorporation. A variation of the basic ^{13}C labeling experiment is the application of a ^{13}C pulse followed by a ^{12}C chase. Analogous to a radioactive pulse/chase experiment, the NMR variant uses more labeled material, but has the advantage of detecting all labeled small molecules in a single experiment. An analysis of solid cell debris can also show flux of label from small molecules in solution to macromolecular structures.

EXPERIMENTAL PROCEDURES

Materials

$^{13}CO_2$ (99.2% ^{13}C)/H_2 (1:4, vol/vol) was obtained from Cambridge Isotopes or from Mound Isotopes. [1-^{13}C]acetate, [2-^{13}C]acetate, [6-^{13}C]glucose, and $^{13}C_2$-acetate were obtained from Cambridge Isotopes. [2,3-^{13}C]pyruvate was obtained from Merck Isotopes. Unenriched CO_2/H_2 (1:4, vol/vol) was obtained unanalyzed from Matheson Gas. ^{13}C-depleted CO_2 (99.96% C-12) was a gift from Los Alamos National Laboratory and was mixed with H_2 to give CO_2/H_2 (1:4, vol/vol) by Matheson Gas. All other chemicals were reagent grade.

Cell Growth

Cells of *M. thermoautotrophicum* strain ΔH were initially grown in pressurized bottles as described previously.[12] Bottles containing cells with $OD_{660} \sim 1$ were used to inoculate anaerobically 1 L of medium in a 2 L fermentor and cells were grown at 62°C, pH 7.1 under continuous purging of CO_2/H_2 and H_2S with stirring. For feeding experiments, anaerobic solutions of ^{13}C labeled molecules were added to bottles with cells containing unenriched CO_2/H_2. In pulse-chase experiments, the cells were grown in the fermentor to $OD_{660} \sim 1$ and the $^{13}CO_2/H_2$ was introduced for 0.5 hr before flushing with unenriched $^{12}CO_2/H_2$. Aliquots of cells were removed from the fermentor at the indicated chase times.

Cell Extracts

Cells were harvested anaerobically by centrifugation (8,000 *g*, 10 min), washed with phosphate buffer, and centrifuged again. The pellet was extracted with 70% ethanol, stirred at room temperature for 30 min, and centrifuged (12,000 *g*, 10 min).

The supernatant was decanted and the residue reextracted with ethanol. Supernatants from both ethanol treatments were combined and lyophilized. The cell debris remaining from the ethanol extraction was dried and used for CP-MASS NMR spectroscopy. Cell wall material was purified from this particulate matter by incubating a suspension of debris with DNase (1 mg) and trypsin (0.1%) at 37°C for 18–24 hr.[18] This removes all protein and nucleic acid and leaves behind the pseudomurein cell wall.[19]

Permeabilized Cells

Cells grown to $OD_{660} \sim 1$ in the fermentor were harvested anaerobically and centrifuged for 10 min at 8,000 *g*. The cell paste was then resuspended in a volume equal to that before centrifugation in anaerobic buffer containing 10 mM potassium phosphate, 10 mM NH_4Cl, 1 mM $MgCl_2$, 3 mM ATP, 5 mM dithiothreitol, and 0.15 mM coenzyme A, pH 6.5. The resuspended cells were placed in 500 ml bottles and sealed. Half of the samples were stored in the refrigerator (controls), while the other half were frozen, then slowly thawed out. Such freezing and thawing produces small breaks in the cell membrane that allow leakage into the supernatant of small molecules normally extracted only with ethanol or perchloric acid.[20] Macromolecules are retained inside and do not leak through the permeabilized membrane.

NMR Spectroscopy

High-resolution spectra of extracts were obtained with a 6.34-T Bruker 270 instrument operating at 67.9 MHz for ^{13}C and 109.3 MHz for ^{31}P, and with a 9.39-T Bruker AM-400 WB instrument operating at 100.6 MHz for ^{13}C. Dry samples were dissolved in 10 mM potassium phosphate, 0.1 mM EDTA, pH 7.2, 50% D_2O. The ^{13}C chemical shifts were referenced to p-dioxane.

CP-MASS NMR spectra were obtained on a home-built pulse spectrometer operating at 79.9 MHz for ^{13}C and 317 MHz for ^1H. ^{13}C spectra were obtained using cross-polarization with a ^1H decoupling field of 100 kHz, ^{13}C rf field of 48 kHz, and a contact time of 2.5 msec. Powdered samples of 25–100 mg were tightly packed into ceramic double bearing rotors (Doty Scientific), and sample spinning rates varied from 2.7 to 3.5 kHz.

RESULTS

$^{13}CO_2$ Labeling of Exponentially Growing Cells

An exponential culture of *M. thermoautotrophicum* exposed to $^{13}CO_2$ for a short time compared to the cell doubling time shows rapid and intense labeling of CPP.[17] Not only is CPP a major carbon pool, but it also turns over rapidly. FIGURE 1 shows *in vivo* ^{13}C spectra for *M. thermoautotrophicum* incubated with $^{13}CO_2$ for 0.5 hr followed by a $^{12}CO_2$ chase for the times indicated. For this experiment, cells were withdrawn from a 2-L fermentor, cooled to 4°C, and examined by both ^{13}C and ^{31}P NMR. The level of ^{13}C incorporation into CPP first increases (as residual $^{13}CO_2$ in the medium is assimilated) then decreases to natural abundance levels. Other observable metabolites do not decrease on this time scale and no new resonances appear. The CPP half-life is about 5 hr compared to a doubling time of about 30 hr. The ^{31}P NMR spectra indicate that the steady-state concentration of CPP does not decrease over this period, but rather increases in accordance with an increased total number of cells. Subsequent repeats of this ^{13}C pulse/^{12}C chase experiment, but using ethanol extracts to improve resolution and signal-to-noise, show much smaller amounts of other ^{13}C enriched molecules with extremely short half-lives (FIG. 2). CPP is again the major labeled species (δ_c = 70.1 [C-3] and 78.6 [C-2] ppm) during a 15-min $^{13}CO_2$ pulse. The maximum enrichment is about 12-fold over the natural abundance background. After 1 hr of the $^{12}CO_2$ chase, the CPP peaks have significantly decreased in intensity. No new resonances appear in the spectrum, indicating that the ^{13}C flux through CPP has been partitioned into large insoluble structures. This $^{13}CO_2$ pulse/$^{12}CO_2$ chase experiment identifies three classes of metabolites with respect to utilization of specifically labeled pools: (1) intermediates in carbon assimilation such as alanine (δ_c = 17.5 ppm [C-3]) and other as yet

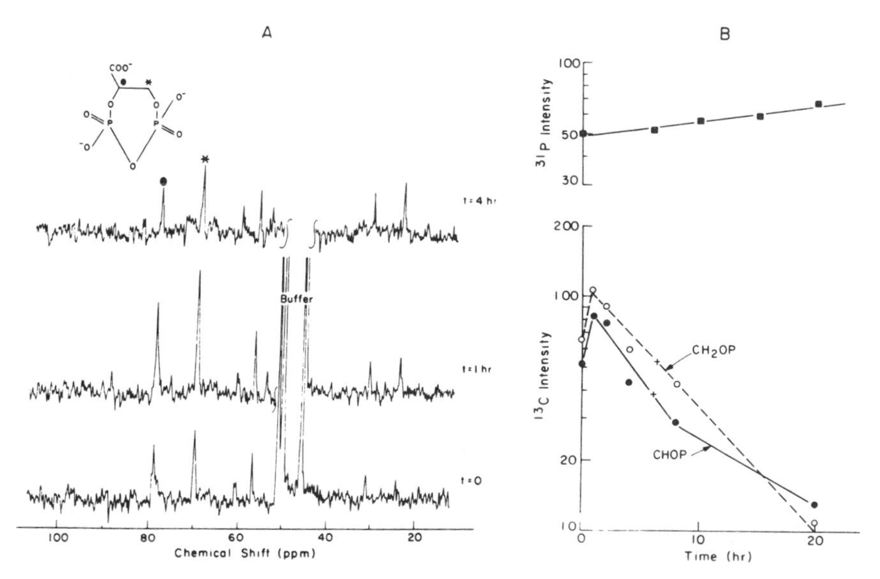

FIGURE 1. H-1 decoupled ^{13}C NMR spectra (67.9 MHz) of *M. thermoautotrophicum* cells exposed to a 0.5 hr $^{13}CO_2$ pulse followed by a $^{12}CO_2$ chase, harvested as a function of time and examined at 4°C. (B) Both the CPP ^{13}C and ^{31}P intensity plotted as a function of the chase times.

unidentified species that turn over very rapidly (~10 min), (2) CPP with a 1-2 hr half-life, and (3) glutamate (δ_c = 27.8 [C-3], 34.3 [C-4], and 57.0 [C-2] ppm) with an even longer half-life (~3-4 hr). Establishing where the ^{13}C from CPP goes during the CO_2 chase should provide a clue to its metabolic role.

^{13}C-CP-MASS Studies of Particulate Fractions

Since in the solution spectra no new resonances appear concomitantly with the decay of ^{13}C–labeled CPP, ^{13}C CP-MASS solid-state NMR spectra of cell debris were obtained. Preliminary CP-MASS experiments showed that solid material from cells given a 15 min $^{13}CO_2$ pulse followed by a $^{12}CO_2$ chase exhibited a small increase in intensity (compared to cells harvested prior to the introduction of ^{13}C) in a region of the spectrum consistent with carbohydrate (~75 ppm). However, the natural abundance ^{13}C background of the cell debris dominated the spectrum. Because CO_2 is the sole carbon source for *M. thermoautotrophicum,* we can dramatically suppress this isotopic abundance by growing the cells on ^{13}C depleted $^{12}CO_2/H_2$. For this cells grown in bottles with $^{12}CO_2$ (99.96% ^{12}C)/H_2 to $OD_{660} \sim 1$, were subjected to a $^{13}CO_2$ (99.2% ^{13}C) pulse for 0.5 hr, then switched back to the ^{13}C–depleted CO_2. Under these conditions the cells have a doubling time of about 30 hr. After 12 hr and 30 hr of the ^{12}C chase, the cells were harvested, extracted with ethanol, and the soluble and pellet

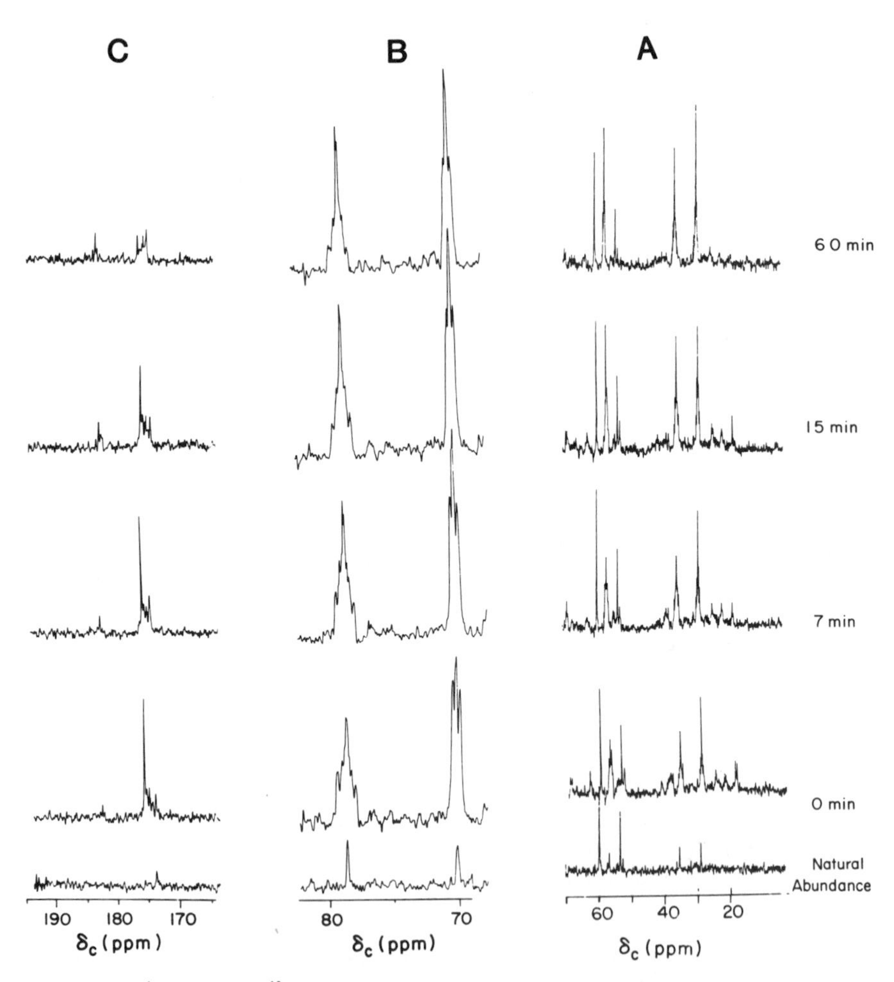

FIGURE 2. ^1H decoupled ^{13}C NMR spectra at 67.9 MHz of ethanol extracts of exponentially growing cells of *M. thermoautotrophicum* exposed to a 15 min. $^{13}CO_2$ pulse, followed by a $^{12}CO_2$ chase and harvested as a function of times after the start of the chase. Three spectra regions are shown: (A) upfield containing glutamate, alanine, acetyl methyls, etc., (B) CPP [C-2] and [C-3], and (C) carbonyl region.

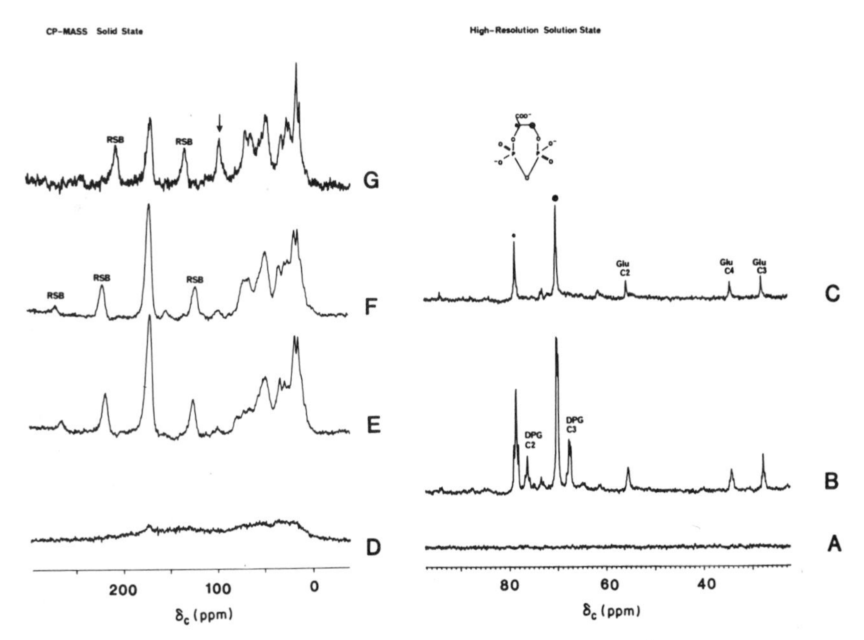

FIGURE 3. ^{13}C CP-MASS NMR spectra (and solution state as well) of ^{13}C depleted cells before and after a $^{13}CO_2$ pulse. High resolution spectra at 100.6 MHz of ethanol extracts are shown for (A) control cells without any $^{13}CO_2$ pulse, (B) 12 hr after a 0.5 hr ^{13}C pulse, and (C) 30 hr post-pulse. CP-MASS NMR spectra at 79.9 MHz of cell debris are shown for (D) cells harvested prior to the ^{13}C pulse, (E) cell debris 12 hr after the initial $^{13}CO_2$ pulse, (F) 30 hr post-pulse, and (G) debris from the 30 hr time point after DNase and trypsin treatment.

fractions examined by ^{13}C NMR. The solution-state spectra indicate an intense CPP label uptake for the 12 hr time point (FIG. 3B), which decays to 75% of its value by 30 hr of the ^{12}C chase (FIG. 3C). Significantly enriched resonances from 2,3-diphosphoglycerate are observed in this experiment; they are not detected if the pulse-chase experiment is performed with cells grown in the fermentor. The incorporation of ^{13}C from CPP into cell macromolecules can now be assessed by CP-MASS (FIG. 3D–G). Note the virually complete elimination of the normal 1.1% ^{13}C background in cell debris from *M. thermoautotrophicum* harvested before the introduction of $^{13}CO_2$. The dramatic incorporation of ^{13}C in FIGURE 4 reflects carbohydrate, protein, and nucleic acids labeled up to 12 hr after the 0.5 hr ^{13}C pulse. Organic solvent extracts of this material do not remove any ^{13}C, suggesting that ^{13}C incorporation into lipids is minimal on this time scale. After 30 hr of the ^{12}C chase, during which time the only change in ^{13}C in solution is the decay of ^{13}C CPP (FIG. 3F), the most significant increase in intensity in the solid phase occurs at 103 ppm, which corresponds to the anomeric carbons of carbohydrates, and at around 75 ppm (also consistent with carbohydrate). Spectra of the residue from trypsin and DNase digests[18] of the 30 hr time point (FIG. 3G) confirm that this arises from carbohydrate in the pseudomurein cell wall structure[15,19] of *M. thermoautotrophicum.*

Thus, the CP-MASS spectra detect the flux of ^{13}C from the soluble CPP pool into particulate matter, primarily carbohydrate. This suggests a link between CPP turnover and carbohydrate biosynthesis.

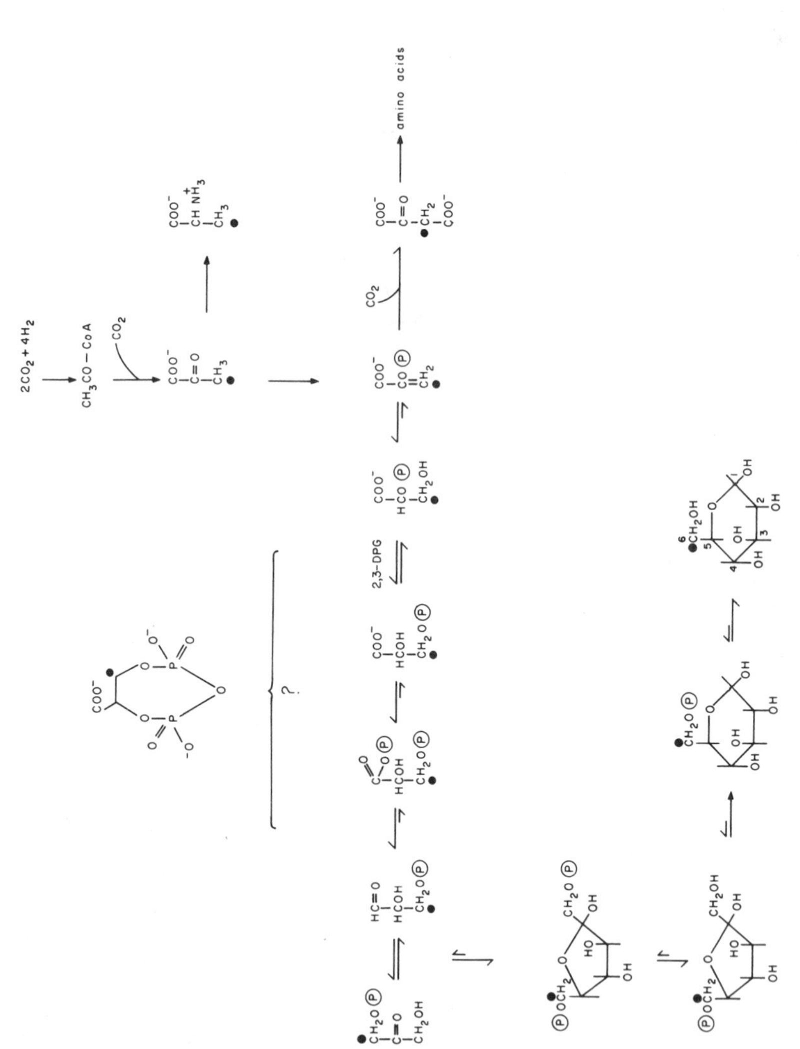

FIGURE 4. Scheme showing the fate of [6-^{13}C]glucose if gluconeogenesis is reversed.

Biosynthesis of CPP

To delineate the pathway for CPP synthesis in *M. thermoautotrophicum,* we have carried out a series of ^{13}C labeling experiments where this methanogen is grown on $^{12}CO_2/H_2$ (for CH_4 generation) and ^{13}C acetate, pyruvate, or other weak acid precursor for assimilation into cell carbon. [1-^{13}C]acetate specifically labels the CHO [C-2] of CPP, while [2-^{13}C]acetate labels the CH_2O [C-3].[17] A similar experiment with [1-^{13}C]pyruvate labels the CPP carboxylate.[17] *In vitro* experiments with cell extracts were also attempted with a variety of combinations of pyruvate, PEP, acetate, ATP, or ADP as carbon and phosphorus donors. Limited (but above background) synthesis of CPP occurred only with PEP and ATP or ADP (the two nucleotides are interconverted in cell extracts). It should be noted that PEP is a branch point for carbohydrate biosynthesis.

An NMR experiment aimed at directly relating CPP and gluconeogenesis is the incubation of ^{13}C glucose with cells and subsequent ^{13}C NMR analysis of CPP for label

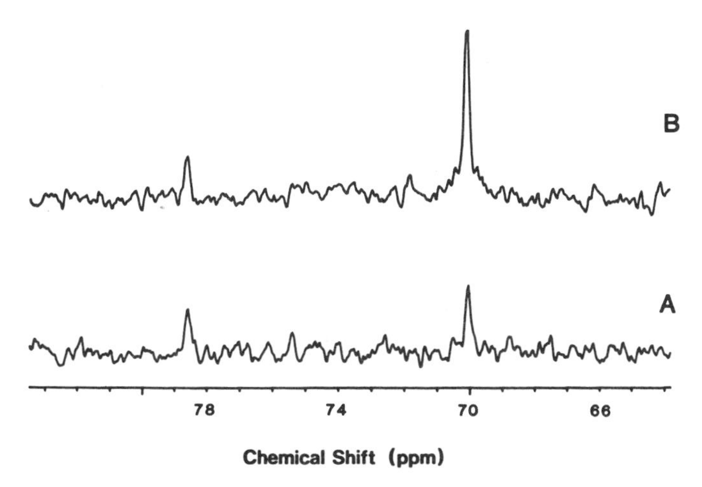

FIGURE 5. ^{1}H decoupled ^{13}C NMR spectra at 100.6 MHz of ethanol extracts of exponentially growing cells fed with (A) unenriched CO_2, and (B) unenriched CO_2 and [6-^{13}C]glucose.

uptake via reverse reactions. The scheme shown in FIGURE 4 predicts that if CPP is involved in the triose conversions, the ^{13}C from [6-^{13}C]glucose will specifically label the [C-3] of CPP. FIGURE 5 also shows the results of such an incubation: [C-3] of CPP is exclusively labeled, which suggests that glucose is cleaved by the reverse of the gluconeogenesis pathway.

Regulation of CPP Biosynthesis and Turnover

The rapid flux of carbon through CPP into an insoluble pool suggested that it is an intermediate in carbon assimilation. Such intermediates are not usually maintained at high intracellular concentrations, so that CPP could have a more immediate function as a storage or regulatory carbon pool. To investigate this we have examined CPP concentrations and labeling efficiency under a variety of metabolic conditions.

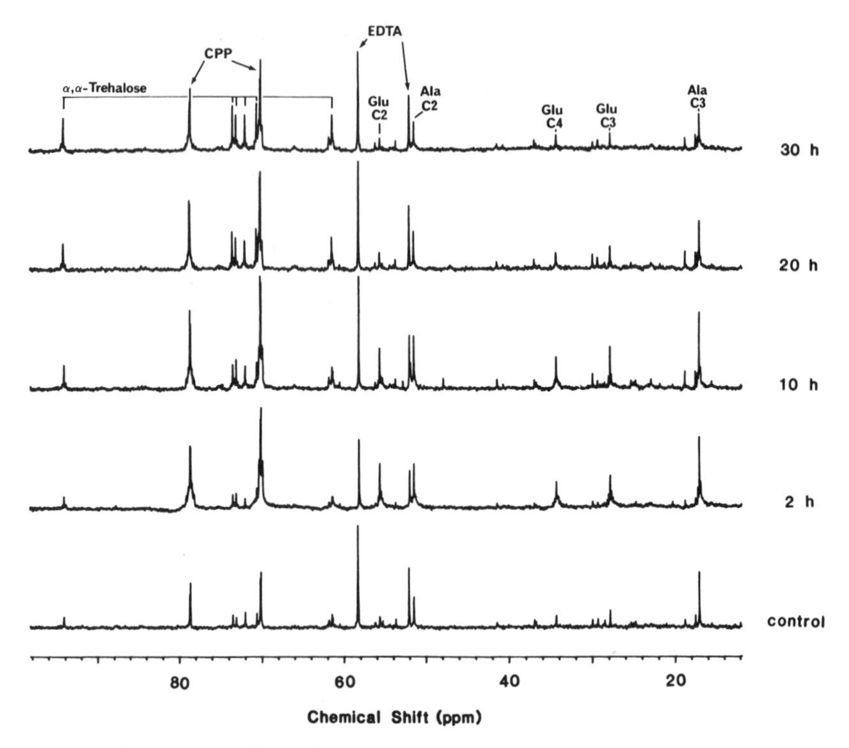

FIGURE 6. ^{1}H decoupled ^{13}C NMR spectra at 100.6 MHz of ethanol extracts of stationary phase cells exposed to a 0.5 hr $^{13}CO_2$ pulse followed by a $^{12}CO_2$ chase and harvested at various time points after the start of the chase.

M. thermoautotrophicum grown to stationary phase were given a 0.5 hr $^{13}CO_2$ pulse followed by a $^{12}CO_2$ chase. Carbon flux varies significantly in these non-growing cells (FIG. 6). CPP is labeled by $^{13}CO_2$ but to a much lower level (2.4-fold above natural abundance) than in exponentially growing cells. The turnover of ^{13}C CPP has a half-life of 9–10 hr and glutamate and alanine have 15–16 hr half-lives. Additionally, the disaccharide α,α-trehalose[21] is detected (δ_c = 94.1 [C-1], 72.0 [C-2], 73.5 [C-3], 70.6 [C-4], 73.0 [C-5], and 61.4 [C-6] ppm) and becomes more labeled as CPP decreases (FIG. 7).

M. thermoautotrophicum cells frozen and then slowly thawed develop leaky membranes that allow passage of small molecules across the cell envelope. Under these conditions no ΔpH or $\Delta\psi$ can be maintained. If enzymes that synthesize CPP are not dependent on the cell metabolic state and sufficient small molecule cofactors or necessary substrates are added for biosynthesis, then CPP should be labeled from incubation with $^{13}CO_2$. Cells permeabilized in this fashion had added buffer components (ATP, coenzyme A, dithiothreitol, $MgCl_2$) shown by Fuchs and co-workers[22] to accelerate acetyl CoA biosynthesis as well as methanogenesis *in vitro*. As shown in FIGURE 8, both the supernatant (FIG. 8B) and ethanol extract of debris (FIG. 8C) of permeabilized cells exposed to $^{13}CO_2$ for 24 hr show very little label uptake into CPP compared to the control ethanol extract of intact cells incubated with $^{13}CO_2$ fu. 24 hr

(FIG. 8A). Incubation of permeabilized cells with $^{13}CO_2$ for only 6 hr fails to label any CPP above natural abundance background levels. The same drop in CPP production also occurs when French Press cell extracts are used. The rate of synthesis of ^{13}C CPP from $^{13}CO_2$ has decreased drastically from that which occurs in intact cells. The supernatant of permeabilized cells shows an 11 to 13-fold decrease in ^{13}C CPP compared to the control ethanol extract. Ethanol extracts of cell debris from permeabilized cells exhibit low amounts of ^{13}C CPP almost equivalent to the amount in the supernatant. This amount of labeling in the debris may be from a few intact cells; it could also imply a fraction of the CPP pool is not freely diffusing despite permeabilization.

An examination of ^{13}C labeled peaks in the spectra of the supernatant from permeabilized cells shows a number of other significant changes. The alanine methyl is clearly labeled and its ^{13}C intensity is now comparable to CPP (in the intact cell extract such a high relative concentration of alanine is not detected except in stationary-phase cells). Because alanine is labeled, $^{13}CO_2$ must be metabolized to pyruvate. Label uptake into glutamate is also observed and is comparable to ^{13}C CPP, a result quite distinct from intact cells. The amount of ^{13}C glutamate implies that phosphoenolpyruvate can be synthesized in the *in vitro* system.

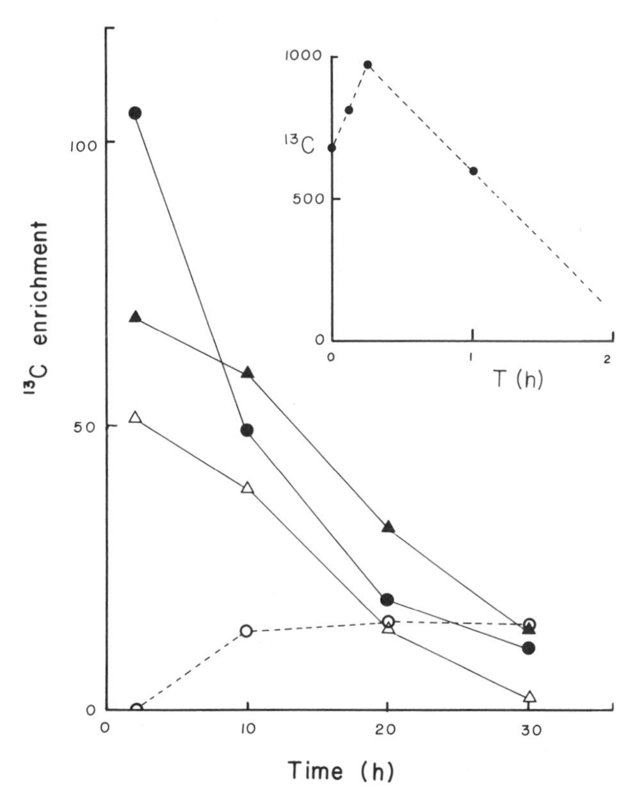

FIGURE 7. ^{13}C integrated intensities of CPP CHO (●), α,α-trehalose (O), alanine C-3 (▲), and glutamate C-3 (△) in extracts shown in FIGURE 6.

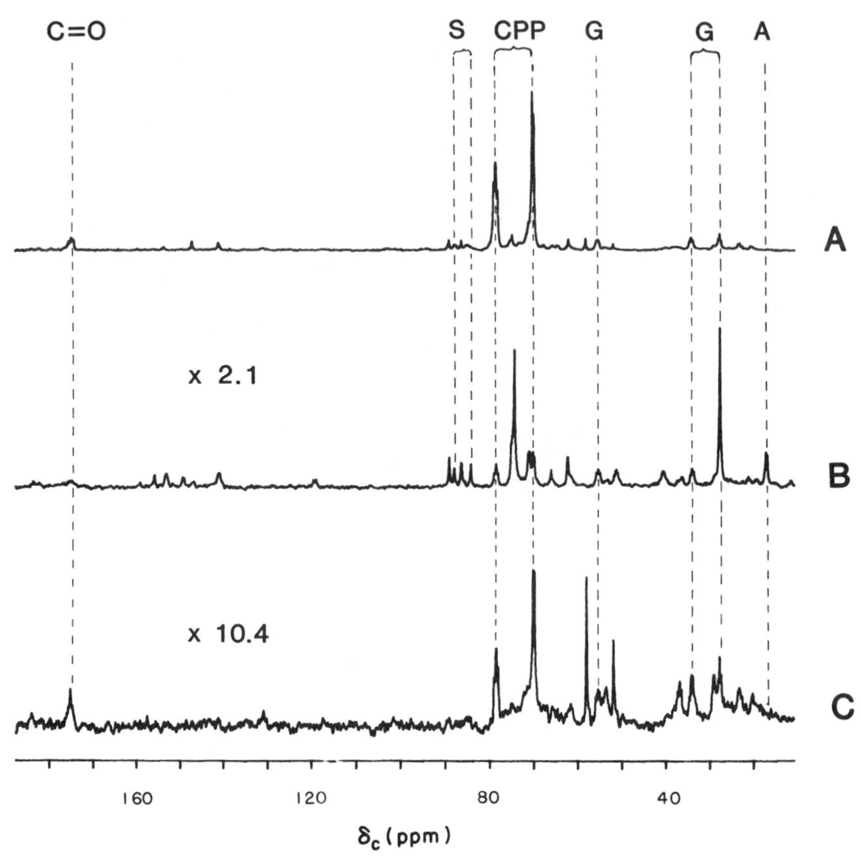

FIGURE 8. ^1H decoupled ^{13}C NMR spectra of (A) ethanol extracts of *M. thermoautotrophicum* cells exposed to $^{13}CO_2/H$ for 24 hr, (B) supernatant of frozen-thawed cells exposed to the $^{13}CO_2$ for 24 hr, and (C) ethanol extracts of cells spun down from the frozen-thawed preparation. The numbers above each spectrum represent vertical expansion from normalized spectra. Dotted lines indicate chemical shifts of peaks of interest (A = alanine methyl, G = glutamate, CPP = 2,3-cyclopyrophosphoglycerate, and S = sugar anomeric carbons).

Assimilation of C Units

Because CPP is rapidly and intensely labeled by $^{13}CO_2$, acetate, and pyruvate, it is a good probe of reactions leading up to C_2 units.[23] CO_2 fixation by the activated acetic acid pathway is indicated in FIGURE 9. The acetate carboxyl carbon is derived from CO_2 via a cyanide-sensitive enzyme (CO dehydrogenase) and the methyl group is derived from part of the methanogenesis pathway. The latter step is inhibited by propyliodide. By using ^{13}C NMR and doubly labeled CPP precursors we can probe any scrambling of C_2 or C_3 units that occurs in *M. thermoautotrophicum,* $^1J_{cc}$ (the coupling constant between CHO and CH_2O of CPP) is diagnostic for integrity of a C_2 unit. If cells are grown on $^{12}CO_2/H_2$ with $^{13}CH_3{}^{13}COO^-$ added, the acetate will be incorporated into CPP. If the acetate unit remains intact, then the CHO and CH_2O carbons will each be a doublet of doublets. The large splitting is due to $^1J_{cc}$, the small splitting to $^2J_{cp}$ (since this is unaltered in these experiments, we will ignore the splitting due to ^{31}P). Degradation of the C_2 unit to C_1 units one or both of which can exchange with $^{12}CO_2$, and subsequent resynthesis of acetate will "scramble" the labeling pattern by introducing ^{12}C species next to ^{13}C. Such cleavage, exchange, and equilibration of the ^{13}C-acetate pool with $^{12}CO_2$ would produce a central singlet for each ^{13}C now next to a ^{12}C. Thus, any deviation from the ^{13}C doublet pattern will indicate scrambling. The relative intensity of the two CPP carbons will also provide information on specificity of the scrambling reactions.

Incorporation of ^{13}C into CPP peaks using $^{13}C_2$ acetate or pyruvate and $^{12}CO_2/H_2$ gives rise to a CPP spectrum, consistent with scrambling (FIG. 10). The CHO appears as a doublet flanking a small singlet (equivalent in intensity to the natural abundance background), while the CH_2O has more total ^{13}C intensity and appears as a large singlet flanked by the doublet. The large singlet is from $^{13}CH_2O$-^{12}CHO units. The result is predicted if the activated acetic acid scheme is operational. Bound ^{13}CO (generated by breakdown of $^{13}CH_3$ ^{13}COY to $^{13}CH_3X$ and [^{13}CO]) becomes exchangeable with the $^{12}CO_2$ pool, presumably via CO dehydrogenase. This reduces ^{13}C uptake into the CHO of CPP. [2,3-^{13}C]pyruvate exhibits the same behavior, only more ^{13}C is incorporated into CPP with pyruvate as the precursor. This is expected since pyruvate is energetically a closer intermediate to CPP than acetate.

KCN inhibits CO dehydrogenase,[24] the primary step in converting CO_2 to the carbonyl of acetate, and hence should prevent bound ^{13}CO (produced from the cleavage of acetate) from exchanging with $^{12}CO_2$. KCN does not inhibit methanogenesis; the latter continues to produce cellular ATP for further biosynthetic reactions. Cells incubated with [1,2-^{13}C]acetate, $^{12}CO_2$, and 50 μM KCN produce CPP with no scrambling of the carbon labels (FIG. 11C). Propyliodide (at concentrations around 50 μM) inhibits the formation of the methyl donor for the acetyl condensation reaction.[21] Therefore, propyliodide affects the other half of the C_2 condensation pathway. Added [1,2-^{13}C]acetate will be converted to CPP, but the exchange of [^{13}CO] with $^{12}CO_2$ will

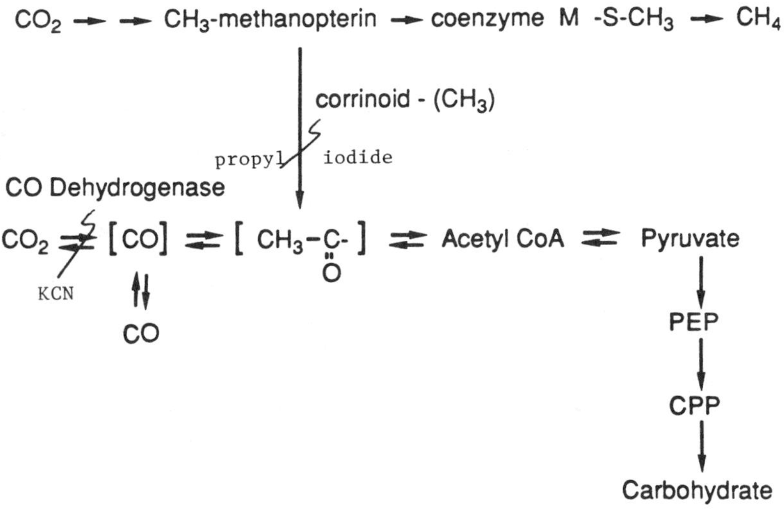

FIGURE 9. Activated acetic acid pathway proposed for methanogens.

not be stopped, hence a CH_2O singlet should be prominent. If the formation of CPP requires metabolic energy (ATP, ΔpH, etc.) then the total amount of ^{13}C incorporated will be reduced. This is in fact observed (FIG. 11D).

The same methodology can be used for examining the effect of other proposed metabolic inhibitors on the activated acetic acid pathway. For example, fluoropyruvate inhibits pyruvate synthetase *in vitro* (D. J. Livingston & C. T. Walsh, personal communication). If it has the same target *in vivo*, then a dramatic reduction in CPP

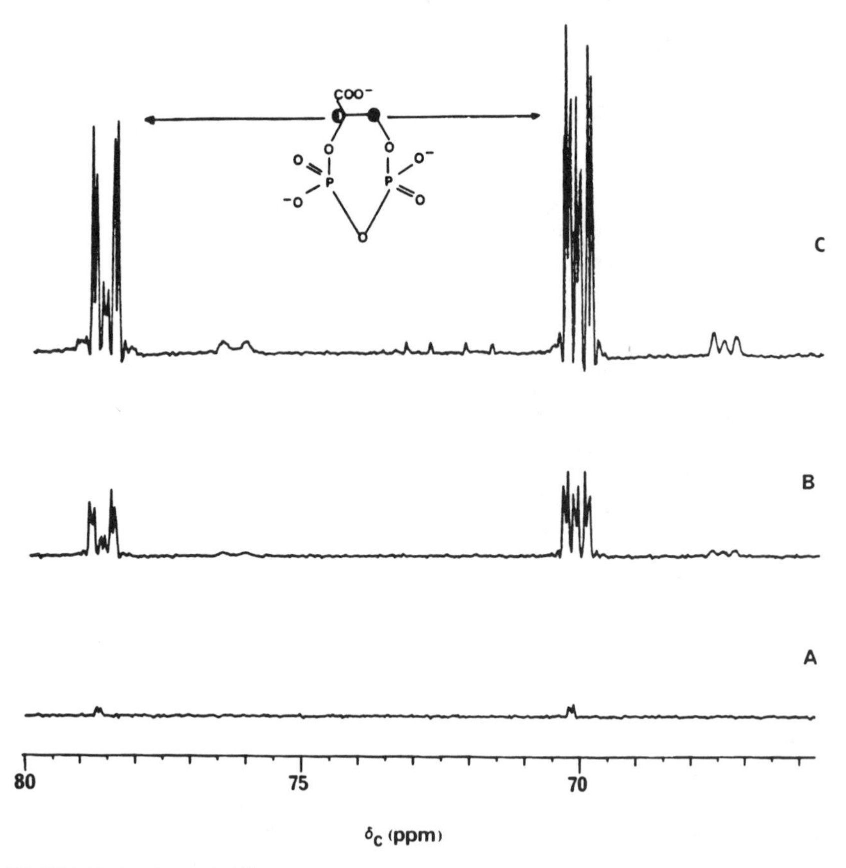

FIGURE 10. ^1H decoupled ^{13}C NMR spectra (100.6 MHz) of ethanol extracts of cells fed with (A) $^{12}CO_2/H$, (B) [1,2-^{13}C]acetate (20 mM) with $^{12}CO_2/H_2$, and (C) [2,3-^{13}C]pyruvate (10 mM) with $^{12}CO_2/H_2$. Spectral intensities are normalized.

intensity (both carbon resonances) should be observed along with a build-up of C_2 units. Cells exposed to 10 mM or 1 mM fluoropyruvate do not grow. The level of ^{13}C incorporated from $^{13}C_2$ acetate into CPP is somewhat reduced but the level of acetate is not increased (FIG. 11E). Furthermore, the scrambling of the CPP ^{13}C-^{13}C unit is enhanced much like what was observed with propyliodide. The turnover rate of CPP in a pulse-chase experiment also decreases,[23] suggesting that this metabolic inhibitor

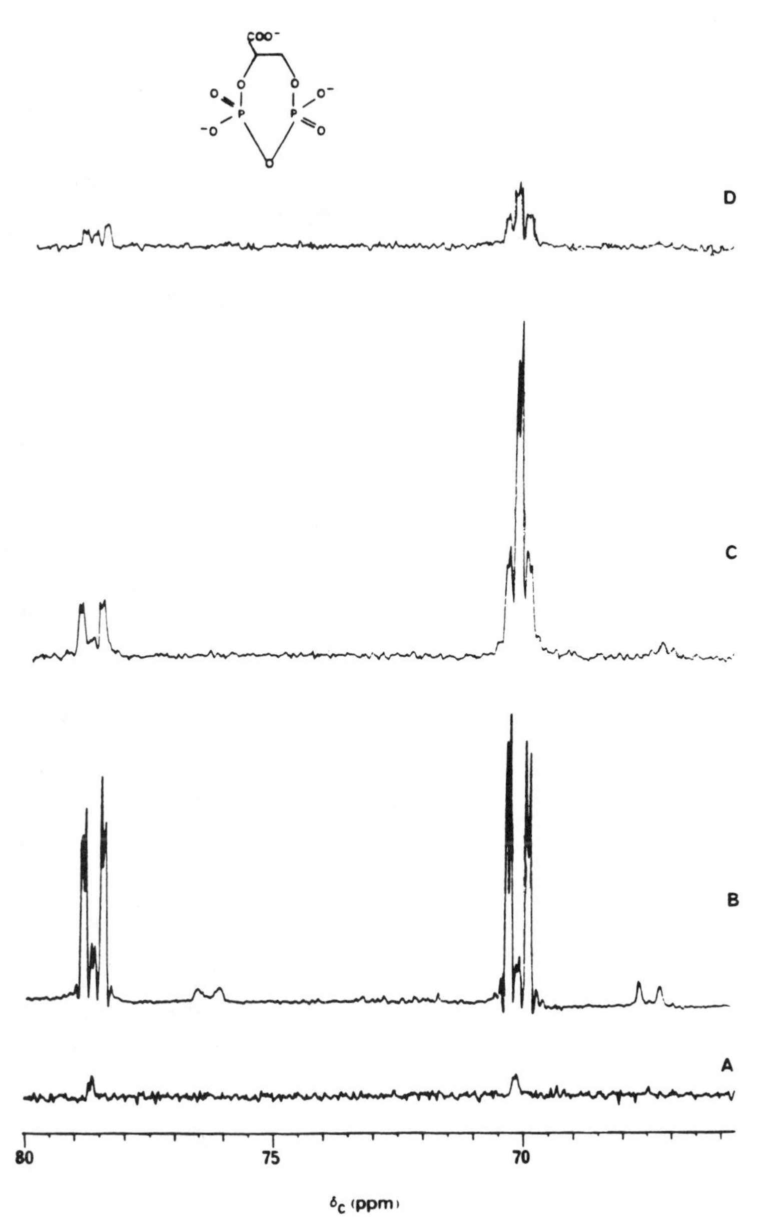

FIGURE 11. ^1H decoupled ^{13}C NMR spectra (100.6 MHz) of ethanol extracts of cells fed for 6 hr with [1,2-^{13}C]acetate (20 mM), $^{12}CO_2/H_2$ and (A) control, (B) KCN, (C) propyliodide, and (D) fluoropyruvate (1 mM).

must be affecting some step in methanogenesis and not inhibiting growth by inactivating pyruvate binding biosynthetic enzymes.

DISCUSSION

These studies of methanogens present a unique case where *in vivo* and *in vitro* NMR spectroscopy has discovered a totally new compound, monitored its dynamics (i.e., turnover) in the intact cell versus in extracts, and identified it as a key component in carbon assimilation from CO_2. The CPP structure is relatively unique—a seven-membered pyrophosphate diester that exists as a trianion at physiological pH. CPP has not been detected in other cells (archaebacteria or eubacteria) and appears to represent a new twist on the standard gluconeogenesis scheme.

^{13}C CPP is not a static reservoir for excess carbon or phosphorus storage in *M. thermoautotrophicum,* but is rapidly metabolized in growing cells. The label is quickly converted into an "insoluble" or particulate species. The use of CP-MASS ^{13}C NMR of ^{13}C depleted cells is the first application of this technique to a preparation of ^{13}C depleted cells. Because CO_2 is the sole carbon source, we can grow the organism on ^{12}C enriched media and more easily monitor the flux of a ^{13}C pulse from solution into macromolecules. The loss of ^{13}C intensity from CPP in solution correlates with the appearance of ^{13}C intensity in the region corresponding to carbohydrate resonances of the solid debris. Such a result connects CPP turnover with carbohydrate biosynthesis. Further evidence for this is provided by the steady-state labeling experiments with [6-^{13}C] glucose. The drop in ^{13}C labeling of CPP in stationary phase cells is also consistent with a role for this compound in carbon assimilation. Without the need for carbohydrate biosynthesis, CO_2 is fixed into pyruvate and then stored as alanine (an amino acid not at high concentrations in exponential phase cells) rather than CPP, and as α,α-trehalose.

Collapse of transmembrane ion or pH gradients (crucial parts of the bioenergetic machinery) in permeabilized cells leads to dramatically reduced CPP labeling. This is not because of low ATP, since these *in vitro* preparations label alanine and glutamate moderately well. For comparison, methanogenesis and acetyl CoA production drop 1,000-fold in frozen/thawed cells or in a French Press extract, and the addition of a variety of soluble compounds only increases this 30-fold at best. Again this suggests careful regulation of CPP enzymes compared to those for other carbon intermediates (e.g., amino acids).

While CPP has a role in carbon assimilation into carbohydrates, it also acts as a probe of other parts of carbon fixation in methanogens. The high concentration of CPP in exponential cultures of *M. thermoautotrophicum* fortuitously allows us to confirm the existence of the activated acetic acid pathway. Label from CPP is rapidly fixed into solids. Therefore, label incorporation into CPP is a probe of reactions leading up to PEP. In particular we can monitor $C_1 \rightarrow C_2 \rightarrow C_3$ reactions. The formation of acetyl CoA from two CO_2 should be partially reversible by CO dehydrogenase. Since this reversibility of one of the CO_2 fixation paths is easily detected in ^{13}C NMR spectra of CPP, we have a sensitive screen for the effect of metabolic inhibitors on this pathway. Future experiments will exploit this probe role for CPP.

An intriguing question is why do *M. thermoautotrophicum* cells maintain CPP at such high intracellular concentrations? It is not known how gluconeogenesis is regulated in this organism. Fuchs has shown that the enzymes that are usually key regulation points for carbohydrate metabolism in eubacteria are not stringently controlled in this methanogen. It is possible that CPP may serve as a regulatory pool for eventual generation of carbohydrate. We suggest that the enzymes that metabolize CPP are the points of control in gluconeogenesis in this methanogen.

REFERENCES

1. DANIELS, L., R. SPARLING & G. D. SPROTT. 1984. The bioenergetics of methanogenesis. Biochim. Biophys. Acta **768:** 113–163.
2. ZEIKUS, J. G. 1977. The biology of methanogenic bacteria. Bacteriol. Rev. **41:** 514–541.
3. CALVIN, M. 1962. The path of carbon in photosynthesis. Science **135:** 879–883.
4. EVANS, M. C. W., B. B. BUCHANAN & D. I. ARNON. 1966. A new ferredoxin-dependent carbon reduction cycle in a photosynthetic bacterium. Proc. Natl. Acad. Sci. USA **55:** 928–933.
5. RUHLEMANN, M., K. ZEIGLER, E. STUPPERICH & G. FUCHS. 1985. Detection of acetyl coenzyme A as an early CO_2 assimilation intermediate in *Methanobacterium.* Arch. Microbiol. **141:** 399–404.
6. LJUNGDAHL, L. & H. G. WOOD. 1982. Acetate biosynthesis. *In* Vitamin B_{12}. D. Dolphin, Ed.: 165–202. Academic Press. New York.
7. STUPPERICH, E. & G. FUCHS. 1984. Autotrophic synthesis of activated acetic acid from two CO_2 in *Methanobacterium thermoautotrophicum.* II. Evidence for different origins of acetate carbon atoms. Arch. Microbiol. **139:** 8–13.
8. HOLLIS, D. 1980. Biological Magnetic Resonance. L. Berliner, Ed. **2:** 1–44. Plenum Press. New York.
9. UGURBIL, K., R. G. SHULMAN & T. R. BROWN. 1979. High resolution P-31 and ^{13}C Nuclear Magnetic Resonance studies of *Escherichia coli* cells *in vivo. In* Biological Applications of Magnetic Resonance. R. G. Shulman, Ed.: 537–589. Academic Press. New York.
10. SHULMAN, R. G., T. R. BROWN, K. UGURBIL, S. OSAWA, S. M. COHEN & J. R. DEN HOLLANDER. 1970. Science **205:** 160–166.
11. BURT, C. T. & M. F. ROBERTS. 1984. P-31 NMR observation of less-expected phosphorus metabolites. *In* Biomedical Magnetic Resonance. T.L. James, Ed.: 231–242. Radiology Research and Education Foundation. San Francisco.
12. KANODIA, S. & M. F. ROBERTS. 1983. A novel cyclic pyrophosphate from *Methanobacterium thermoautotrophicum.* Proc. Natl. Acad. Sci. USA **80:** 5217–5221.
13. TOLMAN, C. J., S. KANODIA, L. DANIELS & M. F. ROBERTS. 1986. P-31 NMR spectra of methanogens: 2,3-cyclopyrophosphoglycerate is detected only in methanobacteria strains. Biochim. Biophys. Acta **886:** 345–352.
14. SEELY, R. J. & D. E. FAHRNEY. 1983. A novel diphospho-P,P-diester from *Methanobacterium thermoautotrophicum.* J. Biol. Chem. **258:** 10835–10838.
15. KANDLER, O. 1979. Zellwandstrukturen bei Methan-bakterien. Zur evolution der Prokaryonten. Naturwissenschaften **66:** 95–105.
16. JARRELL, K. F. & G. D. SPROTT. 1984. Can. J. Microbiol. **30:** 663–668.
17. EVANS, J. N. S., C. J. TOLMAN, S. KANODIA & M. F. ROBERTS. 1985. 2,3-Cyclopyrophosphoglycerate in methanogens: Evidence by ^{13}C NMR spectroscopy for a role in carbohydrate metabolism. Biochemistry **24:** 5693–5698.
18. JACOB, G. S., J. SCHAEFER & G. E. WILSON. JR. 1985. Solid-state ^{13}C and N-15 Nuclear Magnetic Resonance studies of alanine metabolism in *Aerococcus viridans (Gaffkya homari).* J. Biol. Chem. **260:** 2777–2781.
19. KONIG, H. & O. KANDLER. 1979. The amino acid sequence of the peptide moeity of the pseudomurein from *Methanobacterium thermoautotrophicum.* Arch. Microbiol. **121:** 271–275.
20. SOUZU, H. 1980. Studies on the damage to *Escherichia coli* cell membrane caused by different rates of freeze-thawing. Biochim. Biophys. Acta **603:** 13–26.
21. PFEFFER, P. E., K. M. VALENTINE & F. W. PARRISH. 1979. Deuterium-induced isotope shift ^{13}C NMR. 1. Resonance reassignments of mono- and disaccharides. J. Am. Chem. Soc. **101:** 1265–1274.

22. STUPPERICH, E. & G. FUCHS. 1984. Autotrophic synthesis of activated acetic acid from two CO_2 in *Methanobacterium thermoautotrophicum*. I. Properties of *in vitro* system. Arch. Microbiol. **13:** 8–13.
23. EVANS, J. N. S., C. J. TOLMAN & M. F. ROBERTS. 1986. Indirect observation by ^{13}C NMR spectroscopy of a novel CO_2 fixation pathway in methanogens. Science **231:** 488–491.
24. DANIELS, L., G. FUCHS, R. K. THAUER & J. G. ZEIKUS. 1977. Carbon monoxide oxidation by methanogenic bacteria. J. Bacteriol. **132:** 118–125.
25. KENEALY, W. & J. G. ZEIKUS. 1981. Influence of corrinoid antagonists on methanogen metabolism. J. Bacteriol. **146:** 133–140.

DISCUSSION OF THE PAPER

QUESTION: You've mentioned a number of times that methanogens need a very large amount of H_2 to grow and that makes it difficult to do NMR experiments. I'm curious where the H_2 comes in their natural environment.

M. F. ROBERTS: If you look in sludges and swamps, there are layers of these microorganisms with a lot of other aerobes and anaerobes above them. The latter take carbon sources and produce CO_2 and H_2, which get funneled to the methanogens. Then there are a series of methylotrophes that take the methane and convert it to CO_2, making a nice biological cycle. In fact that's one reason why you don't see a lot of usable methane generated in these environments.

R. G. SHULMAN: The cyclic compound looks so much like 2,3-diphosphoglycerate. There are bacterial spores with storage compounds such as 3-phosphoglycerate, which is derived from 2,3-DPG. Could you review your evidence against the cyclic compound being in equilibrium with the 2,3-DPG and acting as a storage compound? Have you varied external feeding and so on?

ROBERTS: If we have the cells growing optimally, and that means in the fermenter where we're bubbling through a lot of H_2, we don't detect 2,3-DPG in ^{13}C NMR experiments. Detectable 2,3-DPG appears as an artifact when we stress the cells, when we limit hydrogen concentration, for example. Certainly the cyclic compound is not generated from 2,3-DPG. In the labeling and pulse/chase experiments, if you look carefully, label flows from the cyclic compound into 2,3-DPG. Now this may be a regulatory path for the cells to control gluconeogenesis by limiting the amount of 2,3-DPG and hence compounds, which may be derived from it. We don't know what the full story is yet.

SHULMAN: It wasn't clear from the one slide you showed, which showed peaks of 2,3-DPG, that the label was flowing from the cyclic compound to 2,3. It looked to me like it was flowing from 2,3 into the cyclic compound, but. . . .

ROBERTS: No, you have to look at many more spectra (I have lots and lots of other spectra . . .). That's the order of decay—the cyclic phosphate first, then 2,3-DPG. Also in *in vitro* experiments we've used 2,3-DPG to see if we can get any biosynthesis of the cyclic compound, and it sits there inertly, even if we add ATP or ADP (or other high energy phosphates). If we use PEP and ATP or ADP, we can actually get some biosynthesis of this compound. Cell extract experiments are very messy for lots of different reasons and we don't get optimal activity (that's an understatement). Yet we know we can transform PEP into this compound, while 2,3-DPG under comparable conditions will not easily convert into the cyclic structure.

Intracellular pH as Measured by ^{19}F NMR[a]

CAROL J. DEUTSCH[b] AND JUNE S. TAYLOR[c]

[b]Department of Physiology and
[c]Department of Biochemistry and Biophysics
School of Medicine
University of Pennsylvania
Philadelphia, Pennsylvania 19104-6085

INTRODUCTION

Intracellular and organelle pH play an important role in cell function, growth, and development.[1-3] Clearly our understanding of the role of pH in such processes is limited by our ability to determine intracellular pH accurately. This led some of us to undertake the development of better methods for measuring pH in cells and organelles.

^{31}P NMR spectroscopy has been widely used to measure intracellular pH in the decade since Moon and Richards[4] reported the first measurement of intracellular pH by ^{31}P NMR. When ^{31}P NMR resonances of endogenous phosphate compounds can be used as pH indicators, they provide a noninvasive means of determining intracellular pH, with concomitant quantitative measurements of ATP and other phosphorylated metabolites. The most reliable endogenous ^{31}P pH indicators are inorganic phosphate and sugar phosphates. However, resonances of endogenous phosphates are often obscured by overlapping resonances (including phosphates in perfusing or suspending media) or are absent because the indicators are present at concentrations below the limit of accurate detection in reasonable data-accumulation times. Human peripheral blood lymphocytes (PBLs), Ehrlich ascites tumor cells, 3T3 cells, and gut and fundus smooth muscle are in this latter category.

When endogenous indicators cannot detect pH differences of interest, exogenous indicators can be supplied. One is then not limited to ^{31}P NMR. The choice of nucleus greatly influences the sensitivity and resolving power of NMR measurements. The obvious NMR nucleus to choose is ^{19}F:^{19}F has a nuclear spin of 1/2, a natural abundance of 100%, and a high detection sensitivity (83% that of ^{1}H). There are no problems with dynamic range or endogenous ^{19}F signals, because biological systems do not normally contain fluorine. ^{19}F has a large chemical shift range and corresponding large changes in shift with protonation of nearby groups. Finally, the F-C bond can be very stable in biological systems. For these reasons, we have developed and tested a set of sensitive, nontoxic ^{19}F pH indicators for measuring cytosolic pH that are applicable to a broad range of problems and easily accessible to interested investigators. The principle is the same as that for ^{31}P determination of pH, that is, the chemical shift of the ^{19}F NMR line is strongly dependent on whether a nearby acidic group is protonated.

[a]Supported primarily by the American Heart Association and National Institutes of Health Grant GM-36433.

DEVELOPMENT OF ^{19}F NMR pH INDICATORS

An ideal NMR pH indicator has the following properties: (1) a large chemical shift difference between conjugate acid and base forms; (2) an exchange rate across biological membranes that is sufficiently slow to meet the slow-exchange criterion; (3) a pK_a that is known accurately for intracellular media and within 1 pH unit of the pH to be measured; (4) no cytotoxicity; and (5) metabolically stable and does not perturb the physiological state of the cells.

In 1981, we published pH measurements at 4°C with the ^{19}F pH indicator, $CF_3CH_2NH_2$ (trifluoroethylamine),[5] which has a shift of 1 ppm per pH unit near its pK_a of 5.7. However, at 25°C its exchange rate across cell membranes is sufficiently rapid to average inside and outside ^{19}F indicator peaks. In order to retain the large chemical-shift dependence on pH of the F-C-C-NH_2 structure, while achieving the desired slower exchange across cell membranes, we turned to more polar molecules, specifically mono-, di-, and trifluoro-α-methylalanines (α-CH_2F-ala, α-CHF_2-ala, and α-CF_3-ala) (FIG. 1). We also reasoned that amino acids are actively transported in most cell systems so that this class of compounds would be good candidates for exogenous pH indicators. The presence of the α-methyl group greatly suppresses the

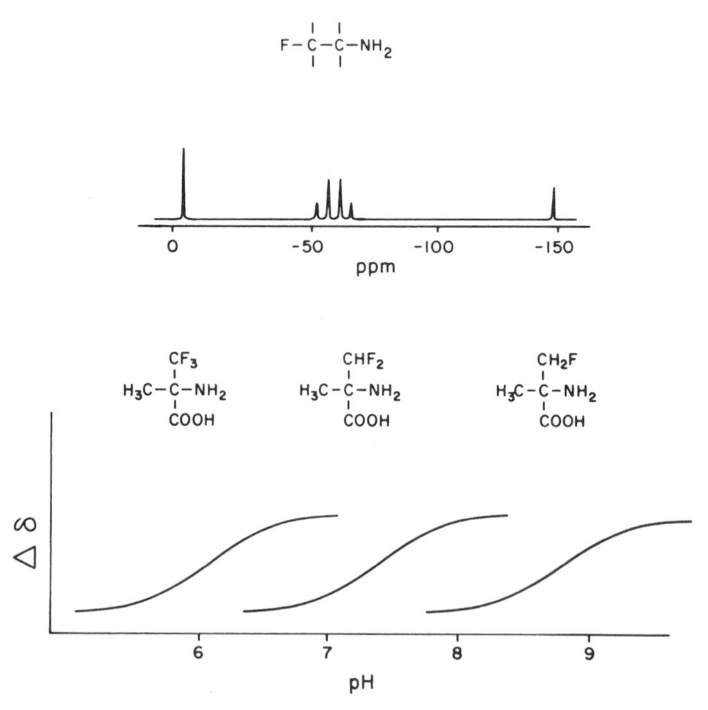

FIGURE 1. ^{19}F NMR resonance frequencies and pH dependence of the chemical shifts (center peak spacing for the difluoromethyl alanine) for trifluoro-α-methylalanine, difluoro-α-methylalanine, and monofluoro-α-methylalanine. The structures of each of these molecules is shown with its corresponding resonance frequency (above) and titration curve (below).

TABLE 1. pK_a and Maximal pH Shifts for Fluorinated Compounds

	pK_a	$\Delta\delta$/pH Unit[a]
α-Monofluoromethylalanine	8.5	1.1
α-Difluoromethylalanine	7.3	0.9
α-Trifluoromethylalanine	5.9	1.2
α-Monofluoromethylalanine methyl ester	6.4	0.8
α-Difluoromethylalanine methyl ester	5.1	1.4
α-Trifluoromethylalanine methyl ester	4	
α-Trifluoroethylalanine	8.0	0.5
N-Methyl α-trifluoromethylalanine	6.0	1.0
α-Monofluorovaline	8.3	1.3
γ-Hexafluorovaline	6.3	0.6
α-Monofluoromethylhistidine	7.7	1.1
α-Monofluoroalanyl-α-monofluoroalanine	6.54	0.35
Fluorophosphate	4.7	0.8

[a] $\Delta\delta$ = change in chemical shift, in ppm. For the difluoromethylalanines, $\Delta\delta$ = change in spacing of the two center peaks of the ^{19}F AB spectrum, in ppm. pK_a and δ values were determined from titration curves at 25°C, 0.5–1 mM indicator; chemical shifts were referenced to trifluoroacetate.

metabolism of amino acids; transaminases do not act on α-methyl amino acids. These compounds have been extensively studied both *in vivo* and *in vitro*. The results show no evidence of any cytotoxicity. In fact, Christensen and Oxender[5] reported no toxicity in mice receiving single injections of α-CF_3-ala or α-CH_2F-ala at 500 mg/kg body weight, and found that over 60% of α-CF_3-ala and 40% of α-CH_2F-ala were excreted unchanged within six hours.

These fluorine-labeled alanines indicate pH by the dependence of the chemical shift of their ^{19}F resonance on protonation of the α-amino group. The α-amino pK_as are approximately 8.5, 7.3, and 5.9, for the α-CH_2F-ala, α-CHF_2-ala, and α-CF_3-ala, respectively. The resonances of these three compounds lie in very different frequency regions, so that α-CH_2F-ala plus α-CHF_2-ala or α-CHF_2-ala plus α-CF_3-ala can be used simultaneously in the same cell suspension to cover a wide range of pH. The observed T_1s of the fluorinated α-methylalanines at 25°C in physiological media range from 1.2 sec for α-CH_2F-ala to greater than 3.0 sec for the di- and trifluoro compounds at 188.2 MHz.

The ^{19}F NMR proton-decoupled spectrum of α-CHF_2-ala is an AB-type spectrum. The center-peak spacing between the center two resonances of the two nonequivalent fluorine resonances is pH dependent, so that this compound has the advantage of not requiring a separate internal standard. The ^{19}F NMR spectra of α-CF_3-ala and of proton-decoupled α-CH_2F-ala consist of single resonance lines.

If the fluorine substituent is moved further from the amino group, the maximal pH-dependent chemical shift is decreased and the pK_a is higher, as shown in TABLE 1. For example, α-trifluoroethylalanine has a shift of 0.5 ppm per pH unit and its amine pK_a is 8.0, compared to a shift of 1.2 ppm per pH unit and a pK_a of 5.9 for the α-CF_3-ala. Hexafluorovaline shows a maximal change in the spacing of its two nonequivalent trifluoromethyl resonances of 0.6 ppm per pH unit. These results indicate that fluorine(s) should be located on the carbon β to the NH_3^+ group in aliphatic molecules, for optimum pH sensitivity.

We found that the fluorinated α-methyl-alanines were the most useful of the molecules in TABLE 1. They have appropriate chemical-shift dependence on pH, exchange rates, pK_as, metabolic stability, plus no cytotoxic effects. Titrations in a variety of ionic media showed that the primary modifiers of pK_a are temperature and K^+ concentrations below 130 mM K^+ (property 3).[16]

THE NEXT GENERATION OF ^{19}F pH INDICATORS

We have recently investigated aromatic fluorinated compounds, in which the fluorine is directly bound to the aromatic ring. We hoped to exploit the resonance interaction between NH_2 and *ortho* or *para* fluorine substituents on the ring. This proved possible for a simple, available class of such aromatic fluorinated amines, the fluoroanilines (FIG. 2). *Ortho* and *para* substitutions give the largest chemical shifts per pH unit, while the *meta* substitution is substantially less; the pH sensitivity of these *o*- and *p*-substituted anilines is at least an order of magnitude larger than the fluoromethylalanines. As shown in TABLE 2, the pK_a of fluoroanilines are in the low acid range, which makes them inappropriate for use in cells or tissues in the physiological range, but useful for lysosomal pH determinations or for gastric cell or organ pH measurements. However, we have been modifying these compounds to increase the pK_a. Our goal is to develop *p*-fluoroaniline analogs of alanine, represented here by N-(2-carboxy-isopropyl)-4-fluoroaniline (pFaa), to measure cytosolic and compartmental pH. The ^{19}F spectrum of this compound is a singlet about 50 ppm upfield from trifluoroacetate; proton decoupling is not necessary and the chemical shift with pH is large. The methyl ester of this compound is taken up and cleaved by human lymphocytes to the indicator shown here. This indicator was stable in cell suspensions for several hours and did not inhibit stimulated DNA replication.

We were prompted to look at fluoroaniline analogs by a recent report from Metcalfe *et al.*,[7] who reported that the fluorinated analog of quene 1, a trans-analog of the calcium indicator quin 2,[14] could be used to measure pH by fluorescence or by

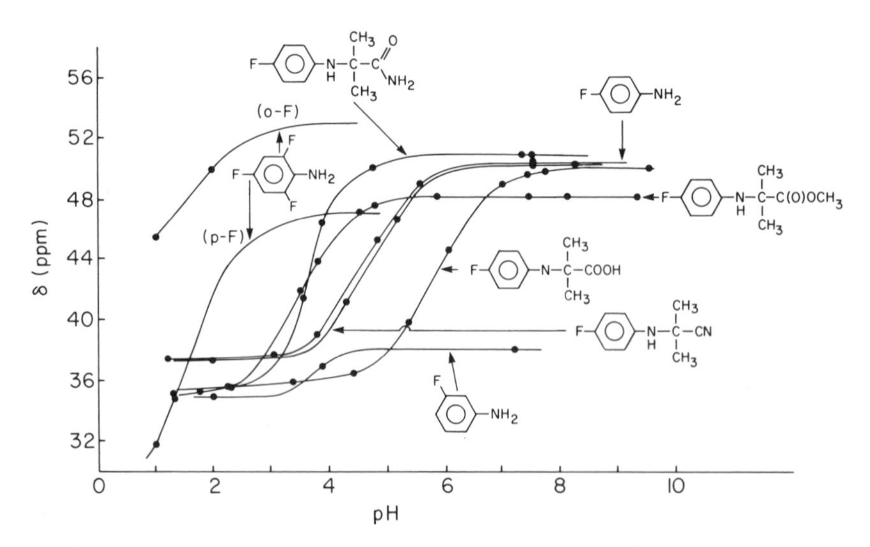

FIGURE 2. pH dependence of chemical shifts of fluorinated anilines and *p*-fluoroaniline derivatives. Chemical shifts were determined at 25°C in Hanks-HEPES (20 mM HEPES, Ca-Mg free) medium for aniline concentrations between 1–5 mM.

TABLE 2. pK_a and Maximal Shifts for Fluorinated Anilines

	pK_a	$\Delta\delta$/pH Unit[a]	Total Shift (ppm)[b]
3-Fluoroaniline	3.5	2.3	3.2
4-Fluoroaniline	4.6	6.0	13.1
2,2,4-Trifluoroaniline	2	*ortho:* 5.0	15
		para: 10.0	15
N-(2-cyano-isopropyl)-4-fluoroaniline	4.75	6.6	13.1
N-(2-amido-isopropyl)-4-fluoroaniline	3.3	6.2	15.8
N-(2-methylcarboxylate-isopropyl)-4-fluoroaniline	3.5	6.8	13.2
N-(2-carboxy-isopropyl)-4-fluoroaniline	5.8	7.0	14.3

[a]Over linear portion of titration curve. $\Delta\delta$ = change in chemical shift, in ppm.
[b]Total change in ppm between conjugate acid and base forms. pK_a and δ values were determined from titration curves at 25°C, 1 mM indicator; chemical shifts were referenced to trifluoroacetate.

NMR spectroscopy. We wondered how much of this molecule was actually necessary for the large chemical shift with pH (4–5 ppm per pH unit). Was the extended conjugation necessary? We found that a simpler system with significant electronic resonance, fluoroaniline, was sufficient for most of the large chemical shift with pH.

There are other promising possibilities: Brown and co-workers,[8] pursuing the work of Chung and Graves[9] on fluorinated pyridoxals, are investigating these compounds as potential ^{19}F pH indicators. The chemical shifts of these compounds are very sensitive to pH, and the pK_as are in the alkaline range. We would also like to cite the work of Korytnyk and Srivastava,[10] who synthesized some of these vitamin B6 analogs some time ago, for different reasons. Gillies[11] is developing a prototype hydantoin structure, a 5,5-dimethyloxazolidine-2,4-dione (DMO) analog, which should also be a sensitive ^{19}F pH indicator.

It is instructive (FIG. 3) to consider some compounds that are not suitable indicators or precursors. The first group does not have the first property of an NMR indicator. We considered 2-F- and 4-F-substituted glucose, since these compounds are known to be taken up and phosphorylated in the 6 position. Our hope was to generate a non-metabolizable indicator with a phosphate-ionizing group. The fluoroglucoses did not have pH-sensitive ^{19}F chemical shifts. However, this is still an attractive idea; these compounds might serve as ^{31}P pH indicators in some systems. The remaining fluorinated amino acids in the first group of FIGURE 3 likewise failed to show pH-sensitive ^{19}F chemical shifts. The second example fails to meet the second criterion of an indicator; it is quite pH sensitive, but exchanges too rapidly across biological membranes, and thus averages signals from two compartments. The last category is dipeptides. The one on the left was not cleaved in the biological systems that we have tested so far (*P. denitrificans,* RINm5F cells). The dipeptides of monofluoroalanine were cleaved to monofluoroalanine, which was rapidly metabolized to pH-insensitive products. However, the use of peptide precursors is still a viable approach to generating sensitive, accurate pH probes *in situ.*

We would also like to indicate some future strategies, in the hope that these strategies will stimulate others to devise probes for their own particular needs. As described above, aromatic fluorinated compounds promise an increase of approximately an order of magnitude in chemical-shift sensitivity to pH. Opella and Reed[12] have suggested the attachment of pH-sensitive fluorinated moieties to a solid support, such as microspheres, to restrict the indicators to extracellular spaces, obviating absorption and possible accumulation and/or toxicity in measurements of extracellular pH in isolated tissues and *in vivo.*

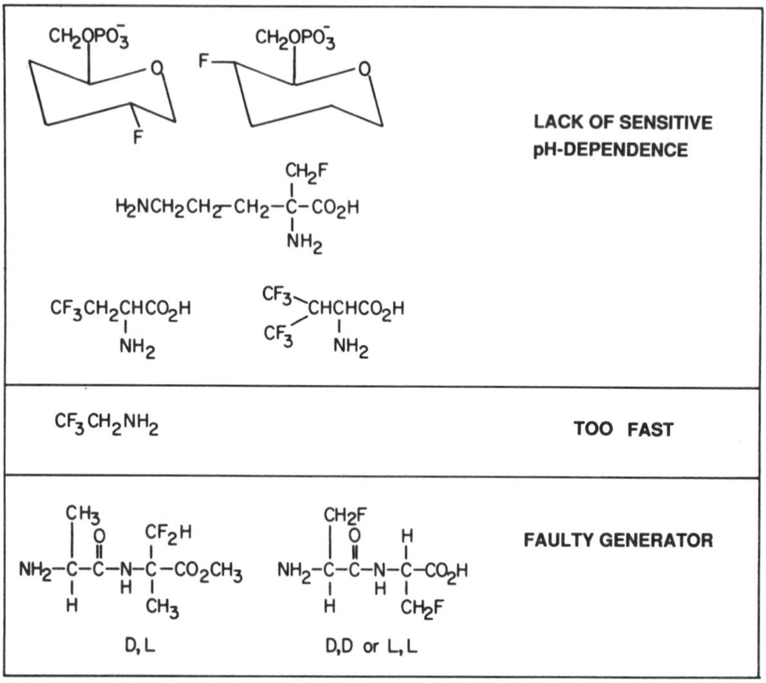

Compounds	
$CH_2OPO_3^-$ (ring with O, F) ; $CH_2OPO_3^-$ (ring with O, F) ; $H_2NCH_2CH_2{-}CH_2{-}C(CH_2F)(NH_2){-}CO_2H$; $CF_3CH_2CH(NH_2)CO_2H$; $(CF_3)_2CHCH(NH_2)CO_2H$	LACK OF SENSITIVE pH-DEPENDENCE
$CF_3CH_2NH_2$	TOO FAST
$NH_2{-}C(CH_3)(H){-}C(=O){-}N(H){-}C(CF_2H)(CH_3){-}CO_2CH_3$ (D,L) ; $NH_2{-}C(CH_2F)(H){-}C(=O){-}N(H){-}C(H)(CH_2F){-}CO_2H$ (D,D or L,L)	FAULTY GENERATOR

FIGURE 3. Compounds that failed as ^{19}F pH indicators or precursors.

We also point out the advantages of incorporating a second fluorine or trifluoromethyl group in the indicator molecule: first, to serve as an internal chemical shift reference (as we have done with α-CHF_2-ala); and second, in some cases, to provide additional pK_a lowering effect (cf. hexafluorovaline). Thus, the fluorine substitution site proximal to the ionized group serves as the pH-sensitive indicating resonance; the second fluorine substitution is several bonds away from the first F and is thus insensitive to pH.

LOADING CELLS WITH EXOGENOUS INDICATORS

With exogenous indicators, there is always the potential problem of getting sufficient amounts of indicator into cells and tissues. Transport of the fluorinated α-methylalanine indicators into cells was too slow to give suitable intracellular indicator concentrations in reasonable times in almost all cells and tissues we studied. We therefore used esters of the fluorinated amino acids as precursors to obtain increased rates of transport into cells and tissues. The less polar esters rapidly diffuse across cell membranes, and the intracellular ester can be cleaved by esterases to generate the desired indicator molecule (the free fluorinated α-methyl amino acid) and free alcohol or phenol. Ester precursors lead to increased intracellular accumulation of indicator, due to the group transport effect.

The methyl esters of α-CH_2F-ala and α-CF_3-ala were the first series of precursor molecules we prepared. The pK_as of the methyl esters (FIG. 4 and TABLE 1) are all less than 6.5, so that the esters are almost completely in the conjugate base form ($-NH_2$) around neutrality. A single ester peak or quartet is observed, representing intracellular plus extracellular ester. The ester resonance lines are separated by 0.6–1.5 ppm from the free acid resonances, so that the esters and acids can be monitored simultaneously. When fluorinated α-methylalanines are incubated with human peripheral blood lymphocytes at concentrations of 1–10 mM, 1–3 hours are required in order to obtain observable intracellular levels. However, the methyl esters of the fluoromethylalanines permeate human lymphocytes and are hydrolyzed to give 1 mM intracellular levels of the free acids within 3–10 minutes at 25°C. The fluoromethylalanines are in fact concentrated in the cells 10–20-fold over extracellular levels by this means.[13] These methyl esters are successful precursors for lymphocytes, hepatocytes, frog skin, and rabbit colon slices (see below).

Methyl esters of fluoroalanines are not hydrolyzed by *P. denitrificans, E. coli,* or RIN m5F insulinoma cells. Cells lacking esterase activity for the methyl esters of CHF_2-ala nevertheless cleave esters with aromatic or larger aliphatic groups esterified to the carboxyl. We explored other precursor molecules, some of which may be useful for applications in particular systems (FIG. 5). Di-, tri-, or tetrapeptides may be useful in cases where peptidase activity is present. More reactive esters may be useful in cases where the less reactive methyl ester is inappropriate. We have recently tested the *p*-chlorophenyl ester of α-CHF_2-ala and found that (1) the ester pK_a is below 5.5, (2) it is hydrolyzed in aqueous salt (300 mOsm) solution with a half-life of 15–30 min at pH 6.5–7.5, and (3) it is transported into cells and is cleaved *in situ.* In RIN m5F cells (see below) and *P. denitrificans,* α-CHF_2-ala *p*-chlorophenyl ester was taken up and cleaved inside the cells at a rate far exceeding the extracellular hydrolysis rate for this ester. Initially, the intracellular α-CHF_2-ala concentration was at least twenty times greater than the extracellular concentration in both cells. Thiol esters are more reactive than oxy esters, so we have studied the benzylthio ester of alanine. The advantage in this case is that a number of cell types and tissues can be screened for esterase activity using simple optical techniques to monitor the generation of benzylthiol. Other labile esters, such as acetoxy, may be worthwhile candidates.[14] Variation of the ester group has allowed us to generate the fluorinated α-methylalanines *in situ* in bacteria and other cells that lack appropriate methyl esterase activity and may enable

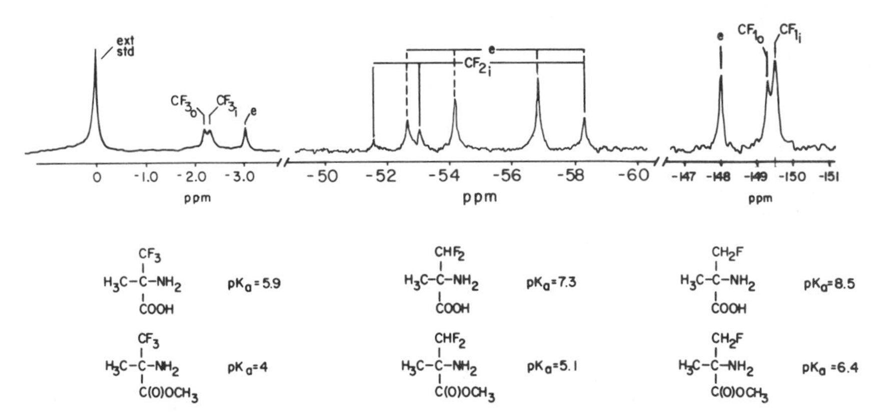

FIGURE 4. ^{19}F NMR spectra and pK_as for (left) trifluoro-α-methylalanine and its methyl ester (e), (center) difluoro-α-methylalanine and its methyl ester (e); and (right) monofluoro-α-methylalanine and its methyl ester (e). Spectra were acquired at 25°C, in Hanks-HEPES (20 mM HEPES, Ca-Mg free) medium.

1 $H_2N-C(CH_3)(H)-C(=O)-NH-C(F_2CH)(CH_3)-CO_2CH_3$

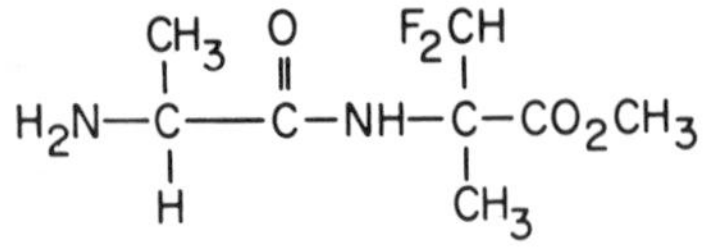

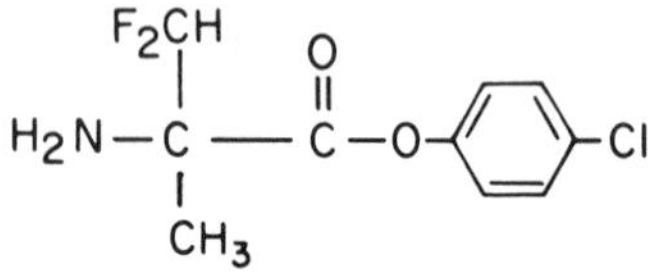

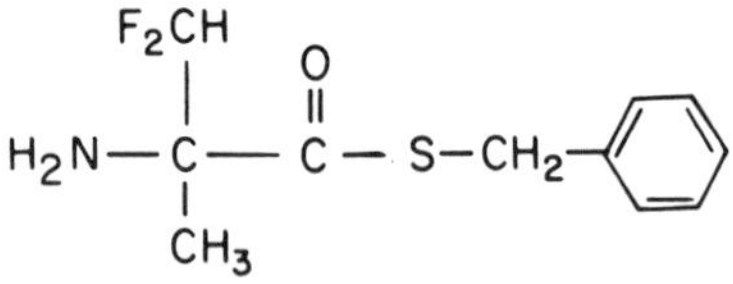

2 $H_2N-C(F_2CH)(CH_3)-C(=O)-O-C_6H_4-Cl$

3 $H_2N-C(F_2CH)(CH_3)-C(=O)-S-CH_2-C_6H_5$

4 $H_2N-C(F_2CH)(CH_3)-C(=O)-OCH_2OCOCH_3$

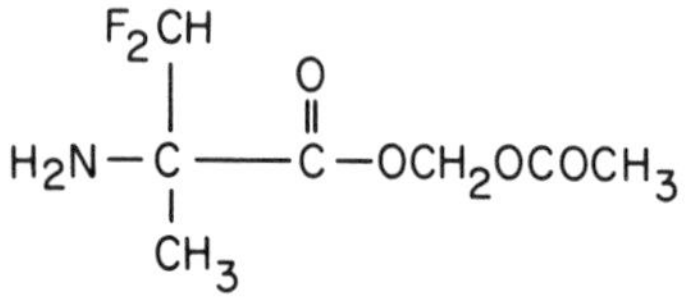

FIGURE 5. Indicator precursors.

us to aim the pH probes at specific cellular compartments or at certain cell types in tissues.

APPLICATIONS TO BIOLOGICAL SYSTEMS

We have successfully determined intracellular pH in a variety of eukaryotic cell suspensions and several tissue preparations with easily available NMR instrumentation and a standard commercial 10 mm ^{19}F probe with proton decoupler coil.[13] Our cell suspensions were run in a simple flow system with external oxygenation chamber, and tissue samples were rolled on fine plastic mesh in a tube sealed to a simple perfusing system. With cell suspensions of 10–20% cytocrit for small cells and 2–5% cytocrit for large cells, signal-to-noise ratios (S/N) of 8–10 can be obtained in 3 to 6 min on the equipment described above. Measurements on cytocrits of 1–5% are possible in longer times. With optimum probe and receiver design and higher fields, a 4–6-fold improvement in S/N could conservatively be expected. This would allow measurements on cell suspensions of $\leq$1% cytocrit, or, alternatively, a time resolution of 6–10 sec per spectrum.

It is necessary for each biological system to verify that the cells or tissues have been maintained in proper physiological condition during the NMR measurements. We have tested the physiological condition of cells and tissues exposed to the conditions of the NMR experiments and flow system and find that cells from the NMR sample at the end of experiments are indistinguishable from healthy control cells in viability as measured by trypan-blue exclusion (for hepatocytes,[15] PBLs,[16] and insulinoma cells[13]); in stimulated DNA synthesis (for PBLs);[16] in glucose and urea synthesis rates and [ATP]/[ADP] ratios (for hepatocytes); and for frog skin,[17] in electrophysiological properties.

Human Peripheral Blood Lymphocytes

For determining the intracellular pH of lymphocytes in media covering a wide range of extracellular pH, α-CHF_2-ala was used in combination with α-CF_3-ala or α-CH_2F-ala. Experiments in which α-CHF_2- and α-CF_3-ala methyl esters were added simultaneously to the same lymphocyte suspension showed that intracellular pH values obtained from the two probes in the same cell suspension agreed to 0.1 pH unit. When used simultaneously, α-CH_2F-ala and α-CHF_2-ala gave intracellular pH values that agreed within 0.07 pH units. FIGURE 6 shows the intracellular pH at various extracellular pHs as determined by α-CH_2F-ala, α-CHF_2-ala, and α-CF_3-ala. Any two of the fluorinated probes allow the construction of a reliable profile of intracellular pH over a 2 pH-unit range of extracellular pH. The precision of these pH measurements was $\pm$0.06 pH unit, for data such as those in FIGURE 6, and is limited by the precision of determining the chemical shifts. All three probes gave results consistent with those reported previously for lymphocytes. In addition, the intracellular pH calculated from equilibrium distributions of the radioactive weak acid DMO is included in FIGURE 6 and this shows reasonable agreement with the ^{19}F NMR results.

As shown in FIGURE 6, quiescent lymphocytes are able to maintain intracellular pH

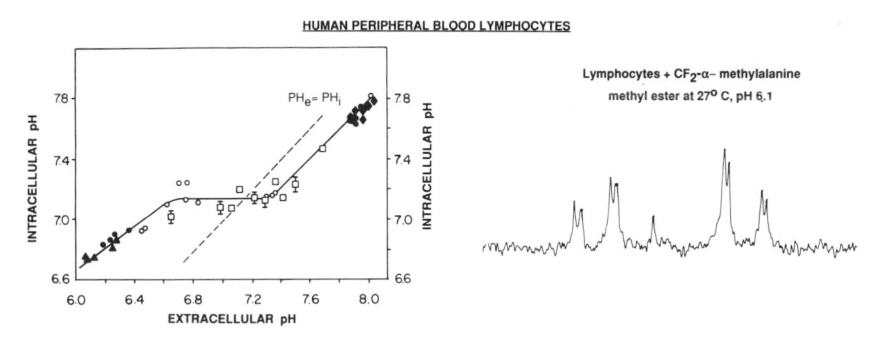

FIGURE 6. (Left) Intracellular pH versus extracellular pH in resting human peripheral blood lymphocytes. Data points are taken from individual ^{19}F NMR experiments using monofluoromethylalanine methyl ester (♦, 1 mM), difluoromethylalanine methyl ester (○, ●, 1 mM), or trifluoromethylalanine methyl ester, (▲, 0.5 mM). The filled symbols represent NMR determinations made with two probes simultaneously present in the same sample. DMO (7–14 μM, □) measurements were carried out as described previously. Error bars represent mean values ± SD; these DMO experiments were carried out on suspensions at 2×10^6 cells/cm^3. Cell water and trapped extracellular water were assayed simultaneously. Temperature ranged from 24–25°C. (Right) 188.2 MHz ^{19}F NMR spectrum of human PBLs +0.85 mM difluoromethylalanine methyl ester in Hanks-HEPES medium (20 mM HEPES, Ca-Mg free) plus 10% D_2O. Temperature, 27°C; medium pH, 6.1.

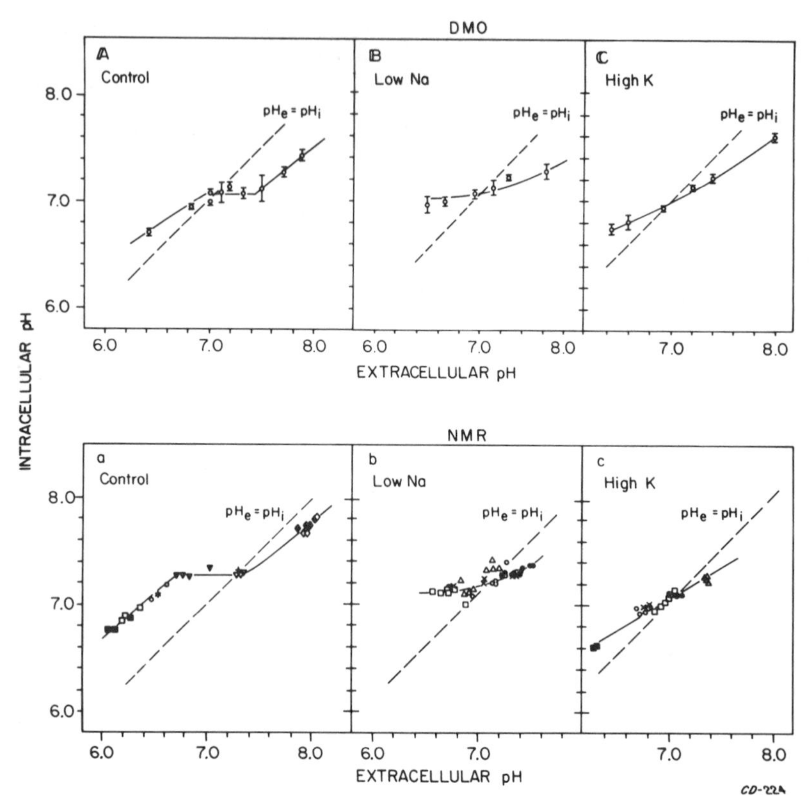

FIGURE 7. Intracellular pH as a function of extracellular pH in resting human peripheral blood lymphocytes. A, B, and C are DMO experiments. The procedure for the determination of pH_i was identical to that described in Cohen *et al.*[20] Values are given as means ± SD for triplicate samples. The bottom panel (a–c) shows ^{19}F NMR pH measurements at 25°C, 188.2 MHz. Lymphocytes were suspended at 20–25% cytocrit in the indicated media, which also contained 12% D_2O, 1–3 mM trifluoroacetate (internal standard), and 0.3–1 mM ^{19}F pH indicator (α-difluoro- or α-trifluoro-methyl alanine methyl ester). Cells were oxygenated as described in Deutsch *et al.*[22] Different symbols represent independent experiments.

at 7.0–7.2 over an extracellular pH range of 6.9–7.4. Secondly, the observed pH gradient changes sign at pH 7.1; the gradient is acid inside with respect to outside for pH_es greater than 7.1, and is alkaline inside with respect to outside for pH_es less than 7.1. Thirdly, neither protons nor hydroxyl ions are in electrochemical equilibrium anywhere in the regulated region, for the measured value of the membrane potential. This regulation requires ATP and it is modified by the ion content of the extracellular medium, as shown in FIGURE 7, and by mitogen. We have also shown that mitogen stimulation leads to little or no change in intracellular pH in stimulated lymphocytes in the first few hours of culture, and that intracellular pH decreased, rather than increased, over the subsequent days in culture.[22] Therefore a shift in intracellular pH is not a necessary or general concomitant of mitogen-stimulated proliferation in lymphocytes.

Isolated Rat Hepatocytes

In hepatocytes, as shown in FIGURE 8, the esters of α-CHF_2-ala and α-CF_3-ala were taken up rapidly and cleaved within minutes to the free acid by endogenous esterases. These studies were carried out over a range of external pH (pH 6–8.2) by varying medium [HCO_3] for a constant 5% CO_2/95% O_2. The intracellular pH of hepatocytes was a nearly linear function of external pH with a slope of 0.45 and $pH_i = pH_e$ at 7.09. The pH_i agrees with a previous measurement on hepatocytes.[18] Other observations on perfused liver[19,20] also show that H^+/OH^- are not in electrochemical equilibrium in this system. With these pH_i data in hand, we were able to determine the (actual) pH dependence of glucose and urea synthesis in hepatocytes and to establish that ΔG for ATP hydrolysis remains essentially constant at -12 Kcal/mol, changing only 0.6% as pH_i goes from 6.7 to 7.5.[15]

Cultured Insulinoma Cells

The RIN m5F cells are insulin-secreting cells derived from a radiation-induced insulinoma from rat pancreas. Unlike PBLs and hepatocytes, these cells do not have detectable methyl esterase activity for methyl esters of fluorinated methylalanines; no free α-CHF_2-ala was observed after approximately one hour incubation of cells with

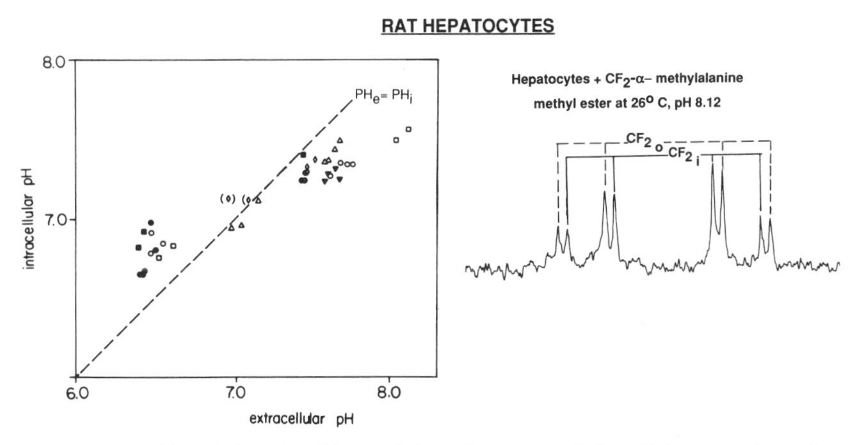

FIGURE 8. (Left) Relationship of intracellular pH to extracellular pH in suspensions of rat hepatocytes. Data points are taken from individual ^{19}F NMR experiments using difluoromethylalanine methyl ester (0.9 mM). Temperature was 25.5–26°C. Linear regression analysis gave a best fit to a straight line with a slope 0.45 and $pH_e = pH_i$ at 7.09. The regression coefficient was 0.96. Different symbols denote different hepatocyte preparations. (Right) 188.2 MHz ^{19}F spectrum of rat hepatocytes treated with α-difluoromethylalanine methyl ester. Hepatocytes were suspended in Krebs-Henseleit saline containing 5 mM NH_4Cl, 2 mM lactate, 2 mM ornithine, 1 mM oleate, and 2% bovine serum albumin, initially at pH 8.46, at 76–87 mg wet weight/ml, with 0.9 mM α-difluoromethylalanine methyl ester, 11% D_2O. Spectra were taken at 5-min intervals (1,000 scans, 4.0 μsec pulse width, 0.35 sec acquisition time). The resonances marked "o" arise from extracellular amino acid while those marked "i" arise from intracellular amino acid. The cps for the intracellular amino acid = 3.62 ppm, corresponding to pH_i = 7.58. The cps for the extracellular amino acid = 3.09, corresponding to pH_e = 8.12.

α-CHF_2-ala methyl ester at 25°C. However, the *p*-chlorophenyl ester of α-CHF_2-ala is rapidly cleaved in Rin cells to give easily observable intracellular α-CHF_2-ala signals. FIGURE 9 shows ^{19}F NMR spectra of a Rin cell suspension containing 1.0 mM α-CHF_2-ala *p*-chlorophenyl ester, taken 8 and 14 minutes after adding ester to the cell suspension. A measurable internal α-CHF_2-ala signal persisted for at least 20 min, long enough to observe the effect of increased extracellular K^+, for example, on intracellular pH. Preliminary results indicate that Rin cells actively regulate their intracellular pH, giving a pH_i versus pH_e curve similar to that of PBLs.

Frog Skin

These exogenous compounds can be used successfully in tissues, as well as in isolated cells. In frog skins (*Rana pipiens*), we determined intracellular pH by both ^{19}F

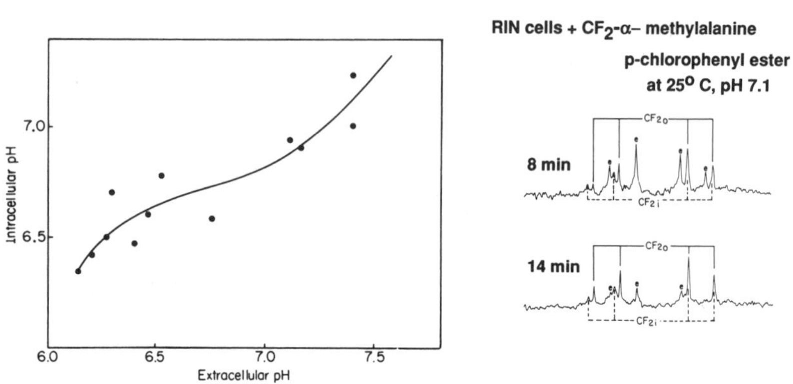

FIGURE 9. (Left) Intracellular pH versus extracellular pH in RINm5F insulinoma cells. Data points are taken from individual ^{19}F NMR experiments using difluoromethylalanine *p*-Cl-phenyl ester (1 mM) at 25°C. (Right) 188.2 MHz ^{19}F NMR spectra of RINm5F cells, 3% cytocrit in RPMI medium containing also 1 mM difluoromethylalanine *p*-chloropheny ester and 12% D_2O. Temperature, 25°C, medium, pH, 7.1. Times after ester addition are shown. (1,000 scans, 3.6 μsec pulse width, 0.2 sec repetition time).

and ^{31}P NMR measurements, using the methyl ester of α-CHF_2-ala. ^{19}F and ^{31}P spectra were obtained consecutively on the same sample in the same 10-mm sample tube (FIG. 10). From four direct comparisons of the two techniques in two experiments, the difference in intracellular pH estimates was less than 0.05. Both NMR approaches indicate that in frog skin, as with the other cells we have studied, acidification of the extracellular medium reverses the sign of the pH gradient present under baseline conditions. An antiport exchange of Cl^- entry for HCO_3^- exit may play a role in the intracellular regulation of pH in frog skin.[21] However, this mechanism would cause intracellular acidification. Therefore, one or more additional processes must be responsible for keeping the cell more alkaline than the external medium at an extracellular pH of 6.9, despite an electrochemical driving force favoring net proton entry and despite the continued metabolic production of acid.

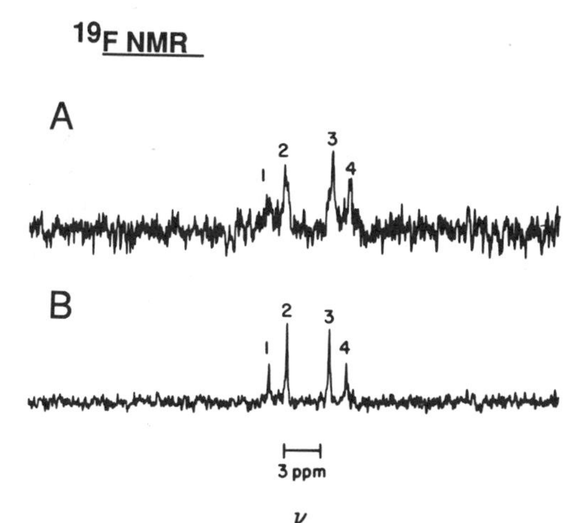

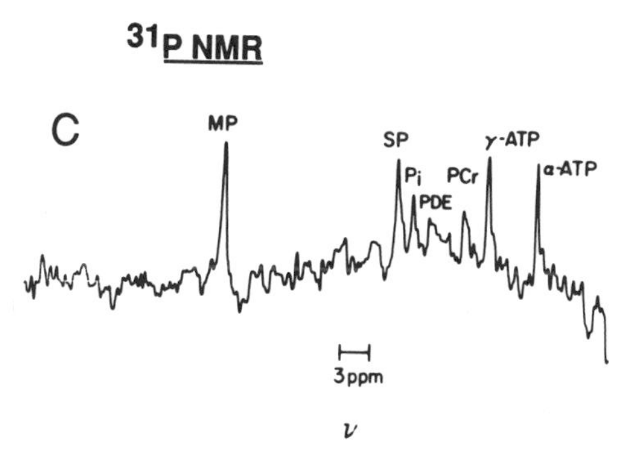

FIGURE 10. 188.2 MHz ^{19}F NMR and 145 MHz ^{31}P NMR spectra of same frog skin. Frog skin had been preincubated in Ringer solution containing 1 mM of ^{19}F probes for 4.7 hr before beginning analysis. (A) ^{19}F spectrum of frog skin, with NS = 14,000 and LB = 7 Hz; cps = 3.92 ppm of cleaved amino acid. This corresponds to calculated intracellular pH of 7.39 at a measured external pH of 7.60. (B) ^{19}F spectrum of free amino acid added to perfusing solution. Number of scans accumulated (NS) was 400, and line broadening (LB) was 7 Hz; Central peak splitting (cps) given by δ_2–δ_3 was 3.5 ppm, corresponding to pH_e = 7.66, at pH measured with meter of 7.63. (C) ^{31}P-NMR spectrum obtained with same sample of tissue after completion of spectrum of B: NS = 565 and LB = 30 Hz; chemical shift of P_i relative to that of PCr ($\delta_{P_i} - \delta_{PCr}$) was 5.31 corresponding to calculated pH_i of 7.34, while relative chemical shift of methyl phosphonate ($\delta_{MP} - \delta_{PCr}$) was 24.82, corresponding to calculated extracellular pH of 7.73.

Rabbit Colon

Rabbit colon slices incubated with the α-CHF_2-ala methyl ester for 1.0 hr took up the ester and cleaved it to the free amino acid.[13] The free amino acid is relatively impermeant; there was no detectable leakage from the rabbit colon tissue over a period of 1 hr, nor was the free amino acid transported into the tissue from the incubation medium. The pH_i showed an alkaline shift (7.35) when Cl^- was replaced with gluconate, and an acid overshoot (6.96) when the medium was changed back to Cl^- medium.

SUMMARY

^{19}F NMR pH measurements with the fluorinated pH indicators we have described are rapid and sensitive, work with readily available commercial instruments, and extend the applicability of nondestructive NMR measurements to systems for which ^{31}P NMR measurements are presently impractical. The family of ^{19}F pH indicators is useful for independent confirmation of pH_i values obtained by ^{31}P NMR, distribution of radioactive weak acids, or other methods. The necessity for using exogenous indicators, which at first appears as a liability, can be turned to advantage also. Our future goal is to direct our measurements unambiguously to compartments (in cells or in tissues) of particular interest, by matching the indicator precursor molecule to the hydrolytic enzyme activities inherent in the target cell or compartment, so that the pH indicator is generated *in situ*.

REFERENCES

1. BUSA, W. B. & R. NUCCITELLI. 1984. Am. J. Physiol. **246:** R409–R438.
2. NUCCITELLI, R. & D. W. DEAMER, Eds. 1982. Intracellular pH: Its Measurement, Regulation and Utilization in Cellular Functions, Liss. New York.
3. ROOS, A. & W. F. BORON. 1981. Physiol. Rev. **61:** 296–433.
4. MOON, R. B. & J. H. RICHARDS. 1973. J. Biol. Chem. **248:** 7276–7278.
5. TAYLOR, J. S., C. DEUTSCH, G. G. MACDONALD & D. F. WILSON. 1981. Anal. Biochem. **114:** 415–418
6. CHRISTENSEN, H. N. & D. L. OXENDER. 1963. Biochim. Biophys. Acta **74:** 386–391.
7. METCALFE, J. C., T. R. HESKETH & G. A. SMITH. 1985. Cell Calcium **6:** 183–195.
8. BROWN, T. Personal communication.
9. CHANG, Y. C. & D. J. GRAVES. 1985. J. Biol. Chem. **260:** 2709–2714.
10. KORYTNYK, W. & S. C. SRIVASTAVA. 1973. J. Med. Chem. **16:** 638–642.
11. GILLIES, R. Personal communication.
12. OPELLA, S. Personal communication.
13. DEUTSCH, C. & J. S. TAYLOR. 1987. NMR Spectroscopy of Cells and Organisms. R. Gupta, Ed. CRC Press.
14. TSIEN, R. 1981. Nature **290:** 527–528.
15. KASHIWAGURA, T., C. J. DEUTSCH, J. S. TAYLOR, M. ERECINSKA & D. F. WILSON. 1984. J. Biol. Chem. **259:** 237–243.
16. TAYLOR, J. S. & C. DEUTSCH. 1983. Biophys. J. **43:** 261–267.
17. CIVAN, M. M., L.-E. LIN, K. PETERSON-YANTORNO, J. TAYLOR & C. DEUTSCH. 1984. Am. J. Physiol. **247:** C506–510.
18. COHEN, S. M., S. OGAWA, H. ROTTENBERG, P. GLYNN, T. YAMANE, T. R. BROWN, R. G. SHULMAN & J. R. WILLIAMSON. 1978. Nature **273:** 354–356.
19. BARON, P. G., R. A. ILES & R. D. COHEN. 1973. Clin. Sci Mol. Med. **55:** 175–181.
20. COHEN, R. D., R. M. HENDERSON, R. A. ILES & J. A. SMITH. 1982. J. Physiol. (Lond.) **330:** 69–80.

21. NUNNALLY, R. L., J. S. STODDARD, S. I. HELMAN & J. P. KOKKO. 1983. Am. J. Physiol. **254:** F792–F800.
22. DEUTSCH, C., J. S. TAYLOR & M. PRICE. 1984. J. Cell Biol. **98:** 885–893.

DISCUSSION OF THE PAPER

QUESTION: In the difluoromethylalanine experiments, where there were equal amounts of material inside and outside, is that a sort of passive diffusion across the membrane? Do you use the D or L acids when you make the materials?

C. DEUTSCH: What I showed you was one point in time. You can put the ester in, watch the intracellular acid peak grow, and then sometime later—it depends on the cell system, the rates are very different from system to system, and in the case of lymphocytes, from donor to donor—but you can watch the difluoromethylalanine leak out, generate the extracellular acid peak, and eventually you'll see it all leak out, unless it's a case like the frog skin, which happens to be relatively impermeable to the free amino acid. What I did was show you one slice in time, coincidently, where they were both present, but you can actually watch the intra and extracellular peaks come up and change sequentially. We use the DL form of the ester.

QUESTION: I have two questions. One relates to the ester loading. How much methanol are you generating in the cells when you load as a methyl ester?

DEUTSCH: In the case of the trifluoro compounds it's less, but it's anywhere from one tenth to one millimolar at maximum.

QUESTION: And the other question is, that there is a potassium dependence of the shift—have you ever come across that as a real problem in cell suspensions?

DEUTSCH: Only in the sense that you have to know both your intracellular and extracellular potassium concentrations, which we take great pains to find out. But that's true in all of your systems—you have to know the ionic strength dependence and you have to know the intracellular ion concentrations.

QUESTION: If your cells are changing volume for some reason and K is moving, how do you account for it?

DEUTSCH: We have to know the magnitude of that change in concentration and we can calibrate that. So we've done extensive work in parallel to monitor what's happening to intracellular potassium concentration and sodium concentration and calcium concentration. So all of those parameters are being monitored.

QUESTION: Is the cation dependence of these pH curves (and that is potassium, sodium, calcium), more sensitive, more of a problem than one finds with the traditional inorganic phosphate titrations?

DEUTSCH: No, it's not more of a problem. There is no dependence on sodium; there is a dependence on calcium and magnesium only when the fluoroalanines are not in a physiological salt medium—that is, if you add 1 mM or 3 mM calcium to a distilled water solution of these probes. In physiological salt solutions there is no dependence on calcium and magnesium. There is a dependence on potassium and that saturates at about 130 millimolar. But it is not a unique or flagrant problem; it's just that we have done a good job to document it in detail. But there is no dependence on sodium; there's no dependence on chloride.

NMR Investigations of Cellular Energy Metabolism

ROBERT S. BALABAN, ALAN KORETSKY, AND
LAWRENCE KATZ

National Heart, Lung and Blood Institute
National Institutes of Health
Bethesda, Maryland 20892

The mechanism for the regulation of energy metabolism within the cell is not entirely understood. For many years it has been established that tissues, such as heart[1–3] and kidney,[4,5] can balance their rate of energy conversion with work output. However, the actual intracellular communication network that results in the coordination of these processes is still an area of active debate and research.

Over the last several years we have been concentrating on the regulation of oxidative phosphorylation within the mitochondria by changes in work output in numerous tissues and observing the biochemical response to this challenge. Since oxidative phosphorylation is a major source of chemical energy (i.e. adenosine triphosphate (ATP)) in most cells, an understanding of the control of this process is key to an overall comprehension of energy utilization by the cell.

Early work on isolated mitochondria led to the notion that alterations in cytoplasmic ADP or P_i levels, due to changes in ATP hydrolysis by the work producing ATPases, are responsible for the appropriate regulation of oxidative phosphorylation.[6,7] The site of respiratory control by these phosphates has been suggested to be a simple Michaelis-Menten type substrate control by ADP and P_i,[8,9] or a near equilibrium relation existed between the phosphorylation potential and the mitochondria redox state.[10,11]

All of the above models concerning phosphates suggest that changes in the adenylate phosphates and P_i should be observed during alterations in work output by a tissue if they are responsible for the observed "feedback" control of respiration. For example, the early[6] and most recent work on the control of respiration by ADP in isolated mitochondria demonstrated a saturable Michaelis-Menten type relationship between extramitochondrial ADP and respiration. The apparent affinity of ADP is approximately 20 μM, which is close to the calculated ADP value in many active tissues.[12] Of interest here is that the slope of the relationship between ADP and respiration is never greater than 1:1. That is, a 50% increase in ADP results in a 50% increase in respiration near the mitochondrial K_m for ADP. Therefore, if this simple mechanism is responsible for respiratory control in the intact cell, a 50% increase in work or respiration should be accompanied by a 50% increase in cytosolic ADP.

The relatively recent demonstration that ^{31}P NMR can be used to follow high energy phosphate contents *in vivo*[12–14] as well as the turnover of these compounds,[15] has permitted the reinvestigation of the role of these phosphates in the regulation of *in vivo* respiration. As a non-destructive technique, each tissue can serve as its own control and not be plagued by freeze-clamp artifacts occurring with these rapidly metabolized substrates.

Several studies using ^{31}P NMR have investigated either directly or indirectly the relationship between high energy phosphates and work output in numerous tissues. The initial ^{31}P NMR studies in animals demonstrated that in skeletal muscle the high energy phosphates do seem to change in accordance with an increase in work output.[16] This has also been confirmed in humans.[8,16] However, several reports in tissues capable of prolonged increases in ATP production via oxidative metabolism have not been as convincing. In the isolated perfused heart, several groups have demonstrated that increases in work output or oxygen consumption are associated with increases in ADP and P_i concentrations.[18–20] However, most have noted that the increase in ADP and P_i alone was not sufficient to explain the increase in oxidative metabolism observed.[18–20] Our own work on the dog heart *in vivo* using a catheter NMR probe[21] demonstrated that moderate increases in work output with pacing (i.e. heart rates from 60 to 180 beats/min), which vary oxygen consumption by at least a factor of two, are not associated with increases in ADP or P_i.[22] This observation has been confirmed even over larger workloads using inotropic agents.[8] In addition, no variation in phosphates was observed throughout the cardiac cycle *in vivo*.[23]

These *in vivo* and *in vitro* data suggested that the simple phosphate models for describing the interaction of metabolism and work in the heart were not adequate to explain respiratory control in the heart. Therefore, it was necessary to hypothesize another mechanism in which mitochondrial respiration may be controlled independently or in concert with the phosphates. Two likely candidates are oxygen delivery and the mitochondrial NAD redox state. Though either or both of these parameters could control respiration, we have concentrated on the mitochondrial NADH redox state as a potential controlling factor over the last several years.

The first objective was to establish the relationship between mitochondrial respiration, phosphates, and the NAD redox state. To accomplish this, we used isolated mitochondria where the rate of respiration, phosphates, and NADH could be controlled as well as accurately monitored. ^{31}P NMR was used to follow the high energy phosphates using a previously described chamber.[24] The NADH redox state was estimated using the NADH fluorescence signal from the suspension.[25] The fluorescence signal was used since it was non-destructive and could be used in conjunction or in parallel with the ^{31}P NMR data *in vitro* or *in vivo*. In these studies the redox state of NAD was adjusted by varying the concentration or type of substrate used while the ATPase rate was varied by adding different amounts of ATPase to the incubation medium. The most intriguing finding of these studies[26] was that the NADH concentration, measured by fluorescence, was linearly related to the mitochondrial oxygen consumption (FIG. 1), with no indication of saturation. This was true for a wide range of ATPase activities including State 3; the maximum respiration rate driven by the phosphates.[6] This result suggests that no matter what the phosphorylation potential is, the rate of respiration would always be influenced by the mitochondrial NADH redox state. Thus, NADH is a powerful regulator of mitochondrial respiratory rate, which can influence respiration with or without changes in phosphates. This suggests that if the cell could regulate the NADH redox state during changes in work output it could increase the rate of ATP synthesis without a large change in the relative concentrations of the phosphates involved, as observed in the *in vivo* and *in vitro* heart.

In earlier studies, we have demonstrated that this linear relationship between NADH and respiration is present in intact cells. In isolated renal tubules, a linear relationship between NADH fluorescence and respiration was found with a wide variety of substrates and work outputs.[27] However, this study was hampered by the lack of ability to measure the phosphates since renal cells do not have large amounts of creatine kinase to permit the determination of ADP concentrations.[28] Therefore, we chose to study the *in vitro* perfused rat heart where we could accurately monitor the phosphates as well as oxygen consumption and NADH fluorescence.

Using a isovolumic Langendorff perfused heart preparation, we monitored the rate of respiration, NADH fluorescence, and phosphates detected by ^{31}P NMR. Respiration was measured by determining the coronary flow and oxygen concentration difference between the efferent and afferent perfusate. NADH fluorescence was monitored using a rapid scanning spectrophotometer in conjunction with an internal fluorescent reference trapped inside the cells.[29] The work jump in these experiments was performed with pacing alone to minimize inotropic effects as in our *in vivo* studies. In glucose-perfused hearts, we found that a 30% increase in respiration induced by pacing was associated with no significant change in high energy phosphates with the ratio of CrP to ATP actually becoming significantly larger, suggesting that ADP decreased during this increase in work. However, the NADH redox state was found to

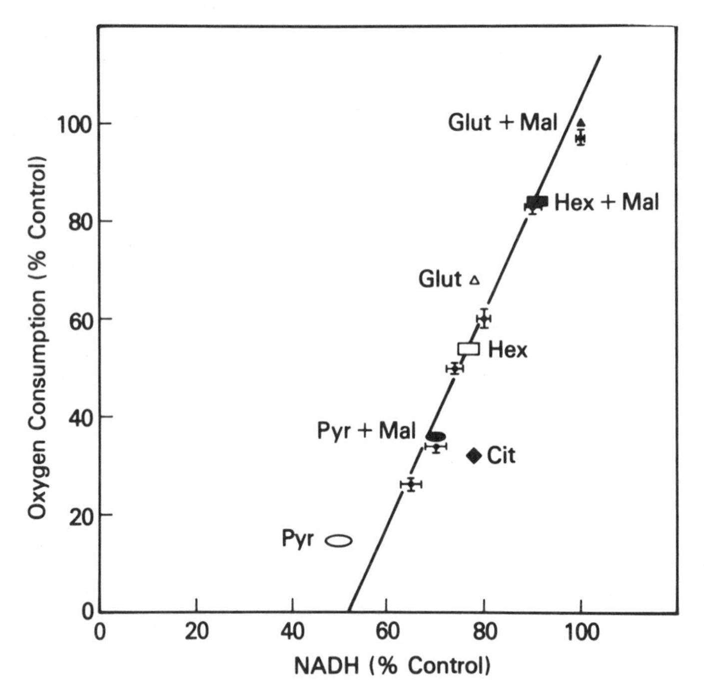

FIGURE 1. A plot of respiration versus NADH fluorescence for the mitochondria respiring in State 3 with different amounts and types of substrates. Abreviations are MAL = malate, GLUT = glutamate, HEX = Hexanoyl-carnitine, CIT = citrate, PYR = pyruvate. (Adapted from Koretsky & Balaban.[26])

increase by 25% with this steady-state increase in work output. This increase in NADH is the opposite effect expected on the mitochondrial redox state if respiration was being driven by alterations in ADP and P_i alone. This result implies that the NADH redox state was the only parameter that changed in the appropriate direction to increase respiration under these conditions.

A probable mechanism for the increase in NADH in response to an increase in work output can be found in the literature. Denton and McCormack[30] have proposed that an increase in mitochondrial calcium will activate mitochondrial dehydrogenases, which in turn will result in an increase in reduced NAD. Hansford[31] has also shown in blowfly mitochondria that this type of activation can result in a net reduction of mitochondrial NAD at State 3. This later result would be consistent with the observations made in the glucose-perfused heart. That is, an increase in work output causes an increase in cytosolic calcium, which then activates mitochondrial dehydrogenases and increases NADH without requiring a significant change in phosphates. To test his hypothesis, we used ruthenium red to block the entry of calcium into the mitochondria.[32] A work jump in ruthenium red–treated glucose-perfused hearts resulted in a large increase in ADP and P_i, in contrast to no change in the absence of ruthenium red. This result is consistent with the notion that the fluxes of calcium across the mitochondrial membrane may be important for the feedback between work and respiration in the intact heart. Another test of the hypothesis was performed using pyruvate-perfused hearts, where the level of some dehydrogenases are already maximized.[32] In the control pyruvate-perfused hearts, the NADH redox state and the phosphorylation state (i.e. $ATP/ADP + P_i$) were higher, consistent with an activation of the dehydrogenases. Upon an increase in work, respiration increased, however, in contrast to the glucose-perfused heart, the inorganic phosphate and ADP concentrations increased sufficiently to entirely explain the observed changes in respiration. Similar results have been found in other laboratories.[20] These latter two results suggest that the modification of dehydrogenase activity may play a key role in the regulation of respiration in the heart under normal workloads.

In summary, ^{31}P NMR has permitted us to test some of our basic concepts concerning the orchestration of energy metabolism in intact tissues. These studies suggest that the high energy phosphates and their hydrolysis products are not solely responsible for the feedback between work and mitochondrial respiration in the heart. In combination with non-destructive optical techniques, information is emerging that suggests that the NADH redox state, controlled by intermediary metabolism, may be one of the important effectors in this basic energy control process of the cell.

REFERENCES

1. Evans, C. L. & Y. Matsuoka. 1915. J. Physiol. **49:** 378–405.
2. Katz, L. N. & H. Feinberg. 1958. Circ. Res. **6:** 656–659.
3. Gibbs, C. 1985. J. Mol. Cell Cardiol. **17:** 727–731.
4. Thurau, K. 1961. Proc. Soc. Exp. Biol. Med. **106:** 714–717.
5. Mandel, L. J. & R. S. Balaban. 1981. Am. J. Physiol. **240:** F357–F371.
6. Chance, B. & G. R. Williams. 1956. Adv. Enzymol. **17:** 65–80.
7. Whittam, R. 1961. Nature **191: 603–604.**
8. Chance, B., J. S. Leigh, B. J. Clark, J. Maris, J. Kent, S. Nioka & D. Smith. 1985. Proc. Natl. Acad. Sci. USA **82:** 8384–8388.
9. Wilson, D. F., C. S. Owen & A. Holian. 1977. Arch. Biochem. Biophys. **182:** 749–762.
10. Jacobus, W. E., R.W. Moreadith & K. M. Vandegaer. 1982. J. Biol. Chem. **257:** 2397–2403.
11. Klingenberg, M. 1961. Biochem. Z. **335:** 263–272.
12. Balaban, R. S. 1984. Am. J. Physiol. **246:** C10–19.
13. Ackerman, J. J. H., T. H. Grove, G. G. Wong, D. G. Gadian & G. K. Radda. 1980. Nature **283:** 167–170.
14. Meyer, R. A., M. J. Kushmerick & T. R. Brown. 1982. Am. J. Physiol. **242:** C1–C11.
15. Alger, J. R. & R. Shulman. 1984. Q. Rev. Biophys. **17:** 83–95.
16. Kushmerick, M. J. & R. A. Meyer. 1985. Am. J. Physiol. **248:** C542–549.
17. Taylor, D. J., P. Styles, P. W. Matthews, D. A. Arnold, D. G. Gadian, P. Bore & G. K. Radda. 1986. Magn. Res. Med. **3:** 44–54.
18. Matthews, P. M., S. R. Willaims, A. M. Seymour, A. Schwartz, G. Dube, D. G. Gadian & G. K. Radda. 1982. Biochim. Biophys. Acta **720:** 163–169.

19. BITTL, J. A. & J. S. INGWALL. 1985. J. Biol. Chem. **260:** 3512–3518.
20. FROM, A. H. L., M. A. PETEIN, S. P. MICHURSKI, S. D. ZIMMER & K. UGURBIL. 1986. FEBS Lett. **206:** 257–259.
21. KANTOR, H. L., R. W. BRIGGS & R. S. BALABAN. 1984. Circ. Res. **55:** 261–266.
22. BALABAN, R. S., H. L. KANTOR, L. A. KATZ & R. W. BRIGGS. 1985. Science **232:** 1121–1123.
23. KANTOR, H. L., R. W. BRIGGS, K. R. METZ & R. S. BALABAN. 1986. Am. J. Physiol. **251:** H171–H175.
24. BALABAN, R. S., D. G. GADIAN, G. K. RADDA & G. G. WONG. 1981. Anal. Biochem. **116:** 450–455.
25. BALABAN, R. S., L. J. MANDEL, S. SOLTOFF & J. M. STOREY. 1980. Am. J. Physiol. **77:** 447–451.
26. KORETSKY, A. P. & R. S. BALABAN. Biochim. Biophys. Acta. (In press.)
27. BALABAN, R. S. & L. J. MANDEL. (Submitted for publication.)
28. BALABAN, R. S. 1982. Fed. Proc. **41:** 42–47.
29. KORETSKY, A. P., L. KATZ & R. S. BALABAN. Am. J. Physiol. (In press.)
30. DENTON, R. M. & J. G. MCCORMACK. 1980. FEBS Lett. **119:** 1–9.
31. HANSFORD, R. G. 1985. Rev. Physiol. Biochem. Pharm. **102:** 2–12.
32. MCCORMACK, J. G. & P. J. ENGLAND. 1983. Biochem. J. **214:** 581–585.

DISCUSSION OF THE PAPER

M. W. WEINER: My question relates to the observation that got you started, the intact animal work which you published in *Science* (Balaban *et al.* 1986. **232:**1121.) some time ago. I have basically two questions about those experiments. The first is that, as I understand it, your catheter coil is, in fact, in the right ventricle, not in the left ventricle, so when you show us a pulse pressure product, talking about blood pressure and work, you're really talking about left ventricular work. My question is, what happens to right ventricular work when you pace? The work output of the right heart may not go up the same extent as the work output of the left heart.

My second question concerns at least the possibility that the phosphorylation potential or the ADP concentration might be changing because of a change in pH. It would be difficult to detect the change in pH because the P_i resonance is obscured by overlying blood resonances and it seemed to me, we know that when we exercise skeletal muscle there's a drop in pH, perhaps an increasing work output in working heart would also cause the pH to drop and that might affect the ADP concentration. . . .

R. S. BALABAN: With regard to pH, an acidification actually takes it the other way, so it would make the data more indicative of an increase in ADP, not the negative result seen here. With regard to the source of the signal, I don't think that's really a very significant issue. If you dissect out the dog heart and look at where we have placed that coil, we are observing the septum between the left and the right ventricle. In that area there is a significant amount of left ventricle, which is going to respond as that work output goes up. I think, you've also seen the work of Dr. Clark where he's actually placed the surface coil on the left ventricle and increased the heart rate pressure product with isoproterenol and indeed found almost no change in those high energy phosphates.

QUESTION: Would you mind clarifying what you found when you use blood pressure or systolic pressure times heart rate as a measure of work, as opposed to pulse pressure times heart rate? Did you find any difference if you use that as a function of heart work?

BALABAN: We haven't analyzed it that way.

J. R. ALGER: I'm wondering about the implications of what you have said about NADH control with respect to what we know from the saturation transfer measurements of heart. If the control is at NADH, one might expect the saturation transfer to show a rapid exchange through the ATPase. This is an apparent P:0 ratio much greater than 3 in the heart and I believe that that has not ever been observed in the heart, although there is some suggestion, maybe in other systems, it has. Could you comment on that?

BALABAN: I agree. In fact, that was one of my original interests in the saturation transfer experiment. I thought it might really establish whether we're talking about a nonequilibrium vs. an equilibrium situation. As you say you would predict that if the mitochondrion was at equilibrium, you would expect a much higher unidirectional flux due to the mitochondrial ATPase than the net flux. I've indicated that if you eliminate the glycolytic flux the ATPase flux does look like it's one way.[12] The question is, where is this reaction occurring and what percentage of the entire pool of ATP is involved in this equilibrium in the mitochondrial matrix space? Can we talk about equilibrium in the cytosol or do we have to talk about equilibrium in the mitochondrial matrix? I don't know if we really have an answer to that, but I agree that's an inconsistency.

Cerebral Lactate Elevation by Electroshock: A ^{1}H Magnetic Resonance Study[a]

JAMES W. PRICHARD, OGNEN A. C. PETROFF, TAKASHI OGINO, AND ROBERT G. SHULMAN

Departments of Neurology and Molecular Biophysics and Biochemistry Yale University New Haven, Connecticut 06510

INTRODUCTION

We showed recently that bicuculline-induced status epilepticus in rabbit brain caused elevations of cerebral lactate detectable *in vivo* by ^{1}H magnetic resonance spectroscopy.[1] The elevated lactate persisted with no sign of recovery throughout 1–2 hour experiments, despite disappearance of intense seizure discharge from the electroencephalogram (EEG) within 30 minutes when the dose of bicuculline was below 1.95 mg/kg. We suggested two possible explanations for the persistence of the elevated lactate: (1) Lactate was trapped in a metabolic compartment less active than the one in which it was produced and (2) bicuculline-induced status epilepticus caused an uncompensated excess of glycolysis over respiration. In order to investigate the matter further, a less intense, more controllable way of producing lactate elevations would be useful. The present study was undertaken to determine whether non-pharmacological activation of rabbit cerebrum by electrical stimulation would cause lactate to rise. It did, and several features of the phenomenon were examined.

METHODS

Animal Preparation

Twenty New Zealand white female rabbits weighing about 2 kg were prepared under enflurane anesthesia as previously described.[1,2] The dura was exposed through a 20-mm circular craniotomy centered 2–4 mm occipital to the bregma in the midline. Silver epidural stimulating electrodes were placed under the lateral margins of the craniotomy for direct stimulation of the cortex. In three animals, a stimulating electrode was placed near each optic nerve via the orbit. The dura was covered with plastic film to prevent drying. The electroencephalogram (EEG) was recorded from silver wires placed subcutaneously mesial to the left orbit and left ear. A catheter was placed in the aorta via the femoral artery to record blood pressure and provide blood samples in which glucose was measured with Dextrostix and an Ames Glucometer.

[a]Supported by grants from the United States Public Health Service (GM 30287 and NS 21708) and the Esther A. and Joseph Klingenstein Fund.

Heart rate, blood pressure, and EEG were monitored continuously on a Grass model 7 polygraph, and rectal temperature on a Yellow Springs Telethermometer. Following surgery, the rabbits were paralyzed with 1 mg/kg pancuronium bromide and 1.5 mg/kg tubocurarine HCl and enflurane was removed from the ventilatory mixture; analgesia was maintained throughout the experiment with 70% N_2O in O_2. Cardiac arrest was induced by KCl injection at the end of each experiment.

NMR Measurements

These were made with a 1.89 Tesla Oxford Research Systems TMR 32/200 spectrometer with a ^{1}H operating frequency of 80.3 MHz. Radio frequency signals were transmitted and received with a two-turn surface coil. A 20-mm coil centered over the craniotomy was used in 8 experiments; the other 12 were performed with a 16-mm coil in contact with the plastic film covering the dura within the skull defect. As in an earlier study[1], ^{1}H spectra were obtained using a combined $1\bar{3}3\bar{1}$-Hahn spin echo pulse sequence ($1\bar{3}3\bar{1}$-τ-$2\bar{6}6\bar{2}$-τ-acquire) adapted for use with a surface coil *in vivo.*[3–6] The sequence was adjusted to deliver maximum excitation at the lactate resonant frequency during all of the experiment except acquisition of the final *in situ* spectrum. Following stabilization of the lactate resonance after cardiac arrest, an additional spectrum was obtained with the excitation power maximum at the resonant frequency of N-acetyl aspartate (NAA). Acquisition parameters were: Time domain points, 2048; filter band width, 1300 Hz; spectra width, 1250 Hz; acquisition time, 0.82 sec; and interval between free induction decays (TR), approximately 2 sec. Small variations in TR across experiments were caused by optimization of signal-to-noise ratio in individual animals. In each animal, this was done by adjusting the duration of the exciting pulse so as to obtain a maximum signal from water with a single pulse and then from NAA by a $1\bar{3}3\bar{1}$-spin echo sequence with the carrier set on the frequency of the water resonance. The τ delays of 100–200 msec partially suppressed signals from lipids co-resonant with lactate because of their shorter T_2s; optimization of τ for this purpose was the major cause of variability in TR. Power spectra were made from the resolution-enhanced average of 32 free induction decays[1] collected in about 64 sec, depending on the exact value of TR.

Shock-induced and terminal lactate increases were measured in difference spectra made by subtracting the scaled average of several control spectra from spectra obtained after cortical stimulation. Stability of resonances during the experiment was judged by the absence of an NAA signal from difference spectra and the unchanged shape of the lactate signal. A terminal lactate/NAA ratio was determined from the areas (weights of paper cut-outs) of the lactate resonance in the final difference spectrum obtained with maximum excitation power on lactate and the NAA resonance in the spectrum obtained with maximum power on NAA. Lactate resonance heights from earlier difference spectra were expressed as fractions of the terminal lactate resonance height and converted to lactate/NAA ratios by multiplication by the terminal ratio.

RESULTS

FIGURE 1 shows ^{1}H spectra obtained by the $1\bar{3}3\bar{1}$ method during the control period and after cortical stimulation with a 10 sec train of 150 V, 150 Hz, 1 msec square pulses, and the corresponding difference spectrum.

FIGURE 2 shows the EEG consequences of a 10 sec shock train with other stimulus parameters the same as for FIGURE 1. Spike-and-wave seizure discharge lasted 20 sec after the end of stimulation; hence intense electrical stimulation of the cortex was present for 30 sec. Rhythmic slow activity waned over the next 3 min; all abnormal

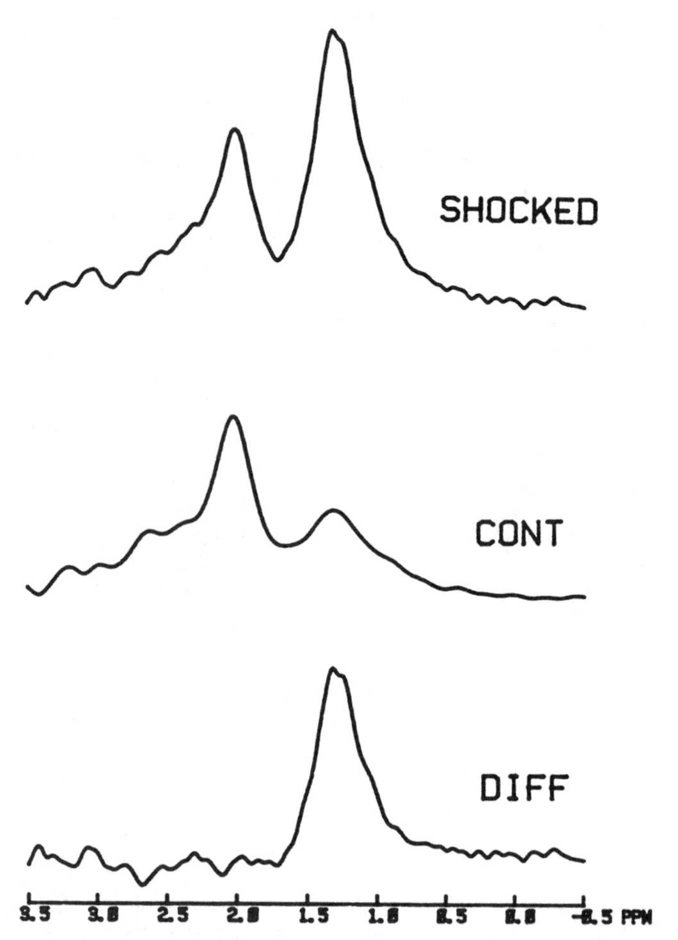

FIGURE 1. [1]H spectra of rabbit cerbrum obtained by the $1\bar{3}3\bar{1}$-spin echo pulse sequence described in METHODS. The control spectrum (CONT) has resonances from NAA and lipids at 2.0 and about 1.3 ppm, respectively. After cortical electroshock with a 10 sec train of 150 V, 150 Hz, 1 msec square pulses, a large increase in the resonance near 1.3 ppm appeared (SHOCKED). Subtraction of CONT from SHOCKED yielded a shock-dependent signal considered to be principally from lactate.

EEG activity had disappeared 4 min after the start of the shock train. Of seven animals that received identical stimulation with this shock train before any other manipulation, the duration of spike-and-wave discharge post-shock varied from 8 to 40 sec, and total electrical disturbance lasted 85–205 sec.

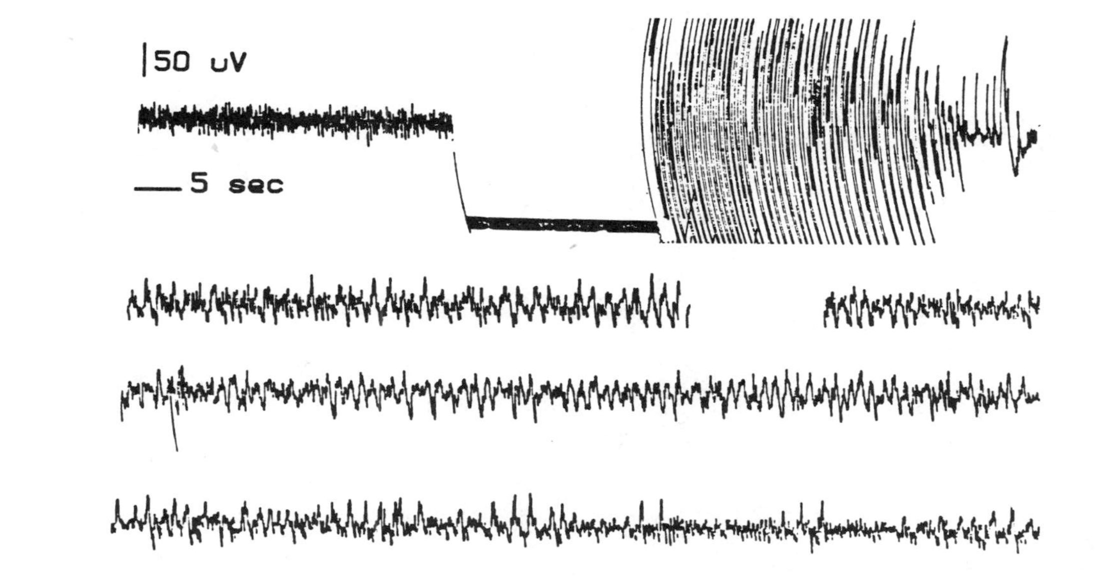

FIGURE 2. Electroencephalogram showing the effect of a 10 sec train of 150 V, 150 Hz, 1 msec square pulses. The top trace shows 25 sec of control EEG followed by the artifact of the 10 sec shock train, which caused high amplitude spike-and-wave discharge lasting about 20 sec. The next two traces show rhythmic slow waves, which decline in the fourth trace, to be replaced by EEG of control appearance in the bottom trace. Traces are continuous; the gap in the second was caused by failure of ink flow. In the amplitude calibration above the first trace, "uV" means μV.

Blood glucose measurements at the time of cortical stimulation were available on 12 animals; the mean value was 212 mg% (standard error of the mean (S.E.M.) 20, range 110–317). Individual values were not correlated with duration of spike-and-wave discharge or total electrical disturbance.

FIGURE 3 shows the time course of lactate elevation in two animals that received a 10 sec train of 150 V, 150 Hz, 1 msec square pulses before any other manipulation. Although the animal with the higher peak lactate/NAA ratio had spike-and-wave discharge lasting 40 sec, compared to 12 sec for the other animal, peak lactate/NAA ratio did not correlate well with duration of spike-and-wave discharge in all seven animals that received the same stimulation or in a total of 17 animals that received cortical shock trains varying in length from 0.1 to 10 sec. Adding the length of the shock train to the duration of spike-and-wave discharge did not improve the correlation. Total duration of electrical disturbance was likewise uncorrelated with peak lactate elevation.

In all animals, lactate elevation persisted far longer than the shock-induced EEG changes did, as FIGURE 3 illustrates. In the seven animals that received 10-sec shock trains, mean time for lactate to reach its control level was 134 min (S.E.M. 14). In all cases, the course of lactate decline was approximately linear with time, with a tendency to level off at late times (FIG. 3).

Selective reduction of stimulus voltage in a 10 sec cortical shock train of 150 Hz, 1 msec square pulses revealed a threshold of 8–20 V for production of a detectable lactate elevation. The corresponding threshold for pulse frequency was 10 Hz when train duration was adjusted to deliver 150 40 V, 1 msec square pulses. When shock train duration was reduced selectively, with other parameters of stimulation held at delivery of 150 V, 100 Hz, 1 msec square waves, lactate elevations were produced by trains of 750, 150, 15, 4, and 2 shocks. Single shocks caused neither spike-and-wave discharges nor detectable changes in lactate in two experiments. FIGURES 4 and 5 show, respectively, the EEG and spectroscopic results of the two experiments in which

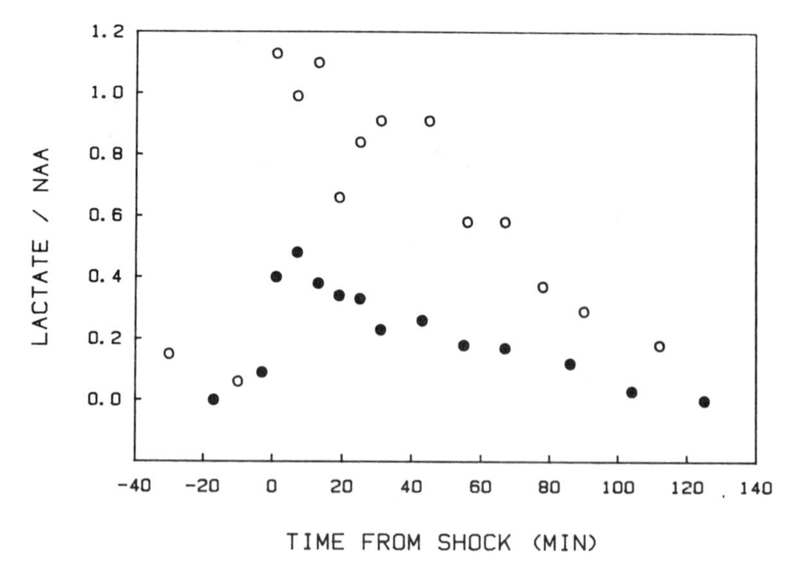

FIGURE 3. Shock-induced changes in lactate/NAA ratios in two rabbits that received 10 sec trains of 150 V, 150 Hz, 1 msec square pulses at time 0.

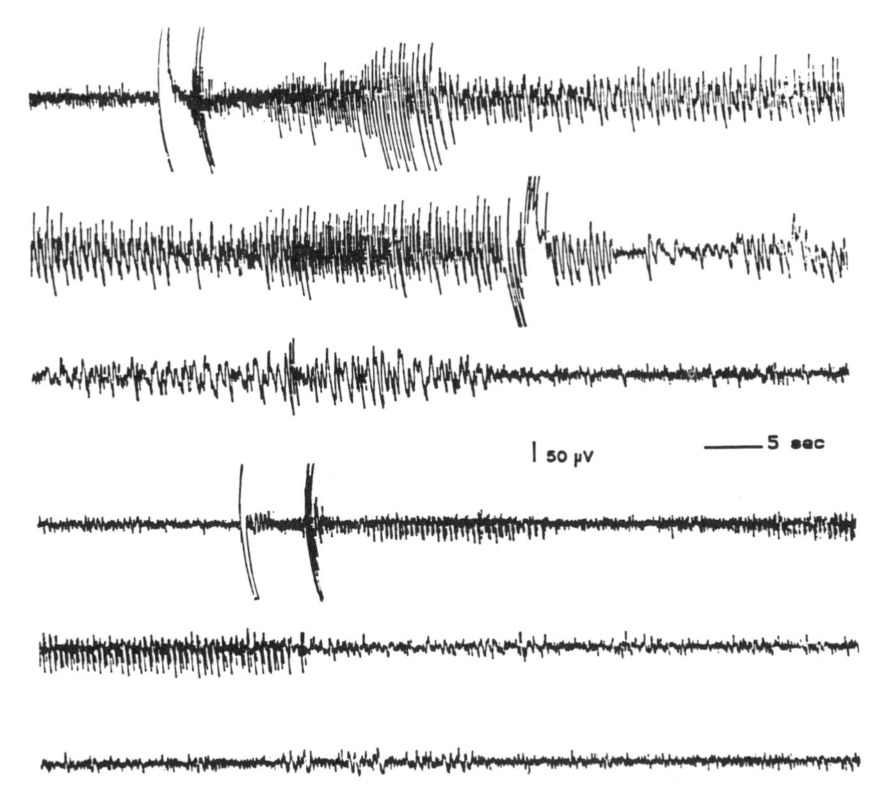

FIGURE 4. Electroencephalographic responses to two shocks in the same rabbits whose metabolic data appear in FIGURE 5 as filled circles (top 3 EEG traces) and open circles (bottom 3 traces). The artifact of the shocks appears near the beginning of the first and fourth traces, after a period of control EEG; the disturbance a few seconds later is an artifact caused by detachment of the stimulating cable to prevent introduction of radio frequency noise into spectra. Rhythmic sharp and slow activity followed the shocks after a brief latency in both animals and disappeared within 3 min; the EEG response was more intense in the animal represented by the upper 3 traces.

two shocks were given. In both cases the electrical response was considerably less vigorous than the one shown in FIGURE 2, and the extent of lactate elevation was less than the ones in FIGURE 3.

In three animals, electrical stimulation was applied to the optic nerves rather than to the cerebral cortex. With sufficiently prolonged stimuli, lactate elevations could be induced repeatedly in all three; in all cases, they returned to control level in 10–30 min. FIGURE 6 shows the time course of lactate elevations in an animal twice stimulated with a 2 min train of 150 V, 150 Hz, 1 msec square pulses.

DISCUSSION

This study shows clearly that cerebral lactate is consistently elevated by brief electrical activation of rabbit brain. The similar result of our earlier study of

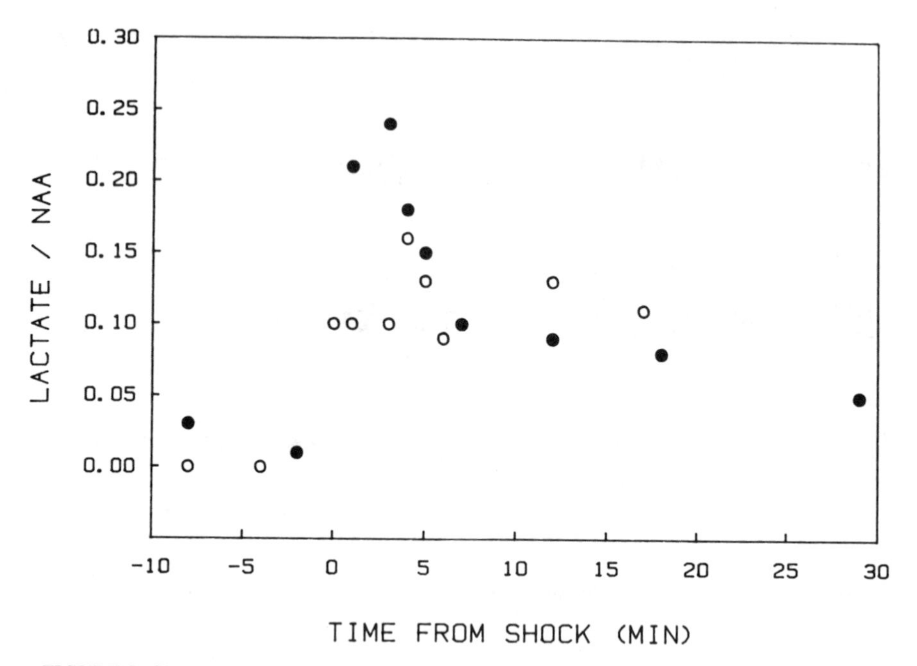

FIGURE 5. Shock-induced changes in lactate/NAA ratios in two rabbits that each received two 150 V, 1 msec square pulses 10 msec apart at time 0. The corresponding EEG responses appear in FIGURE 4.

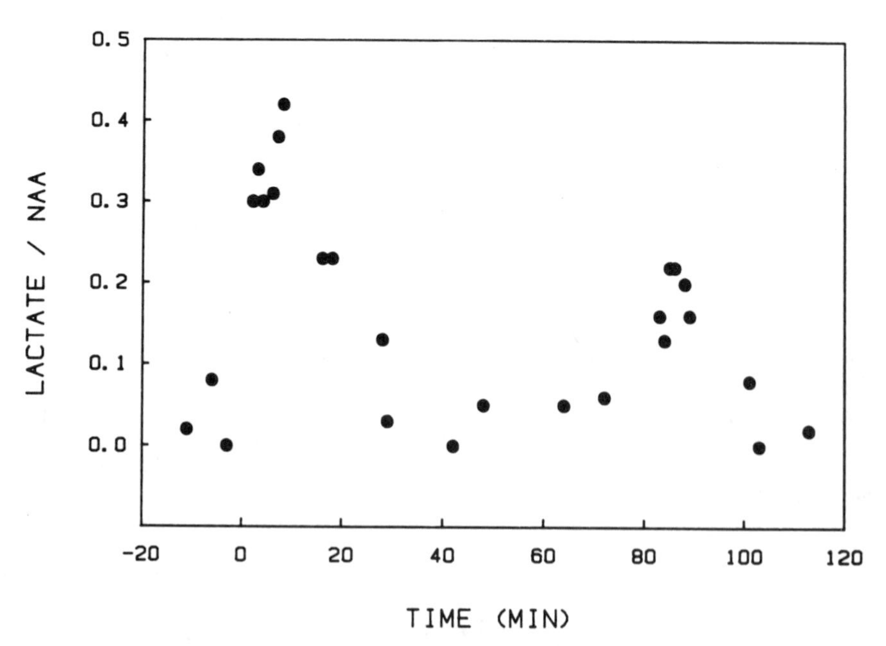

FIGURE 6. Elevation of lactate/NAA ratio in a rabbit stimulated at 0 min and again at 80 min with a 2 min train of 150 V, 150 Hz, 1 msec square pulses delivered via the optic nerves.

bicuculline-induced status epilepticus[1] is therefore unlikely to have been a consequence of some exotic or extra-cerebral action of the drug.

The three experiments in which lactate was elevated by shocks delivered to the optic nerves greatly reduce the possibility that the lactate change was due to tissue damage by electroshock. In these experiments, the stimulating electrodes were more than 1 cm away from the sensitive volume of the surface coil. Current density within the sensitive volume is quite unlikely to have reached levels sufficient to exert direct effects other than a modest degree of membrane depolarization. Most of the metabolic activation of the tissue providing the magnetic resonance signals will have been due to massive release of synaptic transmitters in the visual pathways.

The precise controllability of electrical stimulation allowed preliminary determination of the threshold for elicitation of cerebral lactate elevation. Thresholds of about 10 V, 10 Hz, and 2 shocks for the three parameters of stimulation that we varied show that most of the stimulation we used was much more intense than the minimum necessary to produce observable lactate elevation. This is perhaps why we observed a correlation among stimulus parameters, EEG response, and lactate elevation only when comparing widely different intensities of stimulation, e.g., FIGURES 2 and 3 compared to FIGURES 4 and 5.

One feature of both this study and our earlier one on status epilepticus[1] requires further consideration. Rabbits are prone to develop hyperglycemia after surgery, despite anesthesia, and 12 of our rabbits were indeed hyperglycemic (mean blood glucose 214 mg%) at the time electrical stimulation was delivered. Observation of electrically induced cerebral lactate elevations in other species and in euglycemic rabbits will be necessary before the generality of the phenomenon can be assessed.

Why cerebral lactate remained elevated for 2 hr after a total electrical disturbance lasting less than 5 min is a matter for further investigation. The mean time required for shock-elevated lactate to return to its control level was 134 min (sem 14) in seven animals that received identical stimulation. This tempo of recovery suggests that dissipation of the excess lactate was limited principally by efflux of lactate across the blood-brain barrier, which is much slower than lactate efflux from other tissues. In a study of isolated, perfused dog brain, the maximum rate of post-ischemic lactate efflux was 250 nmol/g/min; although the rate fell to 33 nmol/g/min as brain lactate decreased, the variance of the data was too great to allow distinction between linear and asymptotic regression.[7] Studies of perfused brains in other species obtained lactate efflux values in the same range.[8] Data from several sources suggest that the process can be described by Michaelis-Menten kinetics with $K_m \approx 1.8$ mM and $V_{max} \approx 91$ nmol/g/min.[9] Since the normal lactate concentration of rabbit cerebrum is about 2 mM,[10] the shock-elevated concentration will have been well above 1.8 mM. Lactate efflux at 91 nmol/g/min for 134 min would clear an excess lactate load of 12.2 mM; a decline in efflux rate as cerebral lactate concentration approached K_m would lengthen clearance time or reduce the load clearable in the same time. If such a model and values are appropriate for rabbit brain, the slow lactate dissipation we observed requires no further explanation. However, an additional factor may have affected our results. The shocks must have elevated cerebral glycolytic rate relative to respiration in order to cause the observed lactate rise. For as long as this imbalance persisted after the shocks, it would have produced excess lactate and thereby lengthened the period of lactate elevation.

Why was the excess lactate not metabolized faster? We consider these the leading possibilities: (1) Lactate was trapped in some compartment with less respiratory capacity than the one(s) in which it was produced; this could have been extracellular fluid, cells damaged by the shock/seizure, or healthy cells with normally low respiratory rates into which excess lactate found its way. (2) Mitochondria, though undamaged, never increaed their activity sufficiently to compensate for the period of increased glycolysis, so that the excess lactate was not cleared, even though the entire lactate pool remained in its usual state of rapid exchange with pyruvate.

Experiments with ^{13}C-labeled glucose will be important for further analysis of this problem.

REFERENCES

1. PETROFF, O. A. C., J. W. PRICHARD, T. OGINO, M. J. AVISON, J. R. ALGER & R. G. SHULMAN. 1986. Combined ^{1}H and ^{31}P NMR studies of bicuculline-induced seizures in vivo. Ann. Neurol. **20:** 185–193.
2. PETROFF, O. A. C., J. W. PRICHARD, K. L. BEHAR, D. ROTHMAN & J. R. ALGER. 1985. Cerebral metabolism in hyper and hypocarbia: ^{31}P and ^{1}H NMR studies. Neurology **35:** 1681–1688.
3. BRINDLE, K. M., R. PORTEOUS & I. D. CAMPBELL. 1984. ^{1}H NMR measurements of enzyme-catalyzed ^{15}N-label exchange. J. Magn. Reson. **56:** 543–547.
4. HETHERINGTON, H. P., M. J. AVISON & R. G. SHULMAN. 1985. ^{1}H homonuclear editing of rat brain using semi-selective pulses. Proc. Natl. Acad. Sci. USA **82:** 3115–3118.
5. HORE, P. J. 1983. Solvent suppression in Fourier transform nuclear magnetic resonance. J. Magn. Reson. **55:** 283–300.
6. ROTHMAN, D. L., K. L. BEHAR, J. A. DEN HOLLANDER & R. G. SHULMAN. 1984. Surface coil spin-echo spectra without cycling the refocusing pulsethrough all four phases. J. Magn. Reson. **59:** 157–159.
7. ZIMMER, R. & R. LANG. 1975. Rates of lactic acid permeation and utilization in the isolated dog brain. Am J. Physiol. **229:** 432–437.
8. GILBOE, D. D. 1982. Perfusion of the isolated brain. *In* Handbook of Neurochemistry. A. Lajtha, Ed.: 301–330. Plenum Press. New York.
9. PARDRIDGE, W. M. 1983. Brain metabolism: A perspective from the blood-brain barrier. Physiol. Rev. **63:** 1481–1535.
10. THORN, W., W. ISSELHARD & B. MUELDENER. 1959. Glykogen-, Glucose- und Milchsaeuregehalt in Warmblueterorganen bei unterschiedlicher Versuchanordnug und anoxischer Belastung mit Hilfe optischer Fermentteste ermittelt. Biochem. Zeit. **331:** 545–562.

DISCUSSION OF THE PAPER

G. K. RADDA: I think this is very exciting. Let me ask two questions. Have you considered that maybe the problem in the lactate clearance is that you're generating lactate in one sort of cell which is only a small percentage of the total number of cells, namely the neurons, whereas the glial cells normally are able to mop up lactate. I mean I don't know whether that's true or not. And the second question is, Do you have any information about PDH activity, pyruvate dehydrogenase activity, in these sorts of stimulated brain cells? The two thoughts are that you block the end plate or that there are other cells involved.

J. W. PRICHARD: The second question is easy to answer because the answer is no. However, that is the kind of information that is obviously necessary and it's not impossible to get. The first question is a variation of the trapping idea. The lactate might get made in busy cells, neurons. However all of them would not be equally busy, so they can't all necessarily be considered lactate generators. Whether or not lactate moves from one cell population to another is an open question.

QUESTION: I wonder whether you've repeated any of these experiments in more physiologically relevant systems, namely an initially unparalyzed animal, who was then paralyzed.

PRICHARD: That's a very advanced idea, and that's exactly where we want to go. That is the motivation for bringing the stimulus intensity down to a point where we might have an animal walking around. If we could convince him to lie down on the spectrometer that would be ideal.

PRICHARD: 150 volts, then we let it go down to between 10 and 20.

A. M. HOPKINS: I wonder if you'd care to speculate on whether the cerebrospinal fluid is hiding some of your lactate.

PRICHARD: Yes I would. It's not the cerebrospinal fluid that we normally think of because it would take the lactate too long to get there. In the fluid just outside brain cells, though, there's enough room to hold a lot of lactate, if you take the estimates of extravascular, extracellular fluid volume of about 15%. However, the lactate would have to be pumped out against a gradient.

QUESTION: If you unparalyze the animal you're really going to have to pay attention to the blood gases.

PRICHARD: If you give a seizure to an unparalyzed animal, he makes a lot of lactate in his muscles. That's a well known problem in epilepsy research.

In Situ Brain Metabolism[a]

THOMAS L. JAMES,[b,c] LEE-HONG CHANG,[b] WIL CHEW,[b]
RICARDO GONZALEZ-MENDEZ,[b,d] LAWRENCE LITT,[c,d]
PAMELA MILLS,[b] MICHAEL MOSELEY,[b,c]
BRYAN PEREIRA,[e] DANIEL I. SESSLER,[d]
AND PHILIP R. WEINSTEIN[e]

Departments of [b]Pharmaceutical Chemistry, [c]Radiology,
[d]Anesthesia, and [e]Neurosurgery
University of California
San Francisco, California 94143-0446

As an understanding of cerebral metabolism and circulation may have practical consequences for the treatment of brain injury and for surgery, a considerable body of knowledge has been gathered on the subject over a period of at least twenty years.[1,2] Probably the most striking aspect of the subject is its complexity. The interplay of biochemical and physiological events when cerebral ischemia and hypoxia occur has still not been elucidated. Ischemia, i.e., either partial or total restriction of cerebral blood flow, presents the major medical problem of stroke. Hypoxia (low oxygen levels with normal cerebral blood flow) poses a concern during pulmonary failure, anesthetic malfeasance, and in high altitudes. The effects of ischemia and hypoxia are not identical. These cerebral insults exert various interrelated effects on morphological structure, function, and chemistry that are not simply reversed with reperfusion or restoration of oxygen. Since NMR spectroscopy has the potential for following some metabolic processes noninvasively, there has been some effort made to develop NMR as a technology to examine cerebral metabolism.

Much is known about the mechanism of injury associated with cerebral ischemia. Gross physiological problems include brain edema, increased intracranial pressure, microcirculatory compromise, and post-ischemic recirculation problems, such as the "no-reflow" phenomenon and "loss of reperfusion" syndrome. Biochemical and intracellular physiological aspects include low intracellular pH; the calcium-induced arachidonic acid cascade, excitotoxins, preischemic glucose excess, and oxygen-derived free radical toxicity. Although much is known, optimum clinical stratagems for "brain protection" and "brain resuscitation" remain to be developed. It is suspected that much of the injury secondary to ischemia occurs during reperfusion and that an optimum regimen for reperfusion has yet to be developed. Many patients must also tolerate unavoidable periods of cerebral ischemia during surgery.

Regulation of cellular energy metabolism and the consequences of lack of regulation are central to the problems of ischemia and hypoxia. The important factors for regulation have been reviewed.[3] ATP is the bridge between the metabolic reactions that produce energy (glycolysis in the cytosol and oxidative phosphorylation in the mitochondria) and the energy-requiring functions of the cell including biosynthesis (gluconeogenesis, lipogenesis, protein synthesis, and nucleic acid synthesis), muscle contraction, and ion transport (to maintain ion gradients across cell membranes, transepithelial transport, and nerve conduction). The vast majority of ATP is produced

[a]Supported by grants from the National Institutes of Health (GM34767 and NS22022).

by oxidative phosphorylation in mitochondria. When ATP production by oxidative phosphorylation is limited, e.g., due to anoxia, glycolysis is accelerated increasing ATP production from this pathway. Acceleration of glycolysis increases lactic acid production, which may cause tissue pH to fall. The brain possesses an additional store of energy in the form of phosphocreatine (PCr); this reserve energy acts to maintain the intracellular ATP concentration via the creatine kinase–catalyzed reaction: PCr + ADP + H^+ $\rightleftharpoons$ creatine + ATP, which is in rapid equilibrium in the brain.

Normally, i.e., when oxygen availability and substrate supply are not rate limiting, the rate of tissue metabolism is regulated by the rate of ATP hydrolysis via respiratory control. Most resting (or basal) metabolism is due to ATP hydrolysis (catalyzed by Na-K ATPase) for maintenance of cellular sodium and potassium gradients. If cellular work is increased by muscle contraction, nerve depolarization, or transepithelial transport, this increases ATP hydrolysis and intracellular ADP concentrations, diminishes the cytosolic phosphorylation state, and stimulates mitochondrial respiration. The effects of hypoxia or ischemia on tissue metabolism may be considered in terms of oxidizable substrates and the flow of reducing equivalents from NADH down the electron transport chain to cytochrome oxidase. The redox state, redox potential, and concentration of all the mitochondrial metabolic intermediates are strongly influenced by arterial oxygen tension (pO_2) even at normal levels. As pO_2 falls, the flow of reducing equivalents down the electron transport chain is slowed, resulting in their accumulation along the chain. This increased electron availability compensates for reduced O_2 availability, maintaining essentially constant O_2 consumption at each new steady state of hypoxia. As progressive hypoxia develops, the rate of electron transfer to oxygen is inhibited. If substrate supply is not limited, the dehydrogenases of the citric acid cycle will continue to produce NADH, causing the mitochondrial redox state to become reduced. Localized ischemia presents an even more complex situation than hypoxia because the supply of substrates is reduced and metabolic waste products, such as CO_2 and lactate, also accumulate.

Many of the brain metabolites mentioned above can be monitored directly or indirectly with NMR. Those containing phosphorus can be examined by ^{31}P NMR, which also enables determination of intracellular pH from the chemical shift of the inorganic phosphate (P_i) peak.[4,5] Any lactate produced by glycolysis can be followed with some effort by ^1H NMR.[6] Furthermore, as anesthetics are generally employed with cerebral animal models and, of course, are an important facet of surgery, the effects of anesthetics on cerebral metabolism could potentially be examined via NMR. In addition, the anesthetics themselves can often be followed since many contain fluorine, which is detectable via ^{19}F NMR.

METHODS

Some studies utilized Sprague-Dawley rats (300–400 g) and others utilized New Zealand white rabbits (3.0–4.0 Kg). For study, an animal was anesthetized and mechanically ventilated through an orotracheal tube using an appropriate respirator. A rectal temperature probe was used to control normal body temperature both during surgical preparation and during the NMR experiment as the animals were placed prone on a temperature-controlled cradle in the magnet for the experiment. For most studies, femoral arterial and venous catheters were placed for physiological monitoring and control as well as for fluid infusion. For hypoxic hypoxia studies, the mean arterial blood pressure was maintained within 20% of control by infusion of epinephrine. Depending on the study, some or all of the following monitoring was carried out during the *in vivo* NMR experiment: blood pressure, blood pH, blood glucose, arterial blood gases (P_aO_2 and P_aCO_2), heart rate, temperature, and EEG. Inspired levels of oxygen, nitrogen, carbon dioxide, and anesthetic were controlled using an anesthesia machine. For some experiments, the scalp was retracted and a craniectomy was performed in which a circular piece of cranium was removed such that the NMR surface coil, which was much smaller than the piece removed, could be placed directly on the intact dura.

Experiments with rats were carried out on our home-built NMR instrument, which was configured about a Nalorac 5.6 Tesla, 10-cm diameter horizontal-bore magnet and a Nicolet 1180/293B data system. Experiments with rabbits were performed with our General Electric CSI-II, 2.0 Tesla NMR spectrometer/imager. A two-turn surface coil (size and extent of ellipticity depending on experiment) was tuned to the appropriate nuclear frequency using a balance-matched circuit to minimize dielectric losses.[7] Magnetic field homogeneity inside the brain was optimized by adjusting magnet room-temperature shim currents for optimal linewidth and lineshape on the water proton resonance. For ^{31}P NMR, the broad signal, principally from bone, was eliminated from the spectrum by selective saturation.[8,9]

In some cases the NMR spin-lattice relaxation time (T_1) was measured using the saturation-recovery technique or the inversion-recovery technique.[10] Values of the spin-spin relaxation time (T_2) were measured using the Hahn spin-echo pulse sequence.[10] The Hahn spin-echo sequence, in conjunction with presaturation of the water proton signal, was also used to acquire *in vivo* ^1H NMR spectra. The Hahn spin-echo sequence with an echo delay of 60 msec allowed us to remove fat and short T_2 signals and at the same time to invert the lactate signal. Spectral parameters were 4 K data points, spectral width ± 2,500 Hz, repetition time 1.5 sec, and 168 transients. Chemical shifts were referenced to the N-acetylaspartate (NAA) resonance at 2.02 ppm.

For *in vitro* studies, frozen brains (funnel frozen with liquid nitrogen within 1 min of an *in vivo* experiment) were cut into two hemispheres and brain metabolites extracted by 0.5 N perchloric acid (PCA).[12] After extraction, brain metabolites were dissolved in 0.6 ml phosphate-buffered saline in D_2O (pH 7.2). Lactate concentrations were determined by *in vitro* NMR spectroscopy on our home-built 5.6 T (240 MHz) vertical bore spectrometer and by lactate dehydrogenase (LDH) assays. Quantitative *in vitro* ^1H NMR spectra were acquired using presaturation to minimize the water resonance. Spectral parameters were 8 K data points, 45° pulse width, spectral width ± 2,500 Hz, repetition time 2.8 sec, and 256 transients. Concentrations of each metabolite were referenced to 3 mM sodium 3-trimethylsilyl(2,2,3,4-D_4)propionate (TSP). Duplicate aliquots were used for LDH assays (Sigma technical procedure No. 826-UV).

RESULTS AND DISCUSSION

Hypoxic Hypoxia and Global Ischemia (Cerebral)

Metabolic consequences of an ischemic insult may differ in some regards from those of hypoxic hypoxia (*vide supra*). Conceivably, these differences may be reflected in ^{31}P NMR spectra. FIGURE 1 compares some ^{31}P NMR spectra from an hypoxic hypoxia experiment with those from a global ischemia experiment on rats. The results entailing a 20-min episode of hypoxic hypoxia ($P_aO_2 = 30$ mm Hg), during anesthesia with 1% isoflurane, were quite reproducible.[13] Compared to the control spectrum obtained under hyperoxic conditions ($P_aO_2 \geq 300$ mm Hg) and correcting for saturation, there is a decrease of about 50% in PCr, a three-fold increase in monophosphate (MP), a four-fold increase in P_i signal and no change in ATP. The upfield shift of the P_i peak with hypoxia indicates a decrease in intracellular pH of 0.21 unit.[14] A thorough statistical analysis (repeated measures ANOVA and multiple comparisons tests) supported the qualitative observations. The rats all recovered metabolically and neurologically from their hypoxic experience. In contrast to the changes observed in hypoxic hypoxia experiments with adult rats, the same level of hypoxia in rabbit neonates (10–16 days old) for a longer hypoxic episode (60 min

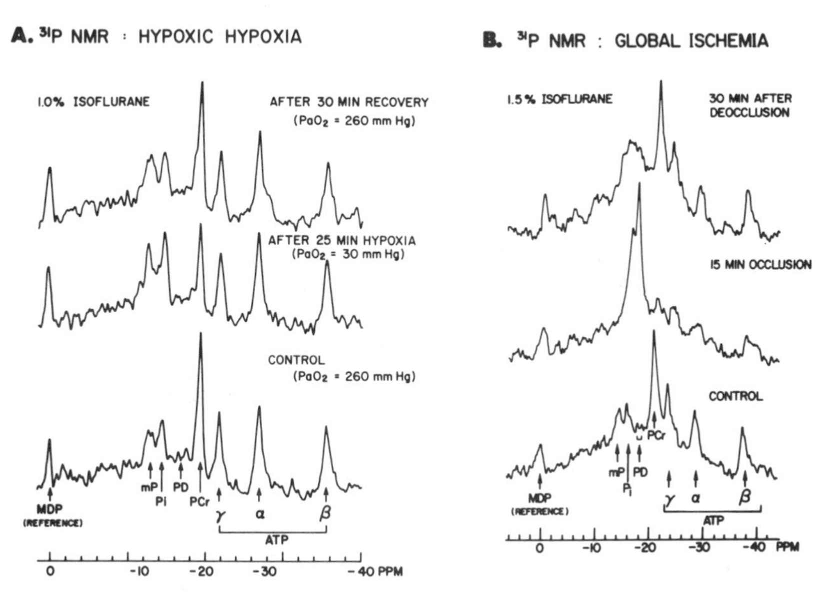

FIGURE 1. Comparison of changes in the ^{31}P NMR spectra (95.9 MHz) of rat brain during isoflurane (1%) anesthesia. Each spectrum was obtained in 5 min with a 2.0 sec repetition time. (A) Hypoxic hypoxia was induced in a rat after recording the control spectrum shown at the bottom. The spectrum in the middle, taken after the arterial oxygen tension was 30 mm Hg for 25 min, exhibits an increase in P_i and a decrease in PCr and pH_i. The recovery of the NMR spectrum is shown in the top tracing, which was obtained 30 min after restoration of 100% oxygen. (B) Global ischemia was induced in a different animal that had previously had both vertebral arteries transected. The control spectrum is shown at the bottom. Remotely controlled suture snares were used to occlude both carotid arteries while the animal remained motionless inside the spectrometer magnet. The spectrum in the center shows the metabolite pattern after 15 min of global ischemia. The top spectrum shows the spectrum that was obtained 30 min after perfusion was restored.

versus 20 min) produced no change in the high-energy phosphorus metabolites, but a 0.2 unit decrease in intracellular pH was observed.[15]

The global ischemia model employed for FIGURE 1B entailed a four vessel occlusion.[16] Considerable variability in the response to that cerebral ischemia insult has been observed in both ^{31}P and ^{1}H spectra. The variability may arise from individual variations in the peripheral vascular system of the rats studied such that global (or nearly so) ischemia is not always achieved. Regardless, for a number of individual rats, a 15-min ischemic episode produced (cf. FIG. 1B) a dramatic decrease in ATP, PCr, and pH_i and an increase in P_i, MP, and lactate (the latter not shown). Some of these animals also recovered metabolically their apparently depleted energy stores.[17] A study using two NMR surface coils, each placed over one hemisphere of the *in situ* rat brain, in a hemispheric brain ischemia model showed that ischemia induced in one hemisphere does not affect the metabolism in the other hemisphere, which can consequently be used as a continuous control.[18]

Lactate Quantitation

Stimulation of glycolysis under hypoxic and ischemic conditions is expected to lead to lactate production, which is believed to account for the drop in pH (*vide supra*). Indeed, this has been observed in biochemical analyses[1] and NMR studies of these cerebral insults.[6,11,17] Recently, the relationship between the lactate concentration measured by *in vivo* ^{1}H NMR and by *in vitro* NMR and enzymatic assay was investigated.[19]

During either hypoxic hypoxia or global cerebral ischemia, ^{1}H NMR spectra of the *in situ* rat brain (with scalp removed) were obtained using the presaturation Hahn spin-echo pulse sequence for the purpose of observing the lactate methyl and N-acetylaspartate (NAA) proton resonances (cf. FIG. 2). Because partially relaxed

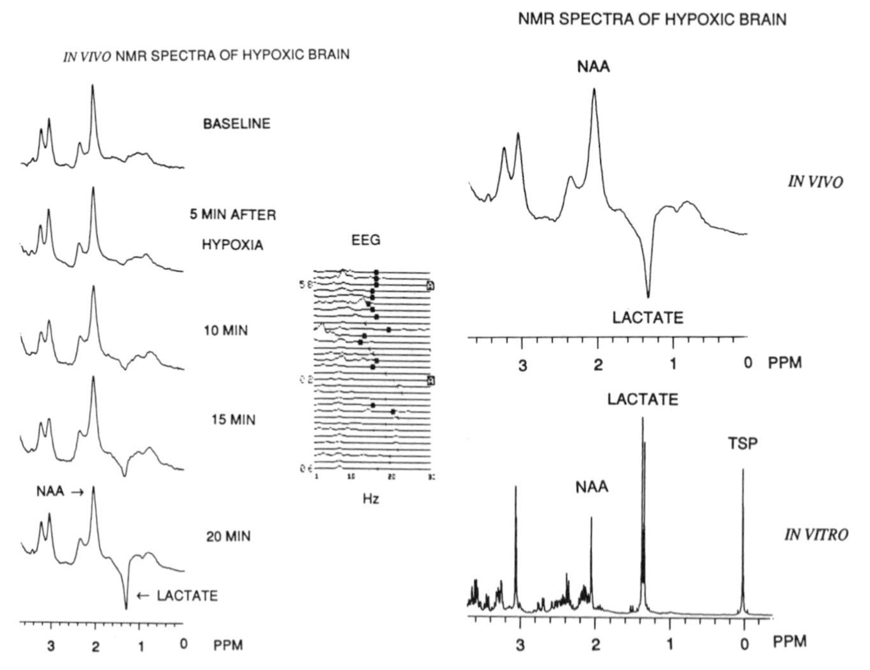

FIGURE 2. ^{1}H NMR spectra (236 MHz) of the hypoxic brain of an anesthetized (1.5% halothane) rat. *In vivo* NMR spectrum of a normal rat brain, 5 min after initiation of hypoxia, 10 min after, 15 min after, 20 min after, quantitative *in vitro* NMR spectrum of the extracted brain metabolites, EEG recording of the brain during hypoxia showing decreased brain activity at the end of the experiment.

TABLE 1. Saturation Factors and T_2 Values of NAA and Lactate in Normal and Injured Rat Brains ($N = 4$)

	Saturation Factor		T_2	
	NAA	Lactate	NAA	Lactate
Normal brain	1.44 ± 0.12	—	220 ± 33	—
Injured brain	1.59 ± 0.09	1.78 ± 0.11	193 ± 38	264 ± 46

and spin-echo spectra were collected, it was necessary to correct for T_1 (differential saturation) and T_2 effects. Consequently, the saturation factors and T_2 values of NAA and lactate were measured (cf. TABLE 1). Following acquisition of *in vivo* spectra, the brain was funnel frozen with liquid nitrogen, the metabolites extracted with perchloric acid, and *in vitro* ^{1}H NMR spectra obtained (following neutralization). Subsequently, lactate was analyzed with the lactate dehydrogenase assay. A number of experiments were run using either hypoxic hypoxia conditions (6% oxygen) or global ischemia conditions (four-vessel occlusion). The results for the hypoxia experiments are summarized in TABLE 2. The LDH assay verifies the *in vitro* NMR results for lactate. Similar results were obtained in ischemia experiments. The last column of TABLE 2 reveals that all lactate is not detected using the Hahn spin-echo sequence *in vivo;* that is probably due to incomplete inversion of the lactate spins because of B_1 field inhomogeneity with surface coils. The factor of 0.28 ± 0.08 may be used to correct for lactate detectability *in vivo* permitting quantitative lactate determinations.

Cerebral Acidosis and Supercarbia

Clearly lactate is produced and the pH_i falls during the cerebral insults of hypoxic hypoxia or ischemia. It has been postulated that the low pH causes, or at least is indicative of, irreversible brain injury.[1,2] As noted above, the animals did recover metabolic and neurologic function following the 20-min hypoxic hypoxia experiment. But that experiment produced a smaller pH drop (from 7.2 to 7.0) than an ischemia experiment, which can yield $pH_i \approx 6.5$. There is another method for lowering the brain pH to test the postulated ill effects of low pH. That method involves conditions of supercarbia. When CO_2 levels are high, there is rapid passive diffusion of CO_2 across membranes, most of which are impermeable to ions, and an increase in the concentrations of intracellular bicarbonate and hydrogen ions. Changes ensue in the PCr concentration and the lactate/pyruvate ratio because of the $[H^+]$ dependence of the equilibrium of the creatine kinase and lactate dehydrogenase reactions. Supercarbia *per se* is not known to cause energy failure.

^{31}P NMR experiments were performed on rats subjected to 15 min of supercarbia in which $P_aCO_2 = 490 \pm 80$ mm Hg, compared with the normal level of 35 mm Hg.[20] As high CO_2 levels induce anesthesia, no further anesthetic was necessary. As shown in FIGURE 3, supercarbia shifts the P_i peak upfield, indicative of a dramatic pH decrease, and causes a 25% reduction in the PCr signal intensity with no discernable effect on the ATP signals. The drop in PCr signal intensity is a consequence of the lower pH, which influences the equilibrium concentrations of species involved in the creatine kinase reaction. With administration of 70% CO_2 discontinued, the spectrum indicates metabolic recovery.

FIGURE 4 shows a comparison of intracellular and intraarterial pH values. The brain cells, which are better buffered than the blood, undergo a smaller decrease in pH_i

TABLE 2. Lactate Concentrations Determined from Hypoxia Experiments

Rat Number	*In Vivo* NMR[a] Lactate/NAA		PCA extracts[b] *In Vitro* NMR Lactate	NAA	Lactate/NAA	LDH Assay Lactate	*In Vivo* Lactate Correction Factor[c]
1	0.60	L	8.7	6.0	1.4	8.3	0.42
		R	8.2	5.8	1.4	7.6	
2	1.10	L	28.6	5.7	5.0	25.4	0.22
		R	26.7	5.2	5.2	25.5	
3	0.91	L	17.0	4.4	3.9	17.5	0.25
		R	22.9	6.5	3.5	19.3	
4	0.69	L	17.7	5.4	3.3	15.7	0.21
		R	16.7	5.3	3.2	15.1	
5	1.32	L	24.9	5.5	4.5	26.1	0.29
		R	25.2	5.5	4.5	28.1	
Average							0.28 ± 0.08

[a]Ratio calculated from peak areas corrected for saturation and T_2 effects.
[b]Concentrations in μmole/g; L and R indicate left and right hemispheres.
[c]The *in vivo* lactate correction factor was calculated by dividing the ratio in column 2 by the ratio in column 6, which represents the fraction of lactate observed by the Hahn spin-echo experiment *in vivo* compared to that by *in vitro* NMR.

and a more complete recovery. The five animals in the NMR study appeared completely normal 5 hr after initiation of the experiment. They were observed to behave normally with no health problems apparent for two months. The complete metabolic and neurologic recovery of the animals in this experiment shows that a 4.3-fold increase in cerebral intracellular hydrogen ion concentration does not, after 15 min, indicate or cause brain injury under conditions of deep anesthesia and adequate oxygenation and perfusion. This is in contrast to hydrogen ion increases of the same magnitude that occur after 15 min of damaging global ischemia.

ADP Concentrations During Hypercarbia: Effect of Anesthetic

The equilibrium constant for the creatine kinase reaction, PCr + ADP + $H^+ \rightleftharpoons$ creatine (Cr) + ATP, is

$$K = \frac{[H^+][PCr][ADP]}{[ATP][Cr]}.$$

From the ^{31}P spectrum, one can obtain relative concentrations of PCr and ATP, as well as $[H^+]$. With the reasonable assumptions that equilibrium obtains and the total known creatine pool ([PCr] + [Cr]) is constant, [ADP] can be calculated. There is evidence that either [ADP] or the phosphorylation potential $[ATP]/[ADP][P_i]$ controls mitochondrial respiration when oxygen and substrate are not rate limiting.[3]

In the presence of a high arterial CO_2 content with no anesthetic other than CO_2, cerebral pH_i and [PCr] decrease while [ATP] is unchanged (*vide supra*).[20] From the observed changes, it was calculated from the creatine kinase equilibrium that in the presence of P_aCO_2 = 490 ± 80 mm Hg, the free intracellular ADP concentration decreases to approximately one third of its control value (P_aCO_2 = 35 mm Hg). The calculated three-fold decrease in ADP and two-fold increase in cytosolic phosphorylation potential indicate that there is increased intracellular oxygenation during supercarbia. However, the calculated decrease in ADP is at variance with an earlier study at $P_aCO_2 = 266$ mm Hg, which concluded there was no change in ADP.[21] In addition to the somewhat lower arterial CO_2 content, that study also employed halothane as an anesthetic.

Consequently, ^{31}P NMR experiments were carried out in a protocol in which the rats ($N = 5$) were subjected to hypercarbia ($P_aCO_2 = 194 \pm 25$ mm Hg) in the presence of 0.5% halothane for 15 min followed by recovery with 100% oxygen and no CO_2 for 45 min, followed by a second similar hypercarbia experiment but with no halothane, and finally a second recovery.[22] ^{31}P spectra were acquired from the brain every 15 min during the experiment. There was complete metabolic recovery between the two hypercarbia episodes and complete metabolic and neurologic recovery after both episodes of hypercarbia. The same protocol was followed for further experiments utilizing 1.0% isoflurane in place of halothane. Blood pressure, heart rate, arterial pH, and blood gas measurements all indicated adequate cerebral perfusion and oxygen availability with the CO_2 and anesthetic levels employed.

TABLE 3 summarizes pertinent results from the hypercarbia experiments. General anesthesia generally causes central nervous system depression with concomitant

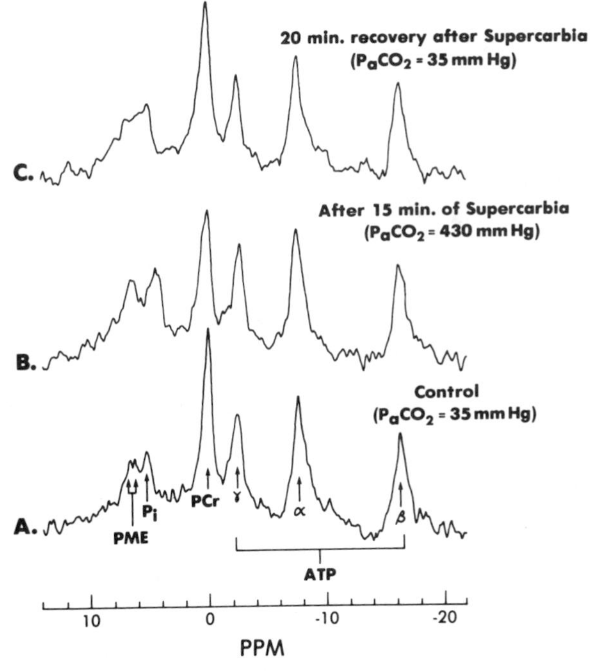

FIGURE 3. ^{31}P NMR spectra (95.9 MHz) of brain from one rat before supercarbia (A), after 15 min of supercarbia (B), and after 20 min of recovery (C). The ATP signal amplitudes are the same in all spectra. The PCr signal during supercarbia (B) is ~75% of that before (A) or after (C) supercarbia. The chemical shift of the P_i peak is closest to that of the PCr peak during supercarbia (B), indicating severe intracellular acidosis. PME, phosphomonoesters, which is identical to monophosphates (MP).

decrease in cerebral oxygen consumption and glucose metabolism. Simultaneous EEG measurements suggested that cerebral metabolic demands decreased during all of the hypercarbia episodes. Anesthesia is not known to change cerebral high energy phosphate concentrations under conditions of adequate tissue oxygenation and normocarbia.[23] This study, however, shows that despite adequate tissue oxygenation and constant ATP levels, the presence of one anesthetic, halothane, causes a significantly greater fall in PCr, with almost no change in ADP, during the stress of hypercarbia. When this stress occurs during isoflurane anesthesia and with no anesthesia, a significant fall in ADP with only a small decrease in PCr results.

The cytosolic creatine kinase reaction evidently stays coupled to ATP metabolism in the same way during hypercarbia with either isoflurane anesthesia or "no anesthesia." The creatine kinase reaction is coupled to mitochondrial reactions such as ATP synthesis, $ADP + P_i + H^+ \rightleftharpoons ATP$, as well as to transport processes and other

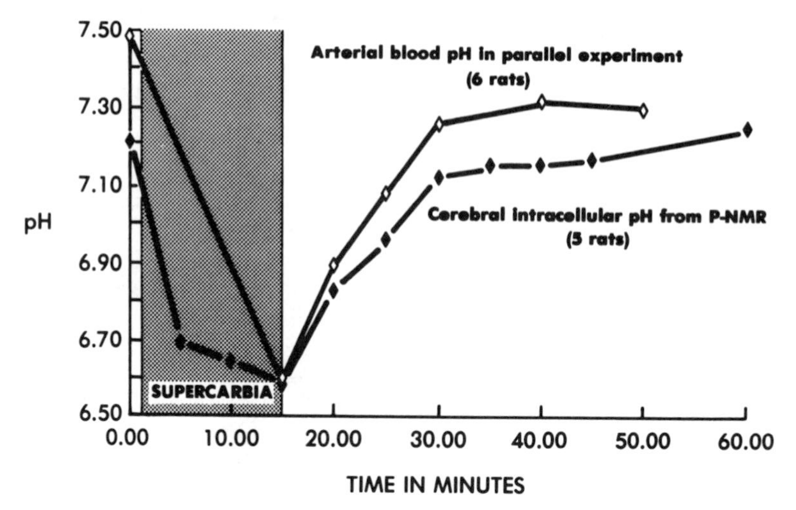

FIGURE 4. The time courses of cerebral intracellular pH (♦) in one set of animals ($N = 5$) and arterial blood pH (◊) in another set of animals ($N = 6$) during supercarbia and recovery. Both sets of animals had supercarbia induced and withdrawn according to the same protocol.

cytosolic reactions that alter ATP and ADP.[3] During hypercarbia with halothane anesthesia, biochemical processes involved in ATP metabolism are perturbed.

Monitoring Cerebral Anesthetics

Pharmacokinetic and metabolic studies of inhalation anesthetics have in the past required invasive techniques, e.g., autoradiography or indirect techniques, e.g., end-tidal gas chromatography. However, Wyrwicz and her colleagues pioneered the use ^{19}F NMR as a means of detecting anesthetic agents and their potential metabolites in the brain.[24] But in contradiction to invasive and end-tidal gas chromatography studies,[25,26] their pharmacokinetic studies resulted in 40% of the ^{19}F NMR signal being observed 7 hr after terminating 30 min of halothane anesthesia. If the elimination of anesthetics

TABLE 3. Effects of Halothane and Isoflurane on Cerebral Metabolic Changes during Hypercarbia ($P_aCO_2 \approx 200$ mm Hg)

Experiments	pH_1	pH_2	ΔpH	ATP_2/ATP_1	PCr_2/PCr_1	ADP_2/ADP_1
Group 1						
Halothane	7.24 ± 0.02	6.75 ± 0.06	0.49 ± 0.06	1.13 ± 0.06	0.58 ± 0.07[a]	0.84 ± 0.15[a]
No halothane	7.30 ± 0.03	6.68 ± 0.07	0.62 ± 0.06	1.09 ± 0.05	0.80 ± 0.05	0.36 ± 0.10
Group 2						
Isoflurane	7.29 ± 0.04	6.80 ± 0.04	0.49 ± 0.03	1.17 ± 0.11	0.91 ± 0.08	0.41 ± 0.12
No isoflurane	7.38 ± 0.08	6.84 ± 0.03	0.54 ± 0.08	1.36 ± 0.18	0.88 ± 0.08	0.44 ± 0.10

Values are means ± SE (statistical errors only). Subscripts 1 and 2 refer, respectively, to control values and values measured after 15 min of hypercarbia. $N = 5$ for each experimental group.

[a]Paired *t* test within group indicates statistically significant difference ($p < 0.01$).

from brain is substantially slower than generally believed, then clinical concerns about long-term anesthetic action following general anesthesia will need to be addressed.

We intended to verify independently the observations of Wyrwicz *et al.*[24] The first experiments were performed on rats measuring the uptake and elimination of halothane from brain via ^{19}F NMR.[27] We were concerned that only halothane from the brain be observed, so the following steps were taken. A relatively small (12 × 8 mm) elliptical surface coil was placed on the rat's head over the brain. The ^{19}F NMR spectra were acquired using Bendall's "depth pulse R,"[28] which improves signal localization and thus should minimize detection of signals from outside the brain. The two-dimensional "topological maps" of signal response from the coil shown in FIGURE 5 demonstrate that the detection volume (the mountains in the "topo maps") for a depth pulse is more sharply defined than that from a typical one-pulse experiment.

Control spectra were obtained following 40 min of 0.25% halothane anesthesia. The inspired halothane concentration was increased to 1% for 60 min at which time the halothane administration was terminated. ^{19}F NMR spectra were acquired every 6 min during the period of halothane administration and for 90 min following its termination. The wash-in and wash-out of halothane from the brain as monitored by ^{19}F NMR is shown in FIGURE 6. After 40 min administration, a steady-state concentration of halothane is reached. Upon wash-out, the signal is decreased to 40% of its maximum value in 34.6 ± 8.0 min ($N = 5$) with no signal detectable past 90 min. To be entirely convinced that signal was being obtained only from brain, in two rats, a bilateral temporo-parietal craniectomy was performed after nearby muscle and scalp tissues were excised; an area of ≈ 2.0 × 2.5 cm of dura was exposed. A 4 × 6 mm two-turn surface coil was placed on the dura and ^{19}F spectra were acquired using the above

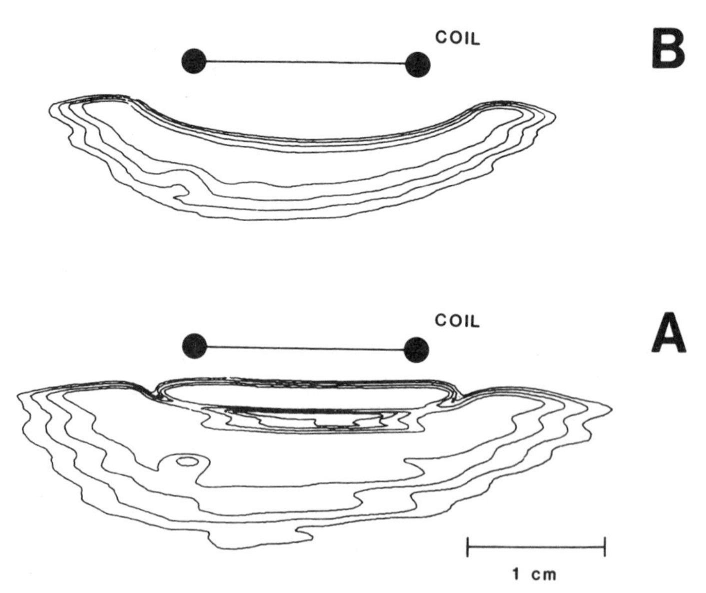

FIGURE 5. Two-dimensional, cross-sectional image showing the magnetic excitation measured in a homogeneous liquid underneath a surface coil. The $\mathbf{B_0}$ field is perpendicular to the plane of the paper. (A) A one-pulse sequence was used. (B) Bendall's depth pulse R[28] was used. The contours topographically define regions of magnetic field strength. A thin horizontal region of substantial signal intensity can be seen in (A) between the coil and the crescent-shaped region of substantial signal intensity.

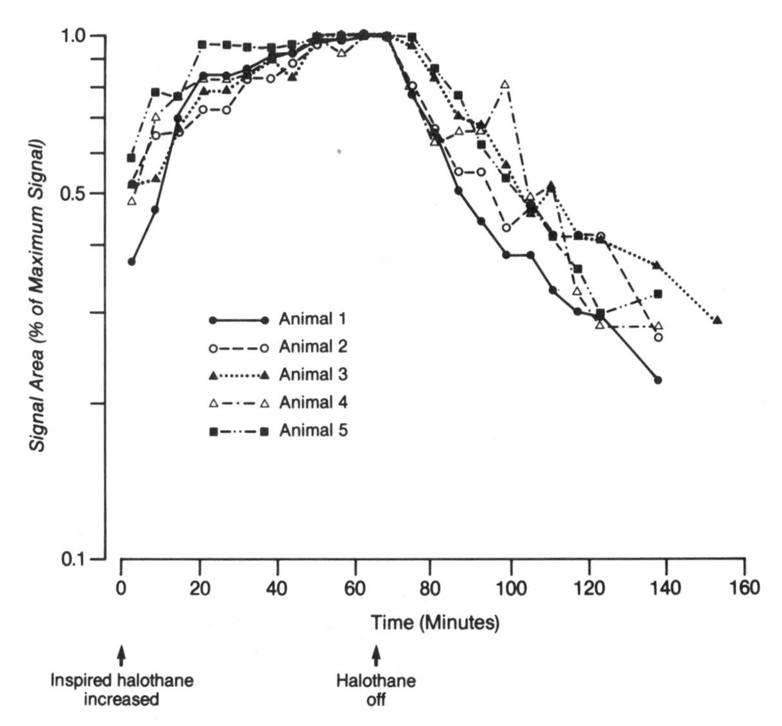

FIGURE 6. A semilog plot of the halothane wash-in and wash-out data from the brain, obtained with the depth-pulse technique, for five animals. No surgery was performed in these animals, which all recovered after anesthesia. The vertical axis (log scale) indicates ^{19}F NMR signal intensity for halothane, while the horizontal axis gives the time in minutes. Arrows indicate when the halothane concentration was increased, and then turned off.

protocol (but with 12 min spectral accumulation). The halothane wash-in and wash-out results confirmed our other results, which are consistent with results from the non-NMR studies.

^{19}F NMR images (spin-echo) have also been obtained showing the *in vivo* spatial distribution of halothane in the rabbit head following respiration for 5 hr with 1.5% halothane in oxygen.[29] Because the halothane concentration is low *in vivo,* and because the measured relaxation times of the ^{19}F resonance peak for halothane were $T_1 \approx 1.0$ sec and $T_2 \approx (3.5–65)$ msec; 1 to 3 hr imaging times were required (pulse repetition time TR = 1 sec, echo time TE = 9 msec) in order to obtain adequate images with a 64 × 256 raw data matrix and a 20-mm slice thickness. A ^{19}F NMR image (coronal) is compared in FIGURE 7 with T_1-weighted ^{1}H coronal images, one being a lipid-only image, of the rabbit head. With this technique, halothane was primarily detected in lipophilic regions of the rabbit head with little or no halothane observed in brain tissue. Because T_2 was shorter in brain tissue than in surrounding fat, a shorter TE than we could obtain is needed for optimal spin-echo imaging of brain halothane.

The uptake and elimination of isoflurane from rabbit brain was also examined via ^{19}F NMR after 90-min exposure to the anesthetic (1.5%). The isoflurane wash-out kinetics we observed with a 1.0-cm coil located directly over the brain (following a craniectomy) differed significantly from the wash-out rate observed with a 3.0-cm coil located over the intact head. This difference arises because the large coil detects tissues

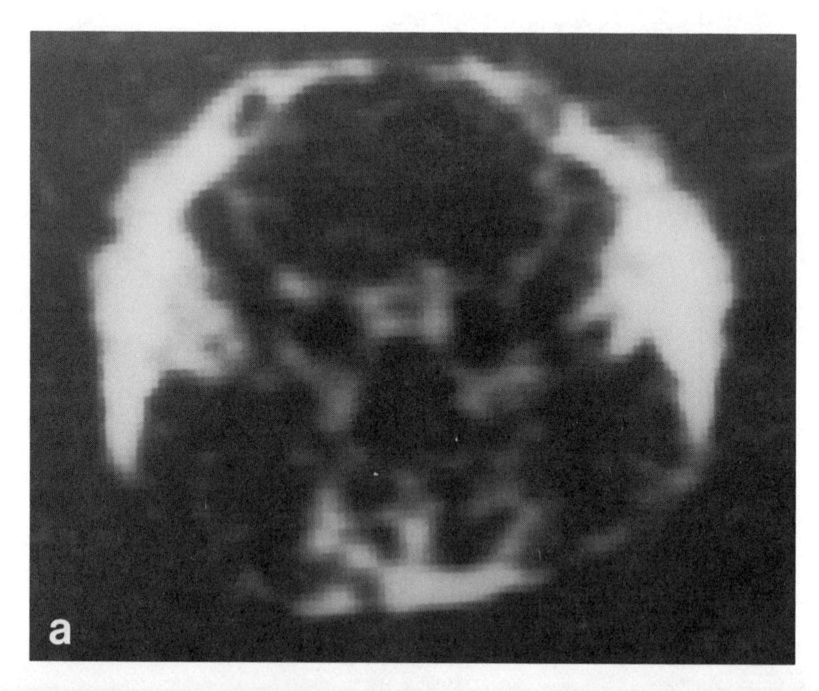

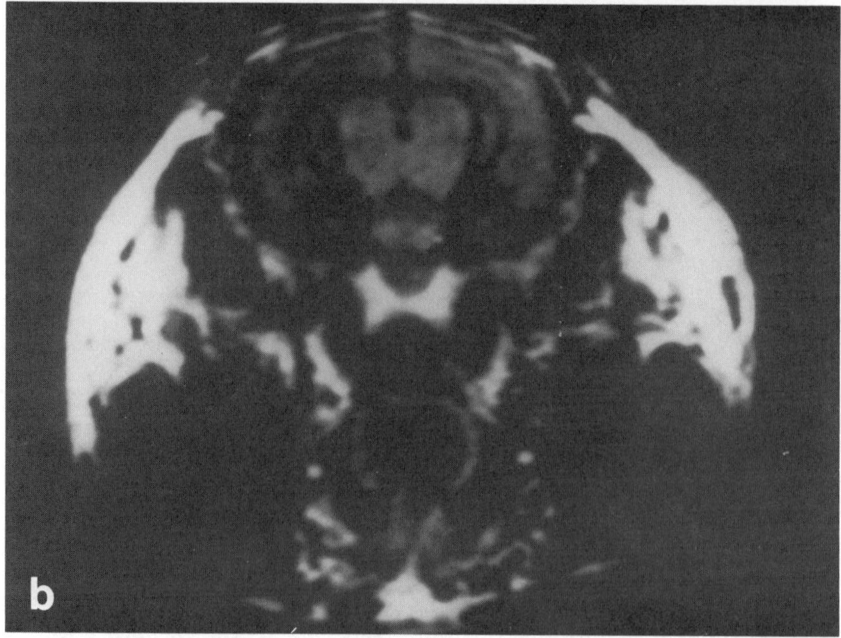

FIGURE 7.

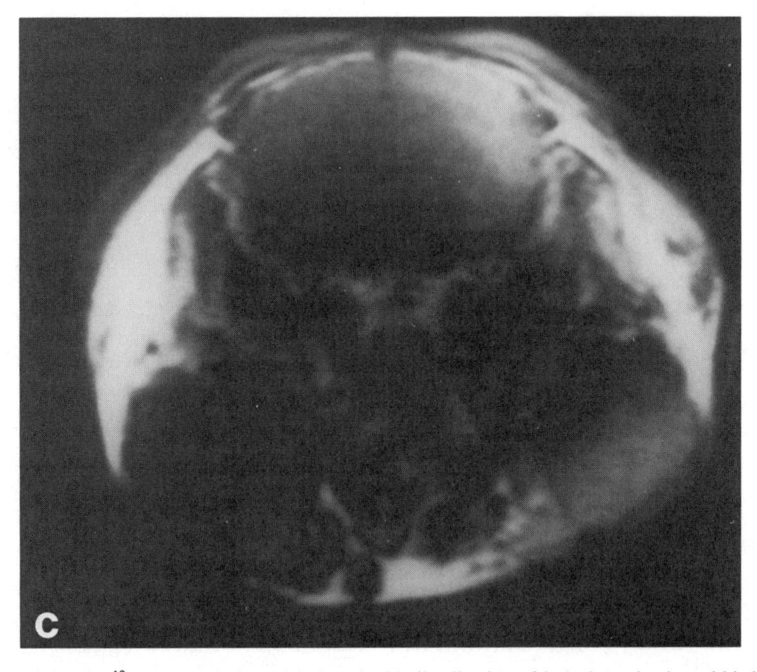

FIGURE 7. (A) ^{19}F coronal image showing the biodistribution of halothane in the rabbit head. The acquisition time was 3 hr, TR was 1 sec, TE was 9 msec, and slice thickness was 20 mm. (B) T_1-weighted ^{1}H coronal image composed by summing three separate 4 mm thick slices from positions corresponding to the front, middle, and rear of the 20 mm thick ^{19}F slice. TR = 300 msec, TE = 15 msec, and 5 min acquisitions for each of the three slices. (C) ^{1}H lipid-only image (water suppressed) composed by summing three separate 4 mm thick slices from positions corresponding to the front, middle, and rear of the 20 mm thick ^{19}F slice.

outside the brain, but in close proximity to it. The isoflurane wash-out kinetics obtained with the small coil agree with previous invasive studies. The 3.0-cm coil resulted in a biexponential wash-out curve with the signal decrease from the faster component being consistent with wash-out from brain (with possible muscle contributions) and the slower-decaying component being consistent with wash-out from fat and possibly bone marrow. The 3.0-cm coil detected substantial ^{19}F signal 7.5 hr after termination of anesthetic inhalation.

ACKNOWLEDGMENTS

We wish to acknowledge the aid and support of Drs. William Hamilton, Max Keniry, Todd Richards, John Severinghaus, and Julia Shuleshko.

REFERENCES

1. SIESJÖ, B. K. 1984. J. Neurosurg. **60:** 883–908.
2. WELSH, F. A. 1984. J. Cerebr. Blood Flow Metab. **4:** 309–316.

3. Erecińska, M. & D. F. Wilson. 1982. J. Membr. Biol. **70:** 1–14.
4. Gadian, D. G. 1982. NMR and Its Application to Living Systems. Oxford Univ. Press. Oxford.
5. James, T. L. & A. R. Margulis, Eds. 1984. Biomedical Magnetic Resonance. Radiology Research and Education Foundation. San Francisco, CA.
6. Behar, K. L., D. L. Rothman, R. G. Shulman, O. A. C. Petroff & J. W. Prichard. 1984. Proc. Natl. Acad. Sci. USA **81:** 2517–2521.
7. Murphy-Boesch, J. & A. P. Koretsky. 1983. J. Magn. Reson. **54:** 526–532.
8. Ackerman, J. J. L., J. L. Evelhoch, B. A. Berkowitz, G. M. Kichura, R. K. Duel & K. S. Lown. 1984. J. Magn. Reson. **56:** 318–322.
9. González-Méndez, R., L. Litt, A. P. Koretzky, J. von Colditz, M. W. Weiner & T. L. James. 19874. J. Magn. Reson. **57:** 526–533.
10. James, T. L. 1975. Nuclear Magnetic Resonance in Biochemistry. Academic Press. New York.
11. Rothman, D. L., K. L. Behar, H. P. Hetherington & R. G. Shulman. 1984. Proc. Natl. Acad. Sci. USA **81:** 6630–6634.
12. Agris, P. F. & I. D. Campbell. 1982. Science **216:** 1325–1326.
13. Litt, L., R. González-Méndez, P. R. Weinstein, J. W. Severinghaus, W. K. Hamilton, J. Shuleshko, J. Murphy-Boesch & T. L. James. 1986. Magn. Reson. Med. **3:** August.
14. Seo, Y., M. Murakami, H. Watari, Y. Imai, K. Yoshizaki, H. Nishikawa & T. Morimoto. 1983. J. Biochem. **94:** 729–734.
15. González-Méndez, R., A. McNeill, G. A. Gregory, S. D. Wall, C. A. Gooding, L. Litt & T. L. James. 1985. J. Cerebr. Blood Flow Metab. **5:** 512–516.
16. Pulsinelli, W. A. & J. B. Brierly. 1979. Stroke **10:** 267–272.
17. Weinstein, P. R., T. Richards, M. A. Keniry, B. M. Pereira, L. Litt & T. L. James. 1985. Society of Magnetic Resonance in Medicine. Fourth Ann. Mtg. London. Abstract p. 293.
18. Chew, W. M., M. E. Moseley, M. C. Nishimura, T. Hashimoto, L. H. Pitts & T. L. James. 1985. Magn. Reson. Med. **2:** 567–575.
19. Chang, L.-H., B. M. Pereira, M. A. Keniry, J. Murphy-Boesch, L. Litt, T. L. James & P. R. Weinstein. 1986. Society of Magnetic Resonance in Medicine. Fifth Ann. Mtg. Montreal. Abstract p. 1071.
20. Litt, L., R. González-Méndez, J. W. Severinghaus, W. K. Hamilton, J. Shuleshko, J. Murphy-Boesch & T. L. James. 1985. J. Cerebr. Blood Flow Metab. **5:** 537–544.
21. Siesjö, B. K., J. Folbergrova & V. MacMillan. 1972. J. Neurochem. **19:** 2483–2495.
22. Litt, L., R. Gonzalez-Mendez, J. W. Severinghaus, W. K. Hamilton, I. J. Rampil, J. Shuleshko, J. Murphy-Boesch & T. L. James. 1986. J. Cerebr. Blood Flow Metab. **6:** 389–392.
23. Siesjö, B. K. 1978. Brain Energy Metabolism: 256–258. Wiley. New York.
24. Wyrwicz, A. M., M. H. Pszenny, J. C. Schofield, P. C. Tillman, R. E. Gordon & P. A. Marin. 1983. Science **222:** 428–430.
25. Cohen, E. N., K. L. Chow & L. M. Mathers. 1972. Anesthesiology **37:** 324–331.
26. Carpenter, R. L., E. I. Eger II, B. H. Johnson, J. D. Unadkat & L. B. Sheiner. 1986. Anesth. Analg. **65:** 575–582.
27. Litt, L., R. González-Méndez, T. L. James, D. I. Sessler, P. Mills, W. Chew, M. Moseley, B. Pereira, J. W. Severinghaus & W. K. Hamilton. 1986. Magn. Reson. Med. **3:** 619–625.
28. Bendall, M. R. 1984. *In* Biomedical Magnetic Resonance. T. L. James & A. R. Margulis, Eds.: 99–126. Radiology Research and Education Foundation. San Francisco, CA.
29. Chew, W. M., M. E. Moseley, P. A. Mills, D. Sessler, R. González-Méndez, T. L. James & L. Litt. 1987. Magn. Reson. Imaging. **5:** 51–56.
30. Mills, P., D. Sessler, M. Moseley, W. Chew, B. Pereira, T. L. James & L. Litt. 1987. Anesthesiology. (In press.)

DISCUSSION OF THE PAPER

P. A. BOTTOMLEY: My first question concerns trying to qualify concentrations of lactate using N-acetyl aspartate: Are you sure that all of the N-acetyl aspartate is NMR observable and that therefore the ratios do give you an accurate absolute concentration of lactate? Would the insults that you're doing on these animals, if done to humans result in a neurologic damage? You mentioned in your animal experiments that there is no neurologic damage, but perhaps animals are different from humans. Would you like to comment?

T. L. JAMES: Yes. Essentially the first question is how confident are we that we have determined the N-acetyl aspartate. All of it *in vitro* and we get the same thing *in vivo*. At the moment there is some reluctance on my part to lay my head on the chopping block to say that we have determined all of it. At present we have no indication that N-acetyl aspartate—some of it—may be tightly bound. However, that's a problem we are concerned with. But on the other hand the results and the numbers we get at least seem to be reasonable.

In terms of the second question: you're right. There is conceivably a difference between human beings and animals in terms of response to some insults, especially if we're to talk about some of these ischemic insults. But with the hypoxic insults, these appear to be very reproducible. They appear to be global in terms of their effects and I suspect that the results will be essentially the same in animals and humans, although the levels of oxygenation we may have to use to achieve the same level of insult may change going from one type of animal to another.

QUESTION: Did you find any correlation between increases in lactate and neurological outcome as you did a dissociation between pH and neurological outcome?

JAMES: There's not a quick answer because . . .

SHULMAN: Yes or no

JAMES: No.

HOPKINS: To confirm your comments, I'm afraid to say that we have done humans under the same conditions, and there was no lasting neurological effect. This refers to CO_2.

JAMES: What insult was this?

HOPKINS: CO_2, 40% with the same pH.

POSTSCRIPT

A. L. HOPKINS: It has been suggested here (from the floor by the previous discussant) and on other occasions, that rats may be unique in their ability to withstand the high levels of CO_2 that were employed by Dr. James and his group. The answer to this is found in a paper published in 1955 demonstrating that both dogs and humans can withstand mixtures containing from 35% to 55% CO_2 provided that all of the precautions that Dr. James used are followed. We found arterial pH to be in the same range as he has reported and can confirm his finding that CO_2 at these levels is an anesthetic. The effect on blood pressure, heart rate, and the EEG is shown in the accompanying figure taken from that work by Clowes *et al.* [Clowes, G.H.A., A.L. Hopkins & F.A. Simeone. 1955. A comparison of the physiological affects of hypercapnia and hypoxia in the production of cardiac arrest. Ann. Surgery. **142**(3): 446–460].

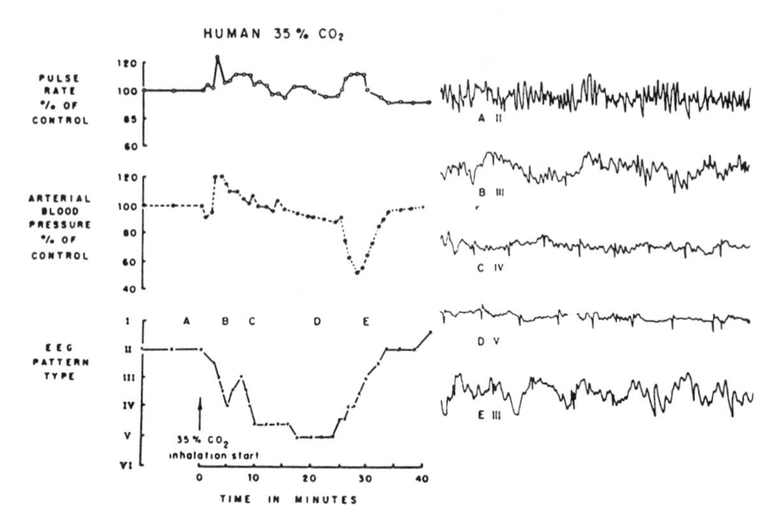

FIGURE 1. The response of pulse rate, blood pressure, and electroencephalographic pattern of man to the administration of 35% CO_2–65% O_2 for 22 min (arterial blood pII 6.82, CO_2: 78 volumes %). (From Clowes *et al.*)

Combined ^{1}H and ^{31}P NMR Studies of the Rat Brain *in Vivo:*

Effects of Altered Intracellular pH on Metabolism[a]

K. L. BEHAR, D. L. ROTHMAN, S. M. FITZPATRICK,
H. P. HETHERINGTON, AND R. G. SHULMAN

Department of Molecular Biophysics and Biochemistry
Yale University
New Haven, Connecticut 06510

INTRODUCTION

A sustained intracellular acidosis may lead to marked changes in the concentrations and turnover rates of cerebral metabolites. Hydrogen ions participate in several enzyme-catalyzed reactions operating near equilibrium (e.g., creatine phosphokinase, lactate dehydrogenase, etc.). Therefore, the steady-state concentrations of substrates linked to these reactions could change according to $[H^+]$.

Changes in $[H^+]$ may also affect rates through metabolic pathways. The effects of pH on tissue glucose utilization have been studied extensively.[1–5] Under steady-state conditions, low intracellular pH decreases glucose utilization, whereas intracellular alkalosis enhances it.[3–5] These effects are ascribed to an influence of $[H^+]$ on the activity of phosphofructo-1-kinase.[6,7] It has been suggested that low pH_i may contribute to the depressed rates of glucose uptake following recirculation after a period of ischemia.[8] Differences may exist, however, between the effects of $[H^+]$ on the glycolytic rate during ischemia and that occurring in the steady state. For example, during exercise in ischemic muscle, acidification (due to glycolytic production of lactic acid) persists to intracellular pH values as low as 6.0.[9]

Changes in pH also have a dramatic effect on the concentrations of TCA cycle intermediates and amino acids linked to the TCA cycle through aminotransferases and dehydrogenases.[10] Changes in the steady-state concentrations of these organic acid anions during acid/base changes has been proposed as a means of achieving intracellular pH regulation.[11,12] The pH-sensitive site(s) that determine the direction and magnitude of these changes are unknown; knowledge of their location would lead to a better understanding of how amino acid levels are altered for those conditions leading to cellular acidosis.

Hypercarbia is an efficient means of inducing intracellular acidosis as carbon dioxide is easily transported through the blood-brain barrier. The diffusion of CO_2 along its concentration or partial-pressure gradient determines net movement into or out of brain cells. Variations in CO_2 tensions in blood will lead, therefore, to parallel changes in brain tissue CO_2 tensions. Following the hydration and rapid dissociation of intracellular CO_2—assisted by the high catalytic activity of carbonic anhydrase—H^+ is liberated, lowering the pH_i according to:

$$CO_2 + H_2O = H_2CO_3 = H^+ + HCO_3^- \quad (1)$$

[a]Supported by National Institutes of Health Grant DK 27121.

Until recently, information concerning the effects of elevated CO_2 tensions on the energy state of brain tissue originated from studies of tissue extracts.[13] Direct measurements by ^{31}P NMR of phosphocreatine, nucleoside triphosphates, P_i, and pH_i during acute and progressive alterations in arterial pCO_2 in rabbit,[14] canine,[15] and rat brains,[16] have duplicated and extended results obtained from other laboratories using metabolite extraction techniques. The NMR method provides, however, the opportunity to determine reaction rates by non-destructive multiple-sampling of substrate levels in single animals—unachievable by conventional extraction techniques.

We present preliminary results of ^{31}P and ^1H NMR experiments designed to elucidate *in vivo* the effect of increased CO_2 on cerebral pH_i, high energy phosphates, and the metabolic pathways of TCA-cycle–linked amino acids in the rat brain.

METHODS

Animal Preparation

White albino male rats, fed *ad libitum* and weighing between 170 to 220 g, were tracheotomized under enflurane anesthesia, paralyzed with *d*-tubocurarine chloride

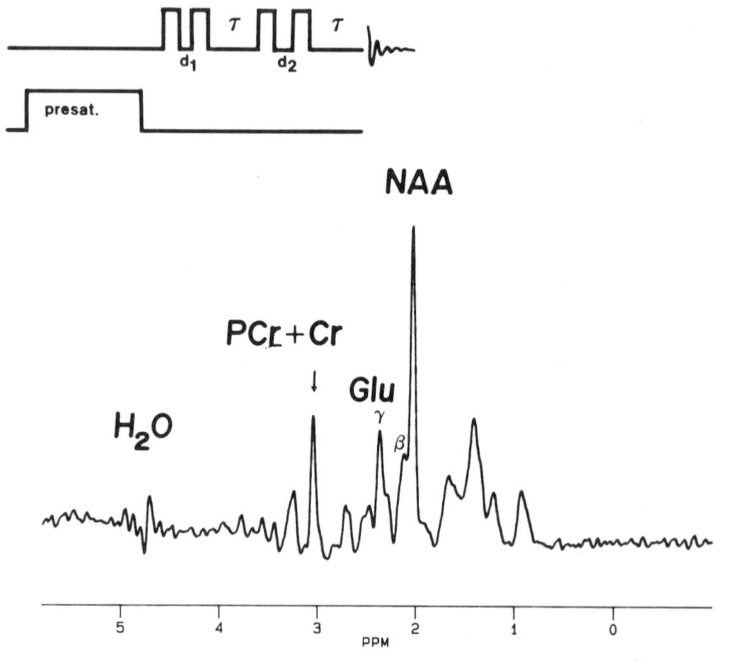

FIGURE 1. ^1H NMR spectrum of rat brain obtained by a semi-selective spin-echo pulse sequence. The pulse sequence depicted consists of semi-selective $1\bar{1}$ pulses (20 μsec each) for both the θ and 2θ refocusing pulse of a spin-echo sequence. The interpulse delays, d_1 and d_2 ($=2d_1$), were adjusted to give a maximum spectral power distribution for the methyl protons of N-acetylaspartate (NAA) at 2.02 ppm. The delay time, τ, was 4 msec. Additional suppression of water was achieved by presaturation (presat). Glu γ and β; glutamate γCH_2 and βCH_2, respectively.

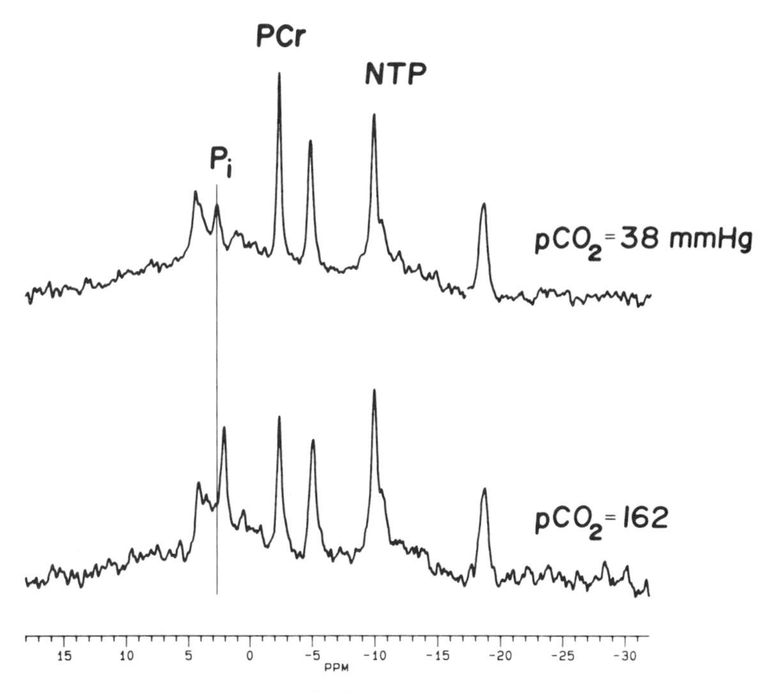

FIGURE 2. Effect of hypercarbia on the ^{31}P NMR spectrum of rat brain. (A) Normocarbia (arterial pCO_2 = 38 mm Hg). (B) Hypercarbia (arterial pCO_2 = 162 mm Hg). The vertical line, centered on P_i in Spectrum A, depicts the upfield shift of the P_i peak position in Spectrum B during hypercarbic acidosis. P_i, inorganic orthophosphate; PCr, phosphocreatine, NTP, nucleoside triphosphate.

(s.c.), and mechanically ventilated with 70% N_2O and 30% O_2. Surgery was performed under 1.5% enflurane anesthesia; its use was discontinued prior to placing the animal in the magnet. Core temperature was maintained near 37°C by a jacket made of Tygon tubing using a temperature-regulated, pump-recirculated water bath. The tail artery was cannulated for sampling of arterial blood. Blood pH, pO_2, pCO_2, and glucose were measured from 100 μl samples. The mean arterial blood pressure was recorded continuously through a polyethylene T-junction connector at the base of the magnet. Following a midline incision through the scalp, the skin was retracted laterally and sutured. The periosteum was removed, leaving the skull clear of tissue between the temporal ridges of the skull. The rat was placed in a Plexiglas cradle and this assembly was attached to the surface-coil probe.

Hypercarbia was produced by replacing N_2O with CO_2 in the gas mixture. A progressive decrease in blood glucose concentration often occurred over the period of hypercarbia. To prevent complications of hypoglycemia, the period of CO_2 administration was not extended beyond 30 min.

NMR Experiments

^{1}H and ^{31}P NMR spectra were obtained at 8.4 Tesla using an AM-360 wide-bore spectrometer (Bruker Instruments) at 360.13 and 145.78 MHz, respectively. The

surface-coil was double-tuned to both frequencies[17] and consisted of an 8 × 14-mm single-turn ellipse. ^{31}P spectra were collected over 5-min periods (512 scans) using a 0.6 sec repetition time and a pulse width of 15 μsec. ^{1}H spectra were collected over 24-sec periods (16 scans) using a 1.5 sec repetition time and a nominal θ pulse width of 20 μsec. Water suppression was achieved both by presaturation and a phase-cycled semi-selective $1\bar{1}$ spin-echo pulse sequence.[18–20] The spin-echo pulse sequence consisted of either a non-selective θ pulse and selective 2θ refocusing pulse or selective θ and 2θ pulses (FIG. 1). The spin-echo delay time, τ, was 4 msec. The intensity maximum of the excitation profile was positioned on the N-acetylaspartate methyl proton resonance at 2.02 ppm. This procedure yielded a spectral intensity profile for resonances extending from about 3.3 to 0.8 ppm (FIG. 1). The relevant peak assignments are given in FIGURE 1 and are based on previous studies.[17,21]

RESULTS AND DISCUSSION

Effects of Hypercarbia on the ^{31}P NMR Spectrum

Following exposure of rats to CO_2, the mean arterial blood pressure and oxygen tension rose while pH fell. The ^{31}P NMR spectrum of a rat brain acquired prior to hypercarbia contains seven discernable resonances (FIG. 2A)—in order of increasing field: phosphorylethanolamine and lesser amounts of phosphorylcholine, P_i, PCr, and NTP γ, α, and β was previously described for rat brain at a comparable field strength and resolution.[17] The intracellular pH calculated from the chemical shift of inorganic

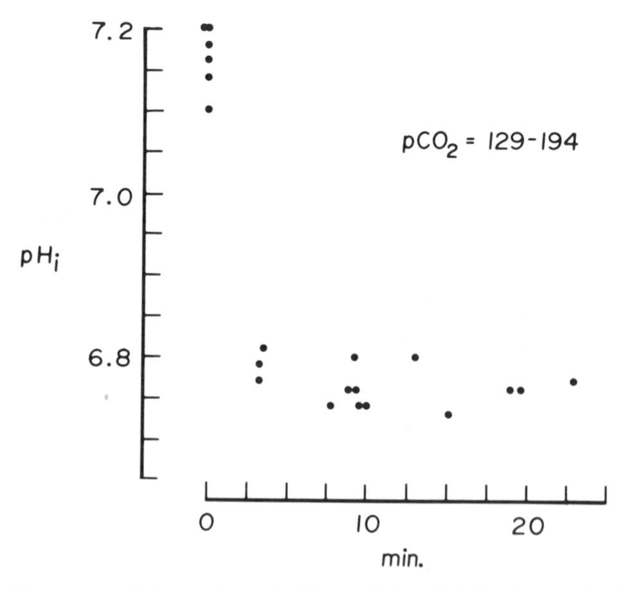

FIGURE 3. Time course of changes in cerebral intracellular pH following sustained hypercarbia in the rat. The CO_2 tensions for the group of animals measured ranged between 129–194 mm Hg. Each point represents the mean time of a 5-min ^{31}P spectrum acquisition. ^{1}H spectra were collected alternately with ^{31}P spectra (see FIG. 5).

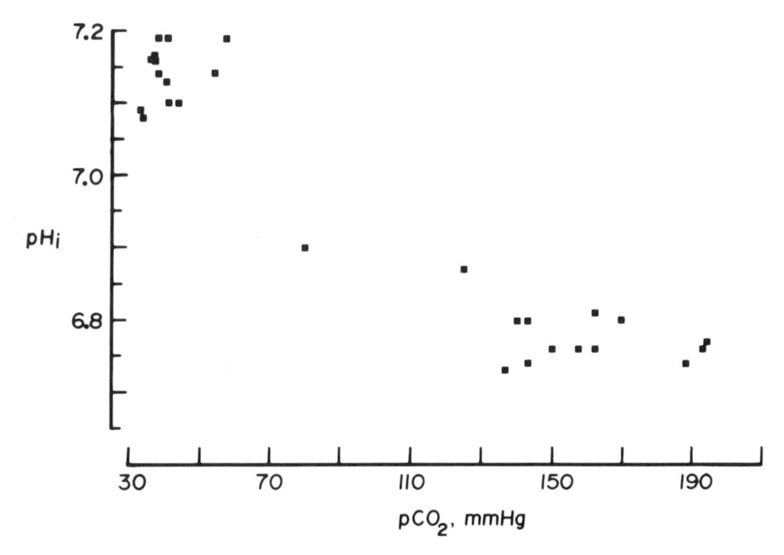

FIGURE 4. Relationship between cerebral pH_i and arterial pCO_2. Intracellular pH values were obtained from ^{31}P NMR spectra during the interval of stable pH_i for a given value of pCO_2.

phosphate and the titration endpoints specified in Petroff *et al.*[22] gave a pH_i of 7.15 for six animals with arterial CO_2 tensions of 36–58 mm Hg. Following hypercarbia (mean pCO_2 = 149 mm Hg; N = 6) of 7–12 min duration, pH_i fell to 6.77; this decrease is represented by a substantial upfield (decreasing δ) shift in the P_i peak position as shown in FIGURE 2B. The acidification was rapid; pH_i fell to values between 6.7 and 6.8 within the first 5 min and then remained stable for at least 20 min (FIG. 3). However, because of the 5-min signal averaging time of ^{31}P spectra, the actual initial rate of acidification was probably greater. As expected, pH_i decreased as the arterial pCO_2 increased (FIG. 4). The greatest decrease in pH_i occurred for CO_2 tensions between 60–130 mm Hg; a lesser incremental decrease occurred for tensions beyond 130 mm Hg.

Concomitant with the fall in pH_i, PCr decreased to 72 ± 7% (S.D.) and P_i increased to 200 ± 30% of their normocarbic peak amplitudes. At no time during hypercarbia did NTP—as measured from the peak height of NTPβ—vary from control levels. These results are summarized in TABLE 1; they are similar, qualitatively, to other ^{31}P NMR studies of hypercarbia. For example, following progressive changes in the CO_2 tension in rabbits,[14] cerebral pH_i fell to 6.9, while PCr decreased by 20% and P_i increased by 50–60% at 130 mm Hg. In unanesthetized rats maintained at a pCO_2 of 200 mm Hg for 15 min,[16] pH_i fell to 6.68–6.84 concomitantly with a 12–20% decrease in PCr.

Effects of Hypercarbia on the ^1H NMR Spectrum

The effect of hypercarbia on the cerebral glutamate level was evaluated in ^1H spectra obtained alternately with ^{31}P spectra. Hypercarbia (pCO_2 = 125–162 mm Hg) of 17–25 min duration, led to a marked decrease in intensity of the γ (2.34 ppm) and β (2.1 ppm) methylene protons of glutamate. The glutamate-to-creatine (Glu_γ/Cr) ratio

TABLE 1. Effects of Hypercarbia on Rat Brain

	pCO_2 (mm Hg)	pH_i	PCr (% of Normocarbic Controls)	P_i	NTP_β	Glu/Cr[a]
Normocarbia	45 (36–58)	7.15				
Hypercarbia[b]	149 (128–170)	6.77	72 ± 7	200 ± 30	100 ± 5	36–72

[a]Changes in cerebral glutamate content were evaluated by calculation of the ratio of the glutamate γCH_2 peak height to that of creatine + phosphocreatine (3.03 ppm) obtained in the ^1H spectrum. The values given were obtained after 17–25 min of hypercarbia.

[b]The values for pH_i, PCr, P_i, and NTP_β represent a 7–12 min period of hypercarbia at the level indicated.

The range of values of pCO_2 is given in parentheses.

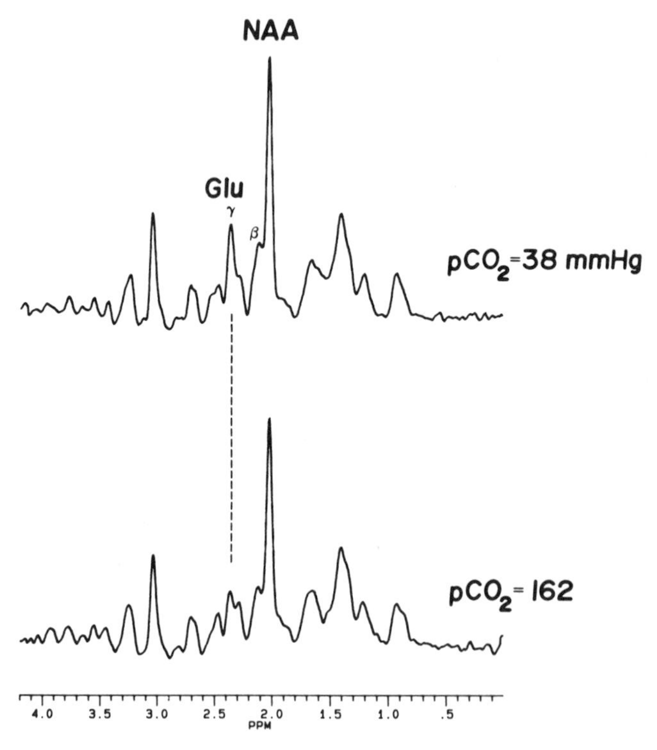

FIGURE 5. Effects of hypercarbia on glutamate levels observed in the ^1H spectrum of rat brain. Upper Spectrum: normocarbia. Lower Spectrum: 20 min of sustained hypercarbia. The spin-echo delay time, τ, was 4 msec. Vertical dashed line depicts position of the glutamate γCH_2 resonance. NAA; N-acetylaspartate methyl protons.

fell to 36–72% of its control value in the brains of four rats (TABLE 1). The effect of hypercarbia *in vivo* on the ^{1}H spectrum of a rat brain is shown in FIGURE 5. The time course for the observed decrease in glutamate in ^{1}H spectra was slower than the time course for changes observed for PCr, P_i and pH_i in ^{31}P spectra. Although the pH_i had stabilized at a lower value after 5 min of hypercarbia (FIG. 3), glutamate decreased progressively throughout the entire period. This observation would suggest that the decrease in glutamate is not the result of an equilibrium-based redistribution, but rather the result of an altered substrate flux. Results from other laboratories have shown that the decrease is not confined to glutamate alone: several amino acids and TCA cycle intermediates undergo a net decrease in concentration during hypercarbic acidosis.[10,12] It has been suggested that an important process regulating cerebral pH_i in response to an imposed intracellular acid load involves the net oxidation of organic acids.[11,12] While much evidence exists for this process, little is known concerning the site(s) in the respective pathways where these fluxes are sensitive to $[H^+]$. Work is now in progress to quantitate accurately the pH_i-dependent changes in glutamate content and determine the underlying cause of this phenomenon.

REFERENCES

1. DOMONKOS, J. & I. HUSZÁK. 1959. J. Neurochem. **4:** 238–243.
2. OPIE, L. H. 1965. Am. J. Physiol. **209:** 1075–1080.
3. DELCHER, H. K. & J. C. SHIPP. 1966. Biochim. Biophys. Acta **121:** 250–260.
4. SCHEUER, J. & M. N. BERRY. 1967. Am. J. Physiol. **213:** 1143–1148.
5. VAN NIMMEN, D., J. WEYNE, G. DEMEESTER & I. LEUSEN. 1968. J. Cerebr. Blood Flow Metab. **6:** 584–589.
6. TRIVEDI, B. & W. H. DANFORTH. 1966. J. Biol. Chem. **241:** 4110–4114.
7. UI, M. 1966. Biochim. Biophys. Acta **124:** 310–322.
8. TANAKA, K., F. A. WELSH, J. H. GREENBERG, R. O'FLYNN, V. A. HARRIS & M. REIVICH. 1985. J. Cerebr. Blood Flow Metab. **5:** 502–511.
9. TAYLOR, D. J., P. J. BORE, P. STYLES, D. G. GADIAN & G. K. RADDA. 1983. Mol. Biol. Med. **1:** 77–94.
10. FOLBERGROVÁ, J., K. NORBERG, B. QUISTORFF & B. K. SIESJÖ. 1975. J. Neurochem. **25:** 457–462.
11. SIESJÖ, B. K. & K. MESSETER. 1971. Factors determining intracellular pH. *In* Ion Homeostasis of the Brain. B. K. Siesjö & S. C. Sørensen, Eds.: 244–262. Munksgaard. Copenhagen.
12. MESSETER, K. & B. K. SIESJÖ. 1971. Acta. Physiol. Scand. **83:** 344–351.
13. FOLBERGROVÁ, J., V. MACMILLAN & B. K. SIESJÖ. 1972. J. Neurochem. **19:** 2497–2505.
14. PETROFF, O. A. C., J. W. PRICHARD, K. L. BEHAR, D. L. ROTHMAN, J. R. ALGER & R. G. SHULMAN. 1985. Neurology **35:** 1681–1688.
15. NIOKA, S., B. CHANCE, M. HILBERMAN, J. S. LEIGH, H. SUBRAMANIAN, J. EGAN, J. MARIS, M. RICHARDSON, E. DONLON & R. S. FORSTER. 1984. 3rd Ann. Meeting Soc. Magn. Reson. Med. New York. Abstract: 555–556.
16. LITT, L., R. GONZÁLEZ-MÉNDEZ, J. W. SEVERINGHAUS, W. K. HAMILTON, I. J. RAMPIL, J. SHULESHKO, J. MURPHY-BOESCH & T. L. JAMES. 1986. J. Cerebr. Blood Flow Metab. **6:** 389–392.
17. BEHAR, K. L., J. A. DEN HOLLANDER, O. A. C. PETROFF, H. P. HETHERINGTON, J. W. PRICHARD & R. G. SHULMAN. 1985. J. Neurochem. **44:** 1045–1055.
18. PLATEAU, P. & M. GUERON. 1982. J. Am. Chem. Soc. **104:** 7310–7311.
19. BENDALL, M. R. & R. E. GORDON. 1983. J. Magn. Reson. **53:** 365–385.
20. HETHERINGTON, H. P., M. J. AVISON & R. G. SHULMAN. 1985. Proc. Natl. Acad. Sci. USA **82:** 3115–3118.
21. BEHAR, K. L., J. A. DEN HOLLANDER, M. E. STROMSKI, T. OGINO, R. G. SHULMAN,

O. A. C. Petroff & J. W. Prichard. 1983. Proc. Natl. Acad. Sci. USA **80:** 4945–4948.
22. Petroff, O. A. C., J. W. Prichard, K. L. Behar, J. R. Alger, J. A. den Hollander & R. G. Shulman. 1985. Neurology **35:** 781–788.

DISCUSSION OF THE PAPER

T. L. James: As I had mentioned a little bit earlier in our hypercarbia experiments, I'd indicated that the ADP concentrations depended on which anesthetic we had present. Also, the phosphocreatine and P_i concentrations are dependent upon which anesthetic we use in the hypercarbia experiments. I apologize, but I did not hear which anesthetic that you were using under the circumstances.

K. L. Behar: We were using a combination of 70% nitrous oxide and 30% oxygen. No additional anesthetic was given beyond that.

James: It's probable that under the circumstances what we've managed to find is the first instance where an anesthetic level actually can change metabolite levels and none of us have bothered to look for these things previously. In the case of your anesthetic I haven't the slightest clue as to whether it can affect it or not.

Behar: Other studies have shown that nitrous oxide has no significant effect on brain high energy phosphates, blood flow, or oxygen consumption, in contrast to barbiturates or other inhaled anesthetics, such as halothane. So, I think it is unlikely that the direction of the changes we observed in the ^{31}P or ^{1}H spectrum during hypercarbia are significantly affected by its presence. However, small changes in the magnitude of putative effects cannot be ruled out.

Deuterium Magnetic Resonance *in Vivo*:

The Measurement of Blood Flow and Tissue Perfusion[a]

JOSEPH J. H. ACKERMAN, COLEEN S. EWY,
AND SEONG-GI KIM

Department of Chemistry
Washington University
St. Louis, Missouri 63130

ROBERT A. SHALWITZ

Department of Pediatrics
Washington University School of Medicine
St. Louis, Missouri 63110

INTRODUCTION

Originally motivated by a need to monitor tissue blood flow concomitantly with metabolic assessment by nuclear magnetic resonance (NMR) spectroscopy, this laboratory has been exploring and developing the use of deuterium (^{2}H,D) as an NMR spin label *in vivo*.[1-4] This laboratory[2] and that of Seelig and co-workers[5,6] have recently demonstrated the feasibility of deuterium spin-imaging after tissue water HOD enrichment through D_2O administration. Brereton *et al.*[7] have recently used deuterium NMR *in vivo* to monitor the turnover of ^{2}H in water and fat in mice after isotopic enrichment through orally administered D_2O. This report will focus on the use of D_2O as a freely diffusible ^{2}H magnetic resonance tracer for measurement of regional blood flow and tissue perfusion.

MAGNETIC RESONANCE PROPERTIES OF DEUTERIUM[8]

Deuterium is a spin 1 nuclide and, as such, possesses an electric quadrupole moment of *ca.* 2.8×10^{-27} cm^2, roughly three hundredths that of ^{23}Na. This quadrupole moment provides an efficient and usually dominant relaxation pathway for ^{2}H. Deuterium linewidths range from fractions of a hertz in specially purified solutions (acetone-d_6, for example) to hundreds of kilohertz in some solid samples. In our hands, at high fields (4.7 and 8.5 Tesla), HOD deuterium linewidths of a few tens of hertz are typical in tissue with spin-lattice relaxation times (T_1) on the order of a quarter of a second. Cogent reviews on the use of ^{2}H as a magnetic resonance probe of molecular structure and dynamics have been published.[9,10]

Deuterium is only about 1% as sensitive as ^{1}H and this combined with a very low natural abundance of approximately 1.6×10^{-2}% results in an overall natural

[a]Supported by National Institutes of Health grants GM-30331 and CA-40411, National Science Foundation instrument grant CHE-8100211, a grant from the American Diabetes Association, and a gift from Mallinckrodt, Inc. (St. Louis). RAS is a fellow of the Daland Philosophical Society.

abundance sensitivity of 1.6×10^{-6} relative to proton. Nevertheless, because of the enormous concentration of equivalent hydrogen in water (111 M), naturally abundant HOD (17 mM) can be readily observed by deuterium NMR in less than a minute of data acquisition. Furthermore, relatively low levels of ^2H enrichment can substantially improve signal-to-noise ratios. For example, assuming 100% NMR visibility, a 1% enrichment of tissue water through administration of D_2O would provide approximately twice the sensitivity of natural abundance tissue ^{23}Na (66% $\times$ 10 mM intracellular, 33% $\times$ 140 mM extracellular). Numerous laboratories have, of course, demonstrated the application of ^{23}Na spin-imaging with humans; the performance of ^2H spin-imaging after D_2O enrichment of body water thus appears feasible.

TOXICITY OF D_2O

There is a substantial body of literature dealing with the toxic effects of D_2O on mammalian systems.[11–29] This literature strongly suggests that D_2O toxicity is not likely to be consequential except at high chronic levels of administration. For example,

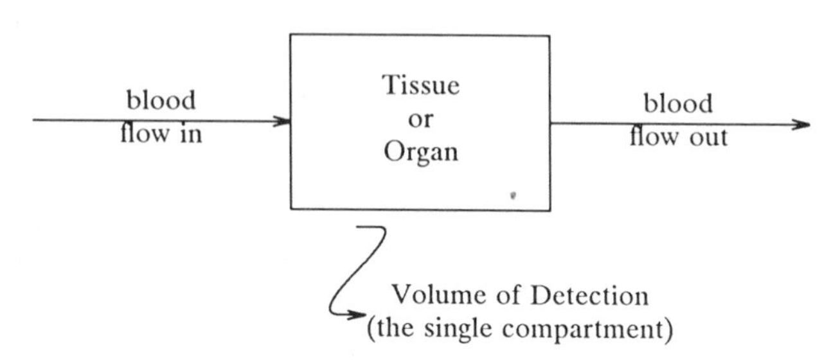

FIGURE 1. Schematic illustration of the single compartment flow model discussed in the text.

Katz *et al.*[23] have shown that mice maintained on drinking water containing 30% D_2O survive for very long periods of time. The only marked effect at this level of deuteration is on the reproductive capabilities of the mice. Chronic levels of 30–40% D_2O in drinking water are required to show a marked decrease in survival time.[23]

Although a more detailed evaluation of the toxicity of D_2O at low acute doses probably will be required prior to clinical use with humans, certainly acute 1–5% D_2O enrichment appears reasonable and should allow safe application of deuterium magnetic resonance techniques for spin-imaging and blood flow studies.

MEASUREMENT OF TISSUE BLOOD FLOW AND PERFUSION

There is an extensive literature on the use of freely diffusible radiolabels such as ^{133}Xe, ^{85}Kr, ^{89}Rb, and $H_2{}^{15}O$ for the measurement of regional blood flow and tissue perfusion. Most methods are based on the classic work by Kety,[30] and excellent reviews are available.[31–33]

Kety[34] was the first to propose that blood flow could be measured by observing the disappearance of a freely diffusible tracer from the tissue of interest. The analysis is most easily described in terms of a simple one-compartment model (FIG. 1) in which one expects rapid tracer diffusion between vascular and extravascular spaces and the tissue compartment of interest is visualized as having the equivalence of single inflow and outflow channels.

A typical measurement protocol attempts to meet the following criteria: (1) at time = 0 the compartment is instantaneously and uniformly labeled with tracer, such as through intraarterial bolus injection (inflow channel) or a direct intratissue injection, (2) at all times > 0 no additional tracer flows into the compartment; that is, there is no recirculation of tracer once it is washed out of the labeled compartment by ongoing blood flow, and (3) the tracer residue in the compartment can be followed closely over time by some mode of external detection, such as deuterium magnetic resonance in the case considered herein. Under these conditions, one expects a first-order kinetic process to govern the disappearance of tracer from the compartment, Eq. (1), resulting in a single exponential decay of tracer residue, $q(t)$, with exponential time constant T, Eq. (2).

$$\frac{dq(t)}{dt} = \frac{1}{T} q(t) \tag{1}$$

$$q(t) = q_0 e^{-t/T} \tag{2}$$

Here q_0 is the tracer residue at $t = 0$.

The central volume principle of tracer kinetics, Eq. (3), describes the relationship between the tracer mean transit time, $\overline{T}$, the single compartment volume of distribution of the tracer, V, and the volumetric flow rate of vascular fluid (blood), F.

$$\overline{T} = \frac{V}{F}. \tag{3}$$

The tracer mean transit time may also be described in terms of the area under the washout curve as Eq. (4).

$$\overline{T} = \frac{\int_0^\infty q(t)dt}{q_0}. \tag{4}$$

Combining Eq. (2) and Eq. (4), it is obvious that for the single compartment model, $\overline{T} = T$, Eq. (5).

$$\overline{T} = \frac{\int_0^\infty q_0 e^{-t/T}\, dt}{q_0} = \int_0^\infty e^{-t/T}\, dt = T \tag{5}$$

Finally, by invoking the central volume principle (Eq. (3)) and changing variables from V, the volume of tracer distribution in the single compartment, to W_{tissue}, the wet-weight of tissue making up the single compartment, Eq. (6),

$$V = \lambda W_{tissue}, \tag{6}$$

one has

$$T = \frac{V}{F} = \frac{\lambda W_{tissue}}{F}. \tag{7}$$

The proportionality factor λ is referred to as the tissue:blood partition coefficient and is determined independently of the flow measurement. For labeled water, λ is the ratio of the water weight of a unit mass of tissue (including vascular spaces) to the water weight of a unit volume of blood. Such water content is readily determined from the wet and dry weights of tissue and blood and the density of blood. The coefficient λ is specific for a particular tissue type and tracer and is typically 0.85–0.95 ml/g for water.

The rate of blood flow and tissue perfusion is then usually expressed as flow per unit mass (100 g) of tissue as given by Eq. (8),

$$\frac{F}{W_{\text{tissue}}} = \frac{\lambda \times 100}{T} \text{ ml blood/100 g tissue/min.} \tag{8}$$

Thus, determination of the exponential time constant governing tracer washout along with independent knowledge of λ allows calculation of the rate of blood flow and tissue perfusion via Eq. (8). The time constant T is readily determined by applying standard curve-fitting routines to the residue washout time course data.

RESULTS AND DISCUSSION

All data presented herein represent *in vivo* surface coil ^{2}H NMR experiments at either 8.5 Tesla (55.5 MHz) or 4.7 Tesla (30.7 MHz). Halothane anesthetized male

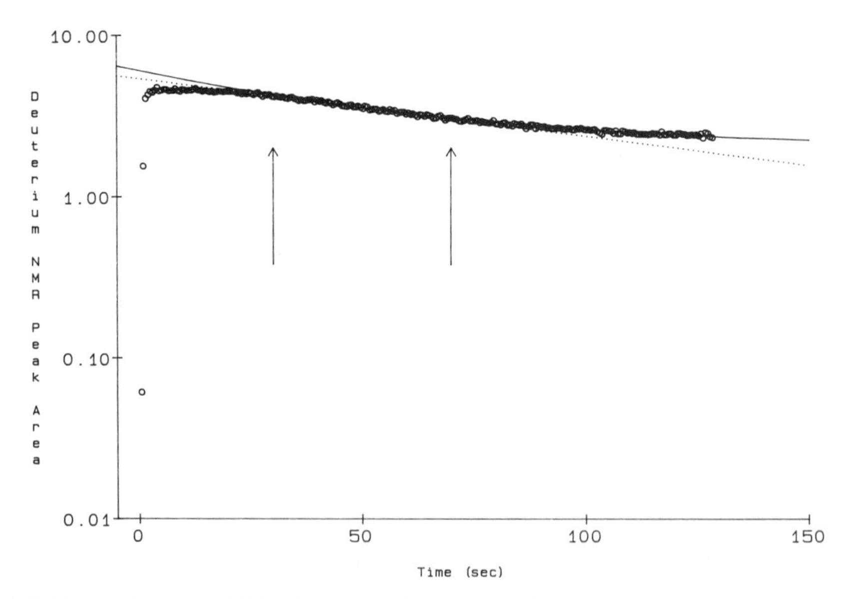

FIGURE 2. Deuterium NMR time course (semilog scale) of the *in vitro* rat hepatic HOD resonance following injection of 0.2 ml D_2O into a branch of the superior mesenteric vein. The open circles represent experimental data points, the dotted line a fit to the single-compartment flow model (between arrows) and the solid line a fit to the two-compartment flow model (between first arrow and end of data) that corrects for recirculation. Calculated hepatic blood flow rates were 41 and 65 ml/100 g/min.

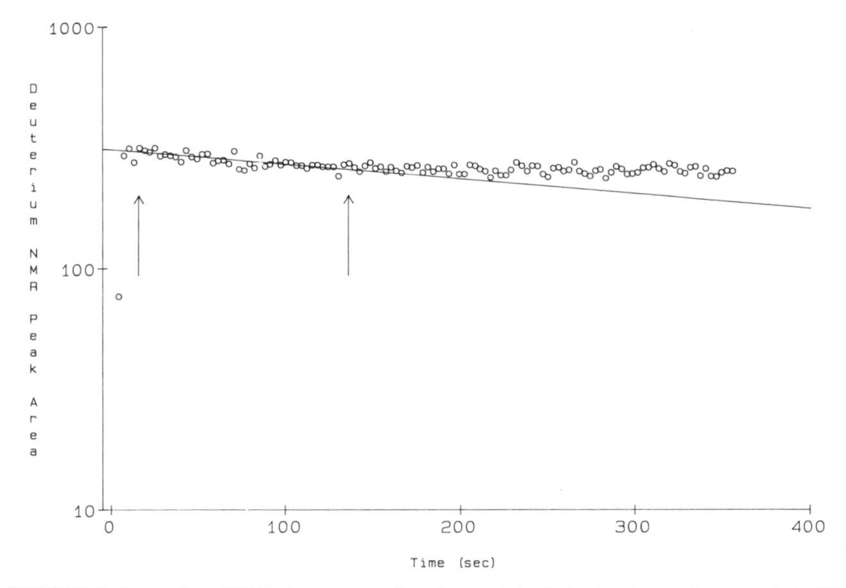

FIGURE 3. Deuterium NMR time course (semilog scale) of the *in vivo* rat leg muscle HOD resonance following intraaortic injection of 0.1 ml D_2O. The open circles represent experimental data points and the solid line a fit to the single-compartment flow model (between arrows). Calculated muscle blood flow was 8.0 ml/100 g/min, respectively.

Sprague-Dawley rats (*ca.* 250 g) and female C3H/HeSnJ and HeJ mice (*ca.* 25 g) were used. During ^2H measurements subjects were maintained in a vertical orientation because of magnet bore orientation.

FIGURES 2–5 show the HOD deuterium NMR washout curves following bolus-type labeling of tissue with D_2O. All plots represent the natural log of the integrated ^2H resonance area of the HOD peak in arbitrary units (ordinate) against time in seconds (abscissa). Therefore, in the absence of background HOD signal, accurate adherence to the single compartment flow model would result in linear plots with slopes equal to $(-T)^{-1}$. However, background HOD is observable in some experiments. This will lead to curvature in the semilog plots at late time points, even if the single compartment model holds, as the deuterium HOD intensity returns to a nonzero background level at $t = \infty$.

Rat Liver Blood Flow and Perfusion

FIGURE 2 shows the HOD residue hepatic washout time course following injection of 0.2 ml D_2O into a branch of the superior mesenteric vein. The superior mesenteric vein forms a natural junction with the portal vein through which about 70% of the liver's blood supply flows. Immediately after D_2O injection, there is a very large rise in ^2H intensity as D_2O enters the hepatic vascular spaces and diffuses throughout the aqueous tissue volume as HOD (proton-deuteron exchange is extremely rapid). After this period of rapid labeling, a linear decrease in ^2H intensity is observed (on a semilog scale), corresponding to the predicted single-compartment washout model. The slope

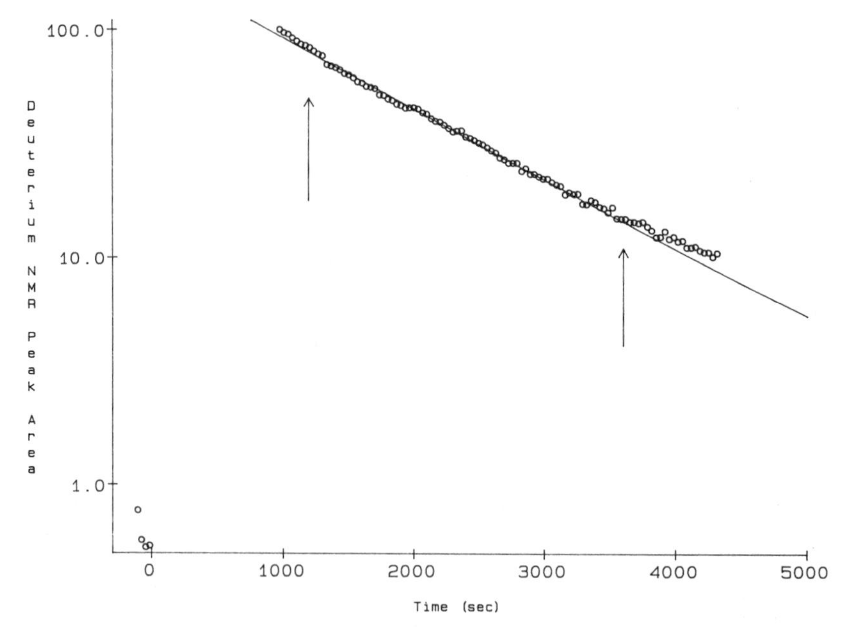

FIGURE 4. Deuterium NMR time course (semilog scale) of the *in vivo* rat leg muscle HOD resonance following single site direct intramuscular injection of 0.15 ml D_2O. The open circles represent experimental data points and the solid line a fit to the single compartment flow model (between arrows). The calculated muscle blood flow was 4.1 ml/100 g/min.

of this linear portion of the plot yields a hepatic blood flow of 41 ml blood/100 g liver/min for this single animal experiment. Inclusion of additional animals yields a mean hepatic blood flow of 61 ± 17 SD ($N = 5$) ml/100 g/min.

At data points late in the time course, behavior not predicted by the simple single-compartment model is observed, namely, curvature in the semilog residue decay plots indicative of nonzero label intensity at $t = \infty$ that is substantially greater than the initial HOD background. This is, of course, due to recirculation of HOD back through the liver after its initial hepatic washout. Ultimately, the ^2H intensity reaches a steady-state level as the HOD is uniformly distributed throughout all aqueous body spaces. On a much longer time scale, ^2H exchange into nonaqueous spaces (proteins, carbohydrates, lipids, etc.) can be observed.[7]

Although space does not permit a full treatment here, recent work from this laboratory suggests that a two-compartment series flow model incorporating conservation of flow and mass will provide a good first approximation correction to the label recirculation problem. With this model, a single exponential washout curve is again predicted but with nonzero tracer residue at $t = \infty$, $q_\infty \neq 0$. Nonlinear, least-squares fitting of the residue washout data extracts q_0, q_∞, and the exponential time constant T. From these three parameters the mean transit time, $\overline{T}$, and, thus, blood flow can be extracted for the compartment of interest.

Application of this two-compartment model to the data in FIGURE 2 yields a blood flow of 65 ml/100 g/min. Inclusion of additional animals where the time course was long enough to give marked evidence of recirculation yields 84 ± 24 SD ($N = 3$) ml/100 g/min. This mean value, as well as that from single-compartment modeling, compares well with the value of 80 ± 35 SD ($N = 17$) determined by the isotope fractionation technique,[35] although the precision of the NMR measurement appears greater.

Rat Muscle Blood Flow and Perfusion

Rat leg muscle blood flow was examined both by intraarterial injection (at the aortic bifurcation) of 0.1 ml D_2O (FIG. 3) or direct intramuscular D_2O injection of 0.15 ml (FIG. 4). In our preliminary experiments, delivery of label solely to the muscle volume under observation by intraaortic injection does not appear to be completely successful. Thus, signal intensity contributed from late arriving label constitutes a significant part of the time course and gives the appearance of substantial early label recirculation. This is not the case with the direct intramuscular injection where a linear semilog plot is observed over a long time course. Single compartment modeling yields muscle blood flows of 6.8 ± 4.0 SD ml/100 g muscle/min ($N = 6$) for the intraarterial injection and 4.1 ml/100 g/min ($N = 1$) for the direct intramuscular injection. These preliminary determinations agree reasonably well with values of 4.97 ± 0.55 SE ml/100 g/min ($N > 10$) determined by ventricular injection of labeled microspheres.[36]

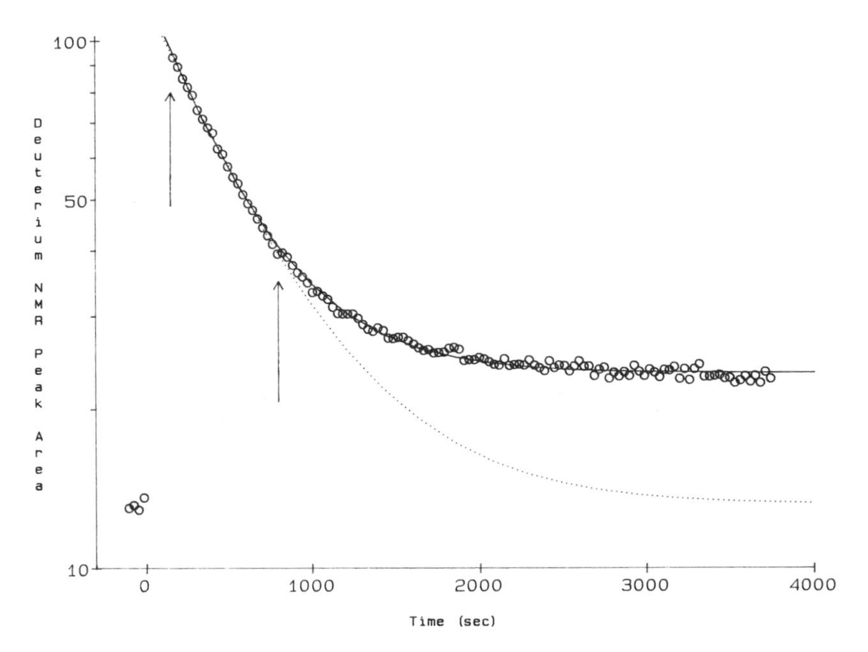

FIGURE 5. Deuterium NMR time course (semilog scale) of the *in vivo* murine RIF-1 tumor HOD resonance following five site direct intratumor injection of 0.05 ml (total) of 20% D_2O. The open circles represent experimental data points, the dotted line a fit to the single-compartment flow model (between arrows) and the solid line a fit to the two-compartment flow model (between first arrow and end of data). Calculated tumor blood flow rates were 10.0 and 11.5 ml/100 g/min, respectively.

Murine Tumor Blood Flow and Perfusion

The blood flow through radiation-induced fibrosarcoma (RIF-1) tumors was examined by multisite direct injection of D_2O into the 1 g tumor mass (FIG. 5). As we found with direct intramuscular injection, a substantial period of linear residue washout (on a semilog scale) was observed before curvature due to tracer recirculation became dominant. Single-compartment analysis yielded a tumor blood flow of 10 ml/100 g tumor/min. Correction for recirculation by two-compartment modeling yielded a tumor blood flow of 11.5 ml/100 g/min. Tumor blood flow is highly dependent on tumor size and, thus, flow comparisons are difficult. However, two compartment modeling of three tumors with an average mass of 1.2 g yielded a mean flow of 11.7 ± 0.5 SD ml/100 g/min, which compares well with a value of 14.9 ± 6.5 ($N = 11$) ml/100 g/min determined by $H_2{}^{15}O$ *in situ* photon activation and washout measurements with tumors of similar size.[37]

CONCLUSIONS

Although the data presented herein must be regarded as preliminary, and rigorous calibration experiments ultimately need to be performed, it is clear that ^2H NMR of *in vivo* systems offers a promising approach to the quantitative measurement of blood flow and tissue perfusion. Even at this early point of development, comparisons with independently measured blood flow values from the literature are quite favorable. Given the widespread availability of magnetic resonance equipment in hospitals and medical research centers, general application of such a technique may be possible. If sensitivity at lower field strengths (0.3–1.5 Tesla) is sufficient, deuterium magnetic resonance spin-imaging of regional blood flow may be feasible in a manner not unlike that of positron emission tomography flow studies with $H_2{}^{15}O$.

ACKNOWLEDGMENTS

Helpful discussions and the assistance of Drs. Jeffrey Evelhoch and Steven Sapareto of Wayne State University, Michigan Cancer Foundation, are gratefully acknowledged.

REFERENCES

1. ACKERMAN, J. J. H. 1986. Methods employing deuterium for enhancing magnetic resonance imaging and spectroscopy. U.S. Patent (Pending). Serial No. 824203. Filed January 30, 1986.
2. EWY, C. S., E. E. BABCOCK & J. J. H. ACKERMAN. 1986. Deuterium nuclear magnetic resonance spin-imaging of D_2O: A potential exogenous MRI label. Magn. Reson. Imaging **4:** 407–411.
3. ACKERMAN, J. J. H.; C. S. EWY & R. A. SHALWITZ. 1987. Deuterium nuclear magnetic resonance measurements of blood flow employing D_2O as an inert diffusible tracer. 1987. Proc Natl. Acad. Sci. USA **84:** 4099–4102.
4. EWY, C. S., D. L. BENNETT, J. J. H. ACKERMAN & R. A. SHALWITZ. 1986. Deuterium NMR measurements of blood flow in normal and tumor tissue. Fifth Annual Mtg. Soc. Magn. Reson. Med. (Montreal, Aug. 19–22, 1986) **1:** 86–87 (abstract).
5. MÜLLER, S., J. SEELIG. 1986. In vivo NMR imaging of deuterium. Fifth Annual Mtg. Soc. Magn. Reson. Med (Montreal, Aug. 19–22, 1986) **2:** 305–306 (abstract).

6. MÜLLER, S. & J. SEELIG. In vivo NMR imaging of deuterium. 1987. J. Magn. Reson. **72:** 456–466.
7. BRERETON, I. M., M. G. IRVING, J. FIELD & D. M. DODDRELL. 1986. Preliminary studies on the potential of in vivo deuterium NMR spectroscopy. Biochem. Biophys. Res. Commun. **137:** 579–584.
8. Magnetic resonance properties taken from NMR tables compiled by Bruker Instruments, Inc. and publicly distributed.
9. MANTSCH, H. H., H. SAITO & I. C. P. SMITH. 1977. Deuterium magnetic resonance: Applications in chemistry, physics, and biology. Prog. N.M.R. Spect. **11:** 211–271.
10. SMITH, I. C. P. & H. H. MANTSCH. 1982. Deuterium NMR spectroscopy in NMR spectroscopy: New methods and applications. Am. Chem. Soc. Symp. Ser. No. 191. Ch. 6. 97–118.
11. AMAROSE, A. P. & D. M. CZAJKA. 1962. Cytopathic effects of deuterium oxide on the male gonads of the mouse and dog. Exp. Cell. Res. **26:** 43–61.
12. BACHNER, P., D. G. MCKAY & D. RITTENBERG. 1964. The pathologic anatomy of deuterium intoxication. Proc. Natl. Acad. Sci. USA **51:** 464–471.
13. BARBOUR, H. G. 1937. The basis of pharmacological action of heavy water in mammals. Yale J. Biol. Med. **9:** 551–565.
14. BENNETT, E. L., M. CALVIN, O. HOLM-HANSEN, A. M. HUGHES, K. K. LONBERG-HOLM, V. MOSES & B. M. TOLBERT. 1958. Effect of deuterium oxide (heavy water) on biological systems. Proc. Int. Conf. Peaceful Uses of Atomic Energy (Geneva) **25:** 199.
15. CZAJKA, D. M. & A. J. FINKEL. 1960. Effect of deuterium oxide on the reproductive potential of mice. Ann. N.Y. Acad. Sci. **84:** 770–779.
16. CZAJKA, D. M., A. J. FINKEL, C. S. FISCHER & J. J. KATZ. 1961. Physiological effects of deuterium on dogs. Am. J. Physiol. **201:** 357–362.
17. HAYES, C. J. & J. D. PALMER. 1976. The suppression of mouse spontaneous locomotor activity by the ingestion of deuterium oxide. Experientia **32:** 469–470.
18. HUGHES, A. M., E. L. BENNETT & M. CALVIN. 1960. Further studies on sterility produced in male mice by deuterium oxide. Ann. N.Y. Acad. Sci. **84:** 763–769.
19. HUGHES, A. M., E. L. BENNETT & M. CALVIN. 1959. Production of sterility in mice by deuterium oxide. Proc. Natl. Acad. Sci. USA **45:** 581–586.
20. HUGHES, A. M. & M. CALVIN. 1958. Production of sterility in mice by deuterium oxide. Science **127:** 1445–1446.
21. KANWAR, K. C. & R. VERMA. 1977. Biologic effects of orally administered deuterium oxide on rat liver. Exp. Pathol. **13:** 255–261.
22. KANWAR, K. C. & R. VERMA. 1976. Oral D_2O administration and enzymatic changes in rat testis. Acta Biol. Med. Ger. **35:** 577–580.
23. KATZ, J. J., H. L. CRESPI, D. M. CZAJKA & A. J. FINKEL. 1962. Course of deuteration and some physiological effects of deuterium in mice. Am. J. Physiol. **203:** 907–913.
24. KATZ, J. J., H. L. CRESPI, A. J. FINKEL, R. J. HASTERLIK, J. F. THOMSON, W. LESTER, JR., W. CHORNEY, N. SCULLY, R. L. SHAFFER & S. H. SUN. 1959. The biology of deuterium. Proc Int. Conf. Peaceful Uses of Atomic Energy, Geneva **25:** 1973.
25. KATZ, J. J., H. L. CRESPI, R. J. HASTERLIK, J. F. THOMPSON & A. J. FINKEL. 1957. Some observations on biological effects of deuterium, with special reference to effects on neoplastic processes. J. Natl. Cancer Inst. **18:** 641–659.
26. PENG, S.-K., K.-J. HO & C. B. TAYLOR. 1972. Biologic effects of prolonged exposure to deuterium oxide: A behavioral, metabolic, and morphologic study. Arch. Pathol. **94:** 81–89.
27. RABINOWITZ, J. L., V. DEFENDI, J. LANGAN & D. KRITCHEVSKY. 1960. Hepatic lipogenesis in D_2O-fed mice. Ann. N.Y. Acad. Sci. **84:** 727–735.
28. TAYLOR, C. B., B. MIKKELSON, J. A. ANDERSON & D. T. FORMAN. 1966. Human serum cholesterol synthesis measured with the deuterium label. Arch. Pathol. **81:** 213–231.
29. THOMSON, J. F. & F. J. KLIPFEL. 1958. Changes in renal function in rats drinking heavy water. Proc. Soc. Exp. Biol. Med. **97:** 758–759.
30. KETY, S. S. 1951. The theory and application of the exchange of inert gas at the lungs and tissues. Pharmacol. Rev. **3:** 1–41.
31. LASSEN, N. A., O. HENDRIKSEN & P. SEJRSEN. 1973. Indicator methods of measurement of

organ and tissue blood flow. Handb. Physiol.; Sect. 2, Vol. 3, Part 1, Ch. 2. Waverly Press.
32. ZIERLER, K. L. 1965. Equations for measuring blood flow by external monitoring. Circ. Res. **16:** 309–321.
33. SHIPLEY, R. A. & R. E. CLARK. 1972. Tracer methods for in vivo kinetics. Academic Press. New York.
34. KETY, S. S. 1949. Measurement of regional circulation by the local clearance of radioactive sodium. Am. Heart J. **38:** 321–328.
35. STEINER, S. H. & G. C. E. MUELLER. 1961. Distribution of blood flow in the digestive tract of the rat. Circ. Res. **9:** 99–102.
36. SONG, C. W., M. S. KANG, J. G. RHEE & S. H. LEVITT. 1980. The effect of hyperthermia on vascular function, pH, and cell survival. Radiol. **137:** 795–803.
37. EVELHOCH, J. L., S. A. SAPARETO, G. H. NUSSBAUM & J. J. H. ACKERMAN. 1986. Correlations between ^{31}P NMR spectroscopy and ^{15}O perfusion measurements in the RIF-1 murine tumor in vivo. Radiat. Res. **106:** 122–131.

DISCUSSION OF THE PAPER

R. G. SHULMAN: I didn't understand why, given that the flow through the tumor was intermediate between the flow rates of the muscle and the liver, why it took so much longer before you had to worry about reflow, 1,000 minutes or so for the tumor and much less time for the other two organs.

J. J. H. ACKERMAN: If one considers only those situations where we were confident in having administered D_2O in a reasonably well-defined pseudo-bolus form *solely* to the tissue of interest at $t = 0$ (i.e., FIGS. 2, 4, and 5), the onset time of what we interpret to be tracer recirculation (i.e., the leveling off of the residue washout curves when plotted on a semilog basis) does scale roughly with measured blood flow rate. The exact relationship between the onset time of tracer recirculation and flow rate will depend, of course, on other factors besides flow rate, such as the ratio of total body mass (e.g., mice *ca.* 20 g, rats *ca.* 250 g) to the tissue mass labeled with tracer at $t = 0$. As mentioned previously, the appearance of very early tracer recirculation under low flow rate conditions (FIG. 3) is, in our opinion, not due to recirculation but rather a poor tracer administration protocol. That is, the intraaortic injection resulted in a non specific tracer distribution to a number of other tissues, thus, effectively presenting the muscle with an ill-defined non-bolus type tracer input function. Late-arriving tracer that originated in other tissue or from the substantial tail of the D_2O "bolus" then gives the false appearance of early recirculation.

NMR Studies of Renal Metabolism:

Regulation of Renal Function by ATP and pH[a]

NANCY SHINE, WILLIAM ADAM, JIAN AI XUAN,
AND MICHAEL W. WEINER

Magnetic Resonance Unit
Veterans Administration Medical Center

Departments of Medicine and Radiology
University of California, San Francisco
San Francisco, California 94121

INTRODUCTION

The function of the kidney is to maintain the volume and composition of the extracellular fluid. This is achieved by glomerular filtration and selective reabsorption of solutes and water by the renal tubular epithelium. Extracellular fluid volume is maintained by tubular reabsorption of sodium. The kidney regulates acid-base homeostasis by reabsorption of bicarbonate and excretion of acid into the urine. It is generally accepted that the energy for active transport by the renal tubules is provided in the form of ATP.[1] Although a variety of physical and humoral factors regulate renal function, inhibition of oxidative phosphorylation has been shown to inhibit sodium transport.[2–11] Furthermore, it has been suggested that the hormone aldosterone stimulates sodium reabsorption by increasing the production of ATP by oxidative phosphorylation.[1,12]

The excretion of H^+ into the urine is largely controlled by varying the production of ammonia produced from the metabolism of glutamine. Despite considerable investigation, the primary signal that regulates renal ammoniagenesis is not completely understood.[13] Nevertheless, previous investigators have suggested that the intracellular pH (pH_i) of the renal tubular cell may be an important regulator of renal ammoniagenesis.[13]

Nuclear magnetic resonance spectroscopy has been used by many investigators to study the metabolism of a variety of organs including the kidney. Ross, Radda, and their colleagues[14] pioneered ^{31}P NMR studies of perfused rat kidneys and the kidney *in vivo.* The goal of the present experiments was to utilize ^{31}P NMR to investigate the role of pH_i in the regulation of renal ammoniagenesis and to determine the importance of renal ATP in the regulation of sodium transport.

METHODS

All methods for these experiments have been previously described elsewhere.[15–18] Experiments were performed using anesthetized Sprague-Dawley rats weighing between 300–460 g. Catheters were inserted into the jugular vein for intravenous

[a]Supported by grants from the National Institutes of Health Grant AM RO1AM33293 and the Veterans Administration Medical Research Service.

infusion, the femoral artery to monitor blood pressure and heart rate, and the renal artery and renal vein to measure renal blood flow and O_2 consumption. The left kidney was exposed by an abdominal incision and the organ placed between a two-turn solenoidal coil tuned to 95.8 MHz for ^{31}P NMR. To minimize the NMR signal for surrounding tissues, a layer of Gortex (Gore Associates, Sunnyvale, CA) was inserted between surrounding muscle and the coil. Animal temperature was maintained and the ureter was catheterized to allow continuous collection of urine. ^{31}P NMR spectra were recorded at 95.8 MHz using a 4 inch horizontal bore magnet (Nalorac Cryogenics) and Nicolet 1180/293B data system. Renal filtration rate and blood flow were measured by inulin clearance using [^{3}H]inulin (25 μcurie) added to 1.3 ml saline infused at 1.3 ml/hr.

In order to increase renal metabolism to maximal rate for saturation transfer studies, the animals were uninephrectomized 7 to 15 days prior to NMR experiments, maintained on a high protein diet consisting of daily cheese supplements for 4 days, and administered daily injections of methylprednisolone (2.5 mg/day) for 4 days. A chronically indwelling renal catheter was implanted for 3 to 4 days prior to NMR experiments. For saturation transfer experiments, the method of Forsen and Hoffman,[19] as described previously by Koretsky *et al.*,[17,18] was followed. The T_1 apparent of the P_i peak was determined using saturation recovery.[20] For all experiments, renal content of ATP was determined by infusing methylphosphonic acid (MPA) *in vivo* as an "internal standard."[21] Briefly, the method involved the acquisition of a kidney spectrum using the two-turn implanted solenoidal coil followed by an intravenous infusion of 4.0 ml of 150 mM MPA, pH 7.45 for 30 min. The animal was then sacrificed in order to insure the maintenance of a steady-state level of renal ATP, and

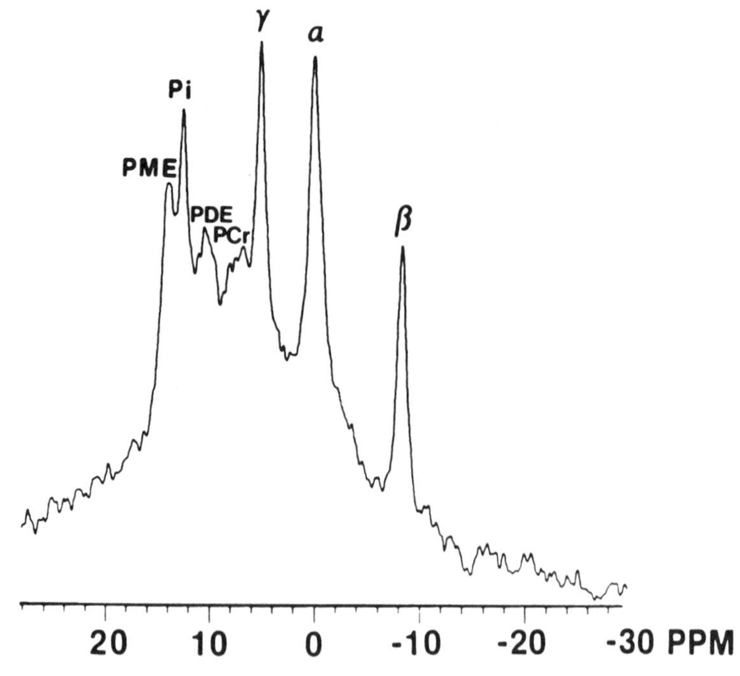

FIGURE 1. ^{31}P NMR spectrum of rat kidney *in vivo*. 90° pulse, 2 sec delay, 128 acquisitions.

TABLE 1. Effects of Acid-Base Disorders on Renal pH_i and Ammoniagenesis

Acid-Base Condition	Renal pH_i	Stimulation of Ammoniagenesis
Control	7.39 ± 0.04	−
Acute metabolic acidosis	7.16 ± 0.03	+
Chronic metabolic acidosis	7.30 ± 0.02	+
Acute respiratory acidosis	7.20 ± 0.05	+
Chronic respiratory acidosis	7.40 ± 0.04	−
K depleted	7.17 ± 0.07	+

another spectrum obtained. From these two spectra the ratio of MPA to ATP was derived. The amount of MPA per gram wet weight of kidney was obtained by placing the excised kidney in a phosphate-calibrated coil. The calibration spectra along with the ratio of intensities for MPA/β ATP were used to calculate μmoles ATP/kidney. Molar concentrations were obtained from the water content of the kidney. No correction was made for the amount of extracellular water.

RESULTS AND DISCUSSION

Measurements of Renal pH_i

FIGURE 1 illustrates a typical ^{31}P NMR spectrum of the rat kidney *in vivo.* The spectrum shows peaks for the three phosphates of ATP, phosphodiesters (PDE), inorganic phosphate (P_i), and phosphomonoesters (PME). Phosphocreatine (PCr) was frequently not present in the ^{31}P NMR spectrum of the kidney. Therefore, experiments were performed on rats in which PCr was clearly visible (but less than 100% of β ATP) to determine if the chemical shift (δ) of α ATP could serve as an appropriate internal standard. pH was varied using acetazolamide, potassium depletion, fructose administration, and acute metabolic acidosis. A roughly linear relationship between δ P_i − δ PCr, and δ P_i − δ α ATP (correlation coefficient 0.90, $p < 0.01$) was demonstrated. This relationship permitted calculation of pH_i from δ P_i − δ PCr in rats in which the PCr peak was not distinct.

The next group of experiments was designed to determine if δ P_i represents pH_i or whether there was a significant contribution from extracellular P_i to the P_i peak. The rationale of these experiments is that the PME peak largely represents intracellular phosphorylated metabolites. Therefore, the change of PME produced by acid-base changes represents changes of tubular cell pH_i without contamination by extracellular (plasma, interstitial fluid, tubular fluid, or urine) P_i. Spectra were obtained before and after administration of acetazolamide, and potassium depletion, and after administration of fructose. The results demonstrated that δ P_i changed in the same direction as δ PME, suggesting that δ P_i indicates pH_i. In addition, δ PME correlated with δ P_i after fructose loading, suggesting that δ P_i primarily represents pH_i in proximal tubular cells because the PME following fructose loading primarily represents fructose-1-phosphate in proximal tubular cells.

TABLE 1 shows the effects of various acid-base maneuvers on renal pH_i. Three maneuvers that had been previously shown to markedly stimulate renal ammoniagenesis were employed: acute metabolic acidosis, chronic metabolic acidosis, and potassium depletion. All of these maneuvers were associated with significant lowering of renal pH_i. Chronic metabolic acidosis produced less acidification of the kidney than acute metabolic acidosis, suggesting adaptation of an intracellular buffering mechanism. Despite the fact that potassium depletion produced extracellular alkalosis (arterial blood pH 7.44), renal pH_i was most depressed by potassium depletion ($pH_i = 7.17$). In contrast to chronic metabolic acidosis, chronic respiratory acidosis did not significantly change renal pH from control values. This is consistent with the fact that chronic respiratory acidosis does not significantly stimulate renal ammoniagenesis. Therefore, the results of these experiments are consistent with the hypothesis that renal pH_i is an important determinant of renal ammoniagenesis.

Importance of Renal ATP for Sodium Transport

In order to investigate the role of renal ATP in the regulation of renal Na transport, three substances were infused, which have been previously shown to alter renal ATP metabolism. First, studies were performed with the classical respiratory inhibitor cyanide, which has been demonstrated to have a profound inhibitory effect on renal Na transport. FIGURE 2 depicts changes in renal ATP, P_i, and fractional reabsorption of sodium before, during, and following intrarenal infusion of sodium cyanide (2 μmoles/min/kg) directly into the renal artery. In order to minimize fluid loss due to the natriuretic effects of cyanide, the rate of i.v. saline infusion was increased to 10 ml/hr. Urine samples were collected at 5 or 10-min intervals during NaCN infusion and recovery. FIGURE 2 shows that cyanide infusions produced a fall of ATP, a rise of P_i, and a fall of fractional sodium reabsorption. By 20 min of cyanide infusion ($T = 40$) ATP had significantly fallen to 86 ± 3% of control and P_i had significantly risen to 136 ± 14% of control while renal sodium reabsorption was significantly reduced to 98 ± 0.6%. After 50 min of cyanide infusion ($T = 70$), renal ATP reached its nadir at 63 ± 5% of control, P_i had risen to 157 ± 12% of control, and fractional renal sodium reabsorption fell to 92 ± 3% of control. Following discontinuance of cyanide, renal P_i fell back to control within 55 minutes. In contrast, recovery of renal ATP was much slower, reaching only 75% of control by the end of the recovery period. This may be attributed to the slow resynthesis of adenine nucleotides. Fractional sodium reabsorption recovered to 96 ± 0.5% 50 min following discontinuation of cyanide and remained constant at this level throughout the remainder of the experiment.

These results clearly demonstrate that intrarenal infusion of cyanide lowers renal ATP levels and similarly inhibits fractional reabsorption of sodium. However, it should be emphasized that the K_m for ATP of the Na-K ATPase is less than 1 mM. Renal ATP concentrations are about 3.6 mM[21]; therefore, a 40% decrease in renal ATP to about 2.2 mM should not be rate limiting for the activity of the sodium pump. These results suggest a number of possibilities. First, the activity of the Na-K ATPase may not be strictly regulated by ATP concentrations, but rather by the phosphate potential, which is affected by concentrations of ADP, P_i, and H^+. Secondly, these measurements were made on whole kidney and do not provide information on ATP concentrations in specific tubular segments. A profound decline in the ATP concentrations of the distal tubule might inhibit reabsorption of sodium by this segment while whole kidney ATP concentrations are less severely affected. The final possibility concerns compartmentation of ATP within the renal tubular cell. Electron microscopy shows that renal mitochondria are placed adjacent to the basal lateral membrane of the proximal tubule. Presumably, the ATP produced by these organelles is directly available to the Na-K ATPase on the basal lateral membrane. Therefore, it is possible that inhibition of mitochondrial ATP production results in a local decrease of ATP concentration and the bulk of cytosolic ATP may not be affected. This raises the possibility that the renal ATP turnover rate is more important than the static concentrations of this metabolite (see below).

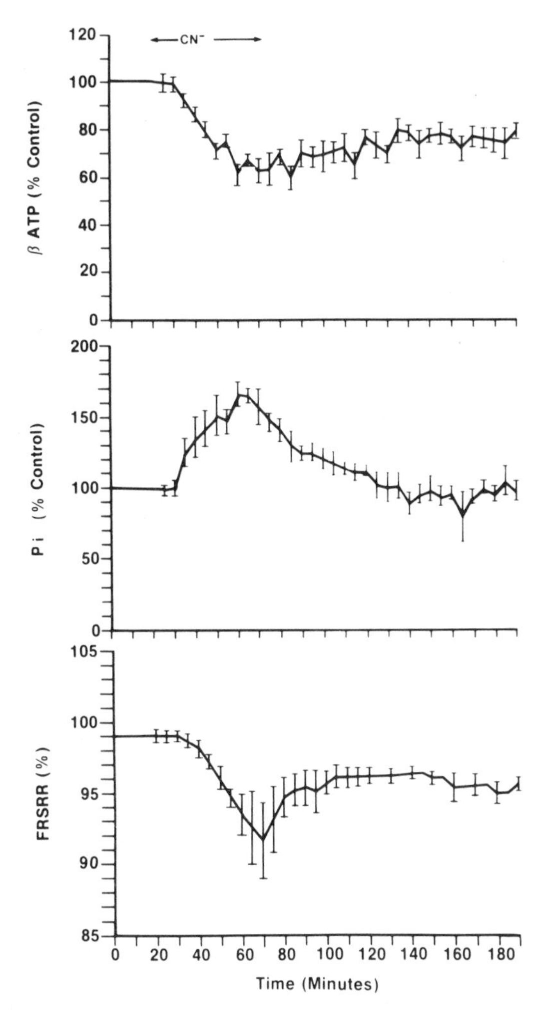

FIGURE 2. Effects of cyanide infusion on β ATP intensity (top), P_i intensity (middle), and fractional sodium reabsorption rates (FSRR) (N.= 4). All data are expressed as the mean ± SEM of the % control. NMR data (N = 6) were obtained from the integrated intensities of ^{31}P NMR resonances. The first 20 min of each experiment were used to acquire control data. Cyanide infusion was instituted at T = 20 min and continued for 50 min as indicated. (P_i, inorganic phosphate; PME, phosphomonoesters; PDE, phosphodiesters; and PCr, phosphocreatine.

Several previous investigators have studied the renal effects of the uncoupling agent 2,4-dinitrophenol (DNP).[2-5,9,10,22] In contrast to the results with cyanide, 2,4-DNP seems to have little effect on renal sodium transport. This suggested to some[3,11] that renal Na transport is not directly linked to ATP, but rather to some high-energy intermediate, which was not affected by uncouplers of oxidative phosphorylation. To investigate this matter further, experiments were performed in which the uncoupling agent 2,4-DNP was infused i.v. while ATP, P_i, and fractional sodium reabsorption were measured. TABLE 2 provides the results of these experiments and compares them with the effects of cyanide and fructose (see below). Infusion of 2,4-DNP (1.5 μmoles/min/kg, pH 7.0 for 1 hr) produced a 30% fall of renal ATP concentrations. In contrast to the effects of cyanide, DNP had little or no effect on fractional reabsorption of sodium. Therefore, despite the finding that both cyanide and 2,4-DNP had a similar effect on ATP levels, cyanide markedly inhibited renal sodium reabsorption whereas 2,4-DNP had no effect. The reasons for this difference are not clear, but the same possibilities discussed above for cyanide must be considered. First, 2,4-DNP may have lowered ATP in the proximal tubule while distal nephron segments were unaffected and maintained high rates of sodium reabsorption. Second, it is possible that 2,4-DNP depleted one pool of cellular ATP while the ATP concentrations adjacent to the Na-K ATPase were maintained. Third, although 2,4-DNP depleted ATP levels, the rate of ATP turnover may have been unaffected.

To investigate further the relationship between renal ATP concentrations and sodium reabsorption, the sugar fructose (previously shown to deplete renal ATP)[17,23] was intravenously infused (16.7 mmoles/kg for 30 min) into rats while ^{31}P NMR measurements were made. TABLE 2 summarizes the results of these experiments, and compares them with cyanide and 2,4-DNP. Despite the fact that fructose infusion produced a profound decrease in renal ATP concentrations, there was no measurable

TABLE 2. Effects of Inhibitors on Renal Transport and Metabolism

Inhibitor	Fractional Na Reabsorption[a] (%)	β ATP (% control)	ATP (mM)[b]
Cyanide ($N = 6$)	95 ± 0.6	76 ± 2.6	2.7
Uncoupler (DNP) ($N = 5$)	99 ± 0.1	70 ± 6.1	2.5
Fructose ($N = 6$)	98 ± 0.8	73 ± 4.0	2.6

[a] Control fractional Na reabsorption = 99 ± 0.1%.
[b] Control β ATP = 3.6 ± 0.6 measured by MPA technique (see Methods).[21]
N = Number of experiments.

effect on fractional sodium reabsorption. These results contrast with those obtained with cyanide and are similar to the findings with 2,4-DNP. There are several possible explanations for this difference. First, is that fructose has a selective effect on ATP concentrations of the proximal tubule because fructokinase is only present in this nephron segment. Therefore, it is possible that diminished ATP concentrations in the proximal tubule led to inhibited sodium reabsorption in this segment while increased sodium reabsorption in more distal segments prevented a decrease in the total fractional reabsorption of sodium. This explanation is unlikely because Morris and co-workers[24] found that massive fructose infusion had no effect on urinary excretion of bicarbonate, phosphate, or amino acids, which are specifically reabsorbed in the proximal tubule. The second possibility is that fructose and cyanide produced depletion of ATP in one portion of the cell while ATP concentrations adjacent to Na-K ATPase were maintained. As discussed above, the respiratory inhibitor cyanide directly inhibits mitochondrial ATP production, which may reduce the size of an ATP pool immediately adjacent to the Na-K ATPase of the basal lateral membrane. In contrast, fructose may deplete a large cytoplasmic pool of ATP, without affecting ATP adjacent to the Na-K ATPase.

In order to investigate the relationship of ATP turnover to renal sodium reabsorp-

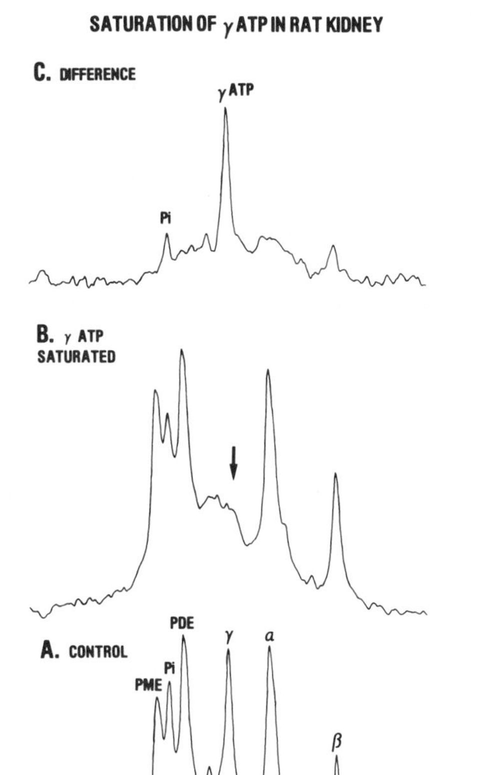

FIGURE 3. *In vivo* saturation transfer measurement of P_i to ATP exchange in the rat kidney. (a) Control spectrum. In order to correct for any stray irradiation, a control spectrum was obtained by selecting rf irradiation at a frequency equidistant from the P_i peak on the opposite side from the gamma resonance (position indicated by arrow). The decoupler channel was used to generate the saturating pulse. The contribution to the spectra from muscle surrounding the kidney was assumed to be insignificant due to the low level of the phosphocreatine (PCr) peak. (b) Spectrum obtained by selectively saturating the gamma phosphate group of ATP (shown by arrow). To minimize spectral changes due to possible biological variations, spectra were obtained by alternately collecting 16 scans of the control and then 16 scans of the gamma-saturated spectrum for a total of 512 scans each. (c) Difference spectrum indicating the amount of intensity decrease in P_i peak on saturation of gamma ATP. Some change in the β ATP peak intensity was also consistently observed. The peaks labeled α, β, γ represent the three phosphate groups of ATP; PDE, phosphodiesters; P_i inorganic phosphate; and PME, phosphomonoesters.

tion, saturation transfer ^{31}P NMR experiments were performed in an effort to quantitate the rate of renal ATP synthesis and to compare this rate with renal oxygen consumption and Na reabsorption. Previous experiments of this type have been reported by Freeman *et al.* for the perfused kidney[25] and the kidney *in situ*[26] and in our laboratory by Koretsky *et al.*[17,18] In contrast to the previous work, the present saturation transfer experiments were performed simultaneously with measurements of renal sodium reabsorption and renal oxygen consumption in the same rat. Renal metabolic rate was enhanced by protein feeding, uninephrectomy, and methylprednisolone (see METHODS). Furthermore, in an effort to determine if the rate of renal ATP synthesis varied with the rate of oxygen consumption and sodium reabsorption, the hormone atrial natriuretic factor was infused to stimulate renal metabolism to an even greater extent.

FIGURE 3 shows a typical experiment in which the saturation pulse was placed on the γ ATP peak of the rat kidney. TABLE 3 shows the data for these experiments and provides previously reported results for comparison. Measurements of T_1 apparent

TABLE 3. ^{31}P NMR Saturation Transfer of the Kidney *in Vivo*

Experiment	$T_{1\ apparent}$ (sec)	[ATP] (μmol/g)	Flux (μmol/min/g)	O_2 (μmol/min/g)	ATP/O
Perfused kidney *in vitro* (Freeman *et al.*[25])	1.2	2.1	16.1	3.27	2.5
Kidney *in vivo* (Freeman *et al.*[26])	1.6	2.4	9.5	2.57	1.7
Kidney *in vivo* (Koretsky *et al.*[17,18])	1.2	1.7	12.0	4–9	0.8–1.7
Kidney *in vivo* (present study)	1.3	2.0	20.1	11.9	0.85

All kidney weights (g) are g wet weight. ΔP_i refers to the change of the P_i peak produced by steady-state saturation of γATP, expressed as μmol/g P_i. In the present experiments these data were obtained from the difference spectrum (control − γATP saturated) using the ATP concentration obtained with MPA infusion.[21] For previous values,[25,26] the values were calculated from the published data for $[P_i]$ and M_a/M_o.

(sec) and ATP concentrations were roughly similar in all studies. The calculated rate of ATP synthesis in the present experiments was 20 μmoles/min/g, similar to the 16 μmoles/min/g previously reported for the perfused kidney.[25] In contrast, Freeman *et al.*[26] calculated the ATP synthesis rate of the kidney *in situ* to be only 9.5 μmoles/min/g. Another difference between the present experiments and those previously reported is the high rate of oxygen consumption, 11.9 μmoles/min/g, which is much greater than that previously reported for the perfused kidney[25] or the kidney *in vivo*.[26] This high rate of oxygen consumption is attributed to the uninephrectomy, high protein feeding, and methylprednisolone injection used to stimulate renal metabolism. Despite the fact that renal oxygen consumption was greatly increased, the net rate of renal ATP synthesis was not very much different than the other studies. Therefore, the efficiency of oxidative phosphorylation, i.e. the ATP/O ratio was only 0.85 in these experiments compared to 1.7–2.5 recorded previously by Freeman *et al.*[25–26] In other experiments, the hormone atrial natriuretic factor was infused producing a further increase in renal blood flow, glomerular filtration rate, and oxygen consumption. In contrast to GFR and O_2 consumption, the rate of ATP synthesis did not change. The results (not shown) demonstrate no relationship between changes in renal O_2 consumption or GFR and the rate of renal ATP synthesis measured by saturation transfer.

It is concluded from the results of these saturation transfer experiments that there is little correlation between the rate of renal oxygen consumption, sodium reabsorption, and renal ATP synthesis. However, these results should be interpreted with caution. The measurement of renal ATP synthesis *in vivo* by saturation transfer is subject to a large measurement error due to a low signal:noise ratio of the P_i peak in the difference spectrum. Furthermore, in the difference spectrum data calculations are based upon the assumption that the T_1 apparent of the P_i peak represents the T_1 apparent of the exchanging P_i pool. If there is a substantial pool of intrarenal P_i (seen in the P_i peak), which is not participating in ATP-P_i exchange, and if the T_1 of the nonexchanging pool is significantly different than the T_1 of the exchanging pool, then the calculated results may not be correct. Notwithstanding these difficulties, the finding that the rate of renal ATP synthesis, measured by this technique, shows virtually no correlation with the renal metabolic rate or sodium transport, suggests one of two possibilities: The first is that renal sodium reabsorption may be energized by a very small pool of ATP, which is rapidly turning over and is not readily detected by the ^{31}P NMR saturation transfer technique. A second possibility is that the rate of renal ATP-P_i exchange measured *in vivo* by saturation transfer does not solely represent net ATP synthesis, but rather represents an "isotopic" exchange catalyzed by other enzymes that participate in ATP metabolism. Recently, Campbell *et al.*[27] reported that the ATP-P_i exchange rate detected in yeast represents "isotopic" exchange catalyzed by 3-phosphoglyceraldehyde dehydrogenase and phosphoglycerate kinase, and not net ATP synthesis energized by oxidative phosphorylation. If this were the case in the kidney, then the present measurements of ATP-P_i exchange would not be expected to bear directly on the energetics of renal sodium transport.

ACKNOWLEDGMENTS

Some of the experiments reported in this manuscript were performed over the last several years by Drs. Samuel Wang and Alan Koretsky in collaboration with the authors. The authors gratefully acknowledge the support and encouragement of Dr. Floyd C. Rector, Jr.

REFERENCES

1. WEINER, M. W. & R. H. MAFFLY. 1978. Energy metabolism and active ion transport. *In* The Physiological Basis for Disorders of Biomembranes. T. Andreoli, J. Hoffman & D. Fanestil, Eds.: 287–308. Plenum Publishing Corporation. New York.
2. STRICKLER, J. C. & R. H. KESSLER. 1963. Effects of certain inhibitors on renal excretion of salt and water. Am. J. Physiol. **205**(1): 117–122.
3. FUJIMOTO, M., F. D. NASH & R. H. KESSLER. 1964. Effects of cyanide, Q_0, and dinitrophenol on renal sodium reabsorption and oxygen consumption. Am. J. Physiol. **206**(6): 1327–1332.
4. MARTINEZ-MALDONADO, M., G. EKNOYAN & W. N. SUKI. 1970. Inhibition of renal tubular sodium reabsorption by dinitrophenol. Am. J. Physiol. **219**(5): 1242–1247.
5. CHERTOK, R. J., W. H. HULET & B. EPSTEIN. 1966. Effects of cyanide, amytal, and DNP on renal sodium absorption. Am. J. Physiol. **211**(6): 1379–1382.
6. KESSLER, R. H., D. LANDWEHR, A. QUINTANILLA, S. A. WESELEY, W. KAUFMANN, H. ARCILA & B. K. URBAITIS. 1968. Effects of certain inhibitors on renal sodium reabsorption and ATP specific activity. Nephron **5:** 474–488.

7. ARCILA, H., H. SAIMYOJI & R. H. KESSLER. 1968. Accentuation of cyanide natriuresis by hypoxia. Am. J. Physiol. **214**(5): 1063–1067.
8. MARTINEZ-MALDONADO, M., G. EKNOYAN & W. N. SUKI. 1969. Effects of cyanide on renal concentration and dilution. Am. J. Physiol. **217**(5): 1363–1368.
9. WEINSTEIN, S. W. 1970. Proximal tubular energy metabolism, sodium transport, and permeability in the rat. Am. J. Physiol. **219**(4): 978–981.
10. WEINER, I. M., L. ROTH & T. W. SKULAN. 1971. Effects of dinitrophenol and cyanide on T_{PAH} and Na reabsorption. Am. J. Physiol. **221**(1): 86–91.
11. URBAITIS, B. K. & R. H. KESSLER. 1971. Actions of inhibitor compounds on adenine nucleotides of renal cortex and sodium excretion. Am. J. Physiol. **220**(4): 1116–1123.
12. SPIRES, D. A. & M. W. WEINER. 1980. Use of an uncoupling agent to distinguish between direct stimulation of metabolism and direct stimulation of transport: Investigation of antidiuretic and aldosterone. J. Pharmacol. Exp. Ther. **214**(3): 507–515.
13. TANNEN, R. L. 1978. Ammonia metabolism. Am. J. Physiol. **235:** F255–F277.
14. RADDA, G. K., J. J. H. ACKERMAN, P. BORE, P. SEHR, G. G. WONG, B. D. ROSS, Y. GREEN, S. BARTLETT & M. LOWRY. 1980. ^{31}P NMR studies on kidney intracellular pH in acute renal acidosis. Int. J. Biochem. **12:** 277–281.
15. ADAM, W., A. P. KORETSKY & M. W. WEINER. 1985. Measurement of renal intracellular pH by ^{31}P NMR. Relationship of pH to ammoniagenesis. *In* Contributions to Nephrology. R. L. Tannen, A. C. Schoolwerth, K. Kurokowa, P. Vinay, S. Karger, Eds. Vol. **47:** 15–21.
16. ADAM, W. R., A. P. KORETSKY & M. W. WEINER. 1986. Measurement of renal intracellular pH in rats *in vivo* using ^{31}P NMR: Effects of acidosis and K^+ depletion. Am. J. Physiol. **251:** F904–F910.
17. KORETSKY, A. P., S. WANG, J. MURPHY-BOESCH, M. P. KLEIN, T. L. JAMES, & M. W. WEINER. 1983. ^{31}P NMR spectroscopy of rat organs *in situ* using chronically implanted radiofrequency cells. Proc. Natl. Acad. Sci. USA **80:** 7491–7495.
18. KORETSKY, A. P., S. WANG, M. P. KLEIN, T. L. JAMES & M. W. WEINER. 1986. ^{31}P NMR saturation transfer measurements of phosphorous exchange reactions in rat heart and kidney *in situ*. Biochemistry **25:** 77–84.
19. FORSEN, S. & R. HOFFMAN. 1963. Study of moderately rapid chemical exchange reactions by means of nuclear magnetic double resonance. J. Chem. Phys. **39:** 2892–2901.
20. MCDONALD, G. & J. LEIGH. 1973. New method for measuring longitudinal relaxation times. J. Magn. Reson. **9:** 358–370.
21. SHINE, N., J. XUAN, A. KORETSKY & M. WEINER. 1987. Determination of renal molar concentrations of phosphorous-containing metabolites *in vivo* using ^{31}P NMR. Mag. Reson. Med. **4:** 244–251.
22. JOHANNESEN, J., M. LIE & F. KIIL. 1977. Renal energy metabolism and sodium reabsorption after 2,4-dinitrophenol administration. Am. J. Physiol. **233**(3): F207–F212.
23. MORRIS, R. C., JR., K. NIGON & E. B. REED. 1978. Evidence that the severity of depletion of inorganic phosphate determines the severity of the disturbance of adenine nucleotide metabolism in the liver and renal cortex of the fructose-loaded rat. J. Clin. Invest. **61:** 209–220.
24. Personal communication.
25. FREEMAN, D. S. BARTLETT, G. RADDA & B. ROSS. 1983. Energetics of sodium transport in the kidney saturation transfer ^{31}P NMR. Biochim. Biophys. Acta **762:** 325–336.
26. FREEMAN, D., L. CHAN, H. YAHAYA, P. HOLLOWAY & B. ROSS. 1986. Magnetic resonance spectroscopy for the determination of renal metabolic rate *in vivo*. Kidney Int. **30:** 35–42.
27. CAMPBELL, S. 1986. Ph.D. Thesis. Yale University.

EDITOR'S NOTE

There was no discussion following this paper's presentation due to lack of time.

^{13}C and ^{31}P NMR Studies of Hepatic Metabolism in Two Experimental Models of Diabetes

SHEILA M. COHEN

Merck Sharp & Dohme Research Laboratories
Animal Drug Metabolism
Rahway, New Jersey 07065

INTRODUCTION

^{13}C NMR spectroscopy provides a useful approach to the study of metabolism in cells, isolated perfused organs, and whole animals.[1–4] After administration of substrates specifically labeled with ^{13}C, signals are measured repetitively from substrate, intermediates, and end products. The regulation of the enzymes of intermediary metabolism must ultimately be understood within the context of the physiology of the whole cell. The nondestructive ^{13}C NMR method is proving to be a particularly versatile way of examining metabolic regulation in intact functioning liver,[5–7] including the kinetics of gluconeogenesis and lipogenesis and the effects of hormonal treatment or fasting on the activity of the phosphoenolpyruvate cycle and on relative fluxes into the tricarboxylic acid (TCA) cycle.

The main precursors of gluconeogenesis in liver *in vivo* (alanine and lactate) enter the pathway as pyruvate. Pyruvate is an important branch point both in the catabolic sequence of glucose and in the gluconeogenic pathway (FIG. 1) from substrates, such as alanine, that enter the pathway below the level of the triose phosphates. Pyruvate can enter the TCA route in two different ways: by carboxylation to form a dicarboxylic acid (oxalacetate) via the activity of pyruvate carboxylase or by oxidative decarboxylation to acetyl-CoA by action of pyruvate dehydrogenase. The other major pathway of pyruvate metabolism is catalyzed by pyruvate kinase. Both pyruvate carboxylase and pyruvate kinase have been implicated in hormonal control of gluconeogenesis from these substrates. Because pyruvate dehydrogenase competes with pyruvate carboxylase for entry of pyruvate into the TCA cycle, its regulation is also pertinent to the control of gluconeogenesis. The ability to examine the control of gluconeogenesis is critical to the understanding of metabolic regulation in models of either insulin-dependent (type 1) diabetes or non-insulin-dependent (type 2) diabetes.

The present study focuses on the use of ^{13}C and ^{31}P NMR spectroscopy in the investigation of metabolism in perfused liver from the streptozotocin-treated rat model of insulin-dependent diabetes and the untreated littermates. In this model, plasma glucose levels are increased threefold and glucagon levels are elevated about twofold over the control values, whereas endogenous insulin levels are very low. Comparison is made with perfused liver from the genetically obese (ob/ob) mouse model of type 2 diabetes and the lean littermates. Obese mice of the age used are both hyperinsulinemic and hyperglycemic. The ob/ob mouse is resistant to both endogenous and exogenous insulin. This resistance correlates with a decrease in the number of insulin receptors in the liver plasma membrane.[8]

Both ^{31}P and ^{13}C NMR are used here to compare the metabolism of ^{13}C-labeled substrates in diabetic liver and in liver from the control littermates. In general, ^{31}P and

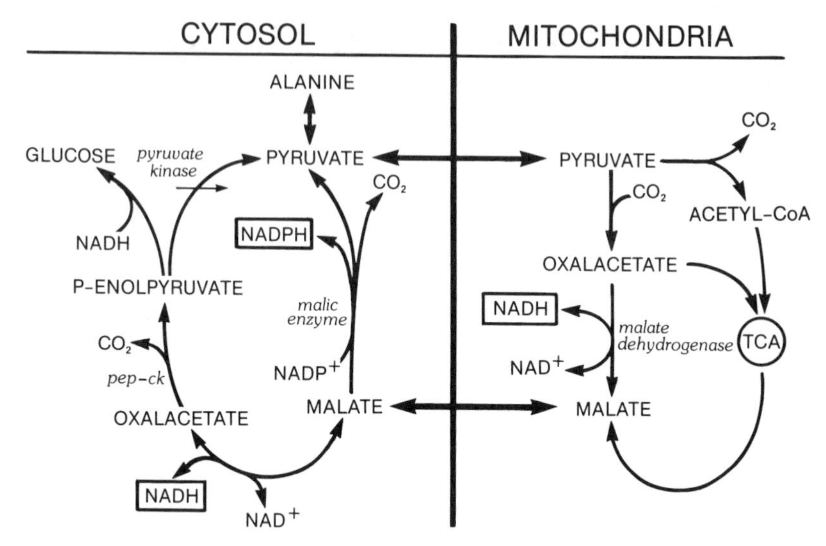

FIGURE 1. Simplified model gluconeogenic pathway emphasizing pyruvate as branch point in hepatic metabolism. *pep-ck* denotes phosphoenolpyruvate carboxykinase.

^{13}C chemical shifts are fairly specific; thus, it is possible to distinguish and measure a large number of metabolites in a single spectrum. Because the information provided by ^{13}C and ^{31}P nuclei is frequently complementary, acquisition of both ^{13}C and ^{31}P NMR spectra of the same perfused liver can give a rather comprehensive view of the regulation of hepatic metabolism.[2]

EFFECTS OF STREPTOZOTOCIN-INDUCED DIABETES ON PHOSPHATE METABOLITES AS STUDIED BY ^{31}P NMR

In spectra of normoxic perfused liver from control rats (FIG. 2a), the three major peaks or resonances arise from the three phosphates of ATP. Additional peaks due to P_i, GPC, and P-choline are also seen in FIGURE 2a. Notably absent from this spectrum are signals from nucleic acids, phosphoproteins, and membrane phospholipids. This circumstance arises because, in general, ^{31}P nuclei in large molecules with decreased mobility give signals that are too broad to be observed by standard high-resolution NMR techniques. Similarly, ADP tightly bound to proteins escapes observation. Thus, for the most part, these ^{31}P NMR spectra measure the free intracellular concentrations of the smaller metabolites. ATP levels were well maintained over 4.6 hr of perfusion in this control liver. From the ratio of the area under the ATPβ peak to the area under the ATPγ peak, which also contains the contributions of the ADPβ peak, an approximate ratio for the free, presumably cytosolic, concentrations of ATP/ADP can be measured. This ratio was 9:1 in control liver. Under the same perfusion conditions, this ratio was only 7.5:1 in liver from the streptozotocin-diabetic rat (FIG. 2b). The ATP/ADP ratios measured in our ^{31}P spectra of liver and the corresponding perchloric acid extracts of the freeze-clamped livers, from either control or diabetic liver, fall within the range measured chemically for cytosol and whole liver, respectively.[9]

The two spectra in FIGURE 2 contrast liver from a streptozotocin-diabetic rat with liver from its untreated littermate. Although it could not have been anticipated, appreciable signal from an unusual P,P′-diesterified pyrophosphate (labeled X in FIG. 2b) was detected in ^{31}P NMR spectra of 25% of the diabetic livers examined ($N = 16$). This metabolite was not observed in the rest of the diabetic livers or in any control liver ($N = 15$). As would be expected for an analogue of NAD^+, the unknown compound appeared as a quartet in ^{31}P spectra of the perchloric acid extract of the liver shown in FIGURE 2b. However, the analogue was distinguishable on the basis of both chemical shifts and J coupling constant from all the familiar compounds of this type that are known to exist in liver.[5]

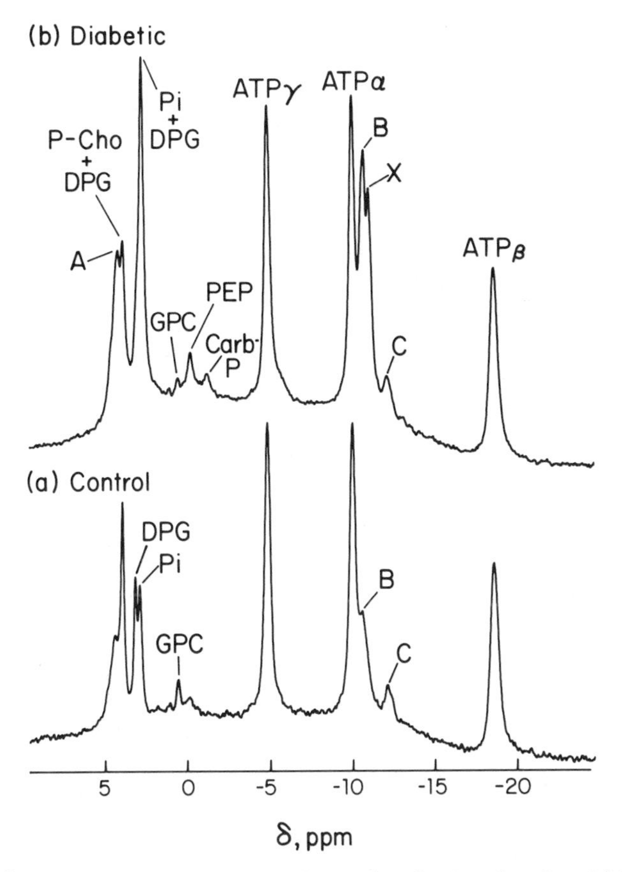

FIGURE 2. ^{31}P NMR spectra at 145.8 MHz of isolated perfused rat liver from (a) 24-hr-fasted control rat and (b) streptozotocin-diabetic rat at 35 ± 1°C. Each spectrum was measured 216–222 min postsubstrate and 16 min after the final NMR spectrum was accumulated in the respective perfusions. Each spectrum is the Fourier transform of 400 FIDs. The labeled ^{31}P peaks include the following: P-Cho, P-choline; P_i, inorganic orthophosphate; GPC, *sn*-glycero-3-phosphocholine; PEP, P-enolpyruvate; Carb-P, carbamoyl-P. Region A includes sugar phosphates, AMP, and 3-P-glycerate. Peak B includes the dinucleotides, predominantly NAD^+. Peak X is an unknown P,P′-diesterified pyrophosphate, and peak C includes the nucleoside diphosphosugars. The peaks labeled DPG are due to D-glycerate-2,3-P_2 in the fresh washed erythrocytes of the perfusate.[5]

The other major difference between ^{31}P spectra of diabetic and control liver is the appearance of carbamoyl phosphate (carbamoyl-P) in the former. With either alanine or pyruvate + NH_4Cl as substrate, carbamoyl-P rose to detectable levels in all diabetic livers studied, whereas no signal for carbamoyl-P was seen in any control liver. Meijer *et al.*[10] have provided strong evidence showing that essentially all the carbamoyl-P present in rat liver resides in the mitochondria. If we assume that this condition obtains under our conditions, intramitochondrial carbamoyl-P levels of 3 to 4 mM are estimated for perfused diabetic liver, in good agreement with the mitochondrial level measured by Meijer *et al.* under ornithine-limited conditions in hepatocytes from normal rats. The ^{31}P chemical shift observed for carbamoyl-P in diabetic liver was reproduced in model solutions (mitochondrial salt conditions) when Mg^{2+} was added and the pH adjusted to 7.5, which is the pH usually estimated for mitochondria. Carbamoyl-P thus appears to be a reliable, and rare, intramitochondrial marker in ^{31}P NMR spectra of liver.

^{13}C NMR OF PERFUSED LIVER FROM STREPTOZOTOCIN-DIABETIC AND UNTREATED CONTROL RATS

As the spectra in FIGURE 3 suggest, our ^{13}C NMR approach makes it possible to determine the metabolic fate of several substances simultaneously. In particular, the main gluconeogenic pathway from [3-^{13}C]alanine into glucose was followed. Because spectra are measured repetitively, the kinetics of glucose synthesis are obtained. These ^{13}C spectra compare liver from a 24-hr fasted normal rat (FIG. 3a) with liver from a diabetic rat (FIG. 3b and c). FIGURE 3 (a and c) contrast spectra of the normal and the diabetic liver at the same period of time (190–200 min) after administration of [3-^{13}C]alanine and [2-^{13}C]ethanol. FIGURE 3 (a and b), on the other hand, contrast spectra of the normal and diabetic liver at different periods of time after the initial administration of substrate, but after addition of the same total quantity of substrate. (Steady-state conditions were approximated by maintaining substrate close to the initial level throughout the perfusion by the addition of small quantities of labeled alanine and ethanol to the perfusion reservoir at frequent intervals. Because of rapid consumption of substrate by diabetic liver, it was more difficult to keep levels constant throughout the accumulation of each spectrum of diabetic liver. However, levels of labeled substrates were monitored frequently during the measurement of each spectrum, so fluctuations were, in general, of short duration. In the low field regions between 60 and 100 ppm, signals for newly synthesized, ^{13}C-labeled glucose are seen. Qualitatively, the ^{13}C enrichment in glucose is similar in both control (FIG. 3a) and diabetic (FIG. 3b and c) liver. Comparison of the spectra in FIGURE 3 suggests that the rate of production of ^{13}C-labeled glucose from [3-^{13}C]alanine was about twice as great in the diabetic liver as in the liver from the 24-hr fasted normal rat. This suggestion is consistent with the greater consumption of ^{13}C-labeled substrate observed for diabetic liver. To test this possibility, aliquots of perfusate were taken throughout the perfusion of livers from diabetic and 24-hr or 12-hr fasted untreated rats, and glucose concentrations were measured by an enzymatic method.[5] The rates of glucose synthesis estimated in this way were 51.7 ± 7.7, 24.2 ± 4.3, and 11.9 ± 1.6 μmol/g liver wet weight/hr for liver from streptozotocin-diabetic, 24-hr-fasted normal, and 12-hr-fasted normal rats, respectively. These rates of glucose synthesis measured for liver from streptozotocin-treated and fasted, untreated control rats are in general agreement with the relative intensities of the glucose resonances in the corresponding ^{13}C NMR spectra. Both measurements correlate well with the elevated glucagon levels observed in this model of chronic diabetes. Endogenous glycogen levels are known to be low in diabetic liver,[11] and in normal liver glycogen levels are almost as low after a 12-hr fast as after a 24-hr fast.[12] Thus, contributions from unlabeled endogenous sources were minimized by both the preparation of the animals and by the 30-min period of preliminary nonrecirculating perfusion in the absence of substrate used here. For these reasons it is assumed that, to a good approximation, glucose present in the liver and perfusate under our conditions was ^{13}C-labeled and was synthesized from the ^{13}C-labeled substrates.

Flow of label is also followed from [3-^{13}C]alanine into C-2, C-3, and C-4 of glutamate and glutamine in all three spectra of FIGURE 3. C-4 of glutamate and glutamine are also traceable to [2-^{13}C]ethanol. Resonances assigned to C-2 and C-3 of aspartate, and to the corresponding carbons in N-carbamoylaspartate, are clearly seen in FIGURE 3a, but not in b and c. The resonance of the randomized C-2 carbon of alanine is in evidence in FIGURE 3a, but is not present at a detectable level in spectra of diabetic liver (vertical lines in FIGURE 3, b and c) under these conditions. Variable levels of β-hydroxybutyrate are seen in all three spectra. The peaks labeled A, C, and D are assigned to base C-5 and ribose moiety C-1 and C-4, respectively, of UTP, UMP, and uridine. Peak B is in the region of glycogen C-1 and CH_2- of PEP. Because a distinct signal from PEP is seen in ^{31}P NMR spectra of diabetic liver under these conditions (FIG. 2b), it seems probable that PEP makes a significant contribution to the intensity of peak B in FIGURE 3, b and c.

An important aid in the assignment of resonances in both ^{31}P and ^{13}C spectra of liver are NMR studies of perchloric acid extracts of freeze-clamped livers.[2,5] The assignment of the resonances of glutathione, in particular, was greatly aided by this approach. First, we noted that the ^{13}C-labeled substrates administered frequently introduced ^{13}C-^{13}C J splitting into the peaks of key metabolites. This J splitting was used to augment chemical shift measurements for identifying metabolites and tracing metabolic pathways. Although the ^{13}C peaks of various metabolites are broadened in spectra of perfused liver (FIG. 3), multiplet structure is resolved in spectra of the corresponding perchloric acid extracts. In spectra of the extracts, the C-3 and C-4 resonances of glutamine and the γ-glutamyl moiety of glutathione showed the same apparent triplet structure as C-3 and C-4, respectively, of precursor glutamate. This information alone was not, however, sufficient to distinguish the reduced and oxidized forms of glutathione, GSH, and GSSG, respectively. A second set of ^{13}C NMR studies based on liver extracts did allow us to make a specific assignment. In this additional approach, the pH-dependent behavior of the ^{13}C chemical shift of C-2 of the γ-glutamyl moiety of either authentic GSH or authentic GSSH, when added to extracts of livers perfused with only unlabeled substrates, was compared with the behavior of the same resonance of glutathione in extracts of livers that had been perfused with the ^{13}C-labeled substrates. These data provided good evidence that the glutathione observed in our ^{13}C spectra of perfused liver (FIGS. 3 and 4) was present in the reduced form GSH.[5]

Two important metabolites, glutathione and N-carbamoylaspartate, were observed and identified in these ^{13}C spectra of liver. N-carbamoylaspartate, which is produced in liver in the second step of the *de novo* pathway of pyrimidine nucleotide synthesis, incorporates the intact aspartate moiety. As anticipated, ^{13}C enrichments observed in liver at aspartate C-2 and C-3 were also observed at the corresponding positions in N-carbamoylaspartate (FIGS. 3 and 4). Appearance of ^{13}C label in the pathway end products, such as uridine, UMP, and UTP (FIG. 3), in spectra of diabetic and control liver suggests that true synthesis of N-carbamoylaspartate, as opposed to turnover, was followed.

Appreciable levels of GSH were observed in ^{13}C spectra of liver from either normal

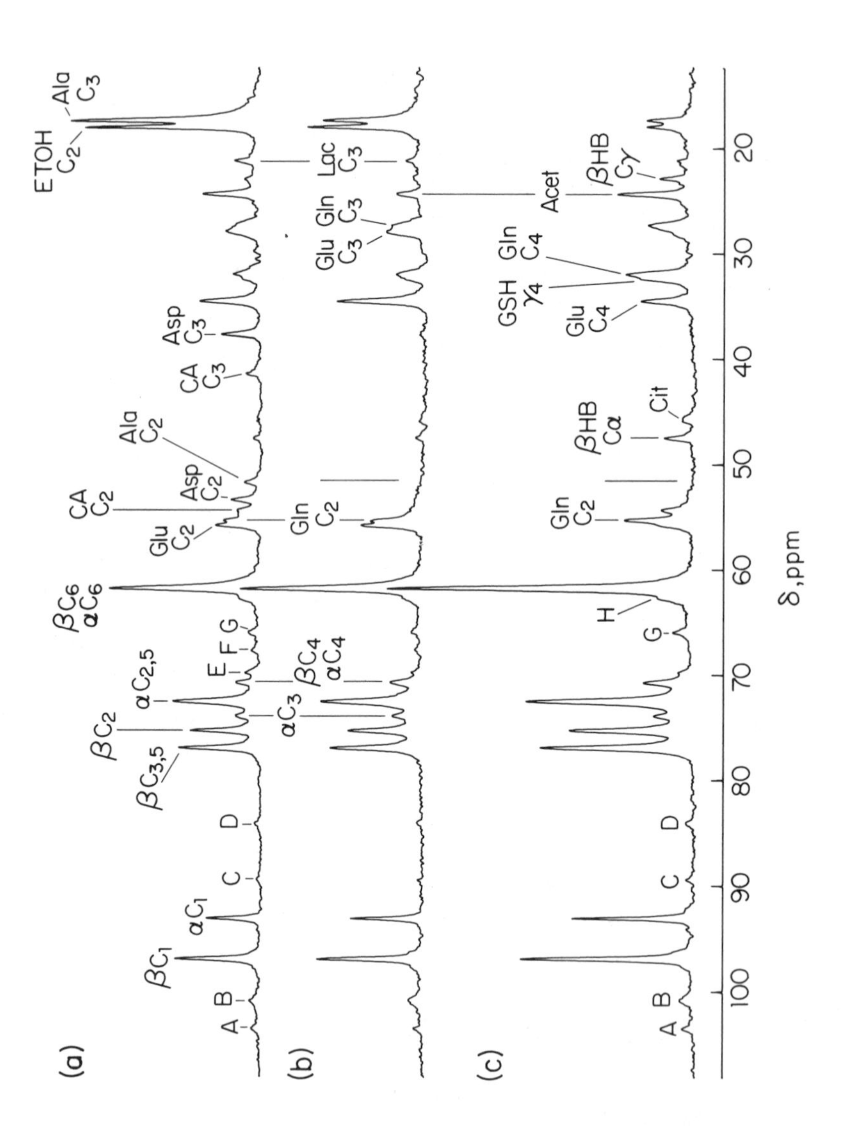
(a)
(b)
(c)
ETOH C2
Ala C3
Asp C3
CA C3
Ala C2
CA C2
Asp C2
Glu C2
βC6 αC6
αC2,5
βC2
βC3,5
αC1
βC1
A B C D E F G
αC3
βC4 αC4
Gln C2
Glu C3 Gln C3
Lac C3
Acet
βHB Cγ
GSH γ4
Gln C4
Glu C4
βHB Cα
Cit
H
δ, ppm
20 30 40 50 60 70 80 90 100

FIGURE 3. ^{13}C NMR spectra at 90.5 MHz of isolated perfused livers at 35 ± 1°C from (a) normal, 24-hr fasted rat and, (b) and (c), a diabetic rat. Each spectrum is from a time sequence of spectra and represents 800 FIDs. Spectrum (a) was measured 190–200 min after the initial addition of 10 mM [3-^{13}C]alanine and 7.3 mM [2-^{13}C]ethanol. Substrate was maintained close to this level by small, frequent (typically every 15 min) additions throughout perfusion period. The ^{13}C natural abundance background spectrum of this liver was measured under identical conditions before substrate was added. For clarity of presentation, this background spectrum was subtracted from the spectrum shown. Spectra (b) and (c) are from a similar series of the same diabetic liver; (b) and (c) were measured 110–120 min and 190–200 min, respectively, post-substrate. This diabetic liver was treated the same as the control liver in (a). Spectra (a) and (b) represent comparison of control versus diabetic for addition of the same total quantity of substrate, whereas (a) and (c) represent comparison after the same total time of perfusion at approximately the same concentration of substrate. The labeled ^{13}C NMR peaks in these spectra include those due to the α and β anomers of D-glucose: βC-1, αC-1 to βC-6, αC-6. Other abbreviations include: Glu C_2, glutamate C-2; Gln C_2, glutamine C-2; Asp C_2, aspartate C-2; Ala C_2, alanine C-2; CA C_2, C-2 of N-carbamoylaspartate; GSHγ_4, C-4 of glutamyl moiety of reduced glutathione (overlaps C-4 of oxidized glutathione); Acet, C-2 of acetate; βHB Cα, C-2 of β-hydroxybutyrate; Cit, citrate-CH_2-; Lac C_3, lactate C-3; and ETOH C_2, ethanol C-2. A–D, see text. E includes lactate C-2 + C-2 of glycerol (ester); F is 3-P-D-glycerate; G is L-glycerol-3-P C-3; and H is C-1, C-3 of glycerol (ester).[5]

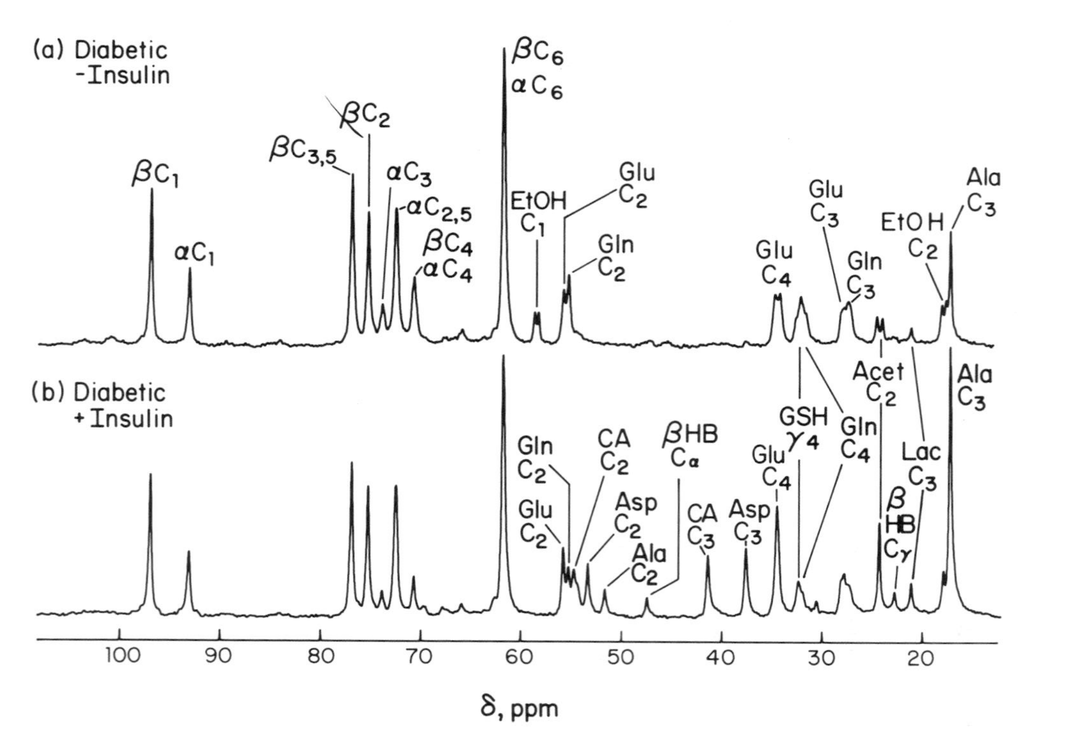

FIGURE 4. ^{13}C NMR spectra of livers from diabetic rats; livers were perfused in the absence, (a), or presence, (b), of insulin. Spectrum (b) is part of a time sequence of spectra and was accumulated during the period 170–180 min after the initial addition of 10 mM [3-^{13}C]alanine and 7.3 mM [2-^{13}C]ethanol. Insulin had been present at 7 nM during a 35-min period before substrate was added; insulin was maintained at 7 nM throughout the perfusion. Spectrum (a) is taken from a similar series of spectra taken of another liver from a diabetic rat and was measured 170–180 min after the initial addition of 10 mM [3-^{13}C]alanine and 7.3 mM [1,2-^{13}C]ethanol. This liver was treated exactly the same as the liver of spectrum (b), except that insulin was absent from the perfusion fluid. Other conditions are as given for FIG. 3. Abbreviations are as given in FIG. 3.[6]

control or diabetic rats (FIGS. 3 and 4). In perfused liver, the C-2 and C-3 signals of the γ-glutamyl moiety of glutathione were not resolved from signals for the corresponding carbons of glutamine. These signals were resolved in spectra of the corresponding perchloric acid extracts. Consequently, observation of the γ-glutamyl C-4 peak, which was resolved, strongly suggests that the peaks labeled, for convenience, simply Gln C_2 or Gln C_3 in spectra of perfused liver include contributions from glutathione.

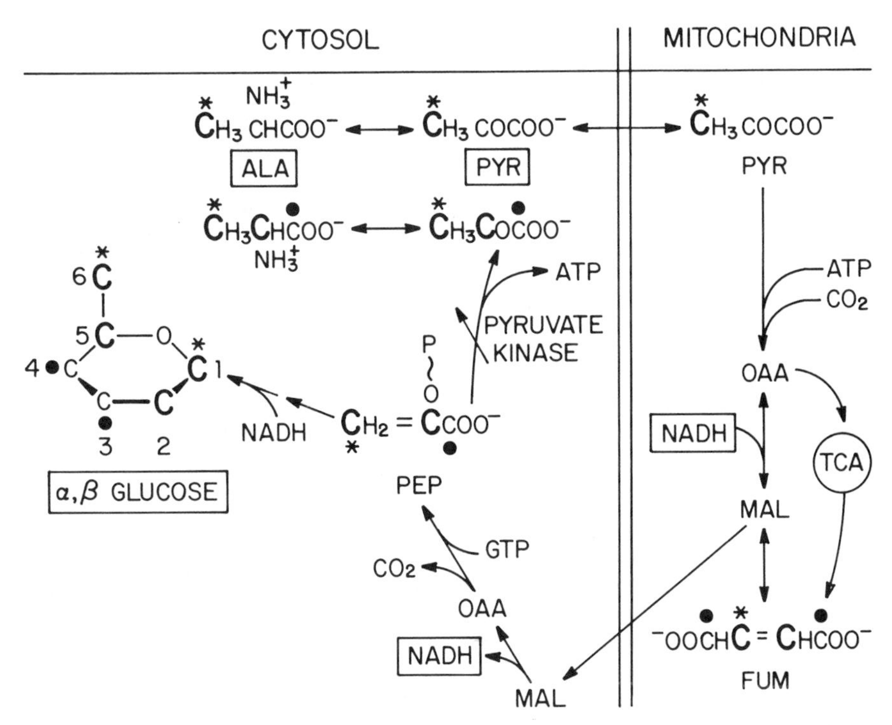

FIGURE 5. Simplified model gluconeogenic pathway. The original label at C-3 of alanine (**C***) is followed into the TCA cycle where randomization by malate dehydrogenase and fumarase exchange occurs; thus, in fumarate (FUM) the label is found with equal probability at either of the two middle carbons (**C**). Further randomization in TCA cycle introduced a small amount of label into the terminal carbons of FUM (C•). MAL is malate and OAA in oxalacetate.[6]

DETERMINATION OF PYRUVATE KINASE FLUX IN PERFUSED LIVER BY ^{13}C NMR

Three futile cycles are known to operate in hepatic glucose metabolism.[13] Cycling that causes ATP hydrolysis without a corresponding change in reactants occurs at irreversible steps where there are two opposing pathways, in which the forward and reverse directions are catalyzed by separate enzymes. In the phosphoenolpyruvate (PEP) cycle (pyruvate → PEP → pyruvate) in rat liver, the path in the gluconeogenic direction from pyruvate to PEP is catalyzed by a complex sequence of enzymes, whereas in the glycolytic direction a single enzyme, pyruvate kinase, catalyzes the reaction PEP → pyruvate (FIG. 5). Although activity is greater in the fed state, there is considerable flux through pyruvate kinase under conditions of active gluconeogenesis in the fasted state.[13]

We have developed a ^{13}C NMR assay of the activity of the PEP cycle in perfused liver carrying out gluconeogenesis.[6] In this assay, livers are perfused under steady-state conditions with a ^{13}C-labeled gluconeogenic substrate ([3-^{13}C]alanine, for example) that enters the pathway as specifically labeled pyruvate. The pathway from [3-^{13}C]alanine into the TCA cycle is followed in FIGURE 5. As shown schematically, label that was randomized in the TCA cycle becomes incorporated into PEP prior to its appearance in either glucose by the usual route or pyruvate by the action of pyruvate kinase. As indicated in FIGURE 5, the middle carbon of PEP is strongly labeled before any label appears at the corresponding carbon C-2 of pyruvate. [A similar circumstance exists for the lesser amount of label found at the carboxylic carbons (C with ● in FIG. 5) of PEP and pyruvate]. This pyruvate is converted to alanine by the active liver alanine aminotransferase. The randomized label will be effectively trapped as alanine if a sufficiently large pool of alanine with the original label ([3-^{13}C]alanine in this example) is present. The flux of PEP through pyruvate kinase as a fraction of the flux of PEP → glucose is estimated from an expression relating these quantities to the relative ^{13}C enrichments at the randomized carbons in alanine as compared with the relative enrichments at the corresponding carbons in glucose. (Measurement of the rate of gluconeogenesis then gives the absolute rate of flux.) Under our conditions of metabolic and isotopic steady state, this relative flux is given by the expression[6]

$$\frac{\text{pyruvate kinase flux}}{(2)\ (\text{gluconeogenic flux})} = \frac{\text{Ala(C-2/C-3)}[1 + (1\text{-}\phi)\text{Glc(C-6/C-5)} + (1\text{-}\phi')\text{Glc(C-4/C-5)}]}{1\text{-}[\text{Glc(C-6/C-5)}][\text{Ala(C-2/C-3)}]} \quad (1)$$

in which all components are readily measured in our ^{13}C spectra. That is, Ala(C-2/C-3) is the ratio of ^{13}C enrichment at C-2 to that at C-3 of alanine; Glc(C-6/C-5) is the corresponding ratio for glucose C-6 and C-5; and $\phi(\phi')$ is the fraction of the ^{13}C-labeled alanine pool in which both C-2 and C-3 (both C-1 and C-2) are labeled in the same molecule.[6] This assay requires the measurement of the ^{13}C enrichments at C-2 and C-3 of alanine in spectra of perfused liver acquired under steady-state conditions during active gluconeogenesis; the alanine intensities must be corrected for T_1 and NOE effects. The ^{13}C distributions in the glucose produced are measured with greater accuracy in spectra of the corresponding perfusates.[5,6] Because ^{13}C distributions in both alanine and glucose are incorporated into this NMR assay, no modifications are required to take into account exchange of label from [^{13}C]ethanol (FIG. 4) entering the TCA cycle as [^{13}C]acetyl-CoA. The assay includes a check on reuse of pyruvate with the randomized label (that is, on the adequacy of the trapping pool); this check uses the ^{13}C enrichment measured at glutamate C-4 and C-5 to estimate the flux from recycled pyruvate into the mitochondrial acetyl-CoA pool.

PYRUVATE KINASE FLUX IN PERFUSED LIVER FROM STREPTOZOTOCIN-DIABETIC AND UNTREATED RATS: EFFECTS OF INSULIN *IN VITRO*

Our earlier ^{13}C NMR investigations demonstrated that the effects of insulin *in vitro* on metabolism in liver from fasted normal control rats were modest.[2,6] In sharp contrast, the spectra in FIGURE 4 demonstrate that treatment of liver from streptozotocin-diabetic rats with 7 nM insulin *in vitro* induced a partial reversal of the differences noted between diabetic and control liver. The spectra in FIGURE 4 compare metabolism in diabetic liver perfused in the absence (FIG. 4a) or presence (FIG. 4b) of insulin; comparison is made at the same time postsubstrate and after addition of essentially the same total quantity of gluconeogenic substrate, [3-^{13}C]alanine. In both cases, as predicted by the main gluconeogenic pathway (FIG. 5), the newly synthesized glucose is strongly labeled at C-1, C-2, C-5, and C-6 and weakly labeled at C-3 and C-4. In the absence of insulin, the rate of ^{13}C-glucose synthesis was high, steady-state levels of certain amino acids were relatively low, and no ^{13}C enrichment was detectable at the randomized alanine C-2 position in diabetic liver (N = 6), as shown in FIGURES 3 (b and c) and 4 (a). However, when perfused diabetic liver was incubated with insulin (FIG. 4b), the rate of ^{13}C-glucose synthesis fell off by 15%, steady-state levels of ^{13}C-labeled aspartate and N-carbamoylaspartate increased sharply, and, most significantly, the relative ^{13}C enrichment at the randomized alanine carbon (C-2) was brought into the range of enrichment observed in liver from 24-hr fasted control rats (FIG. 3a). Thus, the acute regulation of pyruvate kinase by insulin *in vitro* in diabetic liver has been demonstrated by ^{13}C NMR.

An essentially nonketotic form of diabetes is induced by treatment of rats with streptozotocin at the dose level used here (45 mg/kg), whereas at higher dose levels the ketotic form of diabetes is produced. Both [3-^{13}C]alanine and [2-^{13}C]ethanol enter the mitochondrial acetyl-CoA pool as [2-^{13}C]acetyl-CoA, which in turn forms ketone bodies labeled at C-2 and C-4. Labeled β-hydroxybutyrate was observed in ^{13}C spectra of perfused liver (FIGS. 3 and 4), or, more conveniently, in spectra of the perfusion fluid.[5] An average increase of 50% in β-hydroxybutyrate production was measured for diabetic liver compared with control liver in these studies, 7.0 ± 2.0 and 4.6 ± 1.4 μmol/100 g of body weight/hr, respectively. Inter-liver variations suggest that the surprising increase in β-hydroxybutyrate observed in the presence of insulin as compared with the absence of insulin for the diabetic livers shown in FIGURE 4 may not be significant.

The bifurcation of the flux from PEP was measured in control and diabetic liver, in the presence and absence of insulin, by the ^{13}C NMR assay of Equation 1. Results are summarized in TABLE 1. In the absence of insulin, the ratio of pyruvate kinase flux to the gluconeogenic flux showed little variability in liver from 24-hr-fasted control rats (controls 1–4) and ranged from 0.72 to 0.80, independent of substrate. Because of variations in the rates of gluconeogenesis, the absolute rate of pyruvate kinase flux in these livers ranged from 14 to 24 μmol/g liver wet weight/hr. This ratio tended to be higher for liver from 12-hr-fasted rats (0.82–1.20 for controls 8–10), indicating greater relative flux through pyruvate kinase in this state. However, in absolute terms, the rate of flux was lower for the 12-hr-fasted controls, ranging from 12 to 18 μmol/g liver wet weight/hr, because the rate of glucose production was only half of that measured for liver from 24-hr-fasted control rats. Preincubation of liver from 24-hr-fasted rat with insulin for 35 min (control 7) increased the relative pyruvate kinase flux to 0.96. The rate of glucose synthesis from alanine or pyruvate in liver from 24-hr-fasted controls was not significantly depressed by insulin *in vitro,* the rate being 24.2 ± 4.3 and 21.3 ± 3.4 μmol/g liver wet weight/hr in the absence and presence of insulin, respectively. Consequently, the absolute flux through pyruvate kinase was relatively high (25 μmol/g liver wet weight/hr) for control 7. Addition of insulin with substate (controls 5 and 6) had little effect; however, the data on the *in vitro* effects of insulin are limited.

In six livers from streptozotocin-diabetic rats (diabetics 1–6 of TABLE 1) that were perfused in the absence of insulin, pyruvate kinase flux was undetectable by our assay. As a conservative upper limit, we estimate that the relative pyruvate kinase flux in diabetic liver was at least sevenfold lower than that measured in liver from 24-hr-fasted normal control rats.[6] Addition of a physiological level of insulin to the perfusion medium 35 min before substrate increased both the relative flux and the absolute flux through pyruvate kinase in one diabetic liver (diabetic 9) to the level measured in liver from 24-hr-fasted control rats, *viz.*, to 0.74 and 32 μmol/g liver wet weight/hr, respectively. When insulin was coadministered with substrate, pyruvate kinase flux increased to easily observable levels in diabetic liver (diabetics 7 and 8), but the effect was somewhat smaller. Perfusion of diabetic liver under conditions of controlled partial ischemia also brought the pyruvate kinase flux up to detectable levels (data not shown) by an insulin-independent route, presumably through increases in the levels of the

TABLE 1. ^{13}C NMR Measurement of the Effects of Insulin on Relative and Absolute Pyruvate Kinase Flux During Gluconeogenesis in Perfused Liver from Streptozotocin-Diabetic Rats and 12-hr and 24-hr Fasted Untreated Control Rats[a]

Donor Rat	Insulin	Substrate	Pyruvate Kinase Flux/Gluconeogenic Flux	Pyruvate Kinase Flux (μmol/g liver wet wt/hr)
24-hr Fasted Control				
1	−	Ala	0.72	14
2	−	Ala	0.80	24
3	−	Pyr	0.72	—
4	−	Ala	0.72	14
5	+	Pyr	0.80	16
6	+	Pyr	0.72	13
7	++	Pyr	0.96	25
12-hr Fasted Control				
8	−	Pyr	1.20	14
9	−	Ala	1.04	18
10	+	Pyr	0.82	12
Diabetic 1–6	−	Ala or Pyr	ND	ND
7	+	Ala	0.18	12
8	+	Ala	0.52	14
9	++	Ala	0.74	32

[a](Pyruvate kinase flux)/(gluconeogenic flux) and pyruvate kinase flux were calculated using the appropriate form of Equation 1 and the measured rates of gluconeogenesis. "Ala" denotes 10 mM [3-^{13}C]alanine, plus 7.3 mM unlabeled or [2-^{13}C]ethanol. "Pyr" denotes 9.1 mM [2-^{13}C]pyruvate + 5.5 mM NH_4Cl + 7.3 mM unlabeled ethanol. Insulin, when added, was maintained at 7 nM; ++ indicates that insulin was present 35 min before substrate was added; + denotes insulin addition with substrate. ND means not detectable. (Adapted from Cohen.[6])

allosteric activators fructose 1,6-biphosphate and H^+ ion and a decrease in the level of the allostric inhibitor ATP.[6]

Regulation of the activity of pyruvate kinase has been demonstrated previously, by several different approaches, to be important in the hormonal control of gluconeogenesis from substrates that enter the pathway prior to the triose phosphate level.[14–17] An appealing feature of the NMR assay described here is the ability to measure the ^{13}C enrichment at the individual carbons of alanine in whole perfused liver nondestructively, in real time. While comparisons for our exact incubation conditions are not available, the absolute and relative measurements of pyruvate kinase flux given in

TABLE 1 for liver from fasted normal rats are in reasonable agreement with results from several ^{14}C isotopic tracer procedures.[18–20]

The ^{13}C NMR results presented here indicate that insulin exerted its most pronounced effect *in vitro* on liver from the streptozotocin-treated rat, in which model glucagon levels are about doubled. Our ^{13}C NMR data are thus consistent with the current view that insulin's major effect resides in its opposition of the action of glucagon[21] and with measurements on isolated hepatocytes showing that insulin antagonizes the ability of exogenous glucagon to cause the rapid inactivation of pyruvate kinase.[17] Indeed, the most striking effect of insulin *in vitro* observed by our ^{13}C NMR methods was the induction of a huge enhancement (sevenfold or greater) in the flux through pyruvate kinase in diabetic liver, sufficient to bring this flux up to the control level. This effect does not appear to have been demonstrated previously in experimental models of diabetes.

COMPARISON OF ^{31}P SPECTRA AND NATURAL ABUNDANCE ^{13}C SPECTRA OF PERFUSED LIVER FROM GENETICALLY OBESE AND LEAN MICE

Male obese (C57BL/6J ob/ob) mice and their lean littermate controls (Jackson Laboratories, Bar Harbor, ME), 16–20 weeks of age, were used in the present study. All mice were fasted for 20 hr before the start of the perfusion experiment. The body weight of the ob/ob mice ranged from 52–56 g, whereas the leans weighed 24–30 g. All other experimental conditions were as given previously,[5] including NMR conditions and method of liver perfusion, with the exception that here the perfusate contained fresh, washed bovine erythrocytes suspended to a hematocrit of 18% in Krebs' bicarbonate buffer.[2]

Typical ^{31}P NMR spectra of perfused liver from an ob/ob mouse and a lean control mouse are shown in FIGURE 6; in each experiment the liver had been perfused for 44 min under our standard conditions before the spectrum shown was measured. Both spectra show similar intense resonances for the three phosphates of ATP and low levels of intracellular P_i and phosphomonoesters. In contrast to rat liver (FIG. 2), phosphodiesters (0–1 ppm region) are almost undetectable in mouse liver. The wet weight of the liver from the ob/ob mouse was about 2.5 times greater than that of the liver from the lean control. Consequently, the strong similarity between the intensity of the peaks arising from ATP in these two livers was surprising. The ratio of signal to noise of the ATPγ in FIGURE 6b is only about 12% greater than the ratio for the ATPγ peak in FIGURE 6a. A possible explanation for this discrepancy is provided by examination of the natural abundance ^{13}C spectra of these livers (FIG. 7).

The natural abundance of ^{13}C nuclei is only 1.1%. Therefore, the only endogenous metabolites detectable in liver by ^{13}C NMR are those which can accumulate to a high concentration. In liver from fasted rats[7] or mice, the ^{13}C natural abundance background resonances arise mainly from endogenous triacylglycerols. The spectra in FIGURE 7 compare the ^{13}C natural abundance background of liver from an ob/ob mouse with that of its lean control. The assignments of the labeled resonances in FIGURE 7 are listed in TABLE 2.

The level of triacylglycerols in the liver from the fasted obese mouse (FIG. 7b) is about 10 times greater than the level in the fasted control liver (FIG. 7a). The triacylglycerol content of liver from the ob/ob mouse is known to be very high, reaching about 250 μmol/g liver wet weight in the 16-week-old fed mouse.[22,23] Thus a significant fraction of the hepatocyte cytosol is filled with neutral lipids in this model.

It is of interest to note that although peak *i*, which arises from monoenoic fatty acids and the "exterior" olefinic carbons in polyenoic fatty acids, and peak *y*, which is assigned exclusively to polyenoic acids, are of about equal intensity in the leans (FIG. 7a), peak *i* is about three times more intense than the pure polyenoic acid peak *y* in liver from the obese mouse (FIG. 7b).

COMPARISON OF GLUCONEOGENESIS, GLYCOGENESIS, PYRUVATE KINASE FLUX, AND LIPOGENESIS IN PERFUSED LIVER FROM OB/OB AND LEAN MICE

The obese hyperglycemic (ob/ob) mouse is a well-established animal model of obesity and maturity-onset (non-insulin-dependent) diabetes. This model exhibits a number of severe abnormalities associated with carbohydrate metabolism, several of which are age related.[22,23]

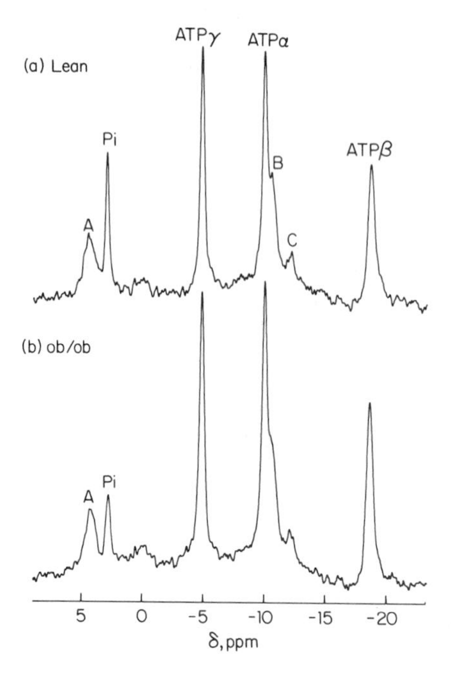

FIGURE 6. ^{31}P NMR spectra of isolated perfused liver from (a) 20-hr-fasted lean mouse and (b) 20-hr-fasted ob/ob mouse. Each spectrum was measured after 44–50 min total perfusion time in absence of added substrate. Labeled peaks are as given in FIG. 2.

Several major pathways are followed simultaneously when ^{13}C-labeled substrate is metabolized by perfused liver from 20-hr fasted mice as the spectra in FIGURE 8 indicate. [2-^{13}C]pyruvate + NH_4Cl was administered to liver from an ob/ob mouse (FIG. 8a) and its lean littermate (FIG. 8b). First, to focus on the spectral region from 60–100 ppm, both livers demonstrated the active production of ^{13}C-labeled glucose and ^{13}C-labeled glycogen. According to the general pathway shown in FIGURE 5, the glucose produced was strongly labeled at C-1, C-2, C-5, and C-6. In both cases our standard perfusion protocol[2] was followed in which each liver is put through a preliminary (35 min) perfusion period during which a substrate-free medium is not

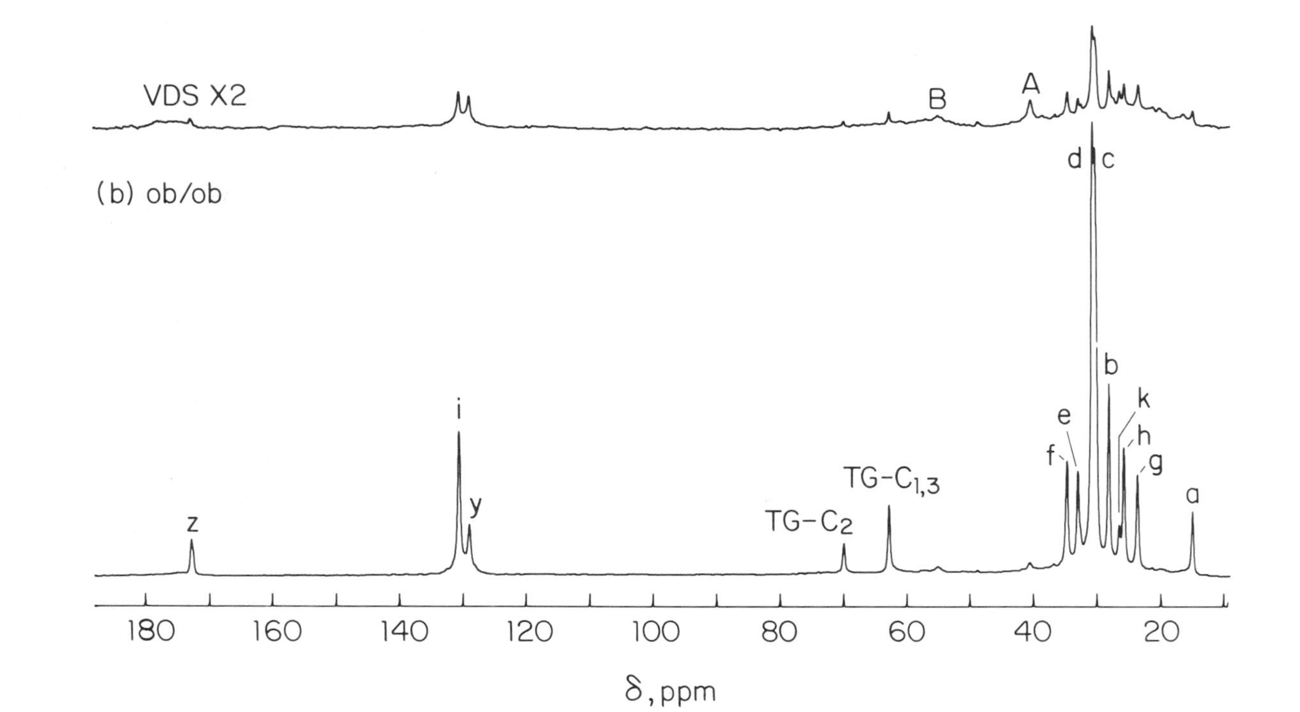

FIGURE 7. ^{13}C NMR spectra showing ^{13}C natural abundance background of isolated perfused liver from (a) 20-hr-fasted lean control mouse and (b) 20-hr-fasted ob/ob mouse before addition of labeled substrate. Labeled resonances are identified in TABLE 2. The notation of VDS × 2 indicates that the gain on the vertical axis was increased twofold when spectrum (a) was read out as compared to spectrum (b).

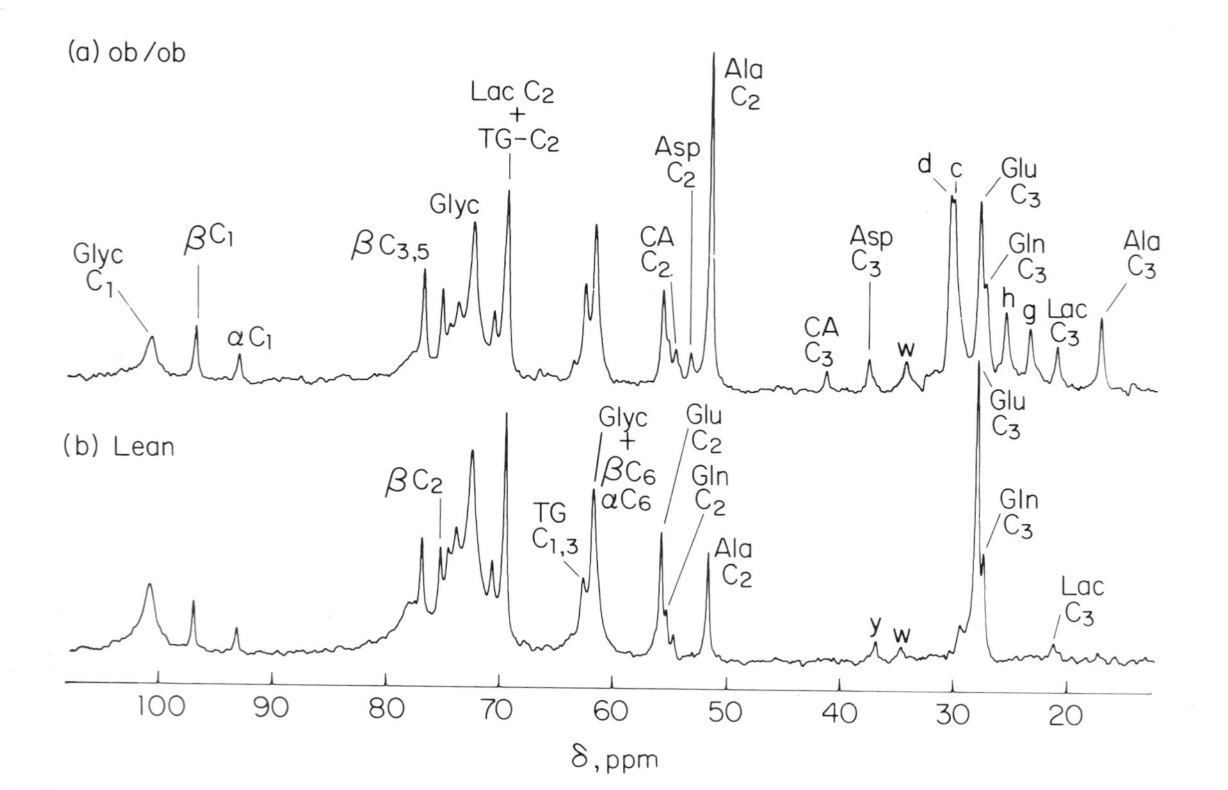

FIGURE 8. ^{13}C NMR spectra of perfused livers from (a) ob/ob mouse and (b) a lean littermate, both fasted 20 hr. Spectrum (a) is part of a time sequence of spectra and was recorded 150–160 min after the initial addition of 7.7 mM [2-^{13}C]pyruvate and 4.6 mM NH_4Cl. The liver from the lean mouse (b) was treated exactly the same as the ob/ob liver shown in (a). Other conditions are as given for FIG. 3. Abbreviations are as given in FIG. 3 with the addition of Glyc, glycogen; TG-C_2, TG-$C_{1,3}$, *c*, *d*, *g*, and *h* as given in TABLE 2; and *w*, *y*, unknown.

recirculated. This preliminary perfusion period was designed to wash out and deplete remaining endogenous substrates. No exogenous glucose was added at any time during the balance of the perfusion period; hence, essentially all glucose present was synthesized by the liver from the ^{13}C-labeled substrate, [2-^{13}C]pyruvate. Consequently, from the time-resolved ^{13}C spectra, such as those shown in FIGURE 8, the kinetics of gluconeogenesis can be followed. ^{13}C NMR has been shown to be uniquely able to follow the kinetics of glycogen synthesis in intact functioning liver.[2] Glycogenesis can be conveniently followed by monitoring the resonance at about 101 ppm arising from ^{13}C at C-1 of the glucosyl units of glycogen (FIG. 8). It is emphasized that under our present conditions, glycogen synthesis must be viewed as strictly a gluconeogenic process, since the liver was first obliged to synthesize the glucose units incorporated into the glycogen molecule from the three carbon precursor, [2-^{13}C]pyruvate. Under these conditions, synthesis of ^{13}C-labeled glycogen was synchronous with the synthesis of ^{13}C-labeled glucose, as was reported earlier for perfused rat liver.[2] Lean mouse liver (FIG. 8b) produced glucose at a rate of 44 μmol/g liver wet weight/hr. After the spectrum shown in FIGURE 8b was measured, a bolus of glucagon (90 nM) was added and the breakdown of glycogen was followed in the subsequent ^{13}C spectra. In this way, the rate of glycogenesis was estimated as 30 μmol glucose units/g liver wet weight/hr, for a total rate of gluconeogenesis of 74 μmol/g liver wet weight/hr for the liver from the lean littermate. For the ob/ob mouse liver shown in FIGURE 8a, the rates of glucose synthesis and glycogen synthesis were 36 and 6 μmol glucose units/g liver wet weight/hr, respectively. Rates of glucose synthesis measured in these ^{13}C NMR studies are comparable to rates reported earlier for a conventional perfused mouse liver preparation with pyruvate as substrate and exceed rates measured with alanine as substrate.[24]

The ^{13}C spectra shown in FIGURE 9 are from a time sequence of spectra of a perfused liver preparation from another ob/ob mouse under the same conditions. These spectra re-emphasize that glycogen synthesis is a very active process in liver from the fasted ob/ob mouse under strictly gluconeogenic conditions with [2-^{13}C]pyruvate + NH_4Cl as the only exogenous substrate (FIG. 9a), and that ob/ob mouse liver is not resistant to the action of glucagon in inducing the acceleration of glycogen breakdown (FIG. 9b). However, acceleration of glycogen breakdown in ob/ob mouse liver required twice the dose of glucagon found to be fully effective for this purpose in lean mouse liver or normal rat liver.[2] Net glycogen synthesis for the liver shown in FIGURE 9 was 129 μmol of glycosyl units, for a total rate of gluconeogenesis of 32 μmol/g liver wet weight/hr.

The second major emphasis in this study is on the activity of the phosphoenolpyruvate cycle in the obese mouse model of non-insulin-dependent diabetes. Completely opposite to the streptozotocin-treated rat model of type 1 diabetes discussed above, the ob/ob mouse displays elevated rates of flux through hepatic pyruvate kinase compared with its lean control as determined by our ^{13}C NMR assay. ^{13}C enrichment at alanine C-2 (FIGS. 8 and 9), which can be traced back directly to the [2-^{13}C]pyruvate substrate, was generally maintained at approximately 8 mM under our conditions. With [2-^{13}C]pyruvate + NH_4Cl as substrate, the principally labeled randomized carbon in alanine is C-3. Thus, to calculate the relative flux through pyruvate kinase with this substrate, Equation 1 must be translated into the appropriate form by interchanging the symbols "Ala C-2" and "Ala C-3" and by interchanging "Glc C-5" and "Glc C-6." Whereas ^{13}C label at the randomized carbon alanine C-3 is barely detectable in ^{13}C spectra of lean mouse liver (FIG. 8b), this carbon is intensely labeled in ^{13}C spectra of ob/ob mouse liver (peak at 17.2 ppm in FIGS. 8a and 9). By this ^{13}C NMR determination, the flux through pyruvate kinase was 25 $\pm$ 11% of the gluconeogenic rate in lean mouse liver ($N = 4$). In ob/ob mouse liver ($N = 6$),

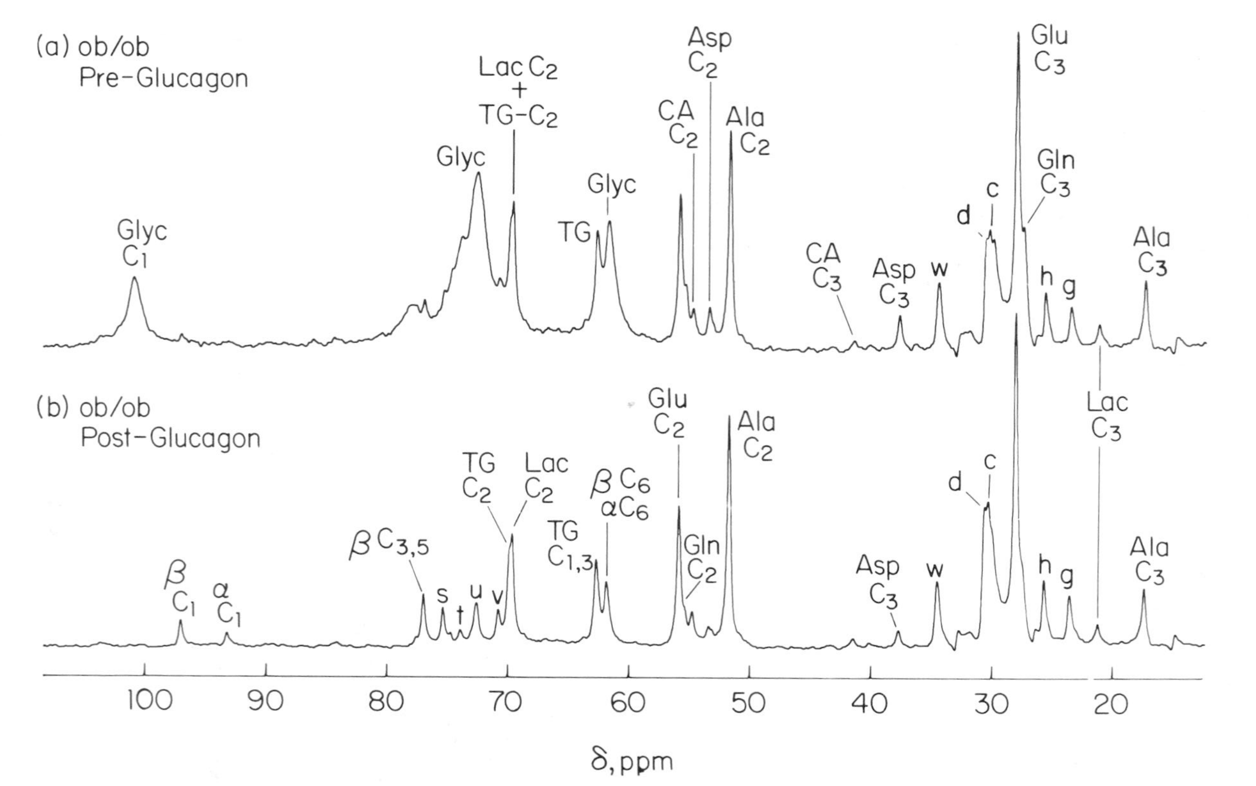

FIGURE 9. ^{13}C NMR spectra of a perfused liver from a 20-hr-fasted ob/ob mouse. Spectrum (a) is part of a time sequence of spectra and was recorded 110–120 min after the initial addition of 7.7 mM [2-^{13}C]pyruvate and 4.6 mM NH_4Cl. Spectrum (b) was measured 30–40 min later, or 30 min after the addition of 170 nM glucagon to the perfusion medium. Other conditions and abbreviations are as given in FIGS. 3 and 8.

pyruvate kinase flux was 102 ± 10% of the gluconeogenic rate. Thus, the relative flux through pyruvate kinase is estimated to be fourfold greater in liver from 20-hr-fasted ob/ob mice than in liver from their similarly fasted lean littermates. The activity of pyruvate kinase has been reported to be three times greater in liver from fed, male ob/ob mice, 10–20 weeks of age, compared with liver from similarly prepared lean littermates.[25] Agreement between the ^{13}C NMR result and the *in vitro* assay of enzymic activity appears to be quite close. It is possible, however, that increased activity of the malic enzyme (FIG. 1), which can interconvert malate and pyruvate, contributed to the apparent flux through pyruvate kinase measured in ob/ob mouse liver. Measurements of the activity of the hepatic malic enzyme indicate that the enzyme is twofold more active in the fed, male ob/ob mouse of 16 weeks of age than in the lean littermate;[22] in overnight fasted mice the increase is reduced to 30%.[26] The contribution of the malic enzyme activity to the apparent flux through pyruvate kinase measured by ^{13}C NMR in these studies is to be tested next by the synthesis and use of the malic enzyme inhibitor, 2,4-dihydroxybutyrate.

TABLE 2. Chemical Shifts and Assignments of Lipid Resonances Observable at the Natural Abundance of ^{13}C in Perfused Liver from ob/ob Mouse[a]

Resonance	Assignment	Chemical Shift (ppm)
a	$*CH_3$—CH_2—	14.73
b	—CH_2—$*CH_2$—CH═CH—	27.95
c	fatty acyl—($*CH_2$)—	30.18
d		30.51
e	CH_3—CH_2—$*CH_2$—CH_2—	32.76
f	—CH_2*CH_2—CO—	34.46
g	CH_3—$*CH_2$—CH_2—	23.47
h	—$*CH_2$—CH_2—CO—	25.60
i	—CH═*CH—CH_2—CH_2—	130.35
k	—CH═CH—$*CH_2$—CH═CH—	26.32
y	═*CH—CH_2—*CH═	128.73
z	—CH_2—CH_2—*CO—	172.43
TG—$C_{1,3}$	glycerol $*CH_2$	62.59
TG—C_2	glycerol *CH	69.74

[a]Letter designations for resonances are as given in FIGURE 7b. *C indicates the specifically assigned type of carbon. Other resonances in FIGURE 7 are A at 40.4 ppm (unassigned) and B at 54.8 ppm (choline $(CH_3)^3N$—).[7]

The third major observation in this study is the demonstration that ^{13}C NMR can be used to monitor the kinetics of lipogenesis via the *de novo* synthesis pathway in ob/ob mouse liver. This approach was developed recently to follow the biosynthesis of fatty acids in perfused liver from fed rats.[7] In this method, livers are perfused under steady-state conditions with labeled substrates that are converted to either [2-^{13}C]acetyl-CoA or [1-^{13}C]acetyl-CoA, which in the *de novo* synthesis pathway label alternate carbons in fatty acids. Fatty acyl carbons labeled by [1-^{13}C]acetyl-CoA (from [2-^{13}C]pyruvate) give rise to resonances distinguishable on the basis of chemical shift from those observed for either endogenous fatty acids or when label is introduced by substrates that are converted to [2-^{13}C]acetyl-CoA in the TCA cycle.[7] As an example of the specificity of this approach, in time-resolved ^{13}C spectra of livers perfused with [2-^{13}C]pyruvate, ^{13}C enrichment at specific positions in the fatty acyl chains gives rise to the resonances labeled *c, d, g, h,* and *z* in TABLE 2. It is important to note that no ^{13}C label is incorporated in the terminal methyl carbon, peak *a* of TABLE 2, under these conditions. Observation of resonances for fatty acyl carbons (*c, d*), *h,* and *g*(and *z,* not shown) in ^{13}C spectra of ob/ob mouse liver (FIGS. 8a and 9) is consistent with the production of ^{13}C-labeled fatty acids by the *de novo* synthesis pathway. The cleanness of the subtraction procedure used to remove the ^{13}C natural abundance resonances (FIG. 7) is demonstrated by the elimination of the background resonances at 14.7 ppm (labeled *a* in FIG. 7), at 40.4 ppm (labeled A in FIG. 7), and at 130.3 and 128.7 ppm (not shown, labeled *i* and *y* in FIG. 7) from the spectra shown in FIGURES 8 and 9. No evidence for the activity of this pathway was found in fasted, lean mouse liver (FIG. 8b). These ^{13}C NMR data demonstrating the activity of the *de novo* synthesis pathway in ob/ob mouse liver are in agreement with the increased activity measured in obese mice of several hepatic enzymes associated with the lipogenic process.[27]

As these investigations may suggest, ^{13}C and ^{31}P NMR spectroscopy can provide useful and novel approaches to the understanding of metabolic regulation and control in the liver in experimental models of diabetes.

REFERENCES

1. COHEN, S. M., P. GLYNN & R. G. SHULMAN. 1981. Proc. Natl. Acad. Sci. USA **78:** 60–64.
2. COHEN, S. M. 1983. J. Biol. Chem. **258:** 14294–14308.
3. NEUROHR, K. J., G. GROLLIN, J. M. NEUROHR, D. L. ROTHMAN & R. G. SHULMAN. 1984. Biochemistry **23:** 5029–5035.
4. CROSS, T. A., C. PAHOL, R. OBERHANSLI, W. P. AUE, U. KELLER & J. SEELIG. 1984. Biochemistry **23:** 6398–6402.
5. COHEN, S. M. 1987. Biochemistry **26:** 563–572.
6. COHEN, S. M. 1987. Biochemistry **26:** 573–580.
7. COHEN, S. M. 1987. Biochemistry **26:** 581–589.
8. KAHN, C. R., D. M. NEVILLE, JR. & J. ROTH. 1973. J. Biol. Chem. **248:** 244–250.
9. SCHWENKE, W.-D., S. SOBOLL, H. J. SEITZ & H. SIES. 1981. Biochem. J. **200:** 405–408.
10. MEIJER, A. J., C. LOF, I. C. RAMOS & A. J. VERHOEVEN. 1985. Eur. J. Biochem. **148:** 189–196.
11. EXTON, J. H., S. C. HARPER, A. L. TUCKER & R.-J. HO. 1973. Biochim. Biophys. Acta **329:** 23–40.
12. SOLLING, H. D., J. KLEINEKE, B. WILLMS, G. JANSON & A. KUHN. 1973. Eur. J. Biochem. **37:** 233–243.
13. KATZ, J. & R. ROGNSTAD. 1976. Curr. Top. Cell. Regul. **10:** 237–289.
14. KRAUS-FRIEDMANN, N. 1984. Physiol. Rev. **64:** 170–259.
15. GROEN, A. K., R. C. VERVOORN, R. VAN DER MEER & J. M. TAGER. 1983. J. Biol. Chem. **258:** 14346–14353.
16. PILKIS, S. J., C. R. PARK & T. H. CLAUS. 1978. Vitam. Horm. (N.Y.) **36:** 383–460.
17. FELIU, J. E., L. HUE & H.-G. HERS. 1976. Proc. Natl. Acad. Sci. USA **73:** 2762–2766.
18. ROGNSTAD, R. 1982. Int. J. Biochem. **14:** 765–770.
19. GRUNNET, N. & J. KATZ. 1976. Biochem. J. **172:** 595–603.
20. FRIEDMANN, B., E. H. GOODMAN, H. L. SAUNDERS, V. KOSTAS & S. WEINHOUSE. 1971. Arch. Biochem. Biophys. **143:** 566–578.
21. UNGER, R. H. 1985. Diabetologia **28:** 574–578.
22. KAPLAN, M. L. & G. A. LEVEILLE. 1981. Am. J. Physiol. **240:** E101–E107.
23. MENAHAN, L. A. 1983. Metabolism **32:** 172–178.
24. ELLIOTT, J., D. A. HEMS & A. BELOFF-CHAIN. 1971. Biochem. J. **125:** 773–780.
25. SEIDMAN, I., A. A. HORLAND & G. W. TEEBOR. 1967. Biochim. Biophys. Acta **146:** 699–603.
26. FRIED, G. H. & W. ANTOPOL. 1966. Am. J. Physiol **211:** 1321–1324.

27. CHANG, H. C., I. SEIDMAN, G. TEEBOR & M. D. LANE. 1967. Biochem. Biophys. Res. Commun. **28:** 682–686.

EDITOR'S NOTE

There was no discussion following this paper's presentation due to lack of time.

Transmembrane Ion Pumping:

High Resolution Cation NMR Spectroscopy[a]

CHARLES S. SPRINGER, JR.

Department of Chemistry
State University of New York
Stony Brook, New York 11794-3400

INTRODUCTION

The alkali metal cations Na^+ and K^+ are ubiquitous in living systems. They are mostly present as free aquo cations and the essence of their biochemistry lies in their unequal distributions across cell membranes and their subsequent transmembrane transport. Sodium is strongly (though not totally) excluded from most cells and the energy stored in the Na^+ gradient is used mostly to drive the active transmembrane transport of small molecules and other ions[1] or, in excitable cells, to transiently diminish or eliminate the transmembrane electrical potential.[2] In an interesting natural reversal of form, K^+ is concentrated in most cells and its gradient is used mostly to maintain this potential, which is exhibited by most cells.[2] Each of these distributions is thus crucial for cell viability and is maintained by an active transport system located in the cytoplasmic membrane of most cells.[3] The sense in which we use the term ion pumping in this paper is just this: the (active) transport of Na^+ and/or K^+ ions across a membrane against their electrochemical potential gradients. In many cases, this is accomplished by the action of the membrane enzyme, $[Na^+ + K^+]$-ATPase, one of the most ubiquitous of all enzymes.[4]

An important goal in physiology and biophysics is to monitor the maintenance and restoration of these ion gradients in whole tissue in real time and as noninvasively as possible. This indicates that a spectroscopic approach is called for. Unfortunately, the closed electronic shells of the Na^+ and K^+ ions preclude the use of the sensitive optical or electron paramagnetic resonance spectroscopies. Fortunately, the most abundant, stable isotopes of sodium (^{23}Na, 100%, $I = 3/2$) and potassium (^{39}K, 93%, $I = 3/2$) have nuclear magnetic moments and are susceptible to nuclear magnetic resonance (NMR) spectroscopy. We have recently reviewed the study of ^{23}Na and ^{39}K NMR of tissue samples.[5] Considering the nuclear magnetic properties of ^{23}Na, along with the abundance of Na^+ in living systems, leads to the realization that ^{23}Na is by far the second most NMR-sensitive nucleus (^{1}H is the first) in biology.[5] This has given rise to considerable tissue spectroscopic activity over a thirty-year period and to a more recent intense interest in medical ^{23}Na NMR imaging.[5]

The early spectroscopic studies revealed two fundamental problems. First, the ^{23}Na signals from the various compartments in tissue are isochronous. Since the Na^+ ions are mostly in the form of the aquo species, irrespective of the compartment containing them, their chemical shifts are effectively indistinguishable. Second, the integrated ^{23}Na spectral intensity often does not correspond to that expected for the total amount

[a]Supported by National Institutes of Health (Grant No. GM 32125) and National Science Foundation (Grant No. PCM 84-08339).

of Na known to be present from chemical analyses: this is the famous "NMR invisibility" problem.[5] Although studies of ^{39}K are much fewer because of its considerably lower NMR receptivity, those that have been conducted demonstrate the same two problems.

The second problem, that of "NMR invisibility," is an interesting and complicated issue. It arises from the electrical quadrupolar nature of the ^{23}Na and ^{39}K nuclei and remains under investigation.[5] It will not be discussed in this paper except to say that, often, a portion of the NMR signal from all of the ^{23}Na or ^{39}K nuclei present in a given tissue compartment is not detected when high-resolution NMR acquisition conditions are employed. Its implications for the transport studies to be described below will be occasionally noted.

THE USE OF SHIFT REAGENTS

The first problem, that of isochronicity, completely stifled early attempts to use NMR to monitor transmembrane ion gradients or transport. One cannot discriminate the intra- and extracellular ^{23}Na and ^{39}K signals in an unadulterated system. An important step toward solving this problem occurred when four groups independently introduced aqueous shift reagents (SR) for the resonances of these, and other, metal cations in 1981 and 1982.[6–9] These are membrane-impermeant anionic complexes of paramagnetic lanthanide ions, which serve to cause a hyperfine shift of the resonance frequency of the metal cation nucleus via a dynamic chemical binding equilibrium with the cation. The chemistry of the equilibria and the SR and their physiological compatibility have been discussed at length.[8,10–12] Their use with suspensions of cells, intact tissue, and *in vivo* has been reviewed.[5]

As examples of the use of SR, FIGURE 1 depicts some ^{23}Na NMR spectra of whole human blood. In FIGURE 1A, the plasma has been made 10.2 mM in the upfield SR $TmTTHA^{3-}$ ($TTHA^{6-}$ represents the triethylenetetraminehexaacetate chelate ligand), while in FIGURE 1B, it was 10 mM in the downfield SR $DyTTHA^{3-}$. The shifted extracellular resonance can be clearly distinguished from the unshifted intracellular signal. The spectrum in FIGURE 1C is of the same sample as in 1B except that an edited spectrum was obtained using an excitation pulse sequence designed to eliminate most of the large signal from extracellular Na^+, Na_o.[13] This can be useful in improving the quantitation of the small signal from intracellular Na^+, Na_i.[14] Since the relative intensities of the two peaks depend on the relative values of the $c \cdot V$ products (c represents the concentration of the cation in a compartment and V represents the cation-accessible volume of the compartment[5]), the Na_o signal is usually much larger than the Na_i signal when the spectrum is obtained under physiological conditions. This is still true in FIGURE 1A and B even though the samples consisted of outdated whole blood whose cells were somewhat Na^+ loaded (c_i ca. 50 mM) due to inhibition of the $[Na^+ + K^+]$ATPase by the low temperature of the blood bank in which the blood had been stored. Of course, the opposite is true for the relative intensities of the K_o and K_i signals obtained from physiological samples.[5]

The shift reagent approach is rapidly gaining in popularity. TABLE 1 presents chronological lists of refereed reports of ^{23}Na and ^{39}K NMR studies of cell suspensions and intact tissue samples using SR. Not all of these studies included transport experiments. When they did, evidence for the activities of various membrane ion transport systems is indicated. The rigor of this evidence covers a wide range.

The remainder of this paper will consist of four examples from our laboratories of

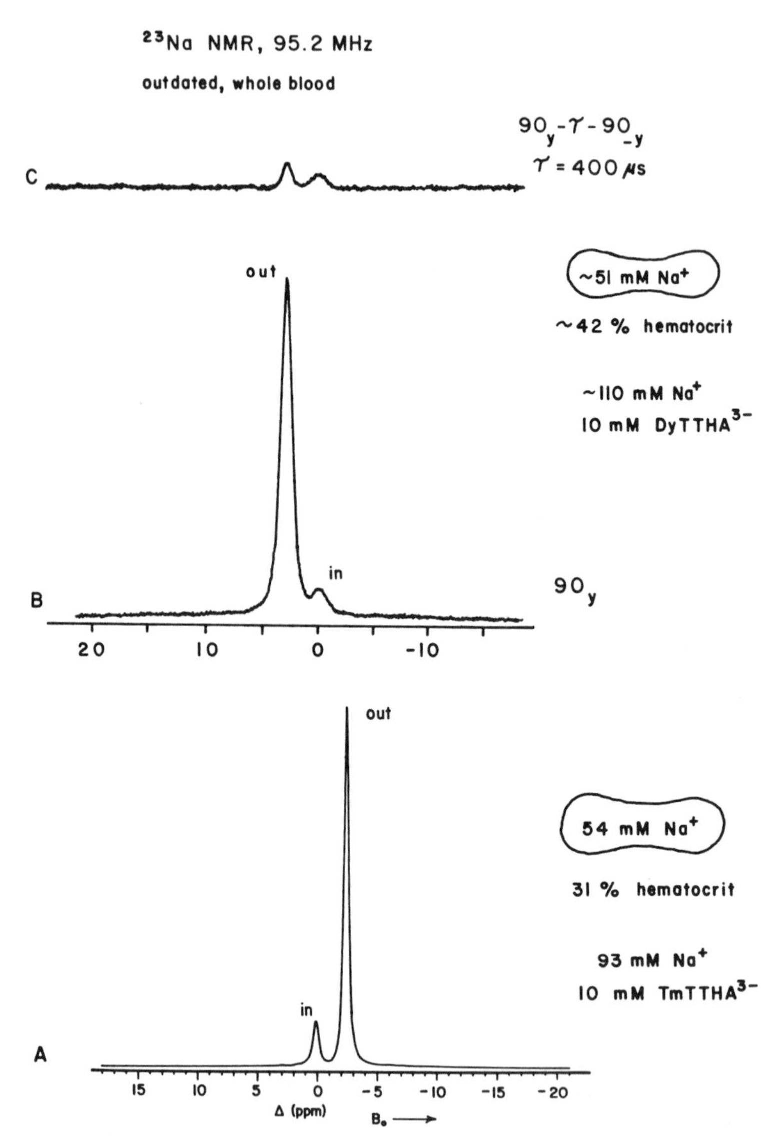

FIGURE 1. ^{23}Na NMR spectra (95.2 MHz) of outdated, whole blood. In the sample for (A), the plasma was 10.2 mM in $TmTTHA^{3-}$ (Fig. 2, Table 2 of Ref. 19). In the sample for (B), the plasma was 10 mM in $DyTTHA^{3-}$ (Fig. 4–16, Table 4-1 of Ref. 13). For (C), the sample was the same as in (B), but the spectrum was obtained with a jump-return pulse sequence designed to mostly eliminate the resonance from the extracellular ions. (From Springer.[5] With permission from the publisher.)

the active transport ("pumping") of Na^+ out of cells. Most of these are not yet published.

STUDIES OF HUMAN ERYTHROCYTES

The first example involves the study of suspensions of human erythrocytes. FIGURE 2 depicts the results from suspensions of fresh cells that have been washed and resuspended (to ca. 24% hematocrit) in low-Na^+ media. The data were obtained by Dr. Pike.[13] The ordinate is the fraction of ^{23}Na NMR signal intensity due to Na_i. Since Na_i and Na_o are 100% NMR visible,[18,19,24,32,33] the ordinate also represents the fraction of Na^+ that is Na_i. Even though these fresh cells contain a normal low level of Na^+ (c_i = ca. 15 mM[33]), there is net efflux because they have been suspended in low-Na^+ media. This is quite obvious in FIGURE 2. The differences in effluxes arise because the cells were given ouabain, a specific inhibitor of $[Na^+ + K^+]$ATPase (the "pump"), glucose, a substrate for the production of ATP to power the pump, both, or neither, as indicated. Mr. Xu has recently begun a project to interpret these and other data with a general mathematical model for the electrodiffusion of ions across cell membranes based on that published by our colleagues Clausen and Moore.[34] The solid curves represent preliminary computer simulations of the data. For all of the curves, the permeability coefficients for the passive transport of Na^+, K^+, and Cl^- were assigned values of 3.45×10^{-9} cm/sec, 6.00×10^{-9} cm/sec, and 500×10^{-9} cm/sec, respectively. In the two cases where ouabain is present, no pump current was allowed in the simulation. In the case where no additions were made, the maximum pump current was assigned the value of 4.67×10^{-9} $Ccm^{-2}sec^{-1}$. In the case where only glucose was added, the maximum pump current was assigned the value of 63.3×10^{-9} $Ccm^{-2}sec^{-1}$. The simulations are quite good; however, we want to stress that they are not fittings. Although Mr. Xu is proceeding to the problem of actually fitting these data sets, this must be approached with great caution because it can become quite multiparametric. Nonetheless, the qualitative nature of the results makes it very clear that the ouabain-inhibitable flux is about an order of magnitude greater for the case where glucose is present than for that where it is not. This is direct evidence for the operation of the $[Na^+ + K^+]$ATPase enzyme.

FIGURE 3 offers even more dramatic evidence from experiments in which fresh erythrocytes were washed and resuspended (45% or 50% hematocrit) in media containing ca. 140 mM Na^+.[13] The ordinate values are much smaller than those of FIGURE 2, because the spectra were not edited as was that of FIGURE 1C. This also gives rise to more scatter in the data. In any case, it is clear that, in most of these experiments, influx of Na^+ occurs. Only in the case where glucose alone was added are the cells able to pump Na^+ out against its electrochemical gradient; and then only until they presumably exhaust the added glucose some time between 10 and 20 hours. The dashed lines are not simulations and are intended merely to guide the eye. The solid curves are computer simulations. The permeability coefficients for the passive fluxes of Na^+, K^+, and Cl^- were assigned values of 1.67×10^{-10} cm/sec, 6.00×10^{-10} cm/sec, and 500×10^{-10} cm/sec, respectively, in all cases except in that for the addition of only glucose, where the value for Na^+ was 3.33×10^{-10} cm/sec. The maximum pump current was assigned values of 4.67×10^{-9} $Ccm^{-2}sec^{-1}$, 0.5×10^{-9} $Ccm^{-2}sec^{-1}$, and zero for the cases with glucose alone, glucose plus ouabain, and ouabain only, respectively. This was possibly necessary because 25 μM ouabain may not be enough to completely inhibit the pump for cells at such a high density and in the presence of glucose. In any case, this is clear evidence for the operation of the pump. Shulman and

TABLE 1. ^{23}Na and/or ^{39}K NMR Transmembrane Discrimination Studies Using SR

	Membrane Transport System Detected						
Subject	$[Na^+ + K^+]$ ATPase	Na^+/H^+ Exchange	Na^+/Ca^{2+} Exchange	$[Ca^{2+}]$ ATPase	$[H^+]$ ATPase	K^+ Channel	Reference
Cell Suspensions							
Yeast cells (*S. cerevisiae*)		X			X	X	Balschi *et al.*[15]
Human erythrocytes							Gupta & Gupta[9]
Human erythrocytes							Brophy *et al.*[16]
Yeast cells (*S. cerevisiae*)		X			X	X	Ogino *et al.*[17]
Human erythrocytes							Pettegrew *et al.*[18]
Human erythrocytes							Pike *et al.*[19]
Amphibian oocytes (*R. pipiens*)		X					Morril *et al.*[20]
Millet cells (*P. miliaceum*)							Sillerud & Heyser[21]

Human erythrocytes						
Dog erythrocytes						Shinar & Navon[22]
Human erythrocytes						
Rat, dog, sheep erythrocytes						Boulanger *et al.*[23]
Human erythrocytes	X					
Duck erythrocytes						Ogino *et al.*[24]
Rat myocytes	X		X	X		Wittenberg & Gupta[25]
Bacterial cells (*E. coli*)		X				Castle *et al.*[26]
Intact Tissue Samples						
Frog muscle						Gupta & Gupta[9]
Frog skin	X	X				Civan *et al.*[27]
Rabbit renal cortical tubules	X					Gullans *et al.*[28]
Rat heart	X		X			Pike *et al.*[14]
Rat muscle						Matwiyoff *et al.*[29]
Corn roots		X			X	Gerasimowicz *et al.*[30]
Rat heart						Fossel & Höfeler[31]

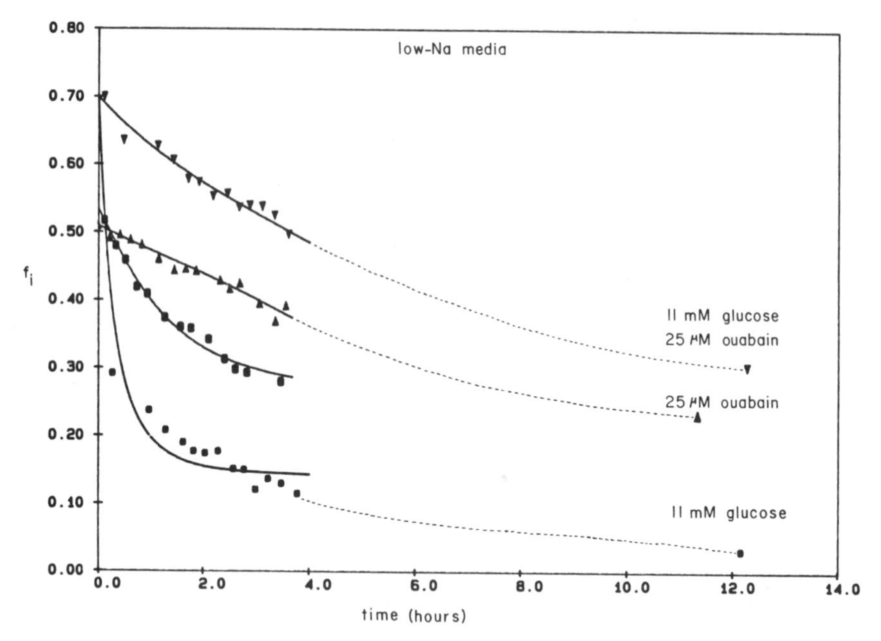

FIGURE 2. Sodium efflux from fresh human erythrocytes suspended in low-Na^+ media. The ordinate is the fraction of total Na^+ that is intracellular. The different additions to the extracellular medium prior to the measurement of Na^+-efflux are indicated. The solid curves are computer simulations (see text). The dashed curves are intended merely to guide the eye.

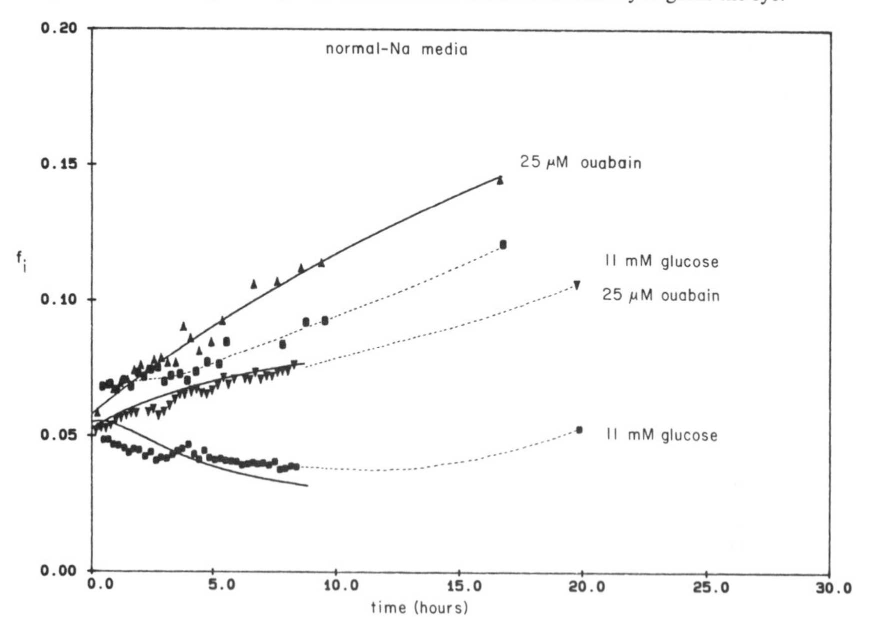

FIGURE 3. Sodium fluxes in suspensions of fresh human erythrocytes in normal-Na^+ media. The ordinate is the fraction of total Na^+ that is intracellular. The different additions to the extracellular medium prior to the measurement of Na^+-flux are indicated. The solid curves are computer simulations (see text). The dashed curves are intended merely to guide the eye.

co-workers have published results somewhat similar to some of those seen in FIGURE 3.[24]

STUDIES OF YEAST CELLS

Our second example involves suspensions of yeast (*Saccharomyces cerevisiae*) cells. We have conducted a number of studies of Na^+ efflux from Na^+-loaded, polyphosphate (P_n)-deficient yeasts suspended in media low in Na^+ and containing phosphorus.[35,36] FIGURE 4 depicts a general overview of these investigations. At the top

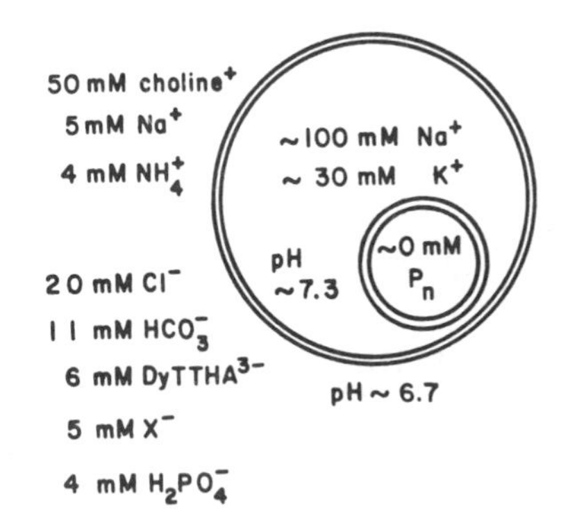

additions to external medium

	N_2	O_2	19 mM KCl	250 mM glucose	
					68 mM MES
N_2	x		x	x	x
O_2		x	x	x	x

FIGURE 4. (Top) The approximate initial concentrations of relevant species common to all Na^+ efflux studies of Na^+-loaded yeasts described in the text. (Bottom) Additions to the extracellular medium just before the commencement of Na^+-efflux measurement for the various experiments.

are shown the extra- and intracellular concentrations of relevant species present at the beginning of each kinetic run. The table at the bottom of the figure lists the varying additions to the external medium just before the different runs begin. Thus the media differ as to whether dioxygen, glucose, or K^+ are present individually or in combinations and as to whether the medium is buffered or not. One should note that as long as the normal membrane potential (ca. 80 mV, inside negative) is maintained, almost all of the Na^+ extrusion is energetically uphill, even for these Na^+-loaded cells. FIGURE 5 shows some of the ^{31}P and ^{23}Na NMR spectra that were simultaneously acquired during the experiment in which O_2 and K^+ were present in the medium. The peaks in the ^{31}P spectra are (1) phosphomonoesters (sugar phosphates, sP); (2) intracellular

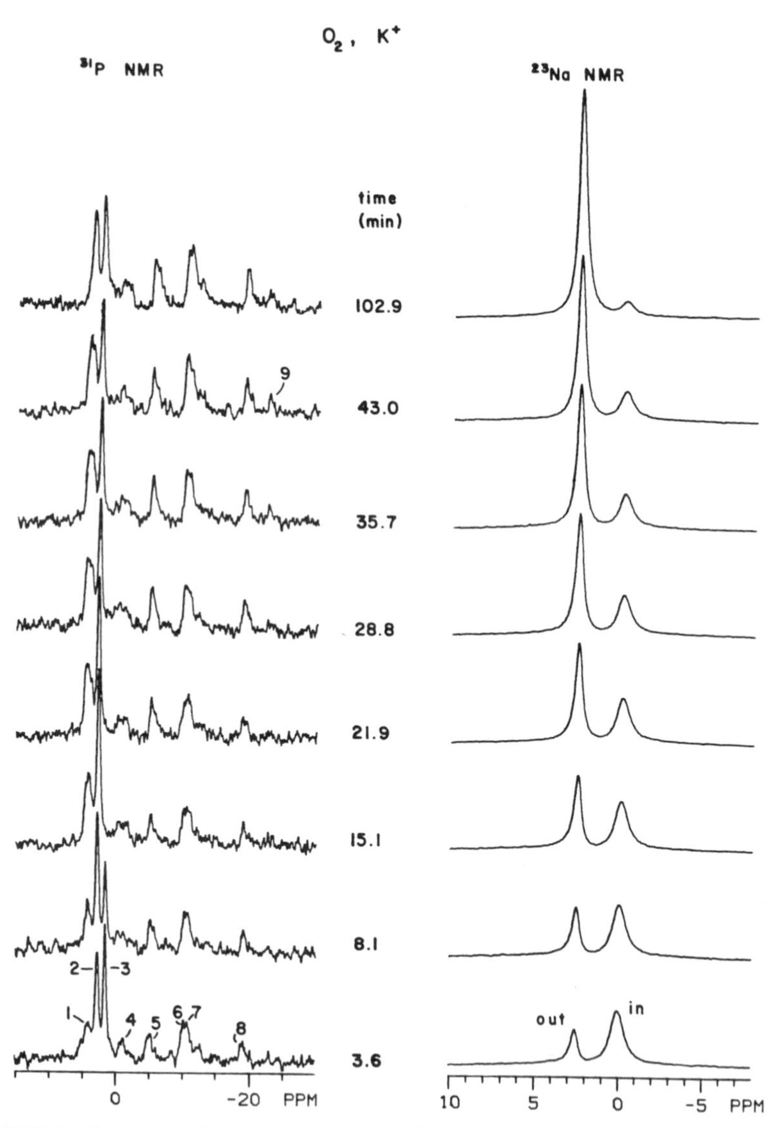

FIGURE 5. Simultaneous ^{31}P (145.75 MHz) and ^{23}Na (95.24 MHz) NMR spectra obtained at 8.5 T on a suspension of Na^+-loaded yeast. The yeast are suspended (23.1%, w/vol) in a medium that is low in Na^+ and contains 3.8 mM $NH_4H_2PO_4$ and 6.3 mM $choline_3DyTTHA \cdot 3cholineCl$, a downfield shift reagent for the ^{23}Na signal. The suspension was bubbled with 95% O_2/5% CO_2, and each pair of spectra is marked with the time of the midpoint of the acquisition which had elapsed since the external medium was made 18.8 mM in KCl. (See Höfeler *et al.*[36] for additional material.)

orthophosphate (inorganic phosphate, P_i), $(P_i)_i$; (3) extracellular inorganic phosphate, $(P_i)_o$; (4) unassigned; (5) γ-P of adenosine triphosphate (ATP) and β-P of adenosine diphosphate (ADP); (6) α-P of ATP and α-P of ADP; (7) P of oxidized and reduced nicotinamide-adenine dinucleotide (NAD); (8) β-P of ATP; and (9) non-terminal P of P_n. The assignments of the ^{23}Na spectra are clear since $DyTTHA^{3-}$ is a downfield SR for $^{23}Na^+$. They are indicated in the figure. These spectra, and others not shown, were analyzed for peak intensities (areas) and the chemical shifts of the P_i peaks. The latter allow an estimation of the intra- and extracellular pH values. The results of such analyses for the experiment where O_2 alone was added to the medium are shown in FIGURE 6. For all of the graphs except those representing pH, the ordinate is the fraction of NMR-visible P or the fraction of total Na. For yeast suspensions, it is known that Na_o is 100% NMR visible but that Na_i is only partially NMR visible.[17,35,36] This is shown in FIGURE 5 by the increase in total ^{23}Na signal intensity as Na^+ efflux continues. This has been taken into account when calculating the ordinate values in FIGURE 6.

Some of the results seen in FIGURE 6 are common to all the experiments of FIGURE 4. Thus, there is always an efflux of Na^+ and an influx of P_i. The initial linear period of Na^+ efflux seen in FIGURE 6 is common to almost all experiments and averages about 15 min. After that, the Na^+ efflux enters a transition period where it either slows (FIGURE 6), ceases, or transiently reverses. This transition period lasts about 10 min on the average and after this the efflux resumes with about the same value as in the initial period (not the case in FIGURE 6). The P_i taken up by the cells accumulates first as $(P_i)_i$, as seen in FIGURE 6, except when glucose is present in the external medium, in which case it accumulates first as sP.

Perhaps the most interesting aspect of the results of these studies is the nature of the production of vacuolar polyphosphates, P_n. These were formed, as seen in FIGURE 6, in almost all experiments. There were three common features attendant to their production. First, there was always an induction period before they were formed. As shown in FIGURE 6, this always coincided quite well with the end of the initial linear period of Na^+ efflux. This is somewhat unusual in that P is entering the cells during this period. In other studies in our laboratories[35] and those of others,[37] the initiation of the formation of P_n has been observed to be coincident with the beginning of plasmalemmal P influx. Second, the P stored in the P_n species appears to have derived mostly from the sP pool. This transformation was very rapid and has also been reported by others.[38] Third, the formation of P_n species was always transient. As also exemplified in FIGURE 6, they were always ultimately consumed and the final distribution of NMR-visible cellular P was usually mostly in the sP pool.

We have offered a detailed interpretation of these results.[36] Two observations are key to our hypotheses. The first is that the vertical extent of the initial linear period of Na^+ efflux corresponds to the extrusion of an amount of Na^+ (about 61 μmol, on the average) almost exactly equal to the amount of NH_4^+ initially present in the extracellular medium (about 59 μmol, on the average). Thus, we propose that the initial phase of Na^+ efflux represents a one-for-one exchange of Na^+ with NH_4^+, which is an active pumping of Na^+ out of the cell against its electrochemical gradient. The yeast cell normally carries out this active transport with K^+ as the external ion exchanging with Na^+. Since the yeast cytoplasmic membrane is not known to have the $[Na^+ + K^+]$ATPase enzyme, it is thought to accomplish this transport by the coupled activities of a probably electrogenic Na^+/H^+ exchange protein, an electrogenic $[H^+]$ATPase, and an electrogenic K^+ channel.[36] In the present case, an NH_4^+ channel is employed in place of the K^+ channel. What is most interesting about this process is that, apparently, its energy demand (for ATP) has precedence over that required for the storage of the incoming P in the form of P_n. Thus, P_n formation does not begin until

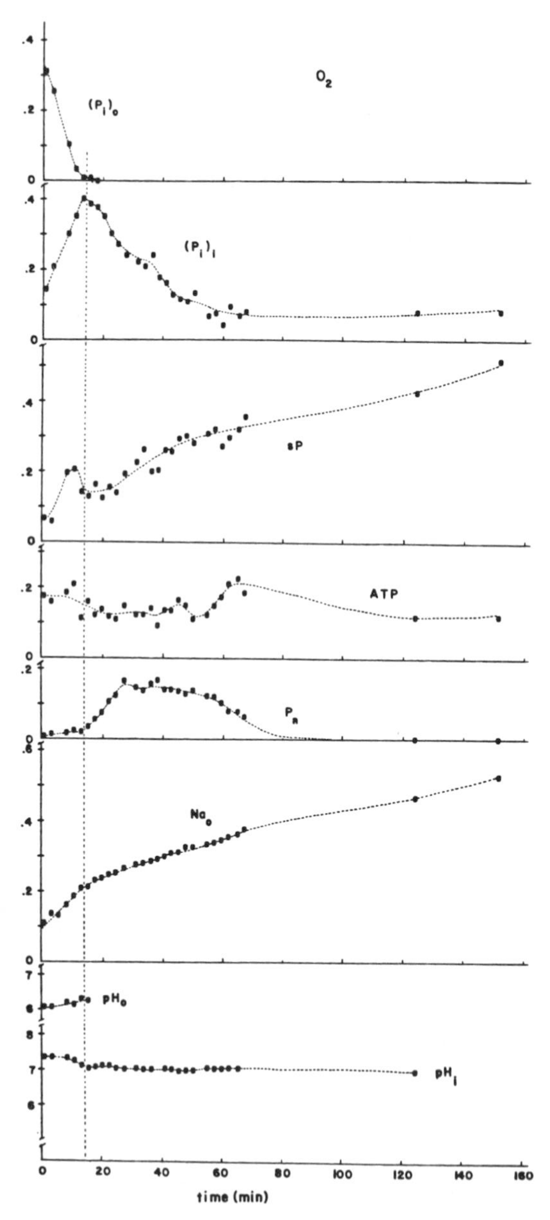

FIGURE 6. Time dependence of spectral data from the yeast experiment with aerobic (only) medium. The ordinates of the top five curves are the fractions of NMR-visible phosphorus. That of the next curve is the fraction of total Na^+ while those of the bottom two curves are pH values. The dashed lines are intended only to guide the eye. The vertical dashed line is drawn to mark the time of end of the initial period of Na^+ efflux. (See Höfeler *et al.*[36] for additional material.)

the NH_4^+ for Na^+ active transport has ceased because of depletion of the extracellular NH_4^+. The remaining Na^+ extrusion, which resumes after the transition period, is known not to be an energy-requiring process. A second observation is consistent with this hypothesis. In the experiment where both O_2 and K^+ were present in the external medium (FIG. 5), the sum of the concentrations of K^+ and NH_4^+ is about six times the concentration of NH_4^+ present in the other experiments (FIG. 4). In this case, the initial linear period of Na^+ efflux persists much longer (probably six times as long) than in any of the other experiments. In addition, no significant amounts of P_n species, compared to other experiments, are ever formed. This was the only one of our aerobic experiments in which this was true. The yeast cells must expend considerably more ATP to pump out an amount of Na^+ equal to the total amount of K^+ and NH_4^+ present in the external medium. Thus, here we see a manifestation of Na^+ pumping not involving $[Na^+ + K^+]$ATPase.

STUDIES OF PERFUSED, BEATING RAT HEARTS

The main advantage of the use of $DyTTHA^{3-}$ as a shift reagent is its relative stability in physiological media.[12] This allows one to study perfused tissue or to conduct *in vivo* experiments with relative impunity. Thus, our third example is a study of active transport in perfused, beating rat hearts. Since we have published these experiments as part of an extensive set of exploratory studies[14] and twice reviewed them,[12,39] we will be very brief here. FIGURE 7 depicts a stacked set of ^{23}Na NMR spectra obtained after the spectral effects of perfusion with a medium containing 10 mM $DyTTHA^{3-}$ had reached steady state. In this situation, the SR has permeated essentially all of the vascular and interstitial spaces and, since this is where the vast majority of Na^+ is located, almost all of the ^{23}Na NMR signal is shifted downfield. The spectra of FIGURE 7 were obtained with an excitation pulse sequence similar to that used for the spectrum in FIGURE 1C. The large extracellular resonance, which is shifted downfield, is not eliminated as well in FIGURE 7 as in FIGURE 1C, and one sees some residual negative intensity that results from phase shifts attendant to the pulse sequence.

At the beginning of the time period exhibited in FIGURE 7, the perfusing medium was switched to one in which the concentration of K^+ was lowered from 5.9 mM to zero and that of Ca^{2+} from 3 mM to 230 μM (all other concentrations remained unchanged, including that of SR). In a very short time, the heart stopped beating. The subsequent spectra are labeled with the time elapsed (to the middle of the acquisition period) since the low-K^+ medium entered the heart. One can clearly see the rise in intensity of a peak representing Na_i (at ca. 0.5 ppm) with time. This represents the leakage of Na^+ into the myocytes because the Na^+ pump is effectively inhibited by the lack of K^+ in the extracellular medium. When the K^+ in the perfusing fluid is restored to its normal concentration (at 23.5 min), the excess Na_i is quickly pumped back out of the heart cells. The intensity data from this experiment are plotted in FIGURE 8. The ordinate represents the intensity of the Na_i peak on an arbitrary scale. It is difficult to be more quantitative because the value of the NMR visibility of intracellular $^{23}Na^+$ is not known at this time. It seems quite likely that the NMR visibility of the extracellular $^{23}Na^+$, in the steady state of SR perfusion, is ca. 40%.[5,12] In addition, it is possible that the visibility of the Na_i peak changes with the degree of Na_i loading. In any case, the effect of the reactivation of the pump with K_o (first arrow in FIG. 8) is clearly seen. When Ca_o was returned to normal (34.7 min, second arrow in FIG. 8), the heart resumed beating. We have interpreted this result to represent a protection of the cells, by low Ca_o, from the effects of the sarcolemmal Na^+/Ca^{2+} exchange protein during

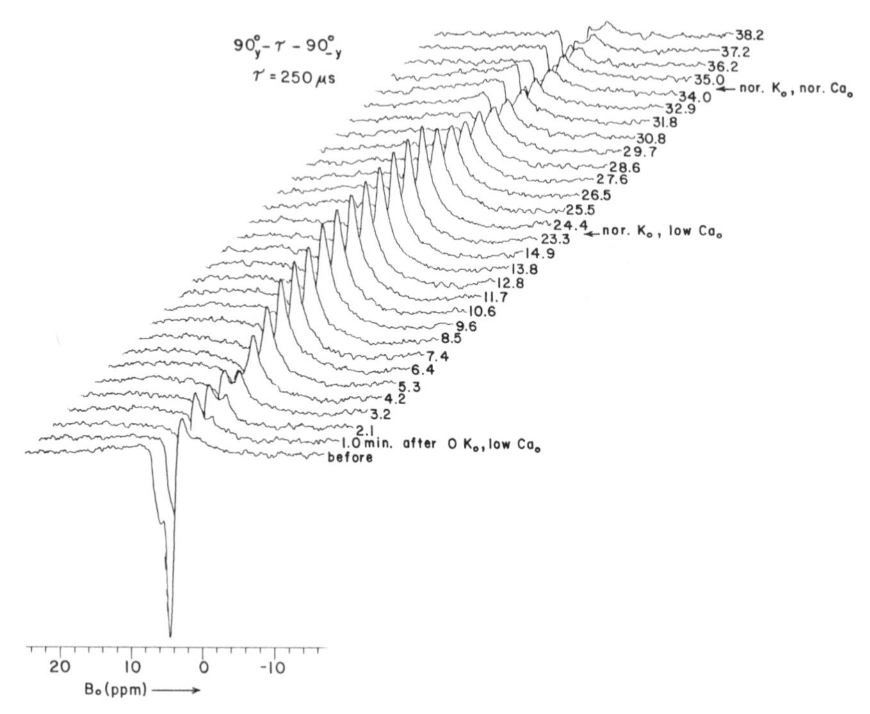

FIGURE 7. ^{23}Na jump-return NMR spectra (95.2 MHz, 8.5 T) of an isovolumic Langendorff-perfused beating rat heart. Before the acquisition of the spectra, the heart had been perfused for 24 min with the standard medium. The heart rate was 290 beats min^{-1}, the systolic/diastolic pressures 150/20 mm Hg, and the coronary flow 23 ml min^{-1}. The first spectrum was obtained at the end of a subsequent 27 min of perfusion with a solution containing 10 mM DyTTHA^{3-} and during which the heart rate was 250 beats min^{-1}, the pressures 150/5, and the coronary flow 22 ml min^{-1}. Subsequent spectra are labeled with the times (midpoints of acquisitions) elapsed since a perfusion medium containing 10 mM DyTTHA^{3-}, no K$^+$, and only 230 μM Ca^{2+} entered the heart. During the 23 min of this perfusion, the pressure was 100 mm Hg. At 23.5 min, a new perfusion medium containing 10 mM DyTTHA^{3-}, normal K$^+$ (5.9 mM) but still only 230 μM Ca^{2+} entered the heart. The subsequent pressure was 20 mm Hg and the coronary flow was 18 ml min^{-1}. At 34.7 min, a new perfusion medium containing 10 mM DyTTHA^{3-} and normal K$^+$ and Ca^{2+} entered the heart. The subsequent heart rate was 230 beats min^{-1}, the pressures were 15/0 mm Hg, and the coronary flow was 16 ml min^{-1}. During all of these perfusions, the heart was bathed with a flowing solution containing 10 mM DyTTHA^{3-}. For each jump-return spectrum, the delay time, τ, was 250 μsec and the number of transients was 128. (From Pike *et al.*[14] With permission from the publisher.)

the period of elevated Na_i. If the Ca_o level was normal during this period, Ca^{2+} would likely enter the cells and the heart would probably not be able to resume beating.

STUDIES OF PERFUSED SHARK RECTAL GLANDS

The shark rectal gland is a very interesting tissue, made up of epithelial-type cells, whose function is the secretion of NaCl from the blood of the shark into the sea.[13,40,41]

Dr. Pike collaborated with Drs. Silva and Epstein, who have pioneered the study of this gland, in conducting ^{23}Na NMR studies while perfusing the gland with $DyTTHA^{3-}$.[13,41]

The process of NaCl secretion turns out to be one of active Cl^- transport. The enzyme $[Na^+ + K^+]$ATPase is involved but, ironically, it is located in the peritubular membrane and thus oriented in the direction opposite from the normal net transport of NaCl. Its activity is coupled (by providing a driving Na^+ electrochemical gradient) to a NaCl-KCl cotransport system, also located in the peritubular membrane and to an electrogenic Cl^- transporter located in the luminal membrane. It is this combination that produces the active secretion of Cl^-.[13] Secretion by the gland is stimulated by the elevation of intracellular cyclic AMP. A question arises as to whether the stimulation derives from a direct effect of cAMP on the Na^+ pump or a direct effect on the electrogenic Cl^- transporter. In the former case, one might expect the level of Na_i to decrease, perhaps only transiently, immediately after elevation of cAMP. In the latter case, one might expect the level of Na_i to increase, perhaps only transiently.

FIGURE 9 depicts the results of an experiment by Drs. Pike and Silva to try to distinguish these mechanisms. It shows a stacked plot of ^{23}Na NMR spectra of a gland obtained after the effects of perfusion with 10 mM $DyTTHA^{3-}$ had reached steady state. It is quite analogous to that of FIGURE 7 except that no attempt was made to eliminate the Na_o peak. The relative intensity of the less-shifted peak is large enough to make this less necessary. A short time, 2.8 min, after the first spectrum was obtained,

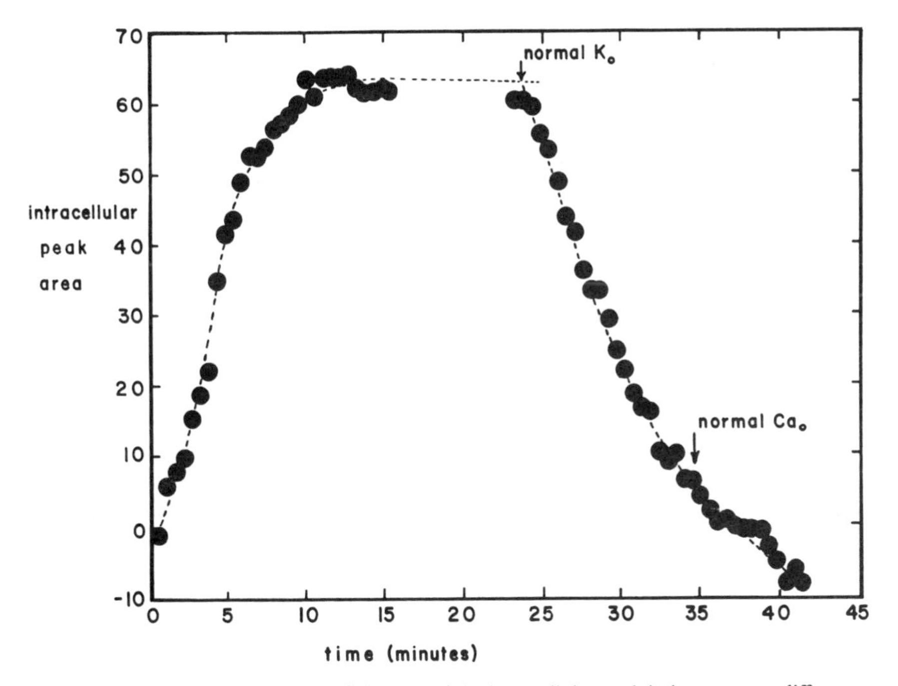

FIGURE 8. The time dependence of the area of the intracellular peak in jump-return difference spectra obtained for the experiment of FIGURE 7. The arbitrary area unit is proportional to the ratio of the area of the intracellular peak in a jump-return difference spectrum to the area under the total heart ^{23}Na resonance in a normal spectrum of the same heart. The dashed line is intended to guide the eye. (From Pike *et al.*[14] With permission from the publisher.)

the perfusing medium was made 50 μM in dibutyryl cyclic-AMP (dbc-AMP), a membrane-permeable form of cAMP, and 250 μM in theophylline, an inhibitor of the phosphodiesterase that degrades cAMP. The intensity of the less-shifted peak is seen to rise thereafter. At 15.8 min, the perfusing medium was made 1 mM in furosemide, a drug known to inhibit the NaCl-KCl cotransport system. After this, the intensity of the

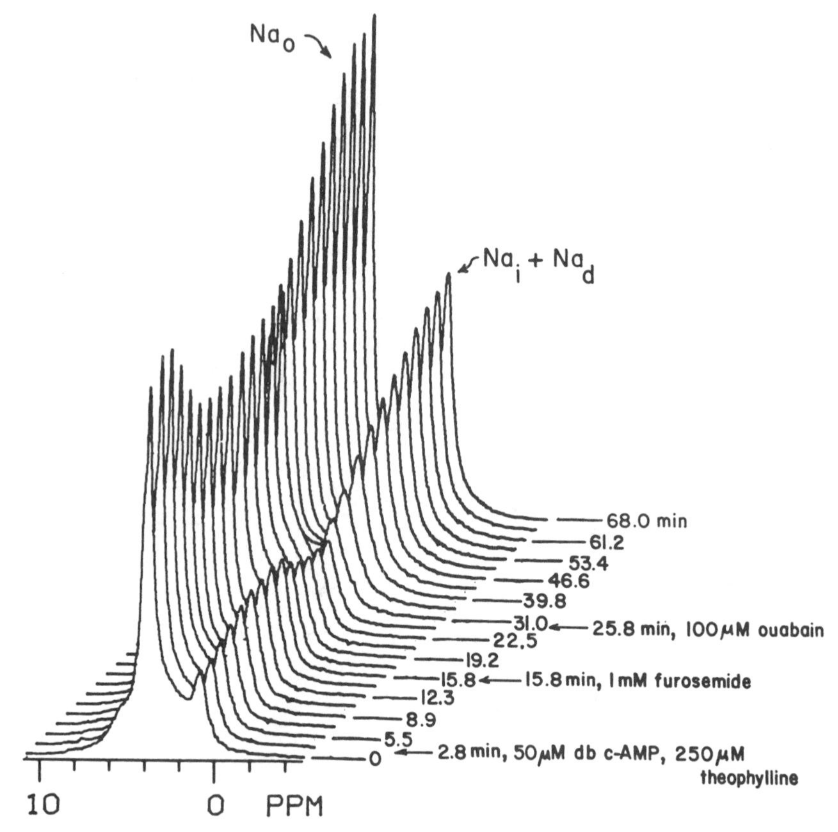

FIGURE 9. Stacked ^{23}Na NMR spectra (95.2 MHz, 8.5 T) of a whole shark rectal gland perfused with shark ringer solution containing 10 mM $DyTTHA^{3-}$, and bathed with sodium-free isotonic mannitol/$DyTTHA^{3-}$ solution. Spectra are plotted with identical vertical scales, and were obtained with 400 transients each, with a total spectral acquisition time of 1.71 min each. A 3 Hz exponential multiplication was used for sensitivity enhancement. The unshifted frequency is defined as 0.0 ppm. Times are reported to the middle of the spectral acquisition period and time zero is defined as the middle of the control spectrum. The arrows indicate when perfusion with the various chemical agents was initiated. (For more information see Pike *et al.*[41])

less-shifted peak falls until the perfusing medium is made 100 μM in ouabain after which the intensity rises again, this time much more dramatically.

These results are consistent with a mechanism by which cAMP acts directly on the electrogenic Cl^- transporter in the luminal membrane. This could cause more Cl^- to leave the cell for the lumen of the gland duct, which could, in turn, cause more Na^+ to enter the cell across the peritubular membrane through the NaCl-KCl cotransport system. The subsequent blockage of this system by furosemide could allow the Na^+ pump to regain dominance and lower Na_i by pumping Na^+ back across the peritubular membrane into the interstitial space. The final addition of ouabain, of course, inhibits the Na^+ pump and allows any Na^+ leakage mechanisms (including the NaCl-KCl cotransport system, now that furosemide is washed away) to become dominant. A problem mollifying these conclusions is that some of the intensity of the less-shifted peak could be contributed by Na^+ in the gland duct. The duct may not be totally accessible to the SR, which was also present in the flowing bathing medium. Duct Na^+ would be expected to rise upon stimulation of the gland. Experiments in which the gland was slit, and the duct therefore much more exposed to the bathing medium, did show a somewhat smaller, less-shifted peak intensity, but they also yielded the same results as seen in FIGURE 9 upon the sequential treatment with the same drugs.[13,41]

Another interesting feature of FIGURE 9 is the downfield shift of the Na_o resonance upon Na_i loading of the cells after treatment with ouabain. Although this was not observed in all shark rectal gland experiments, similar behavior has been observed upon treatment of rat hearts with ouabain and upon Na_i loading of rat heart myocytes induced by global ischemia.[12] This may be an important aspect of perfused tissue studies of this type and possible reasons for its occurrence have been presented.[12]

If the shark rectal gland is subjected to global ischemia by halting the perfusing flow while the bath flow is maintained, the extracellular Na^+ is very quickly washed out of the gland.[13] The same is true for the perfused rat heart.[12] This probably means that in these studies of perfused excised tissue, there is a continuous low flux of ions out of the tissue and into the flowing bathing medium. When the much greater supply flux of ions from the perfusing medium is halted, the low flux to the bath becomes obvious in the NMR spectrum when there is SR in the bath and remaining in the extracellular spaces.

STUDIES *IN VIVO*

There is every reason to anticipate that $DyTTHA^{3-}$ (as well as $TmTTHA^{3-}$, and $TbTTHA^{3-}$, other useful SR) will prove to be relatively non-toxic to the living organism.[12] It is chemically very similar to $GdDTPA^{2-}$ ($DTPA^{5-}$ is diethylenetri-aminepentaacetate), which has become a very popular contrast reagent in the NMR imaging of humans.[42] Balschi, Bittl, and Ingwall have conducted ^{23}Na and ^{31}P NMR experiments on a living rat infused with $DyTTHA^{3-}$.[43] With a surface transceiver coil over a leg gastrocnemius muscle, they were able to observe very interesting changes in the ^{23}Na spectrum as the $DyTTHA^{3-}$ perfused through the muscle. They observed further changes when the muscle was made ischemic. More studies of this type can surely be expected.

ACKNOWLEDGMENTS

I thank all my students and colleagues who have contributed to the aspects of our work which were emphasized here: Paul Allen, Jim Balschi, John Bittl, Simon Chu, Vince Cirillo, Kieren Clarke, Chris Clausen, Dan Dedrick, Marie Dippolito, April Dutta, Frank Epstein, Julie Fetters, Jon Frazer, Herbert Höfeler, Joanne Ingwall, Leon Moore, Marty Pike, Marlene Ricker, John Sachs, Jane Schreiber, Patricio Silva, Tom Smith, and Yan Xu.

REFERENCES

1. SEMENZA, G. & R. KINNE, Eds. 1985. Membrane transport driven by ion gradients. Ann. N.Y. Acad. Sci. **456.**
2. CATTERALL, W. A. 1986. Molecular properties of voltage-sensitive sodium channels. *In* New Insights into Cell and Membrane Transport Processes. G. Poste & S. T. Crooke, Eds.: 3. Plenum Press. New York.
3. CARAFOLI, E. & A. SCARPA, Eds. 1982. Transport ATPases. Ann. N.Y. Acad. Sci. **402.**
4. SKOU, J. C. 1982. The $[Na^+ + K^+]$ATPase: coupling of the reaction with ATP to the reaction with Na^+ and K^+. Ann. N.Y. Acad. Sci. **402:** 169.
5. SPRINGER, C. S. 1987. Ann. Rev. Biophys. Biophys. Chem. **16:** 375.
6. DEGANI, H. & Z. BAR-ON. 1981. Period Biol. **83:** 61.
7. BRYDEN, C. C., C. N. REILLEY & J. F. DESREUX. 1981. Anal. Chem. **53:** 1418.
8. PIKE, M. M. & C. S. SPRINGER. 1982. J. Magn. Res. **46:** 348.
9. GUPTA, R. K. & P. GUPTA. 1982. J. Magn. Res. **47:** 344.
10. PIKE, M. M., D. M. YARMUSH, J. A. BALSCHI, R. E. LENKINSKI & C. S. SPRINGER. 1983. Inorg. Chem. **22:** 2388.
11. CHU, S. C., M. M. PIKE, E. T. FOSSEL, T. W. SMITH, J. A. BALSCHI & C. S. SPRINGER. 1984. J. Magn. Res. **56:** 33.
12. SPRINGER, C. S. 1987. ^{23}Na and ^{39}K NMR spectroscopic studies of the intact, beating heart. *In* Study of Cardiovascular Problems with NMR Techniques. M. Osbakken & J. Haselgrove, Eds. Futura Publishing Co. Mt. Kisco, NY. (In press.)
13. PIKE, M. M. 1985. Development of Aqueous NMR Shift Reagents for Cations and Their Application to Model Membranes, Cells, and Perfused Tissues. Ph.D. Dissertation. State University of New York. Stony Brook, NY.
14. PIKE, M. M., J. C. FRAZER, D. F. DEDRICK, J. S. INGWALL, P. D. ALLEN, C. S. SPRINGER & T. W. SMITH. 1985. Biophys. J. **48:** 159.
15. BALSCHI, J. A., V. P. CIRILLO & C. S. SPRINGER. 1982. Biophys. J. **38:** 323.
16. BROPHY, P. J., M. K. HAYER & F. G. RIDDELL. 1983. Biochem. J. **210:** 961.
17. OGINO, T., J. A. DEN HOLLANDER & R. G. SHULMAN. 1983. Proc. Natl. Acad. Sci. USA **80:** 5185.
18. PETTEGREW, J. W., D. E. WOESSNER, N. J. MINSHEW & T. GLONEK. 1984. J. Magn. Res. **57:** 185.
19. PIKE, M. M., E. T. FOSSEL, T. W. SMITH & C. S. SPRINGER. 1984. Am. J. Physiol. **246:** C528.
20. MORRILL, G. A., A. B. KOSTELLOW, S. P. WEINSTEIN & R. K. GUPTA. 1983. Physiol. Chem. Phys. Med. NMR. **15:** 357.
21. SILLERUD, L. O. & J. W. HEYSER. 1984. Plant Physiol. **75:** 269.
22. SHINAR, H. & G. NAVON. 1984. Biophys. Chem. **20:** 275.
23. BOULANGER, Y., P. VINAY & M. DESROCHES. 1985. Biophys. J. **47:** 553.
24. OGINO, T., G. I. SHULMAN, M. J. AVISON, S. R. GULLANS, J. A. DEN HOLLANDER & R. G. SHULMAN. 1985. Proc. Natl. Acad. Sci. USA **82:** 1099.
25. WITTENBERG, B. A. & R. K. GUPTA. 1985. J. Biol. Chem. **260:** 2031.
26. CASTLE, A. M., R. M. MACNAB & R. G. SHULMAN. 1986. J. Biol. Chem. **261:** 3288.
27. CIVAN, M. M., H. DEGANI, Y. MARGALIT & M. SHPORER. 1983. Am. J. Physiol. **245:** C213.
28. GULLANS, S. R., M. J. AVISON, T. OGINO, G. GIEBISCH & R. G. SHULMAN. 1985. Am. J. Physiol. **249:** F160.
29. MATWIYOFF, N. A., C. GASPAROVIC, R. WENK, J. D. WICKS & A. RATH. 1986. Magn. Res. Med. **3:** 164.
30. GERASIMOWICZ, W. V., S.-I. TU & P. E. PFEFFER. 1986. Plant Physiol. **81:** 925.
31. FOSSEL, E. T. & H. HÖFELER. 1986. Magn. Res. Med. **3:** 534.
32. YEH, H. J. C., F. J. BRINLEY & E. D. BECKER. 1973. Biophys. J. **13:** 56.
33. SCHREIBER, J. K., J. R. SACHS, W. D. ROONEY & C. S. SPRINGER. Manuscript in preparation.
34. LATTA, R., C. CLAUSEN & L. C. MOORE. 1984. J. Membr. Biol. **82:** 67.
35. BALSCHI, J. A. 1984. NMR Shift Reagent Studies of Membrane NH_3 Permeation and Na^+ Transport. Ph.D. Dissertation. State University of New York. Stony Brook, NY.

36. HÖFELER, H., D. JENSEN, M. M. PIKE, J. L. DELAYRE, V. P. CIRILLO, C. S. SPRINGER, E. T. FOSSEL & J. A. BALSCHI. 1987. Biochemistry. (In press.)
37. GILLIES, R. J., K. UGURBIL, J. A. DEN HOLLANDER & R. G. SHULMAN. 1981. Proc. Natl. Acad. Sci. USA **78:** 2125.
38. NICOLAY, K., W. A. SCHEFFERS, P. M. BRUINENBERG & R. KAPTEIN. 1983. Arch. Microbiol. **134:** 270.
39. SPRINGER, C. S., M. M. PIKE, J. A. BALSCHI, S. C. CHU, J. C. FRAZER, J. S. INGWALL & T. W. SMITH. 1985. Circulation **72:** IV–89.
40. EPSTEIN, F. H. & P. SILVA. 1982. Energetics of active chloride transport in shark rectal gland. *In* Chloride Transport in Biological Membranes: 261. Academic Press, Inc. New York.
41. PIKE, M. M., E. T. FOSSEL, P. SILVA, F. H. EPSTEIN & C. S. SPRINGER. Manuscript in preparation.
42. LAUFFER, R. B., T. J. BRADY, R. D. BROWN, C. C. BALIN & S. H. KOENIG. 1986. Magn. Res. Med. **3:** 541.
43. BALSCHI, J. A., J. A. BITTL & J. S. INGWALL. Manuscript in preparation.

DISCUSSION OF THE PAPER

QUESTION: In one of your earlier slides you indicated that the shift reagent concentration you were working with was somewhere in the neighborhood of 10 millimolar. Was that the same for the perfused organ experiments?

C. S. SPRINGER: Yes, for almost every one of the experiments I showed you today the concentration was 10 millimolar.

QUESTION: Can you comment on what the effect of having 10 millimolar concentration of the trivalent anion is on ionic strength?

SPRINGER: As far as ionic strength is concerned, the shift reagent is always substituted for other ions, or other ions are removed so that the general ionic strength and the general osmolarity are roughly the same.

QUESTION: Is this also the case in the sodium-depleted studies where you've got sodium down to about 10 mM?

SPRINGER: Yes.

QUESTION: What kind of K_d are you looking at for the shift reagent, sodium interaction?

SPRINGER: The shift reagent sodium interaction is very, very weak. The binding constants, the reciprocals of the K_d, are less than 10^2 usually. It's a very labile gentle chemical equilibrium—maybe simply some form of ion pairing—so you can also look at the activity of the sodium with electrodes and see that it's not disturbed very much by the presence of the shift reagent.

QUESTION: At that concentration do you have any toxic or any biological effects of the shift reagents?

SPRINGER: No. One of the main reasons for working with the heart was because we could monitor the heart rate and the developed pressure fairly carefully continuously during the experiment. We found that with the perfusion of the shift reagent there is, as long as you're careful to adjust calcium because the shift reagent does also bind calcium, a decrease of the calcium activity. But you can compensate by just a slight amount of added calcium and the heart rate and developed pressure remain quite constant.

QUESTION: Is there any way to distinguish whether you're seeing only the one-half to one-half transition for all the ions or if you're only seeing all the transitions in part? Do you see the same thing for potassium?

SPRINGER: Yes. The question deals with the invisibility issue which I really didn't spend much time on today. A simple statement of the invisibility is that we're only looking at one of the three possible transitions of the sodium, the one-half to one-half. And in most cases in tissue, it has not been studied really thoroughly but in the heart we're pretty sure that in the extracellular case we are only looking at the one-half to one-half transition and probably the intracellular signal reflects only the one-half to one-half transition.

For potassium, that's probably the case too but it's so hard to get spectra with good signal-to-noise ratios that I don't want to say this is absolutely true. In tissue, we're probably going to find that for both extracellular and intracellular signals (at least interstitial and intracellular) we're only seeing the one-half to one-half transition. And therefore, we're only seeing 40% of the intensity. That's one quantum mechanical singularity, there's another one if you have extreme static effects: in a real powder pattern you could only see 20% of the intensity and that's also a possibility.

QUESTION: Do you think that NMR can really help solve problems in ion fluxes in isolated cells and organelles? Most of the problems are dealing with events that happen in seconds or less than a second.

SPRINGER: In the case of fluxes, I don't think so. But I didn't make myself clear. In the heart case, the sodium signal is so strong that each of those spectra were obtained in 12 seconds, which is a best-case kind of time resolution. But then you have a case like a heart, which has a repetitive function, so you can, in principle, gate your signal acquisition to the heart rate. Therefore, each single NMR acquisition only takes milliseconds so that under certain circumstances you can get your time resolution down in the millisecond time scale, but not always.

R. G. SHULMAN: Because electrode measurements always have to be made after jumps, and steady state and equilibrium measurements cannot be made accurately, I think the great advantage of NMR is that it can measure the distribution using the methods described under steady state and equilibrium conditions. These of course can be put together with the transient measurements but from quite a different viewpoint.

A Comparison of Intracellular Sodium Ion Concentrations in Neoplastic and Nonneoplastic Human Tissue Using ^{23}Na NMR Spectroscopy[a]

MELISSA S. LIEBLING AND RAJ K. GUPTA[b]

Department of Physiology and Biophysics
Albert Einstein College of Medicine of Yeshiva University
New York, New York 10461

INTRODUCTION

Because transmembrane ion fluxes and associated changes in intracellular ion concentrations have long been hypothesized as mediators in cell growth,[1–4] the last decade has yielded a wealth of studies eliciting intracellular concentrations of various ions in both neoplastic and nonneoplastic cell lines.[5–14] The sodium ion, in particular, has come to the fore as one such mediator. Cameron *et al.* have observed a direct correlation between the intracellular concentration of Na^+, but not Mg^{2+}, P_i, Cl^-, or K^+, and the proliferation rate of nonneoplastic and malignant animal cell populations.[7,12–14] Similarly, Zs.-Nagy *et al.* and Lukacs *et al.* have found a higher intracellular Na^+:K^+ ratio, due primarily to an increase in intracellular Na^+ concentration, in urogenital and thyroid neoplasias, both benign and malignant, than in their nonneoplastic counterparts.[5,9,10] Koch and Leffert have shown in hepatocyte cultures, as well as in regenerating liver, that sodium influx is a necessary prerequisite for the initiation of cell proliferation.[15] Moreover, Leffert recently noted that the induction of sodium influx by mitogens that act early in the prereplicative phase of the cell cycle is an "apparently 'universal' phenomenon in cultured animal cells."[11] Although the findings of these and other studies strongly implicate the sodium cation to be a mediator in cell growth, the methods employed, such as flame emission photometry, atomic absorption, and electron probe X-ray microanalysis, involve disruption of the cell integrity and/or its content and are, therefore, necessarily unsuitable for *in vivo* studies. Moreover, the measurements obtained from flame emission photometry and atomic absorption, techniques that are relatively simple to use, are not specific to the intracellular compartment and require correction for the presence of contaminating extracellular ions, which introduces sizable errors in the data.

The noninvasive spectroscopic technique of ^{23}Na NMR has enjoyed a growing enthusiasm within the scientific community since the first study of the continuous wave ^{23}Na NMR spectrum of intracellular sodium in muscle, kidney, and brain by Cope.[16,17] Its isotopic abundance of 100%, short nuclear relaxation times, and relatively good detectability at biological concentrations make ^{23}Na an attractive element for NMR spectroscopic studies of biological tissues. Yet, until recently, the information extract-

[a]Supported by National Institutes of Health Research Grant AM-32030 and the National Cancer Institute Core Grant CA-13330.

[b]To whom correspondence should be addressed.

able from a ^{23}Na NMR spectrum of intact cells was significantly limited by an unfortunate spectroscopic characteristic of the hydrated Na^+—the electronic environment of the nucleus remains essentially unaltered and, hence, the nuclear resonance frequency is unchanged in multicompartment systems. However, since the development of membrane-impermeable paramagnetic shift reagents, especially dysprosium tripolyphosphate $[Dy(PPP_i)_2]^{7-}$ discovered by Gupta and Gupta, which shift the resonance frequency of the extracellular sodium ions upfield from the resonance of the intracellular ions,[18-20] there have been numerous reports on the NMR measurement of intracellular Na^+ content of intact cells, both prokaryotic and eukaryotic.[8,18-50]

We have explored the possible involvement in malignancy of intracellular Na^+, an ion strongly implicated as a mediator of cell growth, with a technique that preserves the integrity of the cell and thus possesses the potential for *in vivo* application. Through use of the noninvasive spectroscopic technique of ^{23}Na NMR, we have compared the intracellular Na^+ concentration of uterine leiomyoma and leiomyosarcoma (benign and malignant tumors of smooth muscle) with their nonneoplastic counterpart myometrium, as well as that of colonic adenocarcinoma (a malignant tumor) with its nonneoplastic counterpart colonic mucosa. The specific questions addressed are three fold. First, is there an alteration in the NMR-visible intracellular $[Na^+]$ in neoplastic tissue compared with its tissue of origin? Second, is there a correlation of cell proliferative activity and/or tumor invasiveness with the intracellular $[Na^+]$? Finally, can malignant and benign tumors be distinguished on the basis of their intracellular $[Na^+]$?

MATERIALS AND METHODS

Tissue Preparation

Most of the surgical specimens of human tissue were obtained within three hours of and many immediately after excision from the Department of Pathology of the Jack D. Weiler Hospital of the Albert Einstein College of Medicine. One specimen of colonic tissue, both neoplastic and nonneoplastic, as well as one specimen of leiomyosarcoma were obtained the day following excision with interim refrigerated storage. All specimens, unless obtained immediately after excision, were refrigerated until received for study. The tissue was initially sampled and labeled as neoplastic or nonneoplastic by the resident pathologist by gross observation. The diagnoses of the tissues were obtained upon availability of the final pathological report of the attending physician. The tissue was transported in physiological medium to the NMR laboratory, where it was immediately sectioned into pieces measuring less than or equal to 1 mm in thickness to permit adequate diffusion of oxygen throughout the tissue. The pieces of tissue were then suspended in reagent containing medium at room temperature for a minimum of 30 minutes prior to study. Care was taken to avoid sampling grossly necrotic or hemorrhagic tissue.

NMR Shift Reagent Preparation

The materials $DyCl_3$ and Na_5PPP_i used in the preparation of the highly anionic shift reagent dysprosium bis(tripolyphosphate) were purchased from Bentron, Alfa Division (Danvers, MA) and Sigma (St. Louis, MO), respectively. A solution of 100 mM $DyCl_3$ was titrated with 100 mM Na_5PPP_i to obtain the complex $Dy(PPP_i)_2$. The end point of titration is reached upon disappearance of a white precipitate, which results from the partially insoluble complexes formed at PPP_i levels lower than 2 mole equivalents per mole of Dy^{3+}. KCl, $CaCl_2$, and distilled water were added, as needed, to obtain a final concentration of 145 mM, 5.0 mM, and 2.5 mM of Na^+, K^+ and Ca^{2+}, respectively, in the shift reagent mixture. The shift reagent was diluted with Hanks' balanced salt solution (GIBCO Laboratories, Life Technologies, Inc., Chagrin Falls, OH) to obtain a final concentration of 4.8 mM of $Dy(PPP_i)_2$ in the medium. The final pH of the solution was adjusted with 150 mM HCl and/or 150 mM NaOH to 7.40 ± 0.05. Increasing the concentration of $Dy(PPP_i)_2$ to 7.2 mM, which resulted in better separation of the intra- and extracellular Na^+ resonances, did not have a measurable effect on the intensity of the intracellular Na^+ resonance.

NMR Observation of the Intracellular $^{23}Na^+$ Resonance

Addition of the membrane-impermeable paramagnetic reagent $Dy(PPP_i)_2^{7-}$ to a tissue suspension acts to shift the extracellular $^{23}Na^+$ resonance upfield from that of the intracellular $^{23}Na^+$ resonance, thus permitting direct observation of the intracellular Na^+ (FIG. 1). The upfield shift is caused by pseudocontact interactions of paramagnetic dysprosium with the sodium cation. The large magnitude of the shift arises presumably from the ability of the highly anionic complex to bind the sodium cation in close proximity of the dysprosium moiety. Based on the magnitude of the paramagnetic effect of $Gd(PPP_i)_2^{7-}$, an analogous relaxation agent, on the longitudinal relaxation time (T_1) of ^{23}Na in the $Gd(PPP_i)_2$-Na^+ complex, we have estimated the Dy^{3+} to Na^+ distance to be approximately 4 Å (FIG. 2). This close proximity suggests van der Waals contact.

NMR Measurement of Intracellular Na^+ Concentration

After letting the tissue bathe in the shift reagent medium for a minimum of 30 min, the tissue suspension was placed in the inner tube of a 5 mm/10 mm O.D. coaxial NMR tube combination (Wilmad Glass Company Cat. No. 513-7PP/WGS-10BL). All ^{23}Na NMR spectra were obtained at ambient temperature, on a spinning sample, using an acquisition time of 0.1 sec over an interval of 3 min at a frequency of 53 MHz on a Varian XL-200 Fourier transform NMR spectrometer.

The fractional extracellular (S_{out}) and intracellular (S_{in}) spaces within the sensitive volume of the NMR coil were obtained as shown below in equations 1 and 2[18] by comparison of the extracellular Na^+ resonance intensity (A_{out}) of the tissue suspension with the Na^+ resonance intensity (A_o) of a tissue-free suspension medium in an identical sample geometry (FIG. 3).

$$S_{out} = \frac{A_{out}}{A_o} \tag{1}$$

$$S_{in} = 1 - S_{out} \tag{2}$$

Measurement of the intracellular (A_{in}) and extracellular (A_{out}) Na^+ resonance intensities as well as calculation of the fractional intra- and extracellular spaces in the sensitive volume of the NMR coil yields the NMR-visible intracellular Na^+ concentration via the straightforward equation 3[18] shown below:

$$k[Na_{in}] = \left[\frac{A_{in}\, S_{out}}{A_{out}\, S_{in}}\right][Na_{out}] \tag{3}$$

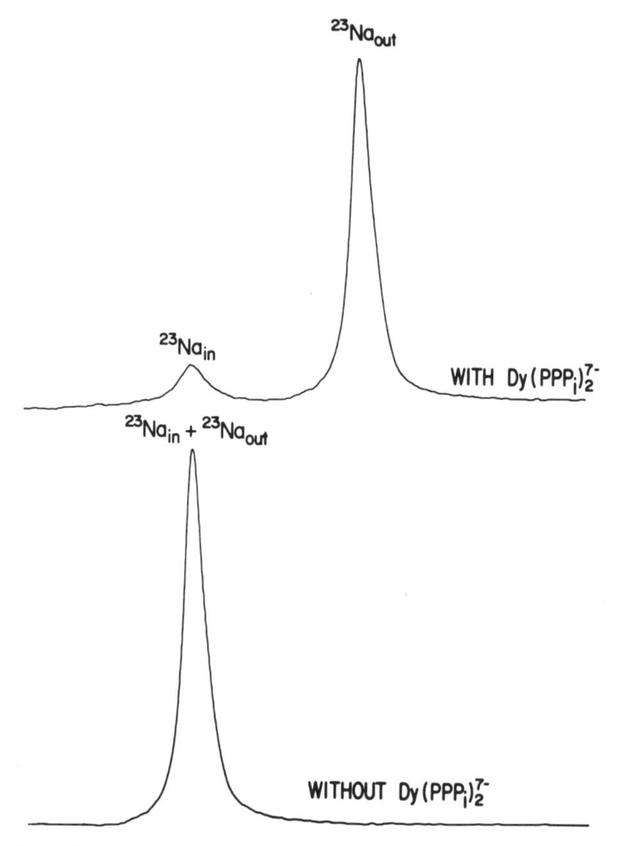

FIGURE 1. ^{23}Na NMR spectrum of a cell suspension with (*top*) and without (*bottom*) 4 mM $Dy(PPP_i)_2^{7-}$ showing spectral resolution of intra- and extracellular Na^+ by the paramagnetic shift reagent. The resonances of intra- and extracellular Na^+ are labeled as $^{23}Na_{in}$ and $^{23}Na_{out}$, respectively.

where $[Na_{in}]$ and $[Na_{out}]$ are the intra- and extracellular Na^+ concentrations, respectively, and k is a constant that takes into account the possibility that first-order nuclear quadrupolar interactions of the intracellular Na^+ with electric field gradients may lead to a 60% reduction in the intensity of their ^{23}Na resonance. We have taken k to be 1 in quantifying our NMR spectra and have thus discounted such interactions. If, at a later date, a reduction in NMR-visible Na^+ due to quadrupolar interactions is found to occur in the tissues under study, the reported concentrations need only be multiplied by a factor of 2.5. When $[Na_{out}]$ is expressed in mM, the units of $[Na_{in}]$ are mmoles/L tissue.

The intensities of the resonances were estimated from the area under their peaks measured using a planimeter.

^{31}P NMR Measurement of Intracellular ATP Concentration

To test tissue viability under our experimental conditions, uterine tissue, both leiomyoma and myometrium, suspended in a 10 mM phosphate-containing physiological medium, was placed in a 10 mm O.D. NMR tube. A ^{31}P NMR spectrum was accumulated at 81 MHz and 10°C with an acquisition time of 0.8 sec following 90° pulses over an interval of approximately 9 hr. The concentration of ATP was estimated by comparing the intensity of the $^{31}P_{\beta}$ resonance of ATP with that of extracellular inorganic phosphate.

RESULTS AND DISCUSSION

When studied through ^{23}Na NMR spectroscopy, the average intracellular [Na^+] observed in five cases of uterine leiomyomas, benign tumors of smooth muscle, was 5.1 ± 1.2 mmoles/L tissue (TABLE 1). As is readily apparent in FIGURE 4, this is almost a factor of three lower than the average concentration of 14.3 ± 1.2 mmoles/L tissue observed in four cases of uninvolved myometrium, the normal smooth muscle layer of the uterus from which the tumor originates. In marked contrast to the sharply lower concentration of intracellular Na^+ observed in leiomyoma, one case of uterine leiomyosarcoma, a rare malignant tumor of smooth muscle, yielded an average intracellular [Na^+] obtained from two separate measurements of 34.6 ± 4.5 mmoles/L tissue (TABLE 1), which represents an almost 2.5-fold increase over the nonneoplastic counterpart (FIG. 5).

In another set of experiments, the average intracellular [Na^+] of four cases of well-differentiated adenocarcinoma of the colon, a malignant tumor that retains a morphological and functional resemblance to its tissue of origin, was 13.9 ± 1.5 mmoles/L tissue, which is almost 1.75 fold higher than the 8.0 ± 1.1 mmoles/L tissue observed in five samples of colonic mucosa, the nonneoplastic counterpart (FIG. 6 and

$Dy(PPP)_2^{7-}$

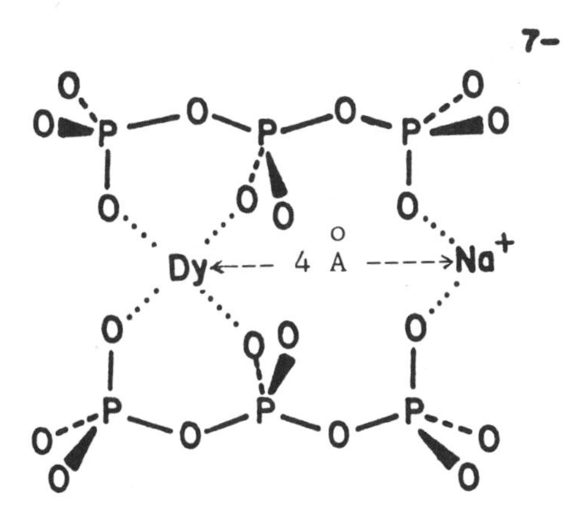

FIGURE 2. A possible structure of the complex of the anionic shift reagent $Dy(PPP_i)_2^{7-}$ with Na^+, consistent with the paramagnetic effects of the relaxation reagent $Gd(PPP_i)_2^{7-}$ on the ^{23}Na nucleus.

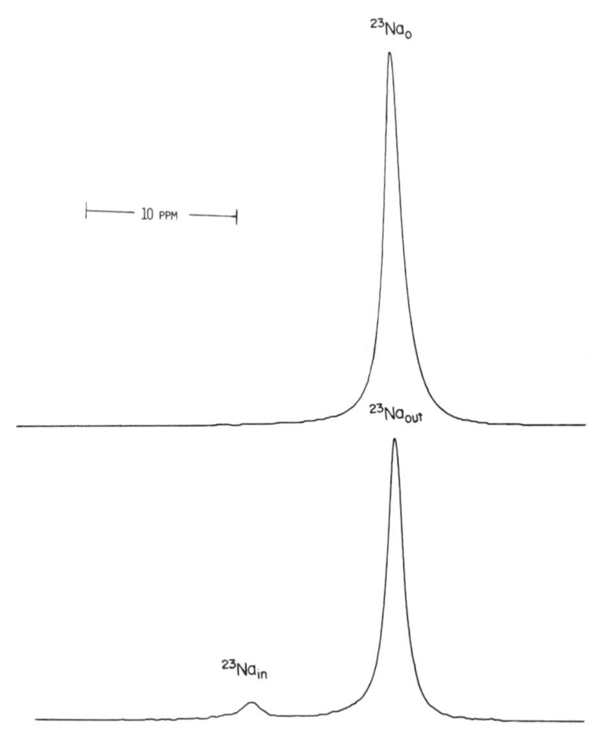

FIGURE 3. Comparison of the ^{23}Na NMR spectrum of a tissue suspension containing 4 mM $Dy(PPP_i)_2$ (*lower trace*) with that of the suspension medium (*upper trace*). The ratio of intensities of the $^{23}Na_{out}$ and $^{23}Na_o$ signals directly gives the fractional extracellular space.

TABLE 2). Two cases of poorly differentiated adenocarcinoma, a more highly invasive malignant tumor that deviates further in structure and function from its tissue of origin, displayed an even higher intracellular [Na^+] concentration of 21.7 ± 0.6 mmoles/L tissue, which represents an almost 2.75-fold increase over the nonneoplastic counterpart and about a 1.5-fold increase over the well-differentiated malignancy.

Several observations support the contention that the tissue was metabolically active at the time of data acquisition. First, the tissue sampled showed no gross evidence of necrosis. Second, the tissue, when suspended in glucose-containing physiological medium for 30 min, altered the color of the dye phenol red from red to light orange. This color change indicates acidification of the medium by lactic acid, which originates from sustained glycolysis in the tissue. Third, the measurements, when repeated twice in succession on the same sample and again at least 30 min later on additional sample,

TABLE 1. NMR-Measured Intracellular Na^+ Concentrations in Uterine Tissue

Tissue	$[Na^+]_i$ (mmoles/L tissue)
Myometrium	14.3 ± 1.2 (N = 4)
Uterine leiomyoma	5.1 ± 1.2 (N = 5)
Uterine leiomyosarcoma	34.6 ± 4.5

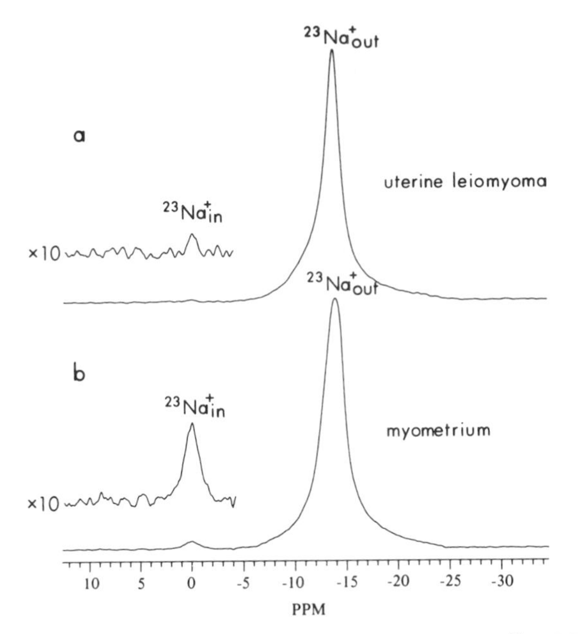

FIGURE 4. Comparison of the resonances of intracellular Na$^+$ in the ^{23}Na NMR spectra of uterine leiomyoma (trace a) and uninvolved myometrium (trace b).

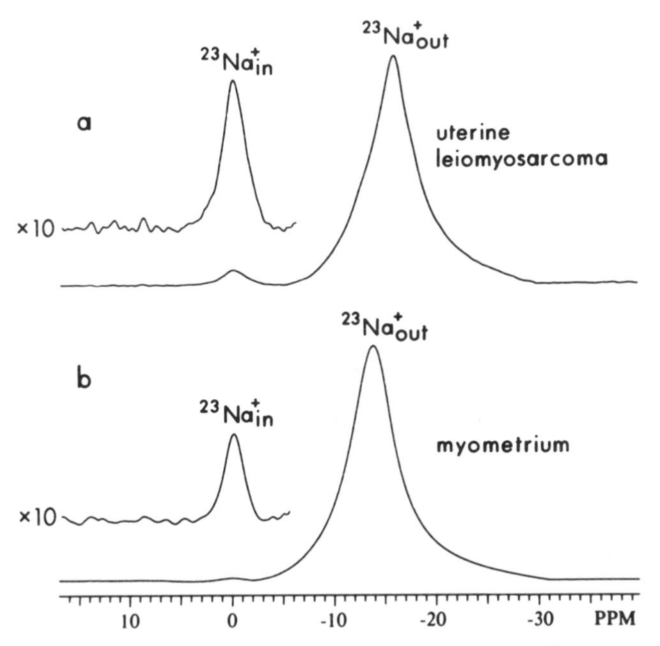

FIGURE 5. Comparison of the resonances of intracellular Na$^+$ in the ^{23}Na NMR spectra of uterine leiomyosarcoma (trace a) and uninvolved myometrium (trace b).

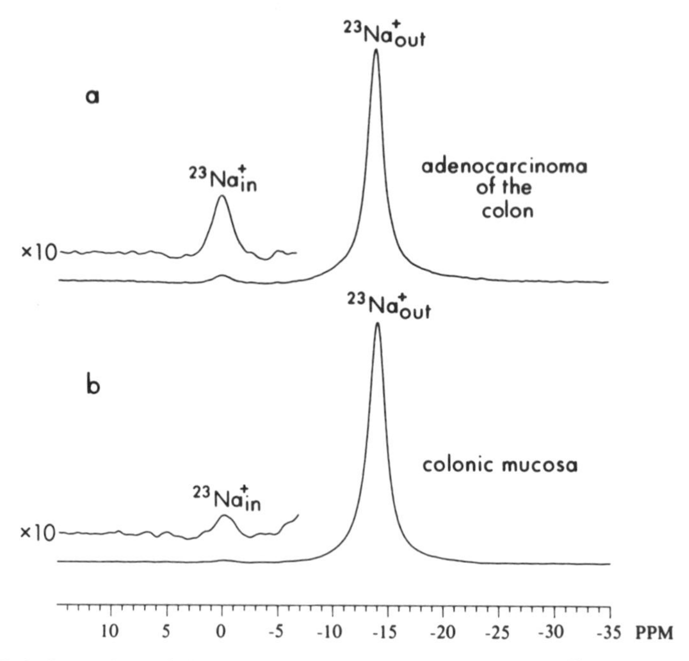

FIGURE 6. Comparison of the resonances of intracellular Na^+ in the ^{23}Na NMR spectra of adenocarcinoma of the colon (trace a) and its nonneoplastic counterpart colonic mucosa (trace b).

yielded results that were identical within experimental error. Fourth, the fact that the intracellular $[Na^+]$ remained well below the extracellular $[Na^+]$ indicates that the ability of the sodium pump to maintain a low level of intracellular Na^+ has not been impaired. Fifth, each measurement required only 3 min of data accumulation during which time the tissue is expected to remain well oxygenated. Moreover, even if the tumor cells are challenged with a reduction in oxygen supply, it is well known that they are able to switch to anaerobic metabolism to maintain their ATP level.[51] As illustrated in FIGURE 7, a ^{31}P NMR spectrum of uterine tissue, both neoplastic and nonneoplastic, obtained over 9 hr at 10°C revealed an intracellular ATP concentration of about 2.0 mM, as calculated by a rough comparison of the intensity of the P_β resonance of ATP with the intensity of the resonance from the 10 mM extracellular inorganic phosphate. This indicates maintenance of significant levels of ATP in the tissue under study. Finally, the intensity of the P_γ resonance of ATP is identical within the noise level to that of the P_β resonance, indicating the absence of ATP breakdown to ADP.

The conclusions to be drawn from the data are fourfold. First, there is an alteration

TABLE 2. NMR-Measured Intracellular Na^+ Concentrations in Colonic Tissue

Tissue	$[Na^+]_i$ (mmoles/L tissue)
Colonic mucosa	8.0 ± 1.1 ($N = 5$)
Well-differentiated adenocarcinoma of the colon	13.9 ± 1.5 ($N = 4$)
Poorly differentiated adenocarcinoma of the colon	21.7 ± 0.6 ($N = 2$)

in the intracellular [Na^+] in neoplastic human tissue relative to its tissue of origin. That the measurement of [Na^+] by NMR spectroscopy is truly representative of the intracellular space is supported by the observation that the intensity of the intracellular $^{23}Na^+$ resonance leveled off after a continual decline over the first 20 min of tissue contact with the shift reagent. This indicates that equilibration of the reagent throughout intercellular space is complete prior to the start of data acquisition. It is possible that the differences observed in the intracellular [Na^+] between tissues may represent an alteration in the relative proportion of intercellular fibrous tissue, which, if inaccessible to Na^+, would dilute the apparent concentration of intracellular [Na^+] as calculated via equation 3. But such a possibility is considered unlikely as the magnitude of observed differences in intracellular [Na^+] would require very large alterations in the volume occupied by such fibers.

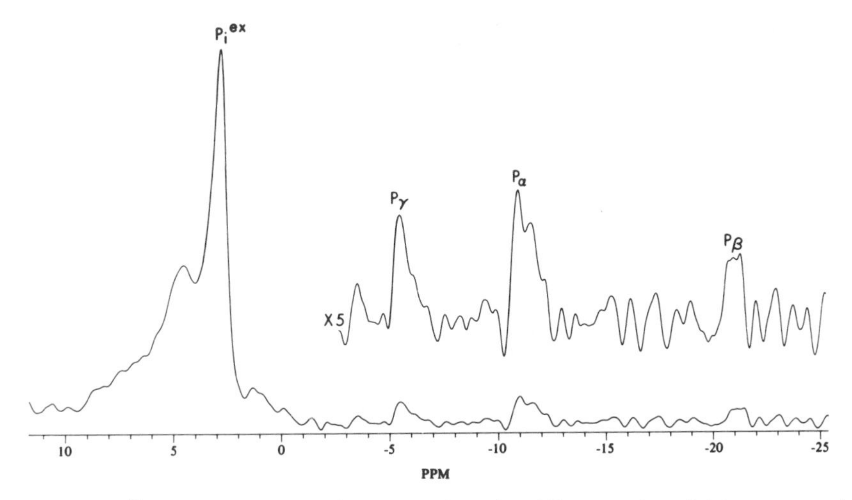

FIGURE 7. ^{31}P NMR spectrum of an approximately 10% suspension of leiomyoma and myometrium at 81 MHz (10°C). This spectrum indicates the presence of significant levels of intracellular ATP in these tissues. The P_α, P_β, and P_γ resonances of intracellular ATP as well as the large resonance of extracellular inorganic phosphate are readily identifiable.

A second conclusion to be drawn from the data is that the degree of cellular proliferative activity and/or tumor invasiveness may have a direct correlation with the concentration of intracellular Na^+. As has been observed with colonic and uterine tissue, the progression from normalcy to neoplasia, from benignity to malignancy, and from well differentiation to poor differentiation are, with one exception, accompanied by increases in the intracellular [Na^+]. Although the intracellular [Na^+] observed in the leiomyomas as compared with the nonneoplastic myometrium appears to be that one exception in the previous statement, most of the excised specimens were taken from peri- or postmenopausal women. Leiomyomas in postmenopausal women tend to regress[52] and, therefore, the lower intracellular [Na^+] is consistent with a regressing tissue possessing little or no proliferative activity. The colonic adenomatous polyp, the benign correlate of colonic adenocarcinoma, would provide a proliferative intermediate between normalcy and malignancy. However, such tissue, although occurring commonly, must be submitted in entirety for histopathologic diagnosis and was, therefore, unobtainable in this study.

A third conclusion inferred from the data is that benignity and malignancy may be distinguishable on the basis of intracellular [Na^+]. That is, once obtained, a measurement of intracellular [Na^+] may represent as good a diagnostic and/or prognostic indicator as histopathologic viewing.

A final conclusion to be drawn is that the technique of ^{23}Na NMR spectroscopy and/or imaging has the potential of becoming a powerful, noninvasive diagnostic tool. While Damadian was first to propose that NMR spectroscopy of biopsy specimens could be used as an adjunct to histopathologic diagnosis, based on his observation that characteristic differences between benignity and malignancy exist in tissue proton spin-lattice (T_1) and spin-spin (T_2) relaxation times,[53] the results to date indicate that proton relaxation time measurements on excised tissue have little diagnostic potential. Once tissue is removed, such measurements offer no additional diagnostic or prognostic value over histopathologic diagnosis. Moreover, Eggleston *et al.* found that the prolongation of T_1 is not specific for malignancy but rather is a more general alteration associated with such tissue disturbances as hyperplasia and inflammation.[54] Although Lauterbur[55] enhanced the usefulness of NMR in oncology by reporting that NMR signals could be spatially encoded to produce images that provide unique anatomic information, the diagnosis of cancer still cannot be made specifically through either *in vitro* or *in vivo* NMR techniques.

The use of ^{23}Na NMR for *in vivo* diagnostic imaging and spectroscopy is still limited by the inability to distinguish the intra- from the extracellular Na^+. Yet, $^{23}Na^+$ remains the next most favorable element for imaging following ^1H. Although ^{23}Na has a sensitivity about an order of magnitude less than ^1H, its relaxation times are about two orders of magnitude shorter than those of ^1H. Therefore, with time-averaging, ^{23}Na produces a signal per unit time comparable to that of ^1H. This imaging characteristic, the 100% natural abundance of the isotope, as well as the finding that NMR-visible intracellular [Na^+] is indeed altered in neoplasia, provide impetus to refine the technique of ^{23}Na NMR to a point where the intracellular compartment can be observed *in vivo*. At least three possibilities exist for such a refinement. The development of nontoxic, membrane impermeable, paramagnetic shift reagents would permit *in vivo* spectroscopy and chemical shift imaging of the intracellular space. Alternatively, it may be possible to use differences in the relaxation behavior of the intra- and extracellular Na^+ to obtain an image representing the intracellular Na^+ compartment.[56] Finally, $^{35}Cl^-$, which possesses a sufficiently broadened and thereby invisible intracellular resonance, could be employed in a difference imaging technique whereby the extracellular signal of $^{35}Cl^-$, after proper normalization to take into account differences in concentration and sensitivity, is subtracted from the signal of $^{23}Na^+$, leaving an image representing the intracellular space. The advent of such technology may provide the medical community with an invaluable technique for the *in vivo* diagnosis of cancer.

ACKNOWLEDGMENTS

We wish to thank the Department of Pathology of the Jack D. Weiler Hospital of the Albert Einstein College of Medicine for their generous cooperation in providing surgical specimens. In particular, we wish to extend special appreciation to Dr. Rachel Morecki for numerous helpful discussions as well as Mr. Angel Rosario for his assistance in obtaining specimens in a timely fashion. In addition, we wish to thank

Miss Julie Gupta for her invaluable assistance in the measurement of resonance intensities.

REFERENCES

1. BOYNTON, A. L., W. L. MCKEEHAN & J. F. WHITFIELD, Eds. 1982. Ions, Cell Proliferation, and Cancer. Academic Press. New York.
2. CONE, C. D., JR. 1969. Electroosmotic interactions accompanying mitosis initiation in sarcoma cells *in vitro*. Trans. N. Y. Acad. Sci. **31:** 404–427.
3. CONE, C. D., JR. 1971. Unified theory on the basic mechanism of normal mitotic control and oncogenesis. J. Theor. Biol. **30:** 151–181.
4. CONE, C. D., JR. 1971. Maintenance of mitotic homeostasis in somatic cell populations. J. Theor. Biol. **30:** 183–194.
5. ZS.-NAGY, I., G. LUSTYIK, V. ZS.-NAGY, B. ZARANDI & C. BERTONI-FREDDARI. 1981. Intracellular Na^+:K^+ ratios in human cancer cells as revealed by energy dispersive x-ray microanalysis. J. Cell Biol. **90:** 769–777.
6. LANNIGAN, D. A. & P. A. KNAUF. 1985. Decreased intracellular Na^+ concentration is an early event in murine erythroleukemic cell differentiation. J. Biol. Chem. **260**(12): 7322–7324.
7. CAMERON, I. L. & N. K. R. SMITH. 1985. Ionic regulation of growth in normal and tumor cells. *In* Water and Ions in Biological Systems. A. Pullman, V. Vasilescu & L. Packer, Eds.: 335–342. Plenum Press. New York.
8. GUPTA, R. K., P. GUPTA & W. NEGENDANK. 1982. Direct observation of the state of $^{23}Na^+$ ions in intact cells and tissues by noninvasive NMR spectroscopy: Intracellular Na^+ ions in human normal and leukemic lymphocytes. *In* Ions, Cell Proliferation, and Cancer. A. L. Boynton, W. L. McKeehan & J. F. Whitfield, Eds.: 1–12. Academic Press. New York.
9. LUKACS, G. L., I. ZS.-NAGY, G. LUSTYIK & G. BALAZS. 1983. Microfluorimetric and x-ray microanalytic studies on the DNA content and Na^+:K^+ ratios of the cell nuclei in various types of thyroid tumors. J. Cancer Res. Clin. Oncol. **105:** 280–284.
10. ZS.-NAGY, I., G. LUSTYIK, G. LUKACS, V. ZS.-NAGY & G. BALAZS. 1983. Correlation of malignancy with the intracellular Na^+:K^+ ratio in human thyroid tumors. Cancer Res. **43:** 5395–5402.
11. LEFFERT, H. L. 1982. Monovalent cations, cell proliferation and cancer: an overview. *In* Ions, Cell Proliferation, and Cancer. A. L. Boynton, W. L. McKeehan & J. F. Whitfield, Eds.: 93–102. Academic Press. New York.
12. CAMERON, I. L. & N. K. R. SMITH. 1982. Energy dispersive spectroscopy in the study of the ionic regulation of growth in normal and tumor cells. *In* Ions, Cell Proliferation, and Cancer. A. L. Boynton, W. L. McKeehan & J. F. Whitfield, Eds.: 13–40. Academic Press. New York.
13. CAMERON, I. L. & K. E. HUNTER. 1983. Effect of cancer cachexia and amiloride treatment on the intracellular sodium content in tissue cells. Cancer Res. **43:** 1074–1078.
14. CAMERON, I. L., N. K. R. SMITH, T. B. POOL & R. L. SPARKS. 1980. Intracellular concentration of sodium and other elements as related to mitogenesis and oncogenesis *in vivo*. Cancer Res. **40:** 1493–1500.
15. KOCH, K. S. & H. L. LEFFERT. 1979. Increased sodium ion influx is necessary to initiate rat hepatocyte proliferation. Cell **18:** 153–163.
16. COPE, F. W. 1965. NMR evidence for complexing of sodium ions in muscle. Proc. Natl. Acad. Sci. USA **54:** 225–227.
17. COPE, F. W. 1967. NMR evidence for complexing of Na^+ in muscle, kidney, and brain, and by actomyosin. J. Gen. Physiol. **50:** 1353–1375.
18. GUPTA, R. K. & P. GUPTA. 1982. Direct observation of resolved resonances from intra- and extracellular sodium-23 ions in NMR studies of intact cells and tissues using dysprosium(III) tripolyphosphate as paramagnetic shift reagent. J. Magn. Reson. **47:** 344–350.

19. GUPTA, R. K., P. GUPTA & R. D. MOORE. 1984. NMR studies of intracellular metal ions in intact cells and tissues. Ann. Rev. Biophys. Bioeng. **13:** 221–246.
20. CHU, S. C., M. M. PIKE, E. T. FOSSEL, T. W. SMITH, J. A. BALSCHI & C. S. SPRINGER, JR. 1984. Aqueous shift reagents for high resolution cationic nuclear magnetic resonance III. $Dy(TTHA)^{3-}$, $Tm(TTHA)^{3-}$, and $Tm(PPP)_2^{7-}$. J. Magn. Reson. **56:** 33–47.
21. CASTLE, A. M., R. M. MACNAB & R. G. SHULMAN. 1986. Coupling between the sodium and proton gradients in respiring *Escherichia coli* cells measured by ^{23}Na and ^{31}P nuclear magnetic resonance. J. Biol. Chem. **261**(17): 7797–7806.
22. KUMAR, A., A. SPITZER & R. K. GUPTA. 1986. ^{23}Na NMR spectroscopy of proximal tubule suspensions. Kidney Int. **29**(3): 747–751.
23. CASTLE, A. M., R. M. MACNAB & R. G. SHULMAN. 1986. Measurement of intracellular sodium concentration and sodium transport in *Escherichia coli* by ^{23}Na nuclear magnetic resonance. J. Biol. Chem. **261**(7): 3288–3294.
24. MONTI, J. P., P. GALLICE, A. CREVAT, M. EL MEHDI, C. DURAND & A. MURISASCO. 1986. Intra-erythrocytic sodium in uremic patients, as determined by high-resolution ^{23}Na nuclear magnetic resonance. Clin. Chem. **32:** 104–107.
25. GUPTA, R. K., A. B. KOSTELLOW & G. A. MORRILL. 1985. NMR studies of intracellular sodium ions in amphibian oocytes, ovulated eggs, and early embryos. J. Biol. Chem. **260**(16): 9203–9208.
26. PIKE, M. M., J. C. FRAZER, D. F. DEDRICK, J. S. INGWALL, P. D. ALLEN, C. S. SPRINGER, JR. & T. W. SMITH. 1985. ^{23}Na and ^{39}K nuclear magnetic resonance studies of perfused rat hearts. Discrimination of intra- and extracellular ions using a shift reagent. Biophys. J. **48**(1): 159–173.
27. RIDDELL, F. G. & M. K. HAYER. 1985. The monensin-mediated transport of sodium ions through phospholipid bilayers studied by ^{23}Na-NMR spectroscopy. Biochim. Biophys. Acta **817**(2): 313–317.
28. GULLANS, S. R., M. J. AVISON, T. OGINO, G. GIEBISCH & R. G. SHULMAN. 1985. NMR measurements of intracellular sodium in the rabbit proximal tubule. Am. J. Physiol. **249:** F160–168.
29. RAYSON, B. M. & R. K. GUPTA. 1985. ^{23}Na NMR studies of rat outer medullary kidney tubules. J. Biol. Chem. **260**(12): 7276–7280.
30. COWAN, B. E., D. Y. SZE, M. T. MAI & O. JARDETZKY. 1985. Measurement of the sodium membrane potential by NMR. FEBS Lett. **184**(1): 130–133.
31. BOULANGER, Y., P. VINAY & M. DESROCHES. 1985. Measurement of a wide range of intracellular sodium concentrations in erythrocytes by ^{23}Na nuclear magnetic resonance. Biophys. J. **47**(4): 553–561.
32. OGINO, T., G. I. SHULMAN, M. J. AVISON, S. R. GULLANS, J. A. DEN HOLLANDER & R. G. SHULMAN. 1985. ^{23}Na and ^{39}K NMR studies of ion transport in human erythrocytes. Proc. Natl. Acad. Sci. USA **82**(4): 1099–1103.
33. WITTENBERG, B. A. & R. K. GUPTA. 1985. NMR studies of intracellular sodium ions in mammalian cardiac myocytes. J. Biol. Chem. **260**(4): 2031–2034.
34. MORRILL, G. A., S. P. WEINSTEIN, A. B. KOSTELLOW & R. K. GUPTA. 1985. Studies of insulin action on the amphibian oocyte plasma membrane using NMR, electrophysiological and ion flux techniques. Biochim. Biophys. Acta **844**(3): 377–392.
35. SHINAR, H. & G. NAVON. 1984. NMR relaxation studies of intracellular Na^+ in red blood cells. Biophys. Chem. **20**(4): 275–283.
36. CIVAN, M. M., H. DEGANI, Y. MARGALIT & M. SHPORER. 1983. Observations of ^{23}Na in frog skin by NMR. Am. J. Physiol. **245**(3): C213–219.
37. OGINO, T., J. A. DEN HOLLANDER & R. G. SHULMAN. 1983. ^{39}K, ^{23}Na, and ^{31}P NMR studies of ion transport in *Saccharomyces cerevisiae*. Proc. Natl. Acad. Sci. USA **80**(17): 5185–5189.
38. BALSCHI, J. A., V. P. CIRILLO & C. S. SPRINGER, JR. 1982. Direct high-resolution nuclear magnetic resonance studies of cation transport in yeast cells. Biophys. J. **38**(3): 323–326.
39. PIKE, M. M., S. R. SIMON, J. A. BALSCHI & C. S. SPRINGER, JR. 1982. High-resolution NMR studies of transmembrane cation transport: Use of an aqueous shift reagent for ^{23}Na. Proc. Natl. Acad. Sci. USA **79**(3): 810–814.
40. FOSSARELLO, M., N. ORZALESI, F. P. CORONGIU, S. BIAGINI, M. CASU & A. LAI. 1985.

^{23}Na NMR investigation of human lenses from patients with cataracts. FEBS Lett. **184:**245–248.
41. SILLERUD, L. O. & J. W. HEYSER. 1984. Use of ^{23}Na NMR to follow sodium uptake and efflux in NaCl-adapted and non-adapted millet suspensions. Plant Physiol. **75:** 269–280.
42. PETTEGREW, J. W., D. E. WOESSNER, N. J. MINSHEW & T. GLONEK. 1984. Sodium-23 NMR analysis of human whole blood, erythrocytes & plasma. J. Magn. Reson. **57:** 185–193.
43. RAYSON, B. M. & R. K. GUPTA. 1985. Steroids, intracellular sodium levels, and Na^+/K^+-ATPase regulation. J. Biol. Chem. **260:** 12740–12743.
44. FOSSEL, E. T. & H. HOEFELER. 1986. Observation of intracellular potassium and sodium in the heart by NMR: a major fraction of potassium is "invisible." Magn. Reson. Med. **3:** 534–540.
45. MATWIYOFF, N. A., C. GASPAROVIC, R. WENK, J. D. WICKS & A. RATH. 1986. P-31 and Na-23 NMR studies of the structure and lability of the sodium shift reagent, bis(tripolyphosphate)dysprosium(III) ($[Dy(P_3O_{10})]^{7-}$) ion, and its decomposition in the presence of rat muscle. Magn. Reson. Med. **3:** 164–168.
46. GUPTA, R. K., A. B. KOSTELLOW & G. A. MORRILL. 1985. NMR studies of the role of intracellular sodium ions in the mechanism of insulin action on an amphibian oocyte. *In* Water and Ions in Biological Systems. A. Pullman, V. Vasilescu & L. Packer, Eds.: 705–714. Plenum Press. New York.
47. MORRILL, G. A., A. B. KOSTELLOW, S. P. WEINSTEIN & R. K. GUPTA. 1983. NMR and electrophysiological studies of insulin action on cation regulation and endocytosis in the amphibian oocyte. Physiol. Chem. Phys. Med. NMR **15:** 357–362.
48. GUPTA, R. K., A. B. KOSTELLOW & G. A. MORRILL. 1985. NMR studies of intracellular sodium ions during the meiotic and early mitotic divisions in amphibians. *In* Magnetic Resonance in Biology and Medicine. G. Govil, C. L. Khetrapal & A. Saran, Eds. Tata McGraw-Hill Publishing Company Limited. New Delhi, India.
49. GUPTA, R. K. 1986. NMR spectroscopy of intracellular sodium ions in living cells. *In* NMR in Living Systems. T. Axenrod & G. Ceccarelli, Eds. NATO ASI Series. D. Reidel Publishing Company. Boston, MA.
50. GUPTA, R. K. 1987. ^{23}Na NMR spectroscopy of intact cells and tissues. *In* NMR Spectroscopy of Cells and Organisms. R. K. Gupta, Ed. **2:** 1–32. CRC Press, Inc. Boca Raton, FL.
51. YUSHOK, W. D. 1971. Control mechanisms of adenine nucleotide metabolism of ascites tumor cells. J. Biol. Chem. **246:** 1607–1617.
52. GOMPEL, C. & S. G. SILVERBERG. 1985. Pathology in Gynecology and Obstetrics. 3rd edit. J. B. Lippincott Co. Philadelphia, PA.
53. DAMADIAN, R. 1971. Tumor detection by nuclear magnetic resonance. Science **171:** 1151–1153.
54. EGGLESTON, J. C., L. A. SARYAN & D. P. HOLLIS. 1975. Nuclear magnetic resonance investigations of human neoplastic and abnormal non-neoplastic tissues. Cancer Res. **35:** 1326–1332.
55. LAUTERBUR, P. C. 1973. Image formation by induced local interactions: Examples of employment of NMR. Nature **242:** 190–191.
56. RA, J. B., S. K. HILAL & Z. H. CHO. 1986. A method for *in vivo* MR imaging of the short T_2 component of sodium-23. Magn. Reson. Med. **3**(2): 296–302.

DISCUSSION OF THE PAPER

QUESTION: I have noticed from the phosphorus spectrum that the tumors you have are largely ischemic. Now let's suppose for a second that they are malignant tumors, which are more active and more likely to be in an ischemic position, will this mean that in your case the high sodium concentration may be because of the more ischemic nature of these particular tumors? Did you rule out this possibility?

R. K. GUPTA: Our ^{31}P NMR spectrum sheds no light on the ischemic nature of the tumor. The large P_i resonance in our spectrum arises mostly from the extracellular P_i and is not an indication of lack of oxygen. In order to permit adequate diffusion of oxygen, the tissue, suspended in oxygenated medium, was sectioned into thin pieces immediately upon receipt. Special care was taken to avoid sampling grossly necrotic or hemorrhagic tissue. I would, however, like to ask my co-author Melissa Liebling to answer your question more fully.

M. S. LIEBLING: Of course that's a very important point. However, that the tissue was metabolically active at the time of data acquisition is suggested by several considerations. (1) The tissue sampled showed no gross evidence of necrosis; (2) the tissue, suspended in glucose-containing physiologic medium, altered the color of phenol red from red to light orange, indicating acidification of the medium from sustained glycolysis in the tissue; (3) the intracellular [Na^+] measurements when repeated twice in succession on the same sample and again at one-half hour later on additional sample, yielded results that were identical within experimental error; (4) the fact that the intracellular [Na^+] remained well below the extracellular [Na^+] indicates that the ability of the sodium pump to maintain a low level of intracellular [Na^+] was not impaired; (5) each measurement required only three minutes of data accumulation during which the tissue is expected to remain well-oxygenated; (6) even if the tumor cells are challenged with a reduction in oxygen supply, it is known that they are able to switch to anaerobic metabolism to maintain their ATP level (Our ^{31}P NMR spectrum indicates maintenance of significant levels of ATP in the tissues under study); (7) the intensity of the P_β resonance of ATP is identical within the noise level to that of the P_γ resonance, indicating retention of ATP; and (8) as long as you have ATP, the sodium pump with its low K_M for ATP will be able to maintain a low level of sodium in the cell.

The actual mechanism of increased intracellular [Na^+] in tumor tissue remains uncertain at this time. It is possible that part of the increase arises from the anoxic nature of the tumor *in vivo*. Whatever the reason for the increased intracellular sodium concentration, our observations remain significant for *in vivo* characterization of tumor tissue by ^{23}Na NMR spectroscopic imaging.

QUESTION: Can you address the point of visibility of interstitial sodium? Because Charles Springer had a very important point that interstitial Na^+ happened to be partly invisible. Do you have any result on that?

GUPTA: We do not. Interstitial sodium in our case appears under the extracellular peak and all of the Na^+ under the extracellular peak seems to be 100% visible, at least in the tissues that we have examined.

C. S. SPRINGER: Whether that is the case in the intact tissue is a different question.

GUPTA: With perfused heart, and we ourselves do not have any direct experience with it, there appears to be a discrepancy in the results from the laboratories of Eric Fossell and Springer. We are puzzled by Springer's results on unperturbed heart where no signal at all is detected from intracellular Na^+, indicating total invisibility of internal Na^+ in this tissue. In contrast to these results, we have observed without any apparent difficulty, an intracellular Na^+ signal in isolated heart cells corresponding to about 8 mM Na^+. Fossell's results on the intact heart agree with our own on isolated heart cells.

QUESTION: There is a whole group of intermediate leiomyosarcomas, particularly in the small intestine but also in the uterus, that is difficult to classify as benign or malignant based on histology. Are you planning to study these borderline tumors or those that are tough to interpret histologically?

LIEBLING: It was just not possible to study a variety of histologic intermediates as these tumors occur exceedingly rarely. In fact, we were fortunate to have obtained even one leiomyosarcoma in our six month study.

QUESTION: It might help getting the small intestine, which has much more of these borderline tumors.

LIEBLING: The leiomyosarcoma occurs rarely enough in the small intestine so that not even one was available during the course of our study. However, the idea of using ^{23}Na NMR prospectively to differentiate benignity from malignancy in those tumors that are difficult to differentiate histologically is an enticing one to pursue. In our study, we separated the histologically well, moderately, and poorly differentiated adenocarcinomas of the colon. Indeed, we did find a direct correlation of the intracellular [Na^+] with the level of malignancy. We would very much like to have studied the adenomatous colonic polyp, which represents the benign correlate of adenocarcinoma of the colon. However, logistically this has been impossible as these lesions must be submitted entirely for diagnostic purposes. Further studies are needed to evaluate ^{23}Na NMR as a diagnostic tool for those lesions that are not clearly benign or malignant by histology.

NMR Studies of Phosphate Uptake and Storage in Plant Cells and Algae[a]

HANS J. VOGEL

Division of Biochemistry
University of Calgary
Calgary, Alberta
Canada, T2N 1N4

INTRODUCTION

Plant cells and algae contain a wide variety of different intracellular compartments. They possess most of the organelles that are normally found in mammalian tissues. However, in addition to these, one may encounter the presence of vacuoles, chloroplasts, amyloplasts, and volutin granules. All these compartments play specialized roles and they may take part in the metabolism. In order to be able to address the question of ion uptake and storage it will be necessary to evaluate the roles of all individual compartments. This requirement almost precludes the use of standard techniques of metabolite analysis. These are generally of an invasive nature, for example, suspensions of intact cells are homogenized and/or extracted with reagents such as perchloric acid. Obviously, information about the intracellular organization of ions and metabolites is lost in this process. Although methods have been developed for the rapid separation of organelles,[1,2] the results are not necessarily always reliable as changes may occur in the course of the separation process. Thus the outcome may be dependent on the buffers that are used, for example. Another second strategy for looking at compartmentalization of ions and metabolites makes use of the selective efflux from several compartments.[3,4] To accomplish this, specific agents are used that will selectively perturb only the plasma membrane, but not the membrane surrounding the vacuoles for example. Although these methods have been used with some success for yeast strains,[3,4] it is immediately obvious that the experimental outcome depends critically on the selectivity of the various agents used for perturbing the membranes.

NMR, because of its noninvasive nature, offers unique opportunities to study intracellular compartmentalization of ions and metabolites. In this paper, some of the reported NMR work with plants and algae pertaining to this question will be discussed. To date, phosphorus-31 NMR has been the most popular NMR method for studies of plant and algae cells. There are several reasons for this choice. First of all the nucleus has a good sensitivity, it is 100% abundant, and many important metabolites are phosphorylated.[5] However, most importantly in this context is that the chemical shift of phosphorylated compounds, such as inorganic phosphate (P_i) and phosphorylated sugars, is a function of the pH in a physiological range.[5,6] Thus, the fact that various intracellular organelles have different pH values has greatly facilitated the studies with plant and algae cells, as the same compound in a different compartment will often

[a]Support from the Alberta Heritage Foundation for Medical Research in the form of a scholarship and funds for the purchase of an NMR spectrometer is gratefully acknowledged. The project concerning metabolism in plant cells is presently funded by the Canadian Natural Sciences and Engineering Research Council (NSERC).

have a different chemical shift. As a result, the various pools of metabolites can be resolved and quantitated independently. Although there are some compounds that can be detected in ^{1}H and ^{13}C NMR spectra of living matter and that also show pH-dependent shifts, most of these titrations lie outside of the physiological range and are thus less useful in separating resonances for the pools in the different organelles.

PLANTS AND PLANT CELLS

In one of the first NMR studies of plant material,[b] Roberts *et al.*[7] made use of ^{31}P NMR to determine the intracellular pH both in the cytoplasm and the vacuoles of maize root tips. FIGURE 1 shows a ^{31}P NMR spectrum of a plant cell suspension as it is nowadays routinely observed in our laboratory. Most of the resonances have been assigned by comparison with standards in extracts of these plant cells.[8] However, the assignment of the two peaks labeled cytoplasmic and vacuolar P_i requires some further comment. First of all it was found that the two peaks merged in whole cell extracts and that they showed a pH titration behavior identical to P_i. At the same time, treatment of the intact cell with proton ionophores also resulted in a collapse of the chemical shift difference.[9,10] These data clearly indicated that both resonances represented separate P_i pools in two compartments with a different pH. The most downfield resonance in the spectrum represents mainly glucose-6-phosphate. This compound is almost exclusively located in the cytoplasm. Its chemical shift in oxygenated plant cells is indicative of a pH of 7.3 in this compartment. Similarly this assignment has been further corroborated by perfusion of the cells with mannose or 2-deoxyglucose.[7,11] Both these compounds can be taken up and phosphorylated by the cell but are otherwise nonmetabolizable analogs for glucose and thus they accumulate in the cytoplasm. Uptake of both of them took place into a compartment with a pH of 7.3. Also, the most lowfield P_i resonance comes from a compartment with a pH of 7.3, thus suggesting that it is the cytoplasmic P_i resonance. This assignment became even more firm when it was shown that turning off the oxygen supply to the cells resulted in a marked drop in pH by 0.7–1.0 pH units in the cytoplasm,[11,13] which was reflected in parallel changing chemical shift values for both the G6P and the cytoplasmic P_i peaks. The other large P_i resonance was assigned to the acidic vacuolar pool, because it resonates at a chemical shift position consistent with a pH of 5.7. It should be noted that this is at the extreme of the titration curve for P_i and thus if the pH were to go down below this value in the vacuole it would be difficult, if not impossible, to observe this in the NMR spectrum. Thus, this pH value should be seen as an upper limit for the vacuolar pH and only changes to higher vacuolar pH can be reliably detected using the P_i chemical shift as a pH indicator. Although attempts have been made to measure vacuolar pH values as low as pH = 4 using the P_i chemical shift,[14] it is the opinion of the author that this is beyond the limits of detectability of the method. In order to measure reliably pH values below pH 5.7 in the vacuole, another pH indicator will be needed that has a pK_a at least

[b] Amongst the first successful NMR studies of living material were the studies on plant material of Schaeffer *et al.*[48] in which they demonstrated that the metabolism of $^{13}CO_2$ to sugar and lipids could be followed in intact soybeans. Since then, a large number of studies have been reported by the same group on the fate of externally supplied isotopically labeled compounds.[49] The questions addressed by these workers required the use of cross-polarization solid-state NMR techniques, which in turn necessitated the use of freeze-dried material. Thus, despite their elegance, these studies are of an invasive nature as the plant material is not kept alive in the course of the NMR experiment. For this reason they will not be discussed here.

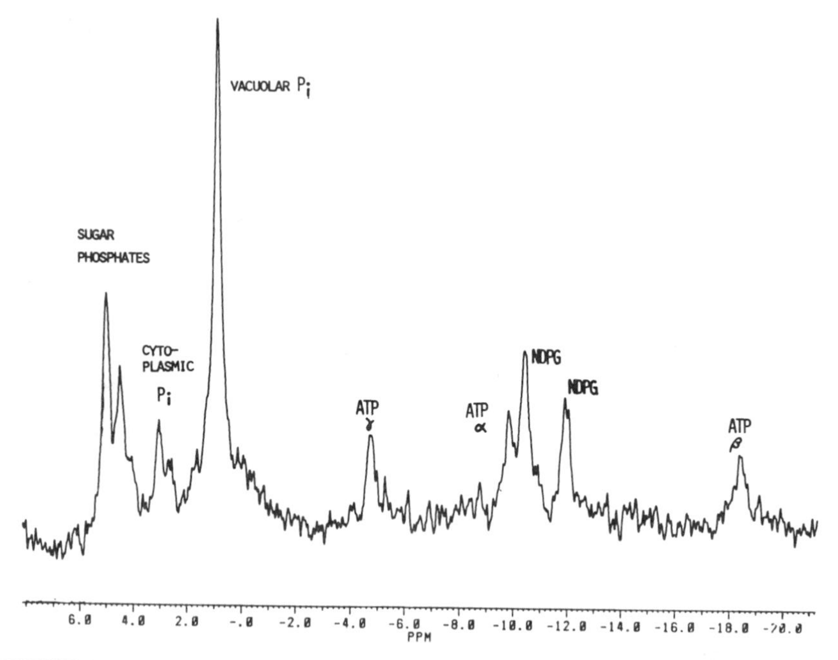

FIGURE 1. Phosphorus-31 NMR spectrum (160 MHz) of *Catharanthus roseus* plant cells perfused with oxygenated medium. The assignment of the resonances is indicated in the figure. All chemical shifts are referenced with respect to 85% H_2PO_4 and upfield shifts are given a negative sign.

two pH units lower than that of P_i (which is 6.8). Some of the compounds that can be observed in ^1H NMR spectra of plant cells[15] may hold promise in that respect.

To date, a wide variety of different root tips and cultured plant cells have been studied by ^{31}P NMR. For example, root tips from maize,[7,16,17] corn,[18,19] and pea[20] have been studied, and cultured cells from *Nicotiana tabacum,*[8,13] *Acer pseudoplatanus,*[9,21] *Daucus carota,*[10] *Catharanthus roseus,*[10,12,22,23] and *citrus* plants[24] have been reported.[25,26] It is surprising to see that all of these give rise to virtually identical ^{31}P NMR spectra. Usually the same resonances are observed (albeit with varying intensities) and the differences in the pH values in the cytoplasmic and vacuolar compartment appear very similar. It has been more difficult to obtain good spectra from stem or leaf tissues. One of the main problems with these tissues appears to be that they are very inhomogeneous, thus leading to variations in the bulk magnetic susceptibility.[27] This renders the resonances extremely broad and difficult to detect. This can be overcome in various ways but most simply by infiltration of water into the tissue airspaces.[27] There is a second reason that cultured plant cells and root tips have been more popular for NMR studies than mature plant cells. The latter usually have large vacuoles and thus the NMR resonances for the cytoplasmic metabolites are weak. In contrast, cultured cells and root tips are less vacuolated and therefore it is easier to obtain a good signal-to-noise ratio for cytoplasmic metabolites in the NMR spectra.

In the majority of the studies reported to date, the NMR spectrometer has been used as an expensive but reliable pH meter. Thus it has been demonstrated that both the vacuoles and the cytoplasmic pH depend on the temperature,[17] but that only the cytoplasmic, but not the vacuolar pH drops considerably during anoxia.[12,13] The latter effect appears to be mainly due to accumulation of CO_2,[13] but in addition accumulation of lactate may play a role.[28,29] Moreover, the cytoplasmic pH appears to be strictly regulated by most cells as it is not much influenced by extremes in intracellular pH[10,13] or various other insults.[16,30] These reports have convincingly demonstrated that the determination of the pH in intracellular compartments by NMR provides an enormous advantage over other methods of intracellular pH determination, such as the distribution of bases. Uncertainties exist in this method because of the corrections needed for vacuolar pH and volume and even because of possible metabolism of the indicator bases, such as dimethyloxazolidine-dione (DMO).[7]

However, NMR spectra as that shown in FIGURE 1 do not only allow the measurement of pH but they also give the possibility to quantitate the P_i pools in the different compartments and to measure the fluxes between them. This became of particular interest since it had been suspected for some time that the vacuole might function as a storage site for P_i.[31] ^{31}P NMR studies with a variety of cultured plant cells quickly demonstrated that this was indeed the case. Upon inoculation of cells into fresh growth media containing P_i, the vacuolar P_i resonance increased dramatically in most—but not all—plant cells.[8-10,24] During this rapid uptake the cytoplasmic P_i pool remained constant. Using ^{31}P NMR combined with ^{32}P radiotracer studies, we could demonstrate that the cells are extremely efficient at totally depleting the growth medium and storing all the P_i in the vacuole.[10] Cells, such as *Daucus carota,* that do not use the vacuole for P_i storage also take up P_i very slowly from the medium.[10] Increasing the level of P_i in the growth medium five times resulted in a five-fold larger vacuolar P_i

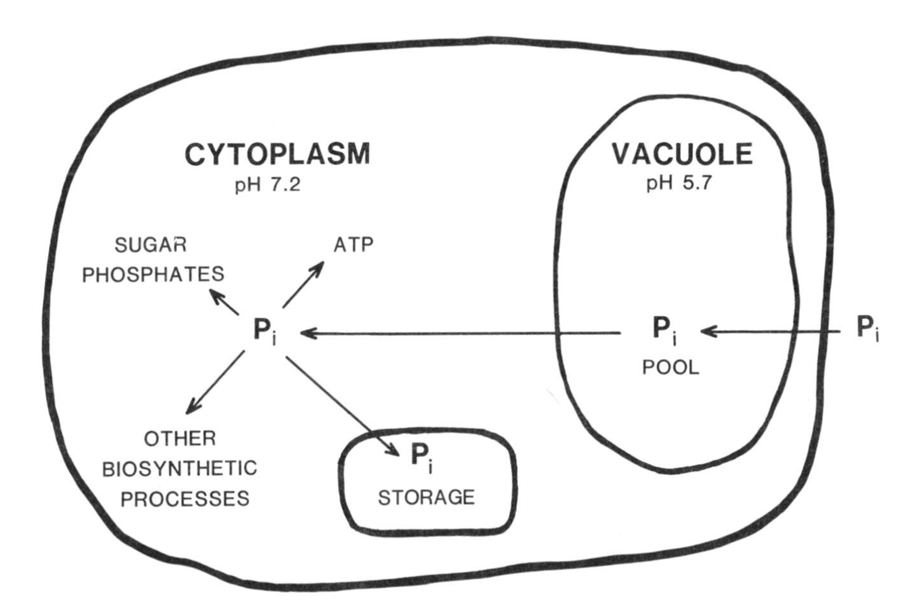

FIGURE 2. Schematic diagram depicting the three different pools of P_i in a *Catharanthus roseus* cell. The pools in the vacuoles and the cytoplasm are NMR visible, the third form probably comprises a metal ion–P_i precipitate in a separate organelle. Upon inoculation in fresh medium, the medium is rapidly totally depleted and all the P_i is stored in the vacuole. The cytoplasmic P_i levels appear to be constant throughout this uptake and during growth.

pool. The rate of uptake was, however, still the same as that with normal P_i levels in the growth medium.[10] It could also be shown that the P_i stored in the vacuole was depleted slowly. One of the most surprising findings in our studies with *Catharanthus roseus* was that the total vacuolar pool became depleted eventually even when five times as much P_i was loaded in the vacuole. Despite the different P_i levels, the same level of growth occurred. From experiments with the radioactive ^{32}P, it became clear that the P_i did not leave the cell and thus it appeared that a separate, NMR-invisible pool was created where P_i was presumably precipitated with metal ions.[10] Indeed, upon making an extract of the cell and treating it with EDTA to chelate the metal ions, a large P_i peak reappeared in the extract spectrum.[10] Thus, these data indicated the existence of a third P_i pool in a separate organelle, presumably a volution granule, as is often observed in yeast or algae (see below). Independent evidence for the existence of such P_i–metal ion precipitates was recently provided by solid-state NMR studies of Alfa-Alfa cells (J. Schaefer, personal communication). The schematic picture in FIGURE 2 indicates the three different compartments and their most likely interrelation. Obviously, there would be an advantage to the plant cell to operate in this fashion. First of all, the rapid total uptake of P_i into the vacuole when this nutrient is available may prepare the cell for times when there is little or no P_i available. Secondly, the cytoplasmic P_i level always remains strictly regulated, which is important as the P_i level of this compartment is known to exert large effects on the activity of numerous enzymatic reactions.[31] Finally, by storing the P_i in a not readily available and only slowly metabolizable metal-ion precipitate form, the vacuole would be available to scavenge for new P_i. In addition, the vacuolar P_i pool does contribute to the cytoplasmic enzymatic activity[27] and thus it may be advantageous that it is regulated to a certain extent itself.

BIOTECHNOLOGICAL USE OF CULTURED PLANT CELLS

Plants have the capacity to synthesize an enormous variety of secondary metabolites. A large number of these are in use or are potentially useful as pharmaceuticals. Well-known examples are compounds such as codeine, opium derivatives, digitoxin, vinblastin, etc. Presently about 25% of the drugs that are in common use are in fact extracted from plants. Thus there is a lot of interest in trying to use cultured plant cell preparations in large scale fermentors to produce some of these pharmaceuticals.[32] There are various questions that need to be considered before such a process could be started: (1) Can the pharmaceuticals be extracted from the vacuole of a cell, without killing it so that the expensive biomass can be used repeatedly? (2) Do the cells survive the harsh environment of a fermentor or would it be better to protect them by immobilization in beads? Would such immobilization change the metabolism in the cell? (3) Can the productivity of cells be improved?

The answer to all these questions is affirmative. In my laboratory, we have been interested in using NMR to look into some of these questions. One of the questions that was addressed was the effectivity of reagents (such as dimethyl sulfoxide) that could be used to extract secondary metabolites. It has been shown that plant cells survived a treatment with 5% dimethyl sulfoxide (DMSO).[33] Looking at the cytoplasmic and vacuolar P_i peaks of the ^{31}P NMR spectrum as indicators of the intactness of both the plasma membrane and the tonoplast (the membrane surrounding the vacuole) we could demonstrate that 5% DMSO perturbed both membranes and that the vacuolar P_i was almost quantitatively released to the external medium.[23] Thus, DMSO is an effective agent for releasing the vacuolar secondary metabolites to the medium while at the same time preserving the viability of the cells.

Whole cell immobilization in agarose and alginate polymer beads has also been studied by NMR.[c] One of the main concerns was that the cells may not be properly oxygenated in these beads. As we discussed earlier, oxygenation of the cells has a marked effect on the cytoplasmic pH. Using this parameter as an indicator, we could demonstrate conclusively that all the cells in a bead are properly oxygenated when the cytoplasmic pH was 7.3 both in perfused immobilized and in control cells.[12] This also demonstrates that products such as CO_2 and lactate can diffuse out of the beads and do not accumulate. The P_i uptake and storage in vacuoles we also studied using ^{31}P NMR. It could be shown that the vacuole was still used as a storage site for P_i. Moreover, the cells were still capable of totally depleting the medium of all its P_i, but the rates of storage into the vacuole and the rate at which the P_i disappeared from the vacuole were considerably slower than in freely suspended cells.[12] This can be explained in two ways. First of all, it is possible that all the cells have a slower P_i uptake because entry is hindered. Secondly, it can not be excluded that only the cells at the outer perimeter of the bead pick up all the P_i and that only a very little reaches the cells at the center of the beads. To differentiate between these two possibilities, ^{32}P-labeled medium was placed on a flat surface of a block of agarose polymer containing plant cells (25%). The results of this experiment are shown in FIGURE 3. Obviously there is a steep gradient in the P_i uptake and the cells closest to the surface take up the majority of the P_i. Thus, it is necessary to use beads that have a diameter of less than 1 mm if one wants to be assured that all cells take up P_i equally.

With respect to the question of improved productivity, NMR has not provided much clues yet. It is well known that certain plant cells show an increased production of alkaloids upon immobilization.[32] It has been suggested that this may be because of changes in intracellular regulatory signals, such as the level of P_i, biochemical energy (ATP), or the pH, for example. Nevertheless, NMR has not shown any large-scale changes in these parameters. Another treatment of plant cells with so-called elicitors is also known to give rise to a dramatically increased production of secondary metabolites.[34] Elicitors are molecules, produced by disease-causing microorganisms, that induce the biosynthesis of antibiotic and antifungal compounds, proteinase inhibitors, and various other substances in the plant as a defense against the invading pathogen. Preliminary experiments in our laboratory with elicitor-treated plant cells have not yet shown any large-scale changes in the regulatory metabolic parameters listed above.

[c] Cell immobilization in agarose beads or threads provides an interesting means of perfusing cell preparations inside an NMR tube. One of the main problems in studying cell suspension by NMR is that high cell densities are necessary to obtain a sufficiently high signal-to-noise ratio in the spectra. This, of course, causes problems with the oxygenation and supply of nutrients to the cells that could be circumvented if the suspensions could be perfused with fresh medium. This is not possible with bacterial or yeast cells as they are too small. However, immobilization in small polymer beads[12] or threads[50] of such cells creates the possibility of perfusing these preparations, thus allowing the measurement of a stable preparation over much longer time spans. This technique has been successfully used to study immobilized suspensions of various bacterial, yeast, plant, and animal cells by NMR.[5,51] It should be realized that plant cells, in contrast to the yeast or bacterial cells, are large enough that they can be retained by a nylon net and thus be perfused in an NMR tube without immobilization. Thus plant cells gave the unique possibility to compare directly perfused preparations of suspended and immobilized cells.[12] Fortunately, these two preparations gave very similar results.

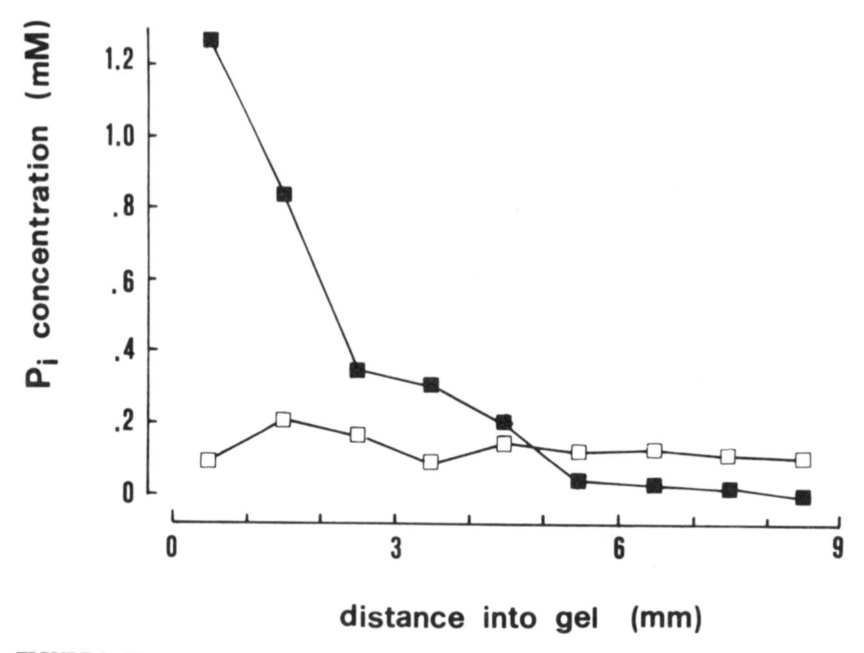

FIGURE 3. The distance into the gel over which plant cells immobilized in agarose medium take up P_i as measured with radioactive ^{32}P. The open symbols refer to a gel without plant cells, the closed symbols are for a gel with plant cells. Note that the cells at the edge of the gel take up a lot of P_i but that this drops off rapidly. (Figure courtesy of Dr. P. Brodelius, ETH, Zurich).

Thus, the regulatory signals giving rise to increased productivity have remained obscure to date.

FRESHWATER AND MARINE ALGAE

FIGURE 4 shows a typical ^{31}P NMR spectrum obtained for the marine alga *Ulva lactuca*.[35,36] It should be noted that there are some unusual resonances in this spectrum that represent polyphosphate, a linear polymer storage form of P_i. The most upfield peaks represent the moieties in the center of the polyphosphate chain, whereas the terminal P_i resonates at -5 ppm. By comparing the integrated intensities of these resonances, one can deduce the average length of the polyphosphate chains. Polyphosphates are found in abundance in almost all microorganisms, but the levels in higher plants or animals are extremely low.[37] The localization of the polyphosphate may vary. In yeast[37] and in the marine algae *Ulva lactuca*[35] the polyphosphates appear to be located in the vacuoles. However, by a combination of ^{31}P NMR and electron microscopy studies, it has been shown that the polyphosphates in the blue-green alga *Cosmarium sp.* are precipitated together with divalent cations in specialized volution granules.[38] In contrast, the polyphosphates in the green alga *Chlorella fusca* are not all NMR visible and they appear to be located both intra- and extracellularly.[39,40] Although the localization of a storage form of P_i in the periplasmic space of algae may seem unusual at first, the same has been observed for some microorganisms,[37,41] so this may be a more general phenomenon. The polyphosphate pools are metabolically active, for example the amount of polyphosphates in or associated with a cell depends on the state of differentiation in synchronized cultures,[39] whereas a shift from anaerobic to aerobic conditions appears to induce the biosynthesis of the linear phosphate polymers.[42]

As was discussed earlier for the higher plant cells, the assignment of the P_i peaks to the P_i pools in the various compartments requires detailed attention. Most authors studying algae have observed several P_i resonances in their spectra and attempted to assign these to the various pools. In studies of *Chlorella,* two P_i peaks were observed by two different groups.[42,43] One group assigned the two peaks to the cytoplasmic and chloroplast pools,[43] but the second group assigned them to cytoplasmic and vacuolar pools[42] and claimed that the P_i pools in the chloroplasts and mitochondria are too low to be detected. Also in *Nitellopsis obtusa* two P_i resonances were observed and these were assigned to cytoplasmic and vacuolar P_i pools based on a comparison of cells with low and high cytoplasm-to-vacuole ratios.[44] Subsequently the effects of illumination and respiration have been studied in these cells,[42–44] and obviously changes in pH do occur under these conditions. One compartment experiences an alkalinization from pH 7.3 to 7.8 during illumination[43,44] while another compartment acidifies.[43] Similarly an 0.4 pH unit alkalinization of the cytoplasm of the photosynthetic cyanobacterium *Synechococcus* was observed upon shifting from dark to light.[45] Shifting from anaerobic to aerobic conditions also shows an increase in pH from 6.8 to 7.4 in one compartment.[42] However, because of the uncertainty in the assignments of the P_i peaks it is difficult to say beyond doubt presently which compartments are affected. Further work will be necessary to obtain firm assignments of these resonances to clarify this situation. Be that as it may, it is encouraging that changes in pH are observed for the different compartments and with more firm resonance assignments, ^{31}P NMR will undoubtedly give a unique insight into the regulation and compartmentalization of pH during different metabolic states, which is indeed an important regulatory parameter in plants.[46]

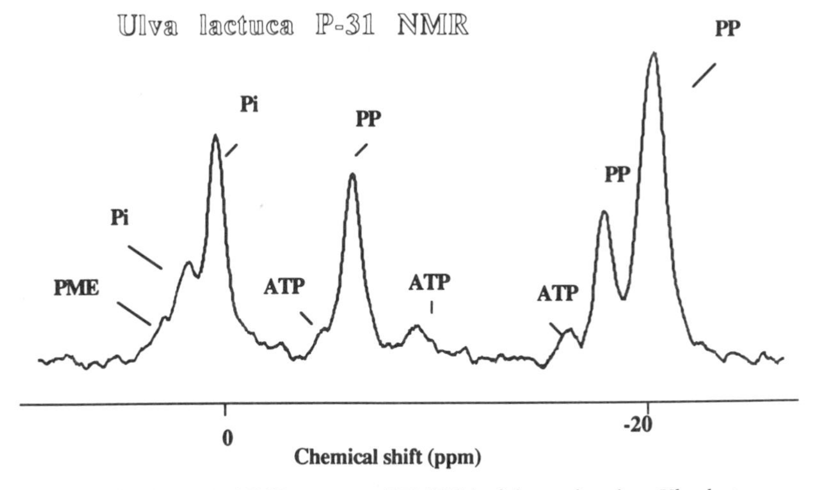

FIGURE 4. Phosphorus-31 NMR spectrum (103 MHz) of the marine algae *Ulva lactuca* grown in fresh seawater containing 0.1 mM P_i. The assignments of the peaks is indicated in the figure. (PP = polyphosphate, PME = phosphomonoesters).

In our studies of *Ulva lactuca,* we observed yet another unanticipated P_i storage form. At P_i values between 0.5–2 mM in the growth medium, a large very broad (2,000 Hz) resonance appeared under the spectrum with the center of the peak at about 0 ppm (FIG. 4). The resonance was of great intensity and the integrated area under this peak was at least 50 times as much as the polyphosphate peak. This peak had a remarkable concentration dependence (FIG. 5), it appeared at maximal intensity around 1 mM P_i in the extracellular medium. Washing of algae, which showed this large peak in the spectrum, with EDTA solutions resulted in the removal of this resonance, without affecting the resonances for the polyphosphates or the other intracellular compounds. From this and subsequent studies with atomic absorption we concluded that an amorphous calcium-P_i precipitate formed on the external surface (periplasmic space)

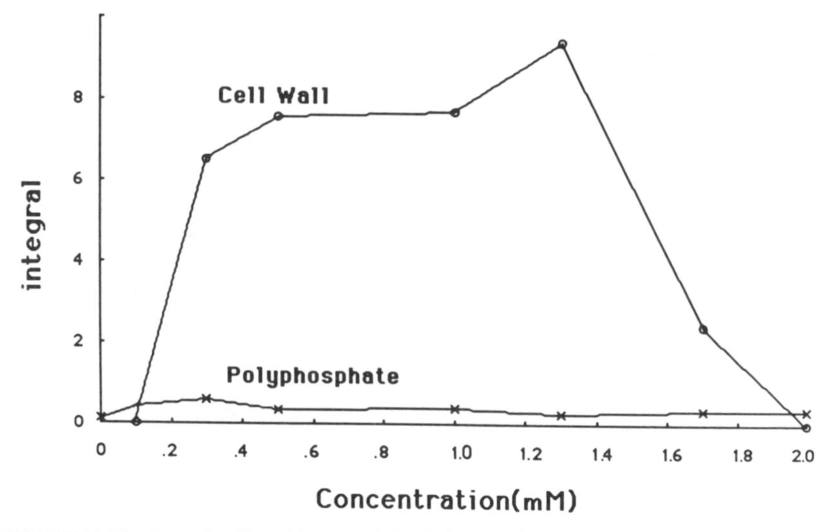

FIGURE 5. The intensity (in arbitrary units) of the broad component and the polyphosphate peak as a function of extracellularly added P_i. Note that the polyphosphate peak remains constant, but that the broad component shows a dramatic increase and decrease between 0.5–2 mM P_i. The intensities plotted here are not corrected for T_1 effects. Experiments with different pulse delays show that the polyphosphate peak as plotted here is at full intensity, but that the broad component is at approximately 40% of its full intensity.

of these algae.[35] The appearance of this extracellular metal ion P_i peak could be modulated by different metabolic conditions such as light, nitrogen source, extracellular pH, etc.[36]

CONCLUSION

The various studies discussed here indicate that ^{31}P NMR can provide unique information about the chemical form and the compartments in which P_i is stored in plant and algal cells. As we have seen, P_i pools have been found extra- or intracellularly and in different organelles. P_i can be stored as P_i, polyphosphates of varying length, metal ion–P_i or polyphosphate complexes, or even as pyrophosphates (as has been observed for *Tetrahymena*).[44] Obviously, it is difficult to generalize and thus it will be necessary to study each species or strain individually to see how it deals with P_i storage and uptake.

ACKNOWLEDGMENTS

The author is indebted to his co-workers, Peter Brodelius, Paul Jensén, Peter Lundberg and Rainer Weich, who collaborated on these projects. The secretarial assistance of Arla Longhurst is greatly appreciated.

REFERENCES

1. SCHLER, M. E., P. HECHT & WILLIAMSON. 1977. Determination of mitochondrial/cytosolic metabolite gradients in isolated rat liver cells by cell disruption. Arch. Biochem. Biophys. **181:** 278–292.
2. AKERBOOM, T. P., H. BOOKELMAN, P. F. ZUURENDONK & J. TAGER. 1978. Intramitochondrial and extramitochondrial concentrations of adenine nucleotides and inorganic phosphate in isolated hepatocytes from fasted rats. Eur. J. Biochem. **84:** 413–420.
3. OKOROKOV, L. A., L. P. LICHKO & I. S. KULAEV. 1980. Vacuoles: Main compartments of potassium, magnesium and phosphate ions in *Saccharomyces* cells. J. Bact. **144:** 661–665.
4. LICHKO, L., L. A. OKOROKOV & I. S. KULAEV. 1980. Role of vacuolar ion pool in *Saccharomyces:* Potassium efflux from vacuoles is coupled with magnesium influx. J. Bact. **144:** 666–671.
5. VOGEL, H. J., P. BRODELIUS, H. LILJA & E. M. LOHMEIER. 1987. Nuclear magnetic resonance studies of immobilized cells. Method Enzymol. **135:** 512–528.
6. MOON, R. D. & J. H. RICHARDS. 1973. Determination of pH in erythrocytes using ^{31}P NMR. J. Biol. Chem. **240:** 7276–7278.
7. ROBERTS, J. K. M., P. M. RAY, N. WADE & O. JARDETZKY. 1980. Estimation of cytoplasmic and vacuolar pH in higher plant cells by ^{31}P NMR. Nature **283:** 870–872.
8. WRAY, V., O. SCHIEL & J. BERLIN. 1983. ^{31}P NMR studies of the phosphate metabolism in cell suspension cultures of *Nicotiana tabacum.* Z. Pflanzenphysiol. **112:** 215–220.
9. MARTIN, J. B., R. BLIGNY, F. REBEILLE, R. DOUCE, J. LEQUAY, Y. MATHIEU & J. GUERN. 1982. A ^{31}P NMR study of intracellular pH of plant cells cultivated in liquid medium. Plant Physiol. **70:** 1156–1161.
10. BRODELIUS, P. & H. J. VOGEL. 1985. A ^{31}P NMR study of the phosphate uptake mechanism of cultured *Catharanthus roseus* and *Daucus carota* plant cells. J. Biol. Chem. **260:** 3556–3560.
11. KIME, M. J., R. G. RATCLIFFE, R. J. P. WILLIAMS & B. C. LOUGHMAN. 1982. The application of ^{31}P NMR to higher plant tissue. J. Exp. Bot. **33:** 656–669 and 670–681.
12. VOGEL, H. J. & P. BRODELIUS. 1984. An *in vivo* ^{31}P NMR comparison of freely suspended and immobilized *Catharanthus roseus* plant cells. J. Biotechn. **1:** 159–170.
13. WRAY, V., O. SCHIEL, J. BERLIN & L. WITTE. 1985. ^{31}P NMR investigation of the *in vivo* regulation of intracellular pH in cell suspension cultures of *Nicotiana tabacum.* Arch. Biochem. Biophys. **236:** 731–740.
14. SCHIBECI, A., R. J. HENRY, B. A. STONE & R. T. C. BROWNLEE. 1983. ^{31}P NMR measurement of cytoplasmic and vacuolar pH in endosperm cells of ryegrass grown in suspension culture. Biochem. Int. **6:** 837–844.
15. FAN, T. W. M., R. M. HIGASHI, A. N. LANE & O. JARDETZKY. 1986. Combined use of ^{1}H NMR and GC-MS for metabolite monitoring and *in vivo* ^{1}H NMR assignments. Biochim. Biophys. Acta **882:** 154–167.
16. ROBERTS, J. K. M., P. M. RAY, N. WADE & O. JARDETZKY. 1981. Extent of intracellular pH changes during H^+ extrusion by maize root-tips. Planta **152:** 74–78.
17. ADUCCI, P., R. FEDERICO, G. CARPINELLI & F. PODO. 1982. Temperature dependence of intracellular pH in higher plant cells. Planta **156:** 579–582.

18. PFEFFER, P. E., S. I. TU, W. V. GERASIMOVICZ & CAVANAUGH. 1986. *In vivo* ^{31}P NMR studies of corn root tissue and its uptake of toxic metals. Plant Physiol. **81:** 77–84.
19. GERASIMOVICZ, W. V., S. I. TU & P. E. PFEFFER. 1986. Energy facilitated Na^+ uptake in excised corn roots via ^{31}P and ^{23}Na NMR. Plant Physiol. **81:** 925–928.
20. ROBERTS, J. K. M. 1984. Study of plant metabolism *in vivo* using NMR spectroscopy. Annu. Rev. Plant Physiol. **35:** 375–386.
21. REBEILLE, F., R. BLIGNY & R. DOUCE. 1982. Regulation of P_i uptake by *Acer pseudoplatanus* cells. Arch. Biochem. Biophys. **219:** 371–378.
22. BRODELIUS, P. & H. J. VOGEL. 1984. Non-invasive ^{31}P NMR studies of metabolism of suspended and immobilized plant cells. Ann. N.Y. Acad. Sci. **434:** 496–500.
23. LUNDBERG, P., L. LINSEFORS, H. J. VOGEL & P. BRODELIUS. 1986. Permeabilization of plant cells; ^{31}P NMR studies of the permeability of the tonoplast. Plant Cell Rep. **5:** 13–16.
24. BEN-HAYYIM, G. & G. NAVON. 1985. ^{31}P NMR studies of wild-type and NaCl-tolerant *Citrus* cultured cells. J. Exp. Bot. **36:** 1877–1888.
25. MCCAIN, D. C. 1983. ^{31}P NMR study of turnip seed germination and sprout growth inhibition by peroxydisulfate ion. Biochim. Biophys. Acta **763:** 231–236.
26. DELFINI, M., R. ANGELINI, F. BRUNO, F. CONTI, A. M. GIULIANI & F. MANES. 1985. Phosphate metabolites in germinating lettuce seeds observed by ^{31}P NMR. Cell Molec. Biol. **31:** 385–389.
27. WATERTON, J. C., I. G. BRIDGES & M. P. IRVING. 1983. Intracellular compartmentation detected by ^{31}P NMR in intact photosynthetic wheat-leaf tissue. Biochim. Biophys. Acta **763:** 315–320.
28. REID, R. J., B. C. LOUGHMAN & R. G. RATCLIFFE. 1985. ^{31}P NMR measurements of cytoplasmic pH changes in maize root tips. J. Exp. Bot. **36:** 889–897.
29. ROBERTS, J. K. M., J. CALLIS, D. WEMMER, V. WALBOT & O. JARDETZKY. 1984. Mechanism of cytoplasmic pH regulation in hypoxic-maize root tips and its role in survival under hypoxia. Proc. Natl. Acad. Sci. USA **81:** 3379–3383.
30. ROBERTS, J. K. M., D. WEMMER, P. M. RAY & O. JARDETZKY. 1982. Regulation of cytoplasmic and vacuolar pH in maize root tips under different experimental conditions. Plant Physiol. **69:** 1344–1347.
31. BIELESKI, R. L. 1973. Phosphate pools, phosphate transport and phosphate availability. Annu. Rev. Plant Physiol. **24:** 225–253.
32. BRODELIUS, P. 1984. Immobilized viable plant cells. Ann. N.Y. Acad. Sci. **434:** 382–393.
33. BRODELIUS, P. & K. NILSSON. 1983. Permeabilization of plant cells resulting in release of intracellularly stored products with preserved cell viability. Eur. J. Appl. Microbiol. Biotechn. **17:** 275–280.
34. LOW, P. S. & P. F. HEINSTEIN. 1986. Elicitor stimulation of the defense response in cultured plant cells monitored by fluorescent dyes. Arch. Biochem. Biophys. **249:** 472–479.
35. WEICH, R., P. LUNDBERG, H. J. VOGEL & P. JENSÉN. 1987. An extracellular calcium-P_i store on *Ulva lactuca* detected by ^{31}P NMR. (In press.)
36. WEICH, R., P. LUNDBERG, H. J. VOGEL & P. JENSÉN. 1987. ^{31}P and ^{14}N NMR studies on the interrelations between nitrogen and phosphate uptake in *Ulva lactuca.* (In press.)
37. KULAEV, I. S. & V. M. VAGABOV. 1983. Polyphosphate metabolism in micro-organisms. Adv. Microbiol. Physiol. **24:** 83–171.
38. ELGAVISH, E., G. A. ELGAVISH, M. HALMANN, T. BERMAN & I. SHOMER. 1980. Intracellular phosphorus pools in intact algal cells. FEBS Lett. **117:** 137–142.
39. SIANOUDIS, J., A. MAYER & D. LEIBFRITZ. 1984. Investigation of intracellular phosphate pools of the green alga *Chlorella* using ^{31}P NMR. Org. Magn. Reson. **22:** 364–368.
40. SIANOUDIS, J., A. C. KÜSEL, A. MAYER, L. H. GRIMME & D. LEIBFRITZ. 1985. Distribution of polyphosphates in cell compartments of *Chlorella fusca* studied by ^{31}P NMR. Arch. Microbiol. (In press.)
41. OSTROVSKI, D. N., N. F. SEPETOV, V. I. RESHETNYAK & A. SIBERLDINA. 1980. Investigation of the location of polyphosphates in cells of microorganisms by high-resolution ^{31}P NMR. Biokhimiya **45:** 517–525. (Biochemistry USSR 392–398.)
42. SIANOUDIS, J., A. C. KÜSEL, W. OFFERMAN, A. MAYER, L. H. GRIMME & D. LEIBFRITZ. 1985. Respirational activity of *Chlorella fusca* monitored by *in vivo* ^{31}P NMR. Eur. Biophys. J. **13:** 89–97.

43. MITSUMORI, T. & O. ITO. 1984. ^{31}P NMR studies of photosynthesizing *Chlorella*. FEBS Lett. **174:** 248–252.
44. MIMURA, T. & Y. KIRINO. 1984. Changes in cytoplasmic pH measured by ^{31}P NMR in cells of *Nitellopsis obtusa*. Plant Cell. Physiol. **25:** 813–820.
45. KALLAS, T. & F. W. DAHLQUIST. 1981. ^{31}P NMR analysis of internal pH during photosynthesis in the cyanobacterium *Synechococcus*. Biochemistry **20:** 5900–5907.
46. SMITH, F. A. & J. A. RAVEN. 1979. Intracellular pH and its regulation. Annu. Rev. Plant Physiol. **30:** 289–311.
47. DESLAURIERS, R., I. EKIEL, A. R. BYRD, H. C. JARELL & I. C. P. SMITH. 1982. A ^{31}P NMR study of structural and functional aspects of phosphate and phosphonate distribution in *Tetrahymena*. Biochim. Biophys. Acta **720:** 329–337.
48. SCHAEFER, J., O. STEJSKAL & F. BEARD. 1975. Carbon-13 NMR analysis of metabolism in soybeans labeled by $^{13}CO_2$. Plant Physiol. **55:** 1048–1053.
49. SKOKUT, T. A., J. MANCHESTER & J. SCHAEFER. 1985. Regeneration in Alfalfa tissue culture. Plant Physiol. **79:** 579–583.
50. FOXALL, D. L. & J. S. COHEN. 1983. NMR studies of perfused cells. J. Magn. Reson. **52:** 346–349.
51. EGAN, W. 1987. The use of perfusion systems for ^{31}P NMR studies of cells. *In* NMR Spectroscopy of Cells and Organisms. R. K. Gupta, Ed. **1:** 135–161. CRC Press. Boca Raton, FL. (In press.)

DISCUSSION OF THE PAPER

QUESTION: We have observed co-transport of phosphate and sodium in other plant cells. Did you find this? [see Reference 24]

H. J. VOGEL: We used a standard growth medium for these *Catharanthus* cells and this is quite low in sodium. We have found in shift-reagent sodium NMR experiments that the intracellular sodium signal is almost undetectable in our spectra, but we do find the extracellular sodium. However, we've never seen it go in together with phosphate.

QUESTION: Regarding the broad signals, did you look for manganese at all in these systems?

VOGEL: We have observed, in the higher plant cells where we believe that we have these intracellular metal ion phosphate precipitates, that we actually need to add quite excessive amounts of EDTA to get a very sharp phosphate resonance in our extracts. From that we conclude that there must be a certain amount of paramagnetic metal ions associated with this. In our studies of marine algae we have not looked specifically for manganese. We do believe that calcium and magnesium are the main counter ions and the atomic absorption data support that as well.

QUESTION: With respect to the *Catharanthus* study showing the phosphate uptake leading to storage of phosphate, there may be a co-transport of divalent metal cations—you did not mention it.

VOGEL: We tried to look at transport of magnesium by magnesium NMR and that has been pretty unsuccessful because in contrast to sodium, potassium, lithium, and ammonium, which are relatively easily studied with the help of NMR shift-reagents, magnesium is significantly broader. The conclusion from this preliminary study was that there was obviously not a large co-transport of this divalent cation. On the other hand, the data were not good enough to exclude that there was some magnesium going in. We've not looked at calcium for the very simple reason that calcium as an NMR nucleus has such a low natural abundance that you have to spend a lot of money to buy all the calcium-43 that you would need to do an experiment like that.

Nuclear Magnetic Resonance Spectroscopy in the Cancer Clinic

JOHN R. GRIFFITHS

The CRC Biomedical Magnetic Resonance Research Group
Department of Biochemistry
St. George's Hospital Medical School
London SW17 ORE UK, United Kingdom

INTRODUCTION

One of the main reasons for the development of biological NMR spectroscopy[1,2] was its promise as a clinical technique. At last it would be possible to detect changes in body chemistry *at the site of the disease,* instead of monitoring their pale reflections in the body fluids. It is now possible to put this promise to the test: whole-body instruments can now be purchased from several manufacturers and many have been ordered for evaluation in hospitals throughout the world. One disease that will be under intense scrutiny by this new method is cancer. Spectra have been obtained from tumors in animals[3,4] and patients[5]; there are already two excellent reviews of the field.[6,7] In this introduction I shall attempt to predict the clinical impact of the method.

CANCER THERAPY

Before cancer treatment begins, the tumor in question must be diagnosed and its type, grade of malignancy, and stage to which it has developed must be assessed, along with the patient's general state of health. These data determine the choice of therapy, which may be surgery, radiotherapy, chemotherapy, endocrine therapy, or any combination of these or other modalities. The orchestration of these various therapies, each with its strengths and weaknesses that vary from one tumor type to another, is the province of the clinical oncologist. His job is further complicated by the tendency of cancer cells to become resistant to new treatment. It is therefore essential to monitor the tumor's regression or progression and to change the treatment if resistance becomes apparent.

How far will it be possible to apply clinical NMR spectroscopy to each stage of anticancer therapy? In this paper I shall consider primary diagnosis, staging, and choice of therapy, while in my other contribution in this volume I shall describe some of our group's work on monitoring treatment.

INITIAL DIAGNOSIS

When a patient has symptoms that suggest a neoplasm, the clinical problem is to find its site, type, and stage. Alternatively, the presenting symptom may be a mass that could be benign or malignant. Magnetic resonance imaging is very suitable for hunting neoplasms, as it can detect soft tissue lesions of a few millimeters diameter. NMR spectroscopy, on the other hand, is not a scanning method: one needs to know where the suspected tumor is before a spectrum can be taken. If the problem is to determine the nature of a lesion, however, NMR spectroscopy provides a unique, non-invasive measurement of its chemical content. Unfortunately, most of the data we have at present come from ^{31}P NMR spectra of animal tumors; these, whether benign or malignant, tend to show identical peaks. The spectrum of the mammary tumor in FIGURE 1(a) is typical, with signals from nucleoside triphosphates (which I shall loosely refer to as ATP, the major constituent), phosphocreatine (PCr), various phosphodiesters, inorganic phosphate (P_i), and phosphomonoesters, all superimposed on a broad hump of signals from immobile phosphate groups, such as those in membrane phospholipids; (a) and (b) are from malignant tumors whereas (c) is benign. The ^{31}P spectrum of a single tumor at various stages in its "life cycle" (see below) can initially resemble that of red muscle or heart, then brain, and later, if PCr has been lost entirely, liver or kidney. The qualitative similarity of all these spectra is not coincidental. It occurs because the high energy phosphate compounds detected by ^{31}P NMR are universal in mammalian cells and thus in the tumors that arise from them.

Despite these problems, there are features in the high energy phosphate spectra of some tumors (peaks 1–4) that may be diagnostically useful. For instance, most tumors contain phosphocreatine, at least at some stage in their life cycle, whereas normal liver does not. Detection of phosphocreatine in the liver would therefore suggest the presence of neoplasm, probably a metastasis. The reason for this assertion is shown in the spectra in FIGURE 3, which are from two Morris hepatomas (well-characterized rat tumors known to derive from liver cells). Neither contains phosphocreatine, presumably because they have developed from hepatocytes that lack the capacity to synthesize that compound. Primary hepatomas are rare, however, at least in the developed countries where NMR spectroscopy facilities are being installed. A more common clinical problem is the differentiation of a solitary metastasis in the liver from a benign lesion. In general, ^{31}P NMR spectroscopy may have a role in distinguishing a suspected tumor from normal tissue or from other disease processes such as hemorrhage, edema, infection, or benign neoplasia. To test this hypothesis, it will be necessary to evaluate the spectra of large numbers of human tumors as well as those of the normal tissues in which they are found and of the benign masses with which they are likely to be confused. It will not be easy to obtain control series, as for ethical reasons one cannot follow an untreated human tumor from its earliest stage to final necrosis. Data on tumors that fail to respond to various forms of therapy may prove a practicable substitute if the therapies do not cause spectral alterations.

The massive peaks of phosphodiesters, inorganic phosphate, and phosphomonoesters (peaks 5–7) in FIGURE 1 (c), and to a lesser extent in (b), seem to be the best tumor markers we have in the ^{31}P spectrum, but as can be seen from (a), they are not always so obvious. This region of the spectrum is prominent in several of the human tumor spectra that have been published.[5,8,9]

STAGING

After initial diagnosis comes staging, which is mainly concerned with establishing the extent to which tumor cells have invaded local tissues or spread metastatically. The data supplied by ^{31}P NMR could give a useful indication as to the development of the primary tumor or of any large metastases.

As animal tumors develop they go through a "life cycle,"[4] shown in FIGURE 2. The high energy phosphate signals in the "young" tumor (peaks 1–4 in FIG. 2 (a)) are

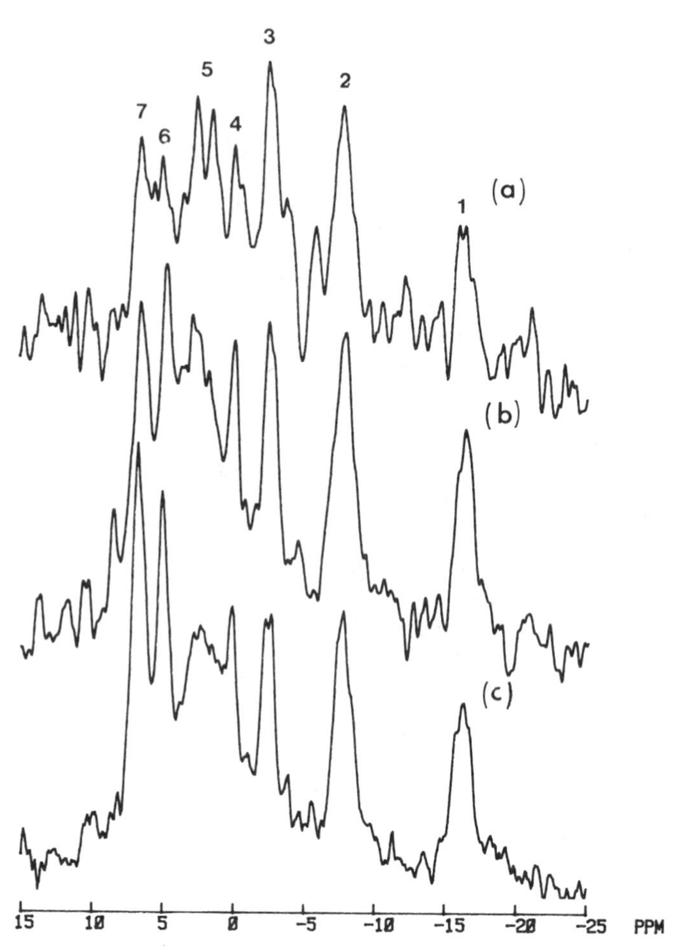

FIGURE 1. ^{31}P NMR spectra of (a) estrogen-sensitive mammary carcinoma induced by 3 injections of nitrosomethylurea in a female, virgin Wistar rat; (b) RIF-1 radiation-induced fibrosarcoma implanted in the flank of a C3H mouse; (c) benign pituitary adenoma implanted in the flank of young, female Wistar rat.[15,16] Animals were anesthetized with intraperitoneal injections of (a) pentabarbitone (30 mg/kg) (b) fentanyl citrate (0.79 mg/kg), fluanisone (25 mg/kg), and midazolam (12.5 mg/kg), or (c) urethane (1.4 g/kg). NMR spectra were obtained at 32 MHz in a 27 cm horizontal bore Oxford Research Systems TMR32-200 spectrometer using 1.1–2.0 cm diameter two-turn surface coils and with 6–15 μsec duration pulses (depending on tumor size); the repetition time was 3 sec (a) and 2 sec (b and c).

smaller a week later in (b) whereas those of P_i, PME, and PDE are more prominent. In the last stage the spectrum of the tumor becomes that of necrotic tissue (FIG. 2 (c)). These changes are probably due to the tumor cells outgrowing their blood supply. In the early stages, the largest fall and rise usually occur in the PCr and P_i peaks, respectively. This suggests that PCr, which is a higher energy phosphate compound than ATP, is donating its phosphate to any ADP that is formed, keeping the ATP peaks almost constant until most of the PCr has been lost, with a more or less stoichiometric rise in the concentration of P_i. Thus, in a growing tumor (as in a normal tissue made ischemic) the main early change is often a fall in the ratio PCr/P_i. The changes that occur in tumors are more complex than simple anoxia, however, and occur over a longer period. In small animal tumors, the surface coil sees an average of the whole neoplasm, and some of the changes in high energy phosphates may be due to severe necrosis in small groups of cells, leading to loss of adenine nucleotides as well as PCr. Indeed, within a single tumor there can be heterogeneity due to local necrosis, cyst formation, hemorrhage, selection of particular clones of cells, and the immunological reaction of the host. These effects will be more evident when large human tumors are investigated, especially when techniques for spatially resolved spectroscopy can be

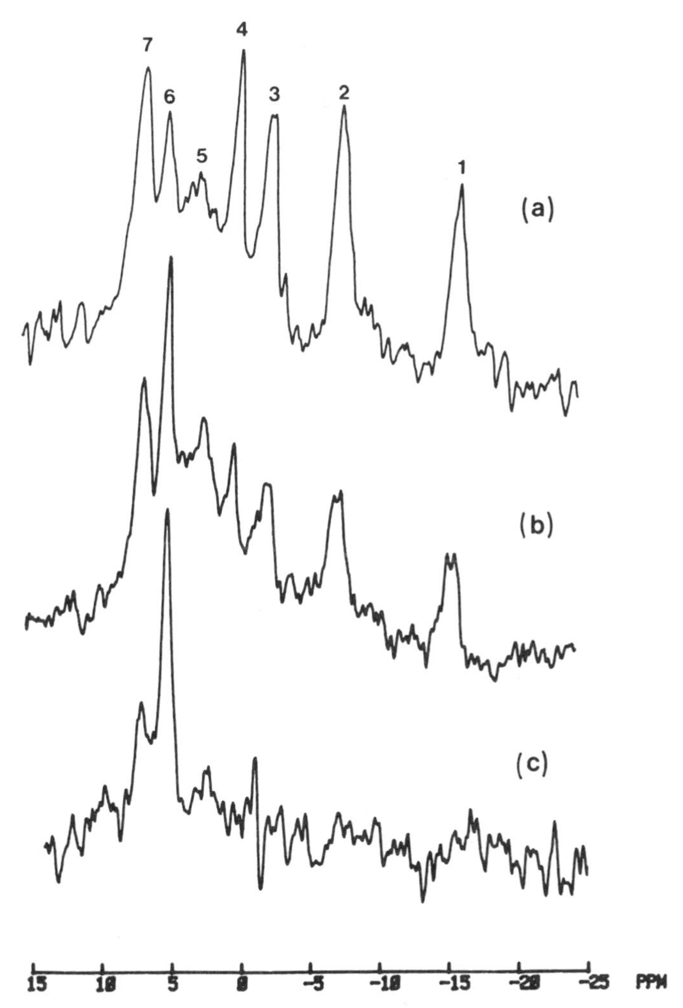

FIGURE 2. ^{31}P spectra of a benign pituitary adenoma at various times after implantation. Conditions as in FIGURE 1 (c).

employed. Further complications are caused by variability between tumors. Even when a single batch of cells is used to raise tumors in a genetically identical batch of laboratory animals, the neoplasms that result often have different spectra and histological characteristics—indeed this can occur if several tumors are raised in the same animal. Clearly the variability will be much greater in clinical practice.

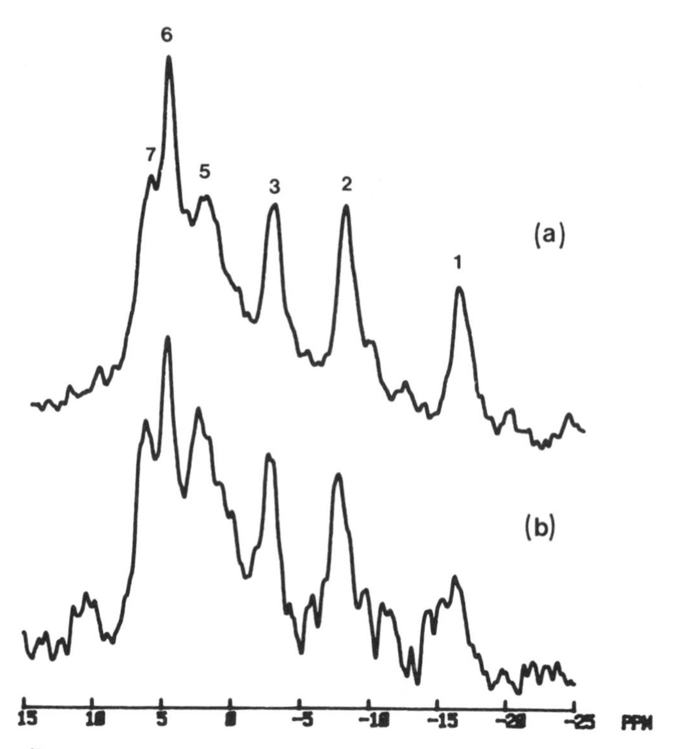

FIGURE 3. ^{31}P spectra of two Morris hepatomas implanted in the flanks of Buffalo rats and studied at about 2 cm diameter. (a) Strain 9618a, well differentiated and (b) strain 7777, poorly differentiated. Spectra were obtained as in FIGURE 1, using a 2 sec recycle time.

^{1}H SPECTROSCOPY

^{1}H NMR spectroscopy has major advantages over the ^{31}P method: it is about an order of magnitude more sensitive and there are many hydrogen-containing compounds of clinical interest. Proton spectra have rarely been obtained from tumors, even in animals, however, for a number of technical reasons. First, special methods are required to eliminate signals from fat and muscle. Secondly, because of the narrow chemical shift range of the ^{1}H nucleus, it is necessary to have an extremely homogeneous magnetic field to resolve the peaks from the small molecules present, i.e. the magnet must be "shimmed" very precisely. Nevertheless, it is possible to obtain spectra of lactate[10] in tumors, and this substance is of some practical importance. We have found that lactate concentration is exquisitely sensitive to changes in tumor energy metabolism,[11] and it may prove a useful clinical indicator in addition to the high energy phosphate spectrum.

When the technical problems have been overcome, ^{1}H spectra might give diagnostically useful information, as the compounds observed are more likely to vary between one tumor and another, and between tumors and normal tissues, than do the ubiquitous metabolites of high energy phosphate metabolism.[12,13] Natural abundance ^{13}C spectroscopy could also provide useful information here.[13]

CHOICE OF THERAPY

The choice of initial treatment is particularly critical as the tumor is likely to spread during ineffective therapy and patients with metastatic cancer are much more difficult to treat. The main considerations are the degree to which the tumor has invaded adjacent tissues or spread its cells to distant organs; these, again, are scanning problems and not amenable to spectroscopic investigation. ^{31}P NMR spectroscopy, however, may be able to provide new parameters that could assist at this stage. In particular, a tumor's response to radiotherapy depends on the concentration of oxygen in its cell water because radiation damage is largely due to free radicals formed from dissolved oxygen. The progressive fall in high energy phosphate compounds in developing tumors probably results from decreasing oxygen supply, so the PCr:P_i ratio could be a useful, indirect index of radiation sensitivity. There are two caveats to be mentioned, however. Firstly, there is evidence from biochemical studies on tumor cells that their decreased oxygen uptake results from an impaired delivery of reducing equivalents to the mitochondrial respiratory chain rather than (or in addition to) an inadequate oxygen supply.[14] Secondly, the low sensitivity of ^{31}P NMR spectroscopy means that one is normally examining the averaged spectra of 10^{10} cells or more. A few deoxygenated cells might not be detected in a neoplasm with a high PCr:P_i ratio but could be sufficient to cause regrowth of the tumor after radiotherapy. There is also a danger that because NMR spectroscopy is confined to large cell masses we may assume that the spectra of these masses are what we need to measure. Unfortunately, tumor regrowth can start from micrometastases that are too small for spectroscopic study and also disseminated around the body to unknown sites. It may be, however, that spectroscopic studies on large primary tumors will give useful insights into the behavior of their micrometastases.

Chemotherapy of tumors often requires competent cellular metabolic pathways, while techniques such as endocrine therapy or the use of tumor-directed drugs depend on the presence of protein receptor molecules on the cell membrane. It is possible that the abnormal spectra seen in late stage tumors are associated with defective metabolism and altered cell membranes, and that such spectra could contraindicate certain therapies.

CONCLUSIONS

NMR spectroscopy has promise as a technique for monitoring tumors at many stages of cancer therapy. Its unique ability to measure cell chemistry non-invasively is likely to be clinically useful for differentiating tumors from benign masses and indicating the stage to which they have developed. It may also be possible to predict the efficacy of some therapeutic modalities from the initial spectrum of the tumor; in particular the high energy phosphate status may indicate tumor oxygen content and hence susceptibility to radiotherapy. ^{1}H spectroscopy, when its technical difficulties have been overcome, could extend the range of metabolites that can be monitored in this way. Even our present ability to monitor tumor lactate could be significant here.

The most promising field of clinical oncology for NMR spectroscopy, monitoring the effects of therapy, has not been considered in this paper. All non-surgical therapeutic modalities are likely to perturb the cancer cell's metabolism and thus the ^{31}P NMR spectrum.

REFERENCES

1. Hoult, D. I., S. J. W. Busby, D. G. Gadian, G. K. Radda, R. E. Richards & P. J. Seeley. 1974. Nature **252:** 285–287.
2. Iles, R. A., A. N. Stevens & J. R. Griffiths. 1982. Prog. NMR Spectrosc. **15:** 49–200.
3. Griffiths, J. R., A. N. Stevens, R. A. Iles, R. E. Gordon & D. Shaw. 1981. Biosci. Rep. **1:** 319–325.
4. Ng, T. C., W. T. Evanochko, R. N. Hiramoto, V. K. Ghanta, M. B. Lilley, A. J. Lawson, T. H. Corbett, J. R. Durant & J. D. Glickson. 1982. J. Magn. Reson. **49:** 271–286.
5. Griffiths, J. R., E. Cady, R. H. T. Edwards, V. R. McCready, D. R. Wilkie & E. Wiltshaw. 1983. Lancet **i:** 1435–1436.
6. Evanochko, W. T., T. C. Ng & J. D. Glickson. 1984. Magn. Reson. Med. **1:** 508–534.
7. Sostman, H. D., I. M. Armitage & J. J. Fischer. 1985. Magn. Reson. Imaging **2:** 265–278.
8. Maris, J. M., A. E. Evans, A. C. McLaughlin, G. D. D'Angio, L. Bolinger, H. Manos & B. Chance. 1985. N. Engl. J. Med. **312:** 1500–1505.
9. Oberhaensli, R. D., D. Hilton-Jones, P. J. Bore, L. J. Hands, R. P. Rampling & G. K. Radda. 1986. Lancet **i:** 8–11.
10. Williams, S. R., D. G. Gadian & E. Proctor. 1986. J. Mag. Reson. **66:** 560–567.
11. Maxwell, R. J., R. A. Prysor-Jones, J. Jenkins, D. G. Gradian, S. R. Williams & J. R. Griffiths. 1986. Proc. Fifth Annual Mtg. Soc. Magnetic Resonance in Medicine **1:** 163–164.
12. Eckel, C & E. T. Fossel. 1986. Proc. Fifth Annual Mtg. Soc. Magnetic Resonance in Medicine **5:** 31–32.
13. Block, R. E. &. B. Parekh. 1986. Proc. Fifth Annual Mtg. Soc. Magnetic Resonance in Medicine **1:** 33–34.
14. LaNoue, K. F., J. G. Hemington, T. Ohnishi, H. P. Morris & J. R. Williamson. 1974. *In* Hormones and Cancer. K. W. McKerns, Ed.: 131–167. Academic Press. New York.
15. Prysor-Jones, A. N., J. J. Silverlight, J. S. Jenkibns, A. N. Stevens, L. M. Rodrigues & J. R. Griffiths. 1984. FEBS Lett. **177:** 71–75.
16. Prysor-Jones, A. N., J. J. Silverlight, J. S. Jenkins, A. N. Stevens, L. M. Rodrigues & J. R. Griffiths. 1985. J. Endocrinol. **106:** 349–353.

Monitoring Cancer Therapy by NMR Spectroscopy[a]

J. R. GRIFFITHS,[b] Z. BHUJWALLA,[c] R. C. COOMBES,[d]
R. J. MAXWELL,[b] C. J. MIDWOOD,[b,d] R. J. MORGAN,[b]
A. H. W. NIAS,[e] P. PERRY,[e] M. PRIOR,[b]
R. A. PRYSOR-JONES,[f] L. M. RODRIGUES,[b] M. STUBBS,[b]
AND G. M. TOZER[c]

[b]*CRC Biomedical Magnetic Resonance Research Group*
[f]*Department of Medicine 2*
[d]*Ludwig Institute for Cancer Research*
St. George's Hospital Medical School
London, SW17 ORE, United Kingdom

[c]*MRC Cyclotron Unit*
The Hammersmith Hospital
London, W12 OHS, United Kingdom

[e]*Richard Dimbleby Department of Cancer Research*
St. Thomas' Hospital
London, SE1 7EH, United Kingdom

MONITORING THERAPY

Cancer therapy is hampered at present by our inability to tell whether a tumor is responding to treatment. The main indicator of success with non-surgical treatments is alteration in tumor size, measured directly or by imaging techniques. Unfortunately, size is a very crude index. Many tumors take months to reach a mass large enough to measure (a gram of tissue contains 10^8–10^9 cells), and extensive metastasis of the resistant cells to distant parts of the body may have already taken place. Also, while waiting for the resistant tumor to grow, one has lost the opportunity to treat it with more appropriate therapy.

NMR spectroscopy offers a unique opportunity for helping to solve this problem. ^{31}P NMR, the most commonly used method, detects the ubiquitous high energy phosphate compounds ATP and phosphocreatine (PCr) and their breakdown product, inorganic phosphate (P_i), in addition to the intracellular pH (pH_i).[1] Almost all non-surgical therapies are likely to alter these parameters,[2] which could therefore be used as indices of therapeutic efficacy. ^{1}H NMR, too, is able to detect substances such as lactate[3] that could be useful indicators. Lastly, it is possible to detect the presence of anticancer drugs themselves in tumors by NMR spectroscopy, particularly those that contain the ^{19}F atom.[4]

Most of the work so far published in this field[5,6] has used animal models. These permit certain simplifications of the clinical situation but suffer from their own artefacts. We will first describe studies on these artefacts and ways of overcoming

[a]Supported by the Cancer Research Campaign, U.K., through a Career Development Award to J.R.G. We thank Roche Products for supplies of floxuridine and doxifluridin.

them and then review progress in our laboratory in monitoring anticancer therapy itself.

^{31}P SIGNALS FROM UNDERLYING MUSCLE

Early studies on human tumors[7–9] used surface coils,[10] and this has also been true of much animal work. Surface coils give spectra with good signal:noise ratios from a wide variety of superficial tumors, but the radiofrequency (B_1) fields they generate are complex and have undefined boundaries. Superficial tumors grown in or under the skin of laboratory rodents are close to muscle, which contains up to 25 mM PCr.[11] A minor extension of the B_1 field into this subjacent muscle can give a PCr signal comparable to the 1–2 mM found in a typical tumor. Worse, a spurious PCr signal may be detected in a tumor that has none or a regressing tumor may appear to develop PCr because adjacent muscle becomes "visible" through it. The PCr:P_i ratio is the most sensitive indicator we have of tumor metabolic status, which makes these potential artefacts particularly annoying. Two general techniques have been adopted to overcome the problem in animal studies.

The first is the Faraday Shield,[12,13] a copper sheet placed around the animal, with a hole through which the coil may be applied to the tumor. When used with surface coils, such shields prevent the side-lobes of the B_1 field extending into muscle around the tumor, but the field may still project through the tumor into subjacent muscle. Glickson[13] has suggested a way of overcoming this problem. Tumors are grown in the loose skin of the rat or mouse; after pulling the tumor away from the body and placing the Faraday Shield closely round the resulting pedicle, the tumor can be placed entirely within a solenoidal coil, maximizing signal-to-noise ratio and minimizing stray field. Lead shields that are essentially similar to Faraday Shields have been used to prevent irradiation of non-tumor tissues in animal radiotherapy studies, but they have a well-recognized disadvantage: the blood supply of the tumor, already precarious, is likely to be further compromised. Tumors given radiotherapy in such a jig are often radioresistant because of anoxia.[14] The problem becomes more serious in NMR studies of tumor metabolism as immobilization of laboratory rodents causes a stress reaction[15,16] that may alter the metabolism of the tumor being studied. Anesthesia is therefore required, causing further cardiovascular stress.[17]

We have studied the effect of such a shield on the ^{31}P spectra of RIF-1 fibrosarcomas in conscious mice. The spectra obtained using a solenoidal coil (FIG. 1, a) are qualitatively identical to those we obtain routinely in anesthetized animals using surface coils although the signal-to-noise ratio is improved about two fold. If the animal is anesthetized while in the shield, the spectrum changes dramatically, with an almost complete loss of high energy phosphate (NTP and PCr) signals; a typical example is shown in FIGURE 1 (b).

Similar, though less marked effects were reported by Rajan *et al.*[18] They found that when an animal bearing an intradermal tumor was anesthetized in a Faraday Shield and a solenoidal coil used to obtain a ^{31}P spectrum, the PCr peak was initially depressed and the P_i peak elevated. A normal spectrum was not established until approximately 30 min after anesthesia. Rajan *et al.*[18] attributed this effect to the anesthetic alone, but our experience suggests that it is due to the combination of stretching the mouse skin in order to place the tumor outside the Faraday Shield and cardiovascular effects due to the anesthetic. We do not find that tumors studied in animals anesthetized with a wide variety of anesthetic agents show initial depression of high energy phosphates if they are studied with a simple surface coil.

Our approach to the problem of contamination by muscle signals has been to define the field of the surface coil by studies with phantoms and to choose an appropriate coil and pulse duration for each tumor. The advantage of this method is that it can be used for studying superficial human tumors with surface coils; Faraday Shields placed between the tumor and patient would hardly ever be feasible. Initially we used a plastic box filled with 150 mM potassium chloride solution in which a glass bulb containing phosphate solution could be moved over a rectangular grid and its signal compared with that of a standard in the center of the coil.[19] The weakness of this phantom is that radiofrequency field penetration may be different in living tissue.

To overcome this objection we have performed further studies with "natural phantoms." Initially, pieces of potato carved to the size and shape of tumors were inserted under the skin of the animal and attempts were made to detect the underlying muscle. Potato, which gives negligible high energy phosphate peaks under these conditions, was used on the assumption that its radiofrequency penetration characteristics would be similar to those of mammalian tissue. If the coil size and pulse duration were correctly matched to the thickness of the phantom the muscle signal was negligible. In another series of experiments we removed a tumor from the animal and kept it at room temperature until it had lost all its high energy phosphate signals. After reimplantation of the tumor into the devascularized bed from which it had been excised, further NMR spectra were collected. FIGURE 2 shows spectra taken before removal of the tumor (a), the excised tumor after 60 min at room temperature (b), and the reimplanted tumor (c) showing no signals from overlaying or subjacent tissues. Assuming that the radiofrequency penetration characteristics of the tumor do not change significantly within an hour of excision, this tumor can be regarded as a natural phantom. All these results suggest that we were able to eliminate spurious PCr signals from underlying muscle by choosing appropriate conditions for the examination.

SIGNALS FROM ANIMAL CUTANEOUS TISSUES

Rat and mouse skin is very different from that of humans: it is loose, and varies in thickness, depending on its location. The rat in particular is able to raise the hairs on its back by using the cutaneous panniculus carnosus muscle. A histological section of rat skin, taken from the upper dorsum, is seen in FIGURE 3. The cutaneous layer on the upper back is thickest and that over the abdomen thinnest. Surface coils are most sensitive to superficial structures, so signals from the muscle layer may interfere with those from subcutaneously or intradermally implanted tumors. The tumor reimplantation study described in the previous section had suggested that contamination from skin signals was negligible as the spectrum in FIGURE 2(c) shows no significant PCr signals, whether from skin or elsewhere. However, we have recently found situations in which this artefact can be important. In a series of Morris hepatomas[20] grown in the flanks of Buffalo rats and in some transplantable hepatomas[21] grown in the upper dorsum of Norwegian hooded rats, positive PCr signals were often seen, whereas enzymatic assays performed on extracts of freeze-clamped tumors[22] showed that they contained negligible PCr and creatine (Cr).

FIGURE 4(a) shows the spectrum of a hepatoma with a strong PCr signal. The skin was then carefully stripped from the tumor and the spectrum shown in FIGURE 4(b) was obtained: all the PCr signal was lost. The tumor did not appear to derive much of its blood supply from the skin but ischemia induced by the process of skin-stripping could, in principle, have contributed to the loss of the PCr signal. However, in a similar experiment with a prolactinoma (FIG. 4, c and d), a tumor known from analysis of

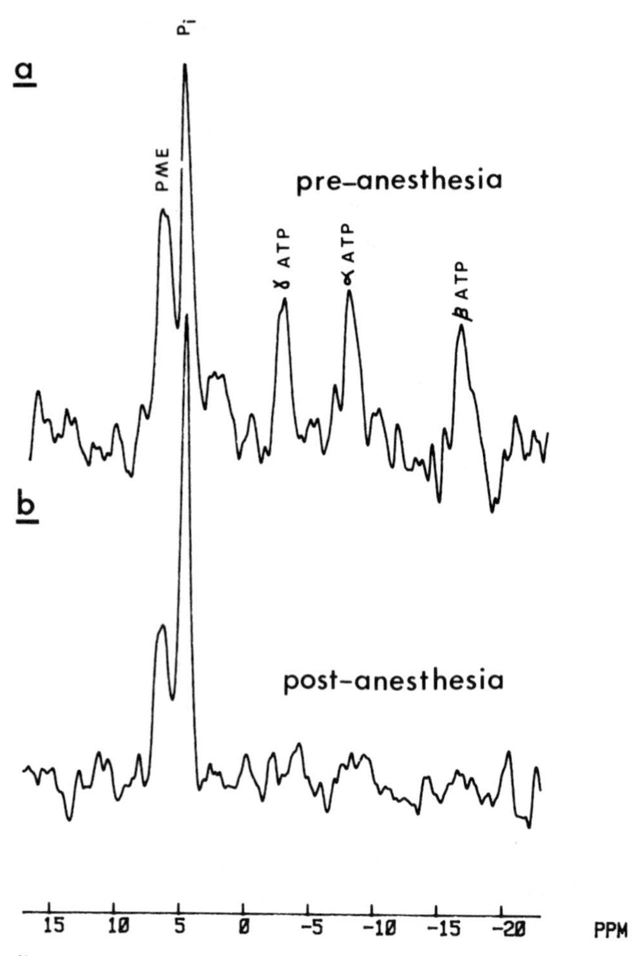

FIGURE 1. ^{31}P spectra of a RIF-1 tumor grown intradermally in a C_3H/HE mouse. In (a) the animal was conscious but restrained in a shield, with the tumor protruding; in (b) it was anesthetized (ketamine 45 mg/kg) while still in the shield. A warm air thermostat was used in the bore of the magnet to keep the rectal temperature of the mouse at 37 ± 0.5°C. ^{31}P NMR spectra were obtained with a 15 mm solenoidal coil using 240 90° pulses at 2-sec intervals in an Oxford Research Systems 1.89 Tesla TMR 32-200 spectrometer operated at 32 MHz.

extracts to contain PCr and also in a mammary tumor (not shown), skin stripping did not remove the PCr signal. Because of the difficulty of rapidly extracting frozen animal skin, we have not been able to assay PCr enzymatically. However, as most of the signal is likely to derive from muscle and as it is known that the PCr:Cr ratio in rat muscle is about 5[11] we can use measurements of total creatine (i.e. PCr + Cr) as an approximate measure of PCr. Enzymatic assays performed on extracts of rat skin show total creatine concentrations of 0.54 ± 0.08 μmol/cm^2 skin, which would be equivalent to a PCr content of approximately 0.4 μmol/cm^2 skin. These data are consistent with the intensity of the signals from PCr in the hepatoma spectra when one allows for the area of skin through which the tumor is observed.

In general, we find that hepatomas quite often have PCr signals attributable to overlaying cutaneous tissue but some other tumors do not. One explanation for this variability may be that some tumors, like the very aggressive Walker carcinosarcoma, invade the skin, while others leave it intact.

The relevance of this artefact to human studies is unclear at present as there is no

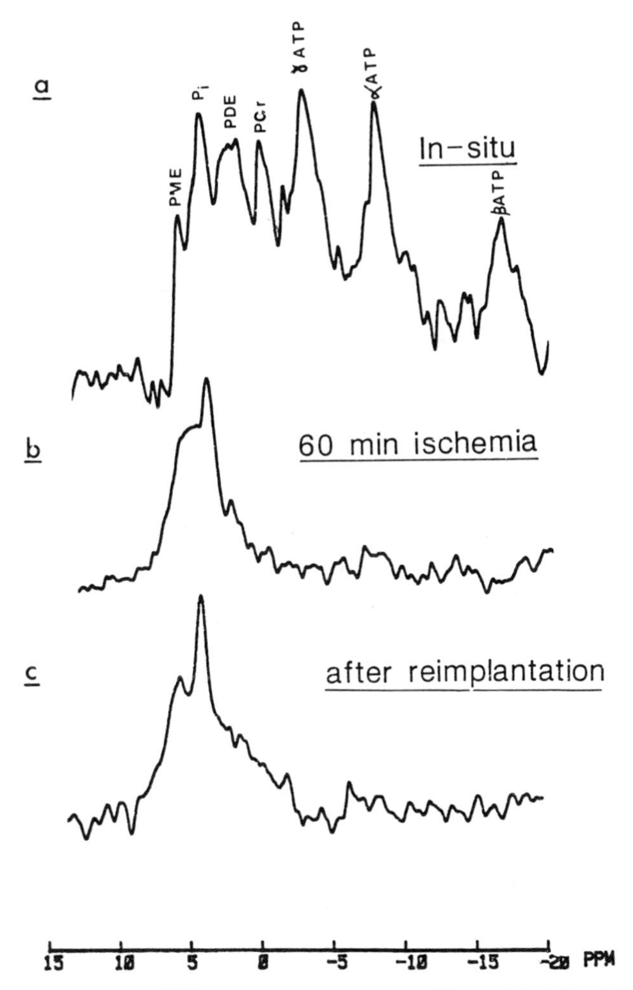

FIGURE 2. An ischemic tumor used as a natural phantom: (a) spectrum of an NMU-induced estrogen-sensitive rat mammary tumor *in situ*. The tumor was then surgically excised and kept for 60 min at room temperature, wrapped in plastic film, before spectrum (b) was obtained. Twenty minutes later, after the tumor had been resutured into the devascularized bed from which it had been excised, spectrum (c) was obtained. In each case 480 8 μsec pulses were given at 2 sec intervals in an Oxford Research Systems TMR 32-200 1.89T spectrometer, operated at 32 MHz.

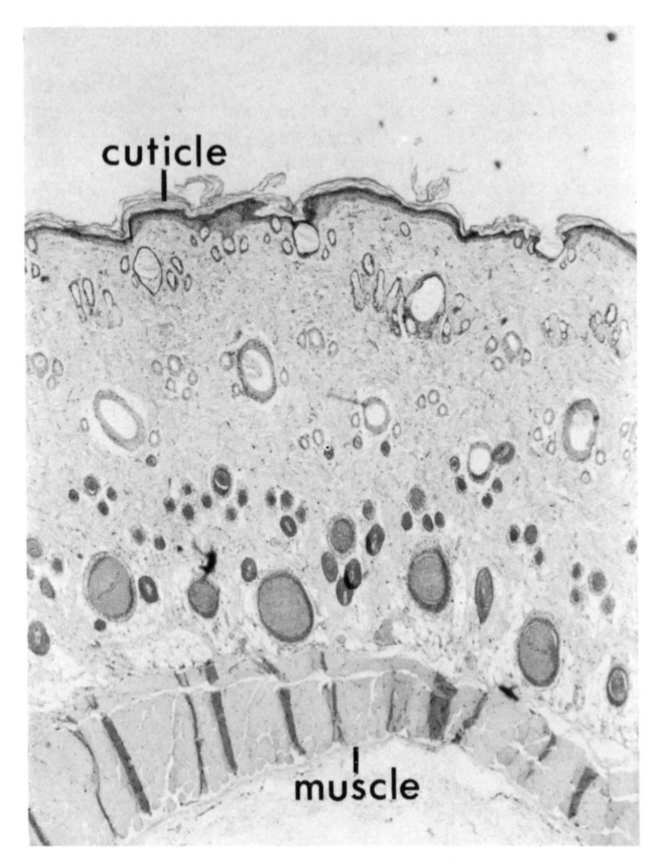

FIGURE 3. Rat skin histology. Section of skin from dorsum of rat, stained with hemotoxylin and eosin; note muscle.

layer of cutaneous muscle in most parts of the human body except for the platysma muscle of the neck and dartos muscle of the scrotum.

TUMOR HETEROGENEITY AND INTRACELLULAR pH

It is obvious from histological studies or even visual inspection that tumors are extremely heterogeneous. Fast growing animal tumors, in particular, tend to become necrotic as they enlarge; classically, in small tumors, a rapidly growing outer shell of tissue should surround a necrotic center, but tumors large enough for NMR studies often show numerous small necrotic foci. In addition to necrosis, areas of hemorrhage or cysts often develop (more in some tumors than in others). Tumors also contain blood vessels induced from the host and a variable proportion of host cells associated with immunological responses. As the tumor grows, the proportions of all these elements change and the tumor cells themselves often undergo variation as one clone predominates over another. Neglecting, for the moment, the additional heterogeneity induced by therapeutic measures, we have a more complex and unstable system than any of the normal tissues studied *in vivo* by NMR. However, surprisingly little evidence of this heterogeneity is evident in the ^{31}P spectra of the tumors studied so far.

Because of its low sensitivity, surface coil ^{31}P NMR is only able to examine large tumors—typically a few hundred milligrams of tissue. Laboratory rodents cannot

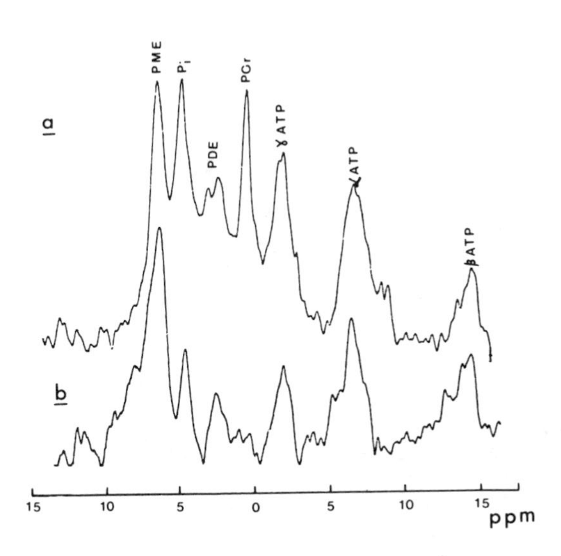

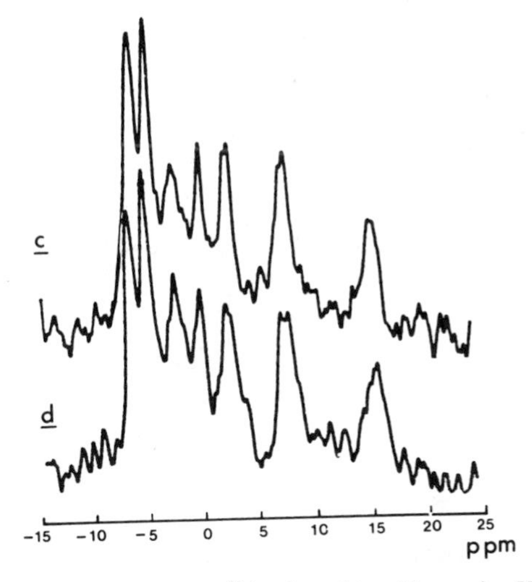

FIGURE 4. (a) ^{31}P spectrum of a hepatoma[21] implanted in a Norwegian Hooded rat. Anesthesia was by i.p. injection of pentobarbitol 60 mg/kg. The spectra represent 480 scans at 5 sec intervals obtained with a 14 mm surface coil, other conditions as for FIG. 2; (b) The same tumor after superficial skin had been reflected; (c) ^{31}P spectrum of a subcutaneous rat prolactinoma; (d) the same tumor after superficial skin had been reflected.

carry tumors of more than a few grams in weight so that ^{31}P spectra represent the average of most of the cells in the tumor. It should be remembered, incidentally, that the inhomogeneous (B_1) fields induced by surface coils mean that some parts of the tumor contribute stronger signals than others. Nevertheless, one would expect to see some evidence of tumor heterogeneity in the ^{31}P spectrum. This is particularly true of the P_i peak, since hypoxic or frankly necrotic regions of a tumor are usually thought to be more acid: either two distinct P_i peaks should be apparent or, more probably, a large number of compartments with slightly different values of intracellular pH (pH_i) would produce a broad P_i peak. In practice, we hardly ever see such spectra. Only one P_i peak is normally present and it is usually as narrow as that of PCr, and narrower than the ATP peaks. Furthermore, the pH_i values observed in animal tumors tend to be close to neutrality.[23] This came as a considerable surprise as, since the days of Warburg,[24] it has been accepted that tumors have a predominantly glycolytic metabolism and therefore produce large amounts of lactic acid. Indeed, measurements of interstitial pH using microelectrodes tend to show that tumors are acid.[25,26] In some tumors we have found acid pH_i values, but usually for a short period, before they become necrotic.[27] Recently, alkaline pH_i values have been reported in human tumor spectra.[9]

How can these paradoxes be resolved? Let us first consider the case of a normoxic cancer cell with a more glycolytic metabolism than that of a healthy cell. As such a cancer cell can exist in a steady state for weeks or months, it follows that the rate at which lactate is lost must be identical to the rate at which it is synthesized. Measurements in various tumors (TABLE 1) show that lactate is at least two-fold higher than in most normal tissues but in seeking to predict the effect of this on intracellular pH we must remember that these tumor cells are chronically adapted to what we would normally regard as a metabolic acidosis. Clearly H^+ ions are lost at the same rate as they are formed (otherwise the cell would rapidly become acidic) so the intracellular pH in the steady state will depend on the passive buffering power of the cellular contents and the nature of any active mechanism for pumping out H^+. If the latter is purely passive diffusion then a rapid rate of H^+ efflux would imply an acidic pH_i, but if an active pump is involved then a homeostatic mechanism could maintain pH_i at or above neutrality.

The simplest hypothesis is that "healthy," well oxygenated tumor cells are able to pump out the H^+ ions they create, producing a neutral or even alkaline cytosol and tending to make the interstitial fluid more acidic. If the tumor blood flow is sluggish then this interstitial fluid will equilibrate slowly with the blood, accounting for the low pH values found with interstitial microelectrodes. As we rarely observe split or broadened phosphate peaks, it seems that little P_i from this extracellular compartment is observed by ^{31}P NMR.

TABLE 1. Lactate in Tumors and Normal Tissues

	μmol/g · wet wt (N)
Prolactinoma, Wistar rat	6.7 ± 1.0 (6)
Hepatoma, Norwegian rat	7.6 ± 1.4 (6)
NMU-induced rat mammary tumor	4.0 ± 1.7 (3)
RIF-1 fibrosarcoma, C_3H mouse	3.2 ± 1.3 (3)
Mean of all tumors studied	6.4 ± 0.9 (26)
Liver, normal[28]	1.36 ± 0.05 (9)
Muscle, normal[28]	0.92 ± 0.08 (6)
Brain, normal[28]	1.35 ± 0.05 (8)
Kidney, normal[29]	1.57 ± 0.65 (9)

Tumors were freeze-clamped *in situ*, extracted with perchloric acid, and assayed for lactate using standard methods.[30]

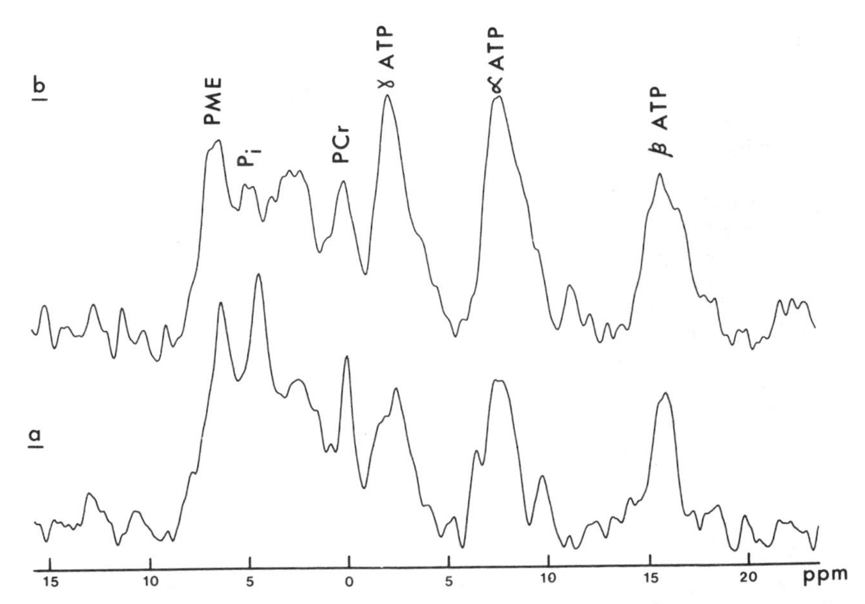

FIGURE 5. ^{31}P spectrum of a RIF-1 tumor implanted subcutaneously (10^5 cells in 0.05 ml Hanks medium) in the flank of a C_3H/KM mouse; (a) before and (b) after 20 Gy radiation (0.92 Gy/min). Anesthesia was by fentanyl (0.79 mg/kg), fluanisone (25 mg/kg) and midazolam (12.5 mg/kg). The magnet bore temperature was maintained, using a warm air blower, to give a mouse rectal temperature of 37 ± 0.5°C. Spectra were obtained using an 11 mm, three turn planar surface coil; 960 6 μsec pulses were applied at 2 sec intervals; other conditions as in FIG. 2.

What, then, of hypoxic or necrotic tumor cells? Again we must remember that a tumor cell becomes hypoxic over a long period whereas most experimental studies on normal tissues are performed acutely. Also, cancer cells rarely contain large glycogen stores, unlike the heart, muscle, or liver cells that are commonly used for studies on ischemia. Thus, a cancer cell that becomes ischemic does not have a large store of carbohydrate to convert into lactic acid and is in any case adapted to a chronic acidosis. The absence of an acid pH_i signal from the hypoxic or frankly necrotic regions that we know to exist in many tumors may therefore indicate that such cells are not in fact acidic. Alternatively, one could postulate that when a cancer cell loses its ability to maintain its pH_i near neutrality, it rapidly loses its P_i pool as well, and therefore gives negligible signal. On the occasions when we are able to observe an acid pH_i it is possible, accepting this hypothesis, that a large number of cells have entered the necrotic phase at the same time, and that their P_i cannot be cleared in the absence of circulating blood.

STUDIES ON TUMOR THERAPY

Radiotherapy

Using ^{31}P NMR we have followed the effect of radiotherapy on RIF-1 fibrosarcomas in mice. When the tumor shown in FIGURE 5 (a) was subjected to radiotherapy (20 Gy) it regressed, and during this regression the P_i peak decreased in size (FIG. 5 b) and the ATP:P_i ratio increased significantly (FIG. 6). This implied that the tumor's metabolic status had been improved by the radiotherapy. Somewhat similar results have been reported from previous NMR studies on RIF-1 tumors[5] whereas Sijens, using murine mammary tumors, found a fall in ATP/P_i.[31] We have now shown a significant rise in PCr/P_i and ATP/P_i with radiation doses as low as 2 Gy, the standard dose fraction in clinical radiotherapy. Two recent reports on radiotherapy of tumors in patients have shown similar results to ours.[32,33]

Why do tumors show this effect? One simple explanation would be that as the tumor reduces in size, the surface coil placed over it acquires progressively more signal from underlying muscle. This cannot be true in the present case as the tumors regress little in the first few days, even after 20 Gy, and after 5 Gy or 2 Gy they are actually continuing to grow. The rapidity with which the effect is observed and the small radiation dose that can elicit it also argue against signals arising from host cells that have invaded the tumor.

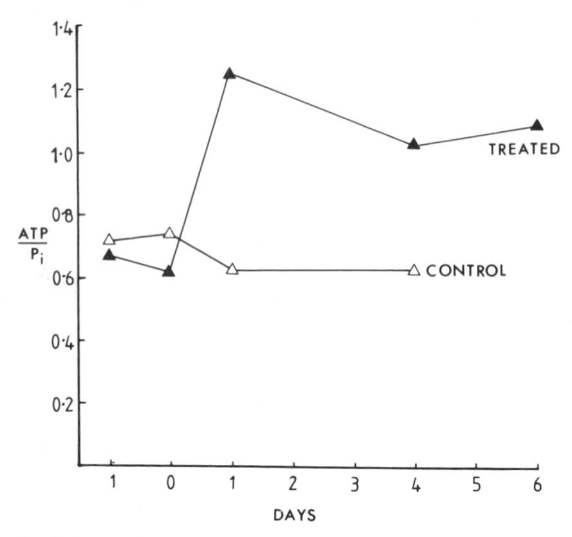

FIGURE 6. Ratio of the integrals of βATP/P_i in the experiments shown in FIG. 6. The data points are the means of 16 experimental animals and 11 controls, which were subjected to sham irradiation.

Evanochko *et al.*[5] studied both the acute and longer term effects of radiotherapy. They found that a dose of 14 Gy to 16/C adenocarcinomas in mice caused an initial fall in PCr followed by a rise, and suggested that the fall might be caused by radiation damage to mitochondria or to the tumor's vasculature. The rise, on the other hand, they felt could be due to an unknown mechanism responsible for the similar rise in PCr they found after chemotherapy or to tumor reoxygenation. We felt that the most probable explanation for the long-term fall in P_i that we observed in RIF-1 tumors after various radiation doses was that the radiation had enhanced the tumor's blood supply, a well known effect in radiotherapy. This hypothesis was supported by measurements of tumor blood flow using 4-iodo[N-methyl-^{14}C]antipyrine; after irradiation there was a progressive and statistically significant rise in blood flow over the period of the experiment.

These results suggest that ^{31}P studies of the effect of radiotherapy could have some clinical applicability. As oxygen enhances the biological effects of radiation, anoxic tumor cells tend to be radioresistant. Thus, if the tumor's blood supply (and hence its dissolved oxygen content) is enhanced following the first dose fraction it may be more sensitive to the next one. If NMR studies of the ^{31}P spectrum are able to predict improved tumor blood flow, we may be able to choose the time of successive fractions to coincide with the improved tumor oxygenation and so enhance their effect.

Endocrine Therapy

Another study on the same lines has been performed on the effect of endocrine therapy on breast cancer. We have examined an estrogen-sensitive mammary carcinoma induced in rats by the carcinogen nitrosomethylurea (NMU). In our first series of experiments, oophorectomy was used to induce tumor regression[34] over a period of weeks. As with radiotherapy, the general effect on the ^{31}P spectrum was a rise in high energy phosphates relative to P_i, but in this case the rise in the ratio PCr:P_i was about two-fold greater than that in ATP:P_i. In some of the tumors, we observed marked changes in the spectrum several days before significant regression occurred.

Recently, we have extended this method to study the effects of drugs on these NMU-induced tumors. Those that regressed in response to the aromatase inhibitor 4-hydroxyandrostenedione (10 out of 16) showed a fall in P_i and a rise in PCr (FIG. 7, b and c), as after oophorectomy. In tumors that carried on growing (4 out of 16) and in 10 controls, there was no effect on the ^{31}P spectrum. Again, NMR changes could often be detected before significant regression.

Endocrine therapy with drugs is an effective treatment of advanced breast cancer and causes less unpleasant side effects than chemotherapy. By measuring the estrogen receptors on the tumor cells one can predict most of the tumors that will respond, but unfortunately a significant percentage of estrogen receptor-positive tumors continue to grow. At present the only way to detect response is to measure the tumor over a period of months and to change the therapy if it fails to regress. The ^{31}P NMR effect observed in the animal tumors suggests that, in the clinic, we may be able to obtain a much earlier indication of failure to respond to endocrine therapy. Such patients could be given a different treatment (e.g. chemotherapy) before the tumor and its metastases had progressed.

^{1}H NUCLEAR MAGNETIC RESONANCE

^{1}H spectroscopy is technically more demanding than ^{31}P because the magnet must be shimmed (i.e. adjusted to give a completely uniform B_0 field) much more precisely and signals from water and fat must be suppressed. One of the most interesting metabolites in this context is lactate, and we have used the spectral editing technique of Williams *et al.*,[35] which eliminates signals other than lactate, to study rat prolactinomas (FIG. 8). This pituitary tumor expends so much energy when it secretes the hormone prolactin that it becomes acid in the same way as a working muscle.[27,36] Using a double-tuned coil, which detects ^{1}H and ^{31}P, we have been able to show that the fall in pH_i occurs at exactly the same time as the formation of lactic acid.[37]

The selectivity for lactate in this pulse sequence depends on the coupling between its –CH and $-CH_3$ protons. Alanine has similar NMR properties but enzymatic assay of tumor extracts has shown that the alanine signal would only be a minor and constant part of the peak assigned to lactate. Signals from the $-CH_2$ protons of tumor

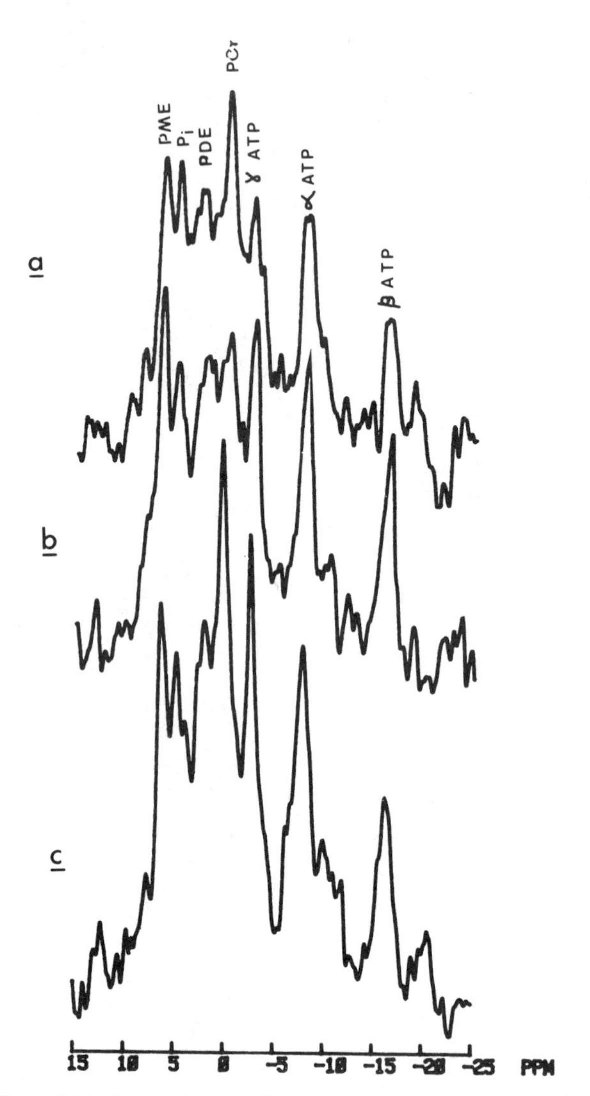

FIGURE 7. Effects of 4-hydroxyandrostenedione treatment (50 mg/kg daily, s.c.) on ^{31}P spectra of estrogen-sensitive NMU-induced rat mammary tumors, (a) before treatment, (b) at commencement of treatment, (c) after 7 days treatment. NMR parameters as in FIG. 2.

triglycerides may also be present in that peak but they too should be constant. We have attempted to assess the significance of this possible contamination of the lactate peak by assaying the lactate in extracts of unstimulated tumors using conventional enzymatic methods and titrating tumor homogenates *in vitro* with alkali. The functional change in the lactate peak integral observed *in vivo* for a given fall in pH_i should correspond to the fractional change in lactate concentration that would induce a similar acidification *in vitro*. This was found to be the case, arguing that contamination of the lactate peak was constant, and not very significant.

^{19}F NUCLEAR MAGNETIC RESONANCE

^{19}F NMR is another spectroscopic technique that has great promise in oncology as it can be used to detect anticancer drugs in tumors and to follow their metabolism.[4] FIGURE 9 shows spectra of the well known drug 5-fluorouracil (5FU) and two analogues, floxuridine and doxifluridin, in tumors. It is thought that all these drugs act by formation of fluoronucleotides, either fluorodeoxyuridine monophosphate, (F-dUMP) an inhibitor of DNA synthesis, or fluoro-UTP. Fluoronucleotides can be seen in the spectrum of the 5FU-treated tumor but not in tumors treated with the other two drugs, although floxuridine and doxifluridin are undoubtedly powerful anticancer

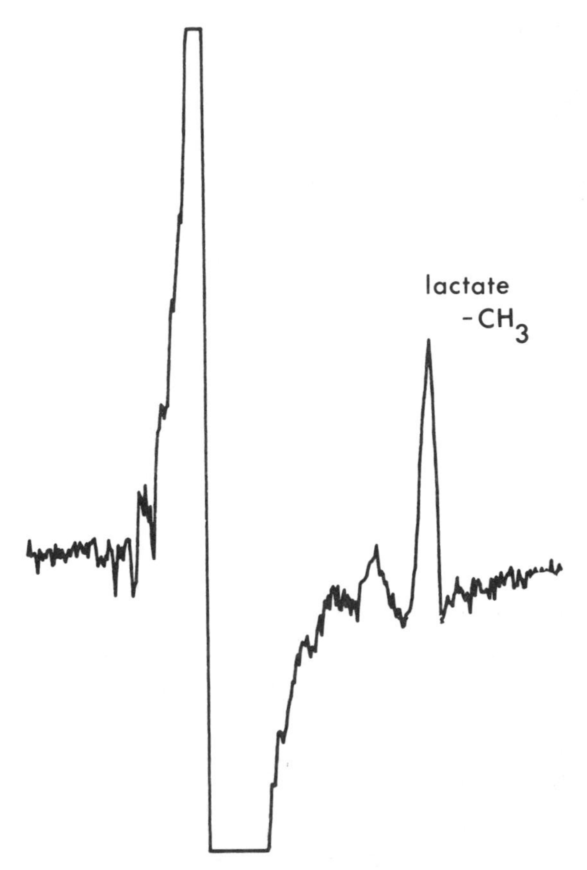

FIGURE 8. Lactate edited ^{1}H spectrum of rat pituitary tumor. A total of 128 acquisitions were collected over a period approximately 7 min using the pulse sequence of Williams *et al.*[36] with a 2 cm surface coil. The large peak is residual water signal.

agents. Indeed, using the drug doses employed in this study, we found that floxuridine was as efficient as 5FU as an inhibitor of the growth of rat prolactinomas. Control tumors took 19 ± 1 days to reach 360 mm^3 in size ($N = 10$), whereas tumors in rats treated with 5FU took 39 ± 3 days ($N = 9$) and those in rats treated with floxuridine took 34 ± 4 days. We have no immediate explanation for this apparent ability of floxuridine and doxifluridin to suppress tumor growth without formation of fluoronucleotides. It may be that the fluoronucleotides detected by ^{19}F NMR consist mainly of

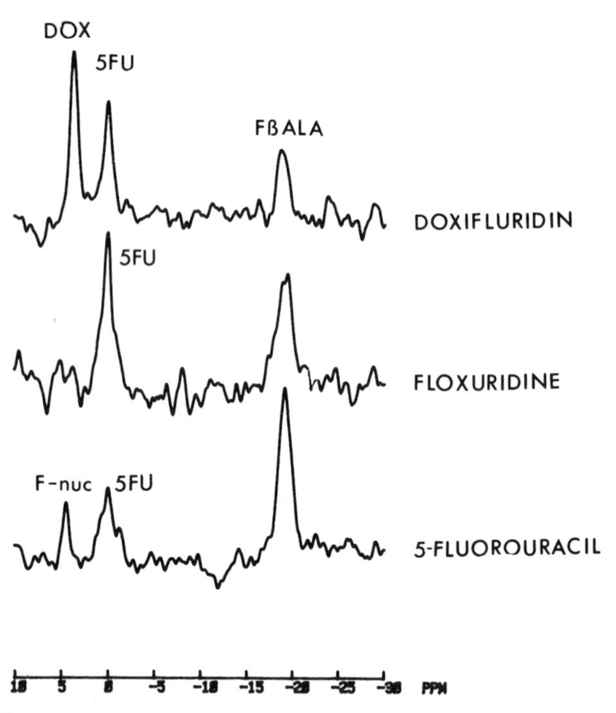

FIGURE 9. ^{19}F NMR spectra of anticancer drugs in rat prolactinomas. F-nuc denotes a peak assigned to fluoronucleotides. ^{19}F spectra were obtained with a 15 mm two-turn surface coil in an Oxford Research Systems TMR-32 200 1.89T spectrometer operated at 72 MHz; 3600 12 μsec pulses were given at 1 sec intervals. The drugs were infused over a 4 hr period at 0.23 mmol kg^{-1} hr^{-1}; the spectra were taken during the last hr of that period.

fluoro-UTP and that small quantities of F-dUMP, sufficient to suppress thymidylate synthetase activity, are still formed.

The ^{19}F NMR tracer method allows us to perform "non-invasive pharmacokinetics" on the drug at the site of its action. In principle, it could be used to determine whether a tumor has taken up an anticancer agent and whether the drug has been metabolized to the active form. Again, therapy could be tailored more precisely to the needs of the individual patient. If ^1H spectroscopy can also be used in this way a wide range of drugs may become accessible to analysis by NMR spectroscopy.

REFERENCES

1. ILES, R. A., A. N. STEVENS & J. R. GRIFFITHS. 1982. Prog. NMR Spectrosc. **15:** 49–200.
2. NG, T. C., W. T. EVANOCHKO, R. N. HIRAMOTO, V. K. GHANTA, M. B. LILLEY, A. J. LAWSON, T. H. CORBETT, J. R. DURANT & J. D. GLICKSON. 1982. J. Magn. Reson. **49:** 271–286.
3. BEHAR, K. L., J. A. DEN HOLLANDER, M. E. STROMSKI, T. OGINO, R. G. SHULMAN, O.A.C. PETROFF & J. W. PRICHARD. 1983. Proc. Natl. Acad. Sci. USA **80:** 4945–4948.
4. STEVENS, A. N., P. G. MORRIS, R. A. ILES & J. R. GRIFFITHS. 1984. Br. J. Cancer **50:** 113–117.
5. EVANOCHKO, W. T., T. C. NG & J.D. GLICKSON. 1984. Magn. Reson. Med. **1:** 508–534.
6. SOSTMAN, H. D., I. M. ARMITAGE & J. J. FISHER. 1985. Magn. Reson. Imaging **2:** 265–278.
7. GRIFFITHS, J. R., E. CADY, R. H. T. EDWARDS, V. R. MCCREADY, D. R. WILKIE & E. WILTSHAW. 1983. Lancet **i:** 1435–1436.
8. MARIS, J. M., A. E. EVANS, A. C. MCLAUGHLIN , G. D. D'ANGIO, L. BOLINGER, H. MANOS & B. CHANCE. 1985. N. Engl. J. Med. **312:** 1500–1505.
9. OBERHAENSLI, R. D., D. HILTON-JONES, P. J. BORE, L. J. HANDS, R. P. RAMPLING & G. K. RADDA. 1986 Lancet **i:** 8–11.
10. ACKERMAN, J. J. H., T. H. GROVE, G. G. WONG, D. GADIAN & G. K. RADDA. 1980. Nature **283:** 167–170.
11. SHOUBRIDGE, E. A. & G. K. RADDA. 1984. Biochim. Biophys. Acta **805:** 79–88.
12. NG, T. C., W. T. EVANOCHKO & J. D. GLICKSON. 1982. J. Magn. Reson. **49:** 526–529.
13. NG, T. C. & J. D. GLICKSON. 1985. Magn. Reson. Med. **2:** 169–175.
14. PHOTIOU, A. 1986. Ph.D. Thesis, London University.
15. KVETNANSKY, R., C. L. SUN, C. R. LAKE, N. THOA, T. TORDA & I. J. KOPIN. 1978 Endocrinology **103:** 1868–1874.
16. TOZER, G. 1987. Some artefacts involved in the radiation response of mouse tumors arising from anaesthesia and physical restraint. *In* Rodent Tumours in Experimental Cancer Therapy. Pergamon Press. Oxford. (In press.)
17. CULLEN, B. M. & H. C. WALKER. 1985. Int. J. Radiat. Biol. **48:** 761–771.
18. RAJAN, S. S., S. J. LI, J. P. WEHRLE, L. DILLEHAY & J. D. GLICKSON. 1986. Proc. Fifth Annual Mtg. Soc. Magnetic Resonance in Medicine **5:** 29–30.
19. HAWKES, D., C. BARTON, L. M. RODRIGUES, M. JENKINS & J. R. GRIFFITHS. 1985. Proc. Fourth Annual Mtg. Soc. Magnetic Resonance in Medicine **2:** 986–987.
20. MORRIS, H. P. & B. P. WAGNER. 1968. *In* Methods in Cancer Research. H. Busch, Ed. **4:** 125–132. Academic Press. New York.
21. REID, E. 1970. Br. J. Cancer. **24:** 128–137.
22. STUBBS, M., L. M. RODRIGUES & J. R. GRIFFITHS. 1986. Abstracts 14th Intl. Cancer Congress. **3:** 842.
23. GRIFFITHS, J. R. & R. A. ILES. 1982. Biosci. Rep. **2:** 719–725.
24. WARBURG, O. 1931. Metabolism of Tumours. R. R. Smith. New York.
25. GULLINO, P. M., H. GRANTHAM, S. H. SMITH & A. C. HAGGERTY. 1965. J. Natl. Cancer Inst. **34:** 857–869.
26. MEYER, J. A. 1974. Ann. Surg. **179:** 88–93.
27. PRYSOR-JONES, A. N., J. J. SILVERLIGHT, J. S. JENKINS, A. N. STEVENS, L. M. RODRIGUES & J. R. GRIFFITHS. 1985. J. Endocrinol. **106:** 349–353.
28. VEECH, R. L., J. W. LAWSON, N. W. CORNELL & H. A. KREBS. 1979. J. Biol. Chem. **254:** 6538–6547.
29. WILLIAMSON, D. H., H. A. KREBS, M. A. PAGE, H. P. MORRIS & G. WEBER. 1970. Cancer Res. **30:** 2049–2054.
30. BERGMEYER, H. U., ED. 1974. Methods of Enzymatic Analysis. 2nd edit. Verlag Chemie. Weinheim.
31. SIJENS, P. E., W. M. M. J. BOVEE, D. SEIJKENS, G. LOS & D. H. RUTGERS. 1986. Cancer Res. **46:** 1427–1432.
32. NG, T. C., S. VIJAYAKUMAR, F. J. THOMAS, A. W. MAJORS, T. F. MEANEY, J. P. SAXTON, M. A. WEINSTEIN & N. J. BALDWIN. 1986. Proc. Fifth Annual Mtg. Soc. Magnetic Resonance in Medicine. **1:** 173–174.

33. BALERIAUX, D., D. A. ARNOLD, C. SEGEBARTH, P. R. LUYTEN & J. A. DEN HOLLANDER. 1986. Proc. Fifth Annual Mtg. Soc. Magnetic Resonance in Medicine. **1:** 41–42.
34. RODRIGUES, L. M., A. N. STEVENS, J. WILKINSON, R. C. COOMBES & J. R. GRIFFITHS. 1986. *In* Magnetic Resonance in Cancer. P. S. Allen, D. P. J. Boisvert & B. C. Lentle, Eds.: 139–140. Pergamon Press. Toronto.
35. WILLIAMS, S. R., D. G. GADIAN & E. PROCTOR. 1986. J. Mag. Reson. **66:** 560–567.
36. PRYSOR-JONES, A. N., J. J. SILVERLIGHT, J. S. JENKINS, A. N. STEVENS, L. M. RODRIGUES & J. R. GRIFFITHS. 1984. FEBS Lett. **177:** 71–75.
37. MAXWELL, R. J., R. A. PRYSOR-JONES, J. JENKINS, D. G. GADIAN, S. R. WILLIAMS & J. R. GRIFFITHS. 1986. Proc. Fifth Annual Mtg. Soc. Magnetic Resonance in Medicine. **1:** 163–164.

DISCUSSION OF THE PAPER

DEUTSCH: This actually is in conjunction with some interesting speculation that Walter Wolf brought up yesterday. It looks like you're generating some of our probes that we made up in our heads. It looks like the fluoro-beta-alanine that you're generating as a catabolic product is now an indigenously generated pH probe and it looked like you were seeing pH shifts and you're measuring pH in these tumors without realizing it.

J. R. GRIFFITHS: Walter mentioned this to me yesterday. As far as I remember, when we actually checked out the various intermediates that we found, that wasn't one of the ones that titrated, in the range we investigated. 5-Fluorodeoxyuridine titrated with a pK of 7.5, but I didn't think the fluoro-beta-alanine did. I guess you'd know which ones titrate and which ones don't.

QUESTION: It looked like when you were administering the 5-fluoro-uracil, you produced F-dUMP but when you gave fluoxuridine you produced 5-fluoro-uracil but then not F-dUMP. Why is that?

GRIFFITHS: I cut a slide because I was going a little slowly. What we think is happening is that the fluoxuridine or doxifluridin possibly are inhibiting the enzyme system that produces F-dUMP from 5-fluoro-uracil. We tried using thymidine as an analog because it's just the same thing with the methyl group in that position, and would thus be expected to inhibit the system too. This is the slide I didn't show you: as you can see if we don't give thymidine you form F-dUMP from 5-fluoro-uracil. If we give a small dose of thymidine, it depresses the detoxification pathway and you don't get any fluoro-beta-alanine. If you give a big dose it also stops the formation of fluoronucleotides. We suspect that those drugs are suppressing the formation of these fluoronucleotides in this particular tumor, which is very interesting. Incidentally, thymidine was actually used in clinical practice as an adjunct for 5-fluoro-uracil in the hope that it would suppress the detoxification to fluoro-beta-alanine. It didn't turn out to be too successful. I wonder whether in fact they might not have suppressed fluoronucleotide formation by giving too much of it. The sort of thing one might be able to check by NMR.

QUESTION: How close are these doses to pharmacological ones that you would use in treating a tumor?

GRIFFITHS: The doses that we're giving in these infusions are similar to the pretty big doses that are used with doxifluridin and floxuridine. They're quite a lot more than would normally be given with 5-fluoro-uracil. We can, in fact, just about get down to the human milligram per kilogram dose of 5-fluoro-uracil and see the signal even in these small animal tumors. I should incidentally have mentioned that Walter Wolf has actually detected the detoxication metabolism of 5-fluoro-uracil in the liver in a patient using a whole body system.

QUESTION: I wonder in view of Digones and Kayes work, whether under conditions where you don't see phosphocreatine, you have looked for creatine kinase activity. It's just conceivable that you are not supplying creatine synthetic pathway when in fact you have the enzyme there. That would be very interesting to know. I wonder if you've studied that.

GRIFFITHS: Yes we have. We've actually assayed phosphocreatine in these various tumors and certainly found creatine kinase activity. The levels are pretty low in the liver-derived tumors and substantially higher in the others. What we haven't done, of course, is a proper time curve of creatine-kinase activity to see whether it falls away at the same time the tumors actually lose their phosphocreatine. That would mean a real mega rat experiment, which we've never tried.

^{31}P and ^{1}H NMR Spectroscopy of Tumors *in Vivo*:

Untreated Growth and Response to Chemotherapy

JANNA P. WEHRLE, SHI-JIANG LI, S. SUNDER RAJAN,
R. GRANT STEEN, AND JERRY D. GLICKSON

The Laboratory of NMR Research
Department of Radiology and Radiological Science
The Johns Hopkins University School of Medicine
Baltimore, Maryland 21205

INTRODUCTION

In the last five years, *in vivo* ^{31}P NMR studies of cancer have become common. Unfortunately, as the number of NMR studies of tumors has increased, a bewildering array of apparently different patterns of growth and response to therapy has been revealed. In order for NMR to become a useful tool for cancer research or clinical management, phenomena must be identified that have a strong correlation with tumor growth or response to therapy. The limits and exceptions to emerging "rules" must be established. In order to accomplish these goals several advances will be required: (1) Well-defined *in vitro* systems must be used to determine which changes occur at the cellular and which occur at the tumor tissue level. An understanding of the mechanisms responsible for NMR-observable changes will greatly facilitate the identification of those changes likely to be useful and reliable, and will contribute to our basic understanding of tumor cell metabolism. (2) The role of tumor type, drug, and dose in determining spectral characteristics and spectral changes must be established by careful comparisons. (3) Studies must be extended to other nuclei. Most previous investigations have used ^{31}P NMR, which allows only a small number of metabolites to be observed. Compounds observed in the ^{1}H or ^{13}C spectrum may show important changes upon tumor growth or response to therapy.

We have been using transplantable tumors grown subcutaneously in mice and rats to study tumor growth and response to therapy. Recently, we have employed ^{31}P spectroscopy to address several basic questions regarding the mouse tumor model: the effects of anesthetics on the *in vivo* ^{31}P spectra of tumors; the relationship between NMR-observable changes following chemotherapy and commonly used therapeutic indices; and the effects of drug dosage on the NMR-observable response to treatment. We have been examining techniques for *in vivo* ^{1}H spectroscopy with suppression of the water and lipid resonances to reveal tumor metabolites. One of the tumor lines studied *in vivo* (radiation-induced fibrosarcoma, RIF-1) has also been grown in a cell culture perfusion system suitable for the NMR spectrometer. This system allows the tumor cells to be observed by ^{31}P and ^{1}H spectroscopy in a highly controlled environment suitable for studies of mechanism.

EXPERIMENTAL METHODS

Tumors and Cells

The RIF-1 tumor was carried according to the protocol of Twentyman *et al.*[1] in C3H/HeJ mice (Jackson Laboratories). Tumors were induced in the right flank by subcutaneous inoculation with 10^5 cells and studied between day 12 and day 24 after inoculation. 9L gliosarcoma was induced in Fischer 344 rats (1 week old) by subcutaneous inoculation with 10^6 cells (work in progress in collaboration with Drs. Henry Brem and Raphael Tamargo). Tumors were examined after they reached approximately 1 cm^3 (day 12–14 following inoculation). Mice were anesthetized for spectroscopy with sodium pentobarbital (65 mg/kg, i.p.). Rats were anesthetized by i.p. injection of a cocktail of ketamine (0.65 mg/kg) plus rompun (0.65 mg/kg).

For cell culture, freshly isolated RIF-1 cells were seeded on collagen-coated microcarrier beads (Cytodex 3, Sigma Chemicals) and allowed to grow for 7 to 10 days in roller bottles with daily changes of RPMI 1640 medium supplemented with 10% fetal bovine serum and 1% penicillin/streptomycin, in an atmosphere of 95% air, 5% CO_2. Immediately prior to NMR spectroscopic examination beads were washed once with perfusion medium and packed in a 20 × 30 mm cylinder with 20 μm nylon filter endcaps. The cell cartridge was perfused inside a modified 25 mm NMR tube at 3–4 ml/min with Earle's Minimal Essential Medium, supplemented with 25 mM HEPES/Na, 10% fetal bovine serum, and antibiotics. The medium was presaturated and continuously gassed with a mixture of 95% O_2/5% CO_2. A combination of reservoir heating, perfusion line insulation, and spectrometer variable temperature control was used to maintain the cells at 37°C.

End Points

In vitro clonogenic cell survival assays for RIF-1 were performed as described by Twentyman *et al.*[1] Volumes of *in vivo* tumors were estimated from caliper measurements of the major and minor axis dimensions parallel to the body surface by assuming the tumor is an oblate ellipsoid.

NMR Spectroscopy

In vivo spectra of subcutaneously implanted tumors were obtained with two- and three-turn solenoidal coils in home-built probes using a Bruker AM 360-WB spectrometer. A Faraday shield[2] was routinely used to prevent the accumulation of signal from body tissue. ^{31}P spectra were obtained at 145.8 MHz, using a flip angle of 70° and a 3 sec interpulse delay. Increasing the delay time further resulted in no alteration of the spectrum. Metabolite ratios were estimated from peak heights. Tumor pH was estimated from the chemical shift of P_i. The *in vivo* spectra contained resonances from phosphomonoesters (PME), inorganic phosphate (P_i), phosphodiester compounds (PDE), phosphocreatine (PCr), and the α, β, and γ resonances of ATP and other nucleoside triphosphates (NTPs). *In vivo* ^{1}H spectroscopy was performed at 360 MHz, using a doubly tuned solenoidal coil (^{31}P/^{1}H). *In vitro* ^{31}P and ^{1}H spectra of perfused tumor cells on microcarrier beads were obtained in a commercial 25-mm broad band probe (Bruker Instruments).

RESULTS AND DISCUSSION

NMR spectra of untreated RIF-1 tumors reflect a high degree of individual variation, even among age and size-matched tumors, as has been reported previously.[3] We have studied a large number (>150) of tumor-bearing mice, attempting to determine the origins and the magnitude of this variability. We have also begun to assess the variability in tumor response to therapy and have accumulated some preliminary data concerning causes of the different response patterns observed under different experimental protocols.

Effects of Anesthesia

Unlike cooperative human subjects, animals require mild anesthesia in order to remain sufficiently immobilized to permit spectroscopic measurements, which may require 15–60 min. The duration of anesthetic effect differs considerably between

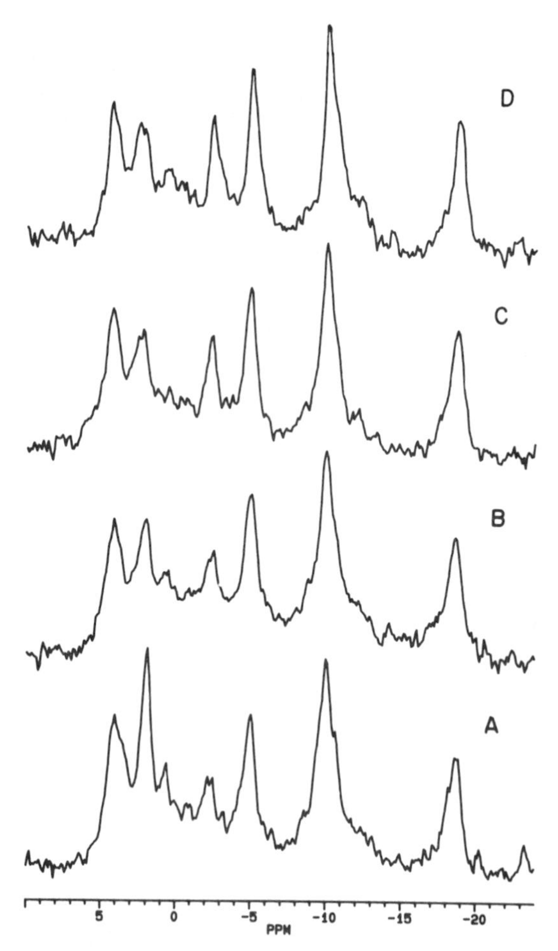

FIGURE 1. *In vivo* ^{31}P NMR spectra of RIF-1 tumor. Spectroscopic conditions are described in METHODS. Spectral resonance identifications (left to right): PMEs, P_i, PDEs, PCr, γ-NTP, α-NTP plus pyridine nucleotides, β-NTP. Spectra were accumulated for 10 min each beginning at 32 (A), 62 (B), 75 (C), and 94 (D) min after injection of sodium pentobarbital (65 mg/kg, i.p.).

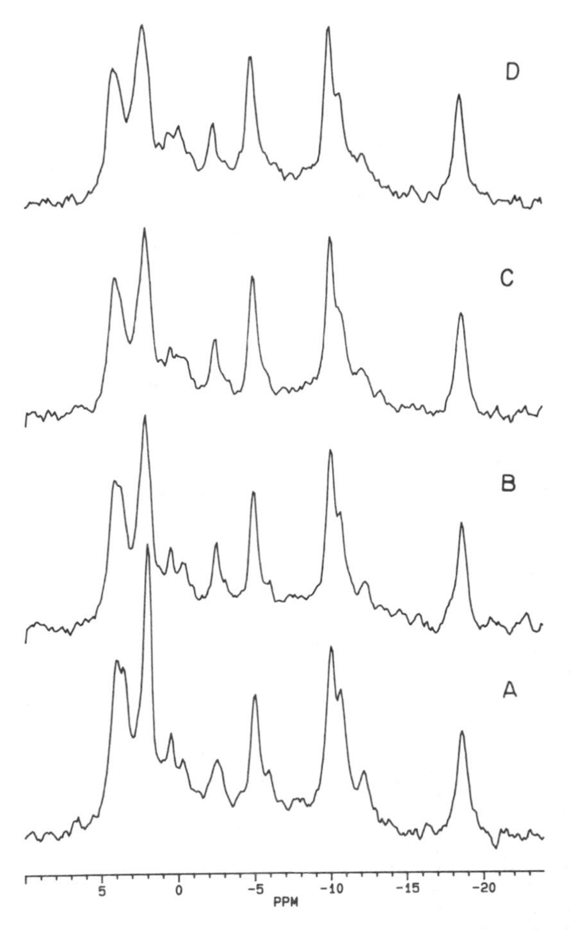

FIGURE 2. As in FIG. 1: 32 (A), 56 (B), 79 (C), and 92 (D) min after injection of ketamine hydrochloride (150 mg/kg, i.p.)

animals, even when doses are carefully adjusted by weight. Studies were performed to determine whether variability in the NMR spectra was being induced by differing response to anesthetics. Mice were anesthetized by i.p. injection, then prepared for spectroscopy as rapidly as possible. Typically, mounting and shimming required 10–15 min. NMR spectra were then collected in blocks of 200 scans (10 min) for the following 90 min, or until the spectrum became constant. In many instances, substantial reductions in the level of phosphocreatine (PCr) and elevations in the level of inorganic phosphate (P_i) were observed in the early spectra, compared to the final, stable spectra obtained after approximately 40 min. Examples of these changes are shown in FIGURES 1 and 2. In one experiment, following anesthesia with pentobarbital (65 mg/kg), 5 of 7 animals showed changes of greater than 10% in the ratio of P_i/NTP-β, comparing the earliest spectrum to the final, stable values. In a similar experiment, 3 of 6 mice injected with ketamine (150 mg/kg) showed changes in excess of 10% over the same one-hour period.

Mouse blood pressure has been shown to fall dramatically following administration of pentobarbital or ketamine.[4,5] Sodium pentobarbital has also been reported to cause a reduction in tumor temperature and a decrease in radiosensitivity in the EMT6 tumor, which was attributed to a reduction in tumor blood flow.[6] The time course of stabilization of the NMR spectra parallels the recovery of blood pressure observed in the physiological studies. ^{31}P NMR spectra of normal tissues are extremely sensitive to conditions that cause hypoxia and ischemia.[7–9] Because subcutaneous tumors are served by the peripheral blood supply, they are likely to be more strongly affected than the central body core by changes in hemodynamic parameters. In order to eliminate artifacts due to these effects, spectra are collected in 10-min blocks until no further change is observed, before collection of the definitive spectrum.

Tumor Response to Chemotherapy

It has been repeatedly observed that the changes in the *in vivo* ^{31}P NMR spectrum during untreated growth of a tumor resemble those of a tissue experiencing increasing degrees of hypoxia.[10–12] The mechanism or mechanisms of these changes remains unknown. Because their magnitude and time-course can vary substantially for different tumors, it is not clear whether the differences are qualitative or quantitative. Some tumors, like the MOPC 104E myeloma,[2] and the 9L gliosarcoma (see below) experience a striking decline in high energy phosphates with growth, with a large increase in P_i and an acid shift of the pH, suggestive of a high proportion of dead, hypoxic, or quiescent cells. Other tumors, such as the mammary 16/C adenocarcinoma[13] and particularly the RIF-1 fibrosarcoma, show more modest changes: an increase in P_i and PME resonances and some decrease in PCr, but relatively high levels of NTP throughout the tumor lifespan. The different growth patterns may result from the inherent metabolic characteristics of the tumors, differences in efficiency of vascularization, host immune response, and many other factors as yet poorly understood.

In many cases, treatment with an effective drug results in spectral changes that are the opposite of those observed during untreated growth. For some tumors these changes can be striking, with increases in PCr and NTP and decreases in P_i. This has been reported previously for the MOPC 104E plasmocytoma treated with cyclophosphamide (CPA)[10] or 1,3-bis(2-chloroethyl)-1-nitrosourea (BCNU),[2] and is also observed in the 9L tumor (FIG. 3). Within 24 hr after treatment with BCNU (10 mg/kg, i.p.) the level of P_i in the 9L tumor is reduced and levels of high-energy phosphates are increased. More moderate changes are observed after treatment of other tumors with the same or with different drugs. Thus, for treatment of mammary 16/C adenocarcinoma with adriamycin,[13] and human breast tumor xenograft MX-1 with adriamycin,[12] as well as for the RIF-1 tumor treated with CPA,[14] chemotherapy resulted primarily in reduction in the magnitude of the P_i and PME resonances, together with an alkaline shift in the pH, reversing the trends of these tumors during untreated growth.

A different response pattern has been reported by Naruse and co-workers, following i.v. administration of high doses of a variety of drugs to tumors of neural origin.[15,16] With these treatments the spectra of the tumors changed to that of severely hypoxic or dead tissue. Because the studies reported to date have employed many different tumor models, drugs, dose schedules, and routes of administration, it has been difficult to determine reasons for the different observations. Using the RIF-1 tumor, we have attempted to determine, for a single tumor-drug pair, whether differences in drug dose result in altered patterns of NMR response.

FIGURE 4 illustrates the spectral changes observed in a RIF-1 tumor during untreated growth between days 12 and 18 after inoculation. The metabolite intensities vary quantitatively from animal to animal, but the changes are not all independent. The P_i/PCr ratios of untreated tumors of different ages and sizes are correlated with tumor pH over a wide range (FIG. 5, $r = 0.8$, $p < 0.005$). It is important to recall that metabolite concentrations and pH values determined by NMR are not necessarily uniform characteristics of each cell in the tumor, but represent average values for a somewhat heterogeneous population.

As we have described previously,[14] treatment of 12–14-day-old RIF-1 tumors with a moderate dose of CPA (150 mg/kg, i.p.) produces a reversal of the changes observed during untreated growth (FIG. 6). A reduction in the level of P_i and PMEs relative to

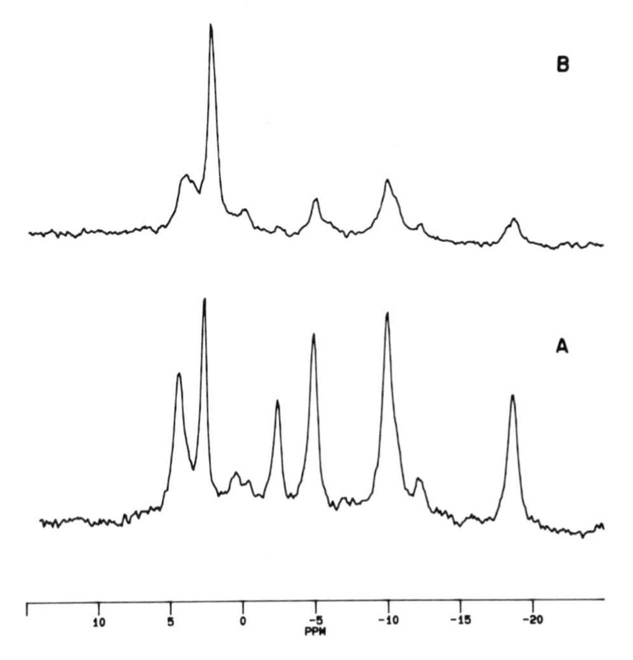

FIGURE 3. *In vivo* ^{31}P NMR spectra of two rats with matched 9L tumors. Spectra were obtained on day 13 after inoculation, 24 hr after injection of a control carrier of 4% ethanol in saline (B), or BCNU, 10 mg/kg, i.p., (A). Spectroscopic parameters were as described in METHODS. Peak assignments as in FIG. 1.

α-NTP and an alkaline shift in tumor pH are observed. Furthermore, after CPA treatment, the population variability of the metabolite ratios was considerably reduced (FIG. 5, crosses). The chemotherapeutic effects of this dose of CPA were measured by tumor growth delay and by clonogenic cell survival (FIG. 7). The clonogenic cell fraction (plating efficiency of treated cells compared to controls) was reduced to $\simeq$0.5% 48 hours after treatment. By this time, NMR changes were already observable, even though acute cell viability (trypan blue exclusion) was still >96%. The clonogenic cell fraction began to increase after day 3, when the tumor volume was constant and the total viable cell number was still decreasing. Externally observable tumor regrowth did not begin until day 7 (FIG. 7A), by which time the NMR spectrum had begun to return to pretreatment patterns.

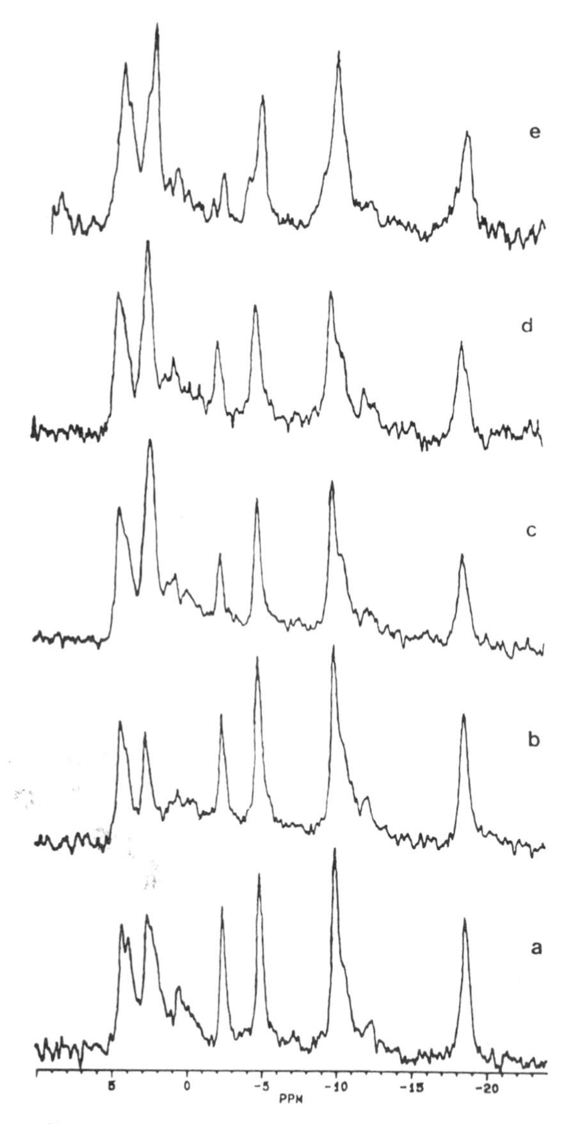

FIGURE 4. *In vivo* ^{31}P NMR spectra of an individual control mouse tumor before (a), and 1 (b), 2 (c), 3 (d), and 5 (e) days after sham treatment with 0.9% saline (0.1 ml, i.p.). Tumor volume increased from 0.7 to 2.0 g during this period. Spectroscopic parameters as described in METHODS. Peak assignments as in FIG. 1.

Administration of a higher dose of CPA (200 mg/kg, i.p.) to mice bearing RIF-1 tumors results in a dramatically different spectral alteration. FIGURE 8B is the spectrum of a 15-day-old tumor, and 8A is a spectrum from a cohort mouse, which had been treated 24 hr earlier with 200 mg/kg of CPA. The reduction in high energy phosphates is striking. This response was observed in 7/19 mice treated with this dose of CPA, while the remainder responded as observed at the lower dose. No difference was observed in cell yield or viability (>96% trypan blue exclusion) 24 hr after the high dose, compared with lower doses or controls. Of the animals given 200 mg/kg, 11/19 mice died within 7 days of drug administration, including all 7 which presented the "dead tumor" spectrum. This may be compared with 0/18 given 150 mg/kg, 1/14 given 100 mg/kg, and 0/18 controls, indicating significantly increased host toxicity of the higher drug dose.

We have shown that in this case, differences in NMR-observable response are a function of drug dose. Naruse *et al.*[15] have reported that treating a human neuroblastoma xenograft with low doses of CPA will induce a "dead tumor" spectrum, but only when drug administration is repeated after 24 hr. Treatment with a single moderate dose did not produce this effect. Thus in this case as well, the response was not specific to the tumor and drug, but was a function of dose and treatment regimen. Considerable additional research will be required to determine what aspects of a therapy protocol result in one or the other type of response, and in what way these two responses relate to ultimate host death, either from the tumor or from drug toxicity.

Although the observation of increased levels of high energy phosphates following effective antitumor therapy appears paradoxical, it has been observed for a wide variety of tumors following chemotherapy[2,10,12–14] and radiotherapy.[17] Several possible explanations may be suggested, but only careful, mechanistic studies will finally determine which, if any, is correct. One possibility is that effects on tumor vasculature result in increased blood flow to the tumor. An increase in tumor blood flow has been reported for RIF-1 tumors following treatment with CPA.[18] Preferential killing of low energy cells or recruitment of quiescent cells into a metabolically more active form would enhance the fraction of highly energized cells contributing to the spectrum. Relatively rapid clearance of dead cells must be occurring, as the level of P_i is reduced

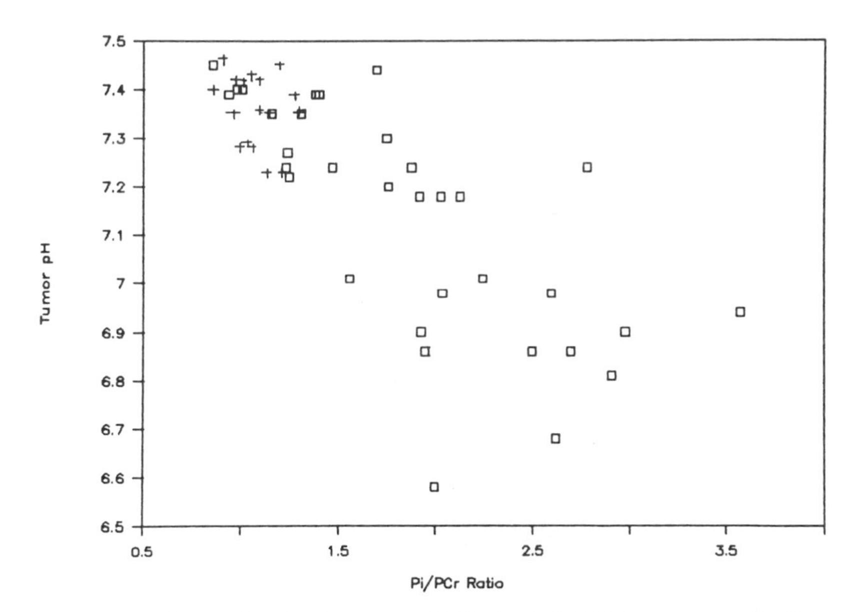

FIGURE 5. P_i/PCr was plotted against tumor pH (P_i chemical shift) for tumors age 12–22 days, for both control (□) and treated (+) animals (150 mg/kg, CPA).

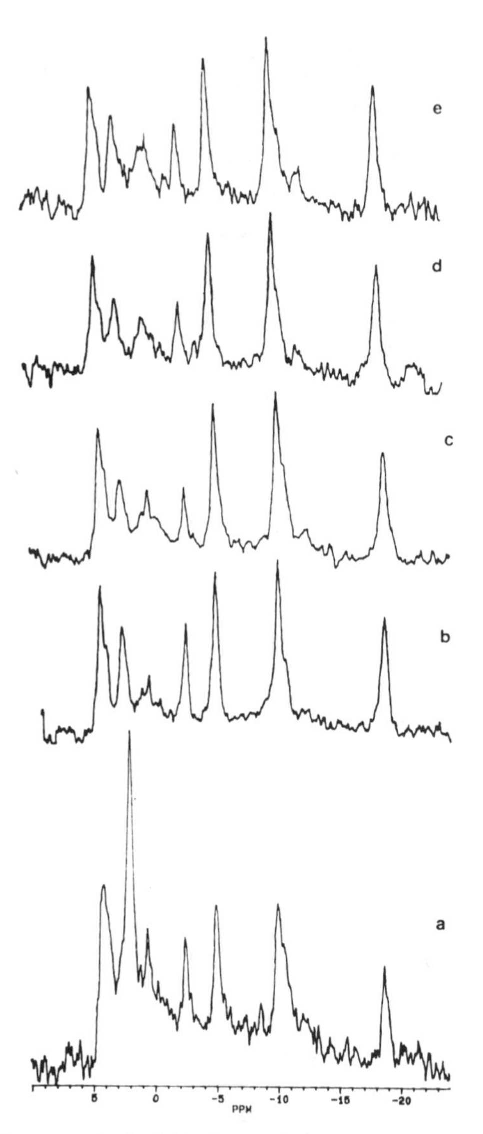

FIGURE 6. *In vivo* ^{31}P spectra of an individual treated mouse tumor before (a), and 1 (b), 2 (c), 3 (d), and 5 (e) days after treatment with CPA (150 mg/kg, i.p.). Spectroscopic parameters as described in METHODS. Peak assignments as in FIG. 1.

in these cases, rather than increased. On the other hand, those protocols that result in the "dead tumor" spectrum must cause a low rate of cell clearance relative to cell death. Significantly, we have observed this type of response after treatment of mammary 16/C adenocarcinoma[13] and Dunn osteosarcoma[19] with hyperthermia, as well as high-dose CPA (see above). In the case of the Dunn osteosarcoma, Lilly *et al.*[19] have shown a linear correlation between the decline in NTP/P_i and reduction in blood flow as measured by ^{133}Xe clearance.[19] Other possible causes for the altered NMR response at higher dose include: cell injury not sufficient to result in trypan blue inclusion, and increased vascular damage or host toxicity resulting in poor perfusion. These last events might result in an *in situ* hypoxia, which would be reflected in the NMR spectrum, but would not be immediately lethal.

In Vivo ^{1}*H Spectroscopy*

^{1}H spectra of tumor extracts[11] and excised tumor tissue[20] suggest the large number of important metabolites that might be visualized by ^{1}H spectroscopy. However, performing ^{1}H spectroscopy *in vivo* has proven considerably more difficult than ^{31}P spectroscopy. Significant magnetic susceptibility heterogeneity has been reported in tumor tissue,[21] which may interfere with water and lipid suppression and spectroscopy. Techniques for the suppression of water and lipid resonances[22–24] all result in some

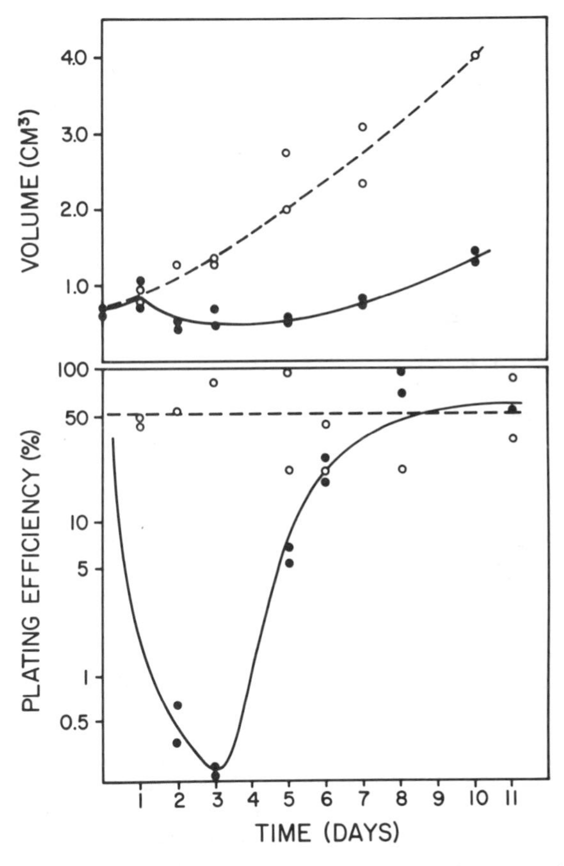

FIGURE 7. Response of the RIF-1 tumor to CPA. Calculated tumor volume (Top) and clonogenic cell fraction (Bottom) on different days following treatment (150 mg/kg, i.p.): ○, control; ● treated.

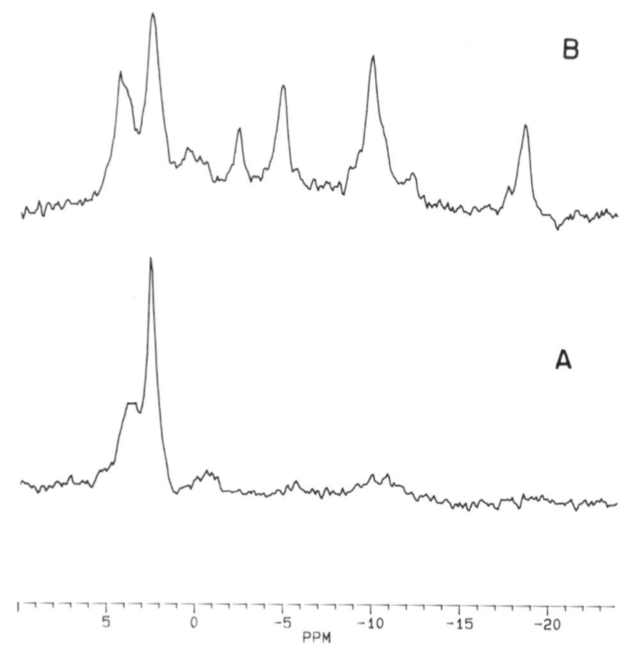

FIGURE 8. *In vivo* ^{31}P NMR spectra of two mice with matched RIF-1 tumors. Spectrum (A) 48 hr after treatment with CPA (200 mg/kg, i.p.); (B) control. Spectroscopic parameters as in METHODS. Peak assignments as in FIG. 1.

restriction of the region that can be observed, and cause some distortion of the quantitative information in the spectrum. We have been evaluating various techniques for application to tumors *in vivo.* Three examples are shown. In each case a solenoidal coil doubly tuned to ^{1}H and ^{31}P was used, together with a Faraday shield around the body of the animal.

The spectrum in FIGURE 9A was obtained using a $1\bar{3}3\bar{1}$ semiselective excitation pulse.[24] Several broad resonances can be seen, including lipid and metabolite contributions. The methyl resonances of lactate and alanine have been identified in this region in high resolution NMR analysis of tumor extracts,[11] and in spectra of excised RIF-1 tumor.[20] Other peaks are as yet unassigned. The spectrum in FIGURE 9B was obtained using a composite binomial–spin echo sequence, designed to excite the entire 1–2 ppm region. Although the echo time (62 msec) has been optimized for lactate, some residual contribution from lipid may still be present. When a particular resonance of interest has been identified, the DANTE pulse sequence[22] can be integrated into the composite pulse to increase selectivity. The spectrum shown in FIGURE 9C was obtained using the homonuclear editing technique described by Hetherington *et al.*[25] On alternate scans, a frequency selective DANTE pulse train[22] is applied during the $2\bar{6}6\bar{2}$ refocussing pulse of the Hahn spin echo sequence. The DANTE excitation is applied at the resonant frequency of the α proton of lactate, inverting the lactate methyl by *j*-coupling, without affecting other resonances. When alternate scans are subtracted, other resonances cancel and a lactate spectrum is obtained.

RIF-1 Tumor Cells in Vitro

Tumor heterogeneity, vascular and immune responses, and other factors complicate the study of tumors *in vivo*. Elucidation of the mechanisms responsible for changes observed in tumor spectra after treatment will require studies in simple, well-defined systems. We have implemented the system of Urgurbil *et al.*[26] for the study of cultured RIF-1 tumor cells. This system, in which cells are capable of replication, promises to be suitable for basic cell biological studies of chemotherapeutic response. Cells growing in monolayer culture on microcarrier beads are packed into a perfusion column that fits inside a standard 25 mm NMR tube and is continuously perfused with growth medium. Levels of ATP and PCr comparable to *in vivo* tumors can be maintained with

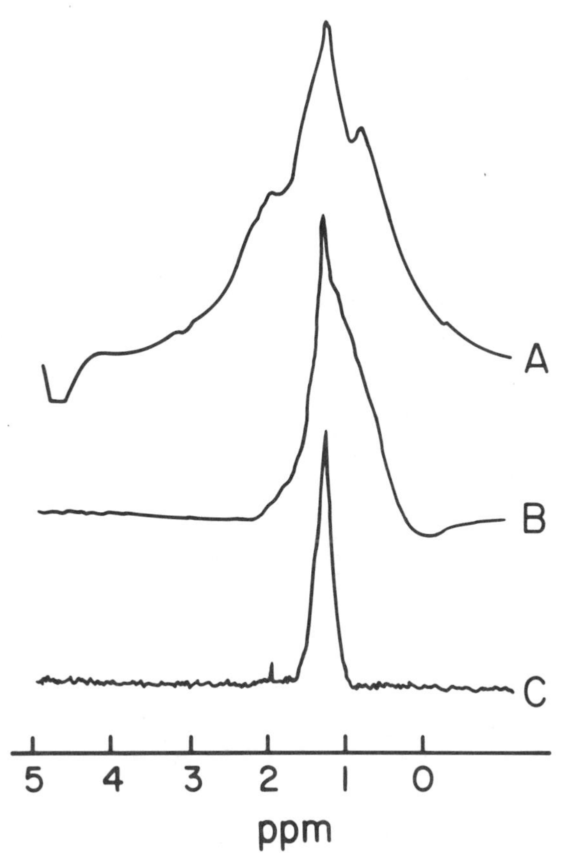

FIGURE 9. *In vivo* ^1H NMR spectrum of RIF-1 tumors. Spectra were obtained using a 1.5 cm solenoidal coil doubly tuned for ^1H and ^{31}P. (a) A 1331 pulse sequence was used, optimized for maximal excitation in the 1–2 ppm region. (b) A 1331-τ-2662-echo sequence, with an echo time of 62 msec was used. The peaks have not been assigned. The resonance position of the lactate methyl is 1.3 ppm. (c) The 1331-τ-2662-τ-DANTE-echo homonuclear editing sequence of Hetherington *et al.*[25] has been implemented to isolate lactate.

proper attention to oxygen supply and temperature (FIG. 10). In this spectrum, intracellular P_i is obscure by extracellular P_i, but it should be possible to distinguish these components using appropriate shift or relaxation reagents. A composite binomial-spin echo sequence has been used in FIGURE 12 to observe the lactate region of the ^1H spectrum during periods of normal flow and ischemia. Because a recirculating perfusion is used, lactate accumulates in the medium (FIG. 11 A). An increase in the lactic acid in the cell chamber during ischemia, and re-equilibration upon reflow are evident (FIG. 11, B and C). Such a system can be used to define precisely the effects of oxygen tension, substrate supply, drug delivery, and toxic waste washout on the tumor NMR spectrum.

CONCLUSIONS

The techniques necessary for the study of cancer chemotherapy by ^{31}P NMR are now well-established. The hard tasks remain: to determine the population variability of the NMR parameters and to select those most useful and reliable in predicting tumor characteristics and drug response; to examine carefully a large number of tumors and drugs, and to identify any common patterns of response; to determine to what extent toxicity to the host influences the changes observed in the tumor after chemotherapy; and to better understand the metabolism of treated and untreated tumors. *In vivo* ^1H spectroscopy of tumors is still in the developmental stage. Resonances in *in vivo* ^1H spectra need to be rigorously assigned. Water and lipid suppression compromise quantitative information; methods to avoid or account for these effects must be developed. Which metabolites visible in the ^1H spectrum are correlated with tumor response to therapy remains to be evaluated. The extension of spectroscopy techniques

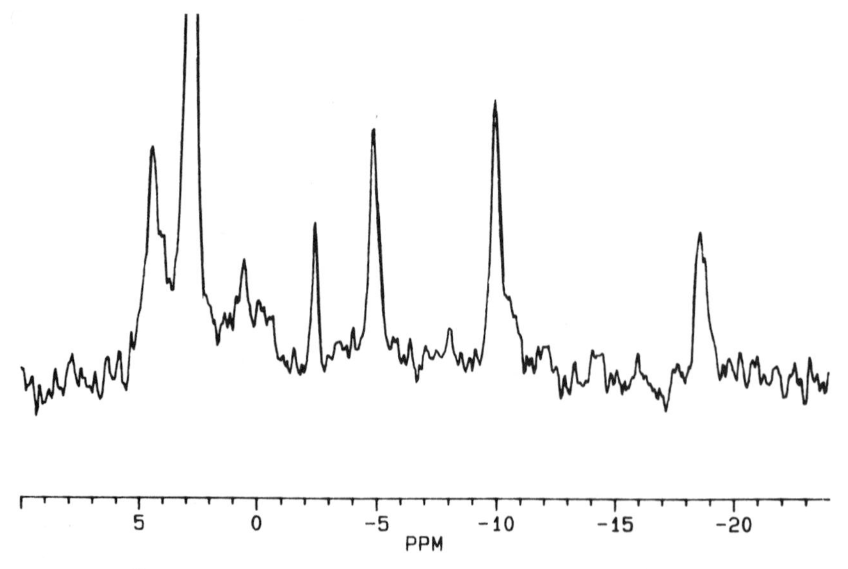

FIGURE 10. ^{31}P NMR spectrum of RIF-1 cells (approximately 5×10^8 in 10 ml) perfused *in vitro* as described in METHODS. 1,500 scans were accumulated, with a repetition time of 3 sec. The large P_i peak originates from medium P_i. Other assignments are as described in FIG. 1.

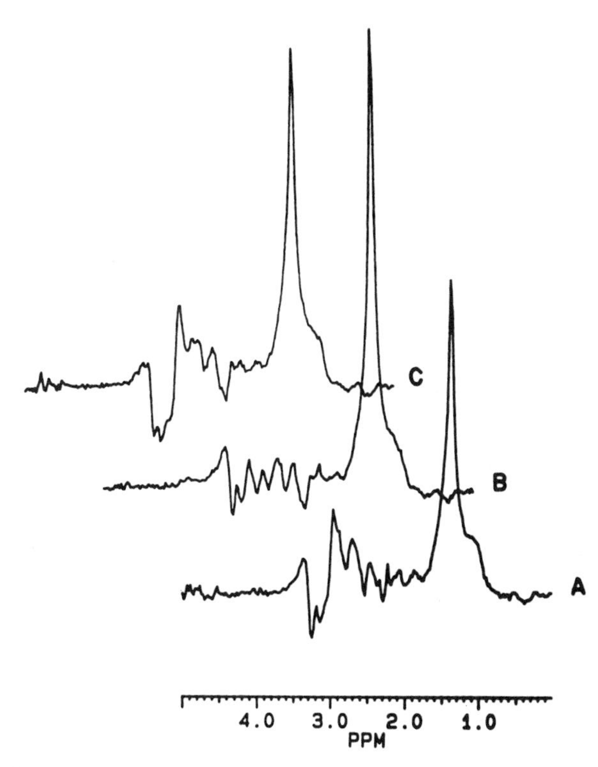

FIGURE 11. ^{1}H NMR spectra of RIF-1 cells shown in FIG. 10. Pulse sequence as in FIG. 9. Sixteen scans were accumulated. Spectra were obtained (A) during recirculating perfusion; (B) after perfusion had been interrupted for 5 min; (C) after flow had been reinitiated.

to the low magnetic field strengths available for human use is already underway, so that information from the laboratory can be translated rapidly to the clinical management of cancer therapy.

REFERENCES

1. TWENTYMAN, P. R., J. M. BROWN, J. W. GRAY, A. J. FRANKO, M. A. SCOLES & R. F. KALLMAN. 1980. J. Natl. Cancer Inst. **64:** 595–604.
2. NG, T. C., W. T. EVANOCHKO & J. D. GLICKSON. 1982. J. Magn. Reson. **49:**526–529.
3. EVELHOCH, J. L., S. A. SAPARETO, G. H. NUSSBAUM & J. J. H. ACKERMAN. 1986. Radiat. Res. **106:** 122–131.
4. JOHNSON, R., J. F. FOWLER & G. D. ZANELLI. 1976. Radiat. Biol. **118:** 697–703.
5. CULLEN, R. M. & H. C. WALKER. 1985. Int. J. Radiat. Biol. **48:** 761–771.
6. ROCKWELL, S. & R. LOOMIS. 1980. Radiat. Res. **81:** 292–302.
7. GADIAN, D. G., D. I. HOULT, G. K. RADDA, P. J. SEELEY, B. CHANCE & C. BARLOW. 1976. Proc. Natl. Acad. Sci. USA **73:** 4446–4448.
8. BAILEY, I. A., A.-M. L. SEYMOUR & G. K. RADDA. 1981. Biochim. Biophys. Acta **637:** 1–7.
9. MATTHEWS, P. M., G. K. RADDA & D. J. TAYLOR. 1981. Biochem. Soc. Trans. **9:** 236–237.

10. NG, T. C., W. T. EVANOCHKO, R. N. HIRAMOTO, V. K. GHANTA, M. B. LILLY, A. J. LAWSON, T. H. CORBETT, J. R. DURANT & J. D. GLICKSON. 1982. J. Magn. Reson. **49:** 271–286.
11. EVANOCHKO, W. T., T. T. SAKAI, T. C. NG, N. R. KRISHNA, H. D. KIM, R. B. ZEIDLER, V. K. GHANTA, R. W. BROCKMAN, L. M. SCHIFFER, P. G. BRAUNSCHWEIGER & J. D. GLICKSON. 1984. Biochim. Biophys. Acta **805:** 104–116.
12. EVANOCHKO, W. T., T. C. NG, J. D. GLICKSON, J. R. DURANT & T. H. CORBETT. 1982. Biochem. Biophys. Res. Commun. **109:** 1346–1352.
13. EVANOCHKO, W. T., T. C. NG, M. B. LILLY, A. J. LAWSON, T. H. CORBETT, J. R. DURANT & J. D. GLICKSON. 1983. Proc. Natl. Acad Sci. USA **80:** 334–338.
14. SCHIFFER, L. M., P. G. BRAUNSCHWEIGER, J. D. GLICKSON, W. T. EVANOCHKO & T. C. NG. 1985. Ann. N.Y. Acad. Sci. **459:** 270–277.
15. NARUSE, S., K. HIRAKAWA, T. HORIKAWA, C. TANAKA, T. HIGUCHI, S. UEDA, H. NISHIKAWA & H. WATARI. 1985. Cancer Res. **45:** 2429–2433.
16. NARUSE, S., T. HORIKAWA, C. TANAKA, T. HIGUCHI, S. UEDA, K. HIRAKAWA, H. NISHIKAWA & H. WATARI. 1985. Magn. Reson. Imaging **3:** 117–123.
17. BHUJWALLA, Z., R. J. MAXWELL, G. M. TOZER & J. R. GRIFFITHS. 1986. Soc. Magn. Reson. Med. (Abs.) **1:** 161–162.
18. SCHIFFER, L. M. & P. G. BRAUNSCHWEIGER. 1985. Proc. Am. Assoc. Cancer Res. **26:** 43.
19. LILLY, M. B., C. R. KATHOLI & T. C. NG. 1985. J. Natl. Cancer Inst. **75:** 885–890.
20. GLICKSON, J. D., W. T. EVANOCHKO, T. T. SAKAI & T. C. NG. 1987. *In* NMR Spectroscopy of Cells and Organisms. R. K. Gupta, Ed. **1:** 99–134. CRC Press. Boca Raton, FL.
21. GADIAN, D. G. 1987. Ann. N.Y. Acad. Sci. (This volume).
22. MORRIS, G. A. & R. FREEMAN. 1978. J. Magn. Reson. **29:** 433–462.
23. PLATEAU, P & M. GUERON. 1982. J. Amer. Chem. Soc **104:** 7310–7311.
24. HORE, P. J. 1983. J. Magn. Reson. **55:** 283–300.
25. HETHERINGTON, H. P., M. J. AVISON & R. G. SHULMAN. 1985. Proc. Natl. Acad. Sci. USA **85**: 3115–3118.
26. UGURBIL, K., D. L. GUERNSEY, T. R. BROWN, P. GLYNN, N. TOBKES & I. S. EDELMAN. 1981. Proc. Natl. Acad. Sci. USA **78:** 4843–4847.

DISCUSSION OF THE PAPER

M. W. WEINER: That was a nice review of the different kinds of changes that you see in tumors following treatment. I thought you were implying that what you call the "Naruse Effect," that is, a fall of high energy phosphates and a rise in inorganic phosphate following treatment, is something associated only with super treatment, which is beyond what is used clinically.

J. D. GLICKSON: I shouldn't have been if I was. It's going to depend on the relative rates of clearance and cell kill. That, and also damage to vasculature.

QUESTION: I'm confused a bit, in that you have a tumor and your NMR signal shows that the region that you are looking at is largely anaerobic and practically dead. You then do something that you call therapy and that makes the tumor look like a nice, live, rapidly growing tissue. Am I missing something?

GLICKSON: No, that's the paradox. That's exactly what Paul Bottomley pointed out at the cancer workshop, and I tried to explain various possible ways in which you could produce that anomalous effect.

C. DEUTSCH: Do you know anything about the pH regulation? The implication is that you can have tumors generating intracellular pH on the order of 6.8 in one case or 6.9, and then they're functioning at an intracellular pH of 7.5 or 7.4. I'm referring to that slide you showed of tumor pH vs. your P_i:PCr ratios. Do you have any insight into what might be changing? I mean you have to invoke a reversal of the pH gradient if you assume this is insight to *in vivo*, and that it's maintaining a totally different intracellular pH in a growing tumor vs. an arrested tumor. That might have some very interesting implications with respect to the regulatory mechanisms. The other question is—I might have misunderstood, but were you talking about the same RIF-1 tumor using two different therapies, in your case as I looked at your slides going by there didn't seem to be any shift in pH.

GLICKSON: In the RIF-1 tumor we have seen alkaline shifts in average tumor pH following either chemotherapy or radiation. Following hyperthermia in the same tumor acidosis is observed. Other tumor lines show very little pH change, either during growth or following treatment.

DEUTSCH: Is there any implication from the different therapies? One results in a very different tumor response in terms of pH regulation and the other shows none.

GRIFFITHS: If I could just answer that as regards the radiotherapy: actually there are some very strange pH changes due to radiotherapy in our hands, which I'd really rather not talk about yet, but basically the tumors go alkaline, some of them go very alkaline indeed. But we're just not too sure whether it's real or an artifact.

GLICKSON: Well, in regard to your first question, what you point out is very true. I mean it's perhaps the most intriguing effect that we've been seeing with this RIF-1 tumor and it's potentially of considerable clinical significance, Dr. Wehrle will be studying this problem I think, in some detail. We saw that there appears—we published and our collaborator presented it here at the New York Academy of Sciences—some data indicating that there is at least a correlation between the pH changes and apparent changes in the cell cycle of the tumor. It appears that when the tumor is going back into a mitotic phase you see the alkaline shifts.

QUESTION: I'd like to comment briefly in regard to the problem of interpreting the pH that you measure in tumors with phosphorous NMR as intracellular pH. There is a large inorganic phosphate peak in many of these spectra especially when the pH is low, it's a very large inorganic phosphate. There is no reason to assume that's intracellular. It could very well be heavily weighted by extracellular pH.

GRIFFITHS: That anesthetic effect you showed where the anesthetic caused quite a big change in the phosphorous spectrum, were they in tumors that were in a Faraday shield, the Faraday shield you showed before?

GLICKSON: I believe they were.

GRIFFITHS: We've got a similar effect, but we don't use Faraday shields routinely. We just use a surface coil on the tumor and rely on tuning the size of the coil to the size of the tumor to minimize contamination. We found the same effect as that, but only when we use a Faraday shield and anesthetic. If you use a Faraday shield just to get the tumor held in position so the animal can't wiggle around, the spectrum looks the same as you get with a surface coil. To give an anesthetic as well, we get the same effect as you.

GLICKSON: The issue that you're alluding to is that in using a Faraday shield (although ours are very lightly applied, we don't have a very small hole on them) in principle you might be pinching the tumor and blood supply off and that could produce some perturbation. But then why should they disappear with time? Except under reduced blood flow you may see a more dramatic effect than of the anesthetic, at least of the peripheral tissues. However, I should mention that back some time ago Mike Lilly was doing work on hyperthermia in Birmingham with us, and he saw effects of anesthetic without any Faraday shield.

Multinuclear NMR Study of the Metabolism of Drug-Sensitive and Drug-Resistant Human Breast Cancer Cells

JACK S. COHEN AND ROBBE C. LYON

Clinical Pharmacology Branch
National Cancer Institute
National Institutes of Health
Bethesda, Maryland 20892

INTRODUCTION

The development of multi-drug resistance[1] is a major problem in cancer chemotherapy and one of the main causes of death. Several differences between drug-sensitive and drug-resistant cell lines have been advanced to account for the phenomenon of multiple (or pleiotropic) drug resistance (PDR).[2–4] Recently, it has been argued that the mechanism of PDR arises from biochemical properties intrinsic to the genotype of cancer cells that have been transformed chemically.[5]

In order to study the biochemical basis of the phenomenon of PDR, we decided to compare the metabolism of drug-sensitive and drug-resistant human breast cancer cells using NMR methods. The drug-sensitive cells are termed wild type (WT) and the resistant cell line was selected for adriamycin resistance (Adr^R), and also exhibits PDR.[6] The levels of major intra-cellular phosphate metabolites, both in cells embedded in agarose gel threads and perfused, and in extracts, have been compared using ^{31}P NMR. The utilization of glucose and production of lactate have been monitored as a function of time using ^{13}C NMR with ^{13}C-labeled glucose. In both studies, major differences were observed between the WT and Adr^R cells. These results may have relevance to *in vivo* studies of tumors. In the treatment of human breast cancer, the levels of phosphate esters may be markers for the clinical diagnosis and monitoring of drug resistance.

MCF-7 human breast cancer cells were chosen for these studies because these cells are well characterized and have been shown to retain most of the endocrinological properties of normal breast tissue.[7] The energy metabolism of cancer cells is quite different from that of normal cells, in that normal cells depend on oxidative phosphorylation to synthesize ATP and cancer cells utilize glycolysis even in the presence of oxygen.[8] It is necessary for NMR studies to be able to perfuse a dense collection of cells trapped in a small volume. We developed a technique to accomplish this with cultured cells by embedding the cells in a fine gel thread.[9] Recently, we improved this technique by reducing the volume required for perfusion so that ^{13}C NMR studies could be carried out with small amounts of ^{13}C-enriched metabolites.[10] We have thus followed the conversion of glucose into lactate, i.e. the process of anaerobic glycolysis, and studied the effect of several drugs and metabolic effectors on this process.

METHODS

MCF-7 human breast cancer cells were grown in IMEM medium supplemented with 5% fetal calf serum (FCS) and penicillin-streptomycin (100 u/ml, 10 mg/l). Adriamycin-resistant cells were obtained by serial passage of the parental WT cells in stepwise-increasing concentrations of adriamycin until cells capable of growing in 10 μM are obtained.[6] These cells were 192-fold more resistant to adriamycin than the WT cells and exhibited high levels of cross-resistance to several other drugs. Prior to biochemical and NMR studies, the cells were grown in drug-free medium for at least 6 weeks; the resistant phenotype was stable when serially passed in drug-free medium for greater than 52 weeks.

Cells were prepared for perfusion by casting in an agarose gel thread (0.5 mm diameter) as previously described.[9] Approximately 3×10^8 cells were used to make ca.

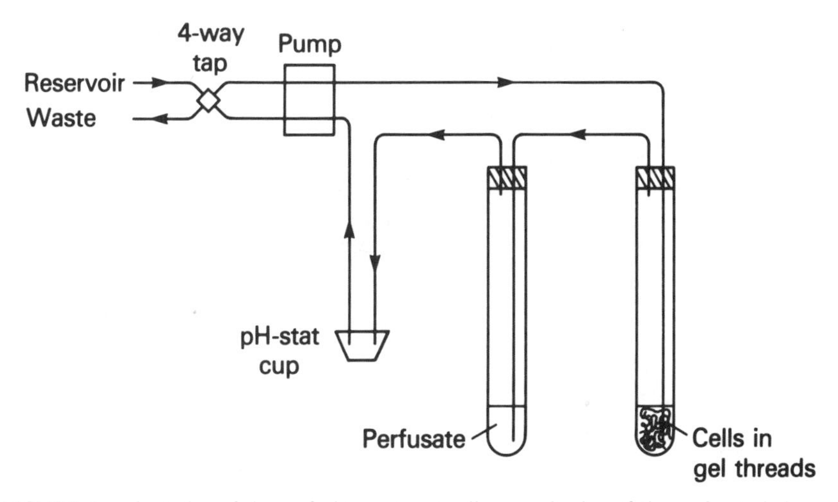

FIGURE 1. Adaptation of the perfusion system to allow monitoring of the perfusate independently of the cells.

2 ml of final gel thread. The gel-cell thread was perfused with RPMI-1640 medium (20 mM HEPES, pH 7.5) in a screw cap Wilmad NMR tube attached to an improved perfusion apparatus[10]; improvements include the capability to close the system from the reservoir, such that the volume of the perfusate in the closed system is ca. 20 ml; the pH can also be monitored and regulated continuously. In the experiments described here, oxygen was always bubbled, except where an inhibitor of oxidative phosphorylation (such as azide) was added.

^{31}P spectra were recorded at 162 MHz on a Varian XL-400 spectrometer. For quantitative spectra, a 40 sec recycle time was used to enable all components to relax completely. Cell extracts were prepared following cell harvest by treatment of cold pellets with 30% perchloric acid in D_2O, following the standard procedure.[10] ^{31}P spectra of extracts were run at 5°C using the deuterium lock. Peak assignments in the cell extracts were confirmed by addition of known phosphates. To quantitate these extracts

1.0 mM diphenylphosphate was added, since its resonance is well resolved from the other metabolites.

^{13}C spectra were recorded at 100 MHz with proton decoupling. [1-^{13}C]-D-glucose was purchased from Merck & Co. Considerations of the ratio of signal to noise (S/N) and compartmentalization of ^{13}C compounds led us to develop the two-tube version of the perfusion apparatus shown in FIGURE 1. The perfusate is monitored immediately after the cell compartment and only extracellular metabolites are observed. This allows much better magnetic homogeneity to be obtained, giving sufficient S/N with 5 mM ^{13}C-glucose and providing more consistent intensity data. Monitoring the tube

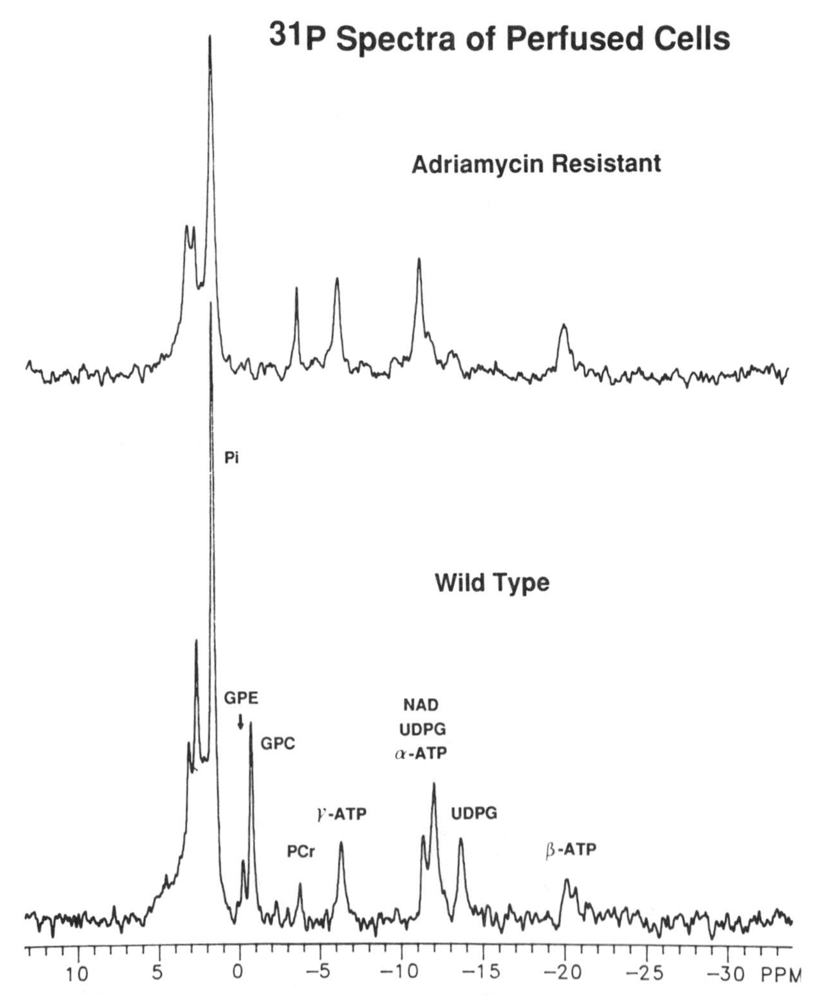

FIGURE 2. Quantitative ^{31}P MR spectra at 162 MHz of perfused MCF-7 WT and Adr^R cells. 200 scans were accumulated with a 40 sec repetition time at 22°C. Perfusion was with RPMI-1640 medium with HEPES buffer (20 mM) at pH 7.5. For abbreviations see TABLE 1.

TABLE 1. Concentration of Phosphate Metabolites in Intact Cells (Relative to β-ATP)[a]

Compound	Wild Type	Adr^R	WT/Adr^R
PME	5.4 (2.0)	2.8 (0.9)	1.9
GPE	0.6 (0.1)	0.08 (0.08)	8.0[d]
GPC	1.3 (0.5)	0.1 (0.06)	13[c]
PCr	0.2 (0.1)	0.6 (0.2)	0.3[c]
γ-ATP[b]	1.2 (0.2)	1.3 (0.09)	0.9
α-ATP, UDPG, NAD[b]	3.0 (0.5)	2.3 (0.4)	1.3
UDPG	1.2 (0.2)	0.5 (0.06)	2.4[d]
β-ATP	1.0	1.0	1.0

[a]Mean of four experiments. Values in parentheses are standard deviations. P_i is not reported since it is not intracellular. Abbreviations are: PME, phosphomonoester, including PE, phosphoethanolamine, and PC, phosphocholine; GPE, glycerophosphorylethanolamine; GPC, glycerophosphorylcholine; and PCr, phosphocreatine.

[b]These peaks could also contain small contributions from ADP and other nucleoside triphosphates.

[c]$p < 0.05$, statistical significance (t-test).

[d]$p < 0.01$.

containing the cells gives contributions from both intracellular and extracellular metabolites (this is not a problem with ^{31}P NMR studies since only P_i is extracellular). ^{13}C spectra were collected automatically at specified time intervals over a period of several hours.

RESULTS

Quantitative ^{31}P spectra of perfused WT MCF-7 and Adr^R cells (FIG. 2) were obtained from eight distinct harvests. The average concentration of each component relative to that of ATP is given in TABLE 1. The Adr^R cells exhibited elevated levels (three fold) of PCr, and depressed levels of GPE (eight fold), GPC (13 fold), and UDPG (two fold). Changes in external pH or temperature (22 and 37°C) caused no significant changes in the ^{31}P spectra. ^{31}P spectra of the cell extracts (FIG. 3) essentially confirmed these results (TABLE 2). As with the perfused cells, PCr was elevated in the Adr^R cell extracts. Creatine kinase was measured in sonicated extracts of cells and the results showed no significant difference between the two cell types (WT, 78 ± 13; Adr^R, 93 ± 4 IU/mg protein).

The signal of [1-^{13}C]-glucose was observed to decrease exponentially with time (FIG. 4). One other prominent ^{13}C signal observed was that of the catabolite [3-^{13}C]-lactate. This is the end-product of glycolysis, and was seen to increase in the perfusate concomitant with the decrease of the glucose signal intensity (FIG. 5). The observed maximum intensity of the lactate signal was ca. 60% that of the original glucose signal; by a comparison of a standard solution, this was found to correspond to ca. 3.5 mM lactate.

The basal rates of glucose utilization and lactate production were about three times faster for the Adr^R cells compared to WT. The rates of glucose utilization and lactate production for WT and Adr^R cells were compared in the presence of various metabolic effectors. Sodium azide (20 mM) had no effect on either rate for both cell lines (FIGS. 6

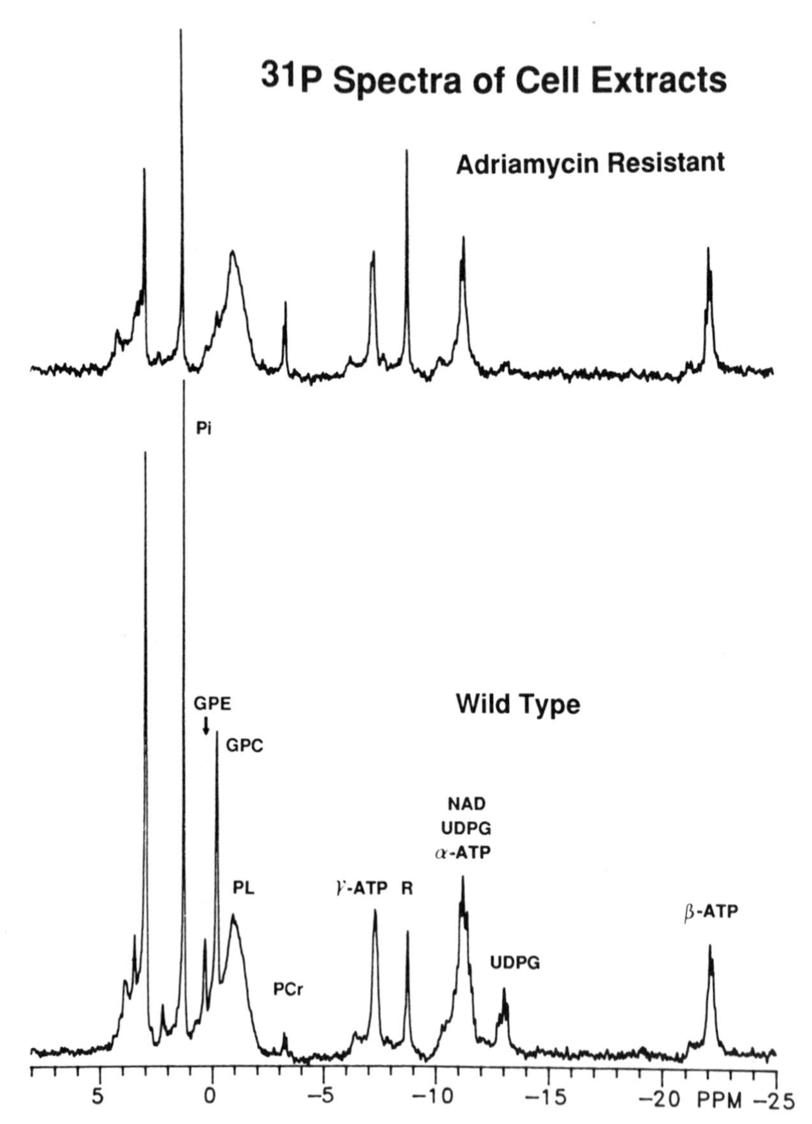

FIGURE 3. Quantitative ^{31}P spectra at 162 MHz of perchloric acid extracts of MCF-7 cells: (A) WT cells (1022 scans); (B) Adr^R cells (1500 scans). Solutions in D_2O at pH 7.0 with 40 sec recycle time at 5°. Peak R is the reference standard diphenyl phosphate (1 mM) added to the extracts.

TABLE 2. Absolute Concentration of Phosphate Metabolites in Cellular Extracts (μmoles per 10^8 cells)[a]

Compound	Wild Type	Adr^R	WT/Adr^R
PME	1.5 (1.5)	0.7 (0.7)	2.1
PE	3.0 (2.9)	1.3 (1.2)	2.3
PC	5.0 (2.4)	2.9 (2.4)	1.9
GPE	1.0 (0.05)	0.2 (0.1)	5.0
GPC	2.8 (1.1)	0.6 (0.5)	4.7
PL	5.9 (2.1)	6.6 (1.9)	0.9
PCr	0.1 (0.09)	0.4 (0.3)	0.2
γ-ATP[b]	2.4 (0.2)	2.5 (0.9)	1.0
α-ATP, UDPG, NAD[b]	4.5 (1.1)	3.3 (1.1)	1.4
UDPG	1.0 (0.8)	0.1 (0.1)	10
β-ATP	1.7 (0.3)	2.2 (0.9)	0.8

[a]Mean of three experiments. Values in parentheses are standard deviations. Abbreviations as in TABLE 1, except PL, phospholipid.
[b]See TABLE 1.

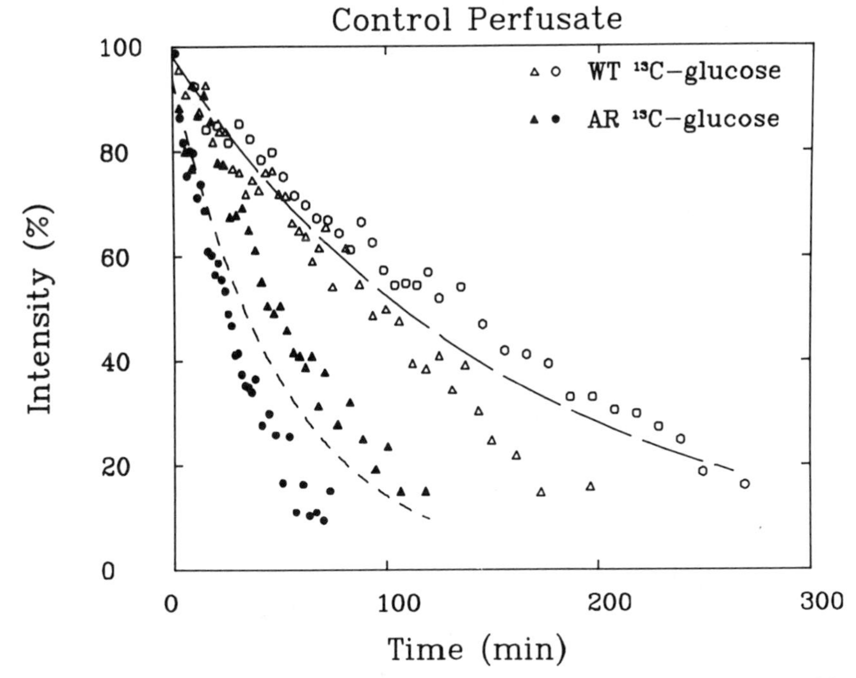

FIGURE 4. ^{13}C NMR signal intensity of $^{13}C_1$-glucose (5 mM) as a function of time corrected for 5×10^8 cells, both WT and Adr^R as indicated.

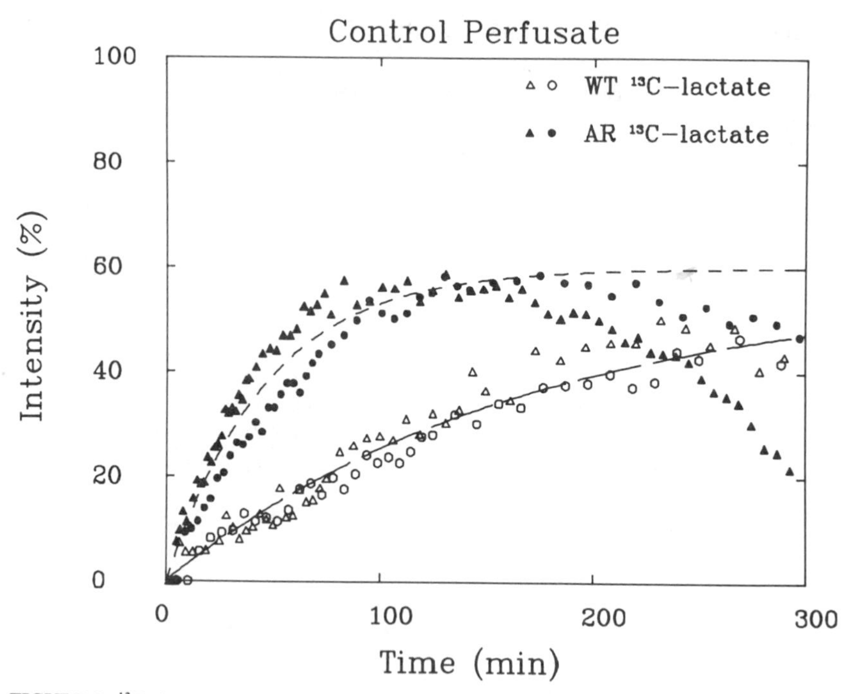

FIGURE 5. $^{13}C_3$-lactate signal intensity as a function of time corrected for 5×10^8 cells, both WT and Adr^R as indicated.

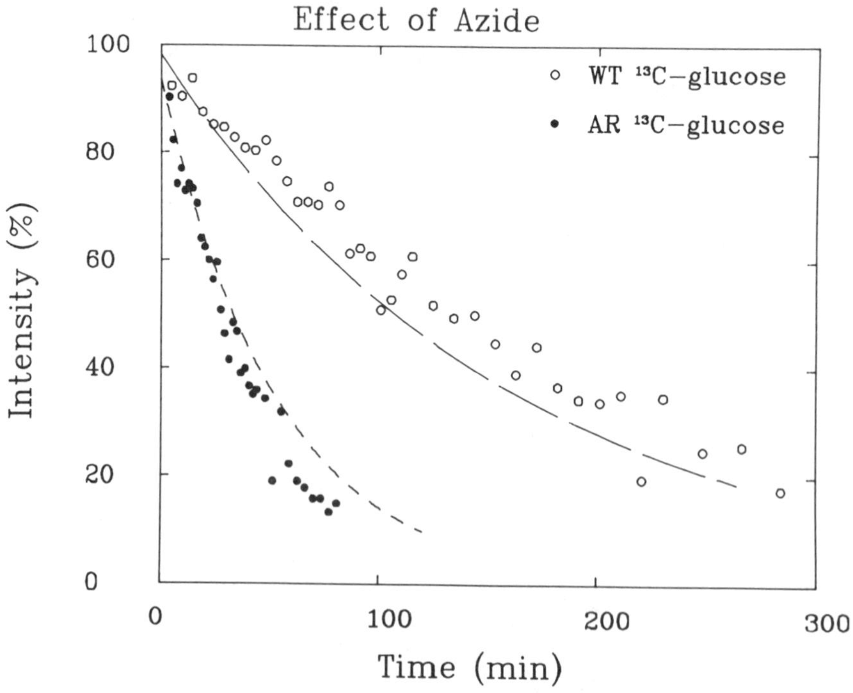

FIGURE 6. ^{13}C-glucose utilization by WT and Adr^R cells as above with the addition of sodium azide (20 mM) to the perfusate.

and 7). The primary effect of azide was the inhibition of lactate re-utilization (FIG. 7). The addition of adriamycin (200 nM) had little effect on the rates of glucose utilization for either cell type (FIG. 8). However, adriamycin reduced the rate of lactate production to 50% in WT cells and 70% in Adr^R cells (FIG. 9).

DISCUSSION

The results described here show significant differences between the levels of intracellular phosphates of the drug-sensitive and drug-resistant cells.[11] These differ-

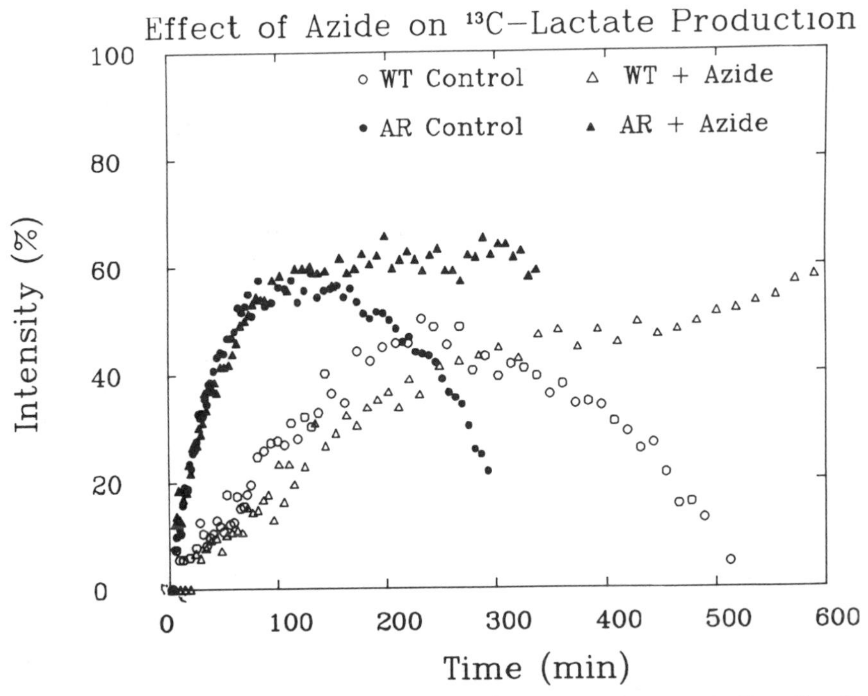

FIGURE 7. ^{13}C-lactate production by WT and Adr^R cells as above, controls (from FIG. 5) compared to the addition of sodium azide (20 mM) to the perfusate.

ences reflect alterations in the control of phosphate metabolism of the cell. Phosphodiester peaks (GPE and GPC) relate to phospholipid (PL) breakdown, UDPG relates to storage mechanisms (FIG. 10), and PCr to bioenergetics. We believe this is the first evidence for such differences related to the phenomenon of PDR. That there are such differences should not be surprising given that mitochondrial hexokinase is responsible for much of the glycolytic activity of tumor cells and one major effect of adriamycin is to produce hydroxyl radicals,[12] which are known to affect the mitochondrial membrane. In the case of heart cells, the effects of adriamycin on the mitochondria lead to cardiotoxicity.[12]

The differences reported here were obtained with MCF-7 cells grown under identical conditions, including the presence of estrogen and insulin. We have carried out experiments with cells grown with charcoal-stripped FCS to ascertain the effects of steroids on the intracellular phosphate components. At present these results are too preliminary to draw any clear conclusions.

The results of the experiments on the uptake of glucose and the production of lactate with the perfused cells showed that the basal rates of glycolysis in the two cell lines are quite different. Our observations are supported by results obtained by enzymatic assay of lactate production in extracts of MCF-7 cells.[13] They also found the glycolysis rate to be three times faster for Adr^R than for WT cells. Notwithstanding the difference in glycolysis rates the ATP levels in the two cell lines were the same.

The Embden-Meyerhoff-Parnas pathway, as well as two subsidiary pathways utilizing glucose, namely the hexose-monophosphate shunt to ribose derivatives and the glucuronyl synthesis pathways, are shown in FIGURE 10. There are several differences in metabolic control known to exist between normal and cancer cells,[8] which lead to increased glycolytic capability of the latter under aerobic conditions. Notably, the presence of a mitochondrial hexokinase leads to increased glucose utilization in cancer cells. Together with other effects, this leads to increased activity of phosphofructokinase, which is the key control point in glycolysis.

Various metabolic effectors were added to detect differences in metabolic regulation between the drug sensitive and resistant cells. Azide, a known inhibitor of respiration via cytochrome oxidase, clearly blocked the metabolism of pyruvate as evidenced by the build-up of lactate. The rates of glycolysis in both cell lines were

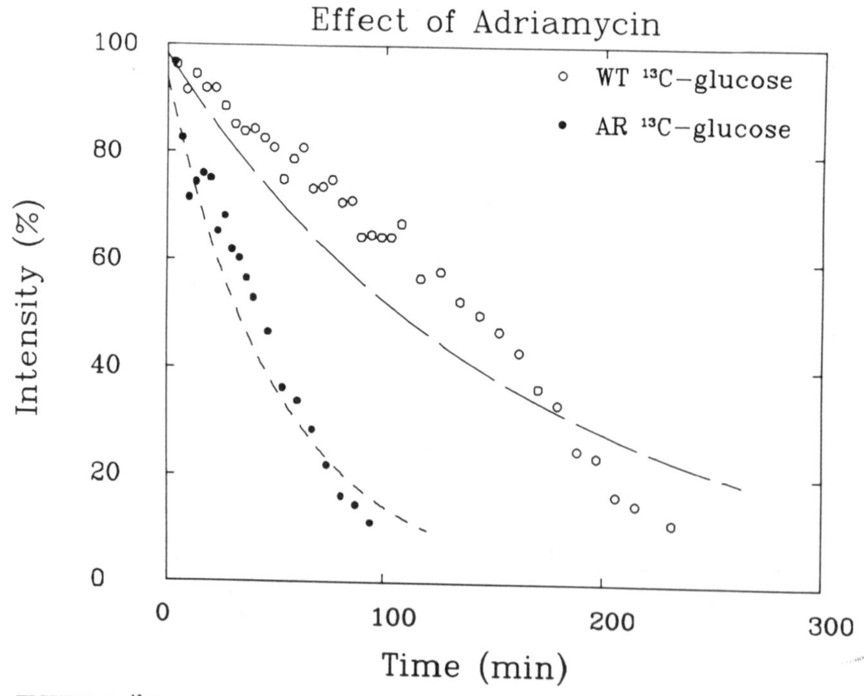

FIGURE 8. ^{13}C-glucose utilization by WT and Adr^R cells as above with the addition of adriamycin (200 nM) to the perfusate.

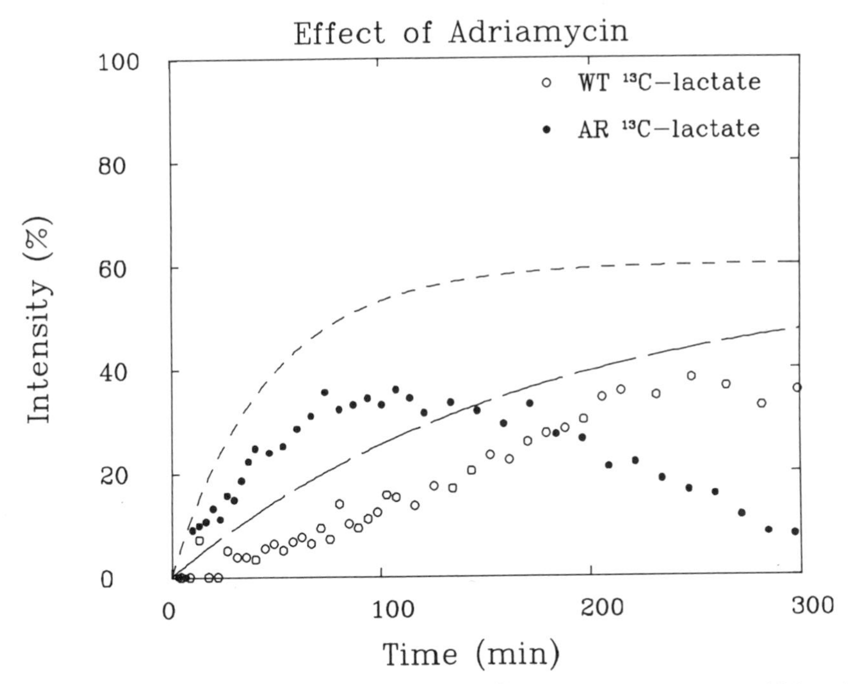

FIGURE 9. ^{13}C-lactate production by WT and Adr^{R} cells as above with the addition of adriamycin (200 nM) to the perfusate.

invariant to the reduction of metabolites provided by the citric acid cycle and oxidative phosphorylation. Adriamycin reduced the rate of lactate production without affecting the rate of glucose utilization. Possibly, adriamycin has enhanced the rate of pyruvate utilization by stimulating oxidative phosphorylation. Another possibility is that more glucose is used by the hexose monophosphate shunt (FIG. 10). This would result in the loss of ^{13}C label from the C-1 position of glucose as CO_2. This seems likely since the shunt is the primary source of NADPH, which is required for metabolic reduction of adriamycin in these cells. We are testing this possibility using $^{13}C_6$-labeled glucose.

These results of controlled *in vitro* cell studies are intended to be both a guide and a feedback mechanism for *in vivo* studies of tumors. We have built an NMR probe specifically designed for the observation of subcutaneous tumors in small rodents. Previous inconsistencies in the results obtained (particularly for PCr) with a variety of tumors under different experimental conditions[14] indicate how important it is to have a controlled *in vitro* system in which the results of any studies on solid tumors can be evaluated and compared. Ultimately, we hope that these NMR spectroscopic *in vitro* and *in vivo* studies in animals will have direct relevance to clinical results on humans.

ACKNOWLEDGMENT

We thank Kenneth Cowan for providing the cell lines, Patrick Faustino for technical assistance, and Charles E. Myers for helpful discussions.

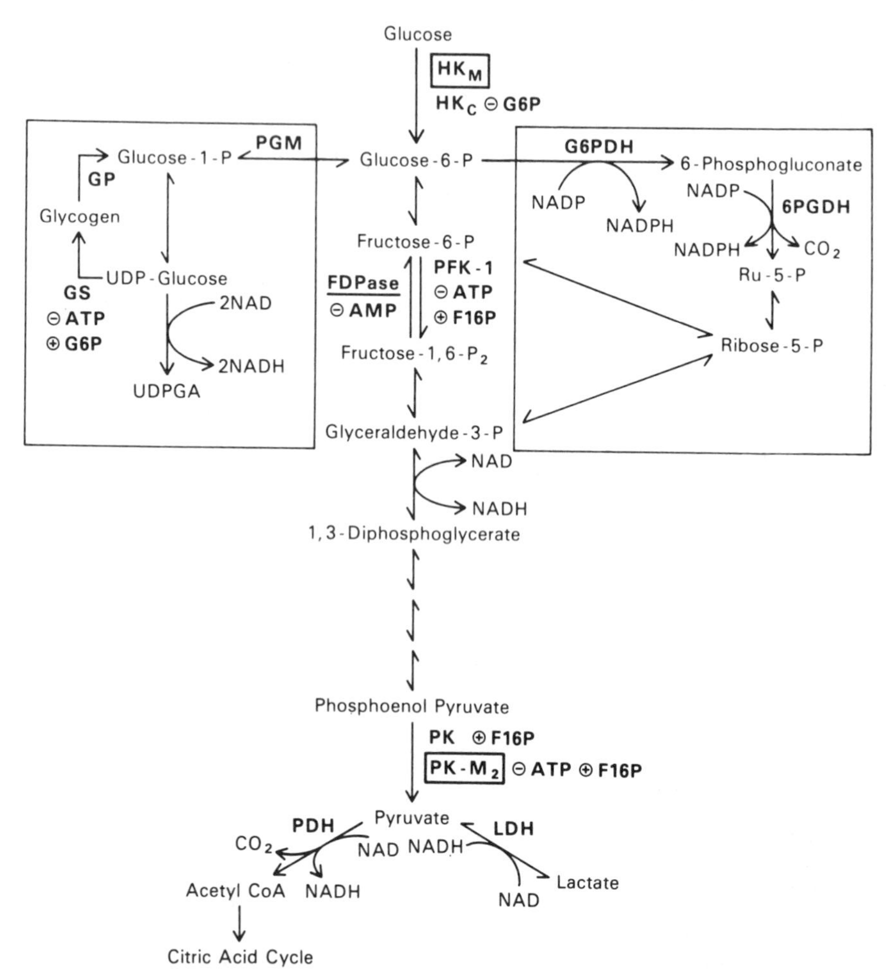

FIGURE 10. Glucose catabolism in mammalian cells featuring the Embden-Meyerhoff-Parnas pathway (center), with the hexose-monophosphate shunt (right box), and glucuronyl/glycogen synthesis (left box). In *normal cells* under "aerobic conditions," pyruvate production is dependent on the acetyl-CoA consumption and mitochondrial production of ATP, which inhibit phosphofructokinase (PFK-1). The subsequent accumulation of G6P blocks its own synthesis by inhibiting the cytoplasmic hexokinase (HKc). Under "anaerobic conditions," respiration and mitochondrial ATP production by oxidation of acetyl-CoA is blocked. The reduction in ATP levels relieves the inhibition of PFK-1. With the inhibition of fructose-1,6 biphosphatase (FDPase) by AMP, the levels of F16P increase and stimulate PFK-1 and pyruvate kinase (PK). The utilization of G6P relieves the inhibition of HKc, allowing glycolysis to proceed at a maximum rate. Pyruvate is converted to lactate in order to re-oxidize NADH formed during the oxidation of G3P. In *tumor cells* ATP inhibits PK-M_2 (tumor type isoenzyme) resulting in an accumulation of F16P until it overcomes the ATP inhibition and stimulates the activity of PK-M_2 and PFK-1. With the high activity of HK_M (mitochondrial bound hexokinase specific for tumor cells, not inhibited by G6P), PFK-1 and PK-M_2, and the de-inhibition of these enzymes, tumor cells maintain a high glycolytic rate under aerobic conditions.

REFERENCES

1. CURT, G. A., N. J. CLENDENNIN & B. A. CHABNER. 1984. Drug resistance in cancer. Cancer Treatment Rep. **68:** 87–99.
2. KESSEL, D. & H. B. BOSMANN. 1970. On the characteristics of actinomycin resistance in L5178Y cell. Cancer Res. **30:** 2695–2700.
3. LING, V. & L. H. THOMPSON. 1973. Reduced permeability of CHO cells as a mechanism of resistance to colchicine. J. Cell Physiol. **83:** 103–106.
4. BIEDLER, J. L. & R. H. F. PETERSON. 1981. Altered plasma membrane glycoconjugates of Chinese hamster ovary cells with acquired resistance to actinomycin D, daunorubicin, and vincristine. *In* Molecular Actions and Targets for Chemotherapeutic Agents. A. C. Sartorelli, J. S. Lazo & J. R. Bertino, Eds.: 453–464. Academic Press. New York.
5. MEYERS, C., K. COWAN, B. SINHA & B. CHABNER. 1987. The phenomenon of pleiotropic drug resistance. *In* Important Advances in Oncology 1987. V. T. DeVita, S. Hellman & S. A. Rosenberg, Eds. J. B. Lippincott Co. Philadelphia.
6. BATIST, G., A. TULPULE, B. K. SINHA, A. KATKI, C. E. MEYERS & K. H. COWAN. 1987. Overexpression of a novel anionic glutathione transferase in multidrug resistant human breast cancer cells. J. Biol. Chem. (In press.)
7. LIPMANN, M. 1976. Hormone-responsive human breast cancer in continuous tissue culture, in breast cancer. *In* Trends in Research and Treatment. J. C. Henson, W. H. Mattheim & M. Rozencweig, Eds.: 111–139. Raven Press. New York.
8. EIGENBRODT, E., P. FISTER & M. REINACHER. 1985. New perspectives on carbohydrate metabolism in tumor cells. *In* Regulation of Carbohydrate Metabolism. R. Beitner, Ed. **2:** 141–179. CRC Press. Boca Raton, FL.
9. FOXALL, D. L., J. S. COHEN & J. B. MITCHELL. 1984. Continuous perfusion of mammalian cells embedded in agarose gel threads. Exptl. Cell Res. **154:** 521–529.
10. LYON, R. C., P. J. FAUSTINO & J. S. COHEN. 1986. ^{13}C NMR studies of the metabolism of ^{13}C-labelled substrates by perfused mammalian cells. Magn. Reson. Med. **3:** 663–672.
11. COHEN, J. S., R. C. LYON, C. CHEN, P. J. FAUSTINO, G. BATIST, M. SHOEMAKER, E. RUBALCABA & K. H. COWAN. 1986. Differences in phosphate metabolite levels in drug-sensitive and resistant cell lines determined by ^{31}P magnetic resonance spectroscopy. Cancer Res. **46:** 4087–4090.
12. MYERS, C. E. 1982. Anthracyclines. *In* Pharmacologic Principles of Cancer Treatment. B. Chabner, Ed.: 416–434. Saunders. Philadelphia.
13. MUINDI, R. J. F., M. SHOEMAKER & K. C. COWAN. 1986. Comparative study of metabolism of sensitive and resistant MCF-7 human mammary adenocarcinoma cells. Proc. Am. Assoc. Cancer Res. **27:** 272.
14. EVANOCHKO, W. T., T. C. NG & J. D. GLICKSON. 1984. Application of *in vivo* NMR spectroscopy to cancer. Magn. Reson. Med. **1:** 508–534.

DISCUSSION OF THE PAPER

R. G. SHULMAN: It seems possible that the adriamycin-resistant cells have some place down the pathway, beyond where you're looking, something which is turning over very much faster, and therefore is reducing the concentration of the glycolytic intermediates, and also turning over the pentose shunt. Is that you're understanding of it?

J. S. COHEN: Well, that's possible but there's a lot known about these cells which I haven't had time to go into, which is not my work. Let me just review for you briefly a couple of those points. It's been found that the resistant cell line has enhanced levels of various conjugating and detoxifying enzymes and in fact, that seems to be the main mechanism of adaptation. So, for example, these resistant cells have very high levels of glutathione-S transferase activity. Detoxification requires reducing equivalents derived from enhanced pentose phosphate shunt activity as a source of NADPH. It's clear that the genetic set up of the cell is now changed. But the energy has to come from somewhere, so the rate of glycolysis is enhanced.

SHULMAN: Yes, but it's further down the line that's creating the demand. I said as my analogy on the work we did on yeast cells, where you support the yeast cells in the starvation medium so there's no place for the energy to go and no real demand for it, you build up high concentrations of these same intermediates. But if you suspend them in growth medium where there's use for the energy, whoosh, the intermediates disappear.

J. R. GRIFFITHS: Could I just interject a question? What do you think the tumors are actually detoxicating when you see this enhanced pentose phosphate shunt activity?

COHEN: They're reducing the drug, adriamycin. One of the points I should have made is that the adriamycin destroys mitochondrial membranes because it's a radical producing . . .

GRIFFITHS: So you've got enough adriamycin in there that it would actually be . . .

COHEN: Oh yes, 500 nanomolar is in the physiologically relevant range. One of the side effects of this drug is that it destroys the heart wall, the heart muscle cells, because it basically destroys the mitochondria. There's almost no oxidative phosphorylation going on in these cells and particularly in the adriamycin-resistant cells.

C. DEUTSCH: There are a number of agents that are putative calcium-channel blockers that have been used in KB cells and CEM cells to reverse multiple drug resistance on specifically adriamycin. Do you see any alteration in the activity of the shunt in the presence of these channel blockers that reverse drug resistance?

COHEN: We haven't looked at that but it's a very good point.

Phosphorylated Metabolites in Tumors, Tissues, and Cell Lines[a]

T. R. BROWN, R. A. GRAHAM, B. S. SZWERGOLD,
W. J. THOMA, AND R. A. MEYER[b]

NMR Department
Fox Chase Cancer Center
Philadelphia, Pennsylvania 19111

[b]*Michigan State University*
East Lansing, Michigan 48824

INTRODUCTION

Several recent studies have attempted to associate changes following chemotherapy, radiotherapy, or hyperthermia in the ^{31}P NMR spectrum of tumors with the response of the tumor to treament.[1–4] Such an ability to monitor metabolic changes in tumors during and following therapy would be quite useful, particularly if it could be done non-invasively by NMR. These initial studies, while encouraging in terms of the signal-to-noise ratio attainable in human and *in vivo* animal studies, have been difficult to interpret in terms of specific metabolic changes. Animal model experiments have shown that the observable phosphorous metabolites in tumors do change following treatment.[4–6] However, the causes of the observed changes have not always been clear, partly due to the complex nature of the whole animal response and partly due to the difficulties associated with localizing the NMR signal to a precise region in the animal. Similar observations on humans have had analogous difficulties.

The recent increase in clinical use of NMR imaging techniques has also led to attempts to correlate the apparent relaxation times, measured from proton images, with various pathological states, particularly in tumors. The relaxation properties of the water protons, the contrast determinant in NMR images, have not, in general, led to unequivocal pathological determinations of various tumors.[7] Here, the problem is not localization but rather the different pathological states that are associated with similar relaxation times. A tumor model that could be used to "calibrate" changes in relaxation times with specific physiological states, such as edema, anoxia, and ischemia, would be quite useful in helping to determine the biological determinants of T_1 and T_2.

From these examples it seems clear that the problems of interpreting *in vivo* studies can be substantial. They range from the complexities of the full range of the responses of the host to the technical problems of localizing the observed signals to the specific tissue of interest. To alleviate these difficulties, we have adapted an arterially perfused tumor preparation developed by Sauer[8] for study by NMR. The preparation uses Morris hepatomas implanted in the inguinal flank of the rat. We were led to this model because of a desire to manipulate various physiological parameters such as flow, oxygen, pH, substrate, temperature, etc. Many of these are difficult or impossible to vary in a whole animal, but can be relatively easily controlled in an isolated perfusion,

[a]Supported by National Institutes of Health grant CA 41078-01.

as has been previously demonstrated in NMR studies with isolated, perfused organs.[9-11] With this preparation, we expect to distinguish among the various effects of different physiological variables, e.g. ischemia, anoxia, therapeutic interventions, etc., on both the ^{31}P spectrum and the proton relaxation times. In addition to the advantages associated with physiological control of this preparation, the NMR signal can only originate from the tumor and high NMR sensitivity is attained.

An early discovery of these studies, which reflects the high NMR sensitivity possible in these isolated tumor perfusions, has been the direct observation of *myo*-inositol, 1,2-(cyclic) phosphate (cPIns), a five-membered cyclic phosphate compound, in an intact Morris 7777 hepatoma at 50–100 micromolar concentrations.[12] This discovery has led to the observation of not only this compound but several other five-membered cyclic phosphate compounds in extracts of other tumors, tissues, and cell lines. In addition to these new cyclic compounds, we have observed and identified three phosphodiesters (PDE) not previously observed by NMR: glycerolphosphoinositol (GroPIns), glycerophosphoglycerol (GroPGro), and bis(glycerophospho)glycerol [b(GroP)Gro]. As has been observed previously,[3,13-18] we also find relatively high levels of phosphocholine and/or phosphoethanolamine in the phosphomonoester region of the ^{31}P spectra in many of the transformed cells.

The full implications of these observations are still unclear. However, the presence of these compounds, many of which are associated with phospholipid turnover in general and phospholipase C hydrolysis in particular, suggests that the tissues and cell lines in which they are observed have high rates of membrane turnover, which may serve to stimulate cellular proliferation.

MATERIALS AND METHODS

NMR

^{31}P NMR spectra were observed at 162 MHz with a Bruker AM400 spectrometer using either a probe designed specially to observe the perfused tumor or the standard 10 mm ^{31}P probe with a Wilmad 1.5 microcell to observe extracts. The coil on the tumor probe consisted of a single turn solenoid with a typical 90° pulse length of 25 μsec with the tumor in place. Chemical shifts are referenced to an internal standard of glycerophosphocholine (GroPCho) at 0.49 ppm. The details of the spectral acquisitions are given in the figure legends.

Extracts

Extracts were typically prepared from up to 5 grams of tissue by the method of Bligh and Dyer[19] as detailed in our previous paper[12] or by a high salt modification of that protocol.[20]

The tissue was clamp frozen between brass plates cooled with liquid N_2 and stored at -80°C until extraction. The frozen material was weighed and then ground in a mortar and pestle under liquid N_2. The powder thus obtained was then added to polyallomar centrifuge bottle containing 30 ml of 0.5 M NH_4HCO_3, 20 mM EDTA, and 84 ml of $CHCl_3$/MeOH (2:5), which formed a monophasic mixture that was then homogenized with a Tissumizer (Tekmar) and then allowed to stand 30 min on ice. After this time, the phases were broken by the addition of 24 ml of ice-cold $CHCl_3$ and 13 ml of ice-cold 2 M NH_4HCO_3. After the mixture was centrifuged at 10,000 *g* for 20 min, the aqueous supernatant was saved, mixed with 30 ml of H_2O, and centrifuged again to remove the residual chloroform and precipitate. The supernatant was immediately passed through a 5 × 22 cm column of Sephadex G25 equilibrated to H_2O. The solvent peak from this column was lyophilized and the resulting powder taken up in 10 ml of H_2O and immediately applied to a 5 × 40 cm column of Sephadex G-10 again equilibrated to H_2O. The fractions containing the metabolites, which emerged just ahead of the solvent peak, were collected and lyophilized. The metabolites were taken up in 1.5 ml of D_2O, 200 μl of Chelex (previously equilibrated to pH 7) was added and the extract was adjusted to pH 7.5 (from about 8.5) with HCl. The Chelex was allowed to settle and the supernatant stored at −80°C. Later extracts included the RNase inhibitor aurintricarboxylic acid (ATA) (10 μM).[21,22]

Extracts of cultured cells grown in tissue culture flasks or on Biosilon microcarrier beads were prepared by appropriate modifications of procedures detailed above.

Chemicals

The 2′,3′ cyclic nucleotides, fructose 1,2-(cyclic) 6-bisphosphate were purchased from Sigma. Other cyclic phosphate compounds were synthesized by one of three methods. The great majority were obtained by a mild alkaline hydrolysis of the appropriate nucleotide-diphosphate precursors, for example, glucose 1,2-(cyclic) phosphate was obtained by incubating uridine-5′-diphosphoglucose at pH 13.0.[23] Glycerol 1,2-(cyclic) phosphate was obtained by methanolysis of phosphatidylcholine[24] while *myo*-inositol 1,2-(cyclic) phosphate was synthesized from *myo*-inositol-2 phosphate by a dicyclohexylcarbodiimide catalyzed cyclization of inositol-2 phosphate.[23,25] No attempt was made to completely purify the synthetic products.

The choline, ethanolamine, serine, and inositol glycerophosphodiesters were purchased from Sigma; others were obtained from the appropriate phospholipid precursors by mild alkaline hydrolysis.[26]

Tumors

Morris (7777 and 5123C) hepatomas (obtained from EG&G Mason Tumor Bank, Worchester, MA) were propagated in a Buffalo rat strain purchased from Harlan/Sprague Dawley. The tumors were serially transplanted after they had grown to 1–3 cm in size, typically in two to four weeks for the 7777 and four to eight weeks for the 5123C. The transplantation and perfusion procedures, which are based on the procedures of Sauer,[8] have been described previously.[12] Approximately 50% of the implants result in tumors suitable for perfusion. The tumors were perfused when they reached from 1 to 2 cm in diameter (2–3 weeks for the 7777, 4–6 weeks for the 5123C).

Extracts of two murine tumors, the EMT-6 mammary carcinoma and the I-347 T-cell lymphoma, were also made. The tumors were harvested about three weeks after implantation.

Tissue Culture

Anchorage-dependent cells were grown either in standard tissue culture flasks or on Biosilon microcarrier beads in agarose-coated tissue culture flasks to prevent the attachment of cells to the surface of the flask. The 3T3, L929 (mouse fibroblast cell lines) and primary rat embryo fibroblasts were grown in a conventional DMEM medium while the MS7 cells were grown in a HAT medium.[27] Typically, 0.5 to 0.75 ml of cells were harvested for extraction.

RESULTS

The upper spectrum in FIGURE 1 shows the ^{31}P NMR spectrum of a Morris (7777) hepatoma obtained in 12 min. The peaks are identified in the figure caption. In addition to the expected peaks in a ^{31}P NMR spectrum (ATP, P_i, phosphomonoesters, and phosphodiesters), there is a new resonance at 16.5 ppm. The initial perchloric acid extractions of the tumor did not isolate the compound responsible for this resonance. Therefore, we investigated neutral extraction procedures based on modified versions of the lipid extraction procedure of Bligh and Dyer.[19] Using these procedures, we were able to extract the compound reliably and reproducibly as shown in the lower spectrum in FIGURE 1. The resonance at 16.55 ppm did not shift as the extract was titrated from pH 4 to 10. At pH 2, it disappeared from the spectrum within two hours. These properties, together with its chemical shift suggested that it was a five-membered cyclic phosphate. Examination of the chemical shifts of a number of compounds in this class (TABLE 1) indicated that *myo*-inositol 1,2-(cyclic) phosphate (cPIns) was the most likely candidate. The proton-coupled ^{31}P NMR spectra of a Morris (7777) hepatoma extract before and after the addition of a small quantity of cPIns, synthesized as described in METHODS, are shown in FIGURE 2a. As can be seen, there is no sign of broadening or splitting of the doublet, confirming the assignment.

Using similar extraction procedures, we have surveyed several tumors, tissue culture lines, and normal tissues to determine whether this or similar compounds could be observed. Thus far, we have observed more than six other resonances in the five-membered cyclic phosphate region (10–25 ppm) in a variety of extracts. Two have been positively identified in a fetal mouse extract to be glucose 1,2-(cyclic) phosphate (cPGlu) and galactose 1,2-(cyclic) phosphate (cPGal). The quality of this identification is shown in the proton-coupled ^{31}P spectra in FIGURE 2b, which compare the extract spectra before and after the addition of the synthesized compound. These two resonances have been reported previously in spectra of extracts from several tissues but were not identified.[28,29] In addition to the cPGlu and cPGal at 11.3 and 10.5 ppm, we have observed resonances at 11.4, 19.1, 19.15, and several near 20.5 (FIG. 3). Comparison with TABLE 1 immediately suggests likely candidates for most of these peaks, however, the identifications cannot be regarded as confirmed until the appropriate extracts have been spiked.

As discussed at greater length below, any extraction procedure has the possibility of introducing artifacts. We have identified one such in our initial preparations and are actively trying to determine if there are more. We think it unlikely, but this caveat should be kept in mind. Obviously, this is not the case for cPIns since it can be observed in the intact tumor.

FIGURE 3 shows both the cyclic phosphate and phosphomonoester spectral regions of several extracts. Although we have not attempted detailed quantitation because of the difficulties in comparing tissue extracts with extracts from cell lines, we have determined that in 3T3 cells (FIG. 3a) the level of cPIns is 50 mM, roughly comparable to that found in the 7777 tumor. The phosphomonoester regions of the spectra (6–2 ppm) were complex and showed considerable variability as observed previously.[3,13–18] These peaks are difficult to identify without spiking with known compounds because of the pH dependence of the resonances. Aside from identifying the large resonances of phosphocholine (PGro) and phosphoethanolamine (PEth), we have not yet made any other definitive identifications in this region.

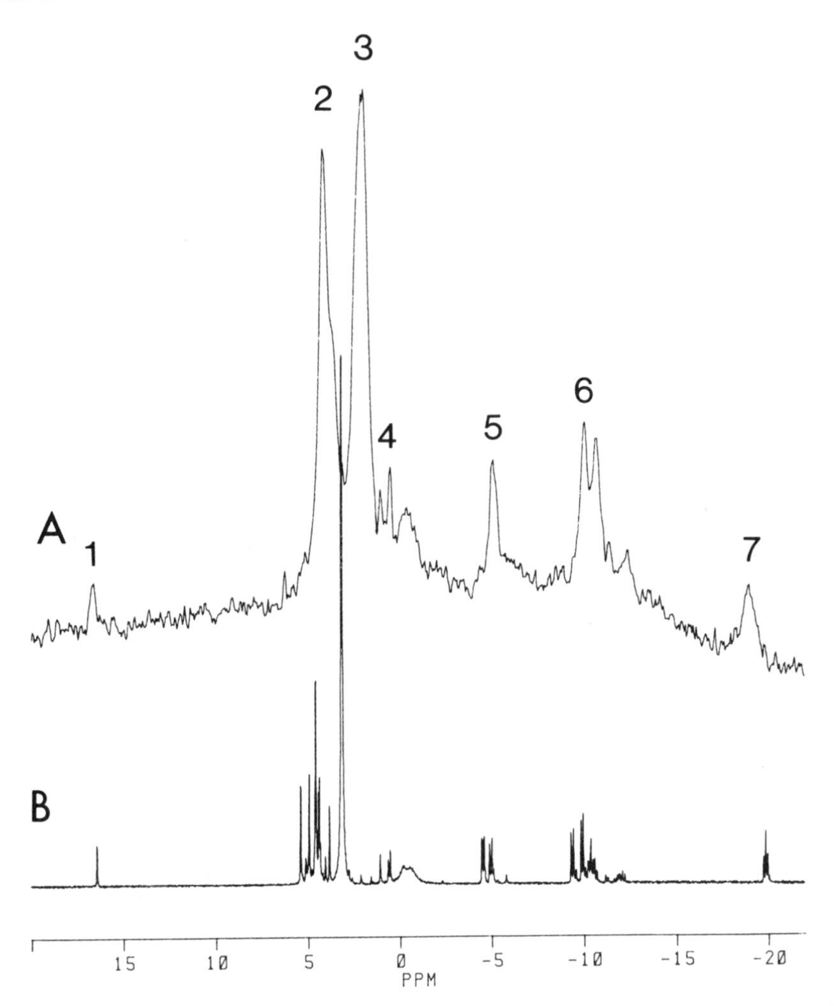

FIGURE 1. ^{31}P NMR spectrum of a perfused Morris (7777) hepatoma (A) and a decoupled spectrum of the aqueous phase of a chloroform/methanol/water extract of a similar tumor (B). (A) 90° pulse, SW = 10,000 Hz, 2K data points, 800 scans with 1 sec pulse interval. (B) 45° pulse, SW = 10,000 Hz, 16K data points, 20,000 scans with 1.4 sec pulse interval. (1) cyclic inositol phosphate, (2) phosphomonoesters, (3) inorganic phosphate, (4) phosphodiesters, (5, 6, 7) γ, α, β, phosphates of nucleoside triphosphates, mostly ATP. The resonances at (5) and (6) also contain contributions from the β and α phosphates of nucleoside diphosphates, mostly ADP.

In the phosphodiester region (2–0 ppm) we have observed peaks at 0.49, 0.56, 0.73, 1.03, 1.41, and 1.54 ppm. These correspond to various glycerophosphodiesters, the chemical shifts of which are presented in TABLE 2. FIGURE 4 shows this region of spectra from four different extracts. They display considerable variation in level from tissue to tissue, although a general pattern of greater complexity seems to be the case in immortalized or transformed tissue. We believe this is the first time GroPIns, GroPGro, and b(GroP)Gro have been observed in ^{31}P spectra of extracts.

TABLE 1. Chemical Shifts of Some Cyclic Phosphates

Compound	PPM
Cytosine 2′,3′-(cyclic)	20.7
Uridine 2′,3′-(cyclic)	20.7
Adenosine-2′,3′-(cyclic)	20.6
Guanosine-2′,3′-(cyclic)	20.6
NADP 2′,3′-(cyclic)	20.5
Ribose 1,2-(cyclic) 5-bis	19.2
Glycerol 1,2-(cyclic)	19.1
Fructose 1,2-(cyclic) 6-bis	17.1
myo-Inositol 1,2-(cyclic)	16.5
Xylose 1,2-(cyclic)	11.4
Glucuronic acid 1,2-(cyclic)	11.3
Glucose 1,2-(cyclic)	11.3
Galactouronic 1,2-(cyclic)	10.7
Galactose 1,2-(cyclic)	10.6

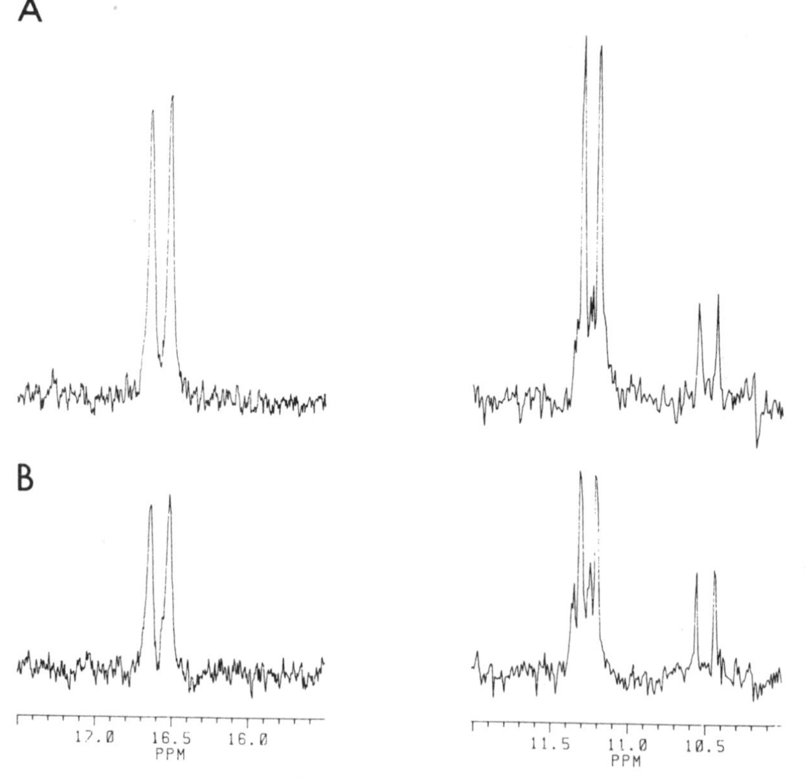

FIGURE 2. Expanded regions of coupled spectra of the aqueous phases of chloroform/methanol/water extracts of a Morris (7777) hepatoma (A) and some near-term mouse fetuses (B) before (bottom) and after (top) the addition of cPIns (A) and cPGlu (B). (A) 90° pulse, SW = 10,000 Hz, 32K data points, 2000 scans with a 3.6-sec pulse interval. (B) 45°V pulse, SW = 11,360 Hz, 16K data points, 10,000 scans with a 1.4-sec pulse interval. A 1 Hz Lorenzian filter was used prior to transformation.

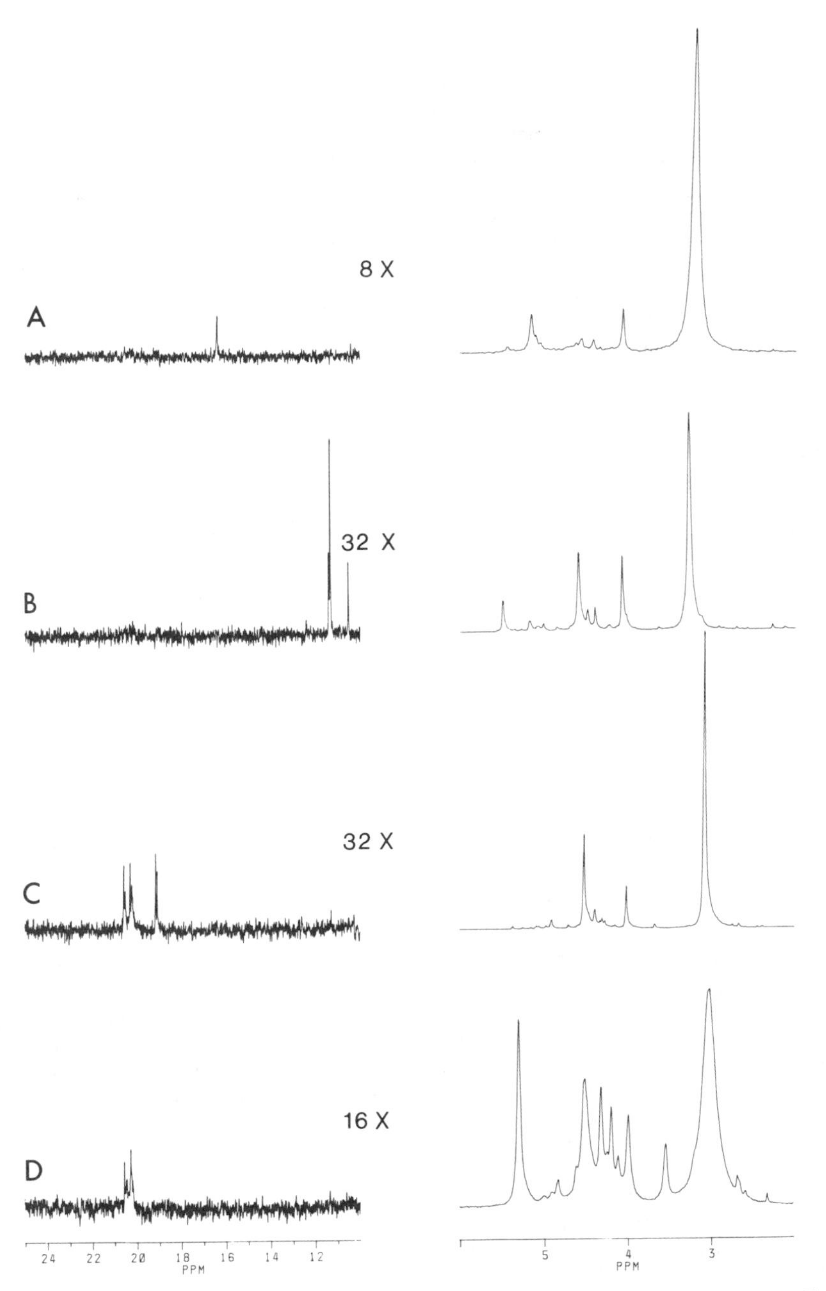

FIGURE 3. Expanded decoupled spectra of extracts of several tissues showing the cyclic phosphate (left) and phosphomonoester (right) regions. The acquisition parameters are as in FIG. 2B, but with 50,000 scans. (A) NIH 3T3 mouse fibroblast cell line; (B) near-term mouse fetus; (C) EMT-6 murine mammary carcinoma; and (D) rat kidney.

DISCUSSION

We have demonstrated the feasibility of examining an isolated, perfused tumor by NMR, thereby ensuring the resultant NMR spectra arose entirely from the tumor. The initial perfusions were not microscopically examined for heterogeneity but on gross examination they were homogeneous. A more detailed study of the tumor histopathology is planned. A detailed understanding of the NMR behavior of these isolated tumors should lead to a better understanding of observed changes *in vivo*.

The high NMR sensitivity of our isolated, perfused preparation is demonstrated by our observation at 16.5 ppm of inositol cyclic phosphate in the intact Morris (7777) hepatoma at a concentration of 50–100 μM. There is no question that in this isolated preparation, the signal of interest is coming from the tumor itself. Such levels of cPIns are presumably an indication of abnormal metabolism in the phosphatidylinositide turnover pathway, although the exact nature of the metabolic defect remains to be determined. We have attempted, in a preliminary series of experiments, to modulate the cPIns levels in these tumors by the addition of lithium chloride to the perfusate[30] and by reversible ischemia. Thus far we have been unsuccessful in causing an observable change, i.e., greater than 10%. Manipulations of the 3T3 cell line have been unsuccessful as well.

TABLE 2. Chemical Shifts of Some Glycerophosphodiesters

Compound	PPM
Glycerophospho:	
-inositol 4,5-bisphosphate	0.39[a]
-choline	0.49 reference set
-inositol	0.56
-serine	0.73
-N,N-dimethylethanolamine	0.78
-N-methylethanolamine	0.91
-ethanolamine	1.03
-glycerophosphorylglycerol	1.41
-glycerol	1.54

[a]The two monoester phosphates of the glycerophospho-inositol-4,5-bisphosphate resonate at 2.84 and 3.80 (pH 7.1).

cPIns is produced from phosphatidylinositols by the activity of phospholipase(s) C and associated enzymes[31] and is degraded to inositol 1-phosphate by a specific phosphohydrolase.[32,33] The high levels of this compound in a Morris (7777) hepatoma could reflect either an increase in the rate of its synthesis or a decrease in the rate of its degradation. If the elevation in cPIns is due to an increase in the rate of its synthesis, then one would expect to find significantly elevated levels in this tissue of the other product of phospholipase C activity, diacylglycerol (DG). On the other hand, if the defect is in the degradative pathway, the activity of the cPIns-specific phosphohydrolase should be lower than in the normal tissue (liver). While we have not yet resolved this very important issue, the discovery of this unusual compound at such relatively high levels in a fast growing tumor stimulated us to investigate the cyclic phosphate spectral region in the extracts of several other tissues, tumors, and tissue cultures. Although we observed CPIns in only some of these, several other cyclic compounds have also been observed. Thus far two other compounds have been positively identified, glucose 1,2-(cyclic) phosphate (cPGlu) and galactose 1,2-(cyclic) phosphate (cPGal)

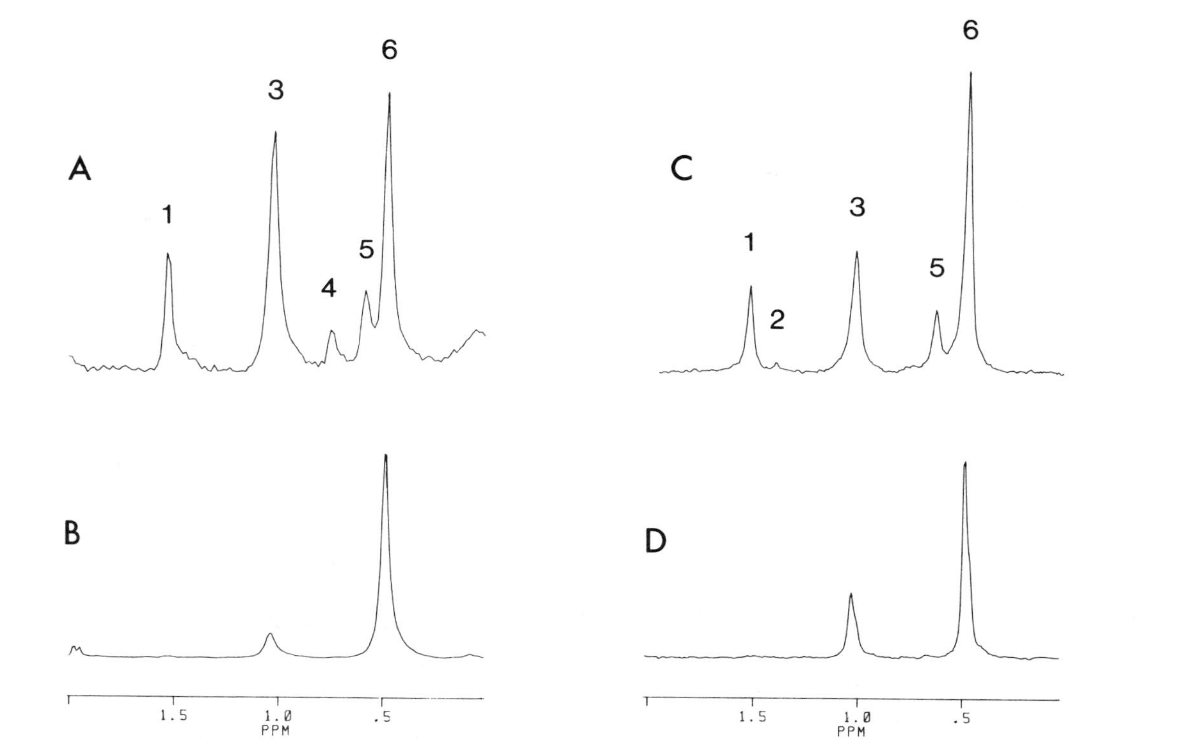

FIGURE 4. Expanded decoupled spectra of extracts from several tissues showing the phosphodiester region. (A) Morris (5123C) hepatoma, (B) rat kidney, (C) NIH 3T3 mouse fibroblast, and (D) Primary rat embryo fibroblast. Acquisition parameters as FIG. 3. (1) GroPGro, (2) b(GroP)Gro, (3) GroPEt, (4) GroPSer, (5) GroPIns, and (6) GroPCho.

in mouse fetus. Several other resonances are very close in chemical shift to known cyclic compounds as can be seen by inspection of TABLE 1: e.g., glycerol 1,2-(cyclic) phosphate (cPGro). Positive identification of these resonances is now in progress.

We have observed levels of PCho or PEth, at least comparable with ATP levels, in virtually all our tumor and cell line extracts, as have many other studies.[3,13–18] We have not attempted to identify the other smaller resonances in this region. The levels of the PDEs in these extracts also appear to be qualitatively higher in tumors and cell lines than normal tissue, although we have not attempted to quantitate this. Of interest is the observation of GroPIns, GroPGro and b(GroP)Gro in these extracts at 0.56, 1.54, and 1.41 ppm, respectively. These have not previously been identified in the ^{31}P spectra of extracts and are probably the result of degradation products of the appropriate phospholipids by various phospholipase A's. Observations of excess PDEs are interesting in view of the recent speculation that the product of the oncogene *src* affects the activity of phospholipase A, which would change the PDE levels.[34]

As mentioned in the results section, the possibility that some of these new peaks are artifacts of the extraction process must be kept in mind. In our early extracts, we observed a number of peaks between 20 and 21 ppm (FIG. 4). These are most probably cyclic 2′,3′ nucleotides resulting from RNase degradation of RNA. We believe they arise from RNase contamination, since we have been able to duplicate the resonances between 20 and 21 ppm by adding RNase to a solution of RNA. Following addition of ATA to the extracts, the peaks were no longer observed in most of the extracts. No such simple explanation for the other cyclic resonances appears possible. Variations of the extraction protocol are now being investigated to determine if any of the cyclic compounds are sensitive to extract conditions.

Intracellular sources of the non-inositol cyclic compounds are difficult to identify at the moment since, as far as we can determine, only cPGro and cPIns have been previously reported in living systems.[35–40] The presence of the other compounds may simply be due to side reactions involving phosphodiester precursors. For example, reactions involving UDPG may give rise to small quantities of cPGlu. Alternatively, these compounds may have substantial, although unknown, biochemical functions. For instance, cPGlu has been reported to be an inhibitor of phosphorylase *in vitro.*[41,42] Further work on these compounds seems indicated.

If the source of high levels of cPIns is a stimulation of phospholipase C hydrolysis of phosphatidylinositol, then there should also be present high levels of the other hydrolysis product, diacylglycerol. The implications of higher than normal levels of DG are substantial since DG is thought to stimulate protein kinase C, a kinase involved in the control of cellular proliferation.[43,44] Elevated DG levels in tumors and cell lines have been observed in a few cases.[45,46] If other cyclic compounds arise from hydrolysis of analogous phospholipids, they too may be indicators of high levels of activity of a phospholipase C and thus high levels of DG. In an analogous way, the presence of PCho and PEth might indicate similarly elevated DG levels from an activated phosphatidylcholine-specific phospholipase C.[47] An exciting, although extremely speculative, possibility is that cells are maintained in an active growing state by the excess activity of diverse phospholipase C's, which generate chronically high levels of diacylglycerol, thereby stimulating persistent cellular proliferation.

ACKNOWLEDMENTS

We would like to thank C. Bergman Meyers, A. Goll, S. Howard, and S. Singh for technical assistance and Dr. D. Brown for supplying us with the EMT-6 and I-347 murine tumors.

REFERENCES

1. GRIFFITHS, J. R., E. CADY, R. H. T. EDWARDS, V. R. MCCREADY, D. R. WILKIE & E. WITTSHAW. 1983. Lancet **1:** 1435–1436.
2. BALERIAUX, D., D. A. ARNOLD, C. SEGEBARTH, P. R. LUYTEN & J. A. DEN HOLLANDER. 1986. SMRM Abst. **1:** 41–42.
3. MARIS, J. M., A. E. EVANS, A. C. MCLAUGHLIN, G. J. D'ANGIO, E. L. BOLINGER, H. MANOS & B. CHANCE. 1985. New Engl. J. Med. **312:** 1500–1505.
4. NARUSE, S., T. HIGUCHI, Y. HORIKAWA, C. TANAKA, K. NAKAMURA & K. HIRAKAWA. 1986. Proc. Natl. Acad. Sci. USA **83:** 8343–8347.
5. NG, T. C., W. T. EVANOCHKO, R. N. HIRAMOTO, V. K. GHANTA, M. B. LILLY, A. J. LAWSON, T. H. CORBETT, J. R. DURANT & J. D. GLICKSON. 1982. J. Magn. Res. **49:** 271–286.
6. ADAMS, D. A., G. L. DENARDO, S. J. DENARDO, G. B. MATSON, A. L. EPSTEIN & E. M. BRADBURY. 1985. Biochem. Biophys. Res. Commun. **131:** 1020–1027.
7. BOTTOMLEY, P. A., T. H. FOSTER, R. E. ARGERSINGER & L. M. PFEIFER. 1984. Med. Phys. **11:** 425–448.
8. SAUER, L. A., J. W. STAYMEAN & R. T. DAUCHY. 1984. Can. Res. **42:** 4090–4097.
9. MEYER, R. A., M. J. KUSHMERICK & T. R. BROWN. 1982. Am. J. Physiol. **242:** C1–C11.
10. COHEN, S. M. 1983. J. Biol. Chem. **258:** 14294–14308.
11. GARLICK, P. B., G. K. RADDA & P. J. SEELEG. 1979. Biochem. J. **184:** 547–554.
12. GRAHAM, R. A., R. A. MEYER, B. S. SZWERGOLD & T. R. BROWN. 1987. J. Biol. Chem. **262:** 35–37.
13. EVANOCHKO, W. T., T. C. NG, M. B. LILLY, A. J. LAWSON, T. H. CORBETT, J. R. DURANT & J. D. GLICKSON. 1983. Proc. Natl. Acad. Sci. USA **80:** 334–338.
14. EVANOCHKO, W. T., T. T. SAKAI, T. C. NG, N. R. KRISHNA, H. D. KIM, R. B. ZEIDLER, V. K. GHANTA, R. W. BROCKMAN, L. M. SCHIFFER, P. G. BRAUNSCHWEIGER & J. D. GLICKSON. 1984. Biochim. Biophys. Acta **805:** 104–116.
15. CARPINELLI, G., F. PODO, M. DIVITO, E. PROIETTI, S. GESSANI & F. BELARDELLI. 1984. FEBS Lett. **176:** 88–92.
16. PROIETTI, E., G. CARPINELLI, M. DEVITO, F. BELARDELLI, I. GRESSER & F. PADO. 1986. Can. Res. **46:** 2849–2857.
17. DESMOULIN, F., J. P. GALONS, P. CANIONI, J. MARVALDI & P. J. COZZONE. 1986. **46:** 3768–3774.
18. COHEN, J. S., R. C. LYONS, C. CHEN, P. J. FAUSTINO, G. BATIST, M. SHOEMAKER, E. RUBALCABA & K. H. COWAN. 1986. Can. Res. **46:** 4087–4090.
19. BLIGH, E. G. & W. J. DYER. 1959. Can. J. Biochem. **37:** 911–917.
20. ISHII, H., T. M. CONNOLLY, T. E. BROSS & P. W. MAJERUS. 1986. Proc. Natl. Acad. Sci. USA **83:** 6397–6401.
21. SARMA, M. H., E. R. FEMAN & C. BAGLIONI. 1976. Biochim. Biophys. Acta **418:** 29–38.
22. GONZALEZ, R. B., R. S. HAXO & T. SEHLEICH. 1980. Biochemistry **19:** 4299–4303.
23. KHORANA, H. G., G. M. TENER, R. S. WRIGHT & J. G. MOFATT. 1957. J. Am. Chem. Soc. **79:** 430–436.
24. MARUO, B. & A. A. BENSON. 1959. J. Biol. Chem. **234:** 254–256.
25. PIZER, F. L. & C. E. BALLOU. 1959. Biochem. J. **81:** 915–921.
26. KATES, M. 1986. *In* Techniques of Lipidology. pp 396–399. Elsevier. New York.
27. JAKOBOVITS, E. B., J. E. MAJORS & H. E. VARMUS. 1984. Cell **38:** 757–765.
28. GREINER, J. V., S. J. KOPP, D. R. SANDERS & T. GLONEK. 1981. Invest. Ophthalmol. Vis. Sci. **21:** 700–713.
29. GLONEK, T., S. J. KOPP, E. KOT, J. W. PETTEGREW, W. H. HARRISON & M. M. COHEN. 1982. J. Neurochem. **39:** 1210–1219.
30. BERRIDGE, M. J., C. P. DOWNES & M. R. HANLEY. 1982. Biochem. J. **206:** 587–595.
31. MAJERUS, P. W., T. M. CONNOLLY, H. DECKMYN, T. R. ROSS, T. E. BROSS, H. ISHII, V. S. BANSAL & D. B. WILSON. 1986. Science **234:** 1519–1526.
32. DAWSON, R. M. C. & N. G. CLARKE. 1972. Biochem. J. **127:** 113–118.
33. DAWSON, R. M. C. & N. G. CLARKE. 1973. Biochem. J. **134:** 59–67.
34. BRUGGE, J. S. 1986. Cell **46:** 149–150.

35. Koch, M. A. & H. Diringer. 1974. Biochem. Biophys. Res. Commun. **58:** 361–367.
36. Ross, T. R. & P. W. Majerus. 1986. J. Biol. Chem. **261:** 11119–11123.
37. Connolly, T. M., D. B. Wilson, T. E. Bross & P. W. Majerus. 1986. J. Biol. Chem. **261:**122–126.
38. Dixon, J. F. & L. E. Hokin, 1985. J. Biol. Chem. **260:** 16068–16071.
39. Sekar, M. C., J. F. Dixon & L. E. Hokin. 1987. J. Biol. Chem. **262:** 340–344.
40. Clarke, N. & R. M. C. Dawson. 1978. Biochem. J. **173:** 579–589.
41. Hu, H. Y. & A. M. Gold. 1978. Biochim. Biophys. Acta **525**–55–60.
42. Kokesh, F. C. & Y. Kakuda. 1977. Biochemistry **16:** 2467–2473.
43. Ball, R. M. 1986. Cell **45:** 631–632.
44. Nishizuka, Y. 1986. J. Natl. Cancer Inst. **76:** 363–370.
45. Preiss, J., C. R. Loomis, W. R. Bishop, R. Stein, J. E. Niedel & R. M. Bell. 1986. J. Biol. Chem. **261:** 8597–8600.
46. Fleischman, L. F., S. B. Chahwala & L. Cantley. 1986. Science **231:** 407–410.
47. Besterman, J. W., V. Duronio & P. Cuatrecasas. 1986. Proc. Natl. Acad. Sci. USA **83:** 6785–6789.

DISCUSSION OF THE PAPER

J. S. Cohen: Just a cautionary note: Nucleotide 2′,3′ cyclic phosphates occur in that region and solutions of ribopolynucleotides produce peaks in that region. I remember a story from Morris Grill who was looking at transfer RNA a few years ago and some other people also were looking at these, and saw peaks in that region. They thought there might be a cyclic phosphate in tRNA but it turns out that over a period in time individual breakages occur and cyclic phosphate forms.

T. R. Brown: The only ones I want to commit myself to identifying are the ones that I've identified. I don't know what the others are. I would be very suspicious about believing anything in the 2′ or 3′ area, but things that are 2, 3, or 4 ppm away are a different matter entirely.

Question: I just want to suggest one area to look for some possible intermediates and that you probably have a very highly stimulated phospholipid metabolism in these cells, particularly a synthetic pathway. I suspect that your chloroform/methanol extraction, unlike the perchloric acid extraction managed to bring into the solution the intermediate phospholipid metabolites which normally would probably fall down to give a pellet.

Brown: Well, one always has to worry about extraction artifacts. That is one of the real nice things about the first peak in that tumor. There is no question it's an extraction artifact since we can see it in the tumor. That's how we came to discover it in the first place.

Some Recent Applications of ^{1}H NMR Spectroscopy *in Vivo*

DAVID G. GADIAN, EDWARD PROCTOR,[a] AND STEPHEN R. WILLIAMS

Departments of Physics in relation to Surgery
[a]Applied Physiology and Surgical Sciences
Royal College of Surgeons of England
35-43 Lincoln's Inn Fields
London WC2A 3PN, England

INTRODUCTION

There is increasing interest in using ^{1}H NMR together with ^{31}P NMR for metabolic studies *in vivo*.[1,2] ^{1}H NMR has the advantages of high sensitivity and potential accessibility to a wide range of metabolites, but is technically more difficult than ^{31}P NMR because of the need to suppress the water and fat signals and because there may be spectral overlap between different metabolite signals. In this paper, we discuss our recent experience in the use of ^{1}H NMR *in vivo*. Most of our work involves the study of small animals at high field strength (8.5 T), but the methods that we describe should generally be applicable at the lower field strengths that are available for studies of human metabolism. We describe some of the technical problems and the ways in which they may be overcome, together with some specific applications to muscle and brain metabolism. In addition, we discuss how some spectroscopic observations have contributed to interesting new developments in the use of magnetic resonance imaging.

MEASUREMENTS OF LACTATE AND pH IN SKELETAL MUSCLE

^{1}H and ^{31}P Measurements of Intracellular pH

We began our ^{1}H studies of rat skeletal muscle by evaluating techniques for suppressing the water and fat signals. We found that the T_2 value of the water signal was very much shorter than for the metabolites, and therefore that selective suppression of the solvent signal could be achieved simply by the use of the Carr-Purcell-Meiboom-Gill (CPMG) multiple echo sequence.[3] FIGURE 1 shows a spectrum obtained using this sequence, in which signals can be detected from creatine + phosphocreatine, taurine, and anserine. Because the chemical shifts of the anserine signals are sensitive to pH, the gradual acidosis that occurs in ischemia can be followed not only by ^{31}P NMR but also by ^{1}H NMR. The data from a single study are plotted in FIGURE 2a. It is apparent that down to a pH value of 6.2–6.3, all three resonances report essentially the same pH. The anserine signal at 7 ppm is the least reliable as it has the smallest chemical shift range (0.42 ppm) compared to 1 ppm for the signal at 8 ppm and about 2.5 ppm for inorganic phosphate. Accordingly, a correlation plot has been constructed (FIG. 2b) in which the pH determined by ^{31}P NMR is compared to that determined by the anserine signal at 8 ppm. All the data from seven experiments are included on this figure; they confirm the good agreement between the ^{1}H and ^{31}P measurements.

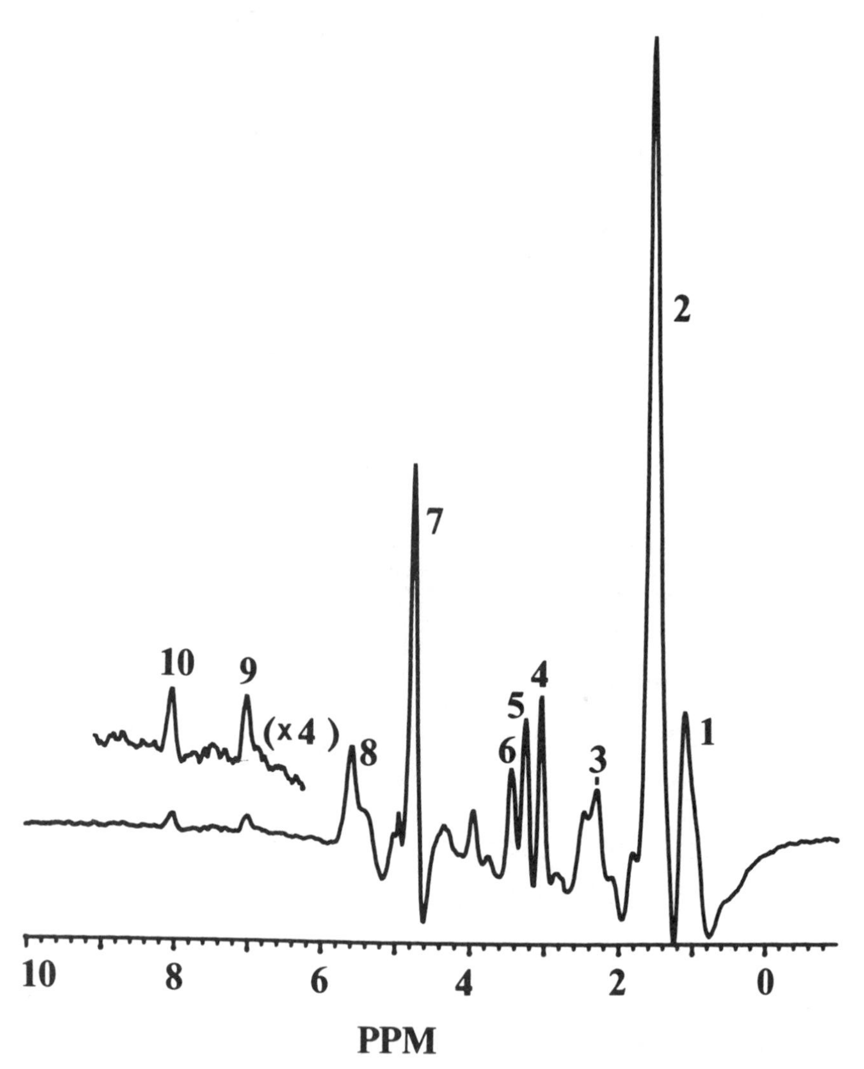

FIGURE 1. ^{1}H NMR spectrum obtained at 360 MHz from hind limb skeletal muscle of an anesthetized rat. Signals were obtained using a saddle-shaped radiofrequency coil of diameter 18 mm and height 17 mm, placed around the thigh of the rat, and doubly tuned to the ^{1}H and ^{31}P frequencies. Data were acquired following a CPMG pulse train of 270 msec duration. The delay between 180° pulses was 1.35 msec with the 180° pulses phase-shifted (±90°) on alternate scans relative to the 90° pulse. 16 scans were accumulated with a 1.7 sec delay between pulse trains. Peaks are assigned as follows: Phosphocreatine (with a small contribution from creatine) 4, taurine 5 and 6, anserine 9 and 10, water 7, various fat protons 1, 2, 3, and 8. Resolution enhancement was accomplished with a trapezoidal window function for the first 20 msec of the free induction decay, and exponential multiplication corresponding to a 10 Hz line broadening. Chemical shifts are measured relative to the phosphocreatine signal at 3.03 ppm.

It is now generally accepted that the ^{31}P chemical shift of the inorganic phosphate signal accurately reflects cytosolic pH to within ±0.1 pH units and that changes in pH can often be measured with a precision of better than ±0.05 units. Nevertheless, the independent confirmation provided by the ^{1}H NMR measurements is of value as it involves the use of a reporter group of different charge to inorganic phosphate. Unlike inorganic phosphate, the dissociation of anserine is from a positively charged to neutral molecule. Therefore, both ionic strength effects and any specific charge interactions (for example with macromolecules) will be different for anserine and inorganic

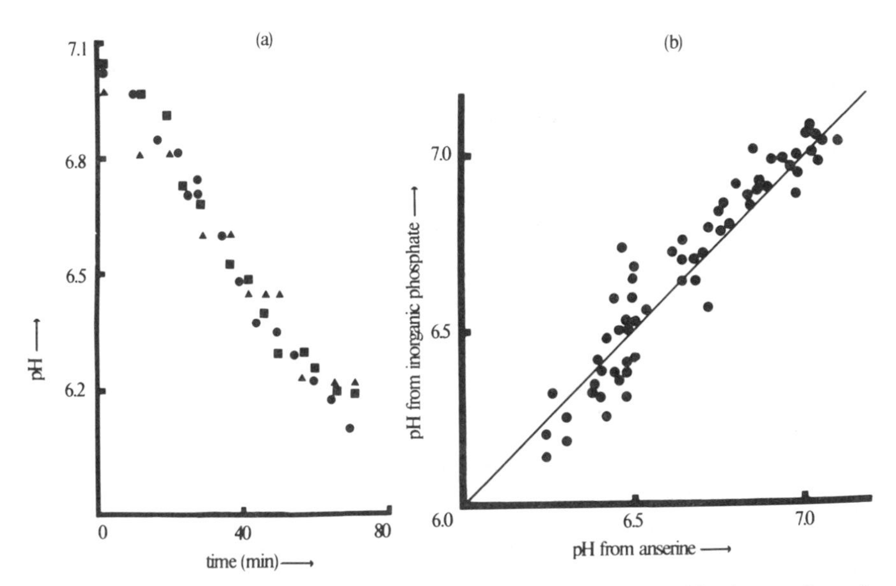

FIGURE 2. The fall in pH during ischemia of the rat leg muscle measured by the anserine and inorganic phosphate resonances. The limb of the anesthetized rat was ligated by remote control from outside the magnet by tightening a snare in place round the top of the thigh. The ligation produced total ischemia with no blood flow to the muscle. On the basis of standard titration curves performed in solution, the pK_a values and the limiting acidic and basic chemical shifts for the anserine resonances were 6.93, 8.59 ppm and 7.59 ppm, and 6.93, 7.26 ppm and 6.78 ppm. The parameters for inorganic phosphate were 6.70 (pK_a) and 3.14 ppm (acidic shift) and 5.68 ppm (basic shift). (a) Data from a single experiment, showing pH determinations from the chemical shifts of the anserine resonances at 8 ppm (■) and 7 ppm (▲), or from the inorganic phosphate resonance (●). (b) Correlation plot of pH determined from the inorganic phosphate resonance against that from the anserine resonance at 8 ppm. The straight line represents the theoretical 1:1 equivalence. Data from seven experiments are included.

phosphate. It is reasonable to conclude, then, that it is indeed pH itself that is the major determinant of the change in chemical shift of these signals as a function of pH.

Correlation of Lactate with Intracellular pH in Ischemic Muscle

The combined use of ^{1}H and ^{31}P NMR (or even, in view of the above results, ^{1}H NMR alone) offers the possibility of correlating the accumulation of lactate with the pH changes that occur in ischemia. However, the ^1H signal from the methyl group of lactate, which occurs at 1.32 ppm, may be masked by the large fat signal that also occurs in this region of the spectrum. As can be seen from FIGURE 1, the fat signals cannot be adequately suppressed with the CPMG pulse sequence, as their T_2 values are comparable with those of the metabolites. Other approaches are therefore necessary in order to detect signals in this region of the spectrum from metabolites such as lactate and alanine. Campbell and Dobson[4] have described a spin-echo double resonance method, which they used to detect certain spin-coupled resonances. This "editing" technique has been used by Rothman *et al.* to resolve specific metabolite signals and to reveal the lactate resonance in the rat brain[5] and in excised muscle.[6] We were able to exploit the technique to monitor lactate production *in vivo* during ischemia[3] and, therefore, to correlate the accumulation of lactate with the decline of intracellular pH. Over the pH range 7.0–6.2, lactate accumulation correlated linearly with pH. With the aid of additional studies of muscle extracts, a value of 63.6 ± 9.1 mmol/pH unit/kg wet wt (s.d. $N = 7$) was determined for the buffering capacity over this pH range.[7]

APPLICATIONS TO BRAIN METABOLISM

Studies of an Inborn Error of Metabolism

The fact that ^1H NMR can be used to study certain aspects of amino acid metabolism suggests that the technique might provide a new approach for studying inborn errors of amino acid metabolism. Phenylketonuria is the most common of these inborn errors, but the precise mechanism by which the systemic enzyme deficiency results in brain disease in this or any of the many other amino acidopathies remains obscure. With a view to assessing the feasibility of monitoring brain phenylalanine in patients with phenylketonuria, we have investigated mutant mice with the inborn error histidinemia.[8]

FIGURE 3 shows a ^1H NMR spectrum obtained at high field strength (8.5 T) from a mutant mouse that had been fed a histidine-enriched diet. The spectrum was obtained using a spin-echo sequence incorporating an initial '$1\bar{3}3\bar{1}$' pulse for solvent suppression. The large (albeit strongly suppressed) signal at 4.7 ppm is from water and the broad signal centered at 1.5 ppm is from fat. Additional narrow signals can be observed from $N\text{-}(CH_3)_3$ protons, for example from choline-containing compounds (at 3.20 ppm), from creatine + phosphocreatine (at 3.03 ppm), and from N-acetylaspartate (at 2.01 ppm). The two signals at 7.83 and 7.08 ppm can be assigned to the C_2 and C_4 protons of histidine. With the aid of various control studies, we were able to quantify the brain histidine concentration, which in a series of 12 animals varied from about 2 to 10 mmol/kg wet wt. These values correlated well with values determined from the same animals by amino acid analysis of the brain extracts, confirming that *in vivo* ^1H NMR can, if all the appropriate controls are performed, give reliable concentration measurements. Further studies at 1.9 T showed that histidine is visible at the lower field strengths that are available for whole body spectroscopy.

The main significance of these studies is that they suggest that it should indeed be feasible to monitor and quantify phenylalanine in brain tissue of humans with phenylketonuria. The phenylalanine signals of interest are from the five protons in the aromatic ring and occur at about 7.4 ppm, i.e. in the same region of the spectrum as the histidine signals. Because of the nature of the spin-spin coupling of these protons, their spectrum simplifies considerably on going from 8.5 T to 2 T, and as a result field strengths of about 2 T should be particularly suitable for monitoring of phenylalanine *in vivo*.

Detection of Brain Lactate

Since detection of brain lactate *in vivo* was first reported by Behar *et al.*[9] in 1983, there has been increasing interest in the non-invasive monitoring of lactate. The main problem with detection of lactate is, as mentioned above in the muscle studies, the large overlapping signal that can be produced by fats. For brain studies, one approach to overcoming the problem is to remove the scalp tissue from which the fat signal

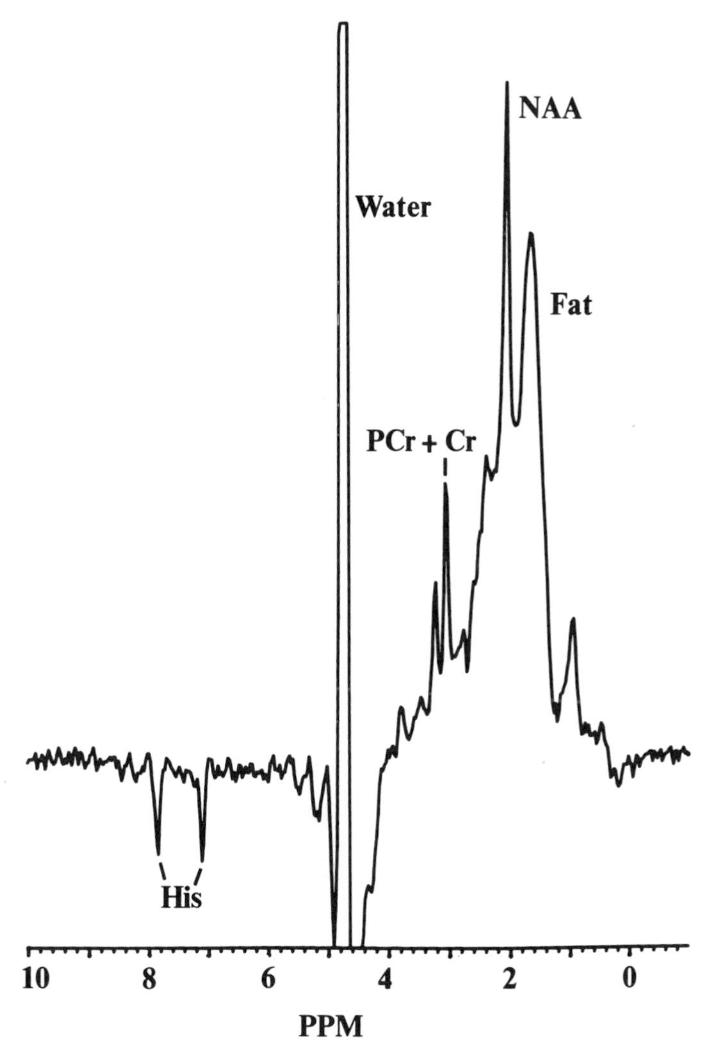

FIGURE 3. ^{1}H NMR spectrum obtained at 360 MHz from an anesthetized histidinemic mouse that had been fed on a histidine-enriched diet. The experimental details are as for FIG. 1a in Reference 8. NAA refers to N-acetyl aspartate, PCr + Cr to phosphocreatine + creatine, and his to histidine. The two histidine signals are negative because of the phase characteristics of the $1\bar{3}3\bar{1}$ pulse.

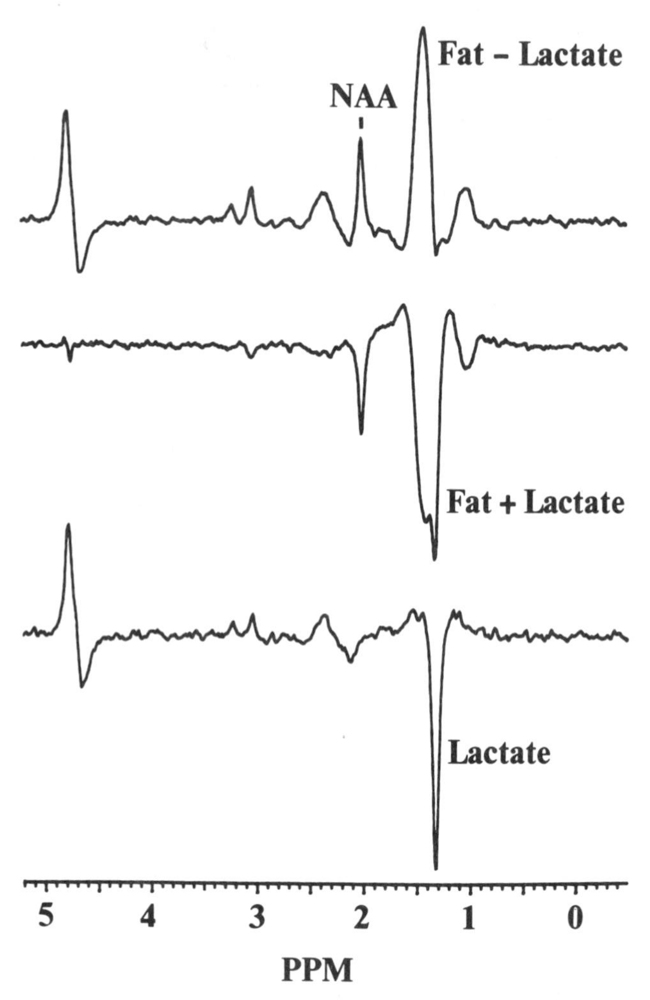

FIGURE 4. ^{1}H NMR spectra obtained at 360 MHz from a rat 1 hr post mortem, showing the detection of lactate from underneath the fat signal. The top spectrum was obtained using the pulse sequence (α-$\overline{3\alpha}$-3α-$\bar{\alpha}$-τ-16α-τ-acquire). The middle spectrum was obtained with the sequence (α-$\overline{3\alpha}$-3α-$\bar{\alpha}$-τ-2α-$\overline{6\alpha}$ -6α-$\overline{2\alpha}$-τ-acquire). The bottom trace represents the sum of these two spectra and shows a large negative signal at 1.32 ppm from lactate. NAA refers to N-acetyl aspartate. (For further details see Williams *et al.*[10])

originates (the brain lipids do not generate high resolution ^{1}H signals). A less invasive approach is to use a pulse sequence for appropriate "editing" of the spectrum, as shown by Rothman *et al.*[5] We recently described an analogous editing scheme that is particularly simple to implement and does not require the use of double irradiation.[10] We demonstrated good suppression of the fat signal for studies of the rat brain *in vivo*. The use of the technique to distinguish lactate from fats is illustrated in FIGURE 4, which shows the detection of a large lactate signal 1 hr post mortem.

The upper two traces in FIGURE 4 show ^{1}H spectra obtained using two different spin-echo pulse sequences (see figure legend for details). For the uppermost spectrum, the sequence was similar to that used for FIGURE 3, except that the echo delay time τ was 68 msec rather than 20 msec. The effect of this sequence is to invert the lactate signal at 1.32 ppm, because of its spin-coupling characteristics. Therefore the fat and N-acetyl aspartate signals are positive, whereas the lactate (which is hidden under the fat signal) is negative. For the middle spectrum, the pulse sequence was modified in such a way that the lactate and fat signals now have the same phase as each other; in fact with identical data processing to that used for the top spectrum, the signals appear negative. On adding the two spectra, the fat signal cancels out, leaving a large negative lactate signal at 1.32 ppm, as shown in the bottom trace.

We are now using this editing method for detecting lactate in a gerbil model of brain ischemia.[11] In this model, we are using the hydrogen clearance technique to measure the changes in regional cerebral blood flow that occur in unilateral or bilateral ischemia. At the same time, we are able to monitor pH, lactate, and phosphorus energy metabolites measured in the same animals with combined ^{31}P and ^{1}H NMR. The NMR measurements are made with two surface coils, one on each hemisphere, so that we can make alternate measurements from the two hemispheres. This combination of techniques allows us to make detailed correlations between regional flow and energy metabolism during ischemia and reperfusion.

The non-invasive monitoring of lactate *in vivo* is of interest, not only because it should provide further information about the biochemical processes that occur during and following ischemia, but also because of its possible impact on spatial resolution in localized NMR studies of disease states. Lactate can accumulate to concentrations of 10–20 mM in ischemia, and this high concentration (10 mM is equivalent to 30 mM in methyl protons) coupled with the high sensitivity of proton NMR means that the achievable spatial resolution for detection of lactate should be much higher than for ^{31}P NMR studies.

FIELD HOMOGENEITY AND THE EFFECTS OF MAGNETIC SUSCEPTIBILITY

^{1}H NMR is far more dependent than ^{31}P NMR on the quality of the static field homogeneity. This is because of the large number of ^{1}H signals in a relatively narrow chemical shift range. In view of our interest in ^{1}H NMR, we have paid particular attention to the optimization of field homogeneity and to the factors that may be important in limiting the homogeneity that can be achieved.

In animal studies, we have found that for the normal brain spectral resolution of better than 0.05 ppm can be achieved.[12] For skeletal muscle, we can achieve better than 0.1 ppm, whereas for liver we have been unable to obtain water linewidths *in vivo* of better than 100 Hz at 360 MHz, corresponding to field inhomogeneity of about 0.2 ppm (taking into account the relatively short T_2 of the liver water). In contrast to the brain and muscle studies, we have experienced difficulty in optimizing the field homogeneity in studies of certain implanted tumors in mice.[13] As a result of this, we became interested in the possibility that variations in tissue magnetic susceptibility (possibly through the accumulation of paramagnetic species) might be a significant factor in determining the local magnetic field profile and, in particular, might be characteristic of diseased tissue.

In agreement with the above spectroscopy studies, imaging studies performed at the NMR Unit of Hammersmith Hospital have shown that susceptibility effects in the normal brain are less than 0.1 ppm over the cerebral hemispheres, becoming larger adjacent to the skull and the major blood vessels.[13] However, marked susceptibility effects have been seen in hematomas and in some tumors.[13,14] These can be interpreted in terms of the accumulation of paramagnetic species, possibly blood breakdown products.

The main significance of these effects is that they could improve the specificity of imaging studies and could provide new biochemical information. However, for spectroscopic studies, the existence of such effects would make it harder to optimize the field homogeneity. Clearly, further research will be needed to evaluate the impact of susceptibility effects in both imaging and spectroscopy.

CONCLUSIONS

Although ^{1}H NMR is technically more difficult than ^{31}P NMR, it offers a number of advantages that make the additional effort worthwhile. Firstly, it provides a means of monitoring a wide range of compounds that are not accessible to ^{31}P NMR, including some of the amino acids, ketone bodies, fats, and lactate. Many additional aspects of intermediary metabolism should therefore be accessible to study. Secondly, the non-invasive monitoring of lactate *in vivo* could enable us to monitor some common disease states with improved spatial resolution. This is of considerable importance as the relatively poor spatial resolution of *in vivo* NMR spectroscopy is a major limitation of the technique.

Another feature of our ^{1}H studies is that the stringent field homogeneity requirements have led us to explore the various factors that influence field homogeneity and signal linewidths. It is of interest that in normal tissue, remarkably narrow linewidths can often be achieved. This should spur us on to developing improved procedures for "shimming," i.e. for optimizing field homogeneity, for this in turn could lead to considerable improvements in the quality of spectra recorded *in vivo.* The fact that susceptibility effects may be significant in certain diseases could generate additional shimming problems in spectroscopy, but could be of considerable interest as an additional contrast parameter in magnetic resonance imaging.

ACKNOWLEDGMENTS

We gratefully acknowledge all our collaborative colleagues, and we thank the Rank Foundation, Picker International, and the Wolfson Foundation for their support.

REFERENCES

1. Avison, M. J., H. P. Hetherington & R. G. Shulman. 1986. Applications of NMR to studies of tissue metabolism. Ann. Rev. Biophys. Bioeng. **15:** 377–402.
2. Williams, S. R. & D. G. Gadian. 1986. Tissue metabolism studied in vivo by nuclear magnetic resonance. Q. J. Exp. Physiol. **71:** 335–360.
3. Williams, S. R., D. G. Gadian, E. Proctor, D. B. Sprague, D. F. Talbot, I. R. Young & F. F. Brown. 1985. Proton NMR studies of muscle metabolites in vivo. J. Magn. Reson. **63:** 406–412.
4. Campbell, I. D. & C. M. Dobson. 1979. The application of high resolution nuclear magnetic resonance to biological systems. Meth. Biochem. Anal. **25:** 1–133.
5. Rothman, D. L., K. L. Behar, H. P. Hetherington & R. G. Shulman. 1984. Homonuclear ^{1}H double-resonance difference spectroscopy of the rat brain in vivo. Proc. Natl. Acad. Sci. USA **81:** 6330–6334.
6. Rothman, D. L., F. Arias-Mendoza, G. I. Shulman & R. G. Shulman. 1984. A pulse

sequence for simplifying hydrogen NMR spectra of biological tissues. J. Magn. Reson. **60:** 430–436.
7. WILLIAMS, S. R., E. PROCTOR & D. G. GADIAN. 1986. Buffering capacity of muscle determined by ^{1}H and ^{31}P n.m.r. Biochem. Soc. Trans. **14:** 1267–1268.
8. GADIAN, D. G., E. PROCTOR, S. R. WILLIAMS, E. B. CADY & R. M. GARDINER. 1986. Neurometabolic effects of an inborn error of amino acid metabolism demonstrated in vivo by ^{1}H NMR. Magn. Reson. Med. **3:** 150–156.
9. BEHAR, K. L., J. A. DEN HOLLANDER, M. E. STROMSKI, T. OGINO, R. G. SHULMAN, O. A. C. PETROFF & J. W. PRICHARD. 1983. High resolution ^{1}H nuclear magnetic resonance study of cerebral hypoxia in vivo. Proc. Natl. Acad. Sci. USA **80:** 4945–4948.
10. WILLIAMS, S. R., D. G. GADIAN & E. PROCTOR. 1986. A method for lactate detection in vivo by spectral editing without the need for double irradiation. J. Magn. Reson. **66:** 562–567.
11. GADIAN, D. G., R. S. J. FRACKOWIAK, H. A. CROCKARD, E. PROCTOR, K. ALLEN, S. R. WILLIAMS & R. W. ROSS RUSSELL. 1987. Acute cerebral ischaemia: concurrent changes in cerebral blood flow, energy metabolites, pH and lactate measured with hydrogen clearance and ^{31}P and ^{1}H nuclear magnetic resonance spectroscopy. I. Methodology. J. Cereb. Blood Flow Metab. **7:** 199–206.
12. GADIAN, D. G., E. PROCTOR, S. R. WILLIAMS, I. J. COX & I. R. YOUNG. 1985. Spectral resolution in metabolic studies of the brain. Abstr. Soc. Magn. Reson. Med. pp. 273–274.
13. COX, I. J., G. M. BYDDER, D. G. GADIAN, I. R. YOUNG, E. PROCTOR, S. R. WILLIAMS & I. HART. 1986. The effect of magnetic susceptibility variations in NMR imaging and NMR spectroscopy in vivo. J. Magn. Reson. **70:** 163–168.
14. YOUNG, I. R., S. KHENIA, D. G. T. THOMAS, C. H. DAVIS, D. G. GADIAN, I. J. COX, B. D. ROSS & G. M. BYDDER. 1987. Clinical magnetic susceptibility mapping of the brain. J. Comput. Assist. Tomogr. **11**(1):2–6.

DISCUSSION OF THE PAPER

QUESTION: I have a question with respect to the proton visibility. In the spectra of the brain was this edited specifically for histidine? We don't see any evidence of the ATP or ADP proton resonances and this seems incongruous given the concentration levels expected.

D. G. GADIAN: I mentioned the compounds that were NMR visible and said that we have no evidence of the NMR invisibility of certain fractions of those. We don't seem to see ATP. I feel that that's due to the T_2 values, which are short for the ATP protons making them invisible in the spin echo experiments of the sort that we're doing.

QUESTION: Two questions. First, with regard to the first set of experiments, halothane causes a significant dissociation of flow and metabolism in brain. Have you looked at any other anesthetics to confirm the observation? Secondly, I didn't catch the duration of ischemia involved in the experiment.

GADIAN: The unilateral ischemia was over a period of an hour and bilateral ischemia half an hour. The anesthesia is halothane/oxygen, we've not tried different types of anesthesia but the Institute of Neurology people have been using pentobarbitol for their previous studies of edema, and we do get the correlation in the critical flow rate. I agree that anesthesia is a problem that one should consider this.

QUESTION: Is there any particular reason you're using the two electrode method for doing the measurements for the hydrogen clearance? I was under the impression that it's well known to give serious errors in the measurement and that four electrode systems are much better.

GADIAN: I'm sorry, we're using four electrodes altogether, two on each side but we're taking the measurements from the two that are inside the coil.

QUESTION: We have a lot of experience with perfused hearts, perfused muscles and they shim up very well. Our perfused tumors shim very well some days and terrible others, it's highly variable. And I thought maybe it's the tumor.

D. I. HOULT: By and large the whole theory of shimming on the FID only works when, (1) you've got a homogeneous $\mathbf{B}_1$ field and (2) you're dealing with a spherical sample. Now basically what I'm saying then is to get a reasonable attempt at shimming, you put a ping pong ball or something in there full of water first and shim on that, hope and pray the system stays constant and then put your sample in. Because shimming on anything else takes forever but you can get false positions in your shimming which I'm sure you're well aware.

In Vivo NMR Studies of Glycogen Metabolism:

A New Look at a Time Honored Problem[a]

EUGENE J. BARRETT, WILLIAM A. PETIT, AND MAREN R. LAUGHLIN

Department of Internal Medicine
Department of Molecular Biophysics and Biochemistry
Yale University School of Medicine
New Haven, Connecticut 06510

INTRODUCTION

Glycogen is the principal storage form of carbohydrate in mammalian species. Its role in the maintenance of cellular homeostasis has been greatly clarified by the large number of studies done over the past 50 years. Though present in most tissues of the body, glycogen concentrations are highest in liver, kidney, and muscle and it is in these tissues that the control of glycogen metabolism has been most extensively studied.[1] In skeletal muscle white fibers, glycogen serves as a ready source of glucose for oxidation and anaerobic glycolysis and, for brief periods, can replace fatty acids as the main energy source for the generation of cellular ATP. In cardiac muscle and in the red fibers of skeletal muscle (both are tonically active), glycogen has a similar role but the activity of the enzymatic apparatus involved with the synthesis and breakdown of glycogen is less active. Liver, which contains glycogen in excess of its own energy needs, will store glycogen during post-cibal periods. That glycogen can then be mobilized in a highly regulated manner during periods of fasting to provide the brain and other tissues (which rely largely or exclusively on glucose oxidation for energy) with an uninterrupted supply of this vital fuel.

Interest in the study of glycogen metabolism derived first from an appreciation of its physiologic importance. However, with the gradual unraveling of the enzymatic control mechanisms involved in the regulation of glycogen synthesis and breakdown, its study has served as a fascinating paradigm for the study of cellular control mechanisms that involve protein phosphorylation.[2] A somewhat simplified schematic of our current understanding of the cascade of enzymes involved in the regulation of glycogen turnover in skeletal muscle is shown in FIGURE 1. Even if we ignore the physiological importance of glycogen, the complexity of its multi-tiered control mechanisms would undoubtedly attract the interest of investigators. Indeed, one thesis of our current work is that as enzymologists have proceeded forward at a gratifying pace in their dissection of the machinery involved in the control of glycogen turnover, our integration of this information into a fully coherent explanation of the physiology of glycogen in the intact organism has lagged.

We propose in this paper that application of NMR spectroscopic methods may further our understanding of glycogen metabolism in two ways; first, they will facilitate the integration of information gained by the enzymologist on the molecular

[a]Supported by U.S. Public Health Service Grants HL 34784 and GM 30287.

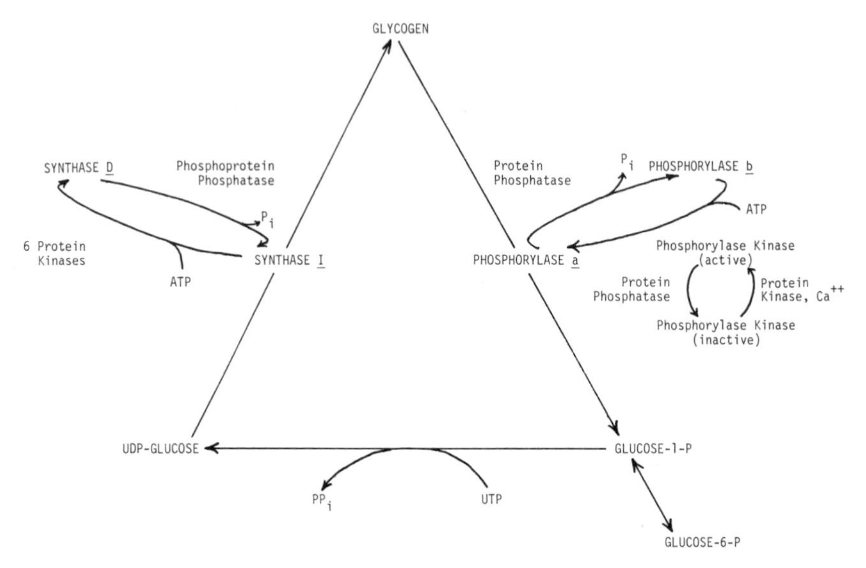

FIGURE 1. Schematic illustration of the pathway of substrate flux in and out of glycogen and of the enzymatic control of this flux.

events involved in the control of glycogen turnover into a unified physiological model and second, as we hope to demonstrate, *in vivo* NMR spectroscopic methods will allow us to address a number of questions relating to the physiology of glycogen synthesis and breakdown that have been difficult to approach using more conventional techniques.

To support this proposal, we will rely principally upon examples from ongoing and planned studies in our laboratory. Where particularly pertinent, information available from other investigators will also be discussed. Our approach has been to develop *in vivo* spectroscopic methods that allow us to quantitate rates of glycogen synthesis and breakdown in absolute terms (i.e. in units of μmol/min or mg/min). We then compare these *in vivo* measurements to the activities of the enzymes involved in glycogen metabolism measured in tissue extracts under conditions designed to approximate the milieu that exists within the cell. These studies have focused almost exclusively upon rat and guinea pig cardiac muscle. However, we will point out that with little alteration these methods can be extended to the study of both liver and skeletal muscle. Before discussing in detail the methods and results of our studies however, some familiarity with the quantitative biology of glycogen in mammalian muscle and liver tissue as derived from classical biochemical studies is needed. This quantitation is particularly important to our appreciation of the power and limitations of ^{13}C-NMR as applied to the study of glycogen metabolism.

QUANTITATIVE ASPECTS OF GLYCOGEN METABOLISM IN MAMMALIAN LIVER AND MUSCLE

In man or animals, in the fed state, glycogen constitutes approximately 0.5–0.9% of the wet weight of skeletal muscle and 4–8% of the wet weight of liver. The glycogen content of heart muscle is similar to that of skeletal muscle (TABLE 1). Skeletal muscle is the main repository of stored glycogen as its mass is approximately 20 fold that of liver and nearly 100 fold that of the heart. On a molar basis, the concentration of glycogen in each of the three tissues can be estimated from the known water content of the wet tissue (75% for heart and skeletal muscle and 80% for liver).[3] Thus, for normal skeletal muscle the concentration of glucosyl monomers present in glycogen averages 100 mmol/L and for liver between 400–600 mmol/L in the fed animal.[4-6] These large substrate concentrations make glycogen an attractive molecule for study by the NMR spectroscopist. However, since only 1.1% of the carbon of glycogen has a ^{13}C nucleus and each of the six carbon atoms of the glucose monomer resonate at slightly different frequencies,[7] the concentrations of the NMR-visible carbons are closer to 1–1.5 mmol/L and 6–8 mmol/L in muscle and liver, respectively. Because concentrations of ^{13}C of approximately 1 mM are required for accurate detection, it should be apparent that detection of the naturally abundant signals from ^{13}C-glucose in muscle glycogen is near this limit of sensitivity. In liver, the situation is better and several laboratories have measured natural abundance ^{13}C spectra clearly demonstrating the presence of glycogen.[8,9]

Observing the presence of glycogen in a spectrum, while no small methodologic feat, affords little physiological information. The power of NMR spectroscopy in

TABLE 1. Glycogen Content of Mammalian Tissue

	Liver	Skeletal Muscle	Heart
	(g glycogen/100 g wet weight of tissue)		
Rat			
Fasting (24 hr)[a]	0.1	0.2	0.2
Fed[b]	4.9	0.5	0.2
Diabetic[b]	1.8	0.5	0.7
Man			
Fasting[c]	0.6	1.6	N.A.
Fed[c]	5.4	N.A.	N.A.

[a] Data from Reference 4.
[b] Data from Reference 5.
[c] Data from Reference 6.
Note: N.A. = not available.

approaching the *in vivo* study of tissue physiology is derived in part from our ability to make serial measurements over time in the face of some perturbation. The question can then be legitimately asked whether we might expect natural abundance ^{13}C-NMR to provide sufficient sensitivity to measure the changes in glycogen concentration that occur in liver or muscle under physiologic or pathologic circumstances. A straightforward answer is available from prior measurements of the rates of glycogen formation and breakdown by these tissues obtained using conventional biochemical assays of either total glycogen content or incorporation and release of radiolabeled glucose in glycogen. In TABLE 2, we summarized the results of such measurements obtained either directly from published studies or calculated from data in those studies.[6,10-13] Since these types of measurements are intrinsically invasive there is little data available from human tissue and estimates based upon it must be considered very rough.

Two facts ought to be obvious from data in TABLE 2. First, the rate of glycogen synthesis in both liver and muscle is slow relative to the rates of glycogenolysis. Second, given these data, it is impossible to measure the rate of glycogen synthesis in liver and muscle on a minute-to-minute basis using natural abundance ^{13}C-NMR. This second point can be illustrated by the following example: the liver (which is capable of synthesizing glycogen more rapidly than muscle) of an animal fed carbohydrate and gluconeogenic precursors synthesizes glycogen at a rate of approximately 0.5 μmol/min/gww. If expressed as a concentration (correcting for the water content of liver), the concentration of glucosyl units in liver glycogen will increase at a rate of approximately 0.63 mmol/L/min. With only 1.1% of this glucose carbon being carbon-13, it requires approximately 2 hr for the ^{13}C-glucose carbon content to change by 1 mM, an amount barely adequate to allow sensitive and precise measurements. This example serves to illustrate the necessity of enriching the ^{13}C content of the compounds that might serve as precursors for glycogen synthesis if sensitive measurements are to be made within a reasonable time frame. Later, we will discuss further the complications introduced by the use of ^{13}C isotope infusions, but would like to first return to the use of ^{13}C-NMR to measure glycogen breakdown in natural abundance spectra. In liver, glycogenolysis can be briefly increased to rates 6–10 fold higher than those encountered in the normal postabsorptive liver by infusing large doses of glucagon.[14] With calculations analogous to those just given for glycogen synthesis we

TABLE 2. Rates of Glycogen Synthesis and Degradation

	Liver	Skeletal Muscle	Heart Muscle
Synthesis	0.4–0.7[a]	.08[c]	0.7[e]
	0.9[b]	.15[d]	
Degradation	2.6[f]	1.2[g]	4.3[h]

[a] Data for perfused liver from Reference 10.
[b] Data for *in vivo* rat liver from Reference 25.
[c] Data for isolated muscle strips from Reference 27.
[d] Data for *in vivo* skeletal muscle from Reference 12.
[e] Data for perfused heart from Reference 26.
[f] Data for *in vivo* liver from Reference 14.
[g] Data for *in vivo* skeletal muscle from Reference 13.
[h] Data for perfused heart from Reference 11.

could expect measurable changes in the natural abundance ^{13}C-NMR spectrum to occur within 10–20 min of the onset of glycogenolysis. In cardiac and skeletal muscle, maximal rates of glycogenolysis, similar to those observed in the glucagon-stimulated liver, can be provoked by anoxia[11] or exercise.[13] We anticipate that the modest glycogen stores initially present would soon be depleted. However, our sensitivity in defining a precise time course to this process will clearly be limited by the small amplitude of the initial glycogen signal.

^{13}C-NMR STUDIES OF *IN VIVO* MYOCARDIAL GLYCOGEN TURNOVER

Description of the Method

Mindful of the limitations of using natural abundance ^{13}C-NMR to quantitatively study the processes of glycogen synthesis and degradation *in vivo*, several years ago we began studies directed at developing a quantitative isotopic method that would allow such measurements to be made.[15,16] As we describe the results of our studies and outline potential future applications, we hope to also illustrate our belief that *in vivo* NMR studies of this type can begin fulfilling its potential in complementing the efforts of the enzymologist to understand more completely the physiological control of glycogen metabolism.

In all our studies described here we used male Sprague-Dawley rats weighing between 400 and 600 g. Animals are anesthetized by intraperitoneal injection of sodium pentobarbital (25 mg/kg). Polyethylene catheters are then placed in the carotid artery and in each internal jugular vein. The arterial catheter is used to continuously monitor blood pressure and the animal is given supplemental infusions of 0.9% sodium chloride if the systolic blood pressure goes below 80 mm Hg. The right jugular venous catheter is inserted at least 5 cm so that its tip is in the inferior vena cava and can be used to obtain mixed venous blood samples. The left jugular catheter is advanced only 1 cm into the jugular vein and is used for infusion of the [1-^{13}C]-glucose. This specific positioning of the venous catheters is chosen to avoid contaminating sampled blood with the infusate prior to the mixing of the ^{13}C-glucose with the systemic circulation. Following placement of the vascular catheters, a tracheostomy is performed and the rat is placed on a respirator. The thoracic cavity is then opened through a sternal incision, the pericardium removed, and the heart gently positioned within a six-turn solenoidal receiver coil. Throughout this surgery and the actual NMR experiment, the animal is maintained in a deeply anesthetized state and its core body temperature is maintained near 37°C with a warming coil. The animal and probe assembly are then placed within the bore of an Oxford Research Systems TMR 32-200 spectrometer. The field is shimmed using the width of the water protons observed at 80.2 MHz. In control studies, ^{31}P spectra of the hearts of animals prepared surgically in this manner were observed for periods of up to 5 hr with maintenance of normal concentrations of phosphocreatine, inorganic phosphate, and ATP.

Once the magnetic field is shimmed (typical linewidths of the water protons ranged from 25–35 Hz), serial 5-min ^{13}C-NMR spectra are obtained over a period of 20–30 min. After collection of these baseline spectra, an infusion of either glucose alone (10 mg/min for 50 min) or glucose and insulin (10 mg/min with 1 unit/min of regular insulin) is begun and serial spectra are taken throughout a 50-min infusion period. During this infusion, blood samples (approximately 1 ml) are obtained at 15, 30, and 45 min for subsequent analysis of [1-^{13}C]-glucose enrichment in the circulating glucose pool as a function of time during the isotope infusion. At 50 min, the animal is removed from the bore of the magnet and the heart rapidly excised and freeze clamped in liquid nitrogen.

The frozen heart is saved for analysis of the activity of glycogen phosphorylases *a* and *b*, and glycogen synthases *I* and *D*, and for the measurement of total glycogen content. In addition, a portion of the frozen heart is used for measurement of the fraction of the glucosyl units of glycogen, which had a ^{13}C-carbon in the C-1 position. For this analysis, samples of the extracted glycogen are hydrolyzed to glucose using amyloglucosidase, and the fraction of the protons bound and therefore J-coupled to a ^{13}C in the 1 position of glucose is determined in proton spectra observed at 360 MHz in a wide bore Bruker NMR spectrometer. Following appropriate correction of the total [1-^{13}C]-glucose of extracted glycogen for that which would be expected from the naturally abundant [1-^{13}C]-glucose, the excess [1-^{13}C]-glucose is considered to have been formed from infused [1-^{13}C]-glucose. Similar proton spectroscopic measurements are made of the [1-^{13}C]-glucose enrichment of the blood samples taken during the labeled infusion.

Since in the heart the only source of glucose monomers for glycogen synthesis is the circulating plasma glucose, the measured enrichment of the plasma glucose with [1-^{13}C]-glucose will provide an accurate measure of the enrichment of the precursor pool being utilized within the heart for the synthesis of glycogen. Stated more explicitly, the enrichment of plasma glucose and of myocardial uridine diphosphoglucose (UDPG) with ^{13}C in the 1-position is essentially the same throughout the labeled glucose or glucose and insulin infusion. From this it should be apparent that the total amount of new glycogen synthesized during the entire infusion period is obtained by simply dividing the excess [1-^{13}C]-glucose found in the glycogen of the excised heart by the time-weighted mean of the enrichment of [1-^{13}C]-glucose in plasma during the infusion period.

Identical measurements could be made more easily using radiolabeled glucose and conventional radioisotope tracer methods. However, using the NMR spectroscopic method described here, the investigator can also use the total amount of [1-^{13}C]-glucose measured in the glycogen obtained from the excised heart to calibrate the signal observed in the TMR spectrum obtained *in vivo* immediately prior to freezing the heart. The peak height of the [1-^{13}C]-glycogen signal will then correspond to an absolute amount of glycogen (in μmol). This peak can then serve as an internal standard, which allows the quantitation of the [1-^{13}C]-glycogen in each of the nine earlier spectra obtained after beginning the tracer infusion. When the total [1-^{13}C]-glycogen content of the heart is thus determined at each time point, the rate of glycogen synthesis between any two times is readily obtained by simply dividing the increase in [1-^{13}C]-glycogen during the time interval by the mean enrichment of the plasma glucose pool with [1-^{13}C]-glucose during that interval.

Therefore, these *in vivo* spectroscopic measurements confer the advantage of having a continuous measurement of the rate of glycogen synthesis throughout the course of the infusion period. The rate of glycogen synthesis in μmol/min estimated in this manner can be expressed mathematically:

$$\text{Synthetic Rate} = \left(\frac{\Delta\ \text{Peak Height}}{\Delta\ \text{Time}}\right)\left(\frac{1}{^{13}\text{C Enrichment of Plasma Glucose}}\right)\left(\frac{\text{Total }^{13}\text{C-Glycogen in Heart}}{\text{Final Peak Height}}\right).$$

The total glycogen concentration of these hearts is determined by digestion of 20–30 mg fragments of frozen heart in 30% KOH at 100°C for 30 min, reprecipitating the glycogen in absolute ethanol, digesting it with amyloglucosidase, and measuring the liberated glucose using the glucose oxidase method. Glycogen synthetase activity is measured using the ^{14}C-labeled UDPG method as described by Thomas *et al.*[17] Synthetase *I* activity is defined as that observed in the presence of 0.11 mM glucose-6-P, and synthetase *D* the activity when the concentration of glucose-6-P was 4.8 mM. Glycogen phosphorylase activity was measured in the direction of glycogen synthesis using ^{14}C-glucose-1-P as described by Gilboe *et al.*[18] Phosphorylase *a* activity corresponds to that observed in the absence of AMP and phosphorylase *b* activity to the activity seen when AMP was present (1.65 mM). Activities of the enzymes measured *in vitro* are expressed as μmoles of glycogen synthesized per minute per gram wet weight of frozen tissue. The radiolabeled UDPG and glucose-1-P were purchased from Amersham Corp. (Arlington Heights, IL), all other reagents used in these assays were purchased from Sigma Chemical Corp. (St. Louis, MO).

Upon first consideration, it might be thought that the rate of glycogenolysis in response to any glycogenolytic stimuli could similarly be quantitated in a continuous manner from the rate of decline of the labeled glycogen signal in the TMR. However, for this to be true it is necessary that the enrichment of the fraction of glycogen being mobilized at any point in time be known. For all practical purposes this requires that [1-^{13}C]-glycogen enrichment be uniform. This can be most easily fulfilled if the following conditions are met. First, the plasma pool from which the glycogen is formed must be at a steady state of [1-^{13}C]-glucose enrichment for most, if not all, of the labeling period. Second, little unlabeled glycogen should be present in the tissue prior to beginning the [1-^{13}C]-glucose infusion (a circumstance readily achieved *in vivo* in the liver by simply fasting a rat overnight but more difficult to achieve in the heart). Third, if the second condition cannot be met it must be known that the glycogen which is mobilized first in response to glycogenolytic stimulus is that which was most recently synthesized (glycogenolysis occurs in an ordered "last in, first out" manner) and thus has a known ^{13}C enrichment. Evidence for the latter occurring in both liver[19] and heart[20] is available and consistent with the known mechanism of action of the phosphorylase and debranching enzymes.

RESULTS

In initial experiments we examined the time course of the increase of the [1-^{13}C]-glucose signal in heart glycogen *in vivo* in the hearts of either rats[16] or guinea pigs[21] before, during, and following an infusion of [1-^{13}C]-glucose. Initially, the size of the guinea pig heart led to its use, but characterization of its glycogen biochemistry is incomplete in comparison to the many such studies of the rat heart both *in vivo* and in isolated perfused preparations. Therefore, all the results reported here were obtained in the rat heart. FIGURE 2 illustrates the baseline *in vivo* ^{13}C NMR spectrum of the heart of a fed rat obtained prior to a 50-min infusion of labeled glucose and insulin and 20 min following completion of the infusion. The difference spectrum contains a single well-defined peak assigned to [1-^{13}C]-glucose in glycogen that resonates at 100.5 ppm. In addition to the presence of the strong glycogen signal, it should be noted that there is no signal apparent from either the alpha or beta anomers of the infused glucose. Strong signals from circulating glucose are present throughout the glucose infusion period but disappear rapidly once the labeled infusion is discontinued. Presumably this reflects the rapid turnover of the circulating glucose pool and either incorporation of glucose into glycogen in body tissues (principally skeletal muscle and liver) or its oxidation and release as CO_2. As we will point out later, this rapid disappearance is helpful in studies of the turnover of the labeled glycogen pool.

In the experiment illustrated in FIGURE 2 the rat received a large dose of insulin (50 units), in addition to whatever endogenous insulin it may have secreted in response to the glucose infusion. As a result, the plasma insulin rises to concentrations far above any that might be encountered physiologically ($>$1,000 μU/ml). Recognizing that the observed glycogen synthesis might be attributed to a strictly pharmacologic effect of this large dose of insulin, we nevertheless used this model in our first series of studies of the measurements of the rates of *in vivo* glycogen synthesis. The enrichment of the plasma glucose pool was estimated as a function of time at 15-min intervals during the infusion period from the fraction of C-1 protons J-coupled to [1-^{13}C]-carbons in glucose (TABLE 3). From this, together with the rate of rise of the glycogen signal over time and the measured quantity of [1-^{13}C]-glucose in heart glycogen at the end of the experiment, we estimated the glycogen synthetic rate using Equation 1 (TABLE 3). In these studies the activities of both the *I* and *D* forms of glycogen synthase were also measured. As seen in TABLE 3 there was an excellent correspondence between the activity of the *I* form of the synthase and the observed rate of *in vivo* glycogen synthesis. This result provides strong evidence that *in vivo* the *I* form of the synthase enzyme is the rate-limiting reaction in glycogen formation in heart during the infusion of glucose and insulin. This result was not unexpected. In light of the large dose of insulin that was given, it might be anticipated that myocyte glucose transport would be maximally stimulated and control of the rate of synthesis would shift to the synthase.

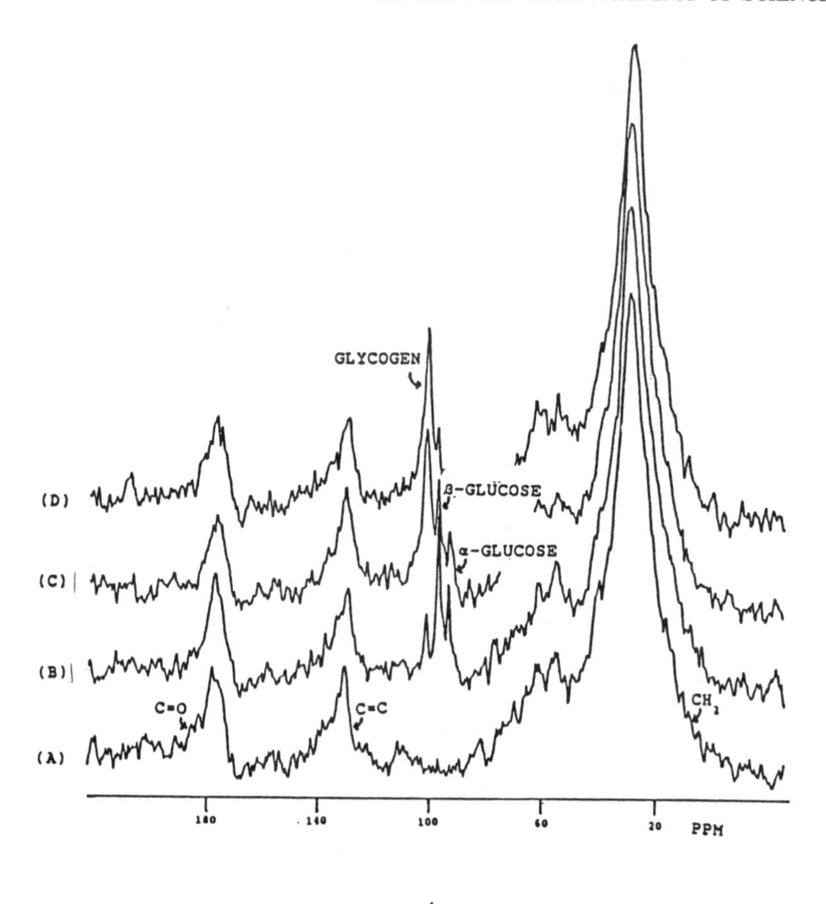

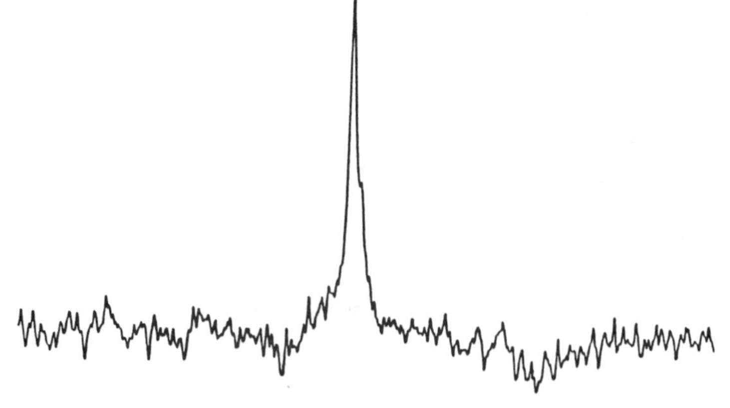

FIGURE 2. The upper panel shows a series of stacked 5 min spectra of the rat heart obtained *in vivo* before (A), at 30 min (B), at 50 min (C), and at 70 min (D) after beginning a 50 min continuous infusion of [1-^{13}C]-glucose and insulin. The lower panel is a difference spectrum obtained by subtracting three spectra obtained during the basal period from spectra accumulated between 65–80 min.

Whether a similar phenomenon occurs when glucose alone is given and only endogenous physiological amounts of insulin available was addressed next. FIGURE 3 indicates the time course of the change in the height of the ^{13}C-glucose signal in heart glycogen after beginning an infusion of [1-^{13}C]-glucose. Note that some incorporation of glucose into glycogen is apparent almost immediately and the intensity of the glucose signal increases steadily throughout the glucose infusion. TABLE 3 indicates the total glucose concentration and the percent enrichment of the plasma glucose signal with [1-^{13}C]-glucose in serial plasma samples. From these values and the calibration of the ^{13}C-glycogen signal in the post-mortem glycogen extract the rates of glycogen synthesis were calculated and these are also indicated in TABLE 3.

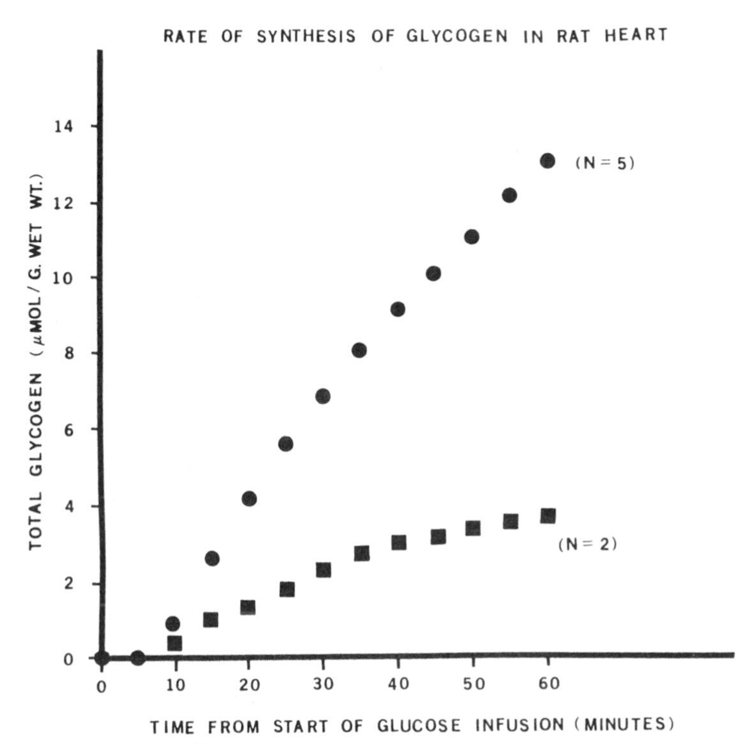

FIGURE 3. The time course of glycogen synthesis on the heart of the intact rat during a continuous infusion of glucose (■) or glucose with insulin (●).

The activities of glycogen synthase *I* and *D* and phosphorylase *a* and *b* were also measured in the extracts obtained from these hearts (TABLE 3). As we had previously seen in the rats infused with both glucose and insulin,[15] the activity of glycogen synthase *I* in these rats corresponded closely to the observed net rate of glycogen synthesis. Again, this suggests that this enzyme is the flux-limiting step for glycogen synthesis *in vivo* in the face of physiologic concentrations of both glucose and insulin.

Examination of the activities of synthase *I* and phosphorylase *a* in these animals reveals a paradox. It is apparent that there is substantially more phosphorylase *a* present than synthase. In spite of this there is net synthesis of glycogen and indeed, from the close correspondence of the *in vivo* net synthetic rate with the initial velocity of the synthase assayed *in vitro* at physiologic concentrations of UDPG, the *in vivo* synthetic rate approximates the unidirectional flux rate. The obvious question then arises of whether phosphorylase *a* is indeed "active" in this circumstance. We had previously observed a similar phenomenon in the rats receiving glucose with large doses of insulin.[16,22] In those studies a "pulse chase" design had demonstrated that when the labeled glucose infusion was replaced by an infusion of [1-^{12}C]-glucose with monitoring of the magnitude of the glycogen signal over an additional hour or more there was essentially no breakdown nor turnover of the labeled glycogen pool, thus indicating an absence of phosphorylase activity *in vivo*. This unidirectional flux of glucose into glycogen in our current study suggests that a similar phenomenon occurs when insulin is present at physiologic concentrations. Currently, we are without an adequate explanation for the lack of *in vivo* activity of phosphorylase *a* despite its apparent abundant activity measured *in vitro* in extracts of the heart. We have discussed elsewhere[16] some of the possible explanations for this: inadequate substrate for phosphorylase (necessarily inorganic phosphate since abundant glycogen is present); artifacts induced by the extraction procedure that activate phosphorylase *a*; and the presence *in vivo* of non-covalent regulators of phosphorylase *a* lost during tissue extraction. Clearly, we view the first and last possibilities as the most exciting, as they require that a portion of the *in vivo* regulation of glycogen turnover reside in mechanisms other than the direct control (by cycles of phosphorylation and dephosphorylation) of either phosphorylase or of the kinase/phosphatase enzyme cascade that regulates these cycles. If these observations are borne out by further investigation, then the information gained by these spectroscopic studies will undoubtedly spur further efforts by enzymologists to provide both a mechanism to explain this regulation and a method to assess the extent to which non-covalent control mechanisms interact with the known phosphorylation-dephosphorylation cycles in the *in vivo* regulation of glycogen turnover.

TABLE 3. Effects of Glucose or Glucose and Insulin Infusion on Heart Glycogen Metabolism

	Glucose and Insulin ($N = 4$)	Glucose ($N = 2$)
Plasma glucose (mg/100 ml)	300	294
Enrichment of plasma glucose with [1-^{13}C]-glucose (%)	35	41
Enrichment of glycogen with [1-^{13}C]-glucose (%)	12	3
Rate of glycogen synthesis (μmol/min · gww)	0.21	0.10
Activity of glycogen synthase *I* (μmol/min · gww)	0.22	0.11
Synthase *D*	0.74	0.48
Activity of phosphorylase *a* (μmol/min · gww)	3.8	5.3
phosphorylase *b*	34.0	44.4

DISCUSSION

We suggested in the introduction that it would be possible to measure the rates of the reactions involved in glycogen synthesis *in vivo* using NMR spectroscopic methods. We further suggested that it would be possible to obtain physiologic information that would either complement our understanding of the regulation of glycogen metabolism as defined by the control of the participating enzymes or perhaps to provide information not readily reconciled to our current constructs of the control of glycogen metabolism. We believe that the work presented here may begin to fulfill both expectations. To date we have measured the rates of glycogen synthesis *in vivo* in rats studied under four separate physiologic circumstances: glucose and insulin infusion;[15,16] glucose infusion alone (see above); glucose and insulin infusion in fasted animals; and in fed Type 1 diabetic animals (unpublished observations). In each of these circumstances, we observed an excellent correspondence between the net rate of glycogen synthesis measured spectroscopically and the activity of the glucose-6-P independent form of the enzyme glycogen synthase. Similarly, in each of these circumstances when net glycogen synthesis was occurring there was a large amount of phosphorylase *a* activity assayable in the extracted heart but this appeared to bear no relationship to the *in vivo* activity of phosphorylase, which was virtually zero.

These findings indicate that in the heart some additional control of phosphorylase activity occurs, the nature of which is not currently understood. The question immediately arises whether this might represent some nuance of glycogen metabolism particular to the heart or whether in fact a similar circumstance might also exist in liver and skeletal muscle. Since these tissues contain much more glycogen than the heart and serve a very important role in the regulation of overall body glucose and fuel metabolism, the answer to this question assumes great importance.

These questions can be addressed using the spectroscopic and biochemical procedures outlined for the heart. In the case of skeletal muscle, no modification of the method is required beyond substituting this tissue for the heart and conducting a similar experiment. For the liver, a different concern arises. One of our assumptions in measuring the rate of glycogen synthesis was that during the infusion of [1-^{13}C]-glucose the enrichment of the precursor pool being used to make glycogen is essentially identical to that of plasma glucose. In the case of the liver, where it has been increasingly appreciated that flux of substrate from gluconeogenesis makes a very significant contribution to the total flux of carbon into UDPG and newly formed glycogen,[23] clearly this assumption is untenable. A different approach however can be applied in this circumstance. Since liver glycogen (unlike heart or muscle) can be virtually depleted by an overnight fast, all glycogen present in liver at the end of an experiment will have been formed during the course of the infusion of labeled glucose. By measurement of the fraction of C-1 protons J-coupled to a [1-^{13}C]-carbon in glucose, the mean enrichment of the glycogen precursor pool with [1-^{13}C]-glucose that prevailed during the course of the infusion can be precisely quantitated. This calculation requires that a correction be made for the contribution of recycled glucose to the [1-^{13}C]-glucose pool. This can be readily done by measurement of the enrichment of the 6-carbon of glucose with ^{13}C.[24] It is also worth noting that by comparing the enrichment of the glycogen synthesized with the average enrichment of circulating [1-^{13}C]-glucose measured during the course of the infusion, an estimate of the relative flux of carbon from gluconeogenesis and from direct incorporation of extracted glucose can be obtained.

From these examples we believe it is quite evident how NMR spectroscopy provides us with a new and powerful tool to study glycogen metabolism in both normal and pathologic circumstances. We would like to indicate two other very specific questions in myocardial glycogen metabolism for which ^{13}C-NMR will be uniquely able to provide us with new information.

The first of these questions is whether the size of the myocardial glycogen pool regulates the rate at which new glycogen is formed, or rephrased, does intracellular glycogen inhibit its own synthesis? Since, as we have outlined above, we can use

^{13}C-NMR to monitor continuously the *in vivo* rate of glycogen synthesis over many hours, it is then possible to examine the rate of glycogen synthesis when the heart is glycogen depleted (by infusion of glucagon) or glycogen replete (via a prolonged infusion of glucose and insulin). We know that these measures will alter the glycogen content of the heart by 50–100%.[11] It then remains only to measure the rate of synthesis using a [1-^{13}C]-glucose infusion at the end of such manipulations to compare the observed rates of synthesis.

The second example relates to our desire to follow-up on the observation that when glycogenolysis is stimulated by rendering an animal anoxic the activity of glycogen synthase *I* is greatly enhanced (50–80% of the total synthase is in the more active *I* form, a greater percent than induced by the stimulation of insulin) within 1 min. Whether this stimulation plays an important role in allowing the heart to replenish glycogen stores depleted by brief periods of anoxia or ischemia is unknown. Such a process could be a potential major benefactor to the heart as it would allow rapid recovery of the heart's anaerobic reserve capacity. A defect in such a mechanism (which could exist in the diabetic heart, known to have an impaired capacity to convert glycogen synthase *D* to the *I* form) could seriously impair the ability of the heart to sustain and recover from brief episodes of either anoxia or ischemia. Our ability to continuously measure the rates of both myocardial glycogen breakdown and synthesis provides us with the opportunity to assess directly whether the apparent activation of synthase *I* by anoxia leads to rapid net replenishment of heart glycogen stores after oxygenation is restored and to examine further whether this process is impaired in pathologic settings.

In summary, we have outlined a new and what we hope will be a very useful method for applying the emerging techniques of *in vivo* ^{13}C and proton NMR spectroscopy to the study of the very complex pathways involved in the regulation of glycogen metabolism. Our studies have emphasized the importance of combining the spectroscopic measurements with more classical measurements of enzyme activity and tissue substrate concentration. We believe that these combined analyses will provide us with a more detailed understanding of glycogen metabolism than was previously possible.

ACKNOWLEDGMENTS

The authors wish to thank Ms. Judith King for expert technical assistance. We also wish to express our deep appreciation to Dr. Robert Shulman for much valued advice and support.

REFERENCES

1. Newsholme, E. A. & A. R. Leech. 1983. Biochemistry for the Medical Sciences. 583–608. John Wiley & Sons. New York.
2. Cohen, P., P. J. Parker & J. R. Woodgett. 1985. *In* Molecular Basis of Insulin Action. M. P. Czech, Ed.: 213–234. Plenum Press. New York.
3. Randle, P. J. & P. K. Tubbs. 1979. *In* The Handbook of Physiology: The Cardiovascular System. 805–844. American Physiology Society. Bethesda, MD.
4. Villar-Palasi, C. & J. Larner. 1960. Arch. Biochem. Biophys. **86:** 270–273.
5. Chen, V. & C. D. Ianuzzo. 1982. Can. J. Physiol. Pharmacol. **60:** 1251–1256.
6. Nilsson, L. H. & E. Hultman. 1974. Scand. J. Clin. Lab. Invest. **33:** 5–10.
7. Canioni, P., J. R. Alger & R. G. Shulman. 1983. Biochemistry **22:** 4974–4980.
8. Alger, J. R., L. O. Sillerud, K. L. Behar, R. J. Gillies, R. G. Shulman, R. E. Gordon, D. Shaw & P. E. Hanley. 1981. Science **214:** 660–662.

9. REO, N. V., B. A. SIEGFRIED & J. J. ACKERMAN. 1984. J. Biol. Chem. **259:** 13664–13667.
10. HEMS, D. A., P. D. WHITTON & E. A. TAYLOR. 1972. Biochem. J. **129:** 529–538.
11. CORNBLATH, M., P. J. RANDLE, A. PARMEGGIANI & H. E. MORGAN. 1963. J. Biol. Chem. **238:** 1592–1597.
12. DANFORTH, W. H. 1965. J. Biol. Chem. **240:** 588–593.
13. COSTILL, D. L., E. COYLE, G. DALSKY, W. EVANS, W. FINK & D. HOOPES. 1977. J. Appl. Physiol. **43:** 695–699.
14. CHAISSON, J. L. & A. D. CHERRINGTON. 1983. *In* Handbook of Experimental Pharmacology. **66:** 361–379. Springer-Verlag. New York.
15. LAUGHLIN, M., E. J. BARRETT, J. ALGER & R. G. SHULMAN. 1985. Proc. Soc. Mag. Res. Med. **2:** 799–800.
16. LAUGHLIN, M. R., W. A. PETIT, J. M. DIZON, R. G. SHULMAN & E. J. BARRETT. 1986. Submitted for publication.
17. THOMAS, J. A., K. K. SCHLENDER & J. LARNER. 1968. Anal. Biochem. **25:** 486–499.
18. GILBOE, D. P. & F. Q. NUTTAL. 1972. Anal. Biochem. **47:** 28–38.
19. SILLERUD, L. O. & R. G. SHULMAN. 1983. Biochemistry **22:** 1087–1094.
20. NEUROHR, K. J., G. GOLLIN, J. M. NEUROHR, D. L. ROTHMAN & R. G. SHULMAN. 1984. Biochemistry **23:** 5029–5035.
21. NEUROHR, K. J., E. J. BARRETT & R. G. SHULMAN. 1983. Proc. Natl. Acad. Sci. USA **80:** 1603–1607.
22. LAUGHLIN, M. R., W. A. PETIT, R. G. SHULMAN & E. J. BARRETT. 1986. Proc. Soc. Mag. Resonance Med. **5:** 129–130.
23. KATZ, J. & J. D. MCGARRY. 1984. J. Clin. Invest. **74:** 1901–1909.
24. SHULMAN, G. I., D. L. ROTHMAN, D. SMITH, C. M. JOHNSON, J. B. BLAIR, R. G. SHULMAN & R. A. DEFRONZO. 1985. J. Clin. Invest **76:** 1229–1236.
25. NEWGARD, C. B., L. J. HIRSCH, D. W. FOSTER & J. D. MCGARRY. 1983. J. Biol. Chem. **258:** 8046–8052.
26. ADOLFSSON, S., O. ISAKSON & A. HJALMARSON. 1972. Biochim. Biophys. Acta **279:** 146–156.
27. WARDZALA, L. J., M. HIRSCHMAN, E. POFCHER, E. D. HORTON, P. M. MEAD, S. W. CUSHMAN & E. S. HORTON. 1985. J. Clin. Invest. **76:** 460–469.

DISCUSSION OF THE PAPER

TORNHEIM: I'm not quite sure exactly how you did your measurement of phosphorylase A vs. B. Was it simple plus and minus AMP?

E. J. BARRETT: Yes.

TORNHEIM: Okay. One thing that may change your phosphorylase A results—that will put it in the other direction and may also account for the lack of activity of phosphorylase A *in vivo*, at least this is what's been observed with the skeletal muscle enzyme—is that there is under some conditions a dependence of phosphorylase A on AMP, but the requirement for phosphorylase A for AMP as an activator is considerably less than that for phosphorylase B. In other words, micromolar levels of AMP may need to be added to measure phosphorylase A *in vitro*. Secondly, very low levels of AMP can perhaps account for the inactivity of the phosphorylase A *in vivo*.

BARRETT: I think your points are good. As you point out, our problem with the *in vivo* measurements is that we see no activity yet we know there's AMP around, so it's not that there's a lack of it in the *in vivo* setting. What we're finding is even the phosphorylase A activity seen in the absence of AMP is not being expressed.

TORNHEIM: Perhaps with the situation in skeletal muscle, where it is thought that perhaps actual levels of AMP is say, if extrapolated based on creatine phosphate, etc.

data. The AMP levels may actually be micromolar or sub-micromolar and that in fact, I mean in skeletal muscle where there's even higher amounts of phosphorylase and phosphorylase A levels, maybe at rest 10 or 15% of total phosphorylase. So there's a whopping amount of phosphorylase A yet there's not glycogen breakdown in the unstimulated muscle and there again the answer may simply be that that enzyme is dependent on either AMP or IMP as an activator.

BARRETT: I think I follow your point. I agree with you in that AMP clearly can be an activator of phosphorylase A as well as phosphorylase B. The puzzling thing about that is that if you take out all of the AMP and there are extracts we've run over an ion exchange column, to try and get rid of all the AMP possible, the activity that we're seeing is still greater than that expressed *in vivo.* Even if AMP's concentration is micromolar in the muscle cell, that's going to give you greater activity than we have, and we've already got too much.

TORNHEIM: That depends on levels of other factors, substrates and so on.

BARRETT: Oh, absolutely.

TORNHEIM: Who's to say exactly what the match is in terms of glycogen levels, phosphate, etc.?

BARRETT: I agree with you. My point is simply that we don't know what those other things are at this point in time, but it doesn't seem to be the phosphorylation state per se, and that's where the emphasis had been.

J. R. GRIFFITHS: The effect of AMP on phosphorylase A at least in skeletal muscle enzyme is a K system in the monoterminology and it's main effect is to alter the K_m's for phosphate and glycogen, which are probably sufficiently high here, so I doubt that that's what's the critical factor. Phosphorylase A is about 10 or 15% of total phosphorylase activity skeletal muscle and that's enought to break down all the glycogen very quickly. One suggestion has been that under *in vivo* conditions phosphocreatine is an inhibitor of phosphorylase A but only when the enzyme is present at the very high concentrations found in muscle cell. This was suggested by Scopes a few years ago and I've never heard of anybody actually refute the suggestion or confirm it.

BARRETT: You're anticipating the directions we're going. Our problem now is to find those likely candidate substrates which may by a noncovalent process regulate this activity of phosphorylase A. There are a number of candidate regulators, phosphocreatine is one, AMP perhaps another, glucose itself has been suggested. Certainly in liver that seems to be the case. The bigger question that arises is how do you go about assaying that. One of the difficulties that I'm sure you're familiar with is that most assays of phosphorylase are run in the reverse direction, in the direction of synthesis. The direction that we're taking is first to set up a good assay in the direction of glycogen breakdown where you can run the assay in the absence of high levels of G-1-P. Any interaction between phosphorylase and either other phosphorylated compounds or glucose itself, will likely be obscured in the presence of high concentrations of glucose-1-P.

^{31}P NMR Studies of the Kinetics and Regulation of Oxidative Phosphorylation in the Intact Myocardium[a]

K. UĞURBIL, P. B. KINGSLEY-HICKMAN, E. Y. SAKO,[b]
S. ZIMMER,[c] P. MOHANAKRISHNAN,
P. M. L. ROBITAILLE, W. J. THOMA, A. JOHNSON,
J. E. FOKER,[b] AND A. H. L. FROM[c]

Department of Biochemistry
and
Gray Freshwater Biological Institute
Navarre, Minnesota 55392

[b]*Department of Surgery*
University of Minnesota
Minneapolis, Minnesota 55455

[c]*Department of Medicine*
Cardiology Division
Minneapolis VA Medical Center
and
University of Minnesota
Minneapolis, Minnesota 55455

INTRODUCTION

In aerobic tissues, the dominant pathway of ATP generation is oxidative phosphorylation. While numerous mechanistic features of this process are understood, there is a paucity of information about the kinetics and regulation of ATP generation in the intact cell under physiological and pathophysiological conditions. The primary cause of this deficiency is the extreme difficulty of obtaining the desired information using classical biochemical methods, which are based on extraction and radioisotope labeling. However, with the relatively recent expansion of nuclear magnetic resonance (NMR) spectroscopy to studies of intact cells and tissues, new methods have become available to investigate the question of metabolic control at the level of oxidative phosphorylation. Accordingly, we have undertaken two separate ^{31}P NMR studies on the problems of respiratory regulation and the kinetics of ATP synthesis in the intact heart. In this article, we present a brief review of the results obtained from our studies.

[a]Supported by National Institutes of Health Grants HL33600, HL26640, and 1K04HL01241, and Veteran's Administration Medical Research funds, and American Heart Association, Minnesota Affiliate.

RESPIRATORY REGULATION IN THE MYOCARDIUM

It is well recognized that increased mechanical activity by the cardiac muscle elicits a commensurately higher rate of oxygen utilization.[1] The mechanism by which this coupling is mediated invokes, by necessity, the question of mitochondrial respiratory control in the intact myocardium. Current concepts of respiratory control, however, are largely derived from studies on isolated mitochondria. Approximately 30 years ago, Chance and Williams reported that the oxygen consumption rate of mitochondria in the presence of non-limiting concentrations of O_2 and carbon substrate was determined by the concentration of exogenous ADP and obtained an apparent K_m of 20–30 μM for ADP utilization.[2] Subsequently, phosphorylation potential (PP), which is equal to $[ATP]/[ADP][P_i]$, and ATP/ADP ratio were also proposed as regulatory parameters in mitochondrial respiratory control.[3–5] In order to investigate whether any of these three potential mechanisms is operative in the intact myocardium we have performed NMR measurements of high energy phosphates in isovolumic, Langendorff-perfused rat hearts concurrently with determinations of myocardial oxygen consumption rate (MVO_2) and mechanical function at six progressively increasing levels of mechanical output. The alterations in workload were achieved by manipulating the heart rate, end diastolic pressure (EDP), and exposure to an inotropic agent.[6] ATP and creatine phosphate (CP) content and intracellular pH were determined from fully relaxed NMR spectra recorded in 10 min at each workload; the cytosolic "free" ADP content was calculated from the creatine kinase equilibrium. The product of systolic pressure and heart rate, the rate pressure product (RPP), was used as an index of myocardial mechanical output. These studies were conducted using three different exogenous carbon substrate conditions, namely, glucose (G), glucose + insulin (GI), and pyruvate + glucose (PG).

FIGURE 1 illustrates typical spectra obtained from hearts under the three substrate conditions at a fairly high RPP (~60,000 mm Hg/min). Over the range of RPP attained, the relationship between MVO_2 and RPP was linear and comparable for the different substrate conditions as previously shown.[1,6] FIGURE 2 illustrates PP and ATP/ADP ratios as a function of MVO_2. The relationship between MVO_2 and ADP concentrations is illustrated as a double-reciprocal plot in FIGURE 3; in this representation, a straight line with a slope of K_m/V_{max} and intercept of $1/V_{max}$ would be expected if ADP availability is the rate-determining factor in respiration.

It is evident from these data that a unique and single value of ADP level, ATP/ADP ratio, of the PP is not associated with a given value of MVO_2. Instead, at any given MVO_2, all three parameters varied significantly with the three different exogenous substrate conditions. This implies that none of these entities serve as a universally applicable parameter that regulates mitochondrial respiration in the intact myocardium.

Inverse correlations observed between PP or ATP/ADP ratio and oxygen consumption rate in mitochondrial suspensions have been the basis for attributing a regulatory role to these parameters in respiratory control. However, it was recently shown that this usual inverse relationship documented between MVO_2 and PP or ATP/ADP ratio in isolated mitochondrial suspensions could be reversed to a direct relationship by altering the means of ADP generation in the suspension.[7] Consequently, it appears unlikely that a causal relationship exists between O_2 consumption rate and PP or ATP/ADP ratio. In agreement with this conclusion, the data on the intact heart did not yield a consistent correlation between MVO_2 and these two parameters. ATP/ADP ratio decreased with increasing MVO_2 only with PG substrate (FIG. 2). In contrast, the ATP/ADP ratio was relatively constant during an approximately three-fold increase in MVO_2 in GI-perfused hearts, and it first decreased and

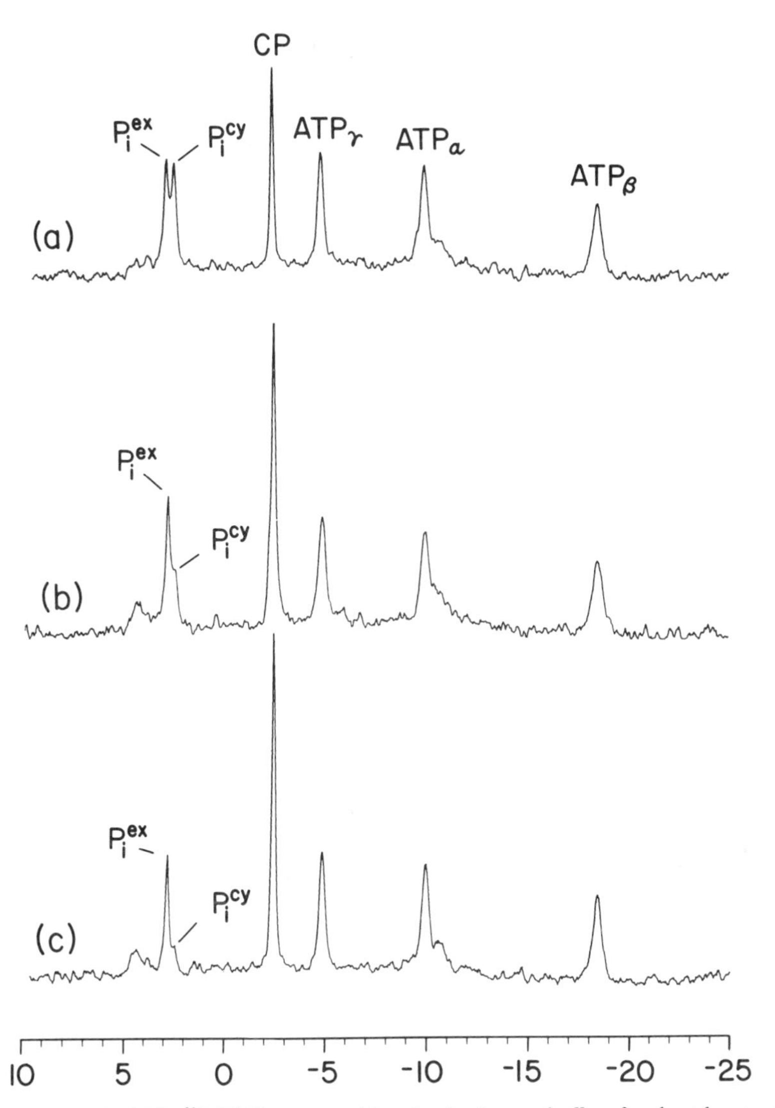

FIGURE 1. 146.1 MHz ^{31}P-NMR spectra of isovolumic, Langendorff-perfused rat hearts at a moderately high workload achieved with HR = 300, EDP = 8, and 40 ng/ml of dobutamine in the perfusate using G (a), GI (b), and PG (c) as the carbon substrate. The spectra are the sum of 40 free induction decays obtained with 90° pulses and 15 sec repetition time.

subsequently increased with increasing MVO_2 during G perfusion. Similarly, PP displayed an inverse dependence on MVO_2 for PG and GI-perfused hearts but not during G perfusion. Even when an inverse correlation was present, all observed values for the ATP/ADP ratio and the PP were at the extremes of the putative regulatory ranges of these parameters when compared to data obtained on isolated mitochondria.[7]

In view of these observations, we conclude that neither PP nor the ATP/ADP ratio controls mitochondrial respiration in the intact myocardium.

For PG substrate conditions, there was a linear relationship between $(MVO_2)^{-1}$ and $[ADP]^{-1}$, which gave an apparent K_m of 25 ± 5 μM and V_{max} of 148 ± 24 μmoles O_2 min^{-1} (g dry weight)$^{-1}$. This K_m is very similar to values obtained from isolated

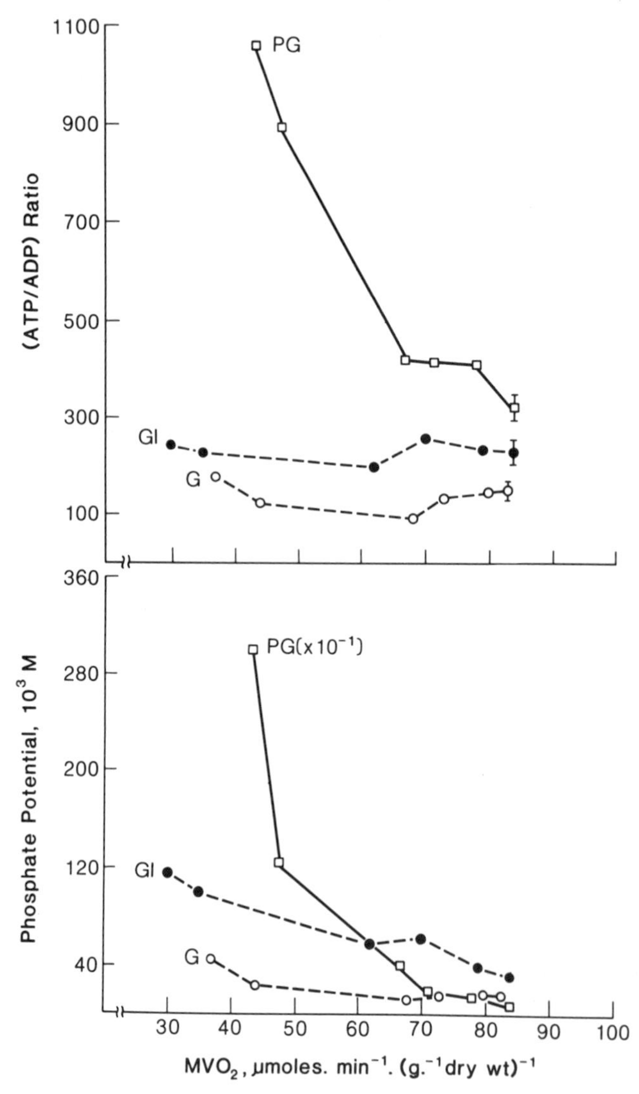

FIGURE 2. ATP/ADP ratio and the phosphorylation potential (PP) obtained for glucose (G), glucose + insulin (GI), and pyruvate + glucose (PG) perfused hearts at six different workloads. Typical error bars are shown for one data point for each perfusion condition. For PP calculations, cytosolic P_i concentration was obtained as described in From *et al.*[6]

mitochondrial studies where ADP availability was shown to be the rate-limiting factor in respiration. The linear dependence between reciprocal MVO_2 and [ADP], and the K_m value of 25 μM suggests that ADP availability may indeed be a rate-limiting step in mitochondrial respiration when exogenous pyruvate is available as the carbon substrate. A similar linear relationship between reciprocal MVO_2 and P_i, which gave a K_m value of 0.38 mM, was also noted in pyruvate-perfused hearts.

The data on the intact heart (FIGS. 2 and 3) can be rationalized if a general substrate limitation is proposed as the regulatory mechanism in respiratory control. Oxidative phosphorylation requires ADP, P_i, and mitochondrial NADH; NADH generation in turn is dependent upon the availability of carbon substrates for the TCA cycle and the activity of the TCA cycle enzymes. It was previously noted that even at very low workloads, levels of cytosolic pyruvate, mitchondrial NADH, and TCA cycle intermediates are at least an order of magnitude lower during GI perfusion relative to those achieved with PG substrates.[1,8,9] In the absence of insulin when glucose uptake rate is reduced, the relative depletion of the TCA cycle intermediates is expected to be even more pronounced. Thus, in the presence of relatively high concentrations of pyruvate, when pyruvate dehydrogenase (PD) is maximally activated[10] and mitochondrial NADH is abundant and non-limiting, the rate of oxidative phosphorylation can be limited by the availability of ADP and/or P_i. The linear relationship observed between $(MVO_2)^{-1}$ and $(ADP)^{-1}$ with a K_m of 25 μM (FIG. 3) and between $(MVO_2)^{-1}$ and P_i^{-1} (not shown) indeed supports this concept. Under G and GI perfusion both the ADP and P_i levels are substantially higher than in PG-perfused hearts at all workloads. Therefore, availability of these metabolites can no longer be rate limiting; in contrast, when glucose is the exogenous substrate, mitochondrial NADH and TCA cycle intermediate levels are known to be lower than in PG-perfused hearts.[1,8,9] This suggests that under these conditions, carbon substrate delivery to the TCA cycle, the activity of the TCA cycle enzymes, and consequent mitochondrial NADH generation may become the rate-determining step in oxidative phosphorylation. These ideas are also consistent with recent observations made in intact dogs[11] and isolated mitochondrial suspensions;[12] in the latter case an increase in MVO_2 was achieved concurrently with a

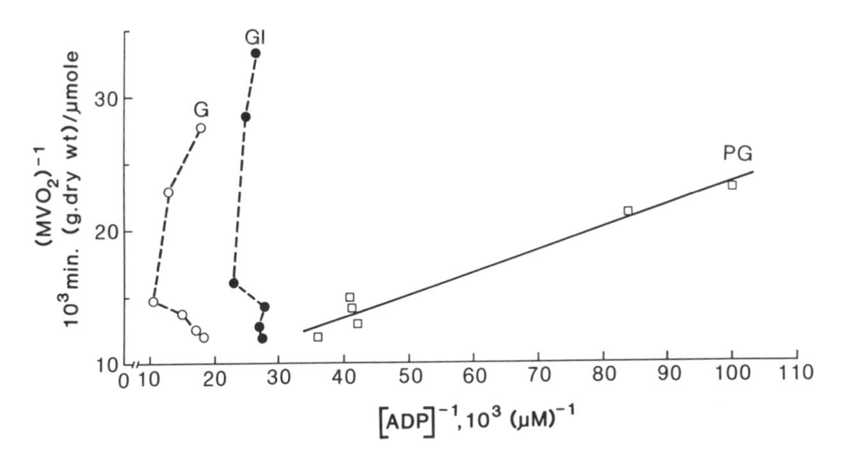

FIGURE 3. Double-reciprocal plot of $(MVO_2)^{-1}$ vs. $[ADP]^{-1}$ in G (○), GI (●), and PG (□) perfused hearts. The solid line was obtained by a linear least-squares fit to the PG data.

decline in extramitochondrial ADP and P_i levels by simply altering the TCA cycle carbon substrate availability supplied to the mitochondria.

If mitochondrial respiratory control is exercised primarily through substrate limitation, then the supply of the requisite substrate(s) must be regulated by the mechanical activity of the cardiac muscle. When ADP and/or P_i is the rate-determining substrate for oxidative phosphorylation (i.e. under PG conditions), mechanical activity can control respiration directly; this follows simply from the fact that the immediate consequence of increased contractile activity is production of ADP and P_i, which in turn can stimulate commensurately higher MVO_2. However, under conditions of abundant ADP and P_i, such as when glucose is the carbon source, a less direct coupling mechanism must exist. One possible point where such control may be exercised is at pyruvate dehydrogenase. The activity of this enzyme is influenced by intramitochondrial Ca^{2+} through regulation by the kinase-phosphatase mediated

TABLE 1. The Effect of DCLA on RPP, MVO_2, CP/ATP Ratio, and ADP Content of Glucose Perfused Hearts (where Endogenous Fatty Acid Utilization Was Inhibited) at Workloads I through IV[a]

	I	II	III	IV
Control				
RPP 10^3 mm Hg · min^{-1}	27.5 ± 1.9	37.7 ± 1.7	49.4 ± 1.9	51.7 ± 2.2
MVO_2 μmoles · min^{-1} · (g dry wt)$^{-1}$	41 ± 5	48 ± 5	60 ± 7.4	59 ± 5
CP/ATP	1.15 ± 0.03	1.07 ± 0.04	1.10 ± 0.06	1.12 ± 0.06
ADP nmoles · (g dry wt)$^{-1}$	185 ± 10	220 ± 14	230 ± 21	235 ± 18
+DCLA				
RPP 10^3 mm Hg · min^{-1}	26.3 ± 1.6	37.7 ± 3.3	58.0 ± 1.2	78.2 ± 3.2
MVO_2 μmoles · min^{-1} · (g dry wt)$^{-1}$	42 ± 2	48 ± 2	70 ± 2	76 ± 3
CP/ATP	1.58 ± 0.12	1.52 ± 0.12	1.39 ± 0.10	1.39 ± 0.05
ADP nmoles · (g dry wt)$^{-1}$	96 ± 11	110 ± 15	149 ± 15	156 ± 11

[a]All values are mean ± SEM, $N = 7$. Endogenous fatty acid utilization is a major energy source in glucose perfused hearts. Bromocrotonic acid was used to inhibit this pathway so that DCLA effect on glucose utilization can be examined.

interconversion system;[13] as is well known, cytosolic Ca^{2+} levels are related to mechanical performance and also determine the intramitochondrial Ca^{2+} levels.[13] It was previously noted that dichloroacetic acid (DCLA) at low concentrations activates PD to its maximum capacity.[10] When we re-examined glucose-perfused hearts (where endogenous fatty acid consumption was inhibited by bromocrotonic acid to drive the metabolism exclusively to glucose utilization), a substantial decrease in ADP level was noted at each workload in the presence of DCLA (TABLE 1). DCLA did not have any effect on pyruvate-perfused hearts where, due to the high pyruvate level, PD is expected to be fully activated.[10] These observations suggest that regulation of PD, possibly through variations in the cytosolic Ca^{2+} levels linked to alterations in mechanical output, may be one of the points where carbon substrate delivery to the TCA cycle and thus NADH generation is regulated. Of course, a similar regulation must also be exercised at other points of the glycolysis pathway and TCA cycle.[13] Not surprisingly, it has previously been shown that the glucose transport system is affected by alterations in workload[14] and the activities of several TCA cycle enzymes are calcium sensitive.[13]

ATPASE KINETICS IN THE INTACT MYOCARDIUM

In a second series of studies, we have investigated in detail the possibility of using magnetization transfer (MT) techniques, in particular, conventional, two-site saturation transfer[15] (CST) and multiple saturation transfer[16,17] (MST) to study ATP synthesis and hydrolysis kinetics in the intact myocardium.

Magnetization transfer techniques provide the unique capability of measuring unidirectional rates of chemical reactions. Their application to studies in intact tissues, however, are complicated. The chemical exchange problems in the intact tissue can be extremely complex due to the partitioning of enzymatic reactions and their reactants among the various subcellular organelles, the possible existence of metabolic compartmentation (e.g., "free" and "bound" pools), and the presence of multiple reactions that utilize the same substrates. Provided that the resonances of the exchanging metabolites are well resolved from each other, the last complication can be eliminated by use of the MST procedure.[16,17] Measurement of creatine kinase (CK) rates in intact hearts is a specific case where conventional saturation-transfer experiments yielded a paradoxical result; subsequent studies with MST showed that the apparent paradox was primarily due to the existence of multiple exchanges that utilized ATP.[17] Despite these complications, however, the potentially high and unique information content of MT measurements justifies efforts to design and perform experiments that examine the validity of magnetization-transfer measurements to study ATP kinetics in intact tissues. No other technique provides the means by which this crucial bioenergetic process can be studied in intact cells.

Until now, the MT studies on myocardial ATP $\rightleftharpoons$ P_i exchange have exclusively been performed in the direction of ATP formation.[18–20] As discussed in greater detail further on, ATP hydrolysis rates are difficult to perform because of signal-to-noise problems and because they require the use of the more complex multi-site exchange methods. The $P_i \rightarrow$ ATP rate can be measured using CST or in principle, the analogous two-site inversion transfer method.[21] The measurements reported so far have employed CST exclusively. Prior to reaching any biochemical conclusions on the kinetics of ATP synthesis from such data, three very fundamental questions must be posed on the use of magnetization transfer methods. (1) What is the origin of the $P_i \rightarrow$ ATP exchange monitored by NMR? (2) If contributions from all pathways other than oxidative phosphorylation are negligible or can be eliminated, is it possible to measure the rate of ATP synthesis due to oxidative phosphorylation by magnetization-transfer? (3) If the answer to the second question is "yes," can one use the measurement to determine the P:O ratio and thus evaluate mitochondrial competence under normal and pathological conditions?

The reasons for the first inquiry are based on the fact that in the intact tissue, the exchange monitored by NMR between cytosolic P_i and ATP is not a simple two-site process (FIG. 4); it can occur potentially in both the $P_i \rightarrow$ ATP and ATP $\rightarrow P_i$ directions through the mitochondria (by means of the mitochondrial phosphate transport, the H^+-ATPase, and the translocase reactions), through catalysis by glycolytic enzymes glyceraldehyde-3-phosphate dehydrogenase (GAPDH) and phosphoglycerokinase (PGK), and potentially through other pathways. The myocardium is primarily an aerobic tissue; most of its ATP requirement is met through oxidative phosphorylation.[1] This strong oxidative inclination, however, is insufficient to exclude

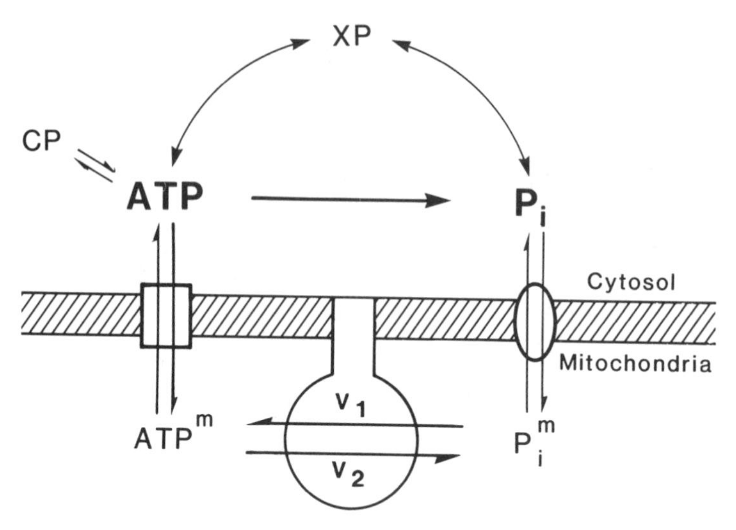

FIGURE 4. Exchange pathways between cytosolic ATP and P_i.

possible non-oxidative contributions to the NMR measurable $P_i \rightarrow$ ATP rate. This simply is a consequence of the fact that magnetization-transfer methods measure unidirectional rates. The net rate of a reaction is the difference between the two unidirectional rates; while the net ATP synthesis rate by a pathway such as glycolysis can be slow, individual enzymes within this pathway can operate with rapid unidirectional rates.

The reason for the second question is based on the fact that the measurement of the $P_i \rightarrow$ ATP rate involves saturation of the ATP_γ resonance. In the intact tissue, however, γ-phosphate resonances of mitochondrial and cytosolic ATP are expected to have approximately the same chemical shift. Consequently, irradiating the ATP_γ resonance position will saturate both the mitochondrial and the cytosolic ATP_γ spins. It should be noted that resonances from a certain spin population need not be detectable in order to saturate them. Thus, in intact cells, even though ATP resonances detected are predominantly of cytoplasmic origin, all ATP pools can be saturated with the appropriate rf irradiation. Therefore, when ATP synthesis occurs predominantly through oxidative-phosphorylation, the rate measured by saturating both cytosolic and mitochondrial ATP_γ spins and monitoring cytosolic P_i resonance is the rate of incorporation of cytosolic P_i into mitochondrial ATP. This rate is mediated by the mitochondrial P_i transport and H^+-ATPase steps. Attributing this rate to the H^+-ATPase alone requires either that the mitochondrial and cytosolic P_i pools are rapidly exchanging relative to their spin-lattice relaxation times and the rate of ATP synthesis, or that the P_i transport operates unidirectionally at all times. In the former limit, the rate measured is the unidirectional rate of ATP synthesis (v_1) by the mitochondrial H^+-ATPase (FIG. 4); this rate is greater than or equal to the net rate of ATP synthesis by oxidative phosphorylation, which would be equal to ($v_1 - v_2$) (FIG. 4). If the latter limit is applicable, then the P_i extrusion from the mitochondria is negligible, and the magnetization-transfer method measures only the rate of P_i transport into the mitochondria; during steady-state, this rate must be equal to the net ATP synthesis rate by oxidative phosphorylation in the tissue. In general, mitochondrial P_i transport is thought to be rapid and not to be the rate-limiting step in oxidative phosphorylation.

Therefore, the question pertinent to MT studies is whether it is sufficiently rapid to average the longitudinal magnetization of the mitochondrial and cytosolic P_i pools so that the rate determined by MT can be ascribed to the mitochondrial H^+-ATPase. If an intermediate exchange condition prevails, the measured rate will represent a complex average of the H^+-ATPase and P_i transport kinetics.

If the potential problems due to P_i transport can be ruled out, the rate determined by CST experiments is the unidirectional rate of ATP synthesis by the H^+-ATPase (v_1). This rate is not necessarily related to the O atom consumption rate (MVO) by the P:O ratio. The exchange process between cytosolic P_i and cytosolic ATP through the mitochondria can occur reversibly. If there exists a non-negligible unidirectional rate (v_2) in the ATP → P_i direction through the H^+-ATPase (FIG. 4), ($v_1 - v_2$) and not v_1 must be proportional to MVO by the P:O ratio. This can be visualized simply by a *gedanken* experiment. Consider a preparation of isolated mitochondria with no leaks and no energy-requiring process such as ion transport, that can dissipate a transmembrane H^+-chemical potential gradient ($\Delta\mu_{H^+}$). If this preparation is supplied ADP, P_i, O_2, and a TCA cycle substrate, it will utilize O_2 and synthesize ATP until $\Delta\mu_{H^+}$ attains a state of equilibrium with P_i, ADP, and ATP levels in the mitochondria (i.e., $\Delta\mu_{H^+} = q$ $(\mu_o + RT \ln ([ATP]/[P_i]\ [ADP])))$; at this point, extra- and intramitochondrial P_i,

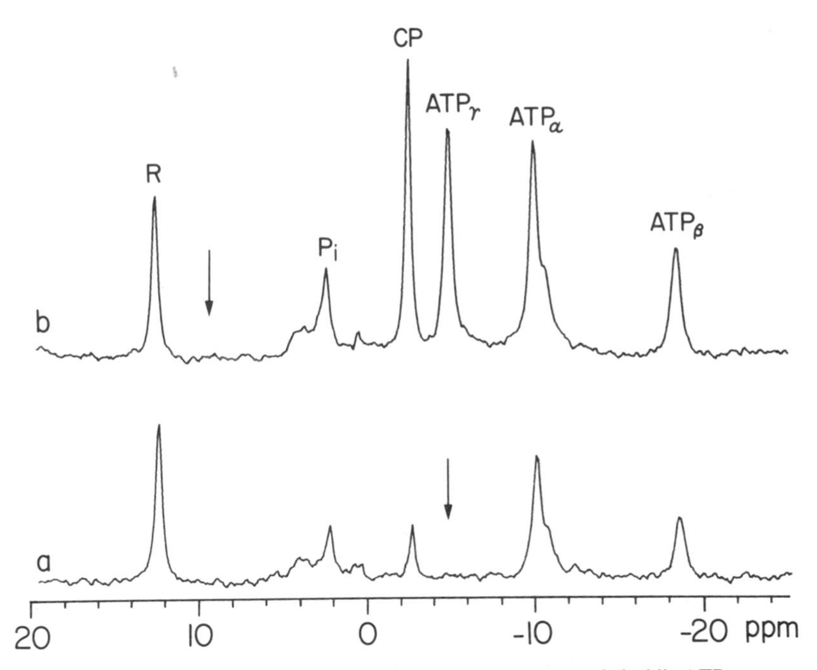

FIGURE 5. 146.1 MHz ^{31}P NMR spectra of a perfused rat heart recorded while ATP_γ resonance was selectively saturated (a), and while the selective irradiation frequency was moved downfield of the P_i resonance symmetrically opposite from the ATP_γ peak (b). Spectra were recorded using 90° pulses, and interpulse delays of 5 sec and 2.8 sec for (a) and (b), respectively. Note that the P_i resonance is fully relaxed in both cases since saturation of ATP_γ reduces the P_i T_1 to less than 0.6 sec in this case. These spectra were taken from a series of spectra accumulated in a time-averaged fashion as described in Kingsley-Hickman *et al.*[20] Peak R is the reference contained in the balloon.

ADP, and ATP levels will also be in equilibrium with each other through the activities of the P_i transport and the translocase reactions. When this point is reached, in the absence of any ATP utilization and any leaks that can dissipate the $\Delta\mu_{H^+}$, ATP and O_2 consumption must cease. However, these mitochondria can still catalyze a back-and-forth exchange between exogenous P_i and ATP even though the net ATP formation does not occur. In other words, neither v_1 nor v_2 is equal to zero but $(v_1 - v_2)$ is. In this domain, the ratio of v_1 (which is the rate determined by magnetization-transfer) to MVO is ∞.

In view of the complexity of the problem outlined above, the ultimate question is whether it is possible to construct experimentally testable criteria for evaluating the fundamental questions raised. The answer is "yes." If mitochondrial P_i transport is sufficiently rapid so that we are measuring v_1, if H^+-ATPase is far out of equilibrium (i.e., $v_2 \sim 0$) so that v_1 is proportional to MVO by the P:O ratio, and if there exist no contributions to the rate measured by magnetization-transfer other than oxidative phosphorylation, then the experimentally determined $P_i \rightarrow$ ATP rate must be equal to K(MVO) at all MVO levels, where K is a constant; if this criteria is satisfied, K is

TABLE 2. Unidirectional $P_i \rightarrow$ ATP Rate at Four Different Workloads

	i (N = 10)	ii (N = 10)	iii (N = 10)	iv (N = 12)
MVO_2 μmoles min^{-1} (g dry wt)$^{-1}$	34 ± 2	49 ± 3	57 ± 2	67 ± 3
$P_i \rightarrow$ ATP rate μmole sec^{-1} (g dry wt)$^{-1}$	7.1 ± 0.8	7.3 ± 0.5	7.5 ± 0.7	6.5 ± 0.9
Rate/MVO	6.4 ± 0.8	4.5 ± 0.2	4.0 ± 0.3	2.9 ± 0.3

All values are mean ± SE. N is the number of hearts on which these measurements were performed. The preparation was isovolumic, Langendorff-perfused hearts supplied with glucose as the primary carbon source. Workloads iii and iv were supplemented with 2 and 3 mM pyruvate, respectively. Under the second, third, and fourth workloads, 50, 80, and 80 μg/L of dobutamine, respectively, was infused into the perfusate, using a side-port in the perfusion line. The mean heart rate was 320, 420, 486, and 600, respectively, and EDP was set at 4–8, 4–8, 16–20, and 24–28 mm Hg, respectively, for the four workloads. Unless the spontaneous heart rate was near the rates given above, hearts were paced to achieve the desired rate.

equal to the P:O ratio. A more complicated dependence on MVO would indicate that one or more of the conditions listed above are not fulfilled. An experimental evaluation of this condition requires simply that the $P_i \rightarrow$ ATP flux is measured at different workloads and MVO_2. There exist two additional experimental checks; one is to perform rate measurements at the same MVO_2 but with different P_i levels; this experiment is relevant to the problems of exchange between cytosolic and mitochondrial P_i exchange, and possible P_i compartmentation. The second is to compare the $P_i \rightarrow$ ATP rate obtained by CST with the ATP hydrolysis rate that can be measured using MST.

All three types of experiments proposed above were conducted on isovolumic Langendorff-perfused hearts. Approximately 50% difference in P_i levels was obtained by adjusting the pyruvate level at a given workload and the $P_i \rightarrow$ ATP rates determined were essentially the same. In a more extensive set of experiments, the $P_i \rightarrow$ ATP rate was examined at four different workloads achieved by pacing, adjusting LV volume and exposure to dobutamine. The details concerning the NMR aspects of the saturation-transfer measurements are given elsewhere.[20] FIGURE 5 illustrates typical spectra recorded during the measurement of the $P_i \rightarrow$ ATP rate at one workload. The

$P_i \rightarrow$ ATP rates together with MVO_2 and mechanical data are given in TABLE 2. Exogenous glucose was the main carbon source for these hearts. The two highest workloads were supplemented with small amounts of pyruvate. Under these conditions, the $P_i \rightarrow$ ATP rate remains virtually unaltered while the MVO_2 changes by a factor of approximately two. Clearly, as judged by the criteria we have established, P:O ratio cannot be derived from these measurements. If we assume that the experimentally determined $P_i \rightarrow$ ATP rate arises exclusively from oxidative phosphorylation, then we must conclude that unidirectional rates v_1 and v_2 of the H^+-ATPase are both non-vanishing and that v_1 is constant while v_2 decreases with increasing workload and MVO_2.

Alternatively, there may exist contributions to the experimentally determined rate from other pathways involved in ATP generation, in particular from glycolysis. Even though the net rate of glycolysis is slow in the myocardium,[1] it is possible that glycolytic enzymes GAPDH and PGK together can catalyze $P_i \rightleftharpoons$ ATP exchange through the phosphorylated intermediate 3-phosphoglyceroyl phosphate (3-PGP) so that the unidirectional rates of exchange exceed the net rate of ATP production by substrate level phosphorylation. The existence of a glycolytic contribution to the NMR-measurable $P_i \rightarrow$ ATP rate was recently demonstrated in yeast.[22,23]

To evaluate experimentally the origin of the saturation-transfer effect in the myocardium, we have repeated these measurements in hearts where GAPDH was either inhibited with iodoacetate (IA) and a small but sufficient amount of pyruvate was supplied to the hearts as the carbon source, or overall glycolytic activity was reduced by eliminating endogenous and exogenous glycolytic carbon sources and supplying the heart with a small but sufficient amount of pyruvate to maintain the required workload.

Inhibition of GAPDH using IA in perfused rat hearts without adverse effects on nucleotide pools and mechanical activity was evaluated in trial experiments. The optimum IA exposure protocol that was finally used in the kinetic experiments was to subject the hearts to 0.15 mM IA in the perfusate for 15 min followed by continuous exposure to a very low dose of IA (0.025 mM) for the duration of the measurements. The left ventricular pressure tracing from a glucose- and a pyruvate-perfused heart exposed to slightly higher IA levels is illustrated in FIGURE 6 (a and b), respectively. FIGURE 7 illustrates the ^{31}P NMR spectra recorded under the conditions of FIGURE 6a. In hearts perfused initially with 11 mM glucose as the only carbon source, within 15 min of IA exposure, the developed pressure and the ATP levels began to decline (FIGS. 6a and 7b) and a large sugar phosphate peak (SP), attributable to fructose 1,6-bisphosphate (FBP), appeared in the ^{31}P NMR spectrum (FIG. 7). If glucose infusion was continued, contractile activity ceased completely. Thus, the GAPDH activity was inhibited to the extent that not even a small amount of work can be supported with glucose as the carbon source. However, if glucose was replaced by pyruvate, heart function gradually recovered (FIG. 6a), further loss of ATP was halted, and FBP slowly diminished in intensity (FIG. 7c).

For the kinetic experiments, hearts were perfused with a low but sufficient amount of pyruvate (0.5 to 3 mM) from the beginning. In the absence of glucose and with pyruvate supplied as the carbon source, exposure to IA did not induce a precipitous decline in the developed pressure (FIG. 6b) nor was there a noticeable accumulation of sugar phosphates or a reduction in ATP levels. However, if pyruvate was replaced by glucose, all contractile activity came to an abrupt halt; alternatively, if the pyruvate-perfused heart was rendered ischemic, they accumulated a large amount of sugar phosphates and experienced a very small pH drop (FIG. 8). In contrast, it is well known that a similar ischemic insult in normal hearts typically results in a large decrease in pH due to production of lactic acid and no detectable accumulation of phosphorylated intermediates of glycolysis. Finally, when an IA-inhibited, pyruvate-perfused heart, depleted of its endogenous carbon sources, was given a brief pulse of glucose, a prominent FBP peak appeared in the ^{31}P NMR spectrum. In the absence of glucose, the FBP disappeared at a rate of 0.5 μmoles/min (g dry wt). This extremely slow rate indicates that any residual GAPDH activity is too slow to be a possible contributor to the MT measurements.

TABLE 3 lists the NMR and functional measurements obtained at two workloads under the conditions listed above. It is clear from these data that there is a large glycolytic contribution since the measured $P_i \rightarrow$ ATP rate is reduced in response to IA. At the high workload, simply eliminating exogenous glucose is sufficient to decrease the rate; the additional intervention with IA inhibition of GAPDH does not cause a further reduction. It is worth emphasizing that upon elimination of the GAPDH/PGK contribution, T_1^*, the spin-lattice relaxation time of P_i measured while saturating ATPγ, increases; this is as expected.

Together with data recorded prior to GAPDH inhibition, FIGURE 9 illustrates the data obtained with either IA inhibition (open squares) or in the absence of exogenous and endogenous glycolytic carbon sources (open circles). Only the data obtained at high workloads are included in the last category of points because only at high workloads was the removal of endogenous and exogenous glycolysis carbon sources sufficient to eliminate the GAPDH/PGK contribution to the $P_i \rightarrow$ ATP rate (TABLE 3).

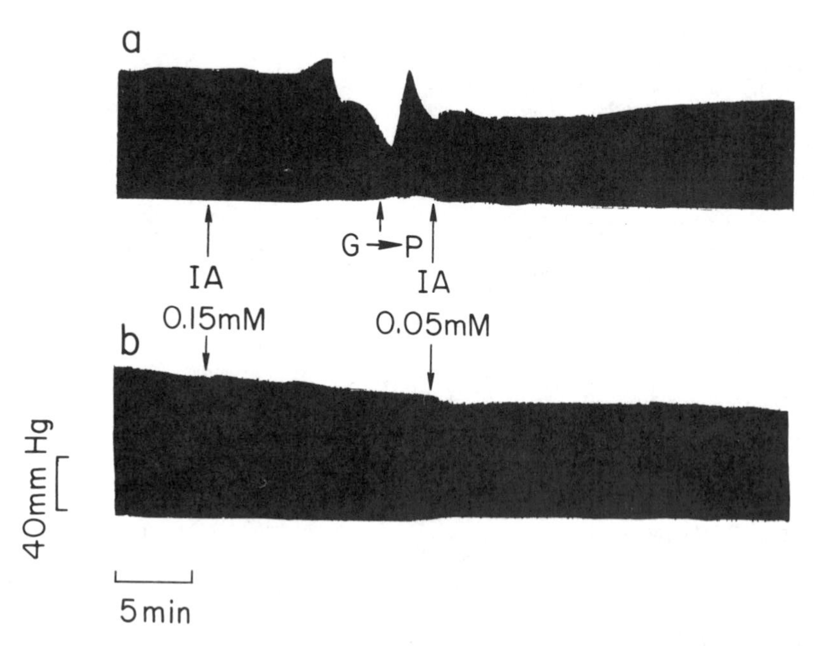

FIGURE 6. Left ventricular pressure versus time, showing the effect of iodoacetate on glucose-perfused (a) and pyruvate-perfused (b) rat hearts. Hearts were initially perfused with a phosphate-free modified Krebs-Henseleit buffer containing either 11 mM glucose or 1 mM sodium pyruvate. At the point designated, 0.15 mM IA was infused. Fifteen min later, IA concentration was reduced to 0.05 mM. In the glucose-perfused heart, carbon source was switched to pyruvate at the point designated by G → P.

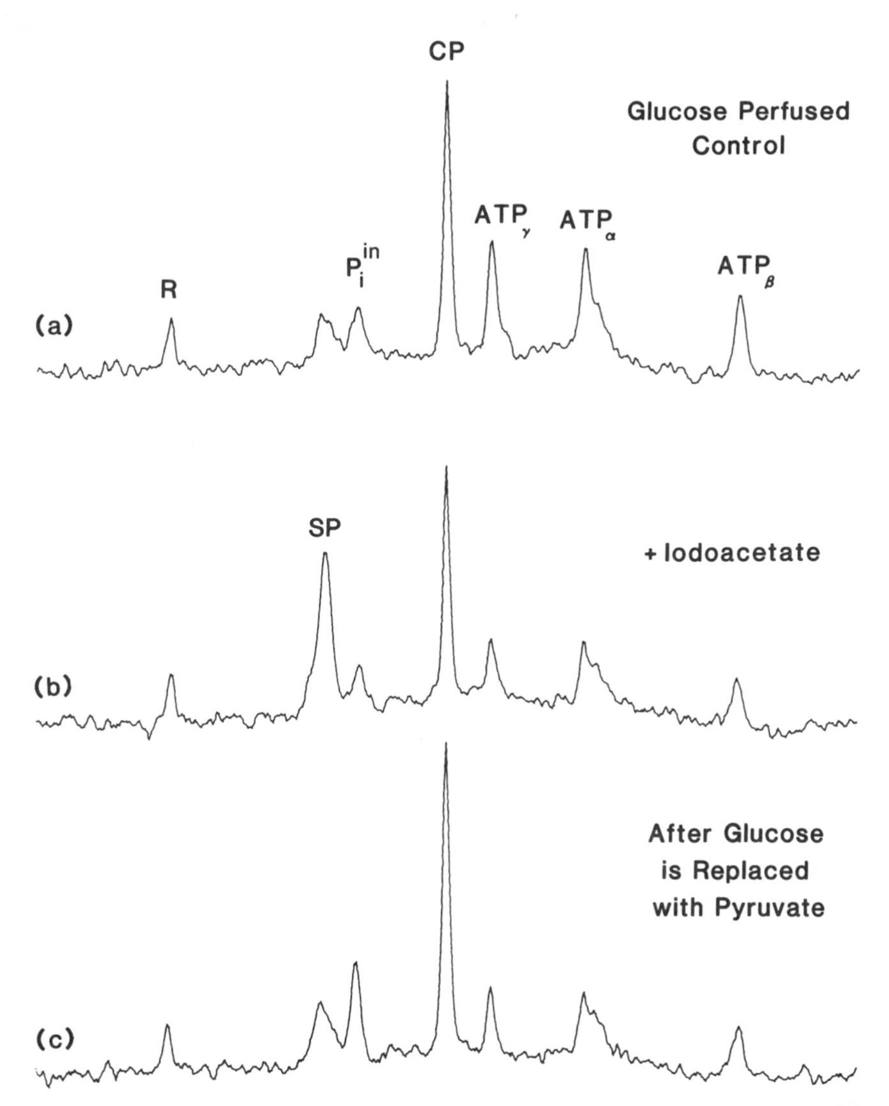

FIGURE 7. 146.1 MHz ^{31}P NMR spectra recorded under conditions specified in FIG. 6a. (a) Glucose-perfused heart, prior to IA infusion, (b) glucose-perfused heart 10 min after onset of IA infusion, (c) subsequent to switching to pyruvate as the carbon source.

As previously discussed, it cannot be assumed that fluxes determined subsequent to explicit elimination of the GAPDH/PGK contribution are free of other potential complications that can influence such kinetic measurements. One criterion we proposed for evaluating whether the net rate of ATP synthesis (and consequently the P:O ratio) can be obtained from these MT studies is that the $P_i \rightarrow$ ATP rate is equal to K(MVO), in which case K would be the P:O ratio. The solid line in FIGURE 9 is the best fit obtained to this equation from all the $P_i \rightarrow$ ATP rates determined subsequent to

GAPDH/PGK inhibition. FIGURE 10 is a plot of ($P_i \rightarrow$ ATP) rate/MVO ratio versus MVO. Two points obtained from ATP hydrolysis measurements (discussed further on) are also included in this figure. The dashed lines correspond to the Rate/MVO ratios of 2 and 3. Except for one point, the data are very tightly clustered near the values of ~2 and ~2.2. The solid line in FIGURE 10 shows the best fit for all data points and corresponds to a constant value of 2.34 ± 0.38 (SD).

The observations outlined above strongly suggest that within the signal-to-noise of these data, our major criteria are fulfilled. However, given the complexities of the MT

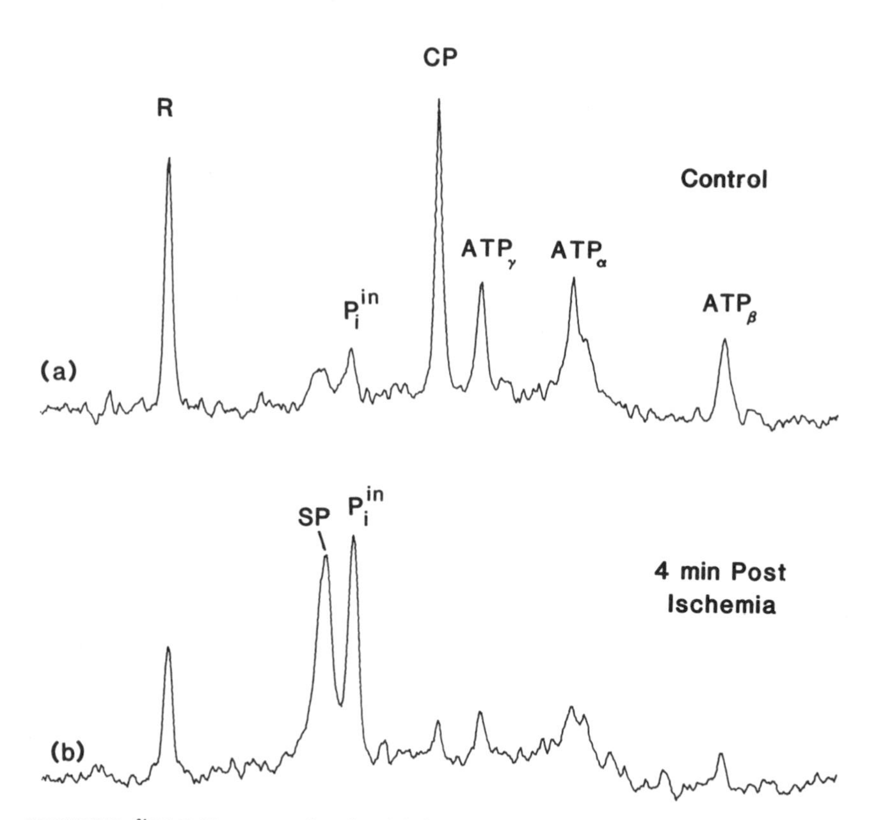

FIGURE 8. ^{31}P NMR spectra of perfused, iodoacetate-treated rat hearts before (a) and after 4 min of global ischemia (b). Individual spectra (each representing the sum of 12 free induction decays recorded with 5 sec interpulse delay and 90° pulses) from four different hearts were added to obtain the spectra shown.

measurements, it is desirable to have alternative experiments for evaluating the validity of these kinetic data. One alternative test that was already mentioned is to determine ATP $\rightarrow P_i$ rates under identical workloads and compare them to the $P_i \rightarrow$ ATP rates. An obvious complication in performing such a measurement is the presence of the CK activity; ATP $\rightarrow P_i$ rate cannot be determined accurately by CST methods in the presence of the rapid CP $\rightleftharpoons$ ATP exchange. The complication introduced by this exchange can be eliminated by use of the MST method. Previously, we have used

TABLE 3. Comparison of MVO_2, RPP, and Kinetic Data Obtained During $P_i \rightarrow$ ATP Rate Measurement in Glucose Perfused, Pyruvate Perfused, Pyruvate Perfused Subsequent to Depletion of Endogenous Carbohydrates and IA Inhibited Hearts at Two Workloads[a]

	$(\Delta M/M_o)$[b]	T_1* (sec)[c]	$P_i \rightarrow$ ATP Rate μmoles sec^{-1} (g dry wt)$^{-1}$	MVO_2 μmole min^{-1} (g dry wt)$^{-1}$	RPP 10^3 mm Hg^{-1} min^{-1}
1st Workload					
Control (N = 10) (glucose perfused)	0.34 ± 0.03	0.59 ± 0.03	7.1 ± 0.8	34.0 ± 2.2	26.4 ± 2.0
Pyr. Perfused (N = 9) (no glucose, depleted)[d]	0.32 ± 0.03	0.84 ± 0.08	5.0 ± 0.9	36.1 ± 1.5	26.5 ± 1.0
IA Inhibited (N = 9)	0.30 ± 0.03	0.73 ± 0.05	3.1 ± 0.4	39.5 ± 1.5	22.1 ± 1.4
2nd Workload					
Control (N = 10) (glucose perfused)	0.32 ± 0.01	0.78 ± 0.03	7.3 ± 0.5	48.8 ± 3.1	57.1 ± 2.6
Pyr. Perfused (N = 10) (no glucose)	0.34 ± 0.04	1.28 ± 0.17	4.2 ± 1.0	53.7 ± 1.5	52.4 ± 3.5
Pyr. Perfused (N = 9) (no glucose, depleted)[d]	0.33 ± 0.04	1.13 ± 0.12	4.3 ± 0.5	55.2 ± 3.3	54.6 ± 3.0
IA Inhibited (N = 9)	0.26 ± 0.02	0.85 ± 0.07	3.7 ± 0.5	50.5 ± 1.6	37.6 ± 2.2

[a]The two workloads correspond to i and ii of TABLE 2; control data are reproduced from TABLE 2. During pyruvate perfusion, pyruvate concentration was low (0.5 and 1 mM for the lower and higher workloads, respectively).

[b]Fractional reduction induced in the P_i magnetization upon saturation of ATP_γ.

[c]P_i T_1 measured while saturating ATP_γ.

[d]Depleted means depleted of endogenous carbon sources by transient perfusion with substrate free media prior to perfusion with pyruvate.

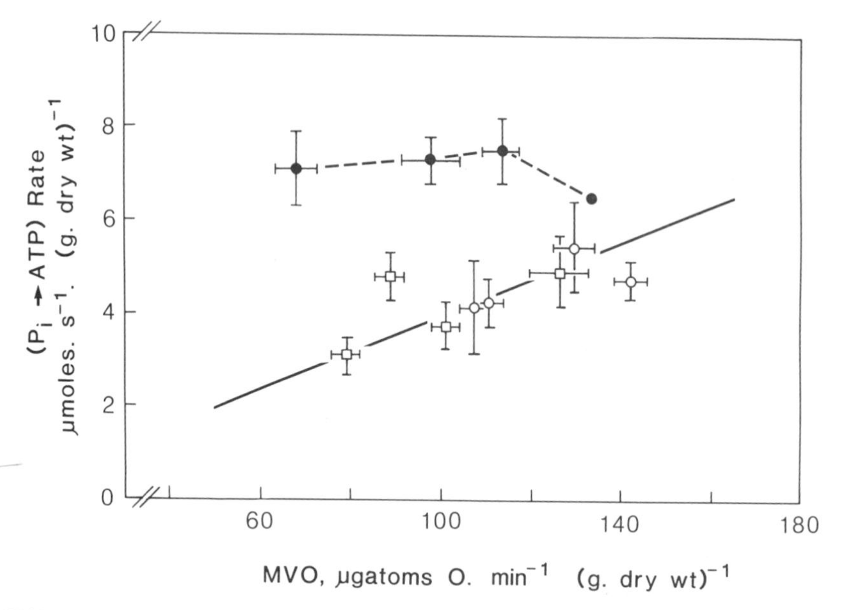

FIGURE 9. $P_i \rightarrow$ ATP rates measured in hearts where glucose was the primary carbon substrate (●), and in hearts where pyruvate (0.3 to 10 mM, depending on workload) was supplied as carbon source either subsequent to depletion of the endogenous carbon sources (○), or in the presence of IA inhibition of GAPDH activity (□). ● points are reproduced from TABLE 2. Solid line is the best fit of both ○ and □ points to the equation, rate = K(MVO).

this method to eliminate the ATP $\rightleftharpoons P_i$ exchange from the problem and measure ATP $\rightarrow$ CP rates.[17]

Since the ATP detected in the NMR spectrum of the rat heart is virtually all cytoplasmic,[17] the ATP $\rightarrow P_i$ rate measured by MST would reflect the sum of all reactions that utilize ATP in the cytoplasm, including GAPDH/PGK, translocase, and the various cytoplasmic ATPases. The GAPDH/PGK problem can be eliminated by procedures already outlined. We know from our studies with respiratory control that in the presence of exogenous pyruvate in amounts sufficient to fully activate pyruvate dehydrogenase, cytoplasmic ADP uptake into mitochondria is the rate-limiting process in oxidative-phosphorylation. This implies that under these conditions, translocase works primarily in the direction of ADP uptake into and ATP extrusion from the mitochondria; consequently, translocase would not contribute to the "utilization" of cytoplasmic ATP. The cytoplasmic ATPases, such as the myosin ATPase and the Na^+/K^+ ATPase, work predominantly, if not exclusively, in the ATP hydrolysis direction; this is the rate we would like to measure.

One possible problem in the MST studies of the ATP $\rightarrow P_i$ rate is the adenylate kinase (AK) reaction. This enzyme catalyzes the reaction ATP + AMP $\rightleftharpoons$ 2ADP and consequently mediates an exchange between ATP_γ and ADP_β phosphates. In principle, the complication introduced by this exchange can also be eliminated by the MST procedure. However, this would require irradiation of the ADP_β resonance, which is not well resolved from the ATP_γ peak. ATP_γ is the resonance we want to monitor while saturating P_i (and CP) for the measurement of the ATP $\rightarrow P_i$ rate. Therefore, in practice, MST cannot be employed in a straightforward fashion to eliminate the AK contribution. However, if the AK activity is a potential source of problems, it would mean that the rate measured by MST for ATP → P_i conversion in the cytoplasm is underestimated. This is acceptable for our purposes, because if mitochondrial H^+-ATPase has a significant reverse rate (i.e., v_2 is not negligible) we would expect the true ATP → P_i rate to be less than the P_i → ATP rates determined by the CST method. Consequently, the ATP → P_i rate measured by MST can be equal to the P_i → ATP rate determined by CST if and only if AK is not a problem and v_2 is negligible.

FIGURE 11 illustrates spectra from the MST measurement of ATP → P_i rate where CK problem was eliminated by a continuous background irradiation of the CP resonance. The experimental details were identical to those reported for the MST measurement of the CK rate. It is seen clearly that there is a measurable saturation-transfer effect on the ATP_γ resonance. TABLE 4 lists both the MST data for the ATP → P_i rate and the CST data for the P_i → ATP rate. At these MVO_2s, in hearts supplied 10 mM pyruvate as the carbon substrate, the ATP → P_i rate did not depend on whether or not the endogenous carbohydrates were depleted, indicating that GAPDH/PGK contribution was insignificant. The ATP → P_i and P_i → ATP fluxes were clearly equal within the error of the determination (TABLE 4). Therefore, we conclude that indeed, the H^+-ATPase is operating unidirectionally in the direction of ATP synthesis under these workload conditions and that AK is not a source of complication in measurements of ATP hydrolysis rates; consequently, the fluxes obtained in both directions can be used to determine the P:O ratio. The P:O ratios obtained from the two MST measurements were 2.08 ± 0.14 (SEM, N = 14) and 2.10 ± 0.28 (SEM, N = 11) in excellent agreement with the value of 2.41 ± .34 (SD)

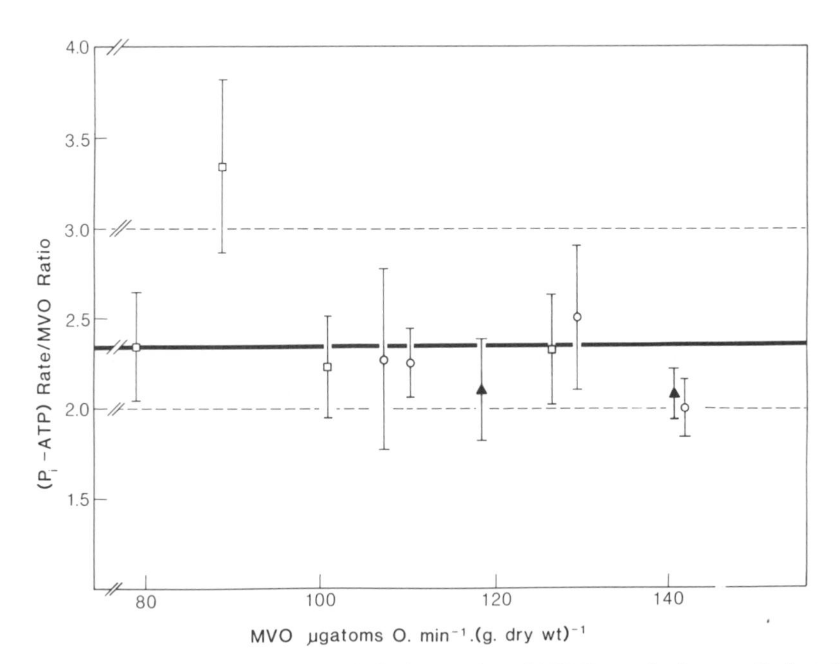

FIGURE 10. P_i → ATP Rate/MVO ratio for hearts where GAPDH contribution was eliminated (O and □) (corresponding to O and □ points in FIG. 9), and ATP → P_i Rate/MVO ratio obtained from the MST measurements (▲) (TABLE 4).

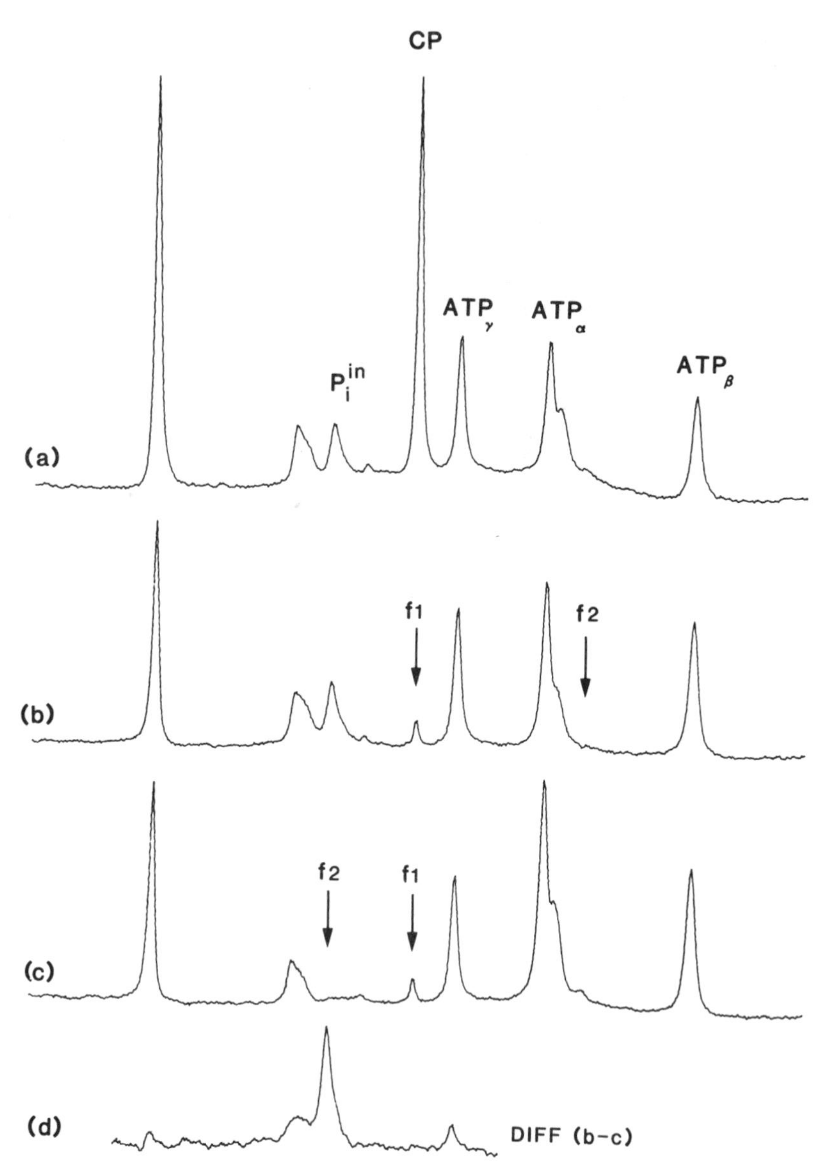

FIGURE 11. ^{31}P MST spectra recorded during the measurement of ATP → P_i rate. (a) no saturation; (b) f1 selectively saturating CP and f2 in control position; (c) f1 and f2 saturating CP and P_i peaks, respectively; (d) the difference between (c) and (b).

obtained from the CST data alone in the opposite direction. As previously mentioned, the average obtained from both the CST and the MST data is 2.34 ± 0.38 (SD). Obtaining the same rates and the same P:O ratios by two different measurements gives us considerably more confidence in these numbers because the problems associated with the two different measurements are different.

CONCLUSIONS

A general conclusion that emerges from these studies is that the regulation of oxidative phosphorylation in the myocardium is based on kinetic rather than thermodynamic[24,25] mechanisms. The rate of oxidative phosphorylation appears to be determined by the availability of its various substrates, and one of the key enzymes involved in this process, the mitochondrial H^+-ATPase, is working unidirectionally.

TABLE 4. Comparison of MVO_2, Rate, and Rate/MVO Data Obtained During Determination of the $P_i \rightarrow$ ATP and ATP $\rightarrow P_i$ Rates at Two Different Workloads Subsequent to Elimination of the GAPDH/PGK Contribution

	Workload A		Workload B	
	$P_i \rightarrow$ ATP	ATP $\rightarrow P_i$	$P_i \rightarrow$ ATP	ATP $\rightarrow P_i$
N	10	11	9	14
MVO_2 μmoles · min^{-1} · (g dry wt)$^{-1}$	55.2 ± 3.3	59.2 ± 1.8	70.9 ± 1.7	70.3 ± 1.1
Rate μmole · sec^{-1} · (g dry wt)$^{-1}$	4.3 ± 0.5	4.1 ± 0.6	4.7 ± 0.4	4.9 ± 0.4
Rate/MVO	2.25 ± 0.19	2.10 ± 0.28	2.00 ± 0.16	2.08 ± 0.14

All data are mean ± SEM. *N* is the number of hearts on which these measurements were performed. Heart rate, EDP, and dobutamine exposure for workload A and B correspond to those used for workloads ii and iv of TABLE 2, respectively. For $P_i \rightarrow$ ATP measurements, the hearts were depleted of endogeneous carbon substrates and perfused with 1 mM pyruvate at the lower and 10 mM pyruvate at the higher workload. For the ATP $\rightarrow P_i$ measurements, hearts were perfused with 10 mM pyruvate. Despite these differences, the P_i, ATP, and CP content were all comparable for each workload (e.g., P_i content for the lower and higher workloads respectively was 9.0 ± 1.0 and 13.1 ± 0.8 μmoles · (g dry wt)$^{-1}$ during the ATP $\rightarrow P_i$ measurements, and 14.9 ± 1.4 and 13.0 ± 2.0 μmoles · (g dry wt)$^{-1}$ during the $P_i \rightarrow$ ATP measurements.

A general conclusion that can be reached from the kinetic studies is that the magnetization-transfer measurements in intact tissues are extremely complicated but experimental criteria exist that can be used to validate the data. Unfortunately, such experimental evaluations must be performed for each system. For example, the extent of GAPDH/PGK activity can be very much dependent on species and the developmental stage of the heart. Similarly, the problem associated with P_i exchange across the mitochondria will depend strongly on the tissue. The rapidity of the exchange between the cytosolic and mitochondrial P_i pools depend on numerous parameters such as the V_{max} of the P_i transport system, the amount of cytosolic and mitochondrial P_i, and the average diffusion distances involved in the cytosol prior to encountering phosphorylation to the $P_i \rightarrow$ ATP measurements. Therefore, careful control studies must be performed in each system under study. Despite these complications and the requirement for extensive control studies, these measurements provide the only means for examining ATP turnover kinetics. In the normal rat myocardium, these studies have provided two biochemically significant contributions. (1) In the MVO_2 range ~30–80 μmoles O_2 min^{-1} (g dry wt)$^{-1}$, which covers a moderate to high workload range for a rat, the mitochondrial H^+-ATPase operates unidirectionally in the ATP synthesis direction. (2) In the intact myocardium, the P:O ratio determined both from MST data in the ATP $\rightarrow P_i$ direction and from the CST data in the $P_i \rightarrow$ ATP direction is significantly different from the canonically used value of 3, and in agreement with the proposal and data of Hinkle and Yu.[26] Finally, it should be noted that unless the possible GAPDH/PGK activity is specifically eliminated, the $P_i \rightarrow$ ATP rates determined by MT methods cannot be ascribed exclusively to oxidative phosphorylation and cannot be used to determine the P:O ratio.

The NMR studies reviewed here have enhanced our understanding of bioenergetics in the normal myocardium. Their application to the compromised cardiac muscle should similarly provide invaluable information on the pathophysiology of this tissue.

REFERENCES

1. Kobayashi, K. & J. R. Neely. 1979. Circ. Res. **44:** 166–175.
2. Chance, B. & G. R. Williams. 1955. J. Biol Chem. **217:** 383–393.
3. Slater, E. C., J. Rosing & A. Mol. 1973. Biochim. Biophys. Acta **292:** 534–553.
4. Klingenberg, M. 1961. Biochem. Z. **335:** 263–272.
5. Holian, A., C. S. Owen & D. F. Wilson. 1977. Arch. Biochem. Biophys. **181:** 164–171.
6. From, A. H. L., M. A. Petein, S. P. Michurski, S. D. Zimmer & K. Uğurbil. 1986. FEBS Lett. **206:** 257–261.
7. Jacobus, W. E., R. W. Moreadith & K. M. Vandegaer. 1982. J. Biol. Chem. **257:** 2397–2402.
8. Williamson, J. R. 1965. J. Biol. Chem. **240:** 2308–2321.
9. Chapman, J. B. 1972. J. Gen. Physiol. **59:** 135–154.
10. Dennis, S. C., A. Padma, M. S. DeBuysere & M. S. Olson. 1979. J. Biol. Chem. **254:** 1252–1258.
11. Balaban, R. S., H. L. Kantor, L. A. Katz & R. W. Briggs. 1986. Science **232:** 1121–1123.
12. Koretsky, A. P. & R. S. Balaban. 1986. Biophys. J. **49:** 207a.
13. McCormack, J. G. & R. M. Denton. 1986. Trends Biochem. Sci. **11:** 258–262.
14. Neely, J. R., R. H. Bowman & H. E. Morgan. 1969. Am. J. Physiol. **216:** 804.
15. Forsen, S. & R. A. Hoffman. 1963. J. Chem. Phys. **39:** 2892.
16. Uğurbil, K. 1985. J. Magn. Reson. **64:** 207–219.
17. Uğurbil, K., M. Petein, R. Maidan, S. Michurski & A. H. L. From. 1986. Biochemistry **25:** 100–107.
18. Matthews, P. M., J. L. Bland, D. G. Gadian & G. K. Radda. 1981. Biochem. Biophys. Res. Commun. **103:** 1052–1059.
19. Bittl, J. A. & J. S. Ingwall. 1985. J. Biol. Chem. **260:** 3512–3517.
20. Kingsley-Hickman, P., E. Y. Sako, P. A. Androene, J. A. St. Cyr, S. Michurski, J. E. Foker, A. H. L. From, M. Petein & K. Uğurbil. 1986. FEBS Lett. **198:** 159–163.
21. Brown, T. R. & S. Ogawa. 1977. Proc. Natl. Acad. Sci. USA **74:** 3627–3631.
22. Campbell, S. L., K. A. Jones & R. G. Shulman. 1985. FEBS Lett. **193:** 189–193.
23. Brindle, K. & S. Krikler. 1985. Biochim. Biophys. Acta **847:** 285–292.
24. Ericinska, M., M. Stubbs, Y. Miyata, C. M. Ditre & D. F. Wilson. 1977. Biochem. Biophys. Acta **462:** 20–35.
25. Ericinska, M. & D. F. Wilson. 1982. J. Membr. Biol. **70:** 1–14.
26. Hinkle, P. C. & M. L. Yu. 1979. J. Biol. Chem. **254:** 2450.

DISCUSSION OF THE PAPER

J. D. GLICKSON: Why can't you use the same method that you use to screen out or take into account the creatine-kinase reaction to take into account the glycolytic pathway? Have you tried saturating fructose 1.6-bisphophate to see if there's an effect?

K. UGURBIL: In principal you can, the problem is that the resonances that you have to saturate are very close to the resonance that you want to monitor, and I also don't have a handle on monitoring whether I am saturating the resonance I should saturate.

E. J. BARRETT: I want to ask if rather than using iodoacetate you were able to simply quantitate the lactate that was being produced or the P_i/ATP exchange at the GADPH or if you can quantitate the extent to which it's exchange vs. flux through the pathway.

UGURBIL: The NMR measures the unidirectional exchange rates and that is not the same as the net rates that you obtain by lactate measurements. In fact, in the heart if you take the rate of lactate production that has been published for example, under maximal work load conditions etc., those are not extremely rapid and so NMR measured rate is an exchange which is going on at a rate that is faster than the net rate of glycolysis.

K. LANOUE: As you know, I'm the one who has done all these isotopic exchanges and we measured v_1 and v_2 isotopically with the isolated mitochondria. They agree quite well with your data without the iodoacetate, but there are some things about the mitochondrial stuff that maybe should be made clear. First of all you don't really expect v_1 to remain constant with work. We have published a paper that says it doubles, maybe a little more than that, maybe triples, in going from no work to the highest work. v_2 goes down to zero and at high work it really does become very unidirectional and our respiratory control ratios are about 2.5, so I'm having a lot of problems. I like the data so much without the iodoacetate, it agrees so well with our *in vitro* studies with the isolated mitochondria. I was looking at your figure with the flat line for the exchange without the iodoacetate which has got to be wrong. It does have a slope and then the difference that you say is glycolysis between that other slope so why does the exchange go up so rapidly as you decrease work? I mean the glycolytic exchange. Why do you think that is?

UGURBIL: Indeed our first interpretation of that relatively flat line that I showed was that exactly. What you are seeing in the isolated mitochondria we were seeing in the entire heart and we went under that assumption until prompted by Sharon Campbells' result. Then we experimentally tried to evaluate the glycolysis contribution. It was nice that it looked like we were in agreement with isolated mitochondria data, but clearly there is a glycolysis contribution. It's not an artifact of iodoacetate because we can just take away glucose and the rate we measure comes down so there's a glycolysis contribution. Now, corollary to this data that you point out is the fact that the glycolytic contribution appears to be work-load dependent. It is high at very low work loads and it seems to decrease at the very high work loads that we are working at. You are approaching the case where almost all of it is coming from oxidative phosphorylation. Not all of it, but almost all of it. I don't know the specifics for this. I can postulate things such as the depletion of GADPH intermediates, the GADPH is becoming more and more unidirectional and that exchange is decreasing. Such mechanisms can be thought about. But why that is happening I don't know.

LANOUE: The flux through glycolosis should be going up.

UGURBIL: The net flux, may be going up, may be staying the same, but that is small compared to the oxidative phosphorylation rate. The net flux of glycolysis may increase while the unidirectional rate through GAPDH/PGK enzymes is decreasing.

R. G. SHULMAN: I think it's important to find out about the GADPH flux, that it's very very rapid and it's cycling, that is you're not measuring just the forward flux through it. If you take the apparent P:O ratio that we first reported in yeast and *E. coli,* the P:O ratio in yeast was 80 and then you bring that down to 3. Sharon brings that down to 3 just by iodoacetate, so the amount of cycling through the iodoacetate is enormous because most of the factor of 77 is just futile cycling.

LANOUE: It's nice that there's apparently no disagreement at the high work load, I mean it is unidirectional, P:O's are what they should be and that should at least put to rest all these stories about ox-phos being thermodynamically regulated.

NMR Spectroscopy for Clinical Medicine Animal Models and Clinical Examples[a]

MICHAEL W. WEINER

Magnetic Resonance Unit
Veterans Administration Medical Center
Departments of Medicine and Radiology
University of California
San Francisco, California 94121

INTRODUCTION

The purpose of this brief review is to acquaint the medical audience with the clinical potential of magnetic resonance spectroscopy (MRS). The interested reader is referred to more complete reviews of this field.[1–5]

CURRENT STATE OF MEDICAL DIAGNOSIS

When a patient goes to a physician for a routine medical examination or because of a specific medical complaint, the diagnostic workup usually consists of (1) a complete medical history; (2) a physical examination; (3) electrical measurements, such as EKG, EMG, and EEG; (4) laboratory tests to evaluate the cellular and chemical composition of blood; (5) endoscopic examination of internal structures through various orifices; (6) imaging studies using conventional X-ray, X-ray computerized tomography, ultrasonography, nuclear medicine scans, and magnetic resonance imaging (MRI); and (7) invasive tissue biopsies.

Once disease is detected, appropriate treatment is instituted using drugs, surgery, radiation, or other therapeutic modalities. To monitor the response to therapy and to determine if diseased tissue is being destroyed or restored to health, the same diagnostic modalities (history, physical, blood tests, scans, etc.) are used to monitor the medical course of the patient.

The last decades have demonstrated impressive advances in the use of technology for medical diagnosis. Clinical laboratory tests of blood composition provide considerable information concerning the chemical composition of extracellular fluid. Scanning studies provide a detailed picture of anatomical structures. Biopsies indicate presence of malignancy, inflammation, or other disturbances, but are invasive and provide little information concerning cell metabolism. All current modalities used for medical diagnosis provide important information concerning anatomy and function, but provide little or no data concerning chemical composition or metabolism. Magnetic resonance spectroscopy (MRS) represents an important new approach to medical investigation and diagnosis because it is uniquely capable of noninvasively monitoring important tissue metabolites in the body.

MRS is unique for several reasons: First, it is a noninvasive technique that uses

[a]Supported in part by the Veterans Administration Medical Research Service and National Institutes of Health Grant R01-AM-33293.

magnetism and radio waves and has few known side effects (aside from the danger of ferrous metal flying into the magnet and heating produced by rf deposition). Secondly, the chemicals detected by MRS are vital to cell energetics, as illustrated in FIGURE 1. Cellular energy is generated by the oxidation of food stuffs, resulting in the production of ATP within mitochondria by the process of oxidative phosphorylation. The ATP is transported out of the mitochondria into the cytosol where it is used for cellular work functions such as muscular contraction, ion transport by the sodium-potassium pump, and biosynthesis. In muscle and brain, phosphocreatine (PCr) serves as a high-energy phosphate reservoir. When the cell is healthy, large amounts of PCr accumulate. When oxidative production of ATP is limited, phosphocreatine is used to synthesize ATP, thus maintaining ATP levels. When ATP is used for biological work functions, it is hydrolyzed to ADP and P_i, which subsequently re-enter the mitochondria for re-synthesis into ATP. Anoxia or ischemia inhibits oxidative phosphorylation and diminish ATP synthesis. If PCr is abundant, this will be used to maintain ATP levels. Once the utilization of ATP exceeds the rate of synthesis, ATP levels will fall. Furthermore, if oxidative production of ATP is slowed, glycolysis will be stimulated, resulting in increased production of lactic acid. This induces tissue acidosis.

FIGURE 2 illustrates typical ^{31}P NMR spectra showing peaks for ATP, PCr, P_i, phosphodiesters, and phosphomonesters. Therefore, the concentrations of these important phosphates can be determined by the NMR technique. Furthermore, tissue pH can be measured by the lateral position (termed the chemical shift) of the P_i peak.[6–8] Finally, lactate can also be monitored in tissues using ^1H NMR with so-called water suppression techniques.[9] Thus, MRS is able to detect noninvasively important

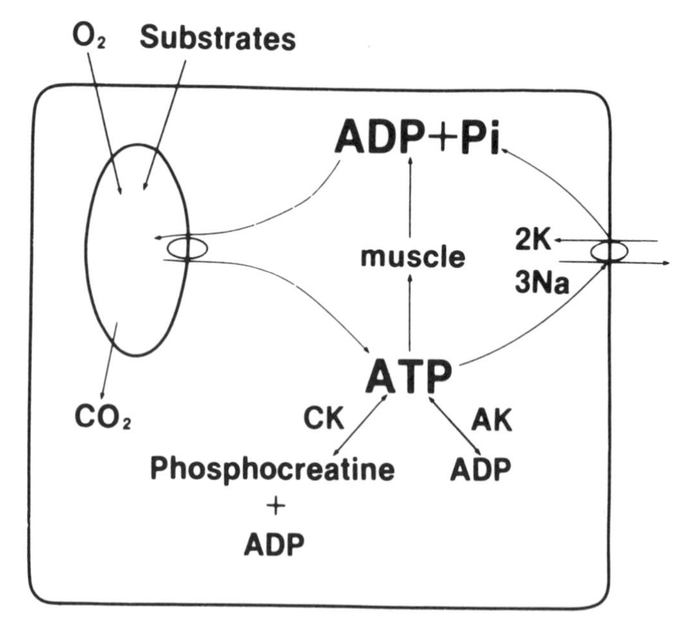

FIGURE 1. A schematic illustration of the high-energy phosphate reactions within the cell. ADP = adenosine diphosphate. ATP = adenosine triphosphate. AK = adenylate kinase. CK = creatine kinase. P_i = inorganic phosphate. The Na-K ATPase at the plasma membrane transports three Na out and two K in for each ATP hydrolyzed. (From Koretsky *et al.*[63] With permission from the publisher.)

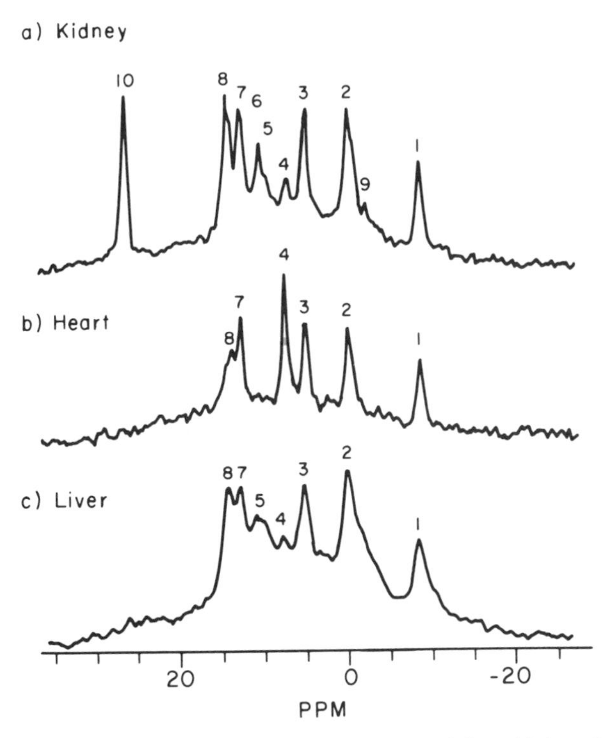

FIGURE 2. Spectra obtained from kidney (a), heart (b), and liver (c) by using chronically implanted coils. Spectra are the sum of 128 acquisitions using a 30-sec pulse (75°) and a 10-sec recycle time. A 30-Hz exponential filter was applied to the free induction decay prior to Fourier transformation. Chemical shifts were referenced to the ATP α-phosphate peak, which was set equal to 0 ppm. Peaks: 1, ATP β-phosphate; 2, ATP α-phosphate/NAD(H); 3, ATP γ-phosphate; 4, PCr; 5, phosphodiester; 6, urine P_i; 7, P_i; 8, phosphomonoester; 9, unknown; 10, MDPA. (From Korestsky *et al.*[51] With permission from publisher.)

compounds that may rapidly change during alterations of cell bioenergetics. This unique feature of MRS has enabled scientists to investigate animal models of human disease and has encouraged early applications of MRS to clinical studies in man.

Muscle

The first use of MRS to study intact tissues involved excised muscle.[10] Since that time many experiments have been performed on animal muscles,[11] and more recently on normal[12] and diseased human muscles.[13] Radda *et al.* have used MRS to investigate a wide variety of metabolic myopathies[14,15] and to detect disturbances in muscle metabolism produced by peripheral vascular disease and congestive heart failure.[16] In collaboration with Dr. Robert Miller, we have been investigating the metabolic basis of fatigue in normal human subjects using MRS.[17,18] Although exercise produces rapid depletion of high-energy phosphates in the form of PCr (ATP does not change very much under usual exercise protocols), results thus far have shown that there is no correlation between muscle PCr concentrations and fatigue. In contrast, there is a relatively good correlation between muscle fatigue and pH.[18] The results also demonstrate a correlation between fatigue and monobasic phosphate concentrations,[18] which confirms animal studies[19] suggesting that monobasic phosphate may be an important determinant of muscle fatigue. It can be expected that MRS will be increasingly used to study many aspects of muscle metabolism in various myopathies and injuries.

Heart

Early studies demonstrating the feasibility of measuring various phosphates in perfused hearts were performed by Jacobus[20] and Radda[21] and their co-workers. It was quickly demonstrated that anoxia or ischemia results in a rapid depletion of PCr accompanied by a rise in P_i and an acid shift of the P_i peak because of tissue acidosis. ATP concentrations are generally well maintained until PCr is depleted. Nunnally and Bottomley[22] demonstrated that localized ischemia of perfused heart produced rapid depletion of high-energy phosphates and a rise of P_i in the ischemic zone. Subsequently, there have been many MRS studies of the perfused heart in order to investigate high-energy phosphate metabolism[23] and the effects of various drugs. MRS studies on the hearts of intact animals have been more difficult to perform, but a number of investigators have reported studies with various laboratory animals.[24-27] Some scientists have found that the static concentrations of phosphates do not change despite wide variations of work load, suggesting that the phosphorylation state may not be the regulator of tissue oxygen consumption.[28] Alternatively, studies in our laboratory[29] have demonstrated that catecholamine infusions produce reversible changes of high-

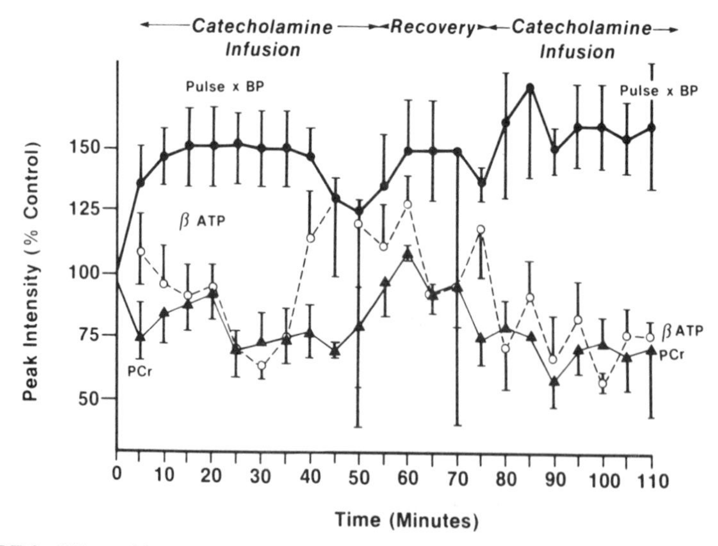

FIGURE 3. Effects of intravenous infusion of isoproterenol and phenylepherine on PCr (▲) and ATP(○) of the rat heart *in situ*. All data are expressed as the mean ± SE % control. Catecholamines were infused for 50 min, followed by 30 min of recovery, followed by a second infusion of twice the original amount.

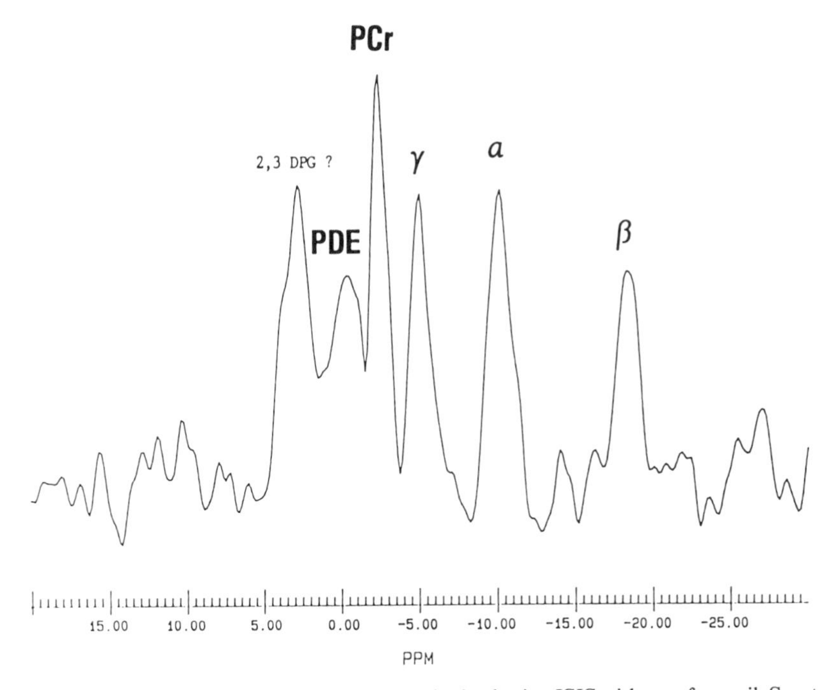

FIGURE 4. ^{31}P NMR spectrum of human heart obtained using ISIS with a surface coil. Spectra obtained at 1.5 T with a Philips Gyroscan by Drs. Karczmar, Matson, Tweig, Gober, Valenza, and Weiner in San Francisco.

energy phosphates (FIG. 3). Several groups have begun to study the effects of acute coronary ischemia *in vivo* with MRS[30,31] and the results show that total occlusion of the coronary artery results in rapid depletion of PCr, a rise of P_i, and tissue acidosis. Using a volume selective localization technique with surface coils (termed DRESS), Bottomley has shown that ^{31}P NMR spectra can be obtained from the dog heart and that the deleterious effects of coronary ischemia may be observed.[31] Chance and co-workers have demonstrated the feasibility of obtaining ^{31}P NMR spectra from hypertrophied baby heart,[32] and both Bottomley[33] and Radda[14] have reported spectra from human hearts using surface coils and volume selective localization techniques (FIG. 4). Bottomley and co-workers[34] reported that myocardial infarction in man was associated with partial depletion of PCr and a prolonged elevation of the P_i peak. Because myocardial ischemia and infarction are such common causes of morbidity and mortality, it can be expected that MRS will be widely used for cardiac studies in experimental animals and eventually human patients.

Brain

Chance *et al.*[35] first obtained ^{31}P NMR spectra from the head of an anesthetized mouse. Subsequently, Ackerman *et al.*[36] used a surface coil to obtain ^{31}P NMR spectra from the brain of a living rat. Thulborn *et al.*[37] first studied the effects of ischemia

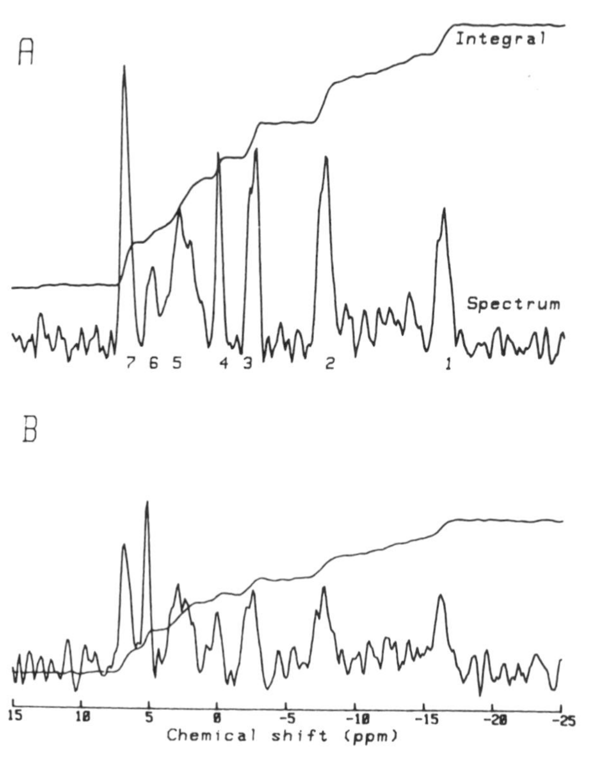

FIGURE 5. ^{31}P spectra from infant 7 at 17 days of age (A) and infant 5 (B) at 6 days (1024 pulses, 2.26 s intervals). (From Cady *et al.*[64] With permission from the publisher.)

produced by carotid occlusion. Ischemia produced the expected loss of PCr and ATP, a rise of P_i, and acidification. Since this pioneering work, many investigators have used MRS to study the effects of ischemia, reperfusion following ischemia, seizures, hypoglycemia, anoxia, anesthetics, metabolic inhibitors, and other agents that affect brain metabolism.[38] Shulman and co-workers[39] pioneered the application of ^{1}H NMR spectroscopy with water suppression to detect changes in concentrations of brain lactate, N-acetyl aspartate, and amino acids. As these techniques are further refined, they should ultimately be applicable to the study of other organs, although the presence of fat makes measurement of tissue lactate concentrations difficult because of overlapping peaks. Bottomley[40] first reported ^{31}P and ^{1}H MRS spectra from the head of normal human subjects. Recently, several investigators have studied patients with cerebrovascular ischemia or infarction using ^{31}P NMR. In general, it appears that patients with chronic, stable strokes do not show very dramatic alterations of brain high-energy phosphates. In contrast, the brains of children with birth trauma or asphyxia often show marked depletion of high-energy phosphates and tissue acidification[41] (FIG. 5). Recent studies[42,43] have indicated that the ^{31}P MRS spectra of brain tumors are different from normal brains, demonstrating high PME peaks. In a number of cases tumor treatment resulted in a decrease in the PME peak, suggesting a metabolic change in response to tumor therapy.[43] Brain tumors appear to be a promising area for application of MRS and it is also expected that this technique will be used to study cerebrovascular disease, brain trauma, and metabolic encephalopathies.

Kidney and Liver

Early studies demonstrated the feasibility of obtaining ^{31}P MRS spectra from perfused kidneys by Ross and co-workers.[44] In particular, the relationship of cell pH to ammoniagenesis[45] and the relationship of ATP to the energetics of sodium transport[46,47] have been studied. MRS studies of cold-stored human kidneys, being preserved for transplantation,[48] suggested that noninvasive measurements of tissue pH and phosphates may provide a predictor of graft function after transplantation. Siegel *et al.*[49] have characterized the metabolic changes produced by acute ischemic renal failure and demonstrated that infusion of Mg^{2+} ATP improved both renal function and the recovery of renal ATP concentrations. It can be predicted from these animal studies that MRS will be useful to investigate normal kidney metabolism. In addition, MRS will serve to investigate and diagnose various forms of acute renal failure (ischemic versus nephrotoxic) and to evaluate the metabolism of renal transplants during rejection, ischemic tubular necrosis, and cyclosporin nephrotoxicity.[50]

The major function of the liver is to regulate carbohydrate and fat metabolism. Animal studies have shown that fructose loading produces reversible depletion of P_i and ATP concentrations within the liver.[51] Shulman and co-workers have pioneered proton and carbon MRS techniques for the investigation of carbohydrate metabolism in isolated cells, perfused livers, and the livers of experimental animals.[52–55] These investigators have already demonstrated that ^{13}C NMR studies of carbon fluxes may be useful to characterize the importance of futile cycling and the rates of glycolysis and gluconeogenesis.[54] The recent demonstration that ^{13}C spectra can be obtained from human liver indicates that these techniques will ultimately be applied to clinical investigations.

Malignancy

Glickson[55] and Griffiths[56] and their co-workers first reported the exciting observation that ^{31}P MRS spectra of experimental tumors show rapid changes after initiation of therapy with a variety of modalities including anti-tumor drugs, radiation, and hyperthermia. These important observations have been repeatedly noted by many investigators.[57] Ross *et al.*[58] reported that the addition of anti-tumor drugs to the perfusate bathing a human hypernephroma produced a rapid increase of the P_i peak. Maris and co-workers[59] have reported spectral changes in a neuroblastoma of human infant (FIG. 6). In this case, a rise of the PME peak seemed to predict tumor growth, while treatment produced an early fall of the PME peak, anticipating response of the tumor to treatment. At the 1986 meeting of the Society of Magnetic Resonance in Medicine in Montreal, several investigators reported abnormal ^{31}P MRS spectra in human tumors. A variety of changes have been noted in human and animal tumors undergoing treatment including decreases in high-energy phosphates, increases in high-energy phosphates, decreases of the PME peak, and changes of the phosphodiester peak. Although a great deal of work needs to be done in this area, it can be predicted that MRS will become widely used for the investigation of human malignancy, particularly in an effort to obtain an early indication of a response to therapy.

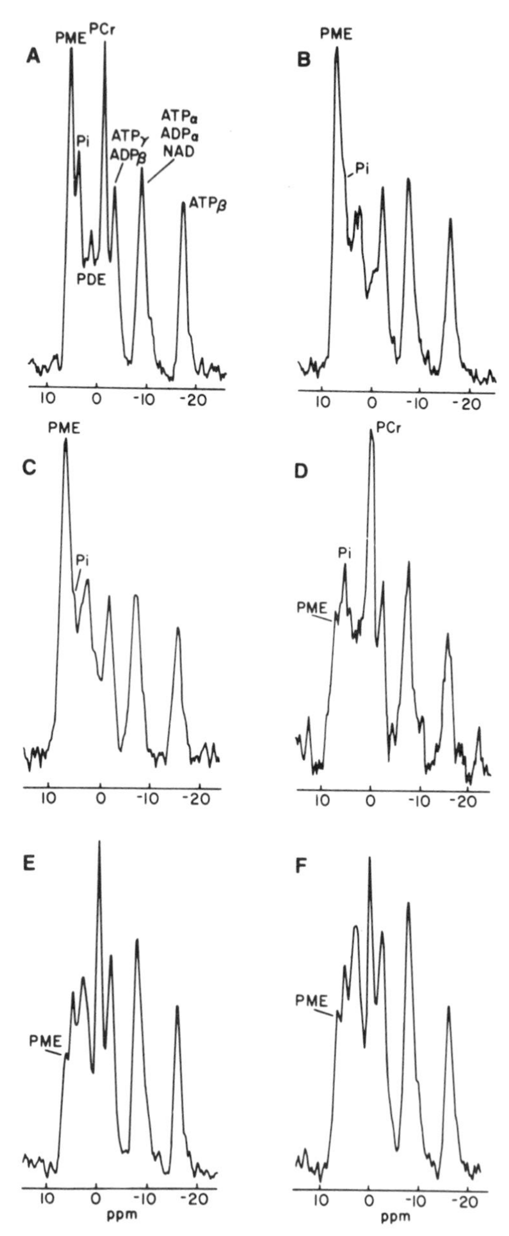

FIGURE 6. ^{31}P NMR spectra in Case 2 as a function of time. (A) shows the spectrum in the first study (October 13, 1984), (B and C) show the spectra 4 and 8 weeks later, (D) shows the spectrum 16 weeks after the initial response to treatment, and (E) and (F) show the spectra 24 weeks after the initial response to treatment at the edge of the liver (E) and anterolateral to the edge (F). (From Maris *et al.*[59] With permission from the publisher.)

PROBLEMS AND FUTURE PROSPECTS

Perhaps the greatest drawback of MRS is the relatively poor sensitivity of the method, which necessitates long run times and the examination of rather large volumes of tissue. However, the sensitivity is improved by higher strength magnetic fields. The increasing availability of whole body magnets with magnetic field strengths greater than 1.5 Tesla increases the prospect of MRS studies being carried out in conjunction with MRI examinations. The major problem is technical, i.e. to develop techniques that provide ^{31}P, ^{1}H, and ^{13}C MRS spectra from specified zones in the body with excellent spatial and spectral resolution. A wide variety of such techniques have been reported and studied in animal systems, but relatively few have been implemented in whole body systems. However, at the time this report, Bottomley[31,33,40] has already demonstrated the feasibility of the DRESS technique in human subjects; Radda, Styles, and their co-workers have demonstrated the use of the rotating frame experiments[60]; den Hollander[61] has developed SPARS; and Ordidge has reported good results with ISIS.[62] It should be emphasized that these are the initial efforts. It can be expected that technical improvements will allow improved spectra to be obtained in shorter periods of time.

In the opinion of this author, MRS will soon establish an important role in clinical investigation. The early results concerning malignancy appear promising and it is expected that a number of medical centers around the world will be actively working in this area. Studies of the brain, heart, liver, kidney, and other internal organs are expected to follow. An important question is whether or not MRS will become part of routine diagnosis. The answer is at present uncertain and will only be answered through the results of clinical scientists working in this area over the next several years. A great deal depends on technical advances and the ability to obtain spectra quickly, because of the high cost of such instrumentation. Nevertheless, this author predicts that before the end of the 1980s, MRS will be used for clinical diagnosis, especially to monitor the effects of therapy, in a number of large referral medical centers.

SUMMARY

Magnetic resonance spectroscopy is able to measure noninvasively a variety of important metabolites involved in cell energetics. These include phosphocreatine, ATP, inorganic phosphate, pH, and lactate. Anoxia, ischemia, and infarction produce rapid loss of high-energy phosphates and accumulation of hydrolysis products. Many animal studies have shown that MRS monitors metabolic changes in various models of human disease. The availability of large, high field magnets and the development of noninvasive localization techniques permits MRS to be performed on selected volumes within the body. It is now clear that MRS in humans will be immediately useful in several areas including studies of malignancy, ischemia, and infarction of various organs and metabolic disorders. It is expected that human MRS will be increasingly used for clinical investigation and eventually for medical diagnosis.

ACKNOWLEDGMENTS

The author wishes to thank Drs. Frank Rico and Gerald Matson for their many helpful suggestions. Secretarial assistance was provided by Lou Thurman.

REFERENCES

1. GADIAN, D. G. 1982. Nuclear Magnetic Resonance and Its Applications to Living Systems. The Alden Press Ltd. Oxford.
2. JAMES, T. L. & A. R. MARGULIS. 1984. Biomedical Magnetic Resonance. Radiology Research and Education Foundation. San Francisco.
3. ILES, R. A., A. N. STEVENS & J. R. GRIFFITHS. 1982. NMR studies of metabolites in living tissue. Progr. NMR Spectroscopy **15:** 49-200.
4. CHAN, L. 1985. The current status of magnetic resonance spectroscopy-Basic and clinical aspects. West J. Med. **143:** 773–781.
5. MATSON, G. B. & M. W. WEINER. 1987. NMR spectroscopy in vivo: principles, animal studies and clinical applications. *In* Magnetic Resonance Imaging. D. P. Stark & W. G. Bradley Jr., Eds. Mosby. St. Louis.
6. MOON, R. B. & J. H. RICHARDS. 1973. Determination of intracellular pH by ^{31}P magnetic resonance. J. Biol. Chem. **248:** 7276–7278.
7. GILLIES, R. J., J. R. ALGER, J. A. DEN HOLLANDER & R. G. SHULMAN. 1982. Intracellular pH measured by NMR: methods and results. *In* Intracellular pH: Its Measurement, Regulation and Utilization in Cellular Functions. R. Nuccitelli & D. W. Deamer, Eds.: 79–104. Alan R. Liss. New York.
8. GADIAN, D. G., G. K. RADDA, R. E. RICHARDS & P. J. SEELEY. 1979. P-31 NMR in living tissue: the road from a promising to an important tool in biology. *In* Biological Applications of Magnetic Resonance. R. G. Shulman, Ed.: 463–535. Academic. New York.
9. BEHAR, K. L., D. L. ROTHMAN, R. G. SHULMAN, O. A. C. PETROFF & J. W. PRICHARD. 1984. Detection of cerebral lactate in vivo during hypoxemia by ^{1}H NMR at relatively low field strengths (1.9 tesla). Proc.Natl. Acad. Sci. USA **81:** 2517–2519.
10. HOULT, D. I., S. J. W. BUSBY, D. G. GADIAN, G. K. RADDA, R. E. RICHARDS & P. J. SEELEY. 1974. Observations of tissue metabolites using Phosphorus 31 nuclear magnetic resonance. Nature **252:** 285.
11. DAWSON, M. J., D. G. GADIAN & D. R. WILKIE. 1979. Mechanical relaxation rate and metabolism studied in fatiguing muscle by phosphorous nuclear magnetic resonance. J. Physiol. (Lond.) **299:** 465–484.
12. TAYLOR, D. J., P. J. BORE, P. STYLES, D. G. GADIAN & G. K. RADDA. 1983. Bioenergetics of intact human muscle: a ^{31}P nuclear magnetic resonance study. Mol. Biol. Med. **1:** 77–94.
13. ROSS, E. D., G. K. RADDA, D. G. GADIAN, G. ROCKER, M. ESIRI & J. FALCONER-SMITH. 1981. Examination of a case of suspected McArdle's syndrome by ^{31}P nuclear magnetic resonance. N. Engl. J. Med. **304:** 1338–1343.
14. RADDA, G. K. 1986. The use of NMR spectroscopy for the understanding of disease. Science **233:** 640–645.
15. ELEFF, S., N. G. KENNAWAY, N. R. BUIST, V. M. DARLEY-USMAR, R. A. CAPALDI, W. J. BANK & B. CHANCE. 1984. ^{31}P NMR study of improvement in oxidative phosphorylation by vitamins K_3 and C in a patient with a defect in electron transport at complex III in skeletal muscle. Proc. Natl. Acad. Sci. USA **81:** 3529–3533.
16. WILSON, J. R., L. FINK, J. MARIS, N. FERRARO, J. POWER-VANWART, S. ELEFF & B. CHANCE. 1985. Evaluation of skeletal muscle energy metabolism in patients with heart failure using gated Phosphorus 31 NMR. Circulation **71(1):** 57–62.
17. MILLER, R. G., D. GIANNINI, H. S. MILNER-BROWN, R. B. LAYSER, A. P. KORETSKY, D. HOOPER & M. W. WEINER. 1987. Effects of fatiguing exercise on high-energy phosphates, force, and EMG: evidence for 3 phases of recovery. Muscle and Nerve. (In press.)
18. MILLER R. G., M. D. BOSK, R. S. MOUSSAVI & M. W. WEINER. 1987. Rule of pH and phosphate in human muscle fatigue. Clin. Res. **35:** 155A.
19. DAWSON, M. J., K. J. BROOKS, D. MCFARLANE & S. J. SMITH. 1986. On the role of $[H_2PO_4^-]$ in determining cross-bridge cycling rate, force production and glycolytic recovery from contraction in skeletal muscle (Abst.). Soc. Magn. Reson. Med. **2:** 447–448.

20. Jacobus, W. E., G. J. Taylor, D. P. Hollis & R. L. Nunnally. 1977. Phosphorus nuclear magnetic resonance of perfused working rat hearts. Nature **265:** 756–758.
21. Gadian, D. G., D. I. Hoult, G. K. Radda, P. J. Seeley, B. Chance & C. Barlow. 1976. Phosphorus nuclear magnetic resonance studies on normoxic and ischemic cardiac tissue. Proc. Natl. Acad. Sci. USA **73:** 4446–4448.
22. Nunnally, R. L. & P. A. Bottomley. 1981. Assessment of pharmacological treatment of myocardial infarction by phosphorus-31 n.m.r. with surface coils. Science **211:** 177–180.
23. Ingwall, J. S. 1982. Phosphorus nuclear magnetic resonance spectroscopy of cardiac and skeletal muscles. Am. J. Physiol. **242:** H729–H744.
24. Ra, J. B., S. K. Hilal & Z. H. Cho. 1986. A method for in vivo MR imaging of the short T_2 component of sodium-23. Magn. Reson. Med. **3:** 296–302.
25. Neurohr, K. J., G. Gollin, E. J. Barrett & R. G. Shulman. 1983. In vivo ^{31}P-NMR studies of myocardial high energy phosphate metabolism during anoxia and recovery. FEBS Lett. **159:** 207–210.
26. Grove, T. H., J. J. H. Ackerman, G. K. Radda & P. J. Bore. 1980. Analysis of rat heart in vivo by phosphorus nuclear magnetic resonance. Proc. Natl. Acad. Sci. USA **77:** 299–302.
27. Kantor, H. L., R. W. Briggs & R. S. Balaban. 1984. In vivo ^{31}P nuclear magnetic resonance measurements in canine heart using a catheter-coil. Circ. Res. **55:** 261–266.
28. Balaban, R. S., H. L. Kantor, L. A. Katz & R. W. Briggs. 1986. Relation between work and phosphate metabolite in the in vivo paced mammalian heart. Science **232:** 1121–1123.
29. Jalles-Tavares, N., G. Karczmar, C. W. Ebert, A. Koretsky & M. W. Weiner. Catecholamines produce reversible depletion of cardiac high energy phosphates. Unpublished.
30. Stein, P. D., S. Goldstein, H. N. Sabbah, Z. Liu, J. A. Helpern, J. R. Ewing, J. B. Lakier, M. Chopp, W. F. LaPenna & K. M. A. Welch. 1986. In vivo evaluation of intracellular pH and high-energy phosphate metabolites during regional myocardial ischemia in cats using ^{31}P nuclear magnetic resonance. Magn. Reson. Med. **3:** 262–269.
31. Bottomley, P. A., R. J. Herfkens, L. S. Smith, S. Brazzamano, R. Blinder, L. W. Hedlund, J. L. Swain & R. W. Redington. 1985. Noninvasive detection and monitoring of regional myocardial ischemia in situ using depth-resolved phosphorus-31 NMR spectroscopy. Proc. Natl. Acad. Sci. USA **82(24):** 8747–8751.
32. Chance, B. 1984. Studies of exercise performance, vascular disease, and genetic disease. *In* Biomedical Magnetic Resonance. T. L. James & A. R. Margulis, Eds.: 187–199. Radiology Research and Education Foundation. San Francisco, CA.
33. Bottomley, P. A. 1985. Noninvasive studies of high energy phosphate metabolism in human heart by depth resolve ^{31}P NMR spectroscopy. Science **229:** 769–772.
34. Bottomley, P. A., L. S. Smith, R. J. Herfkens, T. M. Bashore & J. A. Utz. 1986. Detecting human myocardial infarction with localized ^{31}P NMR: characterization of normal and ischemic tissue (Abst.). Soc. Magn. Reson. Med. **3:** 606–607.
35. Chance, B., Y. Nakase, M. Bond, J. S. Leigh Jr. & G. McDonald. 1978. Detection of ^{31}P nuclear magnetic resonance signals in brain by in vivo and freeze-trapped assays. Proc. Natl. Acad. Sci. USA **75(4):** 4925–4929.
36. Ackerman, J. J. H., T. H. Grove, G. G. Wong, D. G. Gadian & G. K. Radda. 1980. Mapping of metabolites in whole animals by ^{31}P NMR using surface coils. Nature **283:** 167–170.
37. Thulborn, K. R., G. du Boulay & G. K. Radda. 1981. In vivo non-invasive measurements of energy metabolism and pH by ^{31}P NMR in experimental stroke correlated with cerebral oedema. J. Cereb. Blood Flow Metab. **1:** 580–581.
38. Prichard, J. W. & R. G. Shulman. 1986. NMR spectroscopy of brain metabolism in vivo. Annu. Rev. Neurosci. **9:** 61–85.
39. Behar, K. L., J. A. den Hollander, M. E. Stromski, T. Ogino, R. G. Shulman, O. A. C. Petroff & J. W. Prichard. 1984. High resolution ^{1}H NMR study of cerebral hypoxia in vivo. Proc. Natl. Acad. Sci. USA **80:** 4945–4948.
40. Bottomley, P. A., H. A. Hart & W. A. Edelstein. 1984. Anatomy and metabolism of the normal human brain studied by magnetic resonance at 1.5 Tesla. Radiology **150:** 441–446.

41. Hope, P. L., E. B. Cady, P. S. Tofts, P. A. Hamilton, A. M. de L. Costello, D. T. Delpy, A. Chu, E. O. R. Reynolds & D. R. Wilkie. 1984. Cerebral energy metabolism studied with phosphorus NMR spectroscopy in normal and birth-asphyxiated infants. Lancet **2:** 366–370.
42. Younkin, D. P., M. Delivoria-Papadopoulos, J. C. Leonard, V. H. Subramanian, S. Eleff, J. S. Leigh & B. Chance. 1984. Unique aspects of human newborn cerebral metabolism evaluated with 31-P NMR spectroscpy. Ann. Neurol. **16:** 581–586.
43. Baleriaux, D., D. A. Arnold, C. Segebarth, P. R. Luyten & J. A. den Hollander. 1986. ^{31}P MR evaluation of human brain tumor response to therapy (Abst.). Soc. Magn. Reson. Med. **1:** 41–42.
44. Ackerman, J. J. H., M. Lowry, G. K. Radda, B. D. Ross & G. G. Wong. 1981. The role of intrarenal pH in regulation of ammoniagenesis: ^{31}P NMR studies of the isolated perfused rat kidney. J. Physiol. **319:** 65–80.
45. Adam, W. R., A. P. Koretsky & M. W. Weiner. 1987. Measurement of renal intracellular pH in rats in vivo using ^{31}P NMR: Effects of acidosis and K^+ depletion. Am. J. Physiol. (In press.)
46. Freeman, D. S., S. Bartlett, G. K. Radda & B. D. Ross. 1983. Energetics of sodium transport in the kidney: Saturation transfer ^{31}P-NMR. Biochim. Biophys. Acta **762:** 325–336.
47. Koretsky, A. P., S. Wang, M. P. Klein, T. L. James & M. W. Weiner. 1986. ^{31}P NMR saturation transfer measurements of phosphorus exchange reactions in rat heart and kidney in situ. Biochemistry **25(1):** 77–84.
48. Chan, L. *et al.* 1981. Study of human kidneys prior to transplantation, by phosphorus nuclear magnetic resonance. *In* Organ Preservation. Vol. 3. Halasz Jacobsen Pegg, Eds. MTP Press. Lancaster, PA.
49. Siegel, N. J., M. J. Avison, H. F. Reilly, J. R. Alger & R. G. Shulman. 1983. Enhanced recovery of renal ATP with postischemic infusion of ATP-$MgCl_2$ determined by ^{31}P-NMR. Amer. J. Physiol. **245:** F530–F534.
50. Wong, G. C. & B. D. Ross. 1983. Application of phosphorus nuclear magnetic resonance to problems of renal physiology and metabolism. Mineral Electrolyte Metab. **9:** 22–289.
51. Koretsky, A. P., S. Wang, J. Murphy-Boesch, M. P. Klein, T. L. James & M. W. Weiner. 1983. ^{31}P NMR spectroscopy of rat organs, in situ, using chronically implanted radiofrequency coils. Proc. Natl. Acad. Sci. USA **80:** 7491–7495.
52. Cohen, S. M., R. G. Shulman & A. C. McLaughlin. 1979. ^{13}C NMR studies of gluconeogenesis in rat liver cells: Utilization of labelled glycerol by cells from euthyroid and hyperthyroid rats. Proc. Natl. Acad. Sci. USA **76:** 1603–1607.
53. Alger, J. R., K. L. Behar, D. L. Rothman & R. G. Shulman. 1984. Natural-abundance carbon-13 NMR measurement of hepatic glycogen in the living rabbit. J. Magn. Reson. **56(2):** 334–337.
54. Shulman, G. I., J. R. Alger, J. W. Prichard & R. G. Shulman. 1984. Nuclear magnetic resonance spectroscopy in diagnostic and investigative medicine. J. Clin. Invest. **74:** 1127–1131.
55. Ng, T. C., W. T. Evanochko, R. N. Hiramoto, V. K. Ghanta, M. B. Lilly, A. J. Lawson, T. H. Corbett, J. R. Durant & J. D. Glickson. 1982. ^{31}P NMR spectroscopy of invivo tumors. J. Magnetic Reson. **49:** 271–286.
56. Griffiths, J. R., A. N. Stevens, R. A. Iles, R. E. Gordon & D. Shaw. 1981. ^{31}P-NMR investigation of solid tumours in the living rat. Biosci. Rep. **1:** 319–325.
57. Evanochko, W. T., T. C. Ng & J. P. Glickson. 1984. Applications of in vivo NMR spectroscopy to cancer. Mag. Reson. Med. **1:** 508–534.
58. Ross, B., M. Smith, V. Marshall, S. Bartlett & D. Freeman. 1984. Monitoring response to chemotherapy of intact human tumors by ^{31}P nuclear magnetic resonance. Lancet **1:** 641–646.
59. Maris, J., A. Evans, A. McLaughlin, G. D'Angio, L. Bolinger, H. Manos & B. Chance. 1985. ^{31}P NMR spectroscopic investigation of human neuroblastoma in situ. New Eng. J. Med. **312:** 1500–1505.
60. Styles, P. 1987. Spatially resolved ^{31}P-NMR spectroscopy of organs in animal models and in man. Ann. N.Y. Acad. Sci. (This volume.)
61. Luyten, P. R., A. J. H. Marien, B. Sijtsma & J. A. den Hollander. 1986. Solvent-

suppressed spatially resolved spectroscopy. An approach to high-resolution NMR on a whole-body MR system. J. Magn. Reson. **67:** 148–155.

62. ORDIDGE, R. J., A. CONNELLY & J. A. B. LOHMAN. 1986. Image-selected in vivo spectroscopy (ISIS). A new technique for spatially selective NMR spectroscopy. J. Magn. Reson. **66:** 283–294.
63. KORETSKY, A. P. *et al.* 1984. 31 Phosphorous magnetization transfer measurement of exchange reactions in vivo. *In* Biomedical Magnetic Resonance. T. L. James & A. Margulis, Eds.: 231–242. Radiology Research Education Foundation. San Francisco, CA.
64. CADY, E. B. *et al.* 1983. Lancet :1060.

The Biochemistry of Human Diseases as Studied by ^{31}P NMR in Man and Animal Models[a]

GEORGE K. RADDA, ROLF D. OBERHAENSLI,
AND DORIS J. TAYLOR

Department of Biochemistry
University of Oxford
Oxford, England OX1 3QU

INTRODUCTION

In 1981, we reported our first examination of a patient with a muscle disease using ^{31}P magnetic resonance spectroscopy.[1] This patient lacked the enzyme phosphorylase and, therefore, could not break down glycogen during exercise. Consequently, no acidification occurred in his muscle. Since then, we have carried out over 1,200 examinations of patients with a variety of diseases.[2] The majority of our studies (about 800) have been concerned with muscle disorders or diseases where muscle metabolism can help us to understand some of the features of a systemic disorder. Since 1983, when our whole-body (1.9 Tesla) instrument was installed at the John Radcliffe Hospital in Oxford, we have begun a systematic study of other organs, such as the liver, brain, and heart in patients with a variety of diseases. It is the energetics of the tissue, that is, the relationship between the utilization and supply of ATP and how this pathway is controlled, that we can examine by ^{31}P magnetic resonance spectroscopy.

In our clinical investigations we have set up our program of research using a number of guidelines. First, of course, one must select the problem to be studied. Then one needs to establish a protocol for the NMR examination that is appropriate for the patient examination. In our experience, it is essential to devise a test in which metabolism is stressed, e.g. exercise in the case of skeletal muscle, infusion of substrates in the case of liver, and so on. It is then necessary to determine the variability of the metabolic response to such stress in healthy subjects. It is important to look for parameters that are relatively invariant across the healthy population but could be used as a sensitive index for disease. In our muscle investigations, we have outlined before how these parameters tend to be centered around those processes that are involved in the control of the bioenergetic pathways of the tissue.[2,3] Based on such premises, one is then ready for the patient examination. Very often, if the results are unusual, it may be essential to set up an animal model in which such investigations can be extended in a more invasive way and linked to other biochemical observations.

A NEWLY RECOGNIZED MUSCLE DISEASE: AN EXAMPLE

As an example of the type of biochemical information we can derive from ^{31}P MRS, we wish first to describe one case.[4] This study illustrates how, in the light of extensive

[a]Supported by the Medical Research Council, the Department of Health and Social Security, and the British Heart Foundation.

background information on patients with a range of muscle diseases and our control studies, MRS can give an important clue as to the nature of the underlying biochemical defect. This clue can then lead the clinician to the correct tests that usually have to be based on invasive biopsies. The case refers to a 27-year-old man with a 3-year history of exercise-induced muscle pain, myoglobinuria, and elevated serum creatine kinase. Histological examination of a biopsy from quadriceps revealed non-specific myopathic changes with occasional clusters of sub-sarcolemmal mitochondria. Phosphorylase stain was normal and no other abnormality could be detected. Detailed ^{31}P MRS studies of gastrocnemius and flexor digitorum superficialis muscles were then carried out. At rest, no abnormalities were seen. During aerobic dynamic exercise, there was an abnormally rapid decrease in phosphocreatine concentration and the change in intracellular pH was within the normal range. It appeared, however, that the point at which lactate production was turned on was somewhat delayed in comparison to normal controls. This small delay might indicate some kind of block or inhibition of the process of glycogen breakdown. It differs significantly from the observations made in patients with phosphorylase deficiency (McArdle's disease) where there is no acidification during exercise and indeed the tissue goes alkaline.[1,4] During exercise in this patient, we observed that there was a build-up of phosphomonoester, probably glucose-6-phosphate, which could be indicative of a block in glycolysis. The build-up of this sugar phosphate was significantly different from that observed in patients with PFK deficiency where, in any case, there was no acidification of the muscle.[5,6] Recovery from exercise was unusual in that phosphocreatine resynthesis and inorganic phosphate disappearance followed similar and considerably prolonged time courses. In control subjects the rate of inorganic phosphate disappearance is about twice as fast as the rate of phosphocreatine resynthesis. These data indicate that oxidative metabolism is also impaired in the patient, since the slow phosphocreatine recovery is characteristic of some of the mitochondrial diseases we have studied.[7] In the majority of patients with mitochondrial myopathies we also observed an abnormally low phosphorylation potential in the resting muscle. This was absent in our patient. The conclusion has to be that there is some kind of a inhibition of glycolysis as well as some interference with the mitochondrial phosphorylation process. We concluded that the abnormality must be located at a link between the mitochondrial and cytoplasmic compartments. This is achieved by the malate aspartate shuttle or by the α-glycerophosphate shuttle. Since the transport of inorganic phosphate into the mitochondria appeared to be delayed, we felt that the malate aspartate shuttle was the more likely candidate in this disease. In an open muscle biopsy, the properties of isolated mitochondria were then studied. We found that oxygen uptake was similar to control values and we were able to rule out a primary defect in mitochondrial respiration. In contrast, the malate-aspartate shuttle had an activity less than 20% of that in samples from control subjects.[8] This single example illustrates the power of MRS in solving rare and unusual clinical problems. We have seen several cases where MRS has made a significant contribution towards the solution of a clinical problem.[9]

EXERCISE AND MUSCLE METABOLISM IN HEART FAILURE AND PERIPHERAL VASCULAR DISEASE

The study of the relatively rare genetic diseases can contribute a great deal to our understanding of human biochemistry. This, in itself, justifies the use of non-invasive investigations and the expense of magnetic resonance spectroscopy. We can however, do better than this in that we can examine the consequences and changes associated with common disorders such as peripheral vascular disease, heart failure, renal failure, and endocrine diseases, all of which alter muscle metabolism. We can therefore expect to learn something new about the progress of the disease and about the success of therapeutic interventions. For example, heart failure patients are known to suffer from exercise intolerance and this performance does not correlate with hemodynamic abnormalities. We have examined a group of patients in heart failure and have been able to demonstrate that during exercise they show excessive acidification and that there are other metabolic abnormalities in their muscles that cannot be accounted for simply by the reduction of blood flow.[10,11] It appears that there are adaptive metabolic changes to chronic exposure to low cardiac output and some of these changes may result from alterations in the distribution of fiber types. Chance and his co-workers[12] and ourselves[13] have studied many patients with intermittent claudication and have shown that peripheral vascular disease leads to marked changes in metabolism and these changes can be used to indicate the severity of the disease.

MITOCHONDRIAL DISORDERS

We have examined more than 20 patients with mitochondrial myopathies and the pattern that we observed is consistent with the results we reported earlier after our initial 12 investigations.[7] At rest, these patients show a decrease in the phosphocreatine levels and generally the concentrations of metabolites are consistent with a low phosphorylation potential. Most of the patients show slow phosphocreatine recovery after exercise, and this slow recovery indicates very severe inhibition of oxidative phosphorylation. An interesting observation is that although these patients show very high blood lactate levels at rest, and particularly following exercise, the change in their intracellular pH is somewhat less than that of control subjects, and the recovery of the intracellular pH following exercise is considerably faster than in controls. This seems to represent an adaptive mechanism whereby the unwanted and unusable lactate that might be damaging to the cell, can be removed rather more efficiently from their muscles than in controls. Kinetics of the pH recovery indicates that hydrogen ion removal is carrier mediated and is already fully saturated. The increased rate of hydrogen ion removal in the patients therefore is consistent with an increased number of carriers being induced by the chronic conditions.

SOME METABOLIC DISEASES OF THE LIVER

Unlike in skeletal muscle, stressing the liver is not simple. We have, however, developed several protocols in which infusion of substrates such as fructose, glycerol, and glucose can be used to elicit metabolic responses in which normal subjects and those with liver disease can be distinguished. We will first examine the case of fructose handling. We have demonstrated some time ago that ^{31}P NMR spectra from human liver can be obtained rapidly and efficiently either with the use of field-profiling techniques[14] or by Fourier depth selection.[15] We can record spectra in 2–4 minutes of accumulation and follow the time courses of metabolic changes induced by fructose.[2,16] Fructose is handled by the liver initially by phosphorylation using ATP in a reaction catalyzed by fructokinase, and the fructose-1-phosphate is then split by aldolase B into two three-carbon fragments that eventually feed into the glycolytic sequence. These reactions can be followed in healthy individuals if we infuse a relatively large dose of fructose (200 mg/kg body weight) intravenously. In contrast, if fructose is taken orally, a load of 50 g of fructose would produce no observable metabolic changes in normal subjects. It is on this basis that we have investigated, in collaboration with Drs. Leonard and Collins and Prof. Herschkowitz, a number of patients with fructose intolerance and their parents, who are heterozygous carriers of this genetic disease. We have been able to show that, for example in a young boy aged 9, we can observe an increase in the sugar phosphate region following a meal that included a few potatoes (which have sufficient fructose to produce an abnormal response) (FIG. 1). The increase in the sugar phosphate region was accompanied by a very low signal from

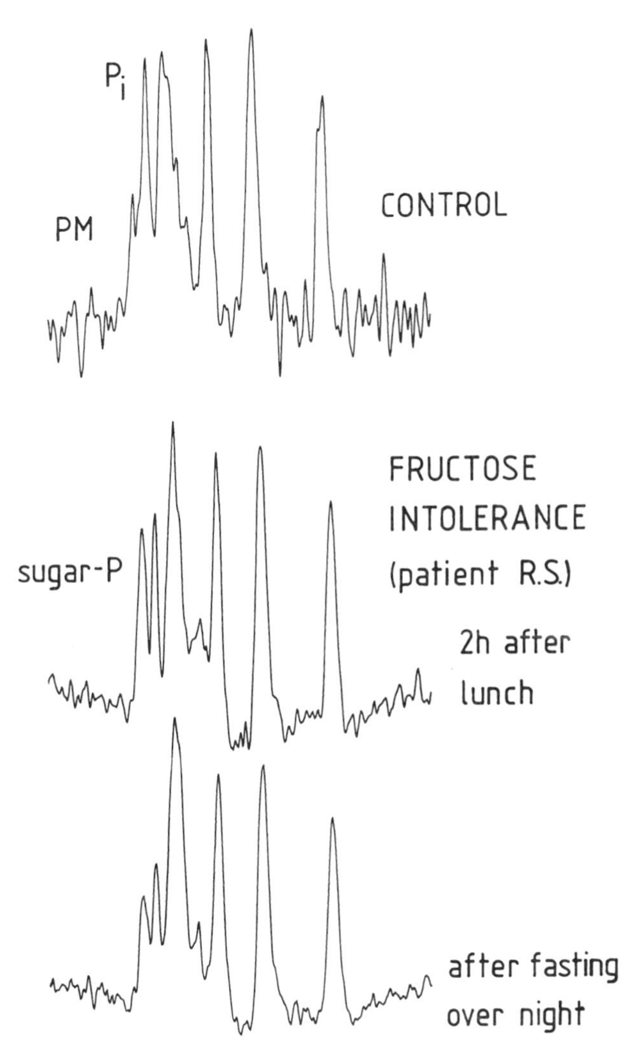

FIGURE 1. ^{31}P NMR spectra of human liver. Control subject and patient aged 10 with fructose intolerance (aldolase B deficiency) after a meal and after a 24-hour fast. Note lower P_i even under fast and elevated phosphomonoester following meal.

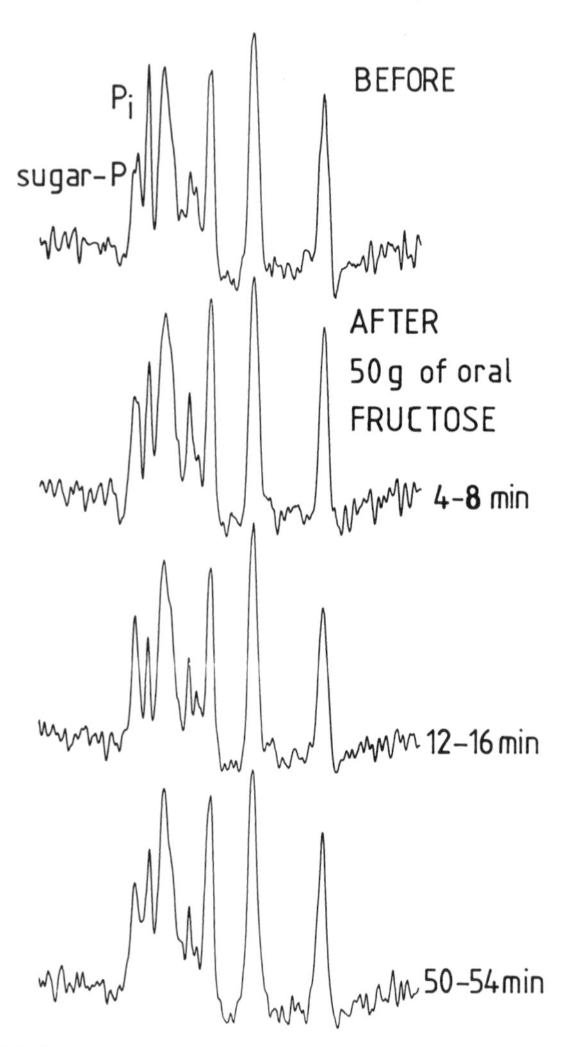

FIGURE 2. ^{31}P NMR spectra of the liver of asymptomatic mother of patient described in FIG. 1. Metabolic response following 50 g oral fructose. Note increase in phosphomonoester and decrease in P_i 5–10 min after fructose.

inorganic phosphate. After a 24-hour fast the sugar phosphate level returned to normal but the inorganic phosphate level was still low. We observed the low P_i signal in several patients, including some older ones, with fructose intolerance even though they were on an apparently completely fructose-free diet. When the parents of such patients are administered with 50 g fructose orally, we can observe an increase in their sugar phosphate signal and a decrease in P_i within 5–10 minutes of the administration. The time course of such changes is shown in FIGURE 2. The carriers are asymptomatic and generally are not on a fructose-free diet, yet quite clearly from the NMR investigation their liver metabolism is abnormal with respect to fructose handling. An important observation that comes out of these studies is that several of the older patients and also several of the male carriers we have studied suffered from gout. We believe our investigations can now rationalize this observation, since the patients run low inorganic phosphate levels and even in the carriers this is induced by ingestion of fructose. Under

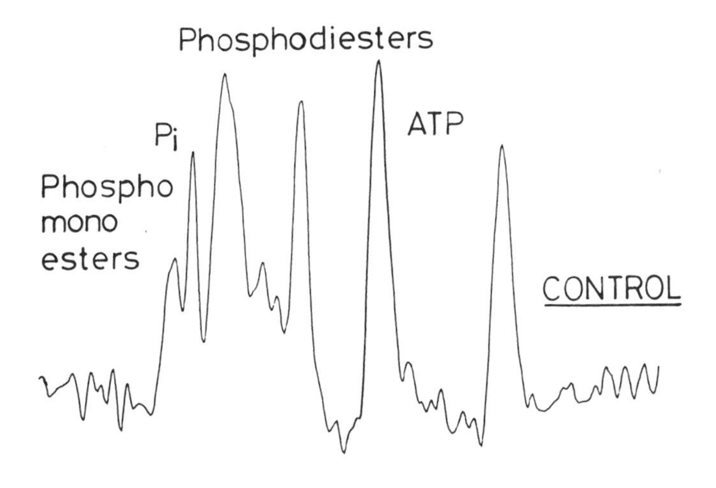

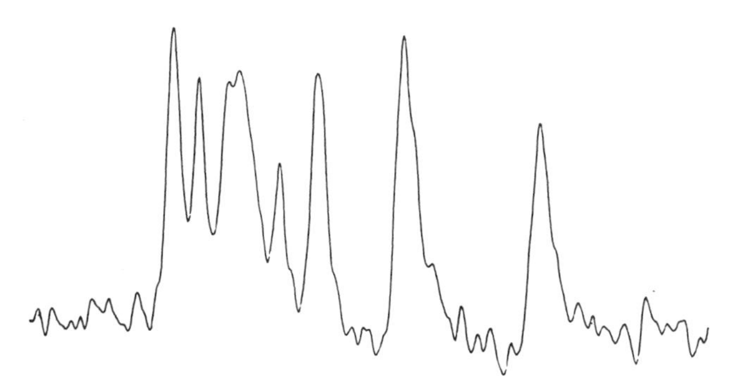

FIGURE 3. ^{31}P NMR spectrum of the liver of a patient with viral hepatitis B compared to control subject.

such conditions AMP-deaminase is activated and there is an increased production of uric acid. In fact, we measured the blood uric acid levels in the carriers following fructose administration and have found it to be elevated compared to those of controls. Thus patients who show no clinical symptoms after fructose ingestion who have lower aldolase B activities in their liver and are carriers for fructose intolerance are likely to suffer from gout more readily than other subjects. This might well be one of the causes of genetically inherited gout.

We have also examined patients with other metabolic liver diseases such as Type 1 glycogen storage disease (glucose 6, phosphatase deficiency) galactosemia, and so on.

INFECTIOUS DISEASES AND CANCER IN THE LIVER

Not a great deal is known about the effect of virus infections on metabolic response. We have studied 6 patients with viral hepatitis (hepatitis B) and have observed significant changes in the ^{31}P NMR spectra of their livers (FIG. 3). In particular, we see a major increase in the signal in the sugar phosphate region. From the chemical shift and T_1 relaxation time of this signal, we conclude that this increase is associated with an increased amount of phosphorylcholine or phosphoethanolamine. We also observe a small decrease in the signal from compounds in the phosphodiester region. When the patient recovers from this condition the spectrum returns to that of the control subjects. It is interesting to speculate that the changes observed in hepatitis B are very similar to those seen in liver cancers such as hepatoblastoma.[17] Since the hepatitis virus, at least in the African continent, often leads to liver cancer this association of spectral changes may well be interesting to pursue. We note that the changes are in those compounds that are involved as intermediates in the biosynthesis or degradation of phospholipids and it could be considered that changes in their concentrations are a result of increased membrane turnover in regenerating or proliferating liver cells.

CONCLUSIONS

In this brief presentation we considered a few examples of the many hundreds of cases we examined by ^{31}P magnetic resonance spectroscopy. The method has already increased our understanding of some diseases, provided new insights through the study of diseases into the control of bioenergetics *in vivo*,[2] and provided information of value to the practicing clinician.[13] The human investigations would not have been possible without many years of extensive work on isolated and *in situ* animal tissues (for a review see Radda & Taylor[9]) and new biochemical and technical developments based on these studies. Some of the techniques used in the work we presented here are also described in this volume.[18] The approach is clearly applicable to the study of many other systems and organs such as the brain[17,19] and heart.[20]

ACKNOWLEDGMENTS

We are grateful to the many colleagues who contributed to this research. Their names appear in the reference list or are mentioned in the text.

REFERENCES

1. ROSS, B. D., G. K. RADDA, D. G. GADIAN, G. ROCKER, M. ESIRI & J. FALCONER-SMITH. 1981. New Engl. J. Med. **304:** 1338–1342.

2. RADDA, G. K. 1986. Science **233:** 640–645.
3. RADDA, G. K., P. J. BORE & B. RAJAGOPALAN. 1984. Br. Med. Bull. **40:** 155–159.
4. RADDA, G. K. 1986. Biochem. Soc. Trans. **14:** 517–525.
5. EDWARDS, R. H. T., M. J. DAWSON, D. R. WILKIE, R. E. GORDON & D. SHAW. 1982. Lancet **i:** 725–730.
6. CHANCE, B., S. ELEFF, W. BANK, J. S. LEIGH, JR. & R. WARNELL. 1982. Proc. Natl. Acad. Sci. USA **79:** 7714–7718.
7. ARNOLD, D. L., D. J. TAYLOR & G. K. RADDA. 1985. Ann. Neurol. **18:** 189–196.
8. HAYES, D. J., D. J. TAYLOR, D. L. ARNOLD & G. K. RADDA. 1986. Biochem. Soc. Trans. **14:** 1208–1209.
9. RADDA, G. K. & D. J. TAYLOR. 1985. Int. Rev. Exp. Pathol. **27:** 1–58.
10. MASSEY, B., M. CONWAY, R. YONGE, P. SLEIGHT, J.G. G. LEDINGHAM, G. K. RADDA & B. RAJAGOPALAN. 1987. Am. J. Cardiol. (In press.)
11. CONWAY, M., B. MASSEY, S. FROSTICK, R. P. YONGE, J. E. G. LEDINGHAM, P. SLEIGHT, G. K. RADDA & B. RAJAGOPALAN. 1987. Eur. J. Clin. Invest. (In press.)
12. CHANCE, B. 1984. *In* Biomedical Magnetic Resonance. T. L. James & A. R. Margulis, Eds.: 187–199. Radiology Research and Education Foundation. San Francisco, CA.
13. HANDS, L. J., P. J. BORE, G. GALLOWAY, P. J. MORRIS & G. K. RADDA. 1986. Clin. Sci. **71:** 283–290.
14. OBERHAENSLI, R. D., G. J. GALLOWAY, D. HILTON-JONES, P. J. BORE, P. STYLES, B. RAJAGOPALAN, D. J. TAYLOR & G. K. RADDA. 1987. Br. J. Radiol. **60:** 367–373.
15. BLACKLEDGE, M. J., P. STYLES & G. K. RADDA. 1987. J. Magn. Reson. **71:** 246–258.
16. OBERHAENSLI, R. D., G. J. GALLOWAY, D. J. TAYLOR, P. J. BORE & G. K. RADDA. 1986. Br. J. Radiol. **59:** 695–699.
17. OBERHAENSLI, R. D., D. HILTON-JONES, P. J. BORE, L. J. HANDS, R. P. RAMPLING & G. K. RADDA. 1986. Lancet **ii:** 8–11.
18. STYLES, P., M. J. BLACKLEDGE, C. T. W. MOONEN & G. K. RADDA. 1987. Ann. N.Y. Acad. Sci. (This volume.)
19. HAYES, D. J., D. HILTON-JONES, D. L. ARNOLD, P. STYLES, G. J. GALLOWAY, J. DUNCAN & G. K. RADDA. 1985. J. Neurol. Sci. **71:** 105–118.
20. RAJAGOPALAN, B., M. J. BLACKLEDGE, W. J. MCKENNA, N. BOLAS & G. K. RADDA. 1987. Ann. N.Y. Acad. Sci. (This volume.)

DISCUSSION OF THE PAPER

B. CHANCE: I had a lot of interest in the fructose stress. I wonder whether the ADP rises high enough to pick it up in the alpha and the beta peaks?

G. K. RADDA: To be quite honest we haven't actually analyzed the difference between the beta and the alpha peaks and to see whether ADP rise is picked up in that particular case. It's true that liver is the only organ where you can actually measure the ADP by NMR. In ethanolic administration in the animal we have seen a substantial ADP rise.

J. COHEN: I'm still worried about the alkalinization of the pH in the tumors that you've reported. You did say it was speculative, but you report your internal pHs to two places in decimals. Now, I think you'd probably agree they're not that accurate given that the lines are very broad and the calibration curve is not really sure. If you did these repeatedly you'd probably find the reproducibility is variable and so the standard error would probably be no more than ± .1. If I look at the values that you've given, that would mean that most of the differences would be not significant.

RADDA: I'm not worried about the alkalinization of the pH like you are. The standard error on the measurement of the normal liver for the intracellular pH is ± 0.02 on 35 different individuals. With the brain, the standard error is slightly higher because the phosphodiester is rather broad next to the inorganic phosphate, but the error is still only in the range of .03 and the pHs that we measure in the tumors are quite often .2 pH units above the corresponding normal tissue. The second point is that rather remarkably and to my surprise, the lines in the tumors are not broad at all, they're rather narrow. Those pHs are pretty good.

K. UGURBIL: If you are making conclusions about the pH in the tumor or any other conclusion, I presume that you're using some sort of a localization technique so that you are sure that you are looking at the tumor and don't have contamination from normal brain tissue. What kind of a protocol do you utilize to insure that this is the case?

RADDA: I deliberately didn't talk about the techniques of localization because Peter Styles is going to talk about this later. We use either field profiling or a combination of field profiling with a rotating frame technique for spatial selection to look at different regions. The details and the limitations of the measurement are discussed in the paper that we published in July. Particular tumors were chosen to be rather large and were characterized by CT scan before the NMR. On that basis we are pretty certain that we are looking at tumor tissue and not surrounding tissue. In any case, the inorganic phosphate peaks in the tumors are considerably higher than it would be in the normal range.

QUESTION: Were the changes that you saw from viral hepatitis unique to type B hepatitis or is that in type A as well and non-A, non-B?

RADDA: We haven't done enough studies to answer that question of other sorts of hepatitis.

QUESTION: You show differences in relative ATP levels for viral hepatitis. I wonder what you use as an intensity reference, say from patient to patient.

RADDA: I showed you on some of the slides but I didn't discuss the absolute intensities. We actually used the ATP as our internal reference, recognizing of course that there might well be conditions where that ATP concentration changes. However, it's a great surprise to us that if we use the same protocol on different people, that is we use the same surface coil, the same setting, and everything else, we can without any external reference just from the peak highs get a reproducibility from individual to individual of those measurements of around 30%, which is quite remarkably good, considering you have no standard and these are the absolute measurements of the peak heights. Now, what that means I don't know, except that we can obviously set up people in the same sort of way and look at the same depth and so on. But we normally refer everything to ATP.

An Approach to the Problem of Metabolic Heterogeneity in Brain:

Ischemia and Reflow after Ischemia[a]

B. CHANCE,[b] J. S. LEIGH, JR.,[b] S. NIOKA,[c] T. SINWELL,[d]
D. YOUNKIN,[e] AND D. S. SMITH[f]

[b]Department of Biochemistry and Biophysics
[c]Department of Physiology
[f]Department of Anesthesiology
University of Pennyslvania
Philadelphia, Pennsylvania 19104

[d]Phospho-Energetics, Inc.
Philadelphia, Pennsylvania

[e]Department of Neurology
Childrens Hospital of Philadelphia
Philadelphia, Pennsylvania

INTRODUCTION

Metabolic heterogeneity has led to great difficulties in interpreting the effects of ischemia or hypoxia on the brain. The salient observation was made initially in the studies of Miyake one of the first applications of phosphorus magnetic resonance spectroscopy (P MRS) to large animal models. In a cat brain model of middle cerebral artery occlusion, they found that the kinetics of phosphocreatine (PCr) and adenosine triphosphate (ATP) decrease were closely linked, a phenomenon that suggests that the creatine kinase equilibrium was not functional in this situation. Experiments by Komatsomoto and collaborators[2] on the same type of model led to the same result. However, similar but detailed studies on a dog model involving bilateral carotid artery occlusion and hypovolemia to 30 to 40 mm Hg have led us to an explanation of the apparent coincidence of PCr and ATP disappearances: namely, steep oxygen gradients and heterogeneous metabolic states of brain cells coexist. Some of these cells are in a stable steady state and others in an unstable state.[3] We hypothesize that the similar disappearance of PCr and ATP is related to the rate of transition from one state to the other and not to disequilibrium of creatine kinase. Thus the loss of PCr and ATP in a brain that still contains substantial amounts of PCr and ATP does not necessarily indicate the assumption of the failure of creatine kinase equilibrium. Instead, we propose the existence of two metabolic states of brain cells, one a normal steady state where $V_{ATPase} \leq V_{m_{ATPsyn}}$ and the other an unsteady state where $V_{ATPase} > V_{m_{ATPsyn}}$. The existence of the two states is confirmed by the observation of two peaks of the P_i, one near normal pH and the other at least one unit more acidic, suggesting at least two separate pH components.

[a]Supported by National Institutes of Health Grants (NS 22881, HL 31934, RR 02305, and HR 34004), The James S. McDonnell Foundation, The Benjamin Franklin Partnership's Advanced Technology Center of Southeastern Pennyslvania, and Phospho-Energetics, Inc.

METHOD

NMR Spectroscopy

A 1.9 Tesla, 30 cm bore Phospho-Energetics 280 MR spectrometer was employed. Proton and ^{31}P were measured at 88.4 and 35.8 MHz. Room temperature shimming resulted in a 0.2 to 0.3 ppm fullwidth at half maximum peak height. A 2.5 cm diameter, two turn, doubly tuned surface coil was employed. The 1331 pulse sequence was used for proton water suppression. For ^{31}P baseline suppression, a presaturation

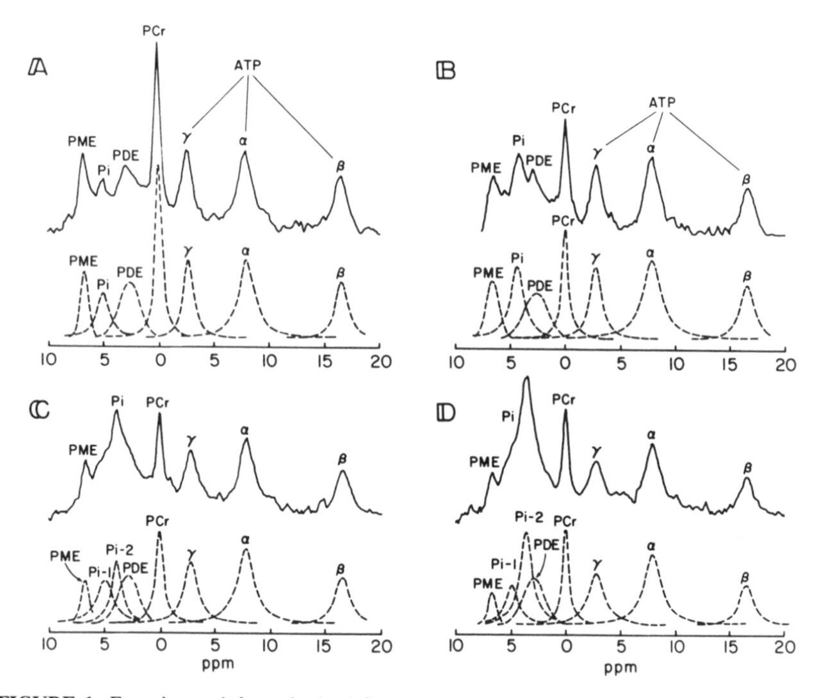

FIGURE 1. Experimental data obtained from the dog brain in hypovolemia and ischemia. Of particular interest is the development of a displaced P_i peak due to acidosis in an ischemic hypoxic compartment. Curve fitting, as indicated in the text, is used in the analysis of these data. Traces A,B,C,D: Stages in the development of two compartments in heterogeneous brain ischemia in dog model. The top trace represents the experimental data and the computer fit while the bottom trace represents the deconvolution. Note the appearance of an additional P_i peak in C and D.

pulse preceded the 90° excitation pulse. The ^{31}P free induction decay (FID) was acquired over an interval of 80 msec with a repetition interval of 4 sec and averaged for convenient intervals of steady-state performance. Typical ^{31}P spectra are shown in FIGURE 1(a–d).

These spectrometric data were fitted by 10 Lorentzian curves. The five peaks corresponding to ATP and PCr, were fitted with single peaks; when necessary, two P_i peaks were used. PME and PDE required special attention because of their asymmetry and were fitted separately on the upfield and the downfield sides. In the case of PDE, the major peak was at 2.45 to 3 ppm, with a minor peak at 1.5 to 2 ppm. In the case of PME, the major peak was at 6.5 to 7 ppm with a minor peak at 5.9 to 6.5 ppm. The concentrations employed were in the range of 200 to 500 μM, which is consistent with analytical data, PEP 50 μM, 1.5 ppm; 2 DPG at 5.725 ppm; phosphorylcholine at 6.24 ppm; 2-phosphoglycerate at 6.0 ppm, a total of 200 μM.

pH_i was calculated according to the formula

$$pH_i = 6.77 + \log \frac{P_i \text{ shift} - 3.27}{5.68 - P_i \text{ shift}},$$

where P_i shift is the chemical shift of P_i with respect to PCr. When two pools of P_i were discernible, the separate pHs were calculated.

In order to convert the areas of ATP, PCr, and P_i into concentrations, we assumed that ATP under control conditions was 2.3 mM, that the total PCr + P_i was 11 mM and that the latter pool was constant during the experiment.

Saturation due to the 4 sec repetition rate was corrected by the factors 1.39, 1.20, and 1.05 for PCr, P_i, and ATP, respectively. With correlation factors, the areas of the MRS peaks would then be proportional to the concentrations of the individual compounds. These factors were obtained from fully relaxed spectra of dog brain.

Surgical Procedure

Five adult Beagle dogs were anesthetized and intubated with isoflurane (5% induction, 2% maintenance) and 70% nitrous oxide in oxygen. Two femoral arteries and veins were cannulated for measuring blood pressure, arterial blood gases, and pH, and for fluid administration. The dogs received intravenous fluids (5% glucose in 0.45% NaCl) at a rate of about 2 ml/kg/hr. Snares were placed loosely around both carotid arteries for later ligation. The dogs were covered with a heating blanket and monitored with a nasopharyngeal temperature probe to maintain the temperature at 37°C. The animals were placed on a conductive cradle and the head was positioned in a stereotaxic holder.[3] To reduce interfering ^{31}P signals, the skin and muscle overlying the skull were widely excised. The top of the skull bone was thinned to improve signal quality. The MRS surface coil was placed on the top of the skull.

Experimental Protocol

At the end of surgery, the nitrous oxide was discontinued and anesthesia maintained with 1.5–2% isoflurane in 30% oxygen, with the balance nitrogen. During the control period, normal arterial blood gasses and pH, and mean arterial blood pressure were maintained for at least 30 min. Bilateral carotid artery occlusion was then obtained by tightening the snares. After occlusion, further baseline data were obtained for another 30-min normotensive period.

Hypotension commenced by draining blood from one of the femoral arteries. The blood was then anticoagulated with ACD solution. Mean arterial blood pressure was progressively lowered to 50, 40, 30, and 20 mm Hg with at least 15 min at each blood pressure level. The MABP was lowered until PCr/P_i decreased to between 1.0 to 0.5. To provide steady state PCr/P_i values, signals from the MRS obtained every 4 min were monitored. Between MAP of 30–20 mm Hg, a rapid drop of PCr/P_i often occurred. The MABP was raised as needed in an attempt to maintain PCr/P_i between 1.0 and 0.5. The steady state for PCr/P_i was maintained for another 1 to 2 hr. During

1) $H^+ + NADH \longrightarrow$ [Flux V/V_M] $\longleftarrow \frac{1}{2}O_2$; outputs: ATP (up); inputs: 3 ADP, 3 Pi

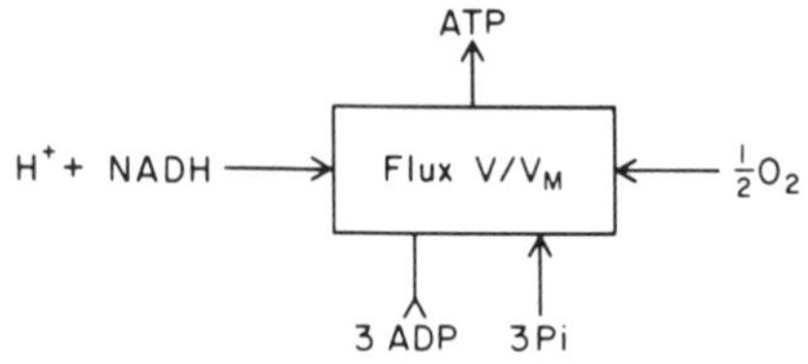

2) $$3ADP + 3Pi + NADH + H^+ + \frac{1}{2}O_2 \longleftrightarrow 3ATP + NAD^+ + H_2O$$

3) The general equation is:

$$\frac{V}{V_M} = \frac{1}{1 + \frac{K^{I}}{ADP} + \frac{K^{II}}{Pi} + \frac{K^{III}}{O_2} + \frac{K^{IV}}{NADH}}$$

4) The specific equation for P NMR studies of ADP control at relatively constant pH, with excess NADH and Pi is:

$$\frac{V}{V_M} = \frac{1}{1 + \frac{0.6}{Pi/PCr}}$$

5) and for ADP and O_2 control is:

$$\frac{V}{V_M} = \frac{1}{1 + \frac{0.6}{Pi/PCr} + \frac{kV}{[O_2]}}$$

where kV takes into account our finding of the effect of V upon K_M for cytochrome oxidase

FIGURE 2. Steady-state model for oxidative metabolism.

this time, the MR spectra were accumulated every 4 min and MABP was altered when necessary to maintain a stable PCr/P_i. After the graded MABP titration and ischemic stress of PCr/P_i at 1.0 to 0.5 for 2 to 3 hr, a recovery period was begun by restoring the shed blood via the femoral vein. This was followed by a 3 hr normotensive (recovery or reflow) period.

THEORETICAL BACKGROUND

The employment of the Michaelis-Menten equations for the representation of oxidative metabolism in the animal model or the neonate brain has enabled us to put forward a hypothesis concerning the relationship between the tissue oxygen concentration and the MRS parameter PCr/P_i. This equation is derived as indicated in FIGURE 2, and it is seen that for a constant V/V_m a decrease of oxygen will cause a proportional decrease of PCr/P_i. In this sense we have identified phosphorus MR spectroscopy as a "tissue oxygen sensor."[4]

It is further possible to take into account heterogeneity induced by steep oxygen gradients in rapidly oxidizing tissues.[5–7] In this case, a microheterogeneous response of the tissue to oxygen lack, particularly when ischemia (as well as hypoxia) is involved. Our underlying assumption is that the presence of "steep oxygen gradients," namely, hypoxia/ischemia is "all or none" and that the tissue is either adequately supplied with oxygen and in a steady state or inadequately supplied and in a nonsteady state. The derivation of an expression that calculates the fraction of tissue volumes in these two states and corrects the PCr/P_i of the steady-state compartment for the contribution to the nonsteady-state compartment is given in FIGURE 3.

The model of tissue O_2 gradients that we have considered is illustrated in FIGURE 4. In this model, there is a sharp border zone between the steady (F) and unsteady (1 − F) state tissue volumes. As hypoxia/ischemia intensifies, the border zone moves to the higher PO_2 region, decreasing F and increasing 1 − F, which in turn decreases the total PCr and ATP and increases the P_i in accordance with Eq. 2. The fact that 1 − F increases shortly after the hypoxic/ischemic stress suggests that the prior bilateral

ENERGY STATE OF SURVIVORS

P MRS estimates the fraction of tissue volume that contains stress survivors. The Pi/PCr value of the survivors can be corrected for the non-steady state cells.

CALCULATION OF Pi

In tissue heterogeneity, Pi will be contributed from three sources.

1) In the non-steady state compartment from the breakdown of ATP (3 phosphates per ATP when the product is adenine, etc.).

2) From PCr breakdown:

$$\therefore (1-F)\left[3ATP_0 + PCr_0\right] \tag{1}$$

3) The third source is from the steady state compartment, $(Pi)_S$. This is unique to this compartment.

The total is:

$$(Pi)_{obs} = (1-F)\left[3ATP_0 + PCr_0 + Pi_0\right] + F(Pi)_S \tag{2}$$

CALCULATION OF PCr

In the case of PCr, only that contribution from the steady state compartment $(PCr)_S$ is observed:

$$(PCr)_S = (PCr)_{obs} \tag{3}$$

Equations (2) and (3) are combined:

$$(Pi)_S = (Pi)_{obs} - \left[(1-F)(3ATP_0 + PCr_0)\right] \tag{4}$$

CALCULATION OF PCr/Pi:

Equation (3) divided by Equation (4):

$$\frac{(PCr)_S}{(Pi)_S} = \frac{PCr_{obs}}{Pi_{obs} - \left[1-F\right]\left[3ATP_0 + PCr_0\right]} \tag{5}$$

Taking reciprocals:

$$\left[\frac{Pi}{PCr}\right]_S = \left[\frac{Pi}{PCr}\right]_{obs} - \frac{\Delta ATP}{PCr_{obs}}\left[3 + \frac{PCr_0}{ATP_0}\right] \tag{6}$$

with $\Delta ATP = ATP_0 - ATP_{obs}$

FIGURE 3. Assay of heterogeneous tissue.

carotid artery occlusion has caused some tissue volumes to operate at the "brink of hypoxia" despite normal P_aO_2.

With the changes in oxygen availability, two distinct values for pH_i can be obtained from the changes in the P_i peak (a split in the peak denotes two separate shifts, see FIG. 1). It seems appropriate to assign these differing values of pH_i to steady- and nonsteady-state cells. One approach is to assign all the acidosis to the latter population. One would expect to find a split in the P_i peak, especially as the quantity $1 - F$ increases to 30%. Indeed, this is observed as indicated by the spectrum of FIGURE 1(d), where P_i peaks at 6.0 and 7.2 are observed. This result suggests that in the

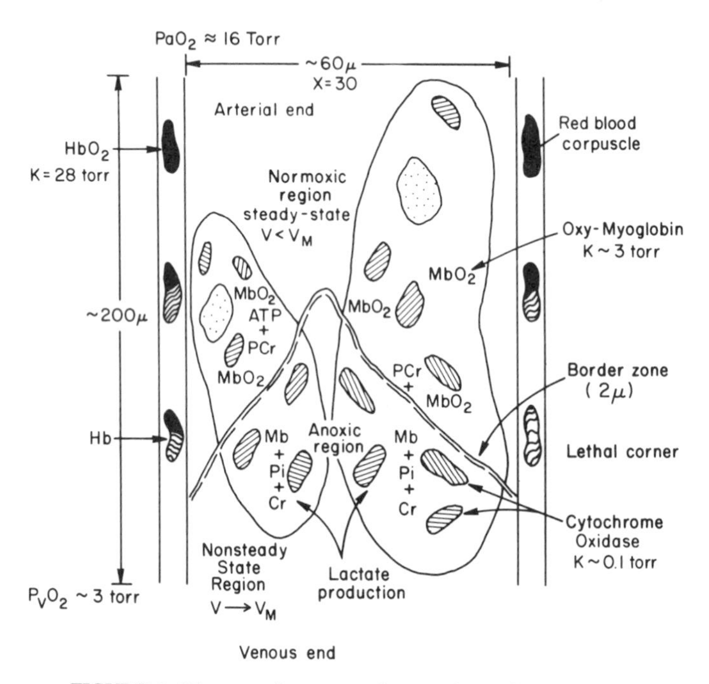

FIGURE 4. Diagram of oxygen and energetic gradients in tissue.

nonsteady-state population nearly all the acidosis and lactogenesis can be attributed to the nonsteady-state population.

EXPERIMENTAL STUDIES

In our animals with bilateral carotid ligation and hypotension, ATP was observed to decline significantly in five of seven animals during the ischemic interval. All of the animals showed a significant loss of ATP during the recovery period. Thus, according to our hypothesis, this model produces a high incidence of significant metabolic heterogeneity. The values of F for the individual animals are included in TABLE 1.

TABLE 1. Summary of Results on Ischemia

	Control						End of the Ischemic Period					End of 3 hr Recovery					
Dog	ATP mM	PCr mM	P_i mM	pH_i	ADP μM	Time of Ischemia	pH_i-1	pH_i-2	ATP	PCr	P_2	pH_i-1	pH_i-2	ATP	P_i	PCr
151	2.3	5.2	2.68	7.10	19	150 min	7.01	NA	2.10	4.56	3.17	7.22		1.42	1.42	2.80
160	2.3	4.48	3.4	7.10	23	200 min	7.04	NA	1.48	2.95	10.0	7.04		1.79	5.0	8.57
161	2.3	4.68	2.82	7.14	25	200 min	7.23	5.4	1.44	3.20	9.0	7.02	5.4	1.97	5.0	3.77
166	2.3	4.12	1.32	7.19	35	105 min	7.17	5.82	1.05	2.08	5.8	7.07	5.85	1.26	3.5	2.26
167	2.3	4.87	1.75	7.08	21	215 min	7.01	5.93	2.01	3.00	6.6	7.12	6.11	1.96	3.5	3.36
169	2.3	4.70	1.65	7.11	17	190 min	7.11	5.59	1.43	2.04	6.91	6.77	5.40	1.26	13.8	1.18
178	2.3	4.86	2.73	7.12	23	70 min	7.19	6.17	1.37	1.69	13.0	7.03	5.68	1.58	22.0	1.0

Time Course of Development of Heterogeneity

The development of heterogeneity is shown by the time course of the increase of P_i/PCr (FIG. 5E). P_i/PCr in both compartments ($P_i - 1$, $P_i - 2$) rises during the hypoxic/ischemic interval at a moderate rate. This is followed by a significantly increased rate of P_i/PCr increase during the recovery interval. At the same time, there is a growing acidity during the ischemic interval in the $P_i - 2$ compartment followed by no recovery during reflow (FIG. 5A). An increase in lactate is shown in FIGURE 5D and this correlates with the $P_i - 2$ compartment pH but *not* with the loss of ATP after reinfusion. In FIGURE 5B, the increase of $1 - F$ shows a linear change (10%/hr) beginning with the onset of hypovolemic stress and a further change following reinfusion, albiet, after a momentary steady state.

We interpret the P_i/PCr increase to be due to the loss of ATP and PCr in the nonsteady-state cells. We further implicate the nonsteady state as the main contributor to the acidity of compartment $P_i - 2$ indicated in FIGURE 5A.

To deduce events in the steady-state compartment, we applied the formula above to obtain the data shown in FIGURE 5E (steady state cells). This suggests that the corrected steady-state ratio (P_i/PCr) is approximately 1 during the ischemic interval. A momentary recovery occurs on reflow, but thereafter no steady state can be obtained. The total brain is then in an "unstable state" as evidenced by the rise of P_i/PCr in FIGURE 5E, the further loss of ATP in FIGURE 5B and the lack of a steady state in FIGURE 5E.

DISCUSSION

Animal Model Data

Two characteristics of this animal model are of significance in our understanding of the effects of hypoxia and ischemia in brain tissue. The first is implicit in FIGURE 5B where the hypovolemia causes a prompt (<10 min) transition of some brain cells from a stable steady state to an unstable state in which both PCr and ATP fall to very low values. Furthermore, the loss of ATP proceeds at a rate of 10%/hr of the total cells under observation by the 2.5-cm diameter MRS probe. Reflow affords a temporary reactivation of ATP synthesis in approximately 10% of the cells indicating that the ischemic insult is reversible in at least a small portion of the cell population. Thereafter, the characteristic effects of reflow into previously ischemic tissues—possibly the effects of free radical damage or progressively impaired cerebral blood flow—predominate and none of the brain cells under observation are in a steady state. The ultimate fate of the ATP and PCr depleted cells has been studied in cat brain ischemia (middle cerebral artery occlusion). After 4 hr of lesion maturation, the neuronal damage was correlated with the ATP deficit.[2]

The corrected value of PCr/P_i over the 2-hr interval of ischemic stress was very close to 1 ± 0.1 as shown in TABLE 1 and calculated by Eq. 6 of FIGURE 3. This is a value at which $V/V_m = 40\%$ and is the level of V/V_m at which the cells are stable.[7] Presumably 10% of this distribution reaches $V = V_m$ and irreversibility every hour. Thus $PCr/P_i = 1 \pm 0.1$ may be a "limit" beyond which the distribution of states has a significant probability of becoming irreversibly unstable under a particular hypoxic/ischemic stress.

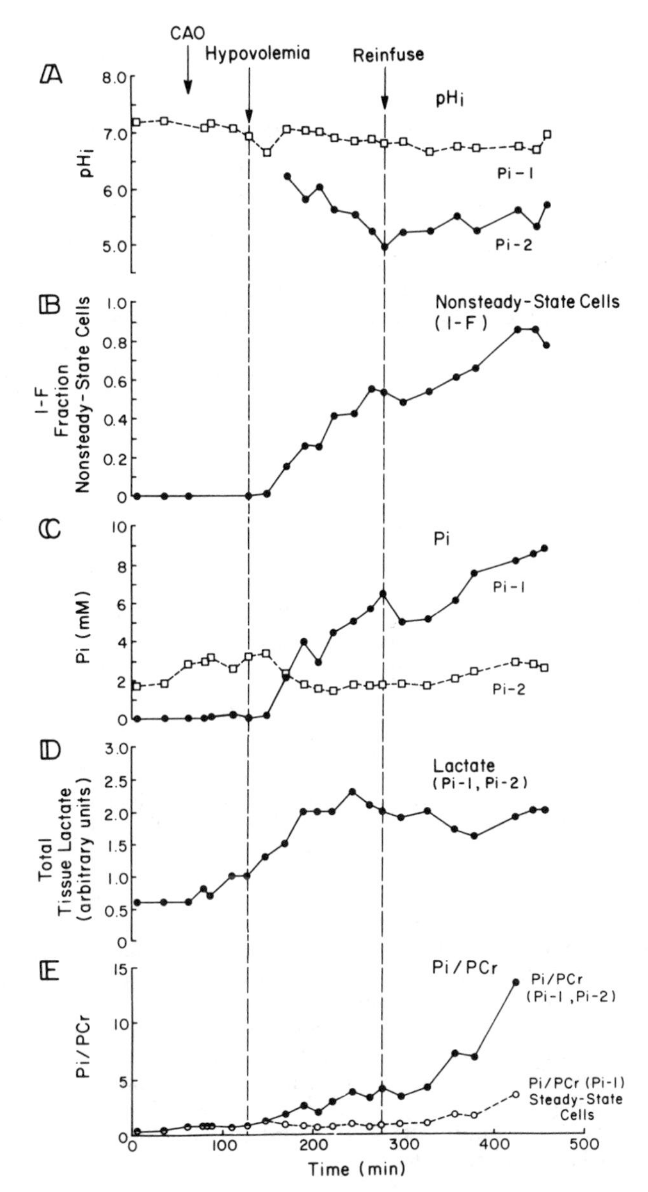

FIGURE 5. Time course of metabolite changes in development of heterogeneity in the adult dog brain. Trace A (pH) scored according to the two compartments, P_{i_1} and P_{i_2} as determined by the curve-fitting procedure. Trace B represents the nonsteady-state cells as calculated above. Trace C represents the content of phosphate in the two pools. Trace D represents the total tissue lactate in both pools, although pH data suggest it refers largely to P_{i_2}. Trace E is the observed P_i/PCr and the corrected value corresponding to P_{i_1}.

Clinical Correlation

The theoretical and experimental study of metabolic heterogeneity in the adult animal brain has significant repercussions on studies of adult humans and neonates. While the latter studies are far from complete in the nearly 500 examinations of pre-neonate brains, we have found four with significant and well documented ATP deficit. An example of hemispheric heterogeneity in a neonate is shown in FIGURE 6 where an ATP deficit is illustrated. This infant experienced severe fetal distress at birth. Focal seizures were noted in the left hemisphere and a 45% ATP deficit was determined using MRS. An analysis by Eq. 6 of FIGURE 3 indicates that PCr/P_i in the steady-state survivor cells is about 1.0 in the left hemisphere. At the present time, a more thorough search is being made for small ATP deficits as a consequence of birth trauma in neonates and it is believed that the frequency of observations of ATP deficits will significantly increase as more precise measurements are made. We may contrast the result obtained with neonates where quantifiable biochemical deficits of brain energy metabolism are found in the early states of the development of brain damage,

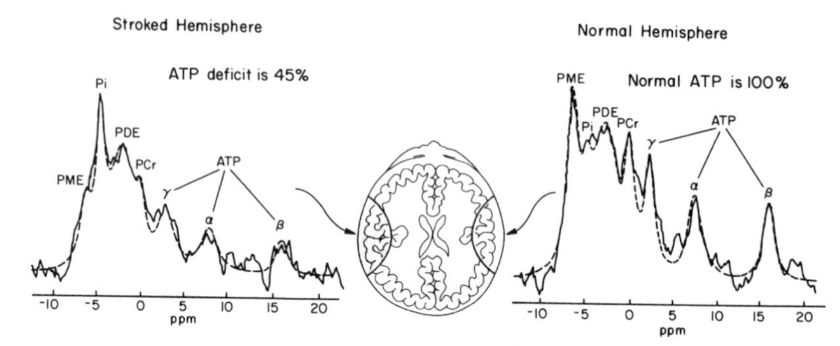

FIGURE 6. Effect of severe respiratory distress on 1-day-old, full-term neonate. MRS hemispheric scans show focal seizure status in the left hemisphere. A later CAT scan confirmed the stroked area. MR shows a 45% ATP deficit in the stroked hemisphere.

and in matured deficits where only CSF is found. This contrasts sharply with P MRS studies of adult brains where little or no biochemical change is found in stroke, and the possibility of silent neurons or replacement by glia as survivor cells may obfuscate the use of P MRS in adult brains to measure nonfunctional tissue volumes.

The correlation between ATP deficit and functional deficits is, at present, a nebulous one since so many factors are interposed between the occurrence of the biochemical lesion and the maturation of the infant to the point where psychological tests can be meaningful. However, we believe that adequate base line data of the type afforded by these biochemical tests have proved useful in the correlation of brain biochemistry and brain development.

SUMMARY

We have proposed that tissue metabolic failure during hypoxia or ischemia is related to the microheterogeneous distribution of tissue oxygen and not to failure of the creatine kinase equilibrium. This theory is based on the concept that sharp oxygen gradients exist in rapidly metabolizing tissue and that shifts in these gradients can place specific cells at risk for metabolic death while relatively adjacent cells escape unharmed; cells that are unharmed meet the steady-state requirements ($V < V_{max}$), those at risk do not ($V > V_{max}$). Though it would seem that confirmation of such a hypothesis would require metabolic delineation at a high resolution, we have shown how ^{31}P MRS provides information supporting this hypothesis. This possible use of MR spectroscopy to define microheterogeneous events suggests further clinical possibilities for this instrument in defining the rate of cell loss and the response to therapeutic interventions.

REFERENCES

1. MIYAKE, H. 1984. Experimental studies on the effects of recirculation on focal cerebral ischemia, following to occlusion of the middle cerebral artery: With respect to regional cerebral blood flow and brain energy metabolism. Arch. Jpn. Chir. **53:** 353–370.
2. KOMATSOMOTO, S., J. H. GREENBERG, S. NIOKA, W. F. HICKEY, K. YOSHIZAKI, V. H. SUBRAMANIAN, B. CHANCE & M. REIVICH. 1986. Is ^{31}P-NMR a predictor of neuropathological damage in focal cerebral ischemia? Soc. Mag. Res. Med. Mtg. **4:** 1115–1116.
3. HILBERMAN, M., V. HARIHARA SUBRAMANIAN, J. HASELGROVE, J. B. CONE, J. W. EGAN, L. GYULAI & B. CHANCE 1985. In vivo time-resolved brain phosphorus nuclear magnetic resonance. J. Cereb. Blood Flow Met. **4:** 334–342.
4. CHANCE B., J. S. LEIGH, JR. & S. NIOKA. 1986. P MRS as a sensor of oxygen in the heart or brain tissue. Soc. Mag. Res. Med. Fifth Annu. Mtg. **4:** 1368.
5. CHANCE, B., S. NIOKA, D. SMITH & J. S. LEIGH, JR. 1986. Biochemical heterogeneity in brain ischemia. Soc. Mag. Res. Med. Fifth Annu. Mtg. **4:** 1372.
6. TAMURA, M., N. OSHINO, B. CHANCE & I. A. SILVER. 1978. Optical measurements of intracellular oxygen concentration of rat heart in vitro. Arch. Biochem. Biophys. **191:** 8–22.
7. CHANCE, B., J. S. LEIGH, JR., J. KENT, K. MCCULLY, S. NIOKA, B. J. CLARK, J. M. MARIS & T. GRAHAM. 1986. Multiple controls of oxidative metabolism in living tissues as studied by phosphorus magnetic resonance. Proc. Natl. Acad. Sci. USA **83:** 9458–9462.

DISCUSSION OF THE PAPER

P. A. BOTTOMLEY: With regard to the results of the infants where we observe that a PCr/P_i ratio of less than 1 is associated with mortality, what fraction of the brain must be involved for this prognosis to be observed? Is it generally true that if you take a surface coil and place it on any part of the skull of a neonate that you observe PCr/P_i of less than 1 and that the baby will have poor prognosis, and for what fraction of the brain is this true?

CHANCE: Studies at Philadelphia and London indicate that PCr/P_i's of less than 1 are warnings of danger, and that values of 0.8 to 0.4 in those pre-term neonates (25 to 30 weeks) having PCr/P_i's between 0.4 and 0.8 are in a population of high mortality. We have attempted to score the fraction of the brain affected by the deficit of ATP either with respect to normal age-matched controls or with respect to an apparently unaffected contralateral hemisphere. When the fractional deficit approaches 50%, the pre-term neonate brain shows many other signs of distress detectable by ultrasound or eventually by CT and MRI. At the same time, I've already shown that localized infarcts can occur in a hemisphere of the brain and can be readily detected by P MRS.

As for prognosis, this is a difficult question to evaluate since the preterm neonate has a highly undifferentiated brain with many of the neuronal connections yet to be made. The brain fluidity is indicated by the high resolution of the ATP spectrum. For this reason, plasticity of neuronal function may occur and an initial insult may be overcome by activation of alternate pathways or the commandeering of existing pathways for a more essential function. It is most important to bridge the gap between brain biochemistry of the pre-term neonate and neurologic dysfunction of the growing child.

QESTION: Is there indeed a generalized decrease of PCr/P_i or are there local regions, which you may have detected by luckily placing the surface coil on the right spot?

CHANCE: Each of the patients is examined by a multiplicity of technologies and EEG is useful in identifying an epilepsy or an epileptic focus. We are likely to be told by the attending physician where to put the coil. Alternatively, we use a large coil and look for ATP deficits as mentioned above.

K. LA NOUE: If I understood what you were implying about the dependence of the rates on oxygen, I think that that means there's tremendous tissue heterogeneity because if you had a bunch of mitochondria well mixed in a sack and you were flowing oxygen past them and you changed the rate of that, then you wouldn't get those results at all. At the low work closer to state 4, those mitochondria won't be able to tell that there was less capacity for oxygen there. What your data evidently mean is that even at low work some of the mitochondria, whether it's part of the mitochondria in the cell or some of the cells, are just completely knocked out. Isn't that right? That's the only way one could anticipate at low work that there was going to be a problem in V_{max}, right?

CHANCE: It's a very good point that you make and surely tissue oxygen gradients are very important. Indeed oxygen may be delivered to the brain tissue parsimoniously, i.e., at a level no more than is necessary than to supply the low work rate, state 4. Although we do not have as extensive data for brain as for skeletal tissues, microvascular control is a key element in the response of tissue to hypoxia and is obviously one of the reasons that the brain function can persist at very low tissue PO_2's. Surely the point you make is that the lower the metabolic rate, i.e., the closer the approach to state 4, the more immune is the system to hypoxia.

Measurement of Phosphocreatine to ATP Ratio in Normal and Diseased Human Heart by ^{31}P Magnetic Resonance Spectroscopy Using the Rotating Frame-Depth Selection Technique[a]

B. RAJAGOPALAN,[b,c] M. J. BLACKLEDGE,[b]
W. J. McKENNA[d] N. BOLAS,[b]
AND G. K. RADDA[b]

[b]The MRC Clinical Magnetic Resonance Facility and
[c]Nuffield Department of Clinical Medicine
John Radcliffe Hospital
Headington, Oxford,
OX3 9DU, England

[d]Cardiovascular Disease
Royal Postgraduate Medical School
Hammersmith, London

INTRODUCTION

Magnetic resonance spectroscopy (MRS) using ^{31}Phosphorus (^{31}P) has been shown to provide useful insights into myocardial energetics in experiments on isolated perfused hearts and in surgically exposed hearts *in vivo* in animals.[1,2]

In particular, the relative concentrations of phosphocreatine (PCr), adenosine triphosphate (ATP), and inorganic phosphate (P_i) together with intracellular pH can yield information about the control of ATP synthesis and utilization. The development of large bore high-field magnets and differential localization techniques have made such measurements possible in deep lying organs in human subjects.[3–7]

Investigation of human heart using ^{31}P NMR requires the achievement of adequate localization as well as the establishment of objective criteria to ensure that the data obtained are representative of cardiac muscle. Any contamination by signal from overlying skeletal muscle will distort the measured PCr/ATP ratio. This is crucial as animal studies show that this ratio in skeletal muscle is very different from that in cardiac muscle.[8,9] We have used a rotating-frame, depth-selection technique[10–12] with a double-concentric surface coil probe[13] to localize signals from flat planar, disc-shaped slices at a series of distances into the chest wall. We then used the relative concentrations of PCr and ATP to confirm the position of the selected slice in the chest. In contrast to previous studies in which a switched B_0 gradient localization method[14] and static field profiling method[4] were used, we use the inherent biochemistry of the

[a]Supported by the Medical Research Council of Great Britain, the British Heart Foundation, the Department of Health and Social Security and, the National Institutes of Health (grant HL/8708).

sample (specifically the PCr/ATP ratio) to provide evidence of localization. This technique has been shown to be reproducible and practical in the study of the human heart and it is now possible to investigate the cardiac metabolism in patients with heart muscle disease.

METHODS

The heart lies directly below two skeletal muscles, the intercostal and transversus thoracis (FIG. 1). The subject lies prone, so that the double-surface coil probe covers the apex and anterior wall of the heart. The position of the apex and interventricular septum of the heart is initially estimated from a two-dimensional echocardiogram. The region of interest is therefore centered on the anterior wall of the left ventricle and includes the left ventricular apex and parts of the right ventricle. Using the rotating-frame depth-selection method, signal is acquired from discs at increasing depths into the chest. The discs interrogated are smaller laterally than the heart, so that there is no contamination from lung or skeletal muscle at the same depth. As deeper slices are interrogated, they will consist of more heart and less skeletal muscle and PCr/ATP will be expected to fall until a plateau is reached. At this point, contamination of the spectra by skeletal muscle will be minimal and the data should represent predominantly heart. As the slices are acquired from progressively deeper regions and pass through the heart wall, the PCr-to-ATP ratios stay constant but the signal-to-noise ratio falls because of the nature of the profile of the receiver coil sensitivity. To determine the plateau region, we first optimized the magnetic field homogeneity using the proton signal. To determine the pulse width at which PCr/ATP reaches a plateau, we then acquired 40 scan spectra (3-minute acquisition time) at increasing depths.

The rotating-frame depth-selection method[12,13]: The experiment relies on a gradient B_1 excitation field in which the nuclear spins experience a tip angle that depends on position of the nucleus and is defined by the B_1 field strength. Magnetization at different field strengths precesses at different frequencies with respect to pulse widths and this difference in frequency distribution can be used to localize signals from a point lying in a particular B_1 field strength by applying a series of pulses that maximally excite this region. The pulse lengths are of the form

$$\theta = (2n + 1)(\pi/2)(-1)^n \; (n = 0,1,2\ldots).$$

Addition of the data files resulting from each pulse width achieves positive coherence signal over a slice of B_1 field gradient with destructive interference of signals from elsewhere in the sample.

We defined the selected region to be Gaussian in shape in the B_1 field direction by applying the following pulses; 20 pulses at $\theta_{\pi/2}$, 14 at 3 $\theta_{\pi/2}$, 5 at 5 $\theta_{\pi/2}$, and 1 at 7 $\theta_{\pi/2}$. We ensured that the B_1 gradient was linear in the region by using a double-concentric surface coil probe. We have shown in phantom experiments that using this system we obtain signals from 6-cm discs of 2-cm width at a depth defined by the length of the pulse. A repetition rate of 3.5 seconds was chosen to remove the possibility of distortion of the sensitive profile due to T1 effects.[12] When the region of the myocardium had been identified by a flat plateau in the PCr/ATP, more scans were required at this pulse length (typically 160) so that spectra with adequate signal-to-noise ratios were available for quantitation by triangulation of peak areas. The peaks are expressed as PCr-to-ATP ratios and the chemical shifts are given with phosphocreatine as the reference peak.

Six normal volunteers were studied using this protocol and in two at least three

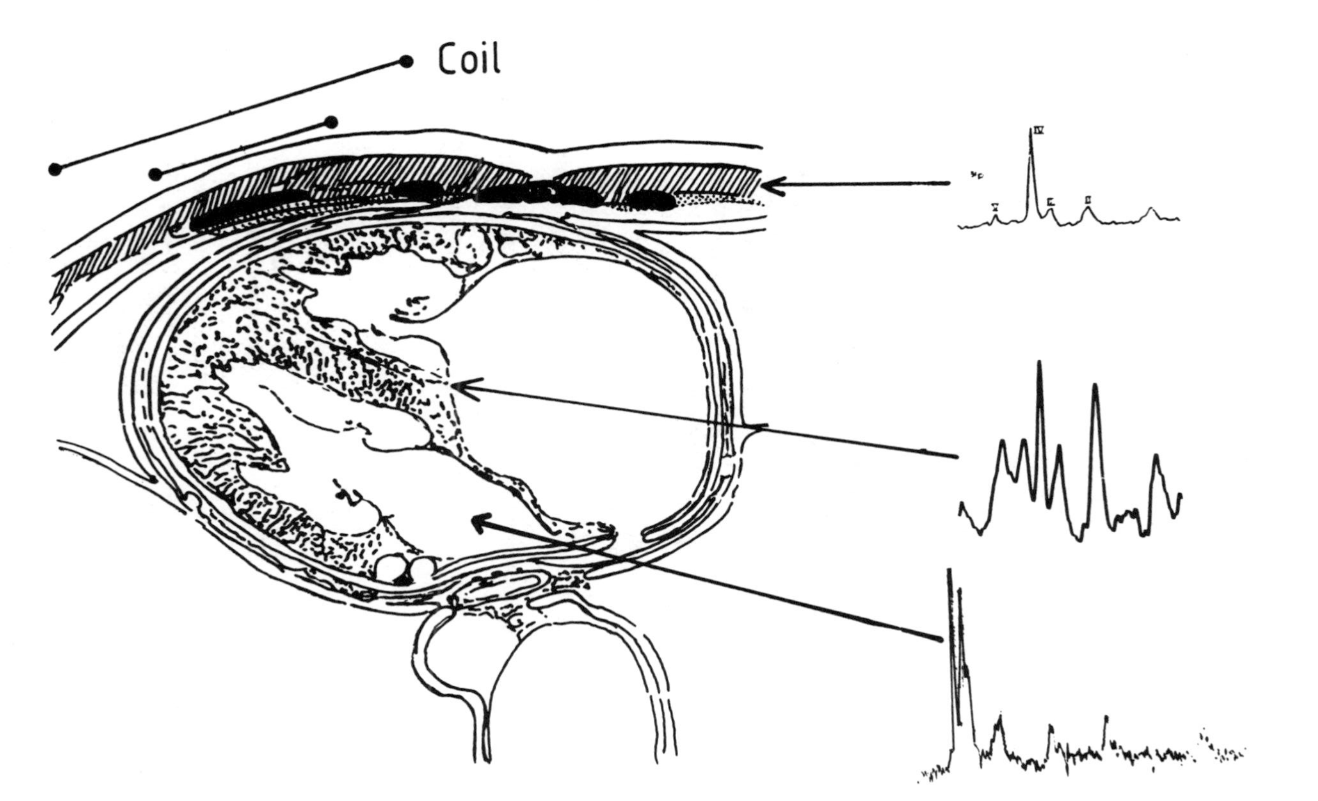

FIGURE 1. Cross section of thorax showing the anatomical proximity of skeletal muscle to the anterior surface of the heart. The figure also shows position of double surface coil. The ^{31}P spectra of skeletal muscle, heart, and blood are from a control subject. The heart muscle spectrum was similar to spectra obtained from surface coils placed on normal rabbit hearts.

studies were done on widely separated days to determine intra- and intersubject variability. In a preliminary study, two patients with hypertrophic cardiomyopathy were also studied. One had disease that was confined to the interventricular septum while the second had global disease that involved both the right ventricular free wall, septum, and left ventricle. Informed consent had been obtained for both the control and clinical subjects.

RESULTS

An example of a series of three-minute depth selection accumulations representing slices from a normal human chest is shown in FIGURE 2. The PCr/ATP starts off at the value of 3–4. As slices are obtained from greater depth, this ratio falls and eventually reaches a plateau value. In the deeper slices two larger peaks appear to the left of the PCr. These have chemical shifts of 5.4 and 6.2 ppm, which at a blood pH of 7.4 would be consistent with them being from 2,3-diphosphoglycerate in red blood cells. There is also a large peak between PCr and P_i in the deeper slices, which probably corresponds to phosphodiesters. Since the 2,3-DPG signal overlaps the region where one would expect to find P_i, measurement of intracellular pH from the spectra is impossible.

FIGURE 3 shows a plot of the PCr-to-ATP ratio against distance. The x axis represents a distance from the coil. The pulse angle dependence of the method allows one to calibrate the distance from phantom studies. As predicted there is a flat plateau in this ratio as spectra are obtained from deeper slices. A typical normal human heart spectrum obtained using a pulse length in the middle of a plateau region is shown in FIGURE 4. In six normal subjects, the PCr:ATP ratio of heart muscle was measured to be 1.55 ± 0.2 (mean ± SD) and in repeat studies in one subject done three times the measured PCr-to-ATP ratio ranged from 1.4 to 1.8.

FIGURE 5 shows a heart spectrum from the patient with septal hypertrophy only. PCr:ATP was similar to that found in controls. Similar ratios were obtained in three separate studies.

FIGURE 6 shows a heart spectrum from the patient with biventricular hypertrophy. PCr:ATP on three different occasions was 0.9 ± 0.2 (mean ± S.D.). In addition, abnormal peaks both in intensity and chemical shift were seen in the diester and monoester regions.

DISCUSSION

Magnetic resonance spectroscopy of ^{31}P has been successfully used to detect changes in high energy phosphate metabolism in isolated perfused rodent hearts[1] and in surgically exposed hearts of animals *in situ*.[9] Changes in the ratio of PCr:ATP or to inorganic phosphate have been seen following ischemia (FIG. 7) and drug intervention. The measurements in surgically exposed hearts were made by placing a surface coil on the heart itself. The use of surface coils to follow phosphorus metabolism in the intact animal or man became a possibility with the development of techniques that could obtain data from deep tissues, such as heart or liver.[3–6] Using these techniques, it has been relatively easy to identify ^{31}P spectrum from liver. This is because liver cells, unlike the overlying skeletal muscle, contain no PCr.[4] Thus the absence of PCr from the spectrum confirms that the spectrum is derived from liver alone. Heart muscle, however, contains the same phosphorus compounds as skeletal muscle but in very different proportions. The PCr:ATP ratio in skeletal muscle in animals and man is

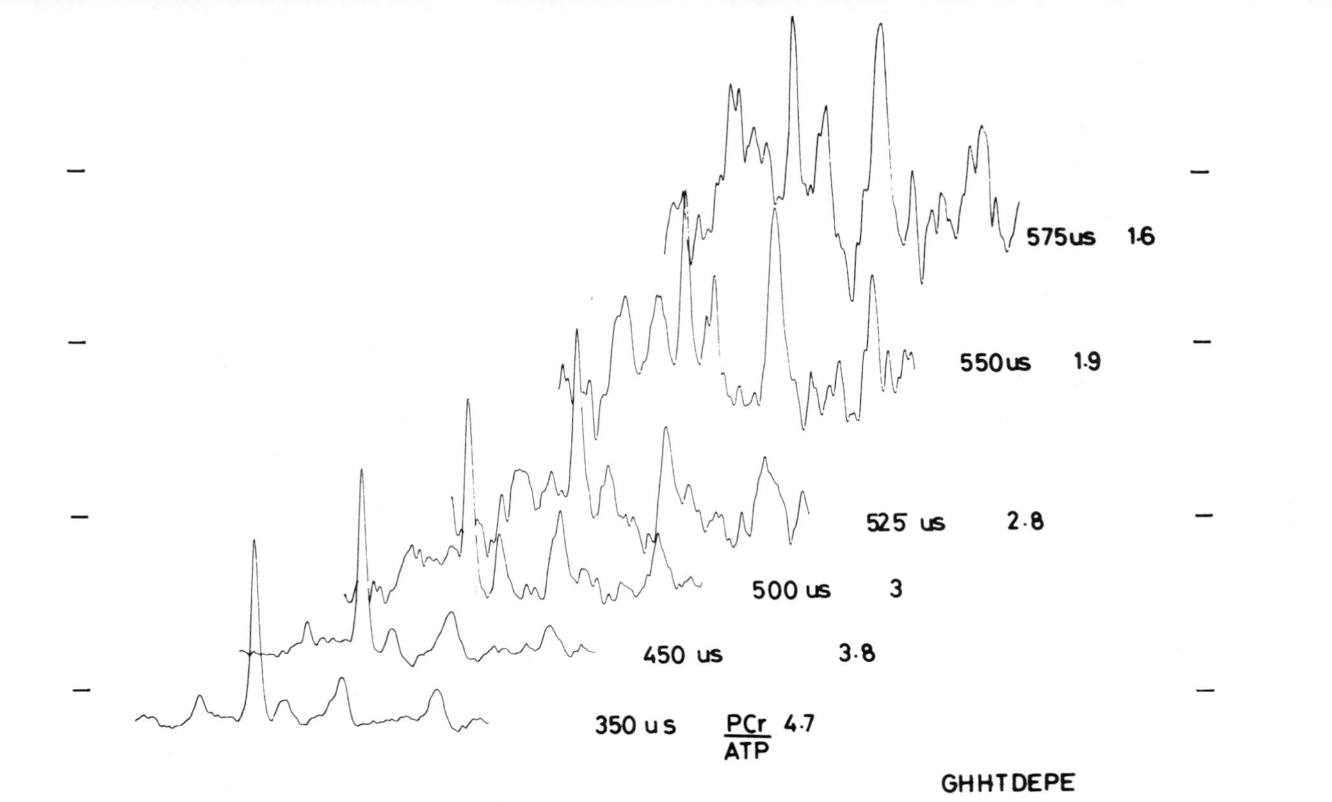

FIGURE 2. ^{31}**P** spectra obtained from slices at increasing depths from the chest wall (3-min accumulations). Note that in deeper slices, PCr:ATP falls, a diester peak is seen, and the peaks of 2,3-DPG appear.

about 4 while in animal heart it is about 1.8–2. It is, therefore, essential to obtain spectra from cardiac muscle with minimal contamination from skeletal muscle. This is made particularly difficult by the close proximity of the anterior surface of the heart to skeletal muscle, which covers the posterior surface of the sternum and lower ribs.

In the case of the first human spectra to be obtained by magnetic resonance spectroscopy, the position of the heart was identified by percussion and spectra were accumulated from a spherical region of interest using the technique of topical magnetic resonance.[3,15] Though this technique enabled one to obtain spectra from depth, while excluding contribution from some superficial tissues, the region of interest cannot be moved in the body to determine metabolite ratios at different depths nor are the edges of the sensitive volume under study clearly defined. Recently, Bottomley has described an imaging technique using field gradients to obtain spectra from human heart.[14] In

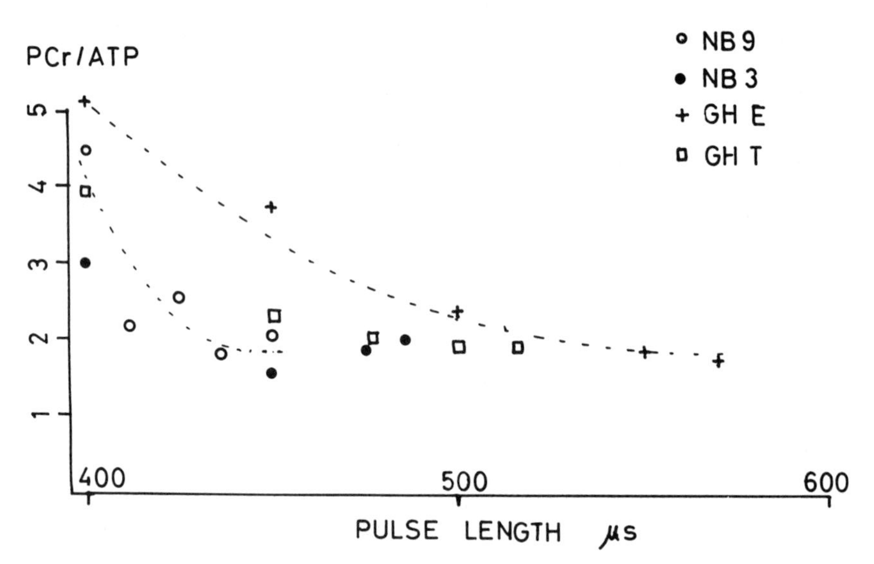

FIGURE 3. Plot of PCr:ATP against increasing depth from four different studies. The ratio reaches a plateau at longer pulse lengths (greater depths) indicating minimal contamination of spectra from skeletal muscle.

this study, only qualitative criteria were used to determine if the spectra were derived from heart muscle. The depth of the heart was estimated from MR images and spectra were qualitatively identified as containing heart muscle by the presence of peaks of 2,3-DPG from red cells.

The depth of the heart from the skin can be estimated from both echocardiography and MR imaging and it is in principle possible to set the parameters of the MR spectroscopy experiment to this depth. Unfortunately, this assumes that the patient can be repositioned for the MRS experiment in exactly the same place as during echocardiographic imaging. In the case of MR imaging, they also assume that the parameters for the phosphorus experiment have not changed from the time that the proton experiment was performed.

The alternative is to use the spectroscopy experiments to determine when the spectra are derived from heart muscle. We argued that once we had located the heart

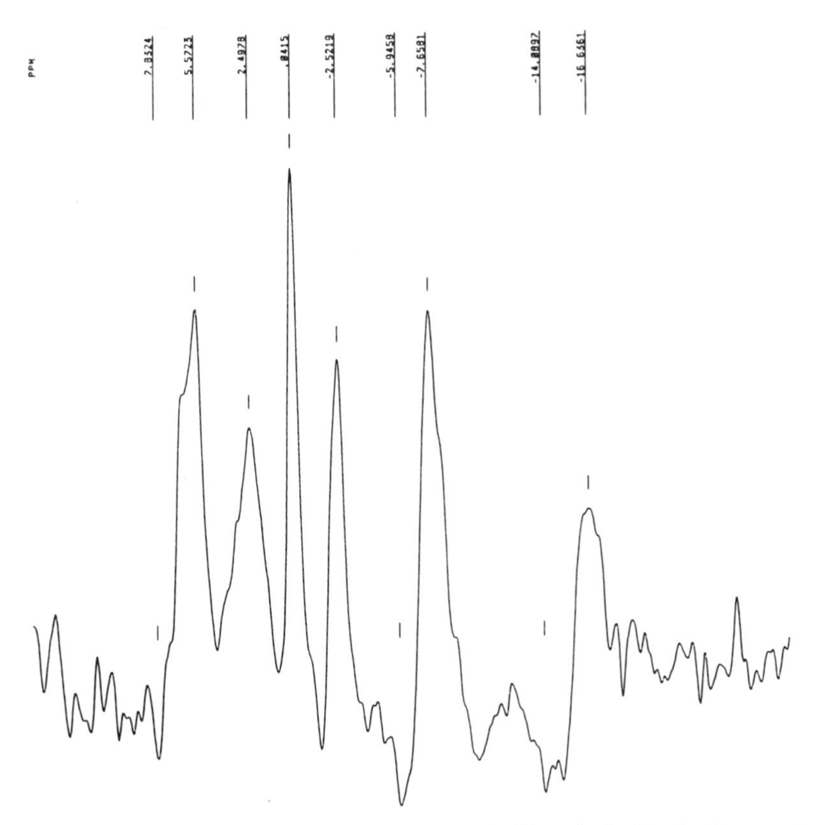

FIGURE 4. Heart spectrum (160 scans) from one control subject obtained in the plateau region as determined by the results shown in FIG. 3.

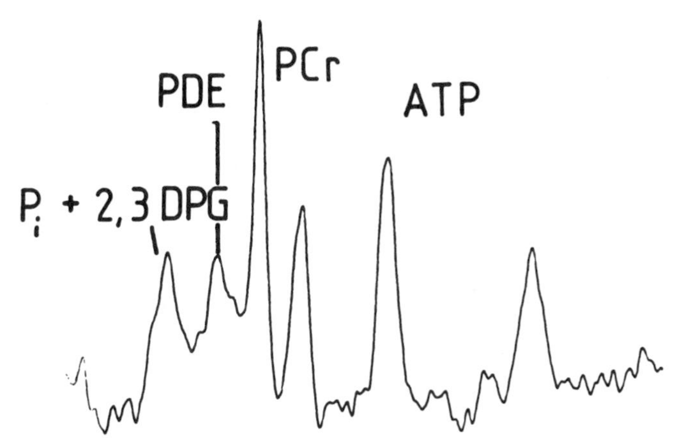

FIGURE 5. Heart spectrum from a patient with hypertrophic cardiomyopathy only affecting the septum. The spectrum is normal.

by two-dimensional echocardiography, by collecting spectra of progressively deeper slices, we should initially obtain signal from predominantly skeletal muscle, then from a mixture of skeletal and cardiac muscle, and finally from cardiac muscle alone. As the slices are moderately thick compared to the thickness of the myocardium, inhomogeneities in cardiac muscle itself would be averaged out. The results support this hypothesis.

Spectra obtained from superficial slices are qualitatively like those from skeletal muscle and the ratios were close to those obtained previously in skeletal muscle. With increasing depth the ratios fell and tended to plateau. These spectra contained additional peaks downfield of PCr not usually seen in skeletal muscle. The chemical shifts of these peaks suggest that they are likely to be 2,3-DPG from red cells. Although this is a qualitative indication that the spectra are from heart, on its own it clearly does not imply that spectrum is free from contamination from skeletal muscle. The ratios we obtained could have been modified by the presence of blood. Fortunately,

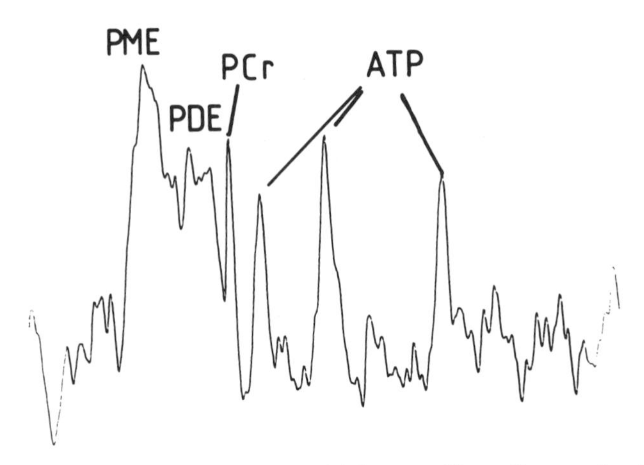

FIGURE 6. Heart spectrum from a patient with hypertrophic cardiomyopathy affecting the whole heart. The PCr-ATP ratio is lower than normal. The region to the left of PCr shows abnormal peaks. The heights of the monoester and diester peaks also appear qualitatively abnormal.

the ratio 2,3-DPG to ATP in blood is about 20:1 so from the height of the 2,3-DPG signal in the spectrum, we can estimate that the contribution from other metabolites from blood to the heart spectra is minimal. Blood contains no PCr.

The peaks of 2,3-DPG lie in the same area as one would expect to find inorganic phosphate. It is therefore currently impossible to measure the relative concentrations of muscle P_i or intracellular pH by ^{31}P MRS.

The spectra obtained by this method and those obtained by the DRESS technique by Bottomley[14] are qualitatively similar. Further experience is necessary before the relative advantages and disadvantages of the two techniques can be determined.

The PCr/ATP is also affected by the relaxation times, T_1 of the ^{31}P nuclei. T_1 determines the time required for the nuclei to return to the B_0 direction after a pulse. This process is exponential so that ideally one should wait at least 4 T_1 in between pulses, otherwise, the intensities of the peaks are reduced as a function of T_1. As T_1 of PCr in animal hearts is about 5 sec, each 40-scan slice would take an unacceptably long

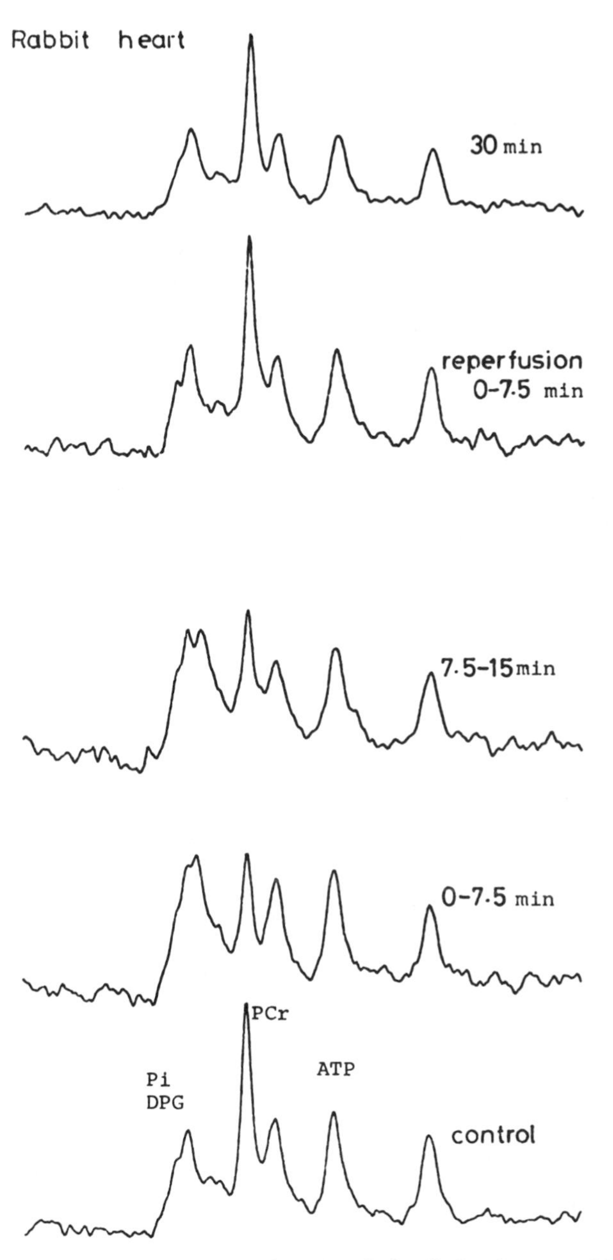

FIGURE 7. Spectra obtained sequentially from a surface coil placed on superficially exposed rabbit heart. Regional ischemia for 10 min was produced by a snare on a branch of the left coronary artery. The region was then reperfused by opening the snare. PCr-ATP ratio falls rapidly during ischemia and recovers with reperfusion.

time. It has been the practice in MRS to pulse faster but obtain correction factors from a limited number of experiments with longer repetition times. In two subjects, PCr/ATP was 2.5 at 5-sec repetition rate and 2.7 at 8-sec repetition rate.

Using the criterion of a plateau in the PCr/ATP ratio, we obtained ratios of 1.55 ± 0.2 in six normal controls (3.5-sec repetition rate; with no correction for relaxation effects). Two of these subjects were studied on three different occasions and the ratios ranged from 1.4–1.7 and 1.5–1.7.

Currently, the limitations of the method are that the subject has to lie flat and prone for about 45 minutes. Though this excludes certain groups of patients, we should still be able to study any patient who can tolerate lying on the catheter table for a similar period of time. We are still limited by the physique of the patient that we can study by this techniques. They have to be slim and the heart has to be within 3–5 cm of the chest wall. This problem is due to the limitation to the power output of the amplifier. A new amplifier, currently being tested, will overcome this problem. This effect is seen as a loss of intensity of the β ATP peak in the heart spectra. As the resonance was placed near the PCr peak, the β ATP being furthest away showed maximum distortion. From phantom experiments, distortion was shown to be minimal of the γ ATP and for this reason we used the PCr to γ ATP ratio for our measurements.

We have made no efforts to allow for respiratory or cardiac movement. By placing the subject in a supine position, we were able to remove the effects of air-filled lungs coming between the heart and the detector coil. We are currently developing methods to trigger the acquisition signal both to the heart beat and respiratory cycle to examine this problem further.

In the preliminary clinical studies, we chose one patient in whom we might have predicted global abnormalities and in one in whom we might have expected to find normal results. As we discussed earlier, the technique only allows us to obtain the signal for a depth of up to 5 cm from the surface coil. The first subject with hypertrophic cardiomyopathy had disease localized to the septum and this was confirmed by two-dimensional echocardiography. The septum from measurements on the two-dimensional echocardiogram was at least 6 cm away from the chest wall so that our chances of being able to see it in the MRS experiment were small. We were therefore examining the right ventricular free wall, the apex, and part of the anterior wall of the left ventricle, all of which were echocardiographically normal. The MRS spectra were also normal on all the occasions that the subject was studied and the PCr:ATP ratios were within normal limits.

The second patient had severe disease of the right ventricle, the septum, and left ventricle. His right ventricular free wall was almost 1.2 cm thick as opposed to the normal of 0.2–0.4. He was in a state of continuous compensated chronic congestive cardiac failure with symptoms of fatigue and raised venous pressure. He required therapy in the form of diuretics. It was in this patient that we found a PCr:ATP ratio lower than any of the values we obtained in our control subjects. The ratios ranged from 0.9–1.2. To ensure that this was not due to a technical problem, we studied the patient on two separate occasions on one day and on a third occasion a week later. On all three occasions we obtained low ratios of phosphocreatine to ATP. This reduced ratio of PCr to ATP implies an imbalance between metabolic demand and energy supply. It has been previously suggested that there may be problems of this kind in severe hypertrophy and hypertrophic cardiomyopathy and that this may be one etiological factor in the development of chronic heart failure as well as the arrhythmias these patients suffer from. Though our observations are preliminary, they are the first to be made in support of this suggestion.

A second abnormal feature in this patient was the presence of the large signal in the diester and monoester region. These peaks from studies in other tissues are known to be contributions of phosphodiesters (such as glycerophosphoryl choline and glycerophosphoryl ethanolamine) and in the monoester region, phosphoethanolamine and phosphocholine. These are components of cell membranes. It is tempting to speculate that the changes we see in the hypertrophic cardiomyopathy are a reflection of the hypertrophic process, i.e. either synthesis or degradation of membrane components. Further studies are clearly required to determine the significance of these observations.

CONCLUSION

We have shown that we can obtain spectra with good signal-to-noise ratio from human heart using rotating-frame depth-selection with a double surface coil probe. The experiments can be carried out in acceptable experimental time for patient investigation. The method relies on internal calibration with the technique, which relies on the available biochemistry of interrogated muscle and makes no assumptions as to what the ratios of metabolites are in cardiac muscle. We have used this to confirm that we are studying predominantly myocardium. We have obtained a normal range for phosphocreatine-to-ATP ratios (without correction for saturation factors) in normal human heart muscle and have compared the preliminary studies with two patients with hypertrophic cardiomyopathy. Abnormalities in PCr:ATP and the diester and monoester region were detected in one patient on more than one occasion. Further studies are warranted to characterize ^{31}P spectroscopic changes in hypertrophy and hypertrophic cardiomyopathy.

SUMMARY

^{31}P magnetic resonance spectroscopy, using surface coils placed on perfused or surgically exposed animal hearts, shows that unequivocal changes in phosphocreatine (PCr) and adenosine triphosphate (ATP) occur during interventions, such as ischemia. Similar measurements seem warranted in man. We have used a modification of the rotating-frame imaging technique to measure PCr-to-ATP ratio non-invasively in human heart. The subject lay prone on a double-surface coil probe with the apex and the anterior surface of the heart covered by the coil in a 1.9 T magnet. ^{31}P spectra were obtained from slices of tissue approximately 6 cm in diameter and 2 cm in thickness. Though skeletal and cardiac muscle contain similar phosphorus metabolites, animal studies show that the ratio in the two are different. We argued that the ratio should start high (skeletal muscle) and plateau at a low value representing cardiac muscle. Using this criterion, which makes no assumption on what the ratio is in heart muscle, the PCr:ATP in six normal subjects was 1.55 ± 0.2. This protocol has been used in a preliminary study in patients with cardiomyopathies.

REFERENCES

1. Flaherty, T. J., M. L. Weisfeldt, B. H. Buckley, T. J. Gardner, V. L. Gott & W. E. Jacobus. 1982. Circulation **65:** 561–571.
2. Radda, G. K. 1983. Br Heart J. **50:** 197–201.
3. Gordon, R. E., P. E. Hanley, D. Shaw, D. G. Gardian & G. K. Radda. 1980. Nature **287:** 736–738.

4. OBERHAENSLI, R. D., G. GALLOWAY, D. HILTON-JONES, P. STYLES, P. BORE, B. RAJAGOPALAN, D. TAYLOR & G. K. RADDA. 1987. Br. J. Rad. **60:** 367–373.
5. ORDIDGE, R. J., A. CONNELLY & J. A. B. LOHMAN. 1986. J. Magn. Reson. **66:**283–294.
6. STYLES, P., C. SCOTT & G. K. RADDA. 1986. J. Magn. Reson. **66:** 402–409.
7. OBERHAENSLI, R. D., D. HILTON-JONES, P. J. BORE, L. J. HANDS, R. P. RAMPLING & G. K. RADDA. 1986. Lancet **ii:** 8–10.
8. BAILEY, I. A. & S. SEYMOUR. 1982. Biochem. Soc. Trans. **9:** 234–236.
9. MALLOY, C. R., P. M. MATTEWS, M. B. SMITH & G. K. RADDA. 1986. Cardiovasc. Res. **20:** 710–720.
10. METZ, K. R. & R. W. BRIGGS. 1985. J. Mag. Reson. **64:** 172–176.
11. GARWOOD, M., T. SCHLEICH, B. D. ROSS, G. B. MATSON & W. D. WINTERS. 1985. J. Magn. Reson. **65:** 239–251.
12. BLACKLEDGE, M. J., P. STYLES & G. K. RADDA. 1987. J. Magn. Reson. **71:** 246–258.
13. STYLES, P., M. B. SMITH, R. W. BRIGGS & G. K. RADDA. 1985. J. Magn. Reson. **62:** 397–405.
14. P. A. BOTTOMLEY. 1985. Science **229:** 769–772.
15. RADDA, G. K. 1986. Control of Bioenergetics: From cells to man by phosphorus NMR. Biochem. Soc. Trans. **14**(3): 517–525.

DISCUSSION OF THE PAPER

B. CHANCE: I'm so glad you've gone into the problems of cardiac phosphorous magnetic resonance studies. We've taken the easy way out by using neonates with a thin sternum and a severe cardiac hypertrophy. I would agree that when the phosphocreatine falls to the ATP level that cardiac failure is imminent as we verified in a couple of cases. Maybe there's a limit to the fall of phosphocreatine. Tell us a little bit about this hypertrophic adult: was he stable?

B. RAJAGOPALAN: He's stable, but he's in a stable form of heart failure. He's limited in terms of exercise tolerance. He can't actually walk very far and he presented initially almost four years ago a very severe form of heart failure that was treated by medical means. He is now stable but can lie flat. But he still has lots of what you might call clinical evidence of heart failure.

CHANCE: Is there a detectable infarct volume?

RAJAGOPALAN: No, he has actually had a cardiac catheterization that shows completely normal coronary arteries. The nearest we can get to image him is a thallium scan, which was done, and that was also normal, and there was no evidence of infarction.

Spatial Localization in NMR Spectroscopy *in Vivo*

PAUL A. BOTTOMLEY

General Electric Corporate Research and Development Center
Schenectady, New York 12301

SUMMARY

Spatial localization techniques are necessary for *in vivo* NMR spectroscopy involving heterogeneous organisms. Localization by surface coil NMR detection alone is generally inadequate for deep-lying organs due to contaminating signals from intervening surface tissues. However, localization to preselected planar volumes can be accomplished using a single selective excitation pulse in the presence of a pulsed magnetic field gradient, yielding depth-resolved surface coil spectra (DRESS). Within selected planes, DRESS are spatially restricted by the surface coil sensitivity profiles to disk-shaped volumes whose radii increase with depth, notwithstanding variations in the NMR signal density distribution. Nevertheless, DRESS is a simple and versatile localization procedure that is readily adaptable to spectral relaxation time measurements by adding inversion or spin-echo refocusing pulses or to *in vivo* solvent-suppressed spectroscopy of proton (^{1}H) metabolites using a combination of chemical-selective RF pulses. Also, the spatial information gathering efficiency of the technique can be improved to provide simultaneous acquisition of spectra from multiple volumes by interleaving excitation of adjacent planes within the normal relaxation recovery period. The spatial selectivity can be improved by adding additional selective excitation spin-echo refocusing pulses to achieve full, three-dimensional point resolved spectroscopy (PRESS) in a single excitation sequence. Alternatively, for samples with short spin-spin relaxation times, DRESS can be combined with other localization schemes, such as image-selected *in vivo* spectroscopy (ISIS), to provide complete gradient controlled three-dimensional localization with a reduced number of sequence cycles.

LOCALIZATION STRATEGIES

The great promise of *in vivo* NMR spectroscopy lies in its ability to provide chemical information about physiologic function, its perturbation by disease, and its restoration to health via therapy. Although proton (^{1}H) NMR spectra from living cells were reported as early as 1955[1] and natural abundance phosphorus (^{31}P) spectra were acquired from blood cells[2] and excised rat muscle[3] nearly two decades later, human *in vivo* NMR spectroscopy was not practical until the advent of larger bore superconducting magnet systems capable of accommodating the human limbs,[4,5] the head,[6,7] and the body[8] only in the last few years. Since human experimentation demanded safe, noninvasive protocols involving intact anatomy, and the body was much larger than the homogeneous volumes that were suitable for spectroscopy within the magnet bore, small circular flat NMR coils positioned on the surface of the body were crucial in providing the first access to spatially localized ^{31}P NMR spectra.[4-7]

Unfortunately, surface coil localization has one major problem: its high sensitivity to surface tissue results in significant contamination of the NMR signals from important deeper lying organs of significant interest such as the brain, the liver, and the heart. The effect is illustrated in FIGURE 1, which shows ^{1}H surface coil NMR images acquired from a single-turn, 6.5 cm diameter surface coil located on the head above the temple.[9] A ^{1}H NMR spectroscopy experiment employing such a coil would collect a total integrated signal composed of approximately equal contributions from the brain and bone marrow, the scalp, and surface musculature (FIG. 1A). Although the proportion of signal derived from brain can be improved somewhat by increasing the NMR flip-angle beyond $\pi/2$ at the surface (FIG. 1B), substantial surface tissue contributions are clearly inevitable.

To address this problem and to improve the control and definition of spatially localized regions for spectroscopic analysis, a battery of spatial localization schemes have been proposed and demonstrated as summarized in TABLE 1.[10] These techniques employ either radio frequency (RF), static, or time-dependent magnetic field gradients to restrict data acquisition to single or multiple selected volumes. They can be combined with either surface detection coils or whole volume detection coils analogous to those used in conventional ^{1}H NMR imaging, depending on the desired depth of the selected volume relative to the surface. At shallow sample depths, surface coils provide dramatic advantages in signal-to-noise ratio over volume coils because they are closer to the signal-generating nuclei and remote from a large fraction of the sample that contributes only noise. The dependence of the signal-to-noise ratio of a 6.5 cm diameter ^{31}P NMR surface coil at 25.7 MHz as a function of depth in the head along the coil axis is shown in FIGURE 2: for depths less than about 5.5 cm the surface coil performance is superior to that of a 27 cm–diameter 30 cm–long cylindrical ^{31}P head coil, assuming that the principal noise source is the sample rather than the coil in both cases.

LOCALIZATION WITH DRESS

Our approach to the elimination of surface tissue contamination in surface coil spectroscopy is to apply an imaging type selective excitation $\pi/2$ NMR pulse in the

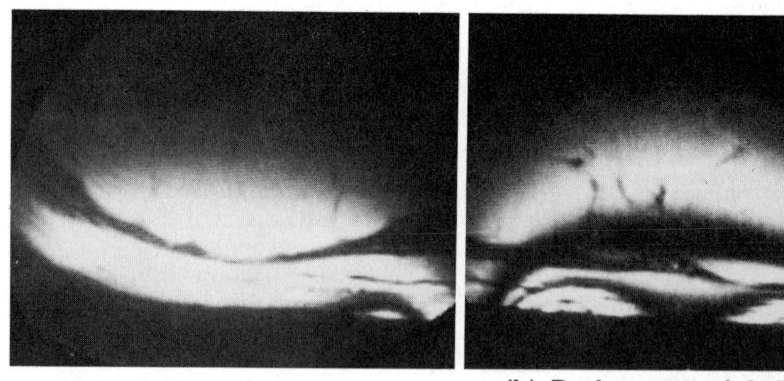

FIGURE 1. ^{1}H NMR images recorded at 64 MHz using a 6.5 cm diameter surface coil located on the human head above the temple.[9] The surface coil was used as both transmitter and receiver with a 2 sec pulse repetition to reduce ^{1}H T_1 relaxation effects. The imaging plane is perpendicular to the plane of the surface coil, which was located at the bottom of the images. In (a), the integrated signal from brain (top) constitutes only 54% of the total signal. This fraction increased to 72% in (b) upon increasing the RF field amplitude 2.5-fold. Dark rings in (b) correspond to the RF field contours of the surface coil at points in the sample that experience integral multiples of a π pulse.

TABLE 1. Methods of Spatially Localizing ^{31}P and ^{1}H Metabolite NMR Spectra

Type	Description/Acronym	Reference
RF gradient		
Single point	Simple surface coil	Ackerman *et al.* 1980. Nature **283:** 167.
	Depth selective RF pulse sequences	Bendall & Gordon. 1983 J. Magn. Reson. **53:** 365.
Multiple points	Rotating frame zeugmatography	Cox & Styles. 1980. J. Magn. Reson. **40:** 209.
Static gradients		
Single point	Topical magnetic resonance/TMR	Gordon *et al.* 1980. Nature **287:** 736.
Time-dependent gradient		
Single point	Single sensitive point	Bottomley, J. 1981. PhysE: Sci. Instrum. **14:** 1081.
	Sensitive disk/DRESS	Bottomley *et al.* 1984. J. Magn. Reson. **59:** 338.
	Volume selective excitation/VSE	Aue *et al.* 1985. J. Magn. Reson. **61:** 392.
	Volume excitation (phase-cycled)/ISIS	Ordidge *et al.* 1986. J. Magn. Reson. **66:** 283.
	Point excitation/PRESS	Bottomley. 1984. US Patent 4,480,228.
	Volume excitation/SPARS	Luyten *et al.* 1986. J. Magn. Reson. **67:** 148.
Multiple point	Selective excitation and projections	Lauterbur *et al.* 1975. J. Am. Chem. Soc. **97:** 6866.
	Projections and weak gradients	Bendel *et al.* 1980. J. Magn. Reson. **38:** 343.
	Other projection methods	Lauterbur *et al.* 1983. Radiol. **149:** 255.
	Phase-encoding gradient—1D–3D	Brown *et al.* 1982. Proc. Natl. Acad. Sci. USA **79:** 3523.
	Phase-encoding gradient—2D	Maudsley *et al.* 1983. J. Magn. Reson. **51:** 147.
	Phase-encoding gradient—3D	Hall *et al.* 1985. J. Magn. Reson. **61:** 188.
	Slice-interleaved DRESS/SLIT DRESS	Bottomley *et al.* 1985. J. Magn. Reson. **64:** 347.

presence of a pulsed magnetic field gradient directed coaxial to the surface coil in a simple, single pulse sequence known as depth-resolved surface coil spectroscopy (DRESS), as depicted in FIGURE 3.[11] The combination of the narrow bandwidth (Gaussian or sinc function modulated) selective excitation pulse and the magnetic field gradient excites a flat plane of nuclei parallel to the plane of the surface coil. The location of the plane relative to the center of the gradient field is $y_i = 2\pi f_i/\gamma G_y$, where f_i is the frequency of the excitation pulse, γ is the gyromagnetic ratio of the observed nucleus, and G_y is the strength of the gradient applied during the RF pulse. f_i can be varied automatically to select planes at different depths y_i relative to the surface coil simply by offsetting the NMR frequency of the selective excitation pulse provided under computer control in a high-field imaging/spectroscopy system with a single sideband transmitter.[8] The slice thickness is $\delta y = 2\pi\delta f/\gamma G_y$ where δf is the spectral width of the selective excitation pulse.

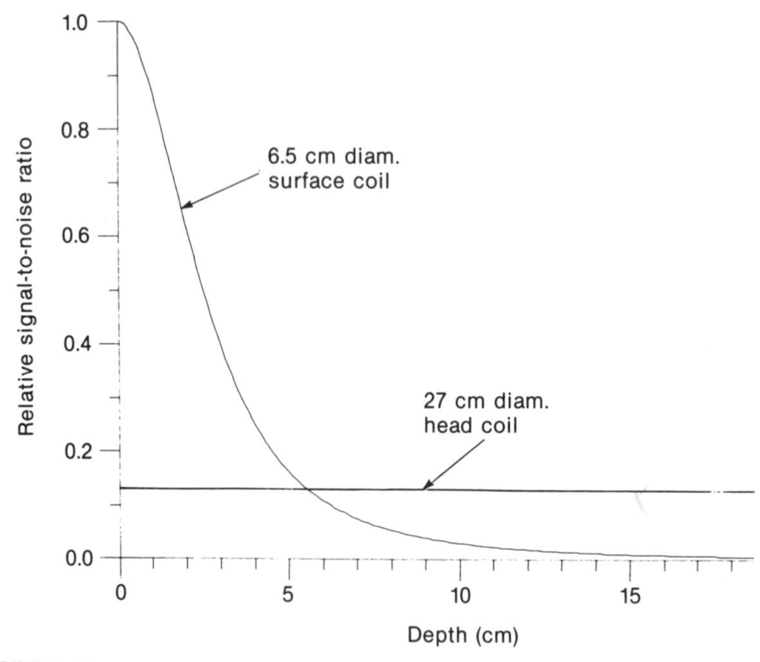

FIGURE 2. The calculated signal-to-noise ratio of a 6.5 cm diameter surface coil as a function of depth along its axis, assuming a constant RF excitation field.[12] At 25.7 MHz and 1 cm depth, the vertical axis corresponds to a measured signal-to-noise ratio for ^{31}P NMR of 1,100 ± 200 ml^{-1} $Hz^{1/2}$ for 15 M H_3PO_4, including all sample losses in the head, but excluding coil losses.[10] The corresponding figure for a 27 cm diameter ^{31}P head coil is 170 ± 20 $ml^{-1}Hz^{1/2}$. The surface coil signal-to-noise ratio becomes comparable to that of the head coil at about one surface coil diameter. (Courtesy W. A. Edelstein.)

The extent of the sensitive volume in the two orthogonal directions parallel to the selected plane in DRESS is determined by the surface coil sensitivity profile, which in turn depends upon the diameter of the surface coil and the distance of the coil from the sensitive plane. The shape of the sensitive volume as a function of depth is thus the same as that of a surface coil at the intersection of its sensitivity profile and the selected plane, as shown in FIGURE 4 for a uniform RF excitation field.[12] The sensitive volume is approximately disk shaped, at least up to the useful depth of about one diameter of the surface coil (FIG. 2), but the diameter of this sensitive disk increases with depth. However, the true size of the selected sensitive volume is not determined by the full-width–half-maximum of the sensitivity profiles (FIG. 4), but rather by the relative contributions of the integrated signal from the sensitive disk and that from the surrounding tissues. The radius of the sensitive disk that represents 50% of the total integrated signal in the selected plane is plotted against depth (in coil radii) in FIGURE 5, assuming a uniform NMR signal density extending infinitely in the selected DRESS plane.

Of course, in real *in vivo* applications of DRESS as well as virtually all other localization schemes that employ relatively coarse volume selection to compensate for the poor sensitivity of ^{31}P and ^{1}H metabolites, the assumption of a homogeneous signal density across the selected volume is often not even approximately true and the real density distribution is usually unknown. Consequently, the shape and size of actual selected volumes can vary dramatically from idealized profiles computed assuming uniform signal distributions. As an example, consider a series of ^{31}P DRESS spectra from a dog myocardium acquired as a function of depth with a 6.5 cm diameter surface coil before and after occlusion of a coronary artery (FIG. 6).[13] Postmortem staining of this heart revealed a 14 g endocardial infarction, consistent with the total depletion of high energy metabolites (phosphocreatine and adenosine triphosphate) apparent in the deepest postocclusion spectrum. This spectrum therefore represents a ^{31}P NMR-visible tissue volume of about 14 cm^3 at most. However, the computed tissue volume representing 50% of the total signal at this depth according to FIGURE 5 is about 80 cm^3, if we divide by a factor of about 2 because the surface coil was used for both transmission and reception in this experiment.[13] Clearly, and fortuitously for heart studies, a large fraction of the sensitive volume here is occupied by tissues such as adipose, lung, and moving blood in the ventricle that contribute negligible detectable ^{13}P NMR signals to the observed spectrum.

DRESS NMR EXPERIMENTS

Simplicity is a key, perhaps unequaled, advantage of the DRESS technique relative to other current spectral localization schemes. This simplicity translates into a versatility of applications employing both conventional NMR pulse sequences, or even incorporating elements of the other localization techniques (TABLE 1) to improve the spatial resolution or enable volume coil detection as opposed to surface coil detection.

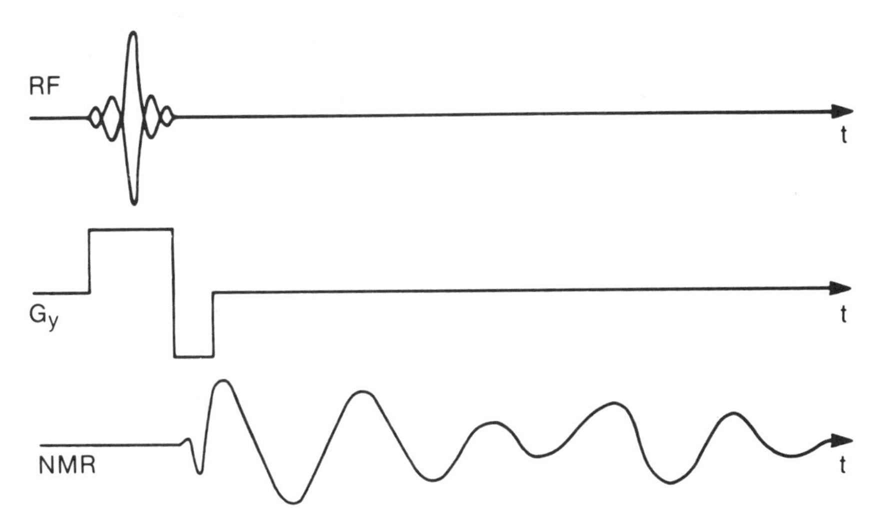

FIGURE 3. Depth-resolved surface coil spectroscopy (DRESS) pulse sequence. A sinc function–modulated $\pi/2$ RF pulse selects a plane parallel to the surface coil when applied in conjunction with a magnetic field gradient ($G_y = \partial B_0/\partial y$, where B_0 is the main magnetic field) directed coaxial to the coil.[11] Data are acquired (NMR) commencing as the nuclei are rephased slightly before the cessation of the negative G_y rephasing lobe.

Such applications are permissible because the entire spatial localization procedure is completed with just a single application of the sequence and because the sequence perturbs the nuclear magnetization from only the selected volume, leaving the NMR signal in all of the remaining space undisturbed. Thus, spatially localized spectral spin lattice (T_1) or spin spin (T_2) NMR relaxation times can be measured by incorporating a conventional square π NMR pulse either at time τ prior to the plane-selective $\pi/2$ pulse of FIGURE 3, or at time τ after the $\pi/2$ pulse, respectively. The sequence is repeated with different τ values and the amplitude of each resonance fitted to $S = S_0 \cdot [1 - 2\exp(-\tau/T_1)]$ or $S = S_0 \exp(-\tau/T_2)$ to yield the individual T_1 and T_2 values in the usual fashion.

Similarly, the DRESS sequence is amenable to solvent suppression techniques, which are pivotal to the *in vivo* detection of millimolar level metabolites, such as lactate in ^{1}H NMR spectra, that are otherwise dominated by water protons at concentrations of around 100 molar.[14,15] A series of four water suppression pulse sequences incorporating DRESS localization are shown in FIGURE 7. In (a) and (b), a long duration, sinc function modulated, $\pi/2$ selective excitation RF pulse of spectral linewidth comparable to and centered on that of the water resonance in the frequency domain, is applied prior to DRESS selection in the absence of any applied field

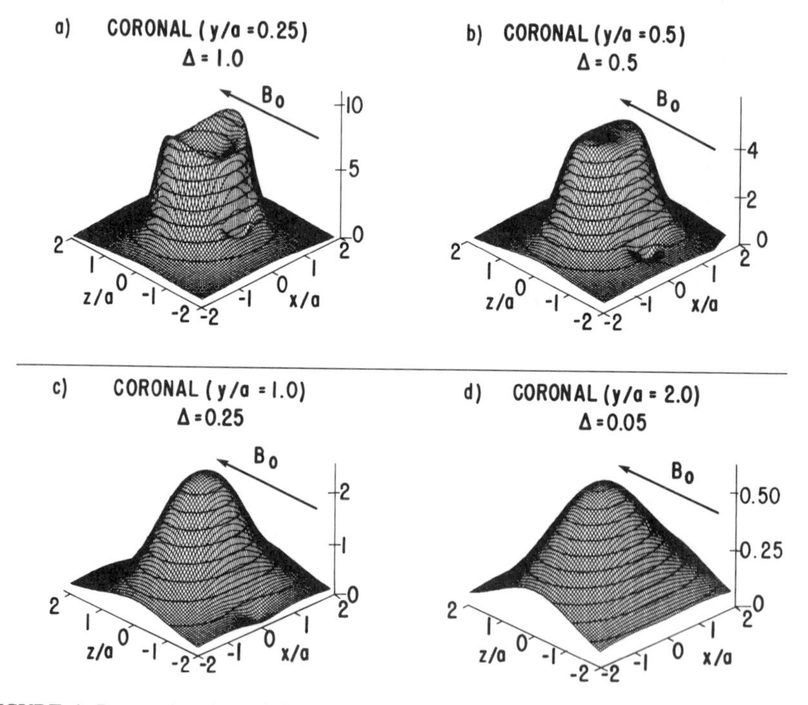

FIGURE 4. Perspective plots of the NMR sensitivity profile of a circular coil of radius a at depths of 0.25 radii (a), 0.5 radii (b), 1.0 radii (c), and 2.0 radii (d).[12] The coil is oriented parallel to the x-z plane with axis at the origin and scales in radii. The vertical axis is proportional to the signal-to-noise ratio depicted in FIGURE 2. The plots are computed from equations for the field transverse to the main field, B_0, given by Smythe (Static and Dynamic Electricity. 3rd edit. 1968. McGraw-Hill. New York.) assuming a uniform RF excitation field, and represent the shapes of the sensitive volume in the selected DRESS planes. (Courtesy J. F. Schenck.)

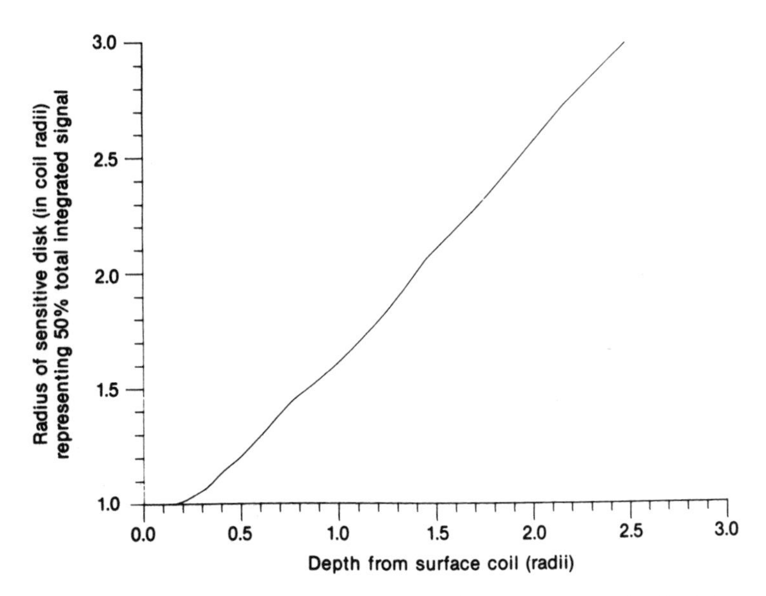

FIGURE 5. The radius of the sensitive volume, assumed circular, representing 50% of the total integrated signal in DRESS selected planes lying parallel to the surface coil as a function of depth along the coil axis. All dimensions are in coil radii. The curve was obtained by iteratively integrating the sensitivity profile until it represented 50% of the total integral at that depth. (Courtesy W. A. Edelstein.)

gradients. Following DRESS selection, all of the nuclei in a selected plane parallel to the surface coil are excited but the H_2O protons have received a net π nutation rendering them, in principle, unobservable.[15] In practice, the amplitude of the H_2O-selective pulse is adjusted for maximum annihilation of the H_2O resonance during data acquisition. The water-suppressed spectrum is then either detected immediately following the slice-selective pulse (FIG. 7(a)) or refocused to a spin echo using a subsequent π pulse applied at time τ later (FIG. 7(b)). The latter sequence provides additional attenuation of tissue ^1H resonances that exhibit shorter T_2 values than the metabolites of interest. Furthermore, if the echo-producing π pulse of FIGURE 7(b) is also a chemical-selective pulse that refocuses only the resonances of interest, the water suppression will be further enhanced.

Discrimination against H_2O signals on the basis of relaxation times is extended in the sequence shown in FIGURE 7(c), wherein the chemical selective pulse is abandoned to be replaced by an initial π inversion pulse applied at time τ_{null} preceding the slice selective $\pi/2$ pulse. τ_{null} is adjusted to eliminate the H_2O resonance at $\tau_{null} \approx 0.69\ T_1$ (H_2O), where T_1 (H_2O) is the water spin-lattice relaxation time. The sequence shown in FIGURE 7(d) is a reduction of that in FIGURE 7(b), in which the initial $\pi/2$ excitation pulse is a chemical-selective pulse tailored to select only the metabolically useful portion of the ^1H spectrum and exclude the H_2O resonance. Slice selection is subsequently performed by the π refocusing pulse applied in the presence of the gradient. In all cases, data are best averaged from two applications of the sequences repeated with the phase of the $\pi/2$ excitation pulse alternated to remove unwanted spurious signals generated by the other RF pulses.[15]

FIGURE 8 shows an example of a normal human brain ^1H NMR spectrum acquired

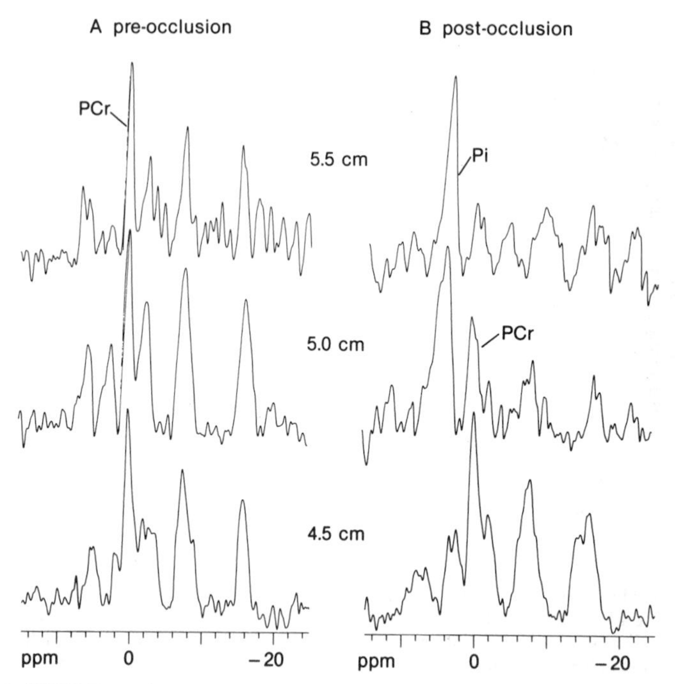

FIGURE 6. ^{31}P NMR surface coil spectra recorded noninvasively *in vivo* at 0.5 intervals through the anterior myocardium of a dog before (A), and 50 min to 70 min after occlusion of the left anterior descending coronary artery (B) using DRESS.[13] Postocclusion spectra at 5.5 cm depth show essentially complete depletion of phosphocreatine (PCr) and adenosine triphosphate, and the appearance of a large inorganic phosphate (P_i) resonance. Less depletion is evident at 5.0 cm and essentially no postocclusion spectral changes are apparent at 4.5 cm. Postmortem staining of the excised heart revealed a 14 g endocardial infarction. Data acquisition periods were 11 ± 1 min per spectrum cardiac gated with a 1.1 ± 0.1 sec pulse-sequence repetition period. Depths are nominally relative to the surface with a 1 cm slice thickness.

in 2 sec at a depth of 5 cm with the water-suppressed DRESS sequence of FIGURE 7(b). A 3-cm diameter surface coil was used to localize to the volume shown in the image. N-acetyl aspartate, phosphorylcholine, and creatine resonances are evident at around 5–10 mM concentrations, but any lactate resonance at 1.3 ppm is sufficiently small as to be obscured by lipid -CH_2- resonances, as might be expected in a healthy brain.

ADVANCED LOCALIZATION SEQUENCES

Adaptations of DRESS that improve its spatial data-gathering efficiency include slice-interleaved DRESS (SLIT DRESS), which enables acquisition of spectra from multiple sensitive disks at different depths in essentially the same time as required for a single DRESS volume acquisition.[16] The idea is to interleave excitation of n different slices during the period τ_r normally allotted for spin-lattice relaxation recovery in the first excited slice. Thus the DRESS sequence of FIGURE 3 is repeated n times faster at intervals of τ_r/n, sequentially advancing the NMR offset frequency f_i to a new depth at each application. To minimize partial saturation effects due to partially overlapping slice profiles, it is prudent to order the offset frequencies nonconsecutively. FIGURE 9 shows a series of six ^{31}P SLIT DRESS spectra as a function of depth in the human liver at 1-cm intervals. The spectra were acquired in a total averaging time of 10 min with $\tau_r = 2$ sec.

Improving the spatial selectivity of DRESS beyond that achieved by surface coil detection in the selected planes parallel to the surface coil (FIG. 4), can be accomplished by adding two echo-producing, spatially selective π pulses at times τ and 3τ following the plane-selective $\pi/2$ pulse, as shown in FIGURE 10.[17] Each π pulse is applied in the presence of a gradient magnetic field pulse directed along each of the two remaining orthogonal axes that are directed parallel to the selected plane, thereby producing two spin echoes. The first spin echo derives from a sensitive line lying at the intersection of the two orthogonal planes selected by the $\pi/2$ pulse and the first π pulse. The second spin echo derives from a sensitive point lying at the intersection of the planes selected by all three pulses, and is collected. The gradient pulses are asymmetric

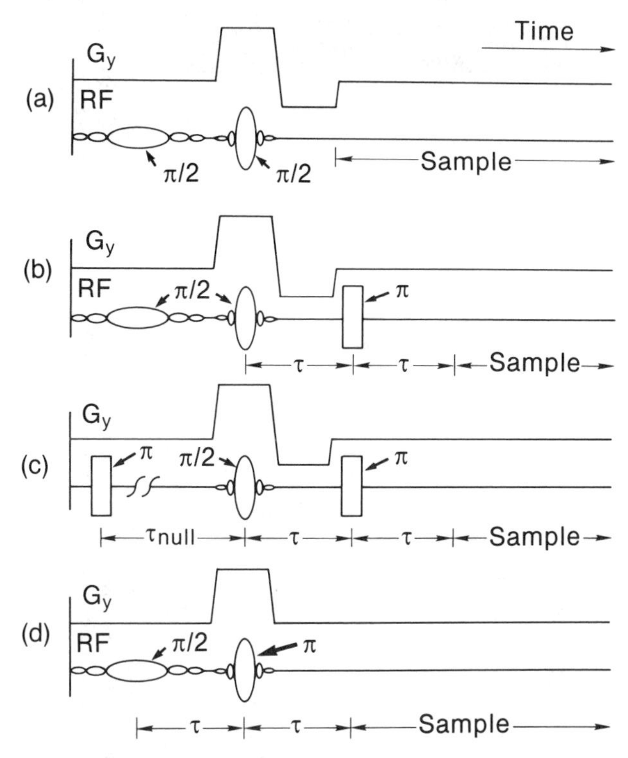

FIGURE 7. Timing diagrams for the gradient (G_y) and the NMR pulses (RF) for four methods of performing solvent suppressed DRESS.[15] Selective pulses are depicted as sinc-function modulated envelopes: they are chemical selective when applied in the absence of a gradient pulse and spatially selective when applied in the presence of a gradient pulse. A negative G_y lobe is not required in (d) if the π RF pulse is symmetrically located with respect to the positive G_y pulse. τ is the time between $\pi/2$ and π RF pulses, τ_{null} is the time between π and $\pi/2$ pulses, and sample is the data acquisition period.

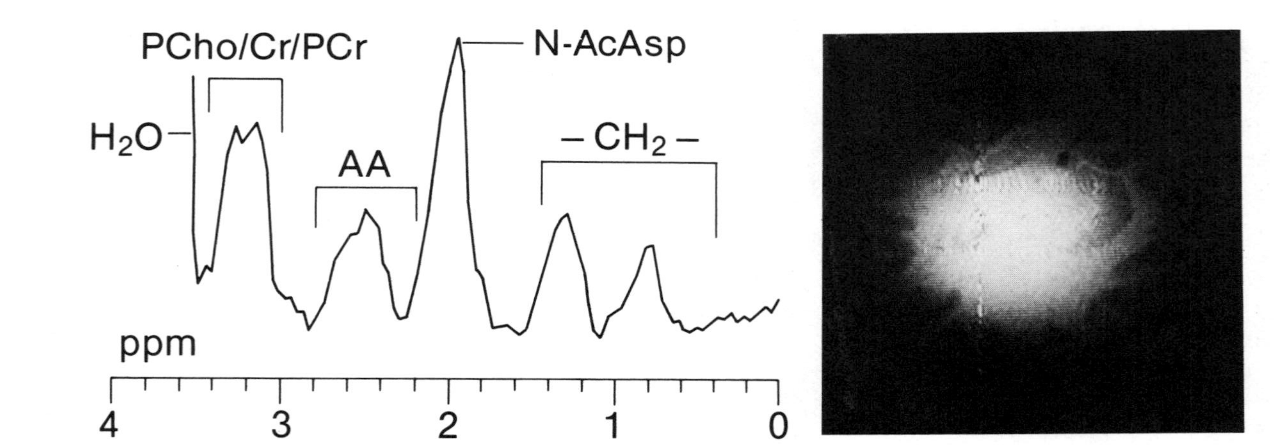

FIGURE 8. H_2O-suppressed ^1H spectrum (left) acquired from a 5-cm deep sensitive disk in the normal human brain imaged at right.[15] The DRESS sequence shown in FIG. 7(b) was used to obtain the spectrum in 2.0 sec. The lactate $-CH_3$ resonance, located at 1.3 ppm is not discernible here (PCho/PCr/Cr, phosphorylcholine and total creatine pool; AA, amino acids including aspartyl, glutamate, and glutamine groups[14]; $-CH_2-$, lipid resonance).

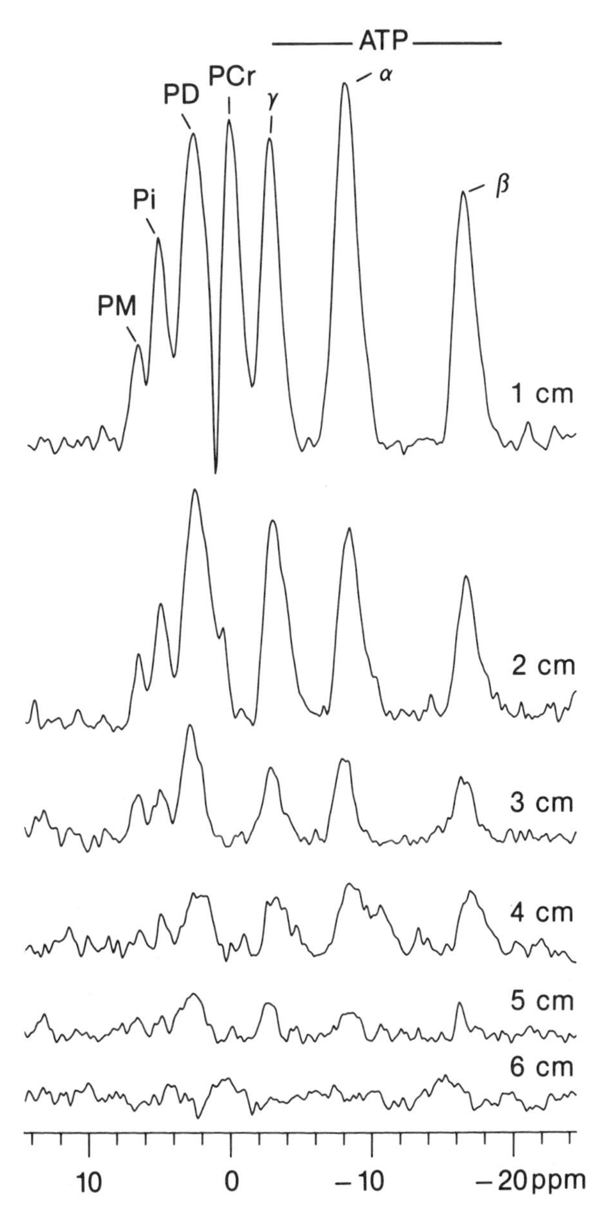

FIGURE 9. *In vivo* ^{31}P slice interleaved DRESS (SLIT DRESS) spectra obtained with a 6.5-cm diameter surface coil located on the human chest above the liver. Depths nominally relative to the surface are indicated. Negligible phosphocreatine (PCr) at depths $\geq$2 cm is consistent with liver metabolism.[8] The complete spectral series was recorded in 10 min with a 2 sec sequence repetition period (PM, phosphomonoesters; P_i, inorganic phosphate; PD, phosphodiesters; γ-, α-, β-ATP, γ-, α- and β phosphates of adenosine triphosphate).

with respect to nuclei that are not selectively rephased by π pulses. Therefore NMR signals excited outside the sensitive point rapidly dephase.

The advantages of this point-resolved spectroscopy (PRESS) technique are that it yields sharply defined volumes localized in all three dimensions in a single application of the pulse sequence, and that it provides automated control of the sensitive volume size and location independent of the size of the detection coil. In fact, surface detection coils are no longer necessary. Its main disadvantage is that the spectral components of interest must possess sufficiently long T_2 values to ensure a detectable NMR spin-echo

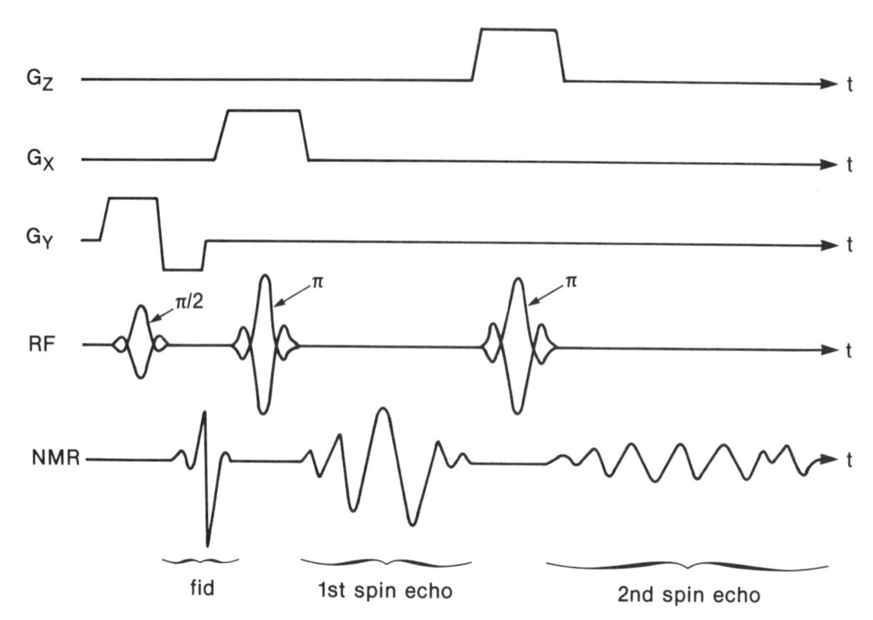

FIGURE 10. Gradient pulse, RF pulse, and NMR signal timing diagram for a point-resolved spectroscopy (PRESS) technique employing DRESS plane localization and two spin echoes.[17] Gradients along the three orthogonal axes, denoted G_x, G_y, and G_z, are applied in the presence of each of the RF pulses to achieve three-dimensional localization of NMR signals persisting to the second spin echo. All NMR pulses are selective and depicted here as sinc function–modulated envelopes. Negative rephasing lobes are unnecessary following the G_x and G_z gradient pulses if these are symmetric with respect to the π pulses. The gradient pulses will also dephase previously excited signals, which are not rephased by the selected π pulses. The sequence generates two spin echoes and one free induction decay (fid).

signal at time 4τ following initial excitation. This is probably unacceptable for *in vivo* ^{31}P NMR since the T_2 of adenosine triphosphate is only about 10 msec,[18] but ^{1}H NMR spectroscopy of lactate and other metabolites or tumor lipids[19] should be accessible.

For spectral components with short T_2 values, and ^{31}P metabolites in particular, relocating the spatially selective π pulses prior to the $\pi/2$ pulse is a viable approach to three-dimensional volume localization with pulsed gradients analogous to the above PRESS technique. However, complete volume localization is no longer possible with a single application of the sequence because there is no way of discriminating between inverted and noninverted signal contributions in one free induction decay following the

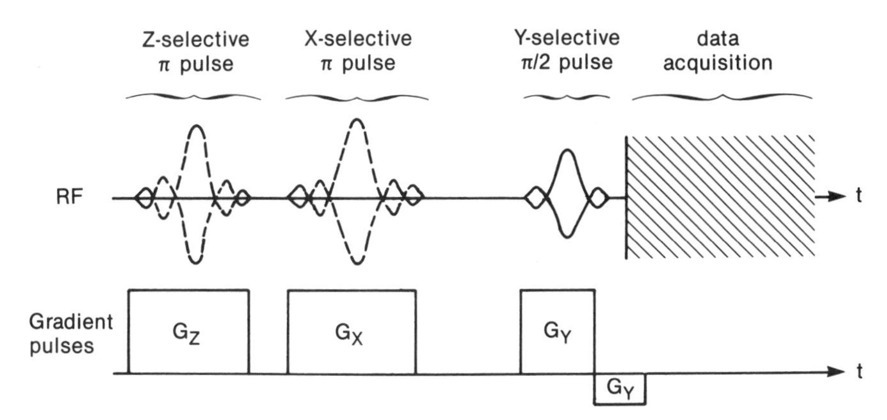

FIGURE 11. RF and gradient pulse timing diagram for combined DRESS/ISIS[20] sequence (CRISIS). The sequence is repeated four times with the three orthogonal gradient pulses G_x, G_y, and G_z applied as shown. However, in each application, the z-selective and x-selective π pulses are either applied or not applied according to the cycle number in TABLE 2. The y-selective excitation and data acquisition is performed as for DRESS (FIG. 3) but the resultant signals are added or subtracted as indicated in TABLE 2. Selective π pulses are depicted as sinc function envelopes, but improved designs are available.[22]

$\pi/2$ pulse. Thus, in the original version of this technique, termed ISIS for image-selected *in vivo* spectroscopy,[20] all three orthogonal gradients are sequentially pulsed prior to a nonselective $\pi/2$ pulse, and the selective π inversion pulses are either applied or not applied during the gradient pulses in $2^3 = 8$ consecutive applications of the entire sequence.

A potential problem of the ISIS approach in regions subject to large physiologic movement such as in the heart, is that motions which occur before the total sequence cycle is completed could introduce substantial volume localization artifacts. A similar problem was encountered in cardiac NMR imaging with an early line-scan technique employing selective inversion.[21] It is aggravated here by the use of a nonselective $\pi/2$ pulse that excites all accessible nuclei in the sample. Such a situation might thus be ameliorated by combining DRESS-selective excitation on the $\pi/2$ pulse for one-dimensional localization, with the ISIS technique in the other two dimensions. This combined strategy would reduce the amount of excited signal present at any point in time by the ratio of the DRESS slice thickness to the sample dimension, that is, by roughly an order of magnitude, and halve the duration of the ISIS sequence cycle to $2^2 = 4$ sequence applications. Its main disadvantage relative to conventional ISIS is the

TABLE 2. Combined DRESS/ISIS (CRISIS) Cycling Sequence for Selective Inversions

Sequence Number	z-Selective Pulse	x-Selective Pulse	Contribution to Total Spectrum
1	off	off	+1
2	on	off	−1
3	off	on	−1
4	on	on	+1

small delay of a few milliseconds required for gradient rephasing following the $\pi/2$ pulse in the DRESS sequence (FIG. 3),[11] during which time some signal might be lost through T_2 decay.[16] In the spirit of pun acronyms in NMR, we might call the combined resolved ISIS sequence, CRISIS. This modified sequence is depicted in FIGURE 11, and the condensed sequence cycle is shown in TABLE 2: sensitive point/volume resolution is obtained by adding and subtracting the resultant signals according to the signs in the last column. Selective inversion pulses that use both amplitude and phase modulation can also be employed.[20,22]

CONCLUSION

DRESS offers a practical means of acquiring spatially localized spectra *in vivo* that can be easily incorporated into a variety of conventional NMR experiments or included as an element of other localization procedures. Its main problems are the poorer spatial localization achieved in directions orthogonal to the slice selective gradient, and possible signal loss incurred during the delay following selective excitation required to refocus the NMR signal. In combination with ^{1}H NMR imaging as a means of locating pathologies or other regions of interest, ^{31}P DRESS studies in brain[23] and heart[24] patients are currently underway to evaluate its clinical utility.

ACKNOWLEDGMENTS

I thank W. A. Edelstein for kindly providing the curves in FIGURES 2 and 5, J. F. Schenck for providing the surface coil plots in FIGURE 4, and L. S. Smith, R. W. Redington and R. J. Herfkens for scientific support.

REFERENCES

1. ODEBLAD, E. & G. LINDSTROM. 1955. Some preliminary observations on the proton magnetic resonance in biologic samples. Acta Radiol. **43:** 469–476.
2. MOON, R. B. & J. H. RICHARDS. 1974. Determination of intracellular pH of ^{31}P nuclear magnetic resonance. J. Biol. Chem. **248:** 7276–7278.
3. HOULT, D. I., S. J. W. BUSBY, D. G. GADIAN, G. K. RADDA, R. E. RICHARDS & P. J. SEELEY. 1974. Observation of tissue metabolites using ^{31}P nuclear magnetic resonance. Nature (London) **252:** 285–287.
4. CHANCE, B., S. ELEFF & J. S. LEIGH. 1980. Noninvasive, nondestructive approaches to cell bioenergetics. Proc. Natl. Acad. Sci. USA **77:** 7430–7434.
5. CRESSHULL, I. D., R. E. GORDON, P. E. HANLEY & D. SHAW. 1980. Bull. Magn. Reson. **2:** 426.
6. CADY, E. B., A. M. DE L. COSTELLO, M. J. DAWSON, D. T. DELPY, P. L. HOPE, E. O. R. REYNOLDS, P. S. TOFTS & D. R. WILKIE. 1983. Noninvasive investigation of cerebral metabolism in newborn infants by phosphorus nuclear magnetic resonance spectroscopy. Lancet **i:** 1059–1062.
7. BOTTOMLEY, P. A., H. R. HART, W. A. EDELSTEIN, J. F. SCHENCK, L. S. SMITH, W. M. LEUE, O. M. MUELLER & R. W. REDINGTON. 1983. NMR imaging/spectroscopy system to study both anatomy and metabolism. Lancet **ii:** 273–274.
8. BOTTOMLEY, P. A. 1985. Noninvasive study of high-energy phosphate metabolism in the human heart by depth resolved ^{31}P NMR spectroscopy. Science **229:** 769–772.
9. BOTTOMLEY, P. A., W. A. EDELSTEIN, H. R. HART, J. F. SCHENCK & L. S. SMITH. 1984.

Spatial localization in ^{31}P and ^{13}C NMR spectroscopy using surface coils. Magn. Reson. Med. **1:** 410–413.
10. Bottomley, P. A. 1986. A practical guide to getting NMR spectra *in vivo*. *In* Medical Magnetic Resonance and Spectroscopy. T. F. Budinger & A. R. Margulis, Eds.: 81–95. Soc. Magn. Reson. Med. Berkeley, CA.
11. Bottomley, P. A., T. H. Foster & R. D. Darrow. 1984. Depth resolved surface coil spectroscopy (DRESS) for *in vivo* ^{1}H, ^{31}P and ^{13}C NMR. J. Magn. Reson. **59:** 338–342.
12. Schenck, J. F., H. R. Hart, T. H. Foster, W. A. Edelstein & M. A. Hussain. 1986. High resolution magnetic resonance imaging using surface coils. *In* Magnetic Resonance Annual 1986. H. Y. Kressel, Eds.: 123–160. Raven Press. New York.
13. Bottomley, P. A., R. J. Herfkens, L. S. Smith, S. Brazzamano, R. Blinder, L. W. Hedlund, J. L. Swain & R. W. Redington. 1985. Noninvasive detection and monitoring of regional myocardial ischemia *in situ* using depth resolved ^{31}P NMR spectroscopy. Proc. Natl. Acad. Sci. USA **82:** 8747–8751.
14. Behar, K. L., J. A. den Hollander, M. E. Stromski, T. Ogino, R. G. Shulman, O. A. C. Petroff & J. W. Prichard. 1983. High resolution ^{1}H nuclear magnetic resonance study of cerebral hypoxia *in vivo*. Proc. Natl. Acad. Sci. USA **80:** 4945–4948.
15. Bottomley, P. A., W. A. Edelstein, T. H. Foster & W. A. Adams. 1985. *In vivo* solvent-suppressed localized hydrogen nuclear magnetic resonance spectroscopy: a window to metabolism? Proc. Natl. Acad. Sci. USA **82:** 2148–2152.
16. Bottomley, P. A., L. S. Smith, W. M. Leue & C. Charles. 1985. Slice interleaved depth resolved surface coil spectroscopy (SLIT DRESS) for rapid ^{31}P NMR *in vivo*. J. Magn. Reson. **64:** 347–351.
17. Bottomley, P. A. 1984. Selective volume method for performing localized NMR spectroscopy. U.S. patent *4 480 228.*
18. Haselgrove, J. C., V. H. Subramanian, J. S. Leigh, L. Gyulai & B. Chance. 1983. *In vivo* one-dimensional imaging of phosphorus metabolites by phosphorus-31 nuclear magnetic resonance. Science **220:** 1170–1173.
19. Mountford, C. E., L. C. Wright, K. T. Holmes, W. G. Mackinon, P. Gregory & R. M. Fox. 1984. High resolution proton nuclear magnetic resonance analysis of metastatic cancer cells. Science **226:** 1415–1418.
20. Ordidge, R. J., A. Connelly & J. A. B. Lohman. 1986. Image selected *in vivo* spectroscopy (ISIS). A new technique for spatially selective NMR spectroscopy. J. Magn. Reson. **66:** 283–294.
21. Hutchison, J. M. S., W. A. Edelstein & G. Johnson. 1980. A whole-body NMR imaging machine. J. Phys. E. Sci. Instrum. **13:** 947–955.
22. Silver, M. S., R. I. Joseph, C. N. Chen, V. J. Sank & D. I. Hoult. 1984. Selective population inversion in NMR. Nature **310:** 681–683.
23. Bottomley, P. A., B. P. Drayer & L. S. Smith. 1986. Chronic adult cerebral infarction studied by phosphorus NMR spectroscopy. Radiology **160:** 763–766.
24. Bottomley, P. A., R. J. Herfkens, L. S. Smith & T. M. Bashore. 1987. Disturbed phosphate metabolism detected in human anterior myocardial infarction by spatially localized ^{31}P spectroscopy. Radiology (In press.)

DISCUSSION OF THE PAPER

D. I. Hoult: I've got compound thoughts for you. You've not addressed several problems. The problem of the eddy currents, the problem with the final technique, the rather poor way in which selective 180 degree pulses for refocusing work, can you just address them?

P. A. Bottomley: Well, we never really had an eddy current problem with DRESS. Perhaps one of the reasons for this is that we used imaging gradient coils of around 0.6 meters or 0.7 meters diameter—much smaller than the 1 meter magnet bore diameter. This means that the gradients do not couple significantly with the magnet structure. A second point is that the bore of the magnet was made of fiberglass so the first metal that the gradient fields hit is cold aluminum, which has a long decay constant. When we actually perform the DRESS experiment we shim on the DRESS sequence and use just the standard static magnet shims to offset any eddy-current effects. The final answer is that we're now even trying to reduce the eddy currents that we do get. This is being done, as was shown in Montreal (SMRM) by Peter Roemer—and Peter Mansfield is doing it as well—by placing shields between the magnetic field gradient coils and the magnet to eliminate any interactions of the coils with the magnet. But I think these problems are really not all that major. The second problem of the poor refocusing ability of the selective 180° pulse, I think that you will probably come up with solutions that will enable us to do that better.

Spatially Resolved ^{31}P NMR Spectroscopy of Organs in Animal Models and Man[a]

PETER STYLES, MARTIN J. BLACKLEDGE,
CHRIT T. W. MOONEN,[b] AND GEORGE K. RADDA

Medical Research Council Clinical Magnetic Resonance Facility
John Radcliffe Hospital
Headington, Oxford, OX3 9DU

INTRODUCTION

It is now more than 10 years since the first demonstration that NMR could be used to obtain spatial maps of chemically shifted compounds.[1] However, the routine acquisition of localized metabolic data from animal and human tissue remains a significant challenge to the spectroscopist. Of the many possible strategies that can be adopted to achieve this goal, one class of experiments relies on a gradient in the B_1 transmitter field to encode the nuclear spins according to position. This idea, first proposed by Hoult[2] as a technique for proton imaging and later demonstrated as feasible for spectroscopy,[3] is known as "rotating frame imaging." Briefly, the experimental protocol involves collecting a set of data accumulations where the length of the transmitter pulse is incremented for each accumulation. Subsequent two-dimensional Fourier transformation (2DFT) yields a matrix where chemical shift is represented along one axis and spatial position along the other. Important developments to the method over recent years have included the use of both single[4] and double[5] surface coil probes for the production of the required B_1 field gradient, permitting the examination of excised organs,[6] animals,[7,8] and man.[5,9] Other related techniques of interest include single volume derivations of the protocol[10–14] and the measurement of localized spin-lattice relaxation times.[15,16]

This paper will describe some of the experiments that we are now routinely using to assess the metabolic energy status of various tissues in both animals and man. In particular we have investigated and are now implementing a phase-encoded version of the rotating frame protocol that offers greater sensitivity than the more usual amplitude-modulated scheme.

ROTATING-FRAME SPECTROSCOPIC IMAGING

The principle of rotating-frame spectroscopy is illustrated in FIGURE 1. Consider a two-compartment phantom with a different chemical species in each compartment, and situated in a gradient B_1 transmitter field as shown in FIGURE 1(a). FIGURE 1(b) shows a set of spectra collected from this arrangement following various durations of the radio-frequency transmitter pulse. It will be seen that the amplitude of each

[a]Supported by the Medical Research Council, the British Heart Foundation, and the National Institutes of Health (Grant HL-18704).

[b]Present address: NMR Research Center, Bldg. 10, Rm. BID-123, National Institutes of Health, Bethesda, MD 20892.

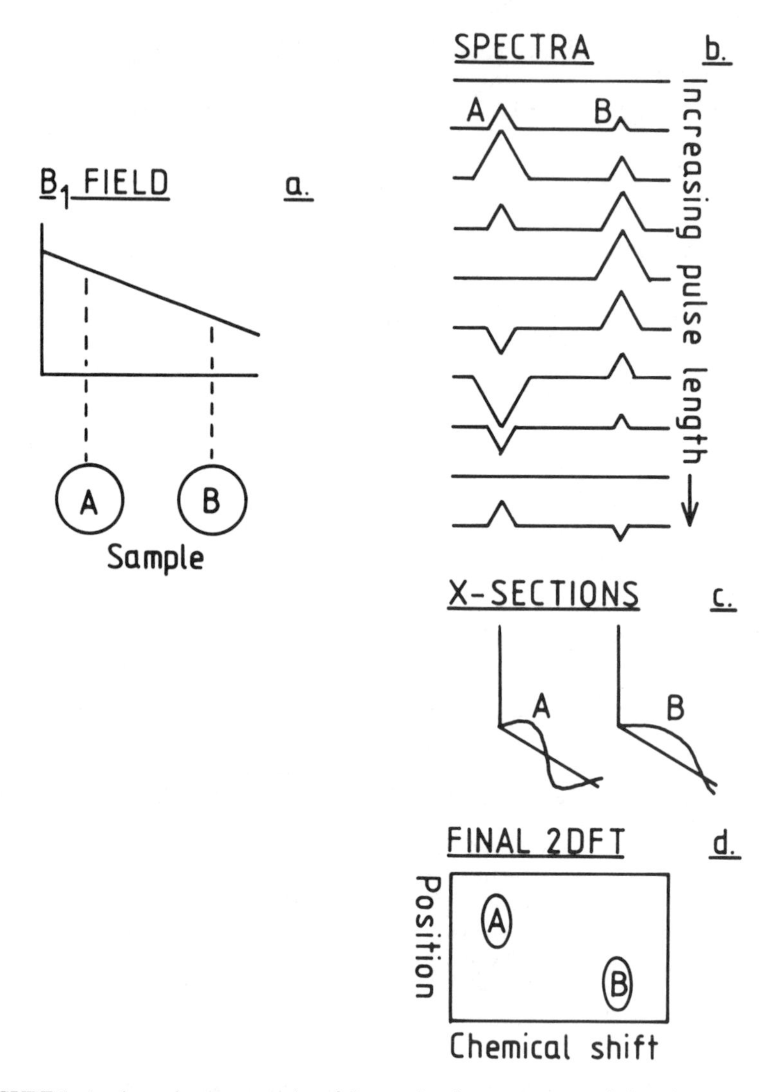

FIGURE 1. A schematic representation of the rotating-frame spectroscopic imaging protocol. (a) A two-compartment phantom placed in a gradient B_1 field. (b) A set of spectra collected at increasing pulse length. (c) Cross sections through the data matrix of (b) at positions corresponding to the centers of the two spectral lines. (d) The final data matrix obtained after a second Fourier transform has been performed on the data in (b).

spectral line varies as a function of the pulse duration, but the rate of change is greater for the sample situated in the higher B_1 field. This is seen from FIGURE 1(c), which plots the amplitudes of the two spectral lines as functions of pulse width. The equation describing this variation in amplitude (A) is:

$$A \propto \text{Sin}\,(\gamma .\, B_1 .\, t_p)$$

where γ is the gyromagnetic ratio, B_1 is the strength of the transmitter field, and t_p is the duration of the transmitter pulse. In order to ascertain the rate with which the signal amplitude changes, the set of spectra are now Fourier transformed with respect to the pulse width. The resulting matrix will have chemical shift on one axis and B_1 on the other, as in FIGURE 1(d). Provided that there is a known relationship between the B_1 strength and position, the B_1 axis will also represent a spatial dimension. This experiment is the amplitude-modulated rotating-frame protocol.

PHASE-MODULATED ROTATING-FRAME SPECTROSCOPIC IMAGING

The amplitude-modulated rotating-frame method suffers from one significant drawback. It will be seen from the several spectra in FIGURE 1(b) that the full signal is only collected when the pulse angle is 90°, 270°, 430°, etc. This is because the nuclear magnetization nutates in the $z'y'$ plane, and any longitudinal magnetization is not detected. In his original paper, Hoult[2] suggested that this loss of sensitivity could be avoided by following the amplitude-encoding pulse with a homogeneous $90°_y$ pulse to convert any z' magnetization into x' magnetization, which would then be detected.

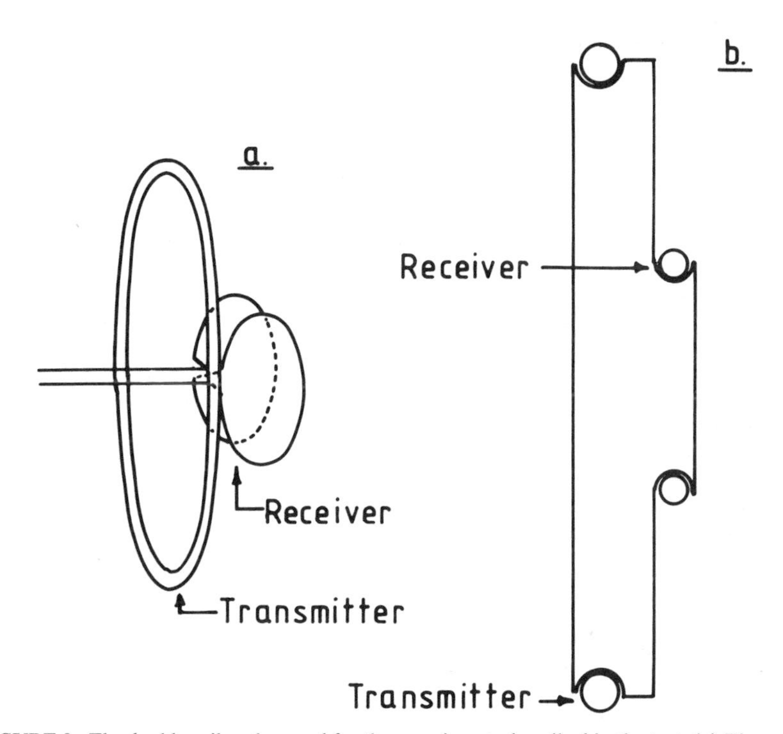

FIGURE 2. The double coil probes used for the experiments described in the text. (a) The probe used for the images of the *in vivo* rat kidney. The single turn surface coil transmitter, diam. 5 cm, is situated 2 cm from the center of the receiver assembly. The latter consists of two elliptical coils, wound in series, with major diam. of 2 cm, minor diam. 1 cm and separated by 1 cm. (b) The probe used for the human studies. The transmitter coil, diam. 15 cm, is situated 2.5 cm behind the receiver coil, diam. 6.5 cm.

This modification to the sequence causes the spatial information to be encoded in the phase of the individual spectra rather than the amplitude.

One further refinement is still necessary. A two-dimensional Fourier transformation performed on a data set acquired in this way suffers from "phase twist," a well known artefact in 2DFT where dispersion and absorption mode signals are mixed, giving a final matrix that cannot be properly phased. Again, a solution has been proposed[2] but not previously implemented. Two sets of data are collected as described, but with the difference that the phase of the $90°_y$ pulse is inverted for one of the collections. These two data sets are processed identically, but then one of the matrices is reversed in the spatial dimension. Subsequent addition of the two matrices produces a data set that can be phased to give pure absorption line shape.

PROBE DESIGN FOR ROTATING-FRAME EXPERIMENTS

The implementation of rotating-frame experiments on non-ideal samples, such as humans, requires special probe design. All of our experiments utilize surface coil transmitters in conjunction with separate receiver coils. By appropriate choice of coil geometry, the sensitive region of the receiver coil can be limited to an approximately linear part of the transmitter field pattern. The data that will presently be shown were obtained using either a Helmholtz type receiver coil in conjunction with a surface coil transmitter or a double concentric surface coil probe. The former was placed around the kidney of a rat, and the latter situated over the required region of the human subjects. Both probes are illustrated in FIGURE 2. Additional components are included to avoid electrical interaction between the two tuned circuits, so that receive and transmit coils operate independently. The circuit that we use to achieve the necessary isolation, which we have described earlier[14] can, with proper choice of component values, be constructed so that there is virtually no induced current in the receiver coil during the transmitter pulse and the region of receiver sensitivity is hardly affected by the presence of the transmitter coil.

It will be seen from the preceding description that the phase-modulated rotating-frame imaging protocol makes particular demands on the probe hardware. The primary transmitter coil must produce the required linear B_1 gradient for the encoding pulse, whereas the $90°_y$ pulse should ideally be homogeneous. These two requirements are difficult to realize when the sample is the size and shape of a human. In the phase-encoded data that we will be presenting, these conditions have not been met —instead the same transmitter coil is used for both irradiating pulses. The reason why this can give acceptable results is that the variation in B_1 field over the sensitive region of the receiver coil in the probe of FIGURE 2(b) is only $\pm 50\%$ relative to the mean value. As it is the sine of the phase-encoding pulse that determines the resulting xy magnetization, a $\pm 50\%$ error in pulse angle results in a much smaller error in phase encoding. The predominant manifestation of this imperfection is that some of the signal appears as a quadrature image along the spatial axis of the two-dimensional matrix.

RESULTS

Several examples of rotating-frame spectroscopic images are shown in FIGURES 3–6. In each case, the individual spectra are representative of approximately planar slices through the tissue.

Spectra from a Rat Kidney, Obtained in Vivo

The spectra shown in FIGURE 3 were obtained by placing the receiver coil FIGURE 2(a) around the exposed kidney of a 150 g male Wistar rat. Subsequent to implantation, the abdomen of the rat was closed over the receiver coil but the larger transmitter surface coil remained outside the animal. FIGURE 3(a) shows data obtained in ≃30 min

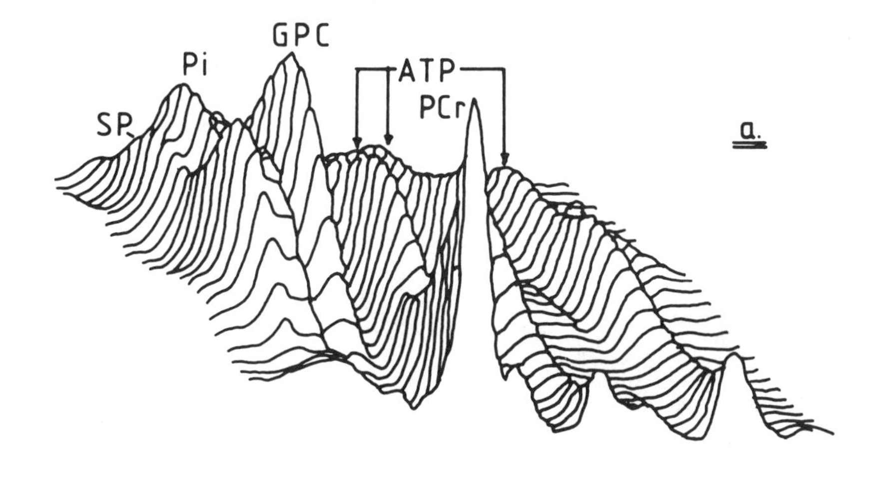

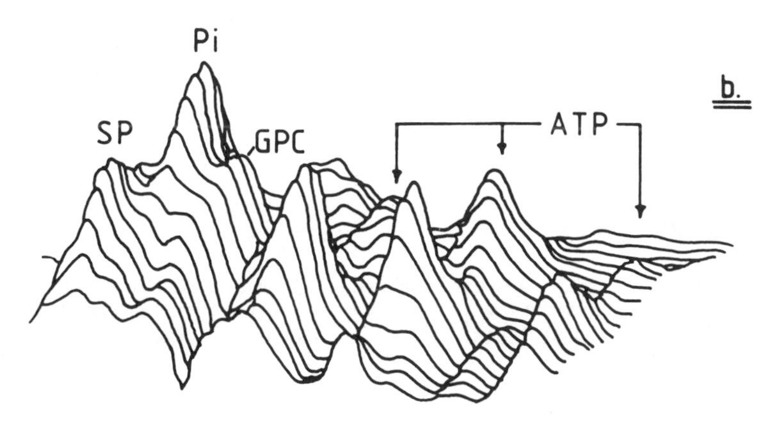

FIGURE 3. Spectra obtained from the *in vivo* rat kidney by the method of amplitude-modulated rotating frame spectroscopic imaging. (a) Spectra from the working kidney prior to hypotensive hemorrhage. (b) Spectra from the same kidney as (a), but following a period of hypotensive hemorrhage. In spectra (a), the complete data matrix is shown, and there is signal from muscle in the abdominal wall as demonstrated by the presence of phosphocreatine. In (b), only that part of the data matrix representing the kidney is shown. Peak assignments: Adenosine triphosphate (ATP), inorganic phosphate (P_i), sugar phosphates (SP), phosphocreatine (PCr), and glycerolphosphorylcholine (GPC).

from the rat kidney. The complete data matrix is shown, and it will be seen that there is a signal from phosphocreatine. This signal comes from skeletal muscle overlying the kidney, and there is clear spatial resolution between the muscle and kidney signals. The animal was then subject to hypotensive hemorrhage. Blood (20 ml) was withdrawn from the femoral vein over a period of 20 min and then this blood was re-injected, again over a period of 20 min. The spectra of FIGURE 3(b) were collected after this insult, and inspection of the data indicates that there is less ATP and more P_i in the central region

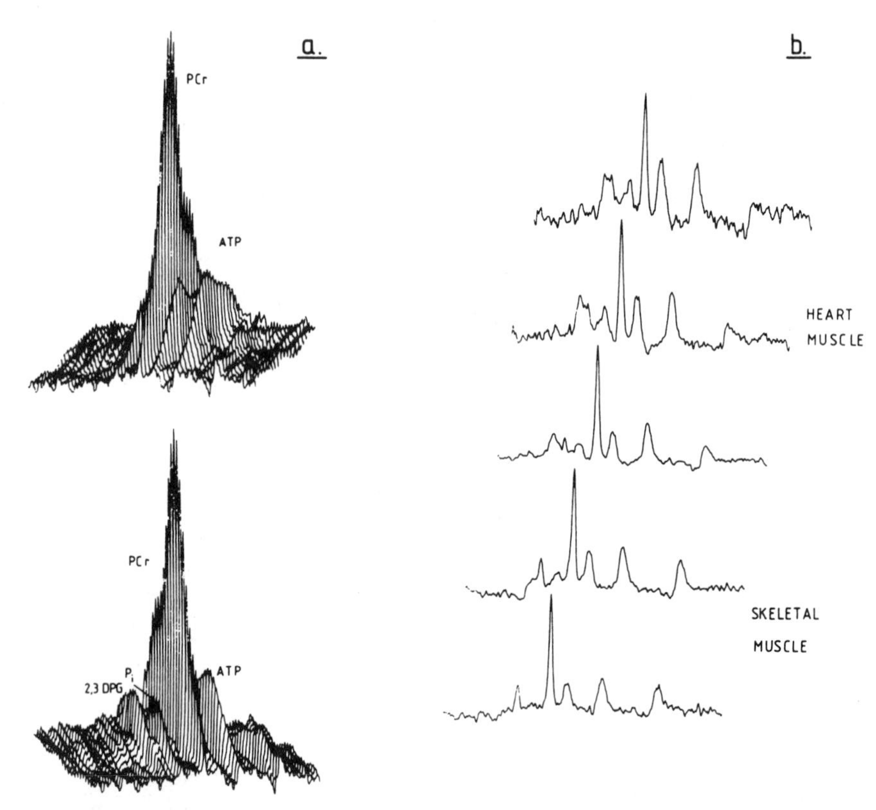

FIGURE 4. Spectra from the chest of a human subject. (a) The data presented as stacked plots. Both plots are of the same data, but shown from different angles. (b) Several individual spectra, representative of increasing depth, from the data set shown in (a). Peak assignments: as FIG. 3 with the addition of 2,3-diphosphoglycerate (2,3 DPG).

of the kidney. This finding is consistent with there being a greater resistance to ischemic damage in the cortex relative to the medulla.

One feature of this protocol is that the receiver coil has an approximately homogeneous sensitivity over the kidney, which simplifies the quantitation of the data in terms of real metabolic concentrations. However, this set of data was obtained using the amplitude-modulated pulse sequence, and suffers from non-optimum sensitivity as a consequence. The spectra in FIGURE 3 were processed using a power spectrum

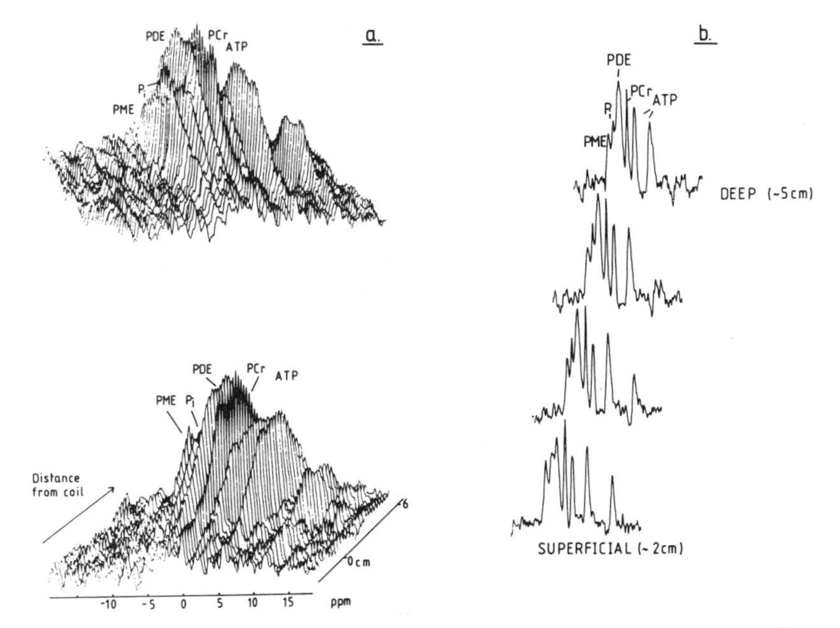

FIGURE 5. Spectra from the head of a human subject. (a) The data presented as stacked plots. Both plots are of the same data, but shown from different angles. (b) Several individual spectra from the data set shown in (a). Peak assignments as FIG. 3 with the addition of phosphomonoesters (PME) and phosphodiesters (PDE).

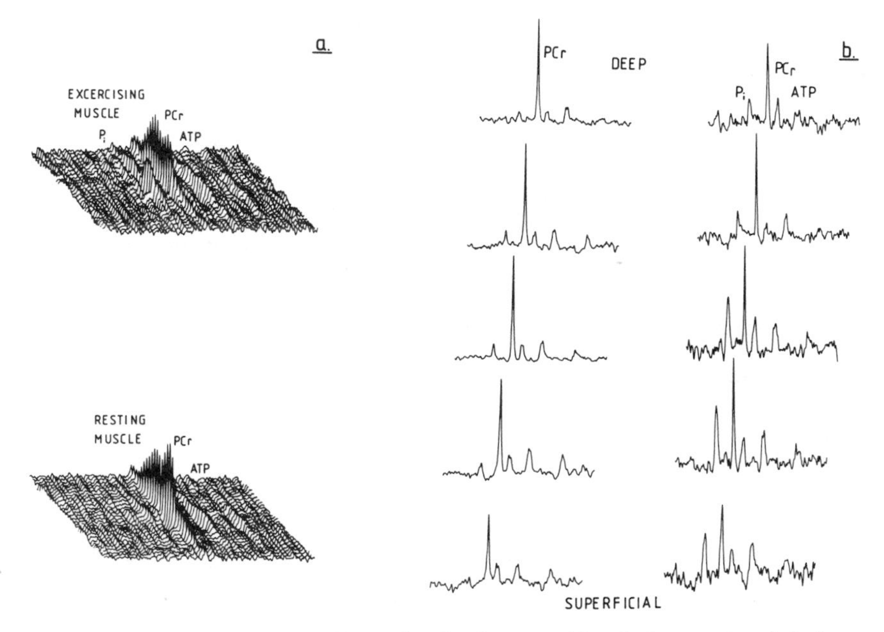

FIGURE 6. Spectra from the calf muscle of a human subject both at rest and during a steady-state exercise regimen. (a) The data presented as stacked plots. (b) Several individual spectra from the data sets shown in (a). The set on the left is from the resting muscle, the spectra on the right from the exercising muscle. Peak assignments as in FIG. 3.

routine. These accentuate the qualitative nature of the result, but distort the relative concentrations in this particular example.

Spectra from a Human Heart

FIGURE 4 is a plot of data collected from a double surface coil probe FIGURE 2(a) placed on the chest of a human volunteer. These spectra were obtained using the phase-encoded rotating-frame protocol. The complete set took ≃30 min to accumulate and FIGURE 4(b) shows selected spectra from the complete data set. It will be seen that each spectrum has adequate signal-to-noise ratio for quantitative analysis. A more detailed discussion of heart NMR will be found elsewhere in this volume,[17] but briefly, the ratio of phosphocreatine to ATP is high (≃4) in the spectra from superficial slices, but falls as the spectra are representative of deeper tissue until the ratio reaches a plateau of ≃1.6. Thus, although the phosphorus spectra of intercostal muscle are broadly similar to those of heart muscle, the rotating-frame imaging method has enabled the two types of tissue to be reliably differentiated.

Spectra from a Human Head

FIGURE 5 shows a set of spectra collected from the double surface coil probe placed over the parietal region of a human head. Individual spectra are shown in FIGURE 5(b) and certain qualitative differences can be seen, in particular in the phosphomonoester and phosphodiester regions of the spectra. There are clear difficulties in the interpretation of such data because of the folded nature of the brain anatomy, but these spectra suggest that there are metabolic differences between grey and white matter. This particular data set was collected in 30 min using the phase-encoded protocol.

Spectra from Human Muscle at Rest and During Exercise

The final example, shown in FIGURE 6, is two sets of spectra obtained from the calf of a human volunteer. In the first, the muscles are at rest, whereas the second data set was collected during an exercise regime involving plantar flexion of the foot against a loaded pedal. The work load was such that the subject experienced no progressive fatigue during the course of the investigation. Examination of the resulting spectra shows that at rest there is no discernible difference between the ratio of metabolites throughout the calf muscles. During exercise, however, there are clear variations with a significant increase in the inorganic phosphate peak in the more superficial tissue, but a much smaller increase in deeper regions. The result suggests that the known fiber type differences between the oxidative soleus muscle and the more glycolytic gastrocnemius are reflected in the metabolic response of the two muscles to exercise, and that spatially resolved NMR can measure these differences in the human.

DISCUSSION

At the present time, several methods for spatial localization have been demonstrated on animal and human subjects, but no single technique has emerged as being optimum for all applications. A critical assessment of the several different approaches is complicated by the fact that machine-specific characteristics can have a dominant effect on the performance of a particular protocol. In addition, the ease of implementation is of vital importance when choosing a method for routine use, and this can be hard to ascertain in the absence of direct experience. Nevertheless, a few observations may help to give some perspective to this rapidly developing field.

In situations where accurate three-dimensional discrimination is required, methods that use selective pulses in conjunction with switched B_0 gradients would seem to be the best choice. An important feature of such methods is that the spectroscopic data can be correlated with a proton image. However, this approach is not without problems, particularly when one or more of the required dimensions is relatively small. In these instances, the strength of field gradient that is needed to suppress the range of chemical shift can be excessive. In addition, data are only collected from one region, which is wasteful, particularly when the study of heterogeneity is itself interesting.

Many of the proposed methods suffer because of instrumental limitations, in particular the time that is needed to apply switched field gradients. Technological improvements will undoubtedly reduce these problems, but at present those methods, which rely on the coherence of transverse magnetization during gradient sequences, have proved to be rather difficult to implement, particularly with short T_2 nuclei such as the phosphorus in ATP.

Given this background, there are both merits and disadvantages in spectroscopic rotating-frame techniques. The encoding process is not compatible with proton imaging as a means of defining the localized region, and the method has only been successfully applied in one spatial dimension. Also, the design of a suitable probe is not a trivial problem for the non-engineer. On the other hand, the method does not require any switched field gradients and is therefore independent of the characteristics of a particular instrument. The applicability of any technique is determined to some extent by the shape of the tissue of interest, and the methods that have been described are especially suitable for the study of planar anatomy, such as the heart wall. The efficiency in terms of sensitivity is excellent, and the phase-encoded version sacrifices only some 15% of signal-to-noise ratio relative to an optimum data collection using a pulse length that satisfies the criterion of the Ernst angle. This is very significant for patient studies where the available time is limited by the tolerance of the subject. The human spectra that have been shown were collected in about 30 min, but the data collection time could have been reduced dramatically by adding several spectra together to give improved signal-to-noise at the expense of spatial resolution. In situations where metabolic heterogeneity is present, imaging protocols have obvious advantages over single volume techniques in that small differences can be observed simultaneously, and this has been demonstrated in the spectra that have been presented.

In summary, rotating-frame spectroscopic imaging is not applicable to the study of all anatomies, but is particularly appropriate and efficient in situations where slice selection matches the anatomy of interest and good spatial resolution is needed in that dimension.

ACKNOWLEDGMENTS

We would like to thank all the members of this laboratory who have contributed to the experiments described here, but in particular Dr. Peter Ratcliffe for his help with the rat kidney preparation, Dr. Bheeshma Rajagopalan who collaborated on the heart project, and Dr. Linda Hands who instigated the human muscle experiments.

REFERENCES

1. LAUTERBUR, P. C., D. M. KRAMER, W. V. HOUSE & C-N. CHEN. 1975. Zeugmatographic high resolution magnetic resonance spectroscopy. Images of chemical inhomogeneity within macroscopic objects. J. Am Chem. Soc. **97:** 6866–6868.
2. HOULT, D. I. 1979. Rotating frame zeugmatography. J. Magn. Reson. **33:** 183–197.
3. COX, S. J. & P. STYLES. 1980. Towards biochemical imaging. J. Magn. Reson. **40:** 209–212.
4. HAASE, A., C. MALLOY & G. K. RADDA. 1983. Spatial localization of high resolution ^{31}P spectra with a surface coil. J. Magn. Reson. **55:** 164–169.
5. STYLES, P., C. A. SCOTT & G. K. RADDA. 1985. A method for localising high-resolution NMR spectra from human subjects. Magn. Reson. Med. **2:** 402–409.
6. GARWOOD, M., T. SCHLEICH, G. MATSON & G. ACOSTA. 1984. Spatial localization of tissue metabolites by phosphorus-31 NMR rotating frame zeugmatography. J. Magn. Reson. **60:** 268–279.
7. BOGUSKY, R. T., M. GARWOOD, G. B. MATSON, G. ACOSTA, L. D. COWGILL & T. SCHLEICH. 1986. Localisation of phosphorus metabolites and sodium ions in the rat kidney. Magn. Reson. Med. **3:** 251–261.
8. BLACKLEDGE, M. J., D. J. HAYES, R. A. J. CHALLISS & G. K. RADDA. 1986. One dimensional rotating frame imaging of phosphorus metabolites *in vivo*. J. Magn. Reson. **69:** 386–390.
9. BLACKLEDGE, M. J., B. RAJAGOPALAN, R. D. OBERHAENSLI, N. M. BOLAS, P. STYLES & G. K. RADDA. 1987. Quantitative studies of human cardiac metabolism by ^{31}P rotating frame NMR. Proc. Natl. Acad. Sci. USA. **84:** 4283–4287.
10. PEKAR, J., J. S. LEIGH & B. CHANCE. 1985. Harmonically analyzed sensitive profile. A novel approach to depth pulses for surface coils. J. Magn. Reson. **64:** 115–119.
11. METZ, K. R. & R. W. BRIGGS. 1985. Spatial localization of NMR spectra using Fourier series analysis. J. Magn. Reson. **64:** 172–176.
12. GARWOOD, M., T. SCHLEICH, B. D. ROSS, G. B. MATSON & W. D. WINTERS. 1985. A modified rotating frame experiment based on a Fourier series window function. Application to *in vivo* spatially localized NMR spectroscopy. J. Magn. Reson. **65:** 239–251.
13. GARWOOD, M., T. SCHLEICH, M. R. BENDALL & D. T. PEGG. 1985. Improved Fourier series windows for localization in *in vivo* NMR spectroscopy. J. Magn. Reson. **65:** 510–515.
14. BLACKLEDGE, M. J., P. STYLES & G. K. RADDA. 1987. Rotating frame depth selection and its application to the study of human organs. J. Magn. Reson. **71:** 246–258.
15. BOEHMER, J. P., K. R. METZ & R. W. BRIGGS. 1985. One dimensional spatial localization of spin-lattice relaxation times using rotating-frame imaging. J. Magn. Reson. **62:** 322–327.
16. BLACKLEDGE, M. J., R. D. OBERHAENSLI, P. STYLES & G. K. RADDA. 1987. Measurement of in vivo ^{31}P relaxation rates and spectral editing in human organs using rotating frame depth selection. J. Magn. Reson. **71:** 331–336.
17. RAJAGOPALAN, B., M. J. BLACKLEDGE, W. J. MCKENNA, N. M. BOLAS & G. K. RADDA. 1987. Measurement of phosphocreatine to ATP ratio in normal and diseased heart by ^{31}P magnetic resonance spectroscopy using rotating frame depth selection techniques. Ann. N.Y. Acad. Sci. This volume.

DISCUSSION OF THE PAPER

P. A. BOTTOMLEY: Would you like to make a comment about the biological variability and detectability of P_i in normal brain?

P. STYLES: What I would say is that no matter what technique we use, we see variations in the level of P_i in brain. George Radda looks like he's got something to say.

G. K. RADDA: If you look at the standard variation and the standard error of the various ratios in the various components, there is no doubt that you get the biggest variation in P_i and there are two reasons for that. It's a small peak to start with and you know it varies a bit, but the other is that this is quite close to an extremely broad and very large phosphodiester signal and sometimes it's not terribly well resolved from that one. You also have various data manipulations that all of us use in one form or another, deconvolution or other techniques, where you in fact might artificially lose some of the P_i.

Non-Invasive Localized NMR Spectroscopy *in Vivo:*

Volume Selective Excitation

WALTER P. AUE

NMR Unit of the Medical Faculty
University of Berne
CH-3012 Berne, Switzerland

INTRODUCTION

In vivo nuclear magnetic resonance (NMR) spectroscopy has gained an incredible popularity among many scientists all over the world. The ability to observe metabolic processes inside a living system with a noninvasive tool is indeed fascinating and opens up innumerable possibilities for biochemical and medical investigations. For larger living systems like animals and humans, undifferentiated chemical information from the whole object will in general be of little interest. Rather, the measurement should be able to focus on a region of interest, like a certain organ, without contributions from surrounding tissue.

In order to allow the observation of a selected volume, animals have been surgically prepared and the exposed organs measured by conventional NMR techniques. This procedure, however, stands in strong contrast to the noninvasive character of NMR and can not be applied to humans. Therefore, many different spectroscopic techniques have been proposed to acquire information from a well delineated volume inside a living system without using surgery. In a recent comprehensive review,[1] the majority of the existing localization schemes has been classified and evaluated. This contribution will focus on one of these methods, namely volume selective excitation (VSE)[2] with a short explanation of its principle, an illustration of its performance with a localized NMR spectrum of a rat liver *in vivo* and a discussion of the main features of the technique with respect to the most relevant points of interest for practical applications.

PRINCIPLE OF VOLUME SELECTIVE EXCITATION

The VSE sequence of radiofrequency (rf) pulses and pulsed gradients of the static magnetic field is shown in FIGURE 1. Actually, its elements are borrowed from the field of NMR imaging, where they are used to excite nuclei in a planar slice of an object, exclusively.[3] In simple terms, the effect of the VSE sequence applied to an object containing NMR active nuclei is as follows.

At first, a linear gradient Gx is superimposed onto the otherwise perfectly homogeneous static magnetic field, in which the sample is placed. By this means, a specific strength of the static magnetic field is assigned to every cross-section of the sample, which is orthogonal to the direction of the gradient, or in other words, is parallel to the y- and z-axes. Then, a frequency selective 45° rf pulse is applied to the whole sample, which, however, affects the nuclei in the one slice of the object, only, in which the strength of the static magnetic field fulfills the nuclear magnetic resonance condition. At the end of the 45° rf pulse, the nuclei in the selected slice of resonance are deflected by +45° from their equilibrium orientation along the *z*-axis, whereas all the other nuclei have not been affected.

The next step is a very short and intense −90° rf pulse. This pulse makes all the nuclei precess by −90° around the axis of the rf field, because its spectrum is wide enough to fulfill the resonance condition of the nuclei for the whole sample, irrespective of the applied gradient. Now, the nuclei in the selected slice are deflected −45° from the *z*-axis. All the remaining spins are orthogonal to the *z*-axis.

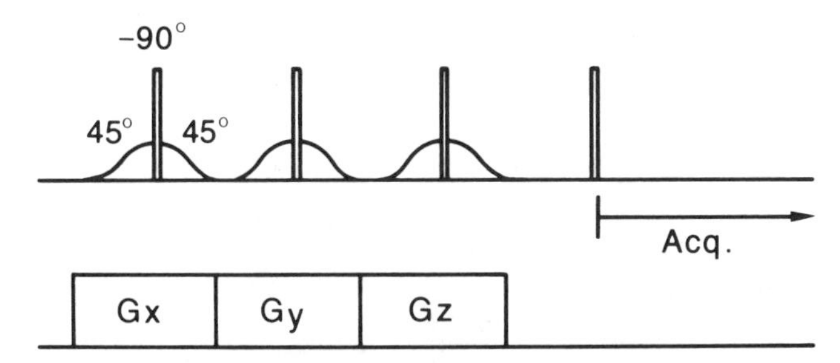

FIGURE 1. Volume-selective excitation sequence of radiofrequency pulses and pulsed gradients of the static magnetic field. It is composed of three selection elements, each of which preserves macroscopic magnetization in a plane orthogonal to the direction of the gradient. The effect of the three elements in sequence is to preserve magnetization in the volume common to the three planes, exclusively, whose signal can be acquired with a fourth strong radiofrequency pulse.

The last step of a VSE element is, again, a frequency selective 45° rf pulse, which takes the nuclei in the selected slice back along the *z*-axis. The macroscopic magnetization of the nuclei in the remainder of the sample, which are not affected by the selective rf pulse, vanishes very quickly in the presence of the gradient due to destructive interference of the magnetic moments of the individual nuclei precessing at different frequencies.

In conclusion, the net effect of a +45°, −90°, +45° rf pulse sequence in the presence of a gradient of the static magnetic field is to preserve macroscopic nuclear magnetization in a selected slice and eliminate magnetization elsewhere. If we are interested in the spectrum of the selected plane, we can apply a conventional broad-band rf pulse and acquire the induced NMR signal with standard techniques. On the other hand, we can choose from the previously selected plane a column of nuclei in *z* direction by applying a second VSE element with a gradient Gy. The selection of a cube of nuclei, finally, is achieved by applying a total of three VSE elements with mutually orthogonal gradients, as it is shown in FIGURE 1. Again, NMR data from the volume of interest can be acquired using standard techniques.

For didactical reasons, it has been assumed, that the rf field applied to the sample is very homogeneous and very intense. In practice, however, these assumptions might not be fulfilled. Therefore, in cases with inhomogeneous rf fields, like for example measurements with surface-coils, and/or reduced rf power, two slightly different VSE sequences have to be run in a compensated experiment in order to obtain a clean selection of the sensitive volume.[4,5]

RESULTS

In order to illustrate the performance of VSE on a living system, we ran a combined NMR imaging and VSE experiment on an anesthetized rat employing a 12 inch horizontal bore spectrometer at 1.9 Tesla (Bruker Analytische Messtechnik GmbH, D-7512 Rheinstetten 4, FRG). In the upper half of FIGURE 2, the proton NMR image shows a cross-section through the rat's abdomen at the level of the liver. The main anatomical features can easily be seen and are marked, accordingly. The lower part of FIGURE 2 displays two phosphorus NMR spectra, obtained on the same rat using the same rf antenna: The spectrum to the right was acquired from the whole rat abdomen

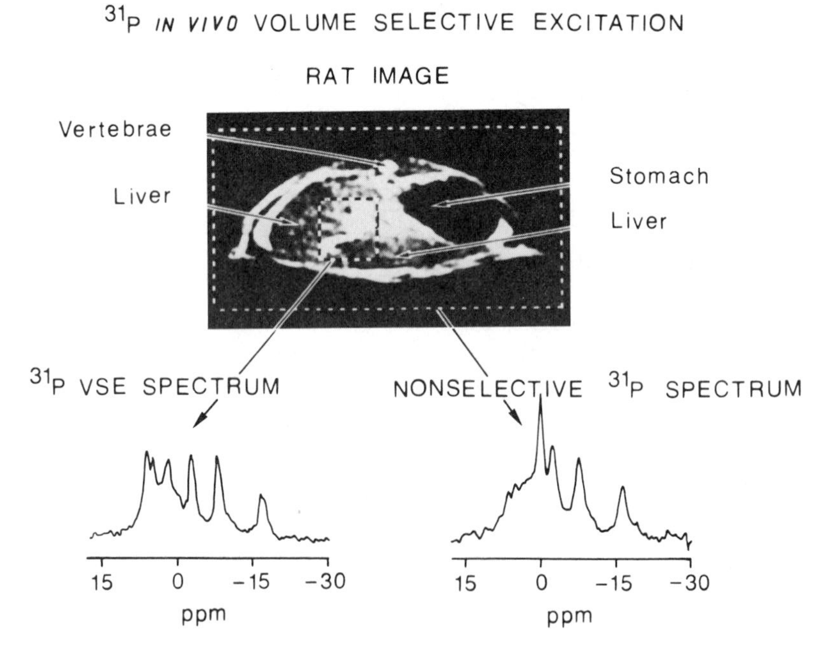

FIGURE 2. *In vivo* NMR experiments performed on an anesthetized rat at a magnetic field of 1.9 Tesla. (Top) Proton image of a cross-section through the rat's abdomen at the level of the liver. Light areas indicate high proton density. (Bottom right) Phosphorus NMR spectrum of the whole abdomen with the dominant peak of phosphocreatine at 0.0 ppm. (Bottom left) Phosphorus spectrum obtained from the rat's liver (inner dotted frame) by volume-selective excitation. The absence of the phosphocreatine peak emphasizes the clean selection for liver tissue.

without any localization and shows the three signals from adenosine triphosphate (ATP) at −2.4, −7.7, and −16.2 ppm, the predominant peak from phosphocreatine (PCr) at 0.0 ppm, inorganic phosphate at 5.1 ppm, and phosphomonoester at 6.4 ppm.

The spectrum to the left was acquired from the small framed region of the rat's liver, exclusively, by application of the VSE technique. Again, we recognize the signals from ATP, inorganic phosphate, and phosphomonoester. The additional signal at 2.1 ppm from phosphodiester and the absence of the PCr signal emphasize the clean volume selection for the rat's liver, which does not contain any PCr.

DISCUSSION

In this section, the performance of VSE will be discussed with respect to the most relevant points of interest for practical applications. In essence, they are related to the main aim of the investigation, the localization, limitations imposed by the sample, and limitations imposed by the equipment.

A short acquisition is of prime importance in three cases: First, if the dynamics of physiological processes are to be investigated, the data acquisition must be faster than the process under investigation. Second, motion in the sample could render measurements useless, if the data acquisition is long compared to the time scale of the motion. Third, quick measurements are necessary, if the result of the acquisition is needed for immediate interaction. This is the case, for example, if the homogeneity of the magnetic field is to be optimized on the strong proton signal from within the region of interest. In practice, two acquisitions per measurement are a maximum, since, otherwise the optimization process will last too long. VSE can achieve accurate localization in a single acquisition with favorable equipment. With inhomogeneous rf fields and little rf power, a compensated experiment with two acquisitions has to be performed.

The sensitivity per unit time per volume will always be of main interest, since NMR investigations in principle are rather insensitive. This is a direct consequence of the low energy of the applied radiation, which, on the other hand, makes *in vivo* NMR noninvasive. In general, maximum sensitivity can be achieved, if all the possible loss mechanisms are avoided. The only loss mechanism inherent in the compensated VSE sequence is a vanishing effect on samples with very fast longitudinal relaxation.[4] Otherwise, maximal sensitivity per unit time per volume is retained.

Spectral resolution is crucial whenever many inherently narrow spectral lines with small differences in the chemical shift need to be distinguished reliably. A good example is the proton spectrum on a low field machine, where many lines crowd within a narrow range of frequencies. In practice, spectral lines are broadened by magnetic field inhomogeneity for two reasons. First, it is technically demanding to generate a strong magnetic field with an inhomogeneity of approximately 10^{-7} over a large volume. Second, a homogeneous magnetic field is distorted upon insertion of an object by its bulk magnetic susceptibility, which can hardly be compensated by correction fields, and therefore spectral lines are broadened considerably. This broadening can be kept at a minimum, if small regions of interest are chosen, within which magnetic field homogeneity can be optimized more easily. With its short acquisition time, VSE is well suited to optimize homogeneity on the actual selected volume.

Besides static magnetic field inhomogeneity, there is also dynamic inhomogeneity caused by eddy currents induced by the switching of the gradients. This phenomenon affects resolution whenever magnetic field gradients are switched rapidly shortly before the data acquisition. In the VSE sequence, the macroscopic magnetization in the region of interest is oriented parallel to the magnetic field whenever the gradients are switched and is therefore not affected by the eddy currents. The observation pulse is only applied after the switching transients have vanished, and therefore, the spectral resolution is unaffected.

The accuracy of the position of the sensitive volume with VSE is determined by the equipment. For a magnet of 2 Tesla with strong gradients of 20 millitesla/m, volumes as small as 0.5 cm^3 can be selected with an accuracy of 1 mm for protons and 2 mm for phosphorus. Since the accuracy is proportional to the strength of the gradients, a clean localization can not be achieved on equipment with weak gradients.

VSE is very flexible with respect to the position and the extension of the sensitive volume. By proper choice of the rf, the sensitive volume can be positioned at any position within the sample and the rf antenna, and with the application of one, two, or three selection steps, the sensitive volume has the shape of a slice, a column, or a cube, respectively.

During the VSE sequence, macroscopic magnetization outside the sensitive volume is destroyed. Therefore, the technique does not permit time shared multivolume sampling.

In some compounds like for example phosphorus in ATP in certain tissues, the nuclei have a very short transverse relaxation time T_2. Since transverse relaxation can only affect transverse magnetization, a short T_2 does not affect the sensitivity and the quantitative information of a VSE experiment, because the magnetization is oriented parallel to the magnetic field prior to the application of the observation rf pulse.

In conclusion, VSE is one of the few NMR methods capable of producing high resolution spectra of a well delineated volume of interest at any position inside a living system. In addition, it features short minimum acquisition time, optimum sensitivity per volume and time, and quantitative spectra, which are not affected by fast relaxation processes in the sample.

REFERENCES

1. AUE, W. P. 1986. Localization methods for in vivo nuclear magnetic resonance spectroscopy. Rev. Magn. Reson. Med. **1:** 21–72.
2. AUE, W. P., S. MÜLLER, T. A. CROSS & J. SEELIG. 1984. Volume-selective excitation. A novel approach to topical NMR. J. Magn. Reson. **56:** 350–354.
3. POST, H., P. BRUNNER & D. RATZEL. Personal communication.
4. MÜLLER, S., W. P. AUE & J. SEELIG. 1985. Practical aspects of volume-selective excitation (VSE). Compensation sequences. J. Magn. Reson. **65:** 332–338.
5. MÜLLER, S., W. P. AUE & J. SEELIG. 1985. NMR imaging and volume-selective spectroscopy with a single surface coil. J. Magn. Reson. **63:** 530–543.

DISCUSSION OF THE PAPER

QUESTION: In biological samples, are the T_1 sufficiently long so that you don't start picking up regions outside of your region of interest?

W. P. AUE: If you ever start picking up regions of non-interest you use a compensated experiment. The one with the two shots takes care of it. This is cancelled because it changes the phase of the *last hard pulse* which inverts the signals of non-interest. We went down to T_1s as short as 70 milliseconds, thing works fine.

J. R. ALGER: In my experience with rats, to obtain a spectrum of liver one really had the large problem with the breathing motion, if that were true I would expect your image to be somewhat more blurred than it was and that you would have a more difficult time obtaining the very beautiful spectrum that you did have. Can you comment on how that animal was handled in such a way to eliminate that?

AUE: Well, first of all the image has been terrible so you might not have seen the artifacts there. Second, we chose a cube small enough so even though you had breathing motion it would not slip out of the liver tissue. Nothing special was done to the animal.

QUESTION: In your phantom of spheres have you tried filling that phantom box with a physiological electrolyte solution? I'm supposing that air was in interstitial spaces between spheres.

AUE: I could try filling with a physiological solution, sure. I don't think it makes a difference with respect to the selection schemes.

QUESTION: I thought you might get a little bit more lossy imaging that way.

AUE: Okay that lengthens the 90 degree pulse some, I don't think this gives us basic trouble.

K. UGURBIL: I don't understand your claim that there are no T_2 losses. Even though there's quite a bit of time that you're spending with your gradient on but your magnetism in the z direction during your soft pulses, there is a time when the spins are going through the transverse plane and they will be decaying due to T_2 even during the pulse.

AUE: If you compare the length of these pulses to the 20 millisecond T_2 which you have for ATP in tissue if I'm right (I have learned these figures from someone else) I think it's absolutely negligible. There is some portion, I agree, where the thing goes through some deviation from z but there again I assume the soft pulse is strong compared to T_2.

Spatial Localization by Rotating Frame Techniques

D. I. HOULT, C-N. CHEN, AND L. K. HEDGES

Biomedical Engineering and Instrumentation Branch
Division of Research Services
National Institutes of Health
Bethesda, Maryland 20892

INTRODUCTION

The rotating frame method of imaging[1] has recently received considerable attention[2-5] as a technique for spatial localization of spectra. While in true imaging experiments, the implementation relies on the production of various irradiating field (B_1) gradients with carefully crafted radio-frequency coils, most spectroscopy experiments have utilized the gradients inherent in the fields produced by surface coils. However, whatever the experimental constraints may be, it is worth looking in some detail at the various facets of the fundamental experiment, so that the desired data are not compromised, either in terms of signal-to-noise ratio or by the presence of undesirable artifacts.

The simple experiment performed by Hoult[1] "to demonstrate feasibility" has a number of serious drawbacks, which are described in his article. However, they are corrected in the more sophisticated phase-encoding scheme he proposes, which is the one the authors strongly recommend. In the basic experiment, the B_1 transmitting field has a strength that depends upon spatial position, but which is also very much greater than ΔB_0—any offset field caused by chemical shift or a gradient in the static field. Thus if B_1 is applied along the $-\tilde{y}$ axis in the rotating frame the magnetization M precesses about B_1 from its starting position (presumably the $\tilde{z}$ axis) down toward the $\tilde{x}$ axis and continues on its way, always lying in the $\tilde{x}\tilde{z}$ plane as shown in FIGURE 1. However, the angle θ through which it flips depends not only upon the pulse length, say t_1, but also upon the strength of the B_1 field, which in turn is a function of spatial position. Now the signal strength ($\propto M_{\tilde{x}}$) following the pulse is proportional to the sine of the flip angle, and so if we collect the free induction decay (f.i.d.) following a pulse of length t_1, and Fourier transform in the usual way, the amplitude of the absorption spectrum will vary as sin ($\gamma B_1 t_1$) where γ is the gyromagnetic ratio of the nuclear species. If we now repeat the experiment, but each time incrementally increase the length of the pulse, the stacked plot of FIGURE 2 shows clearly that the amplitudes of the absorption spectra (we, of course, discard the dispersion spectra) vary sinusoidally with t_1, but that the oscillatory variation in amplitude is more rapid for one chemical shift species than for another. Thus the B_1 field at species A must be greater than that at species B, and so A and B must be at different locations. To quantify the oscillatory behavior, knowing that the variation is as sin ($\gamma B_1 t_1$), we now take the **sine** Fourier transform of each column of absorption mode data (the cosine transform gives a dispersion spectrum and must not be used) and the resulting two-dimensional plot gives a graph of concentration versus chemical shift in one dimension, and B_1 field strength (i.e. spatial distance) in the other. Note that one must not simply take an "off-the-shelf" two-dimensional Fourier transform (2DFT) program; in general, unwanted dispersive components will appear in the real part of the final data.

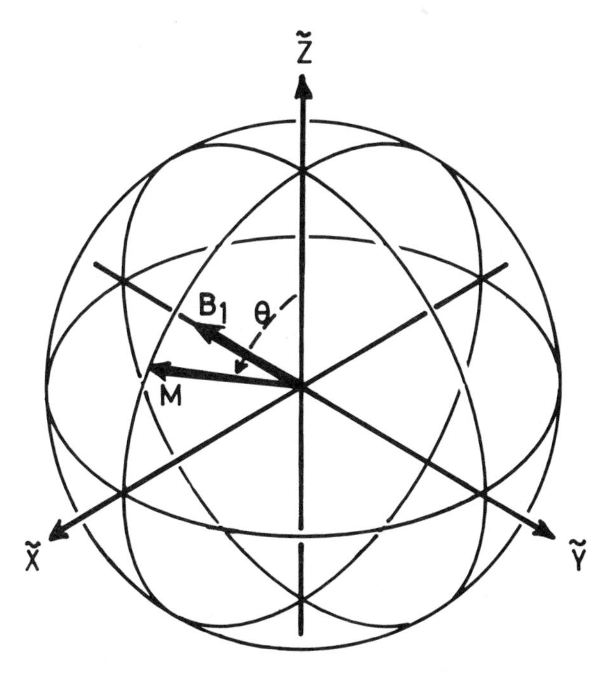

FIGURE 1. The application of a strong B_1 field along the $-\tilde{y}$ axis in the rotating frame causes magnetization to nutate in the $\tilde{x}\tilde{z}$ plane. However, the instantaneous position of the magnetization depends not only upon time, but also upon the strength of B_1, which is a function of spatial position. The NMR receiver detects $M_{\tilde{x}}$, and $M_{\tilde{y}}$, which is zero during the pulse.

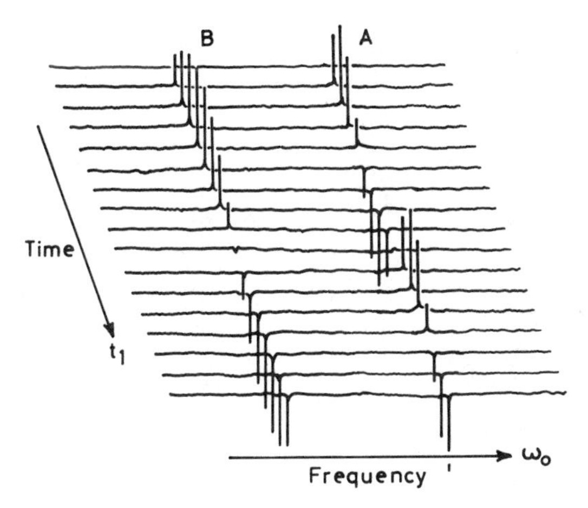

FIGURE 2. A stacked plot showing the evolution of the absorption spectra from two small samples (A and B) as the pulse length t_1 is varied incrementally from experiment to experiment. (A) clearly oscillates in amplitude more rapidly than (B), and we may therefore conclude that it experiences a stronger B_1 field, and is at a location different from that of (B).

The obvious drawback to this simple method is that it only uses the $\tilde{x}$ component of magnetization, $M_{\tilde{x}}$, and ignores the magnetization in the $\tilde{z}$ direction. Thus, there is an immediate loss of $\sqrt{2}$ in signal-to-noise ratio. However, after a pulse, magnetization can be anywhere in the $\tilde{x}\tilde{z}$ plane, including in the $-\tilde{z}$ direction; therefore one must also wait approximately five longitudinal relaxation times ($5T_1$) between pulses if the spin system is not to retain an imprint of its previous travels. More rapid repetition will produce artifacts in the data.[3] It follows that this very basic experiment loses considerable sensitivity. In an attempt to recoup some of the loss, Garwood *et al.*[3] have talked of putting the magnetization into the $\tilde{x}\tilde{y}$ plane after data accumulation so that T_1 recovery at all spatial locations is essentially that following a 90° pulse. This technique certainly helps the recovery problem, but still does not allow one to sample the $\tilde{z}$ magnetization. To do that, one must abandon the simple experiment and proceed to the full protocol proposed by Hoult.[1]

PHASE-ENCODED ROTATING FRAME IMAGING

The experiment is shown in FIGURE 3, and the 90° pulse following the gradient pulse, having a 90° phase shift, essentially flips the "pancake" of magnetization from the $\tilde{x}\tilde{z}$ plane to the $\tilde{x}\tilde{y}$ plane. Thus, if the B_1 field of the 90° pulse lies along the $-\tilde{x}$ axis, magnetization which prior to the pulse had a flip angle of θ, will have a phase of $-(90-\theta)^0$. On the other hand, reversal of the phase of the 90° pulse in a second experiment gives a signal phase of $+(90 - \theta)°$, as shown. The 90° pulses convert amplitude modulation of the signal into phase modulation, and the use of two opposing 90° pulse phases, in pairs of experiments with the same encoding period t_1, creates the

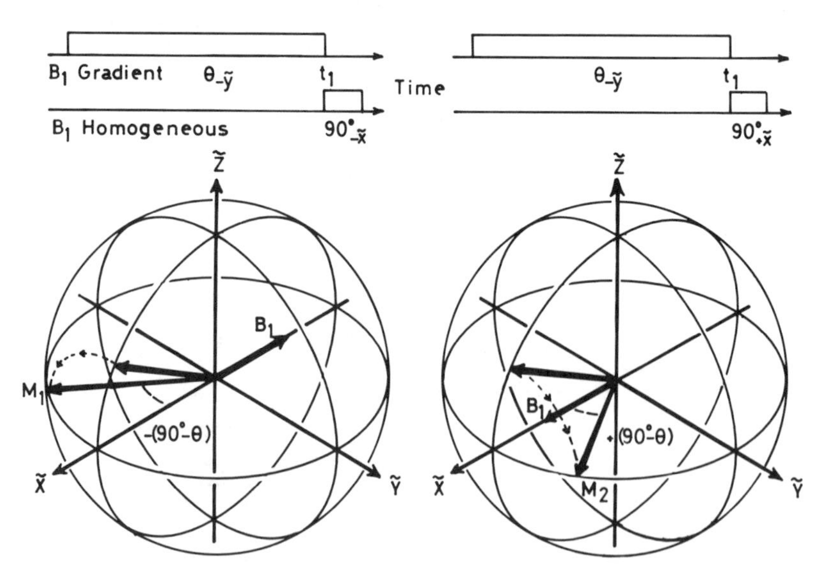

FIGURE 3. The phase-encoded rotating-frame experiment. The 90° pulse, following the period of B_1 gradient application, places the magnetization in the $\tilde{x}\tilde{y}$ plane. However, repetition of the experiment, with the phase of the 90° pulse inverted, generates data that are the complex conjugate of those previously obtained.

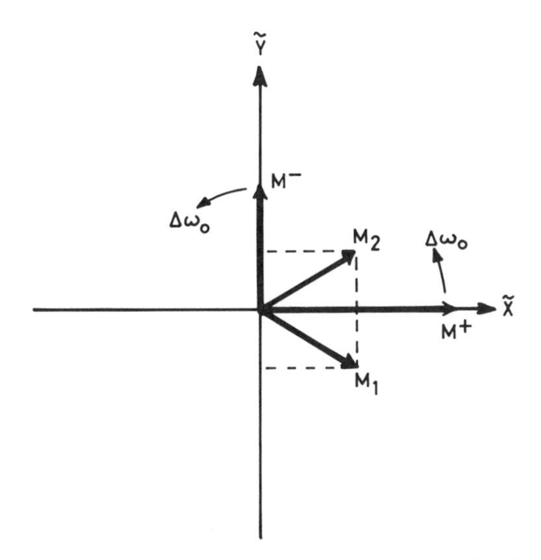

FIGURE 4. The sum and difference of the data generated in FIGURE 3 lie along the $\tilde{x}$ and $\tilde{y}$ axes, respectively. The phases of the two sets of data are therefore precisely defined, enabling absorption mode spectra to be obtained from each.

conditions necessary for the prevention of dispersive components in the final data. There are several ways of understanding this fact. An adroit explanation is that as $\theta = \gamma B_1 t_1$, and as there are two signals with phases that include respectively $\pm\theta$, t_1 can be considered to have symmetrically positive and negative values. Thus the second Fourier transformation is of symmetrical data, which avoids the generation of dispersive components. A more prosaic approach is to consider the sum and difference of the two free induction decays (f.i.d.'s) generated, remembering that the only discrepancy between the two lies in their starting phases—i.e. where in the $\tilde{x}\tilde{y}$ plane they commence their data-generating existence. In general, the phase of one of the f.i.d.'s is arbitrary—it is say $(90-\theta)°$, whatever θ may be—so in general, the Fourier transform of the f.i.d. contains dispersive components. However, the sum of the two f.i.d.'s always commences on the $\tilde{x}$ axis, while the difference always commences on the $\tilde{y}$ axis, as shown in FIGURE 4. As the phases of the sum and difference are precisely defined, it is easy to ensure that we always have an absorption signal: we select the real part only of the FT of the sum, and the imaginary part only of the FT of the difference (we assume the instrument is correctly phased), and the two ensuing spectra form one of the rows of a complex absorptive matrix, whose evolution with increasing t_1 is as shown in FIGURE 5. Comparison of FIGURES 2 and 5 shows that we have now a second data set where the evolution is as $\cos \gamma B_1 t_1$. In effect, we have quadrature data in the t_1 dimension with their concomitant $\sqrt{2}$ signal-to-noise ratio improvement and ability to distinguish between positive and negative nutational frequencies $\pm\Delta\omega_1$. ($\omega_1 = -\gamma B_1$.) We now perform a complex transform, in the t_1 direction, of our complex data and the real part of the transform is the absorptive two-dimensional spectrum or image. As usual, we throw away the imaginary component, which is dispersive. Zero-filling equal to the number of data points is employed when transforming in either direction, to retain resolution and information in the absorption mode.

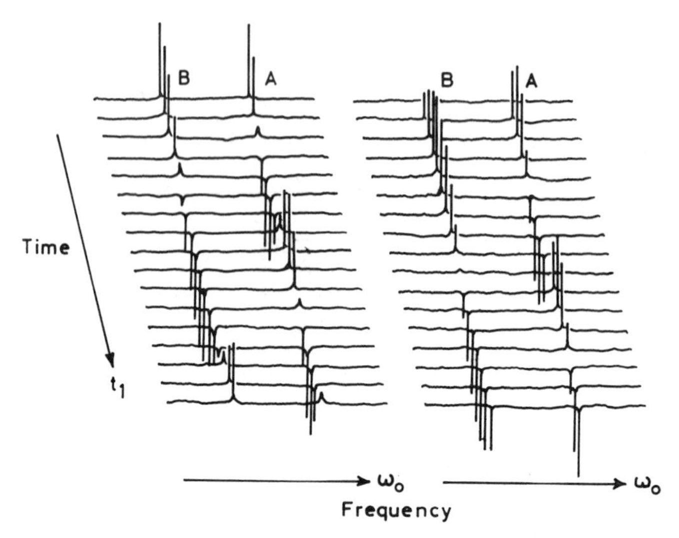

FIGURE 5. The evolution of the two sets of data of FIGURE 4, following Fourier transformation, with increasing pulse length t_1. Both co-sinusoidal and sinusoidal variation of amplitude are now present, allowing quadrature Fourier transformation in the t_1 direction.

PRACTICALITIES

The experiment works very well, but demands that the 90° pulses be reasonably good throughout the volume of interest, and therefore an additional transmitter coil having a uniform field may be needed. This is perfectly feasible; indeed, our equipment has three r.f. coils that are electronically decoupled to render their interaction negligible. Several schemes have been proposed[6-8] to cope with the problems of inductive coupling between coils in a multi-coil probe, but the effective methods all rely essentially on the same principle: induced current in a coil is prevented from flowing by placing a large impedance in series with that coil. The inductive coupling is thereby largely eliminated. An implementation of this principle for a transmitter coil is shown in FIGURE 6, and the blocking can be switched in and out at will.

Other practical problems that may have to be dealt with are $B_1 \not\gg \Delta B_0$ over the entire sample, limited computer memory, excessive r.f. power deposition, and r.f. power instability. The latter annoyance is purely a question of radio-frequency power amplifier and probe design, and if the system is designed from the outset for high-power continuous wave (c.w.) operation, few difficulties should be encountered. For example, with our equipment, which includes an Amplifier Research model 100A15 100W power amplifier, we can detect neither droop nor phase shift in a 64 msec 100 W pulse, when feeding the signal from a search coil into the NMR receiver to monitor the B_1 field's amplitude and phase.

When B_1 is not much greater than ΔB_0, the magnetization moves in a cone about the effective field, $B_{1\text{eff}}$, as shown in FIGURE 7, and so, resolving the motion of the magnetization onto $B_{1\text{eff}}$ and the plane perpendicular thereto, it is clear that there is a static component aligned with $B_{1\text{eff}}$ that does not change with increasing t_1. The final 90° pulse will tip this component close to the z axis, but its presence will still be highly visible in the final data as a spurious signal about the zero frequency line in the ω_1

dimension. If the B_1 field is so small that $B_1 < \Delta B_0$ pertains for certain spectral lines or spatial positions, then the experiment no longer functions properly. It is also worth noting that when B_1 is not much greater than ΔB_0, $B_{1eff} > B_1$, and so the frequency in the ω_1 direction does not bear a linear relationship to B_1. Thus in an image, for example, there is quadratic distortion, for which the computer can quickly correct as it is predictable.

One way of reducing the impact of the zero frequency artifact, while conserving computer memory, is to employ data aliasing in the t_1 phase-encoding dimension, again a trick mentioned in the original article.[1] For example, suppose B_1 varies in strength across the sample by a ratio of two to one. The Nyquist criterion for the t_1 dimension is that the phase spread across the sample should not increase by more than 360° with each increment of pulse length t_1. Thus the first pulse would flip magnetization 360° (=0°) at one side of the sample and 720° (=0°) at the other. However, in the middle of the sample, the flip would be 540° (=180°). Thus, as t_1 is varied, the signal from the middle of the sample oscillates as ±1, i.e. it is at the Nyquist frequency, while signals from the two extremities appear not to evolve, they stay constant in phase at effectively 0° with increasing t_1. Following Fourier transformation, data from the extreme edges of the sample are therefore in the middle of the field of view, along with the zero-frequency artifact, while the signals from the center of the sample are removed to the edges of the field of view. This undesirable state of affairs is easily rectified with the computer by "rolling over" the data so that, for example, the top row of the data matrix moves down to the center, while the bottom row "rolls round" the back of the matrix, comes back in at the top and then also moves down, etc. (This is exactly analogous to the aliasing phenomenon that one experiences in a simple one-dimensional quadrature Fourier transform experiment.) The field of view is now representative of the sample, but the zero-frequency artifact is at the edge of the data matrix, where it is less troublesome. To reduce the artifact's size, the magnetization should always be

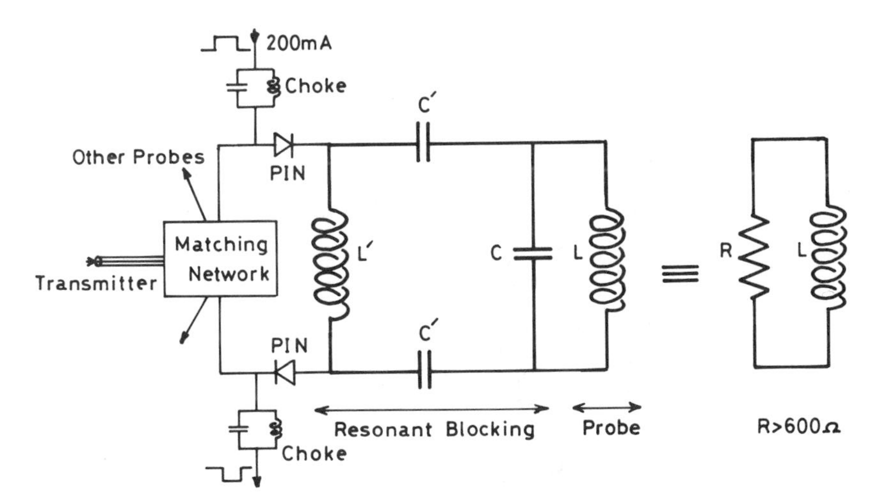

FIGURE 6. An example of the principle of current blocking as applied to a transmitter coil. The PIN diodes are r.f. switches controlled by a d.c. bias current. L′, C′, and C are chosen such that they form a resonant circuit that presents a large current blocking resistance to the probe coil when the PIN diodes are open. In addition, L′, C′, and C are chosen such that a convenient impedance (typically a lossy inductor) is presented to the matching network when the diodes are closed.

perpendicular to B_1 during the phase-encoding gradient pulse, which implies that at the start of the experiment, M should lie along the $\pm\tilde{x}$ axes. This is more easily said than done, and if a preparatory pulse is needed to perform this feat, it will normally need to be a composite pulse[9] with a homogeneous B_1 field. However, we shall see in the next section that there exists another method of producing magnetization in approximately the correct direction.

SLICE OR VOLUME SELECTION

In discussing rotating frame techniques so far, we have considered spatial resolution in only one dimension, that of the B_1 field gradient. At the expense of chemical shift information, a static, non-switched, field gradient may provide information in a second dimension, but in general, three B_1 field gradients in orthogonal directions will be needed with the appropriate electrical isolation and coil design. It is also often desirable to limit the volume from which data are obtained, for example, to get a slice in an image, to obtain a spectrum from a localized region or to restrict signals to a volume where $B_1 \gg \Delta B_0$. Such selection may be accomplished in the presence of a B_1 field gradient[10] (in exact analogy to techniques used when a gradient in B_0 is present) with the aid of a selective pulse, B_2. Just as, to be effective, B_1 must be perpendicular to B_0, so B_2 must be perpendicular to B_1, i.e. in the $\tilde{y}\tilde{z}$ plane if B_1 is

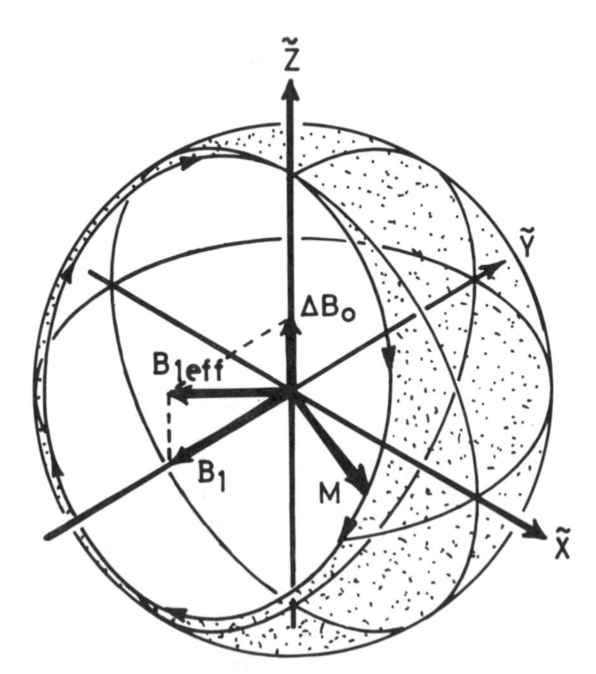

FIGURE 7. The effects of being off-resonance are to lift B_1 out of the $\tilde{x}\tilde{y}$ plane and to cause the magnetization to nutate in a cone about B_{1eff}. If the magnetization at the start of the experiment is in the $\tilde{z}$ direction, there is a component along B_{1eff} that is static and therefore does not evolve with t_1. Hence, its projection on the $\tilde{x}\tilde{y}$ plane is manifest in the data as a zero-frequency artifact. However, if the magnetization at the start of the experiment has been placed along the $\tilde{x}$ direction with the aid of a composite 90° pulse, all the magnetization nutates.

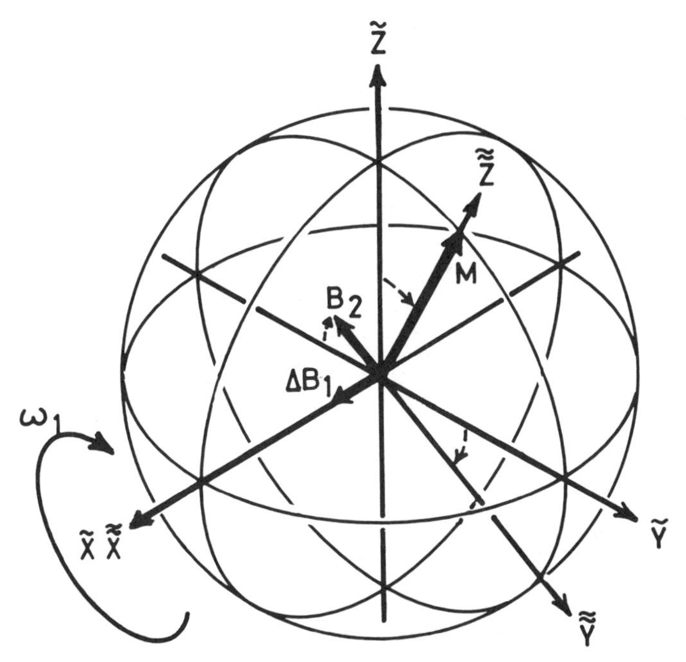

FIGURE 8. A second rotating frame, which spins about B_1 (here shown along the $\tilde{x}$ axis) with angular frequency ω_1. When $\omega_1 = -\gamma B_1$, the B_1 field is effectively reduced to zero. When $\omega_1 \neq -\gamma B_1$, there remains an offset field ΔB_1. An oscillating field in the $\tilde{z}$ and/or $\tilde{y}$ directions can be decomposed into two counter-rotating components, one of which moves with the frame. The latter component's position in the $\tilde{y}\tilde{z}$ plane is determined by the phase of the oscillation, and is therefore controllable. The initial position of the magnetization in the frame is determined by its orientation at the start of the experiment, which, as shown, was along the $\tilde{z}$ axis. With B_2 lying along the $-\tilde{\tilde{y}}$ axis, on-resonance ($\Delta B_1 = 0$) magnetization is flipped onto the $\tilde{\tilde{x}}$ axis. Off-resonance ($\Delta B_1 \gg B_2$) magnetization is not affected.

directed along the $\tilde{x}$ axis. Further, B_2 must alternate at the Larmor frequency, in this case γB_1, so that it can be decomposed into two counter-rotating components, one of which precesses with the magnetization. We have two possible mechanisms for creation of B_2. The first is main field modulation: B_0 is varied as $B_{00}+2B_2 \sin(\gamma B_1 t + \phi)$, where ϕ is an adjustable phase constant. The second is B_1 phase modulation, where B_1 comprises the main field in the $\tilde{x}$ direction and a small oscillating field, $2B_2 \cos(\gamma B_1 t + \phi)$, in the $\tilde{y}$ direction. If $B_2 \ll B_1$, we may analyze the effect of B_2 by utilizing a second rotating frame ($\tilde{\tilde{x}}, \tilde{\tilde{y}}, \tilde{\tilde{z}}$), which rotates about the $\tilde{x}$ axis with frequency ω_1. This frame, coincident with the first rotating frame at time $t = 0$ and at times $2\pi n/\omega_1$, where n is an integer, is shown in FIGURE 8. There are many possible ways of using B_2 to excite selectively a region within a sample. For example, a preparatory 90° pulse may align the magnetization with B_1 along the $\tilde{\tilde{x}}$ (i.e. $\tilde{x}$) axis prior to the application of B_2. B_2 then behaves exactly as B_1 would in the usual slice selection experiment and a selective 90° B_2 pulse produces dephased signal in the $\tilde{\tilde{y}}\tilde{\tilde{z}}$ plane, with the need for echo formation by the equivalent of gradient reversal (inversion of B_1 phase) or Hahn echo (a 180° B_1 pulse along $\tilde{y}$). In order to obtain signal just from the slice, the unperturbed $\tilde{x}$ magnetization is then returned to the $\tilde{z}$ direction at the appropriate time by a 90° pulse (B_1 in $+\tilde{y}$ direction) and coherent magnetization from within the slice is left in the $\tilde{x}\tilde{y}$ plane, the phase having been determined by, among other factors, ϕ. A far simpler slice selection experiment applies a B_1 gradient field of which the $\tilde{x}$ component ($B_{1\tilde{x}}$) is constant in time, and the $\tilde{y}$ component ($B_{1\tilde{y}}$) varies in time as:

$$B_{1\tilde{y}} = -2B_2 \frac{\sin at}{at} \cos \gamma B_{10} t; \; t \geq 0 \tag{1}$$

where a is a constant, and B_{10} is the $B_{1\tilde{x}}$ field strength at the center of the slice to be selected—in other words, a single, selectively phase-modulated pulse is applied. Then, as shown in FIGURE 8, in the second rotating frame M_0 is along the $\tilde{\tilde{z}}$ axis where, left to its own devices, it decays with time constant $T_{2\rho}$ given by:

$$\frac{1}{T_{2\rho}} = \frac{1}{2}\left[\frac{1}{T_1} + \frac{1}{T_2}\right] \tag{2}$$

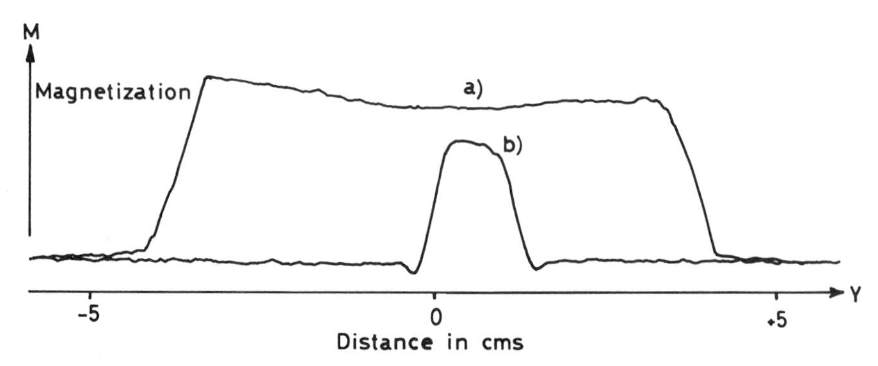

FIGURE 9. The results of application of a selective pulse using phase modulation of B_1, as shown in FIGURE 8. Using an 11 msec pulse, signal was obtained selectively from a 1-cm thick region of an 8-cm long tube of water. The plots are (a) a graph of equilibrium magnetization versus distance and (b) $\tilde{x}\tilde{y}$ magnetization versus distance following the selective pulse. The loss in signal amplitude is almost fully explained by relaxation, for $T_{2\rho}$ was 75 msec. Plot (b) is the difference between two experiments with opposing B_2 phases. This technique eliminates unwanted signals from outside the selected region.

However, B_2, being so phased, is along the $-\tilde{\tilde{y}}$ axis, and tips the magnetization from a well-defined region onto the $\tilde{\tilde{x}}$ axis where it remains spin-locked to B_1, decaying with time constant $T_{1\rho}$. Thus magnetization within an approximately rectangular slice is left at the end of the experiment coherently aligned along B_1 in the $\tilde{x}$ direction (FIG. 9), while magnetization outside the range of influence of B_2 ($\Delta B_1 \gg B_2$) is unperturbed and lies dephased in the $\tilde{\tilde{y}}\tilde{\tilde{z}}$ ($\tilde{y}\tilde{z}$) plane. The experiment may then be repeated in the same f.i.d. with the appropriate phases of B_1 and B_2 (90° different) for the other two gradient coils, and coherent magnetization is finally obtained from a box of sample. Such an experiment may take many milliseconds, depending upon the strength of B_1 and the width and sharpness of the defined region, and its duration in comparison to $T_{2\rho}$ is then of concern. However, it is worth noting that in many biological experiments $T_2 \ll T_1$, and so $T_{2\rho} \sim 2T_2$, giving somewhat longer than expected to perform the experiment—an advantage shared by many rotating frame experiments, including the imaging experiment described earlier. Needless to say, there is residual signal from unwanted regions where B_1 is not much greater than ΔB_0, but the experiment may be repeated, with the phase of B_2 inverted, and the difference in the two sets of data taken, to remove the spurious signals. Thus, if three-dimensional box selection were required, eight experiments would be needed to obtain a "clean" signal. Note that in a one-dimensional experiment, the coherent signal from the selected slice lies along B_{1eff}, which is close to the $\tilde{x}$ axis (in the $\tilde{x}\tilde{z}$ plane in the presence of ΔB_0)—the required condition for reduction of the zero-frequency artifact mentioned earlier.

CONCLUSION

Rotating frame methods have been slow to gain acceptance because of misconceptions as to how the technique is best implemented, and concern over r.f. power deposition. There can be no doubt that they deposit much more power in the sample than does a simple 90° pulse, the exact amount depending, of course, on the details of the experiment. However, they do overcome the cost of, and difficulties associated with, switched field B_0 gradients, and so have a place in the armory of techniques available to the biomedical NMR researcher. As the NMR system only responds to B_1 fields in the xy plane, there is a degree of freedom (z) that may be exploited in the design of appropriate gradient coils, and at the time of writing, we are exploring designs for various applications. The presence of a homogeneous B_1 field, in addition to gradient fields, is a valuable asset, provided the phases of the various fields bear a reasonably constant relationship over the volume of interest. Finally, one should not forget that B_2 can be applied by modulation of the main field, as well as by modulation of the phase of B_1, and this attribute leads to some intriguing possibilities where the amplitude and phase of B_2 vary spatially.

REFERENCES

1. HOULT, D. I. 1979. J. Magn. Reson. **33:** 183–197.
2. COX, S. J. & P. STYLES. 1980. J. Magn. Reson. **40:** 209–212.
3. GARWOOD, M., T. SCHLEICH, G. B. MATSON & G. ACOSTA. 1984. J. Magn. Reson. **60:** 268–279.
4. HAASE, A., C. MOLLOY & G. K. RADDA. 1983. J. Magn. Reson. **55:** 164–169.
5. GARWOOD, M., T. SCHLEICH, B. D. ROSS, G. MATSON & W. D. WINTERS. 1985. J. Magn. Reson. **65:** 239–251.
6. HEDGES, L. K. & D. I. HOULT. 1985. Abstracts, Fourth Annual Meeting, Society of Magnetic Resonance in Medicine. (London, August 19–23), pp. 1096–1097.
7. EDELSTEIN, W. A., C. J. HARDY & O. M. MUELLER. 1986. J. Magn. Reson. **67:** 156–161.
8. HAASE, A. 1985. J. Magn. Reson. **61:** 130–136.
9. COUNSELL, C., M. H. LEVITT & R. R. ERNST. 1985. J. Magn. Reson. **63:** 133–141.
10. HOULT, D. I. 1980. J. Magn. Reson. **38:** 369–374.

DISCUSSION OF THE PAPER

T. R. BROWN: You didn't say how long it takes to do the whole sequence though.

D. I. HOULT: It's entirely up to you and it depends on your coil design and what Rf you've got. That particular sequence—this is for samples 10 cm in size, I'm using 100 watt power amplifier to 15 milliseconds. But I'm using a very modest fire, for samples 10 cm in size I could very easily at 5 MHz, which was the frequency I was working with, have gone on up to 5 kilowatts.

Volume Selection Using Gradients and Selective Pulses

R. J. ORDIDGE AND P. MANSFIELD

Physics Department
Nottingham University
Nottingham NG7 2RD, England

J. A. B. LOHMAN AND S. B. PRIME

Oxford Research Systems
Nuffield Way
Abingdon, England

INTRODUCTION

Localized measurement of NMR spectra by the use of the Image Selected *In-vivo* Spectroscopy technique (ISIS) is discussed in section 1, and examples are given of phosphorus spectra from the human head.

The problems associated with gradient switching in superconducting magnets during application of ISIS and related methods are discussed in section 2, and two methods of overcoming these difficulties are described. Section 3 describes the relevance of cube selection techniques for the investigation of tissues with a complicated and intricate spatial distribution.

1. MEASUREMENT OF A SELECTED CUBE

Several techniques are currently available to achieve spatial selection of well-defined volumes of tissue for the purposes of NMR spectroscopy.[1-4] The Image Selected *In-Vivo* Spectroscopy (ISIS) technique demonstrates the accuracy and flexibility currently achievable with localization schemes that employ selective radiofrequency (RF) pulses in combination with linear magnetic field gradients. The details of the ISIS method have been presented elsewhere.[2] The main advantage is that it allows an NMR spectrum to be obtained from a well-defined cube of material positioned by reference to a standard MR image of the subject. The cube, which can be varied in size and position, is selected by the use of a sequence of eight experiments and employs three selective pulses in the presence of three orthogonal field gradients. The spatial localization is achieved through selective spatial encoding of the z spin magnetization by the use of pre-pulses. A spin-echo experiment is thus avoided, removing the need for a 180° RF pulse or gradient reversal. Measurement of the NMR spectrum thus gives a true representation of the chemical content of the tissue without distortion due to variations in the spin-spin relaxation time (T_2). This feature is particularly important for *in vivo* NMR phosphorus investigations of the human brain, where several components have a particularly short T_2 value, notably the adenosine triphosphate (ATP) resonances.

EXPERIMENTAL RESULTS

All experiments were performed on a 2.1 Tesla whole body superconducting magnet with a BRUKER (BIOSPEC) spectrometer. The NMR probe consisted of two concentric coils tuned to the phosphorus and proton resonance frequencies. Both of these were incorporated into a former with an access sufficiently large to accommodate a human head. The coils were constructed by employing the principles of the slotted tube resonator,[5] and each gave a uniform RF field over a larger volume than required by the head. The coils produced a linearly polarized RF field along a single axis of the sample. The proton coil was used to perform shimming of the proton lineshape for the desired sample, and also could be used to produce MR images of the subject. These show sufficient detail to allow accurate positioning of the ISIS cube. The phosphorus head coil was then used as both transmitter and receiver to obtain the localized spectrum from the cube.

The phosphorus spectrum in FIGURE 1A was obtained from a human head without spatial localization. The data are a result of 32 averages with a repetition delay of 2 sec. The spectral assignments (chemical shift in ppm) are: (I) sugar phosphate region (6.0), (II) inorganic phosphate (4.9), (III) phosphodiester region (3.0), (IV) phosphocreatine (0), (V) γ-adenosine triphosphate (γ-ATP, -2.5 ppm), (VI) α-ATP (-7.6), and β-ATP (-15.9). There is also a very broad peak from relatively immobile phosphorus nuclei, which are contained within bone and may be present in other body tissues and fluids.

FIGURE 1B shows an ISIS phosphorus spectrum from a 9 cm cube positioned in the center of the head. The cube position is indicated on the corresponding cross-sectional proton head image. The phosphorus spectrum is a result of 256 averages with a repetition delay of 2 sec. The experimental time was therefore 8.5 min and the tissue volume investigated was 730 cm^3. The ISIS experiments were performed by using a sinc-shaped RF pulse for selective inversion, which had a spectral bandwidth of 1.6 kHz. The spectral assignments are the same as in FIGURE 1A, however the selected spectrum shows a reduction in the relative size of the phosphocreatine peak compared to ATP, and a corresponding increase in the phosphodiester peak. The proton image was from a sixteen-plane, multislice series of images, which were obtained in an experimental time of approximately 8.5 min. The slice thickness was 6 mm and the in-plane resolution is approximately 1 mm.

The ISIS phosphorus spectrum shown in FIGURE 1C was obtained from a 6 cm cube positioned centrally in the brain as shown by the corresponding proton image. The experimental parameters are the same as in FIGURE 1B apart from the increased gradient strength required to select a 6 cm cube. The selected tissue volume was 220 cm^3. However, apart from a reduction in signal-to-noise ratio, the spectral detail is similar to that in FIGURE 1B.

FIGURE 1D shows a phosphorus spectrum obtained from the 3 cm cube shown in the corresponding proton image. The spectral detail is almost obscured by the noise, however the data illustrate the minimum size of selected volume that is observable using a head receiver coil, and in an experimental time of less than 10 minutes.

DISCUSSION

These results demonstrate that usable phosphorus spectra can be obtained using coil designs that surround the subject, in the manner employed in MR imaging machines. Indeed, whilst surface coils are useful for the investigation of tissue volumes

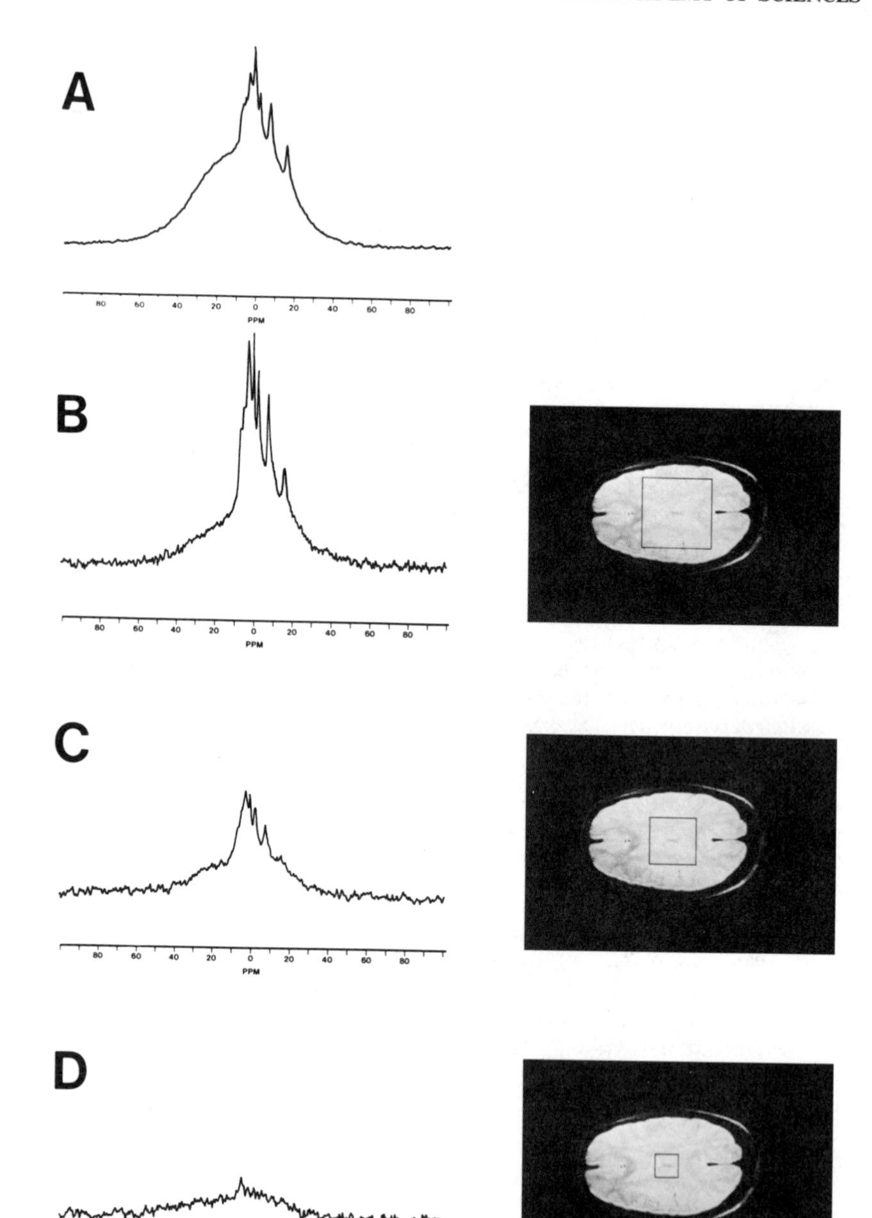
A
80 60 40 20 0 20 40 60 80
PPM
B
80 60 40 20 0 20 40 60 80
PPM
C
80 60 40 20 0 20 40 60 80
PPM
D
80 60 40 20 0 20 40 60 80
PPM

on the periphery of the subject, it is unlikely that there is any advantage in the use of these coils for the investigation of deep-seated tissue of the type shown in FIGURE 1. Furthermore, if spatial localization techniques such as ISIS are to be used to their full advantage, then the spectroscopist must be able to choose the volume of interest from a complete two- or three-dimensional MR image of the subject. The corresponding receiver coil for NMR spectroscopy must therefore detect signal with equal efficiency over the complete volume depicted in the proton images.

A feature common to most of the spatial localization methods is the use of pulsed magnetic-field gradients. The problems associated with this requirement are discussed in the next section.

2. OVERCOMING THE PROBLEMS ASSOCIATED WITH PULSED MAGNETIC FIELD GRADIENTS

The main problems associated with switching magnetic field gradients in superconducting magnets occur following a gradient pulse. The spatially localized NMR spectrum is usually measured following a sequence of gradient pulses. The first effect is a residual magnetic field gradient caused partly by the inductance of the gradient coil system, but mainly through the induced eddy currents in the magnet and probe structure. This effect can be removed by cancellation, which is achieved by overdriving the gradient waveform with a negative overshoot upon gradient removal. The overshoot is tailored to have a time constant identical to that of the magnet system, so that the net effect is approximately cancelled during decay of the eddy currents, which typically takes tens of milliseconds. However, this cancellation does not remove the second main effect, which is a time-dependent variation of the main magnetic field (B_0) caused by circulating eddy currents induced in the magnet bore tube during gradient removal. The eddy currents are initially induced in order to counteract the effect of localized fields produced by the gradient coils. However, the decaying eddy current distribution upon gradient removal is not constrained to be in the local vicinity of the gradient coils. Field variations in addition to B_0 are also produced by the eddy currents, however, these fields decay rapidly to leave a B_0 term that decays approximately exponentially with a much longer time constant.

An uncorrected FID following a gradient pulse produces spectra with major distortions, which can totally obscure all high resolution detail. One method of overcoming this problem is to measure the B_0 field variation as a function of time and

FIGURE 1. (A) The total phosphorus spectrum from the whole head. The spectral assignments (chemical shift in ppm) are: sugar phosphate region (6.0), inorganic phosphate (4.9), phosphodiester region (3.0), phosphocreatine (0), γ-ATP (−2.5), α-ATP (−7.6), and β-ATP (−15.9). There is also a very broad peak from relatively immobile phosphorus nuclei contained within bone and may be present in other body tissues and fluids. (B) An ISIS phosphorus spectrum from a 9 cm cube positioned in the center of the brain. The spectrum shows a reduction in the relative size of the PCr peak compared to ATP and a corresponding increase in the phosphodiester peak. Both features are characteristic of brain tissue. The volume of tissue selected is 730 cm^3 for this size of cube. (C) ISIS phosphorus spectrum from a 6 cm cube centered in the same location as the cube in (B). The selected tissue volume is now 220 cm^3. The spectral detail is similar to (B), but with a corresponding reduction in signal-to-noise ratio. (D) ISIS spectrum from a 3 cm cube centered as in (B). The selected tissue volume is now only approximately 30 cm^3 and spectral detail is almost obscured by the noise.

correct the FID accordingly. The simplest means of achieving this is to measure the phase of an on-resonance signal from a single component spectrum as it evolves following a gradient pulse. This phase angle evolution can then be subtracted from all spectra obtained under similar conditions in order to remove the effect of B_0 variation. This technique is described in more detail in reference 6.

FIGURE 2A shows the proton spectrum from a 10 cm sphere filled with methanol, and was obtained using a 1.5 Tesla/1 meter bore BRUKER (BIOSPEC) spectrometer. The uncorrected spectrum from the same sample but measured 4 msec after three simultaneous gradient pulses of 0.2 Gauss/cm applied along orthogonal axes is shown in FIGURE 2B. The same data are shown in FIGURE 2C following correction with a time-dependent field term previously measured under the same conditions using a water sample. The field term was evaluated as a time-dependent phase shift, which was simply subtracted from the phase of the methanol signal. As predicted, the spectral detail has been mostly restored by the correction.

The effect of eddy currents can therefore be removed by data processing and compensation. An alternative approach is to avoid the initial generation of the eddy currents by modification of the gradient coil design.[7] This technique involves the addition of an active magnetic screen to the gradient coil system. Metal surfaces at the boundary of the screened region, i.e. the magnet bore tube, would normally have currents induced within the metal in order to produce a "mirror image" of the gradient current distribution. This prevents penetration of the magnetic field into the metal. These same currents can therefore be eliminated by creating the same mirror field distribution by use of an appropriately designed active screen. The active screen is pulsed with a current waveform identical to the gradient waveform. The design of the screen is chosen to produce a magnetic field distribution outside the screened volume which is the exact reverse of the field that would have existed in the presence of unshielded gradient currents.

The advantage of this approach is that since the eddy currents can be completely eliminated, all terms in the erroneous eddy field distribution are removed. Gradients can therefore be switched rapidly, immediately prior to the accumulation of NMR spectra. The active screen results in a reduction of the gradient field strength and degradation of linearity of the gradient field. However this can be effectively compensated by using a double-active screen design.[8]

Several of the existing spatial localization schemes (e.g. ISIS, SPARS, and VSE) have been specifically designed to include a time delay in order to allow the induced eddy current distribution to decay. The emergence of these new screening and compensation techniques should result in better spectra and reduced errors in spatial localization that may arise when measuring nuclei with a short spin-lattice relaxation time.

3. INVESTIGATION OF FRACTIONAL TISSUE COMPONENTS USING CUBE SELECTION TECHNIQUES

Most of the proposed spatial localization schemes can select accurately defined cubic volumes of tissue that may be positioned by reference to an MR image. A cubic volume is useful for the investigation of volumes of tissue that are mainly homogeneous, e.g. muscle, liver, etc. However, spectroscopists are also interested in analyzing specific types of tissue rather than the mixture of components that may be present in a cube positioned within a complicated organ, such as the brain. Gray matter in the human brain may have a different phosphorus spectrum from white matter; however, it

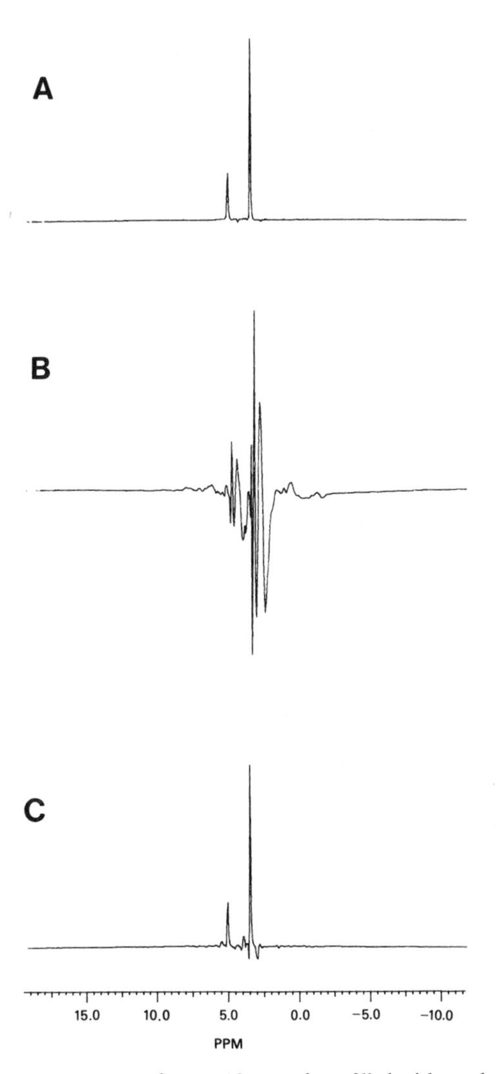

FIGURE 2. (A) A proton spectrum from a 10 cm sphere filled with methanol. (B) The same spectrum as (A), but sampled 4 msec after a gradient pulsing sequence consisting of 3 simultaneous gradient pulses applied along orthogonal axes with a strength of 0.2 Gauss/cm. (C) The spectrum of (B) after correction using the measured time-dependent phase function.

TABLE 1. Fractional Mixtures for Components A and B in the Six Simulated Experiments

Experiment Number	1	2	3	4	5	6
Fraction of spectrum A	0	0.75	0.5	0.25	0.25	0.25
Fraction of spectrum B	0.25	0.25	0.25	0.25	0.5	0

is unlikely that these two components can be separately measured using existing spatial localization techniques. The advantage of the ISIS, VSE, and SPARS techniques is that through an accurate knowledge of the size and location of the selected volume, the amount of each tissue component can be assessed by reference to the corresponding MR image. If the two volumes can therefore be analyzed with a significant difference in the proportion of two main tissue components within each of the selected cubes, then simple mathematics should provide the individual spectra of each tissue component.

THEORY AND THEORETICAL RESULTS

This simple idea may be extended to the investigation of tissue volumes that consist of multiple tissue components, provided the relative concentrations of the components vary in a known manner. If there are m different tissue components, we require n linearly independent measured spectra, in order to derive the individual spectrum of each component, where $n \geq m$.

A spectrum may be represented as a set of coefficients $A_i(f_k)$ defined over a discrete frequency domain of p points (f_k; $k = 0 \ldots p - 1$), where the subscript i refers to the component for which A_i is a spectrum. For a mixture of m components, the kth point in the jth composite spectrum may be expressed as:

$$G_j(f_k) = \sum_{i=0}^{m-1} C_{ij} A_i(f_k), \tag{1}$$

where the coefficients C_{ij} are the total volume of the ith component present in the jth composite spectrum. The only assumption we make is that the amount of the ith signal component in the jth composite mixture is proportional to the volume of the ith tissue component in the jth selected cube. This information can be determined from a standard MR image showing reasonable spatial detail within the volume of each selected cube. It may be as simple as counting the number of picture elements occupied by each tissue component.

Equation 1 represents a set of simultaneous linear equations that can be solved independently at each frequency for the unknown component spectrum $A_i(f_k)$.

A theoretical simulation was performed on a BRUKER (Aspect 3000) computer using the NAG Pascal library subroutine FØ4EDC. This routine finds the minimal

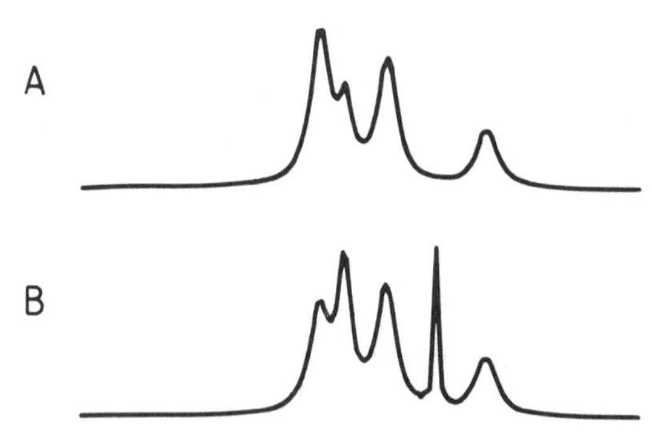

FIGURE 3. Two simulated spectra used to represent tissue components (A) and (B), respectively.

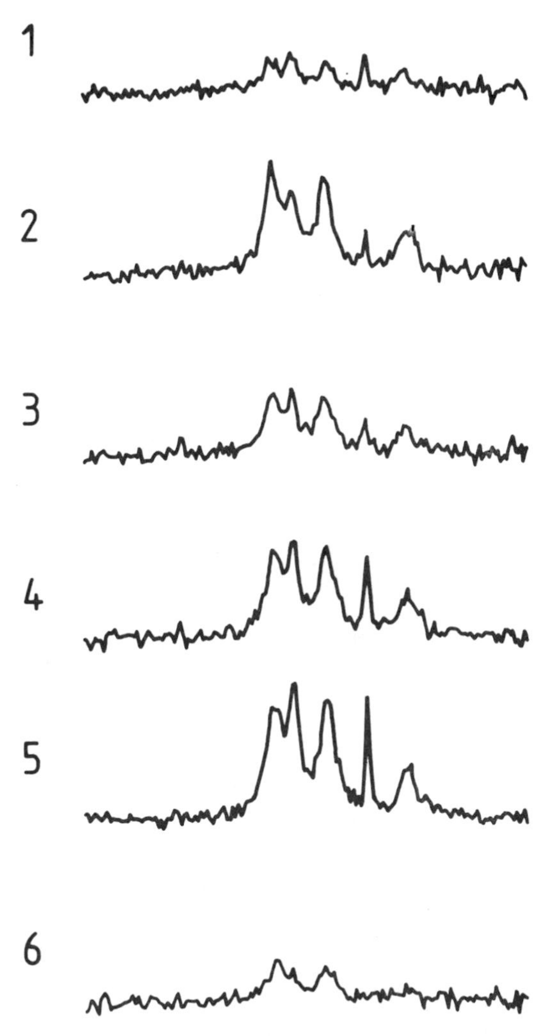

FIGURE 4. Six combinations of spectra 3A and 3B as determined from the mixture coefficients of TABLE 1. Random noise of equal amplitude has been added to all data in order to simulate spatial localization experiments with the correct mixture of tissue components.

least-squares solution to an overdetermined set of linear simultaneous equations (i.e. where n is greater than m). The simulated spectra of two tissue components were chosen to be significantly different in detail, as shown in FIGURE 3. These two components were then mixed in six different proportions in order to represent the spectra that might be derived through measurement of six different cubic volume elements. The six mixtures are defined in TABLE 1. Random noise was added to each of the six composite spectra to represent normal experimental noise, and the result is shown in FIGURE 4.

The final result of the least-squares fit to determine the two individual component spectra is shown in FIGURE 5 and demonstrates correlation with the initial spectra of FIGURE 3. Since the total measured amount of component A is greater than component B in the six experiments, the final signal-to-noise ratio of component A is greater than that of component B, as indicated by FIGURE 5.

An inherent assumption in this technique is that all six cubic volumes have been measured by the same technique and with an equal signal efficiency. This implies the use of a homogeneous receiver coil system as used in MR imaging.

4. CONCLUSION

Spatial localization to a cubic volume element has been demonstrated by the ISIS technique, with examples of phosphorus spectra from the human head. The problems associated with the use of gradient pulses in localization techniques for NMR spectroscopy have been discussed and two solutions have been proposed to overcome

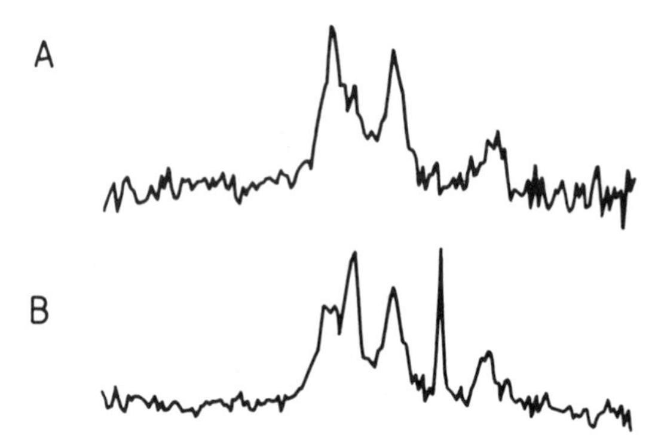

FIGURE 5. The result of a minimal least squares fit to determine the true spectra for tissue components (A) and (B), from the set of six spectra shown in FIGURE 4.

these problems. Selection to a cubic volume element does not allow spectroscopists to analyze individual tissue components within a cube positioned in an intricate organ such as the brain. However, by utilizing the accuracy in spatial selectivity and positioning provided by techniques such as ISIS, mathematical processing can subsequently be applied to extract this information. This technique is more efficient than simply reducing the size of the cube until it fits neatly within the desired volume of tissue. The method therefore might find applications in the extraction of the individual spectra of gray and white matter from the human brain. It may also prove equally useful for the examination of other complicated and intricate organs within the human body.

ACKNOWLEDGMENTS

The phosphorus head spectra were obtained in cooperation with the Department of Molecular Biophysics and Biochemistry of Yale University. The authors would also like to thank members of the development team of Oxford Research Systems for their assistance with hardware and software.

REFERENCES

1. BOTTOMLEY, P. A., T. B. FOSTER & R. D. DARROW. 1984. J. Magn. Reson. **59:** 338.
2. ORDIDGE, R. J., A. CONNELLY & J. A. B. LOHMAN. 1986. J. Magn. Reson. **66:** 283.
3. AUE, W. P., S. MUELLER, T. A. CROSS & J. SEELIG. 1984. J. Magn. Reson. **56:** 350.
4. LUYTEN, P. R., A. J. H. MARIEN, B. SIJTSMA & J. A. DEN HOLLANDER. 1986. J. Magn. Reson. **67:** 148.
5. SCHNEIDER, H. J. & P. DULLENKOPF. 1977. Rev. Sci. Instrum. **48**(1): 68.
6. ORDIDGE, R. J. & I. D. CRESSHULL. 1986. J. Magn. Reson. (In press.)
7. MANSFIELD, P. & B. CHAPMAN. 1986. J. Phys. E: Sci. Instrum. **19:** 540.
8. CHAPMAN, B. & P. MANSFIELD. 1986. J. Phys. D: Appl. Phys. **19:** L129.

DISCUSSION OF THE PAPER

T. R. BROWN: I do not believe that that last technique that you described will work in three dimensions and I'd like to hear you explain it. Clearly you can modulate the k space vectors adequately in one dimension; but the problem is if you want to do a four-dimensional experiment and that is to say take chemical shift imaging information, and you want to have a reasonable spectral band with it then your sampling time has got to be, let's say, a millisecond or some number like that if you want one KHz. That means you have to go through every k space point every millisecond in order to get a full 3-dimensional image. I do not think any conceivable set of wave forms is going to manage to do that on the gradient coils.

R. J. ORDIDGE: Due to the limitations in time I could not of course explain that last technique adequately. Perhaps I should not have tried because of the complicated principles involved. Each experiment however produces two dimensions of information, one representing chemical shift and the other a spatial dimension. Three or four dimensional images require a series of experiments. Therefore, because only two dimensions of information are captured in each shot of the experiment, it is relatively easy to sample one line of k space along a spatial direction in each millisecond.

Image-Guided Localized ^{1}H and ^{31}P NMR Spectroscopy of Humans

JAN A. DEN HOLLANDER AND PETER R. LUYTEN

Philips Medical Systems
NL-5680 DA Best, The Netherlands

INTRODUCTION

NMR spectra of human tissues *in situ* are usually obtained by means of surface coils.[1] Surface coils are the method of choice for studying superficial tissues, such as muscle and adipose. They provide an optimal sensitivity for these tissues and a variety of depth-selection techniques are available[2–4] to suppress interfering signals from areas that overlie the tissue of interest. However, if one wants to study specific lesions, such as brain tumors, then a more flexible approach is needed. It should be possible to define a volume of interest located anywhere in the human brain and a precise control over location and size of that volume is required. In order to be able to examine the entire brain, RF coils are needed that surround the head. This sacrifices some NMR sensitivity, but it does provide added flexibility and convenience as compared to the usual surface coils approach.

To obtain localized NMR spectra from well defined areas within the brain, we have taken an approach that integrates imaging and spectroscopy. The basic idea is that one first obtains a ^{1}H NMR image of the subject, and from this image one selects a volume of interest (VOI). The next step is to optimize the spectrometer settings for this particular VOI, including the RF power and magnetic field homogeneity. The last step is a spectroscopic examination over the VOI; one should be able to use a variety of spectroscopic techniques to obtain as much spectroscopic information as possible from the VOI. One wants to be able to study different nuclei and to use techniques such as water suppression, spectral editing, Carr-Purcell-Meiboom-Gill (CPMG) sequences, and inversion recovery.

The techniques used for the localized spectroscopic examination need to be based upon the approach used to obtain the NMR image; this assures that imaging and spectroscopy are correlated, and that the spectrum is obtained from the volume chosen on the basis of the image. A variety of techniques have been described for this volume selection; these are derived from volume-selective excitation (VSE), introduced by Aue *et al.*[5] We have been using pulse sequences adapted for use on a whole body NMR imager; different pulse sequences are used depending upon the particular requirements posed by the nucleus being investigated.

IMAGE-GUIDED LOCALIZED ^{1}H NMR SPECTROSCOPY

Spatially resolved spectroscopy (SPARS) has been employed to obtain localized ^{1}H NMR spectra.[6] SPARS is a preparation step; its purpose is to provide the spatial selection. SPARS selection can then be combined with different observation sequences; in FIGURE 1 it is combined with a 1331-2662 sequence used for solvent suppression. It may, however, equally well be combined with different observation

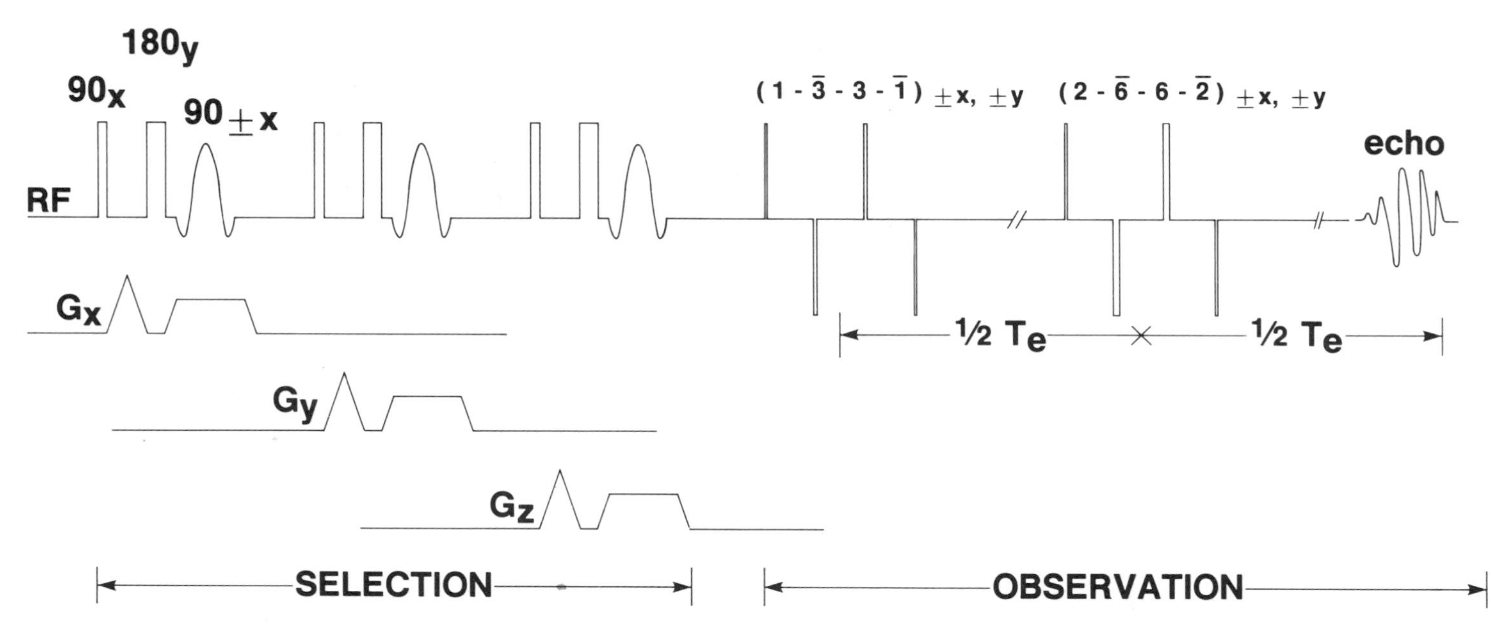

FIGURE 1. The SPARS selection pulse sequence. SPARS is combined with a 1331-2662 spin-echo sequence used for water suppression. A detailed explanation of SPARS is presented in the text.

sequences, such as a simple 90 degree observation pulse, inversion recovery, spin-echo or CPMG, and sequences used for NMR imaging. A few examples will be discussed below in more detail.

Volume Selection by SPARS

The SPARS selection sequence as presented in FIGURE 1 consists of three steps, one for each spatial coordinate. Each of these steps is made up of a non-selective spin-echo sequence, followed by a slice-selective 90 degree reset pulse. The non-selective 90 and 180 degree pulses are given in the absence of any applied magnetic field gradients. This is necessary because the head- and body-coils used in a whole-body NMR scanner give rise to 90 degree pulse-lengths, which exceed 100 μsec. If magnetic field gradients were applied during these non-selective pulses their limited bandwidth would not allow uniform excitation of the entire sensitive volume. Following the 180 degree pulse, a magnetic field gradient is switched on for the slice-selective reset pulse; a correction gradient is included between the non-selective 90 and 180 degree pulse to correct for dephasing by the selection gradient. The net result of the sequence is that within the selected slice, nuclear magnetization is reset along the longitudinal direction, whereas outside this particular slice magnetization is left in the transverse plane. Transverse magnetization is then dephased by the selection gradient, and also relaxes away by transverse relaxation processes. The selective reset pulses are phase-alternated, such that in successive scans the magnetization is either reset along the direction of the applied magnetic field or against that direction. A NMR spectrum from the selected slice can now be obtained after an additional non-selective 90 degree excitation pulse; successive scans are added or subtracted according to the applied phase-cycling scheme. The basic selection sequence can be repeated for the other two spatial coordinates; in this way a block-shaped VOI is obtained. A phase-cycling scheme of eight successive scans is required to combine all possible phase alternations of the three reset pulses used for the volume selection process. The three dimensions of the selected volume can be varied independently by choosing the appropriate gradient strength, and selection pulse bandwidth.

As the VSE experiment, SPARS is based upon a logical separation between the selection and the observation periods. A common problem in the use of switched field gradients for spatial localization is that eddy currents are induced in the cryostat. These eddy currents, which may have relatively long time constants, may cause loss of spectral resolution in the measured NMR spectrum. This effect of the eddy currents may be reduced by introducing a delay time between the selection and the observation period; this delay time should be kept short compared to T_1 of the measured NMR signals.

Shimming the Volume of Interest

Spectral resolution may also be improved by shimming the magnetic field over the VOI. The SPARS sequence can be of help in doing this. As mentioned, the full SPARS experiment requires averaging of eight successive scans for optimal suppression of unwanted signals. However, each individual scan does already provide a localized NMR signal from the VOI. This can be used for interactive shimming of the local magnetic field. To do this the ^{1}H NMR water signal from the VOI as measured by the SPARS sequence is continuously monitored, and the shim currents are adjusted until the linewidth of that signal is optimal. We have followed this procedure routinely; ^{31}P NMR studies are also preceded by a localized ^{1}H measurement using SPARS to shim the VOI.

An illustration of the application of SPARS is presented in FIGURE 2. These ^1H NMR spectra were obtained from the lower leg of a normal volunteer;[7] the leg was put through a 30 cm diameter RF coil normally used for head imaging. The spectra were obtained on a 1.5 T whole body Philips NMR scanner, operating at 63.86 MHz for ^1H

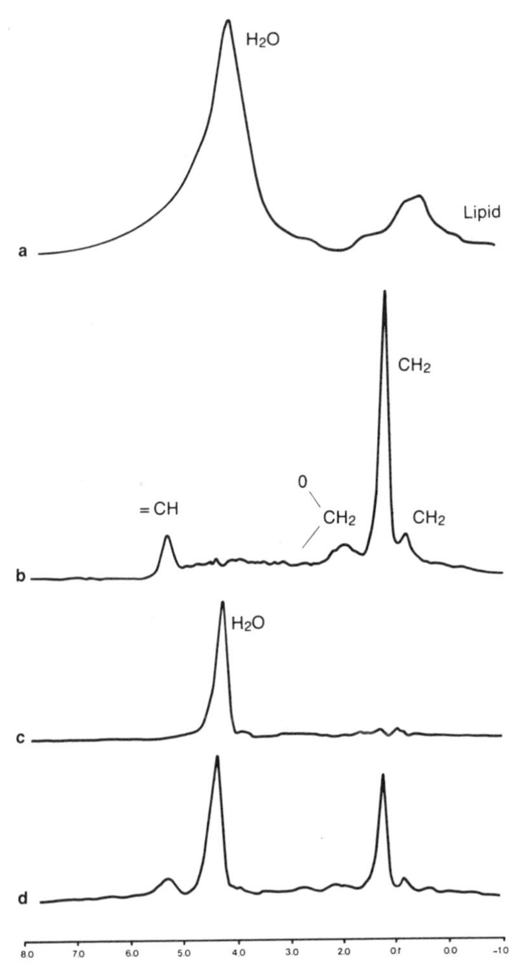

FIGURE 2. 63.89 MHz ^1H NMR spectra obtained of the lower leg of a normal volunteer, employing a 30 cm diameter RF coil normally used for head imaging. (A) Spectrum of the whole lower leg. (B) SPARS localized ^1H NMR spectrum from a 1.5 × 1.5 × 1.5 VOI located in the bone marrow of the tibia. (C) Spectrum from a VOI located in the muscle. (D) Spectrum from a VOI that contained subcutaneous adipose and muscle tissue. Units are ppm.

NMR. The upper spectrum is that of the whole leg, without any spatial localization. It shows the rather poorly resolved water and fat signals usually observed in human tissues. The second spectrum (B) is that of a 1.5 × 1.5 × 1.5 cm VOI located in the bone marrow of the tibia. The resonances in this ^1H NMR spectrum are assigned to saturated and unsaturated fats: it shows signals from CH_2, CH_3, C=CH, and glyceryl protons. This spectrum is nicely resolved (linewidths of about 0.2 ppm), because an excellent magnetic field homogeneity can be achieved over these small VOIs. The third spectrum (C) is that of a VOI located in muscle tissue; it shows a large signal from water and some small resonances from fats. The last spectrum (D) comes from a VOI that contained both muscle and subcutaneous adipose. Signals are observed from water (mainly from muscle) and fat signals from adipose.

Target Imaging by SPARS

Because of the logical separation between the selection and the observation period, the SPARS sequence can be easily combined with different observation sequences. One of the first of these we explored was the combination of SPARS with imaging;[8,9] this provides images that highlight the VOI. The significance of this experiment is that it gives feedback of how well the spectral localization step works. It may also serve to confirm that the VOI chosen for a spectroscopic investigation is properly selected by the SPARS sequence. This procedure is not part of routine spectroscopic investigations because of the time pressure when examining patients. A useful application of this sequence is the possibility of obtaining target images of high spatial resolution from certain areas within the head or body.[8]

SPARS Relaxation Studies

An experiment of more practical implications is the combination of SPARS with inversion recovery, spin-echo, and CPMG sequences.[10,11] This makes it possible to perform detailed T_1 and T_2 measurements over localized VOIs within the body; in this way spectroscopically resolved relaxation data are obtained. The SPARS/inversion recovery experiment is implemented by starting with a non-selective 180 degree inversion pulse and a variable delay time; these are followed by the SPARS selection sequence and a non-selective 90 degree excitation pulse. In FIGURE 3, a SPARS/inversion recovery measurement is shown that was performed on the human lower leg. A $1.5 \times 1.5 \times 1.5$ cm VOI was located within the bone marrow of the tibia was selected, and a SPARS/inversion recovery experiment was done over that volume. As shown in the figure, T_1 relaxation can be followed in detail for each of the resolved resonances. Localized T_2 experiments have been implemented by first applying SPARS, and then the CPMG or Hahn spin-echo sequences. Experiments have been performed to measure separate T_2 relaxation times for each of the resolved spectral resonances of human adipose and bone marrow.

Contrast in NMR images depends upon differences in T_1 and T_2 relaxation; therefore, detailed relaxation studies of human tissues *in situ* have clinical significance. The spectroscopic T_1 relaxation measurement of FIGURE 3 clearly shows that the resolved resonances differ in their T_1 relaxation times. The T_2 experiments have not only revealed different T_2s for each of the resolved resonances, but also that adipose resonances show different decay rates with CPMG as compared with single-echo experiments. These differences between CPMG and single-echo experiments have been interpreted as a J-modulation effect, present in the adipose resonances.[11]

We have applied the SPARS relaxation method to examine a number of patients suffering from brain tumors.[12] T_1 and T_2 relaxation times have been obtained that are more accurate than those obtained by imaging; furthermore non-exponential T_2 relaxation could be detected. A variability was found in T_1 and T_2 relaxation times between tumors diagnosed to be of similar type, but the systematic differences in T_2 relaxation times between tumors of different type exceeded this non-specific variability. This fact may be important for tumor identification on the basis of ^{1}H NMR relaxation times.

^{1}H NMR Observation of Metabolites in Human Tissues

The two main obstacles in observing low concentration metabolites in ^{1}H NMR spectra of tissues are the small chemical shift range of these spectra and the intense water and fat signals. An excellent spectral resolution is a prerequisite for observing

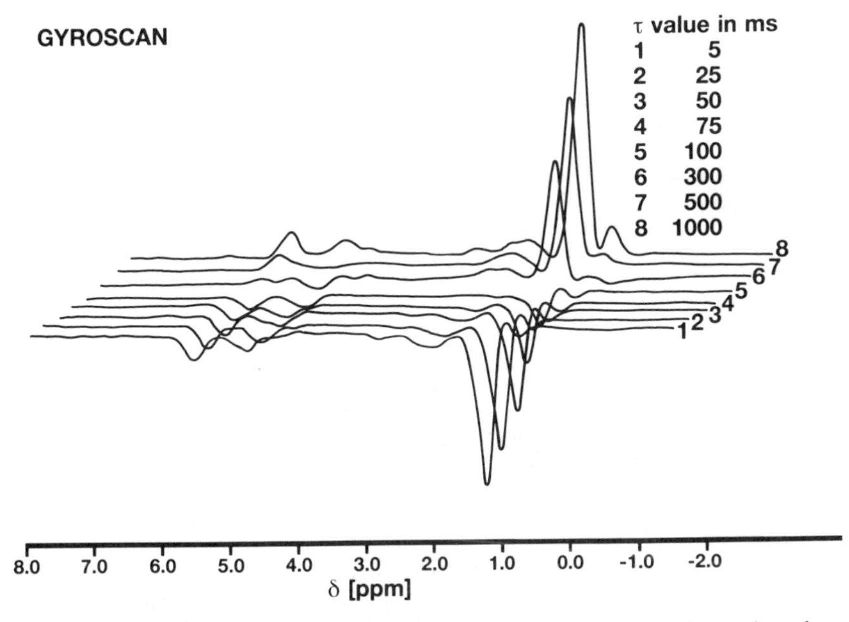

FIGURE 3. A SPARS/inversion recovery measurement was performed on the human lower leg. SPARS was used to select a 1.5 × 1.5 × 1.5 cm VOI in the bone marrow of the tibia, and an inversion recovery experiment was done over that volume. T_1 relaxation could be followed for each of the resolved resonances.

metabolites by ^{1}H NMR. The fact that we are able to shim the magnetic field homogeneity over the VOI by using SPARS helps to improve spectral resolution. A higher magnetic field strength is also important for spectral resolution. At this point, the magnetic field strength available for whole body applications is limited to about 2 tesla. Because 1.5 tesla is the field strength used in many clinical imaging systems, it is important to establish whether or not observation of metabolites in humans is feasible at these relatively low field strengths.

Metabolites are present in concentrations of a few mM, while tissue water and fat protons have a concentration of up to 70 M. It will be necessary to suppress these intense water and fat signals as much as possible to uncover the ^{1}H NMR signals of the

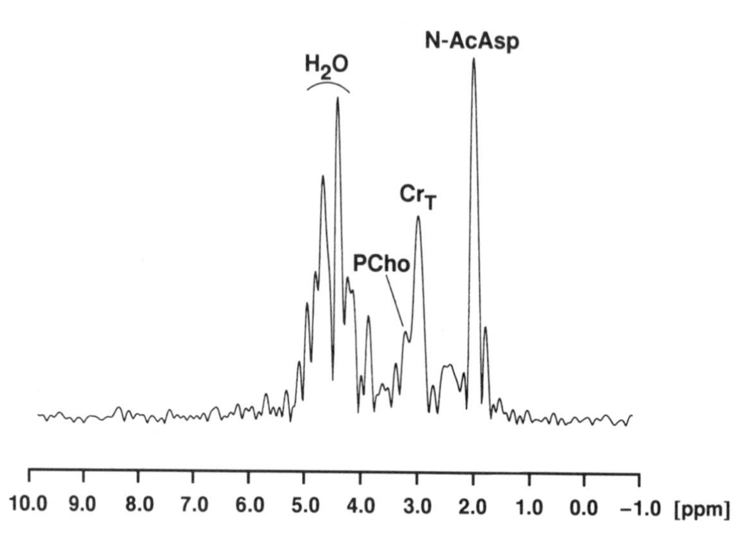

FIGURE 4. Water-suppressed 63.89 MHz ^{1}H NMR spectrum of a 3 × 3 × 3 cm volume located in one of the brain hemispheres of a normal volunteer. SPARS was used to select the volume, and water suppression was done using a 1331-2662 sequence; 128 acquisitions were collected; spin-echo time used was 408 msec. The spectrum shows signals from the methyl group of N-acetylaspartate (N-AcAsp), creatine and phosphocreatine (CrT), and phosphocholine (PCho). The pattern around 5 ppm is due to the residual water resonance.

metabolites. SPARS by itself offers a significant initial step in this suppression process. As discussed before, in contrast to some other spectral localization techniques, SPARS actually suppresses unwanted signals coming from outside the VOI. Since the VOI is usually much smaller than the NMR-sensitive area, this is important in overcoming limitations in dynamic range.

The fat signals are particularly annoying in that they coincide in chemical shift with some of the interesting metabolites. It has been found in animal model studies[13] that brain contains few NMR-observable fat signals. For whole body applications, it means that fat signals arise from surrounding tissues, such as bone marrow and subcutaneous adipose, but not from the brain itself. For this reason the brain is the most suitable organ to successfully observe metabolites by ^{1}H NMR spectroscopy. By selecting a VOI located within the human brain by means of the SPARS sequence, we have overcome the problem of overlapping fat signals.

The last step to be taken is to suppress the intense water signals. We have combined the SPARS selection sequence with the 1331 pulse for water suppression.[14] The sequence show in FIGURE 1 has been used to obtain a water-suppressed ^{1}H NMR spectrum of the human brain. FIGURE 4 presents the water-suppressed 63.89 MHz ^{1}H NMR spectrum of a 3 × 3 × 3 cm volume located in one of the brain hemispheres of a normal volunteer. Water suppression was done using a 1331-2662 sequence. Full EXORCYCLE phase cycling was employed, leading to a total of 128 acquisitions. Whole echos were collected; spin-echo time used for FIGURE 4 was 408 msec. The spectrum shows signals that have been assigned to N-acetylaspartate, creatine and phosphocreatine, and phosphocholine. In the spectral region between 2 and 3 ppm, a small signal is observed that presumably comes from glutamate. The complex pattern around 5 ppm is due to the residual water resonance.

T_1 and T_2 relaxation times can also be measured for the metabolites observed in the water-suppressed ^1H NMR spectrum of the human brain. The spectra of FIGURE 5 were obtained for different spin-echo times, using the pulse sequence of FIGURE 1. These spectra show that the N-acetyl aspartate methyl group has a long T_2 relaxation time. The resonance at about one ppm, presumably coming from proteins or fats, has a short T_2. These spectra show that ^1H NMR can be applied to observe metabolites in the human brain at 1.5 Tesla, although less information can be obtained than at the much higher magnetic fields available for small animal studies. Experiments are in progress to investigate whether or not water-suppressed ^1H NMR spectroscopy can be successfully applied to examine brain pathologies.

IMAGE-GUIDED LOCALIZED ^{31}P NMR SPECTROSCOPY

Localized ^{31}P NMR spectra were obtained using a modification of the sequence for image-selected *in vivo* spectroscopy (ISIS), introduced by Ordidge *et al.*[15] ISIS is similar to VSE and to SPARS in that it consists of a selection sequence, followed by an observation sequence. It works by using slice-selective inversion pulses, one for each of the spatial coordinates. Two measurements are needed for the slice-selection process. The first measurement consists of a slice-selective inversion pulse in the presence of a switched field gradient, followed by a non-selective 90 degree observation pulse; in the second measurement the selective inversion pulse is omitted. Subtraction of the results of these two measurements yields a localized NMR signal from the selected slice. To achieve volume selection, this selection process is extended to the other two coordinates. The three slice-selective inversion pulses are cycled through all possible on/off combinations, leading to a total of eight experiments. The acquired NMR signal is added or subtracted such that the signal from the VOI adds coherently, but signals from outside the VOI subtract out. This cycling is similar to the phase-cycling scheme

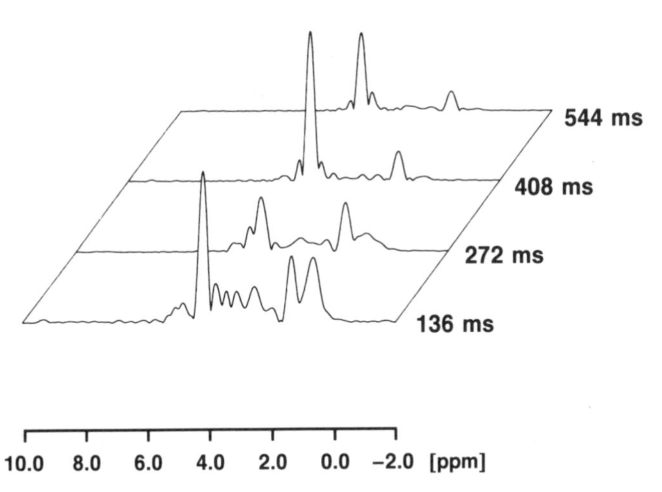

FIGURE 5. Water-suppressed ^1H NMR spectra from a VOI selected in the brain of a normal volunteer. Spectra were obtained for different spin echo times, using the pulse sequence of FIGURE 1. These spectra show that the N-acetyl aspartate methyl group at 2.02 ppm has a long T_2 relaxation time, while the resonance at about one ppm has a short T_2.

used in SPARS; the difference is that ISIS spatial localization depends upon this addition/subtraction scheme, while in SPARS it is used to correct for pulse imperfections and off-resonance effects. ISIS is therefore more susceptible to motion artifacts, cannot easily be used for interactive shimming, and does not help to overcome dynamic range problems when observing small signals in the presence of intense water and fat signals. We have been using this sequence for localized ^{31}P NMR spectroscopy because it avoids the use of spin-echos, allowing measurement of short T_2 signals.

For an optimal definition of the localized volume, and perfect inversion in spite of RF inhomogeneity, we implemented frequency-modulated (FM) inversion pulses of the hyperbolic secant type.[16,17] We used a pulse of 5 msec with a bandwidth of 1600 Hz. The FM pulse has a superior inversion profile compared to the conventional sinc-gauss pulses, resulting in better defined localized volumes. Furthermore, the FM pulses have the property that above a certain threshold spin inversion is perfect, independent of RF power used. Thus by choosing a RF power level well above this threshold, an optimal inversion is achieved, eliminating the need of tedious optimization procedures.

Localized ^{31}P NMR Spectra of the Human Brain

We have used the techniques described above to obtain localized ^{31}P NMR spectra of the human brain.[18] A small ^{31}P NMR coil was inserted into the standard ^{1}H NMR head coil used for imaging. A ^{1}H NMR image was obtained to determine the location and size of the localized volume from which the ^{31}P NMR spectrum should be obtained. The ^{1}H NMR water signal was monitored for interactive shimming of the

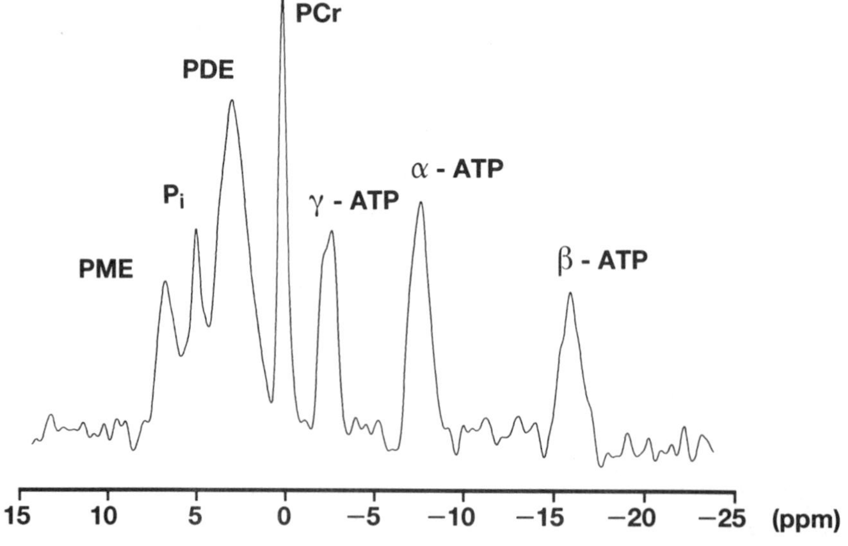

FIGURE 6. Localized 25.9 MHz ^{31}P NMR spectrum of a 6 × 6 × 6 cm VOI centered in one of the brain hemispheres of a normal volunteer. The spectrum was obtained using a 1.5 tesla whole body NMR instrument, and is the averaged result of 256 measurements, acquired with 3 sec pulse intervals. Signals are observed from ATP, phosphocreatine (PCr), phosphomonoesters (PME), inorganic phosphate (P_i), and phosphodiesters (PDE).

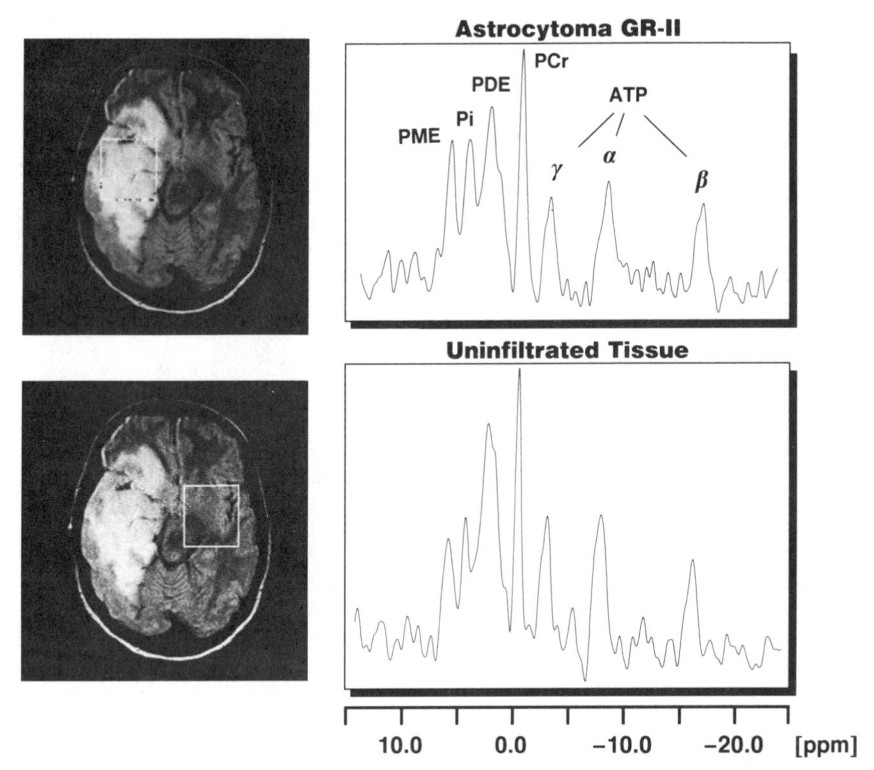

FIGURE 7. ^{1}H NMR images and localized ^{31}P NMR spectra of patient with a large astrocytoma grade II of the right temporal lobe. The VOIs (65 cc volume each) selected for spectroscopy are indicated in the images; one was centered in the tumor and the other in the uninfiltrated left hemisphere. Each of the two ^{31}P NMR spectra were obtained in 512 acquisitions, taken with 3 second pulse intervals (25 minutes measurement time). The tumor spectrum shows an elevated PME signal.

field homogeneity over the volume of interest; this was done by single-scan SPARS localization. The NMR instrument could be switched quickly from ^{1}H NMR imaging to ^{31}P NMR spectroscopy, since there was no need to move the patient or change the RF coils. The three dimensions of the localized volume could be adjusted independently by choosing the appropriate selection gradient strength. FIGURE 6 shows the 25.9 MHz ^{31}P NMR spectrum of a $6 \times 6 \times 6$ cm volume centered in one of the brain hemispheres of a normal volunteer. The spectrum was obtained on a 1.5 tesla Philips whole body instrument, using 3 sec pulse intervals, and a total accumulation time of 13 minutes.

FIGURE 7 shows ^{1}H NMR images and localized ^{31}P NMR spectra of patient who was diagnosed as suffering from a large astrocytoma grade II of the right temporal lobe, extending into the frontal and parietal regions. The NMR examination started with imaging, and two VOIs (dimension of $43 \times 43 \times 35$ mm, 65 cc) were selected, one centered upon the tumor and another VOI, which was centered in the left hemisphere. FIGURE 7 shows the ^{31}P NMR spectra measured of these two VOIs, whose position are indicated in the images. Each of the two spectra are the averaged result of 512 acquisitions, taken in 3 second pulse intervals (25 minutes measurement time). The two spectra were measured without changing coils or patient position. The ^{31}P NMR spectrum from the uninfiltrated left hemisphere is very similar to that obtained of a normal volunteer (FIG. 6). Comparing the two spectra from the tumor patient shows that the main difference between the two is that the phosphomonoester (PME) signal is elevated in the tumor spectrum. There was no significant difference in tissue pH (tumor pH was 7.02, uninfiltrated tissue pH 7.00).

A number of brain tumor patients have been examined by localized ^{31}P NMR spectroscopy.[19] The results show a significant variation, but often an elevated PME signal, and a reduced PCr signal has been observed in the tumor spectra. For a few patients, we performed a ^{31}P NMR examination before and after tumor treatment.

CONCLUSION

NMR spectroscopy of humans requires a precise and flexible definition of the sensitive volume for the examination. The approach we have taken integrates imaging and spectroscopy; the volume of interest is selected on the basis of an image. We have used volume coils instead of surface coils for studies of the human brain. This makes it possible to examine lesions located anywhere within the brain. The more homogeneous RF field from the volume coil provides a uniform NMR response over the VOI, which is necessary for obtaining well defined sensitive volumes. The volume coil allows measurement of several VOIs, without moving the patient or the coil. This facilitates direct comparison of spectra from a tumor and from non-infiltrated tissue of the same patient. Loss of NMR sensitivity as compared with surface coils is partially compensated by a more efficient localizing procedure, so that more time is available for spectral averaging within the same total examination time.

To maximize spectral information from human tissues different spectroscopic techniques should be combined with spectral localization. Using the spectral localization techniques discussed in this paper we have obtained target images, performed localized T_1 and T_2 measurements, and obtained water-suppressed ^1H NMR spectra of human tissues. This can be easily extended to other experimental techniques, such as spectral editing and saturation transfer. In this paper, we discussed the combination of SPARS localization with different spectroscopic techniques; however, ISIS is equally suitable for making such combinations.

The localization sequence used for ^1H NMR spectroscopy, SPARS, allows a signal from the VOI to be obtained in a single scan. This can be applied to monitor the water signal from the VOI for the purpose of interactive shimming to optimize spectral resolution. This procedure has become part of both ^1H and ^{31}P NMR spectroscopic examinations.

The results presented here show that image-guided localized spectroscopy provides an efficient approach to whole body spectroscopy. ^1H NMR images and spectroscopic information are obtained in a correlated way, such that the tissue examined by spectroscopy can be inspected on the image. Questions concerning tissue identification, heterogeneity, and partial volume effects can be addressed by imaging. That may provide information essential for the interpretation of spectroscopic results.

ACKNOWLEDGMENTS

We thank A. J. H. Marien, B. Sijtsma, J. P. Groen, L. J. Oosterwaal, and J. Bunke, who have been involved in various aspects of the work presented here. Drs. W.

Steinbrich (Department of Radiology, University of Cologne School of Medicine), C. M. Anderson, and M. W. Weiner (Department of Radiology, University of California San Francisco, and the S. F. V. A. Medical Center), have contributed to the localized ^{1}H NMR T_1 and T_2 relaxation studies of tumor patients. Drs. D. Balériaux, C. Segebarth (Hôpital Erasme, Université Libre de Bruxelles, Brussels, Belgium), and D. L. Arnold (Montreal Neurological Institute and Hospital, McGill University, Montreal, Canada) have collaborated on the localized ^{31}P NMR studies of tumor patients. All NMR experiments were performed using a Philips S15 Gyroscan MR imager, at Philips Medical Systems, Best, The Netherlands.

REFERENCES

1. Ackerman, J. J. H., T. H. Grove, G. G. Wong, D. G. Gadian & G. K. Radda. 1980. Mapping of metabolites in whole animals by 31P NMR using surface coils. Nature (London) **283:** 167–170.
2. Styles P., C. A. Scott & G. K. Radda. 1985. A method for localizing high resolution NMR spectra from human subjects. Magn. Reson. Med. **2:** 402–409.
3. Bendall, M. R. & R. E. Gordon. 1983. Depth and refocussing pulses designed for multipulse NMR with surface coils. J. Magn. Reson. **53:** 365–385.
4. Bottomley, P. A., T. H. Foster & R. D. Darrow. 1984. Depth-resolved surface-coil spectroscopy (DRESS) for *in vivo* ^{1}H, ^{31}P and ^{13}C NMR. J. Magn. Reson. **59:** 338–342.
5. Aue, W. P., S. Mueller, T. A. Cross & J. Seelig. 1984. Volume-selective excitation. A novel approach to topical NMR. J. Magn. Reson. **56:** 350–354.
6. Luyten, P. R., A. J. H. Marien, B. Sijtsma & J. A. den Hollander. 1986. Solvent-suppressed spatially resolved spectroscopy. An approach to high resolution NMR on a whole body MR system. J. Magn. Reson. **67:** 148–155.
7. Luyten, P. R. & J. A. den Hollander. 1986. ^{1}H MR spatially resolved spectroscopy of human tissues *in situ*. Magn. Reson. Imaging **4:** 237–239.
8. Luyten, P. R. & J. A. den Hollander. 1985. Target imaging using volume-selective excitation. Abstr. Soc. Magn. Reson. Med. 4th Annu. Meeting: 1021–1022.
9. Luyten, P. R. & J. A. den Hollander. 1986. Observation of metabolites in the human brain by MR spectroscopy. Radiology **161:** 795–798.
10. Luyten, P. R. & J. A. den Hollander. 1986. Spatially resolved high resolution ^{1}H NMR relaxation measurements of human tissues *in situ*. Biophysical J. **49:** 471a.
11. Luyten, P. R., C. M. Anderson & J. A. den Hollander. 1987. ^{1}H NMR relaxation measurements of human tissues *in situ* by spatially resolved spectroscopy. Magn. Reson. Med. **4:** 431–440.
12. Steinbrich, W., C. M. Anderson, M. W. Weiner, J. Bunke, P. R. Luyten & J. A. den Hollander. 1986. Measurement of ^{1}H MR relaxation rates in human brain tumors *in situ* by spatially resolved spectroscopy. Abstr. Soc. Magn. Reson. Med. 5th Annu. Meeting: 585–586.
13. Behar, K. L., J. A. den Hollander, M. E. Stromski, T. Ogino, R. G. Shulman, O. A. C. Petroff & J. W. Prichard. 1983. High-resolution ^{1}H nuclear magnetic resonance study of cerebral hypoxia *in vivo*. Proc. Natl. Acad. Sci. USA **80:** 4945–4948.
14. Hore, P. J. 1983. Solvent suppression in fourier transform nuclear magnetic resonance. J. Magn. Reson. **55:** 283–300.
15. Ordidge, R. J., A. Connelly & J. A. B. Lohman. 1986. Image-selected *in vivo* spectroscopy (ISIS). A new technique for spatially selective NMR spectroscopy. J. Magn. Reson. **66:** 283–294.
16. Baum, J., R. Tycko & A. Pines. 1985. Broadband and adiabatic inversion of a two-level system by phase-modulated pulses. Phys. Rev. A **32:** 3435–3447.
17. Silver, M. S., R. I. Joseph & D. I. Hoult. 1985. Selective spin inversion in nuclear magnetic resonance and coherent optics through an exact solution of the Bloch-Riccati equation. Phys. Rev. A **31:** 2753–2755.

18. LUYTEN, P. R., J. P. GROEN, D. L. ARNOLD, D. BALERIAUX & J. A. DEN HOLLANDER. 1986. ^{31}P MR localized spectroscopy of the human brain *in situ* at 1.5 Tesla. Abstr. Soc. Magn. Reson. Med. 5th Annu. Meeting: 1083–1084.
19. BALERIAUX, D., D. L. ARNOLD, C. SEGEBARTH, P. R. LUYTEN & J. A. DEN HOLLANDER. 1986. ^{31}P MR evaluation of human brain tumor response to therapy. Abstr. Soc. Magn. Reson. Med. 5th Annu. Meeting: 41–42.

DISCUSSION OF THE PAPER

T. R. BROWN: I wondered what the echo time was that you were using on the protons. I'm still concerned that we don't see any ATP in the proton spectrum. I think we all agree that it's a T_2 problem, but I just want to make sure I know your echo time. And the other thing is what was the field used for the T_1 measurements?

J. A. DEN HOLLANDER: We have been using different echo times for different measurements. The SPARS sequence itself has a rather short echo time, there's a 5 msec delay between the first 90 excitation pulse and then the 90 reset pulse so that gives you a 15 msec total T_2 relaxation period. However, for the water suppressed spectrum, that is probably the one that you are referring to, the total echo time in that case was 272 msec, that's probably enough to make the ATP go away. All measurements were done at 1.5 tesla.

NMR Microscopic Studies of Eyes and Tumors with Histological Correlation

JAMES B. AGUAYO, STEPHEN J. BLACKBAND,
JANNA P. WEHRLE, AND J. D. GLICKSON

Division of NMR Research
Department of Radiology
Wilmer Ophthalmological Institute
and
Department of Biological Chemistry
The Johns Hopkins University School of Medicine
Baltimore, Maryland 21205

MARK A. MATTINGLY

Bruker Medical Instruments
Manning Park
Billerica, Massachusetts 01821

INTRODUCTION

Recent reports have demonstrated the feasibility of performing NMR microscopy on small biological samples at magnetic fields ranging from 1.5 to 9.4 Tesla.[1-6] The highest in-plane resolution obtained to date is 10 × 13 microns on single cells.[1] At a resolution of 50 microns, which has been achieved with a commercial 1.5 T instrument,[2] noninvasive detection of tissue histological patterns should become possible. In many cases, these tissue patterns will be sufficient to characterize the type or stage of a disease process in a manner similar to conventional histological evaluation. Unlike conventional histologic techniques, however, NMR microscopic examination does not require tissue fixing, staining, or sectioning, which can produce artifacts and prohibit *in vivo* examination. Furthermore, NMR microscopy can be used to sequentially follow tissue changes and to provide dynamic information (T_1, T_2, and diffusion coefficients) and a limited degree of chemical information (e.g., lipid/water distinction).

This study is directed at exploring the utility of *in vivo* NMR microscopic examination of tissues in animal models and in pathological specimens. We have conducted high field ^{1}H NMR microscopic studies and comparative histological examinations of a number of murine tissues (subcutaneous tumors and various intrathoracic organs, which have been examined by NMR *in situ*) and excised eyes.

EXPERIMENTAL METHODS

NMR Microscopy

All measurements were performed on a conventional Bruker AM 400-WB spectrometer equipped with an 89 mm bore superconducting magnet (Bruker Medical Instruments, Billerica, MA) and with imaging hardware and software. The imaging accessories consisted of gradient coils (50 mm in diamter), gradient power supplies, three programmable waveform memories, and a radiofrequency (rf) modulator. The rf probe assembly utilized a 10 mm or a 25 mm diameter slotted tube resonator or a 5 mm diameter horizontal solenoidal rf coil.

A spin-echo pulse sequence was employed,[7] and the 2D Fourier transform method[8] was used for image reconstruction. For routine images, echo times (TE) were kept very short (17 msec) to minimize both imaging times and self-diffusion effects. Images along the three principal axes were obtained with either single- or multi-slice programs, and image display was achieved by a 512 × 512 matrix display processor in conjunction with a high resolution black-and-white monitor. All experiments were conducted at ambient temperature. The use of small-sized gradient coils was necessary for achieving gradients between 5 and 20 G/cm, which were adequate to dominate the water linewidth.[9] A matched bandwidth of 12.2 to 25 kHz was used. Image profiles and scout images were employed to locate the region of interest and to determine the frequency offsets and bandwidths required to use the optimal square matrix that could encompass the sample.

S/N ratios were calculated using standard Bruker software, which computes S/N according to following equation:

$$S/N = I_{tissue}/RMS$$

where I_{tissue} represents the signal intensity of a particular region of tissue. RMS is the root mean square of the background noise signal intensity. Signal averaging and/or short TEs and long TRs were used to maximize image signal intensity. No experiments required more than four averages.

Image contrast (C) was determined from the following equation:

$$C = (I_{tissue1} - I_{tissue2})/I_{tissue1}$$

where $I_{tissue1}$ and $I_{tissue2}$ are the relative signal intensities of two tissue regions in an image.[10] In the case of ocular studies, $I_{tissue1}$ was the relative signal intensity of vitreous and $I_{tissue2}$ was the signal intensity of the lens nucleus.

Field Homogeneity and Spatial Resolution

Field (B_1) homogeneity of the slotted tube resonator and solenoidal rf coils was determined by imaging a cylindrical water phantom. The uniformity of the image indicated that rf field inhomogeneity and penetration artifacts were absent. Image resolution (voxel dimension) was calibrated by imaging glass capillary phantoms of known diameter (FIG. 1). The internal diameter of the glass NMR container used in the eye and cell studies also served as a standard for calibration of spatial resolution and determination of gradient linearity.

Excised Tissues

Human eyes were obtained from the International Medical Eye Bank within 24 hours of expiration. Animal eyes were obtained from albino rabbits (1 lb) and from CD albino mice, which were killed with pentobarbital prior to the enucleation of the eyes.

All images were obtained within 6 hours of enucleation. The mouse eyes were placed in a 1.1 mm (i.d.) tube with buffer solution to minimize tissue deterioration. The buffer solution also served as an internal standard of image intensity.

Mouse Studies

C3H/HEJ female mice were purchased from Jackson Laboratories (Bar Harbor, ME). Studies on the thorax were performed on an 8 week-old mouse euthanized with 400 mg/kg pentobarbital immediately prior to imaging. For tumor studies, 10^5 RIF-1 cells maintained by the protocol of Twentyman *et al.*[11] were subcutaneously implanted in the scalp of a 4-week-old mouse and allowed to grow for 4 weeks prior to examination, at which point the mice were sacrificed (*vide supra*). Images were obtained with a 25 mm diameter slotted tube resonator.

FIGURE 1. An axial image through a 4.1 mm tube filled with five 1.5 mm (o.d.) capillary tubes surrounded by water. The internal diameters of the tubes from largest to smallest are 0.9, 0.6, 0.4 mm, and (two) 0.2 mm. Imaging delays were a TE of 17 msec and a TR of 500 msec, and one average was obtained. The S/N of the water is 47:1, an image matrix size of 256 × 256 was used, and voxel dimensions are 16 × 16 × 500 microns. The image was obtained using a 5 mm solenoid rf coil.

Histological Studies

Immediately after imaging, tissues and whole mice were fixed in 10% neutral buffered formalin. Samples were embedded in paraffin blocks and sectioned at 6 to 8

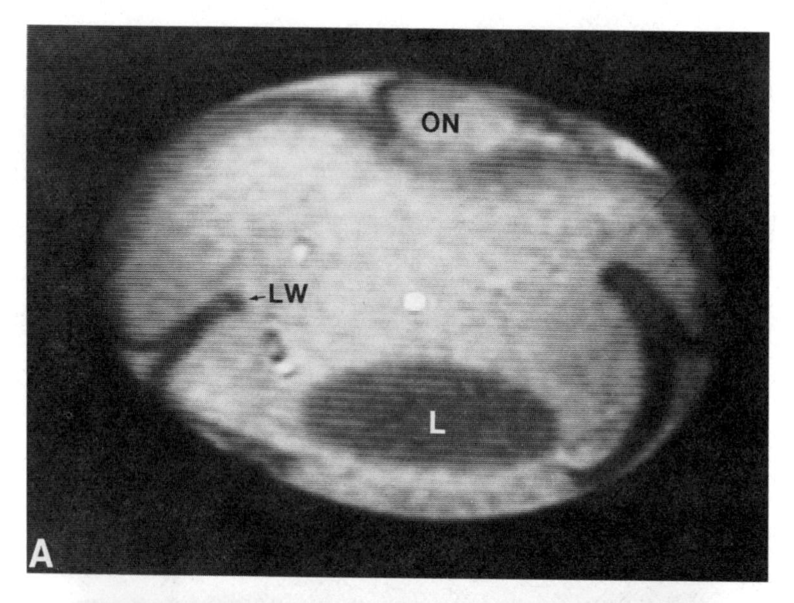

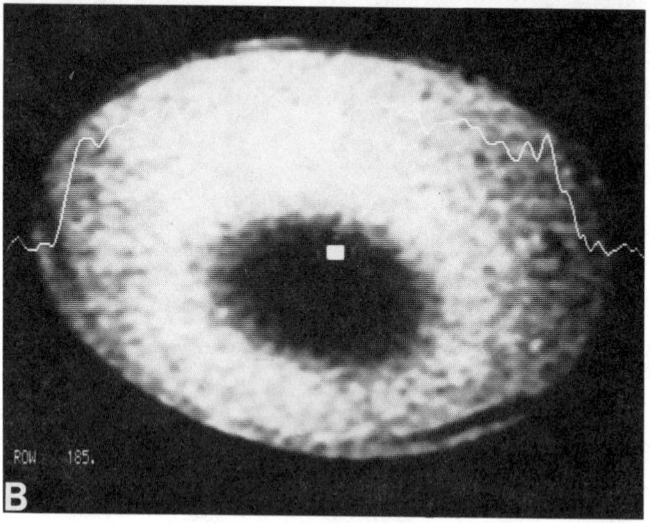

FIGURE 2. (a) Image of a human eye using a 25 mm slotted tube resonator. The imaging delays were a TE of 17 msec and a TR of 1 sec, which produced the 256 × 256 data array in 4 min and 16 sec. The in-plane resolution of the image is 120 × 230 μm. The outer eye wall, which represents the retina-choroid-sclera complex (EW), is folded into the vitreous cavity because the eye was slightly larger than the NMR container (25 mm) and thus had to be compressed. The lens (L) has a low signal intensity secondary to the lack of protons in the lens nucleus. The optic nerve (ON) indents the posterior section of the eye and has an intermediate signal intensity. The bright spot in the center of the image is an artifact produced during image processing. (b) Image of a rabbit eye using a 10 mm slotted tube resonator. A TR of 2 sec and a TE of 17 msec was used.

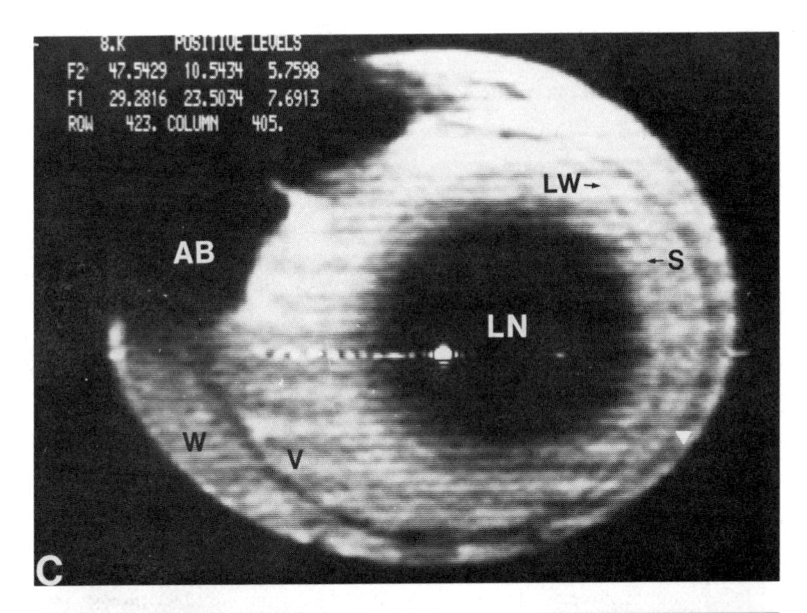

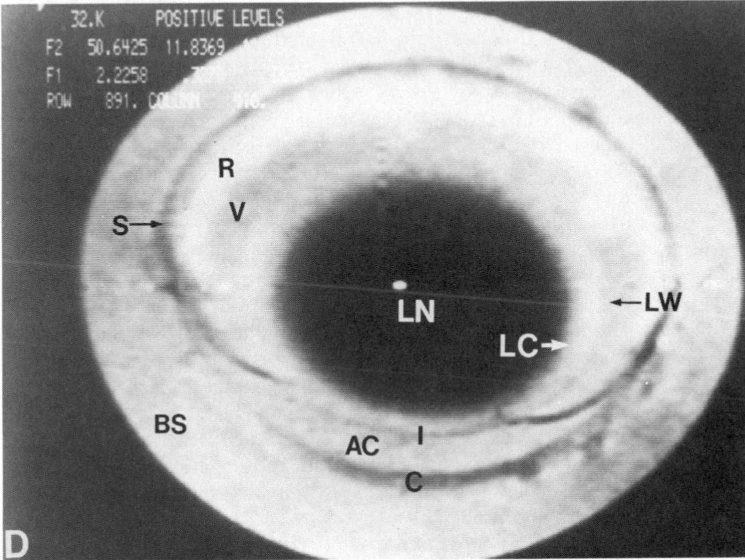

FIGURE 2. Continued. (c) Image of a rat eye (3 × 3 × 3 mm) using a 5 mm solenoid. The matrix size is 256 × 256. The slice thickness is 250 μm. AB = an air bubble, W = water bubble, LW = lens wall, V = vitreous, LN = lens nucleus, and S = sclera. (d) An image of a mouse eye in buffer solution in a 4.1 mm diameter capillary tube (5 mm solenoid). The 256 × 256 data array has an in-plane resolution of 16 × 27 microns and a slice width of 500 microns. A TR of 4 sec and a TE of 20 msec was used, resulting in a total imaging time of 17 min. The bottom-to-top of the image corresponds approximately to the front-to-back of the eye. C = cornea, AC = anterior chamber, I = iris, LN = lens nucleus, LC = lens cortex, LW = lens wall, V = vitreous, R = retina, S = sclera, and BS = buffer solution.

microns. Sections were mounted on slides and stained with hematoxylin and eosin (H&E). The samples were oriented so that the sectioned planes corresponded approximately to the imaging planes.

RESULTS AND DISCUSSION

Excised Tissue Studies

The eye is a particularly interesting organ for microscopic studies because it contains a diverse number of cell types and structures having different spin densities and spin relaxation characteristics. It is effectively a closed system and is subject to little deterioration after excision. Because eyes can be obtained in different sizes from different species, coils of different diameter (5 mm to 25 mm) were employed, as shown in FIGURE 2. All of these images (human, rabbit, rat, and mouse eyes) are primarily spin density–weighted; thus, the signal intensity reflects water content of the different ocular tissues.[12] The lens is known to have the lowest percentage of water of the body tissues (40%) and sclera has both a low water content and a short T_2.[12] Much of the water present in the lens nucleus is believed to be bound,[13] which would lead to further loss of signal intensity in the region of the lens nucleus. In contrast, vitreous (99% water) and aqueous humor (100% water) have high signal intensities. The vitreous cavity is composed of a homogeneous mixture of water, hyaluronate, and collagen.

These images demonstrate that in-plane resolutions between 16 and 32 microns can be obtained on excised tissues of less than 4 mm in diameter.[4] For objects between 5 and 10 mm and between 10 and 25 mm in diameter, resolutions on the order of 39 to 100 microns and 88 and 200 microns, respectively, can be obtained in similar imaging times. The S/N on these samples is of the same order of magnitude because of the smaller rf coils used at the higher resolutions. To obtain better resolution on the larger samples without the use of micro-surface coils or zoom imaging methods, prohibitively long imaging times and larger imaging matrices than are presently available would be required.

The highest resolution obtained on an excised tissue sample is $16 \times 23 \times 250$ microns (FIG. 2d) These dimensions correspond to a volume of 1×10^{-4} ml. Because of the increased resolution and tissue contrast when compared to FIGURE 2a–c, a significant amount of fine structure is distinguishable. The lens capsule, which is the basement membrane of the lens epithelial cells, is clearly resolved. Its thickness is known to be between 10 and 40 microns.[14] Individual cell layers of the ciliary body are also discernible. FIGURE 3 shows a histological section of a similar mouse eye.

Even though the resolution of light microscopy (0.2 microns) is far greater than that obtained by NMR microscopy, the level of structural clarity obtained on tissue samples such as the mouse eye approaches that seen in low-magnification histologic examination. However, the sectioning process introduces significant distortions in the image of the eye obtained from the light microscope, particularly for soft tissues such as the eye wall. Because of the compact nature of the lens nucleus, which leads to shearing and increased resistance to the sectioning blade, the image of the lens is significantly distorted. The absence of sectioning requirements is a major advantage of NMR microscopy and partially compensates for its poorer resolution when compared to light microscopy.

Mouse Thorax

FIGURE 4 shows a multi-slice image study through the thorax of a mouse. A number of anatomical structures, such as the pulmonary bronchi, thoracic vessels, and heart, can be seen. There is a substantial chemical shift artifact originating from the fat layer on the thoracic wall. It is displaced 13 pixels with respect to the water image, which corresponds to a 3.2 ppm shift. This chemical shift, in turn, corresponds to the actual spectral separation of the water and lipid resonances.[15] At this frequency, structures to the right of the fat pockets are difficult to visualize. Although the artifact does create problems for viewing adjacent structures, techniques already described[15,16]

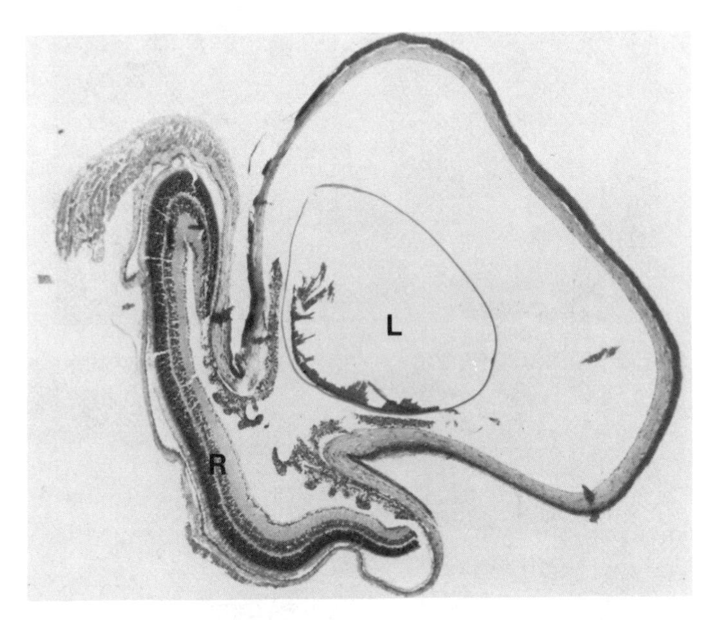

FIGURE 3. Histologic section through a mouse eye. The structures observed in the image of FIG. 2d are discerned. The eye is distorted due to sectioning artifacts, which can result in displacement of tissue, as demonstrated in the lens. Note, however, that there are no distortions in the NMR micrograph.

for removing it and for separately detecting water and lipid images (or sums and differences of these images) can be implemented on our instrument.

The chemical shift information can be utilized to monitor lipid metabolism at various times.[1] On small objects such as a frog egg,[1] the shift is large enough to generate distinguishable lipid and water images without the use of selective detection techniques. The displaced fat image in the mouse thorax is both broader than expected and smeared. The broadening occurs because the linewidth of fat is greater than that of water; although the gradients used in this image dominated the water linewidth, they did not dominate the lipid linewidth. Thus, the given resolution applies only to the water image. Image smearing also exists from volume-averaging effects because of the high signal intensity of fat.

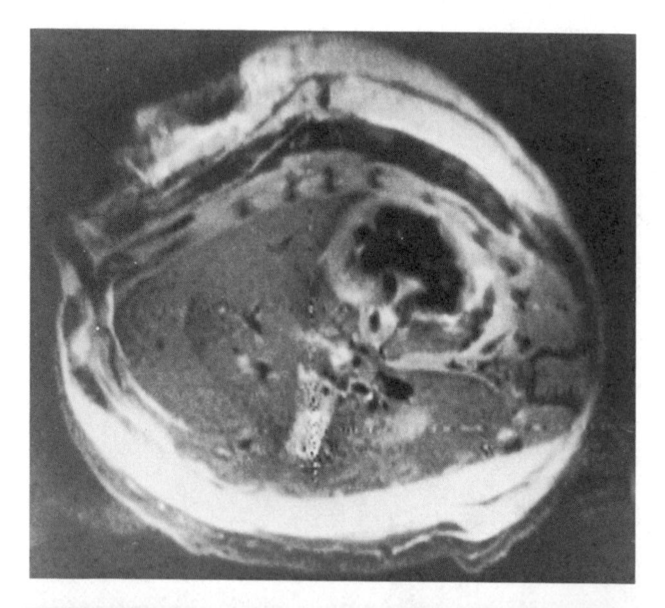

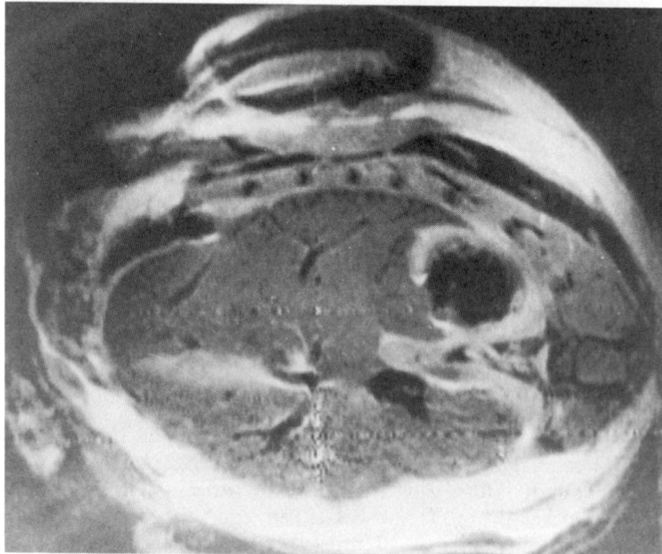

FIGURE 4. An axial section through the thorax of a mouse. The image matrix is 256 × 256. The in-plane resolution is 88 × 97 microns with a 500 μm slice width. A TR of 2 sec was used and two averages were obtained. The heart, mediastinal, and pulmonary vessels can be seen. The fat on the rib cage is markedly shifted.

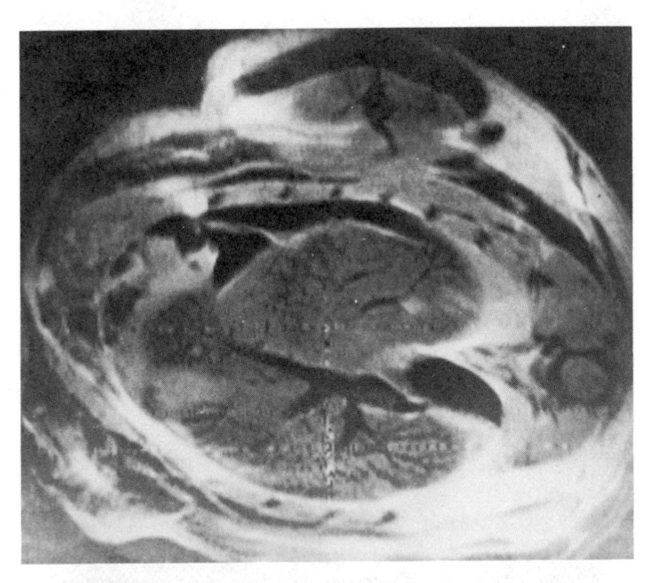

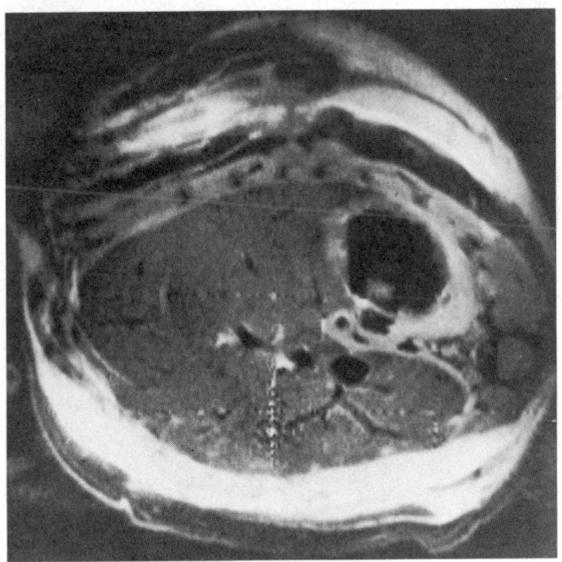

FIGURE 4. Continued

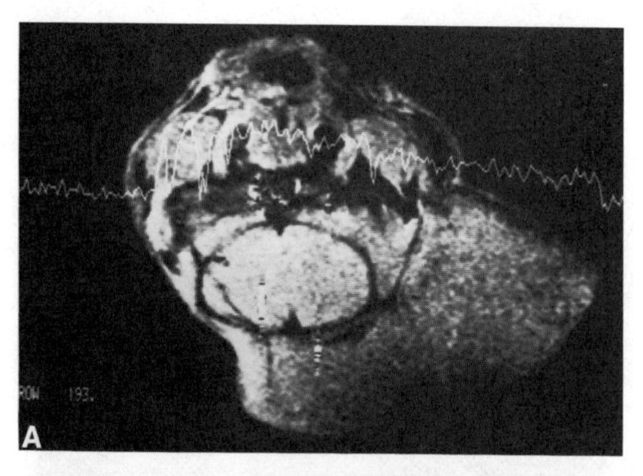

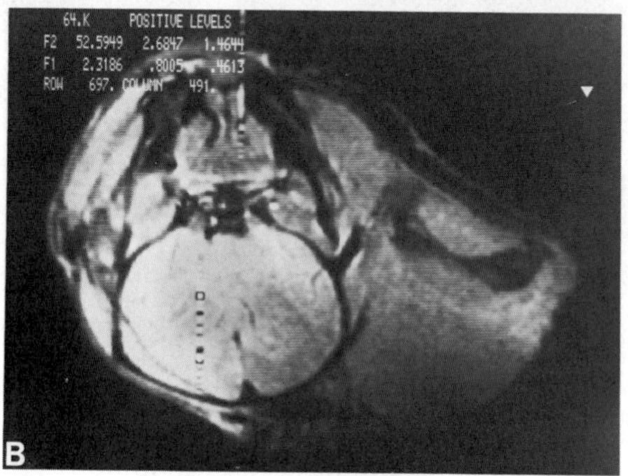

RIF-1 TUMOR

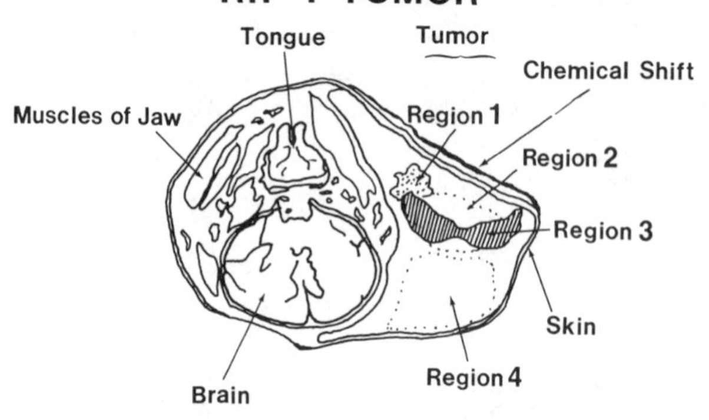

FIGURE 5. (a) An axial section taken through the head of a mouse with a subcutaneously implanted RIF-1 tumor. Some structural details can be seen in the brain of the mouse. The large mass to the right of the brain is the large, subcutaneously implanted tumor, which appears to be homogeneous in this image. (b) Image of mouse tumor with area of necrosis obtained as part of the multi-slice sequence used to produce (a). Some heterogeneity can be observed in the tumor, such as the large lesion in the shape of a figure eight. Five different region regions are clearly seen within the tumor. (c) Illustration of (b) indicating the anatomy of the mouse.

Mouse Tumor

Images of a subcutaneously implanted RIF-1 tumor are shown in FIGURE 5. The chemical shift effect of fat observed on the outside edge of the tumor does not pose a problem since there is little, if any, fat in the mouse head. With a resolution of 97 × 88 microns and a slice width of 500 microns, a tremendous amount of structural detail is observed. The muscle compartments of the mandible and head, as well as regions in the oral- and naso-pharynx, are easy to discern. Although the cortical and hippocampal convolutions of the brain are evident, little tissue contrast is seen because of the small amount of gray matter in the brain of the mouse, as is verified in the corresponding

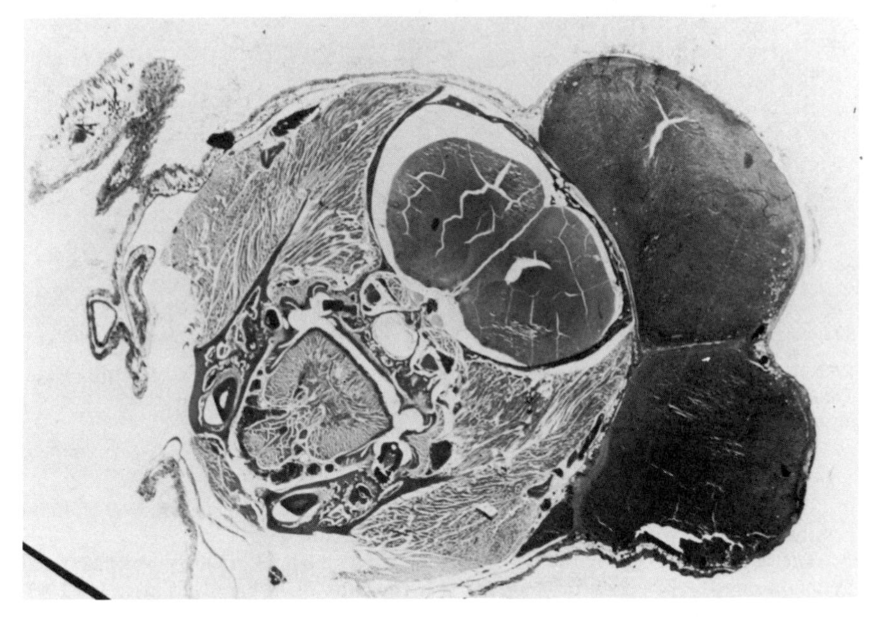

FIGURE 6. Histological section of the same mouse imaged in FIG. 5. Although the exact sections do not correspond, a general correlation is evident.

histological section (FIG. 6). The large mass to the right of the brain is the subcutaneously implanted RIF-1 tumor. We noted no infiltration in either the NMR image or the histological sections. Heterogeneous regions were observed in different NMR sections through the mouse tumor (FIG. 5B). Although the precise nature of these heterogeneous regions is unknown, some are believed to originate from regions of hyaline necrosis, which were observed in histological sections (FIG. 7) in some subcutaneously implanted RIF-1 tumors.

To examine the dependence of transverse relaxation (T_2) on image signal intensity, TE was varied while all other imaging parameters were kept constant (FIG. 8). The S/N from five regions within the images were examined. As the TE was increased from 17 msec to 77 msec, an approximately exponential decrease in S/N was observed

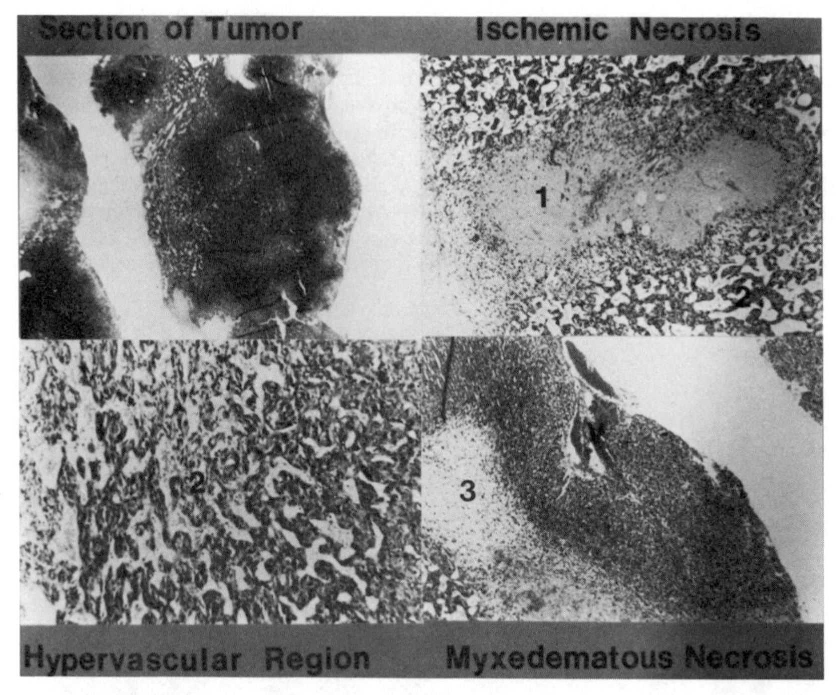

FIGURE 7. A histological sections of a RIF-1 subcutaneous grown tumor in a C3H/HeJ mouse. Three distinct regions are noted (1) ischemic necrosis, (2) hypervascular region, and (3) myxedematous necrosis.

in all of the regions, as expected (FIG. 9). Further studies are required to evaluate possible differences in relaxation rates between the tissue types.

The dependence of longitudinal relaxation time on signal intensity was examined by increasing the TR from 0.1 to 4 sec at a constant TE value of 17 msec (FIG. 8). Image signal intensity increased at longer TRs, a trend illustrated in FIGURE 9, exhibiting the expected exponential form. Additional studies are required to evaluate possible differences between the tissue types.

CONCLUSION

In this paper we described the adaptation of a conventional high field, wide bore spectrometer with imaging hardware and software for NMR microscopy studies. Capabilities of the instrument are sufficient to resolve cyto-architectural features in excised tissues and small animals. These results suggest that NMR microscopy should be immediately applicable to the examination of excised tissues, such as those encountered in medical pathology, and for the examination of lesions in experimental animal models.

Medical diagnosis is based on microscopic examination of diseased tissues, whereby structural and staining characteristics of a tissue make it possible to determine disease states and processes. Current techniques require that the sample be fixed in formalin, imbedded in a firm matrix, sectioned, mounted onto a slide, and stained for microscopic evaluation. Not only does this process take several days before the tissue can be examined, but it can also introduce artifacts such as those demonstrated by the histological preparation of the eye. NMR images of multiple sections through a tissue, on the other hand, can be obtained within 30 minutes. Although less than that obtained by optical methods, the resolution of NMR microscopy is sufficient to resolve cell clusters and structures as small as basement membranes and is free from the artifacts that can occur in histological examination.

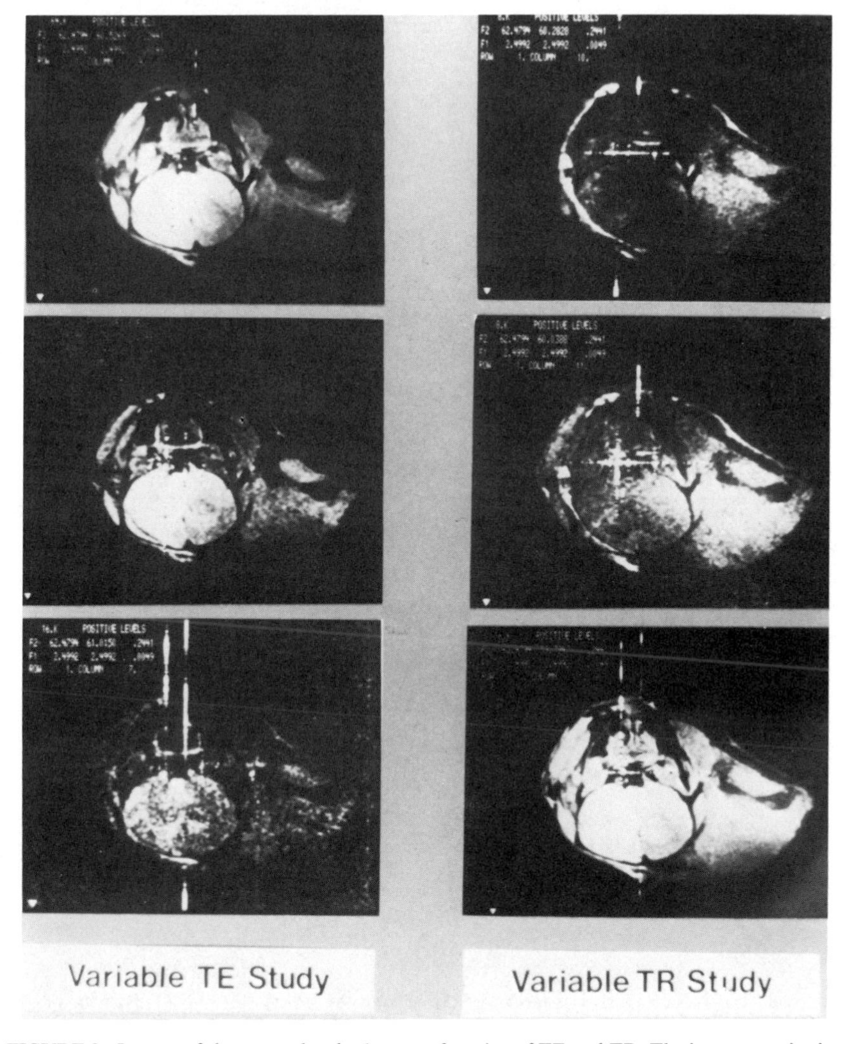

FIGURE 8. Images of the mouse head taken as a function of TE and TR. The image matrix size is 256 × 256. Two averages were obtained. For the TE-weighted images, the TR is fixed at 2,050 msec. From top-to-bottom the images have a TE of 17, 37, and 77 msec. For the TR-weighted images, TE was fixed at 17 msec. The TR values were 0.5, 1, and 4 sec.

The information obtained by NMR microscopy reflects both the distribution of water in a biological system and its movement and interaction with its cellular matrix. NMR microscopy can also provide chemical (lipid), flow, and diffusion information, which are not readily obtained by other techniques. Still, the most important advantage of NMR microscopy over histologic methods is that live samples can be examined, making it possible to examine a lesion in its natural environment and follow changes as the lesion evolves over time. Because NMR does not disrupt tissue, NMR and histological examinations can serve as complementary techniques. As a result, NMR microscopy offers a new and unique view of tissue to the pathologist and medical researcher. NMR microscopy may also be enormously useful to the cell biologist as a means to investigate the properties of the cell nucleus and cytoplasm during the processes of malignant transformation or embryogenesis.[1] These applications of NMR microscopy appear likely as resolution continues to improve via smaller or more sensitive rf coils and by imaging at higher fields.

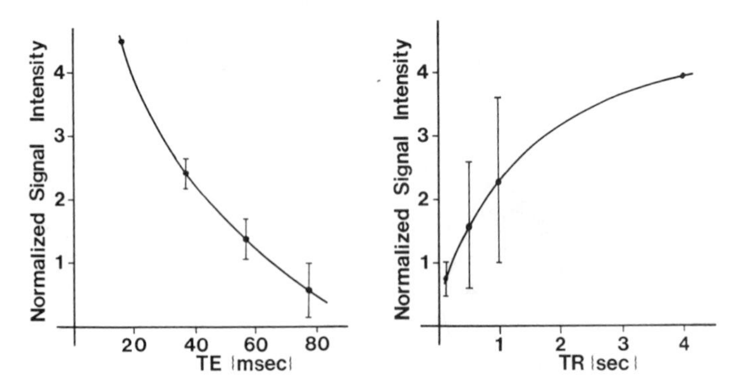

FIGURE 9. Graphs illustrating the dependence of the signal intensities in the images as a function of TE and TR. The data points are an average of several regions within the mouse head. More data will be required before observations on the different relationships between the tissues can be made, but an exception is seen in the lipid signal, which has a shorter T1 than the rest of the tissues on the TR-weighted images.

ACKNOWLEDGMENTS

Many thanks to C. Paella Martin, Joseph Schoeniger, H-M. Cheng, and J. Eggeleston for their assistance.

REFERENCES

1. AGUAYO, J. B., S. J. BLACKBAND, J. SCHOENIGER, M. MATTINGLY & M. HINTERMANN. 1986. NMR imaging of a single cell. Nature **322:**190.
2. JOHNSON, G. A., M. THOMPSON, S. GEWALT & C. HAYES. 1986. Nuclear magnetic resonance imaging at microscopic resolution. J. Magn. Reson. **68:** 129.
3. MANSFIELD, P. & P. GRANNELL. 1975. Diffraction and microscopy in solids and liquids by NMR. Phys. Rev. B. **12:** 3629.

4. HALL, L. D. & S. SUKUMAR. 1982. Chemical microscopy using a high-resolution NMR spectrometer. A combination of tomography/spectroscopy using either ^{1}H or ^{13}C. J. Magn. Reson. **50:** 161.
5. MEYER, P. & T. R. BROWN. 1985. Proton imaging at 400 MHz. 4th Annu. Mtg. Soc. Magnetic Resonance Med. **2:** 1105.
6. LAUTERBUR, P. 1984. Spectroscopic imaging of microscopic objects. IEEE Trans. Nucl. Sci. NA-31 No. 4.
7. EDELSTEIN, W. A., J. M. S. HUTCHINSON & G. JOHNSON. 1980. Spin warp NMR imaging and applications to human whole body line-scan imaging. Phys. Med. Biol. **25:** 751.
8. KUMAR, A., D. WELTI & R. R. ERNST. 1975. NMR fourier zeugmatography. J. Magn. Reson. **18:** 69.
9. MANSFIELD, P., MORRIS, P. G. 1982. NMR imaging in biomedicine. Adv. Mag. Res. Suppl. 2, Academic Press: New York.
10. ROSE, A. 1974. Vision: Human and Electronic. pp 1–17. Plenum Press. New York.
11. TWENTYMAN, P. R., J. M. BROWN, J. W. GRAY, A. J. FRANKO, M. A. SCOLES & R. F. KALLMAN. 1980. A new mouse tumor model system (RIF-1) for comparison of end-point studies. J. Natl. Cancer Inst. **64:** 595.
12. HUGGERT, A. & E. ODEBLAD. 1959. Proton magnetic resonance studies of some tissues and fluids of the eye. Acta Radiol. **51:** 385.
13. NEVILLE, M. C., C. A. PATTERSON, J. RAE & D. WEOSSNER. Nuclear magnetic resonance studies and water "ordering" in the crystalline lens. Science **184:** 1072.
14. HOGAN, M. J., J. ALVARADO & J. WENDDELL 1971. Histology of the human eye. W. B. Saunders Company. Philadelphia.
15. DIXON, W. T. 1984. Simple proton spectroscopy. Radiology **153:** 189.
16. YEUNG, H. & D. KORMOS. 1986. Separation of true fat and water images by correcting magnetic field in homogeneity separation by method of *in situ*. Radiology **159:** 783.

Decoupler Heating in Phantoms at 1.9 T Using Surface Coils

JEFFRY R. ALGER[a]

Departments of Diagnostic Imaging and Molecular Biophysics and Biochemistry
Yale University School of Medicine
New Haven, Connecticut 06510

^{1}H decoupling will be helpful for further extension of the present ^{13}C NMR experiments on animal metabolism to human investigation. However, heating associated with absorption of the decoupler power is the anticipated risk factor associated with decoupling. Here calorimetric measurements of the decoupler heating produced in a phantom sample under conditions that simulate those of *in vivo* ^{13}C spectroscopy are performed in an effort to see how severe the heating might be.

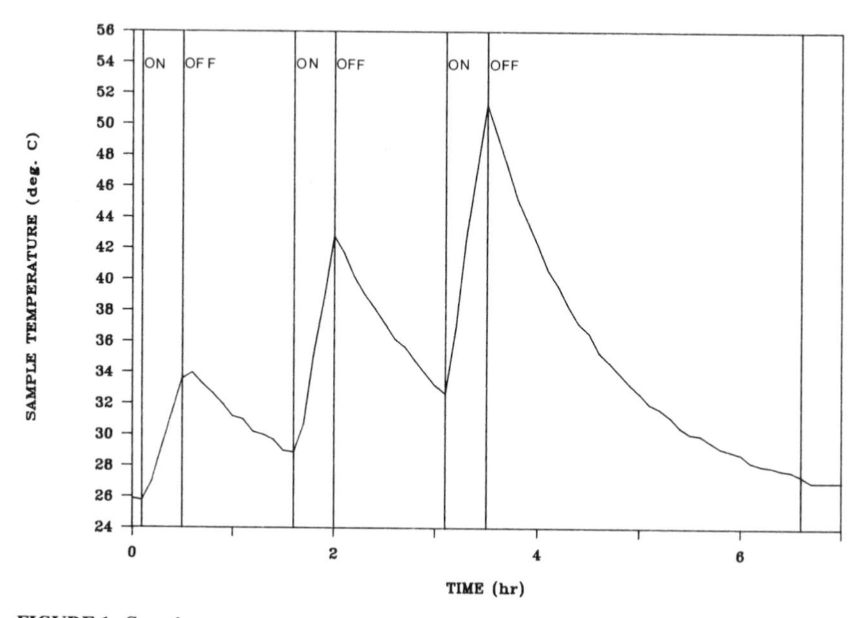

FIGURE 1. Sample temperature versus time. The decoupler is on during the periods labeled "on" at time average power levels of 27 W during the first period, 47 W during the second, and 65 W during the third period. The decoupler is off during the interspersed time periods where the cooling rate is measured.

[a]Present address: National Institute of Neurological and Communicative Disorders and Stroke, National Institutes of Health, Bethesda, MD 20892.

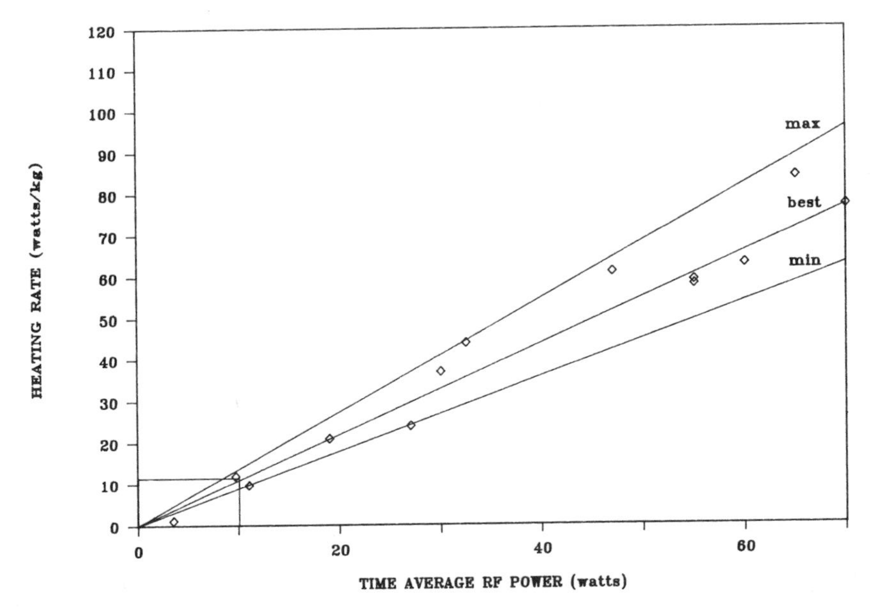

FIGURE 2. Total heating rate (corrected to account for the effects of self-cooling) versus the time average RF power level. The observed data are shown as squares, together with a best fit line, and estimates of the maximum and minimum error in the slope due to data scatter are also shown. The slope of the best-fit line is 1.1 W/kg heat per watt of RF. The sample mass is 0.75 kg, so that 0.82 W of heat appear for each watt of RF. In lower left corner, the figure shows that 11.6 W/kg will be produced by broadband decoupling using an average RF power level of 10 W.

The phantom sample consisted of a plastic box of dimensions 120 mm × 174 mm filled to a height of 38 mm with 750 ml of 135 mM potassium chloride. The bottom of the box was located about 7 mm above the plane of a 80.9 MHz single-turn proton surface coil having a diameter of 80 mm. Experiments were performed with an Oxford Research Systems TMR-32 spectrometer. The sample temperature was measured during and after decoupling with a Luxtron fluoroptic thermometer.

A typical series of measurements are shown in FIGURE 1. Decoupling increases the temperature during the periods labeled "on." The existence of a substantial cooling capacity is also apparent in the periods labeled "off." The instantaneous rate of cooling was determined to be a linear function of the difference between the sample temperature and surroundings.

The total heating rate was calculated by measuring the temperature increase during decoupling and adding this to the cooling rate at the same temperature as evaluated after turning the decoupler off. The total heating rate was determined at a number of incident RF power levels. The results plotted in FIGURE 2 show that 82% of the applied RF power appeared as sample heat.

Heat deposition resulting from application of power sufficient for single frequency decoupling (determined to be in the neighborhood of 1.5 W, resulted in a heating rate of about 1.6 W/kg), which is close to the U.S. Food and Drug Administration's guideline limits for RF exposure in NMR studies. Broadband decoupling conditions (about 10 W of applied RF) substantially exceeded the guidelines depositing about 11 W/kg. Nevertheless, the temperature increases induced by these decoupling conditions are small. Broadband decoupling conditions would produce a temperature change of about 9°C/hr in a perfectly insulated 1.0 kg sample. Single frequency decoupling conditions would produce less than 1.3°C/hr. In fact, the rate of temperature increase in the phantom during single frequency decoupling is so small that it is difficult to measure. At 37°C the passive cooling of the phantom sample is sufficient to maintain the temperature constant even during broadband decoupling. Given the presence of flow in the living system and the fact that only a small part of the body is subjected to the decoupling field, the observed increase in tissue temperature caused by decoupling is expected to be even smaller than it was in this phantom study.

^{31}P NMR Evaluation of *in Vivo* Renal Ischemia and Reperfusion Injury in Rabbits and Dogs

LOUIS F. MARTIN,[a] DAVID M. FEHR,[a]
ANASTASIUS O. PETER,[a] JOSEPH B. SANFORD,[a]
IDA N. GORMAN, AND RICHARD W. BRIGGS

Departments of [a]Surgery and Radiology
Milton S. Hershey Medical Center
Pennsylvania State University College of Medicine
Hershey, Pennsylvania 17033

Despite extensive investigation of post-ischemic acute renal failure, the relationship between renal adenosine triphosphate (ATP) concentration changes and recovery after various ischemic periods remains ill-defined. Phosphorus-31 nuclear magnetic resonance (^{31}P-NMR) has the potential of providing a noninvasive, continuous assessment of phosphate metabolite concentrations throughout preischemic, ischemic, and recovery periods; and thus the potential of quantitating the time course of ATP concentration changes during ischemia, reperfusion, and subsequent events. We have examined two different animal models of *in vivo* renal ischemia to better define the relationship between acute ischemia and ATP concentration

MATERIALS AND METHODS

Animals

New Zealand white female rabbits were anesthetized using 1% endotracheal halothane. Intravenous 0.9% NaCl was used to replace third space losses (5–10 ml/kg/hr) and to provide maintenance fluid replacement (4 ml/kg/hr). The left kidney was exteriorized through a midline abdominal incision and the renal artery and vein were surrounded by a snare that could be used to reversibly occlude them. An NMR surface coil was placed over the exposed kidney.

Mongrel dogs were similarly anesthetized. The NMR surface coil was permanently implanted by attaching it to the lateral aspect of the left kidney by sewing a piece of woven Dacron around the coil and the kidney. An inflatable arterial occluder was placed around the renal artery.

NMR Spectra

The animals were placed in an Oxford Instruments magnet with a 26 cm diameter bore and a 1.89 Tesla field strength interfaced to a Nicolet 1280 computer and a 293C pulse programmer. Experiments were performed at 32.5 MHz with radiofrequency pulses of 36 μsec with a one-sec delay between pulses. High energy phosphates were analyzed during a 40-min control period. Data are reported as percent of the β-ATP peak recorded during this control period. Rabbits had renal blood flow occluded for 40 min while dogs had it occluded for 90 min. Renal blood flow was confirmed in rabbits using radioisotope-labeled microspheres. After reversal of occlusion, β-ATP levels were monitored for recovery periods of 90 min in rabbits and 140 min in dogs.

Additionally, ^{23}Na and ^{31}P NMR Fourier series images (FSI) were collected on each dog kidney to localize sodium ion and phosphorus metabolites within the organ. A total of nine pulse increments were used to generate the spatial maps.

RESULTS

FIGURE 1 presents observed changes in kidney ATP concentration in rabbits and dogs. Blood flow in the left kidney of the rabbits went from almost 4 ml/min/g to undetectable levels during ligation, then returned to control values within 5 min of reperfusion. The blood flow in the right kidney (no ischemia) did not vary significantly during these three periods. ATP levels fell to about 30% of control levels during ischemia in both rabbits and dogs. In the dogs, after 40 min of occlusion, ATP levels rose gradually, probably representing a return of arterial flow due to a malfunction of the arterial occluder. Therefore, ATP levels appear to have returned to 90–100% of control values by 65–85 min after reperfusion in the rabbits. Dog renal ATP values

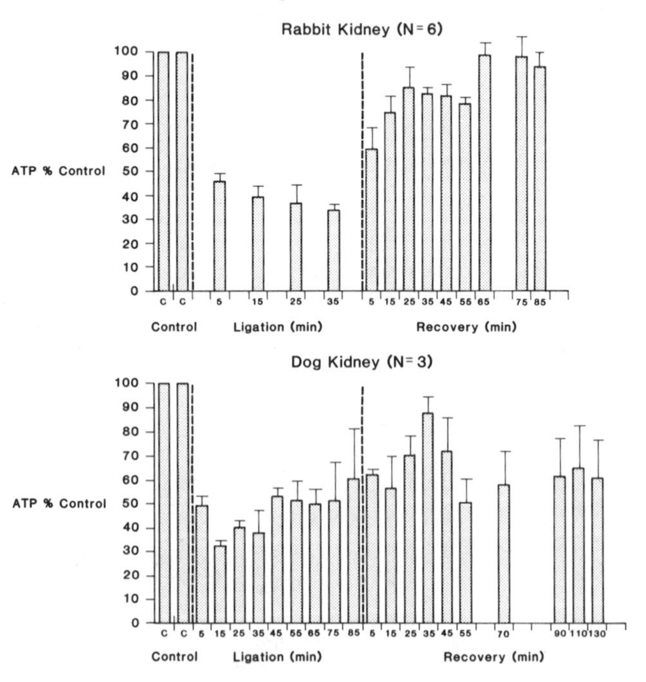

FIGURE 1. The graphs demonstrate the time course of changes in renal ATP concentration from spectra obtained before, during, and after 40 minutes of total ischemia for rabbits and 90 minutes of total ischemia for dogs. The x axis for the graph from each group of animals is scaled so that the control, ligation, and recovery periods are similarly spaced.

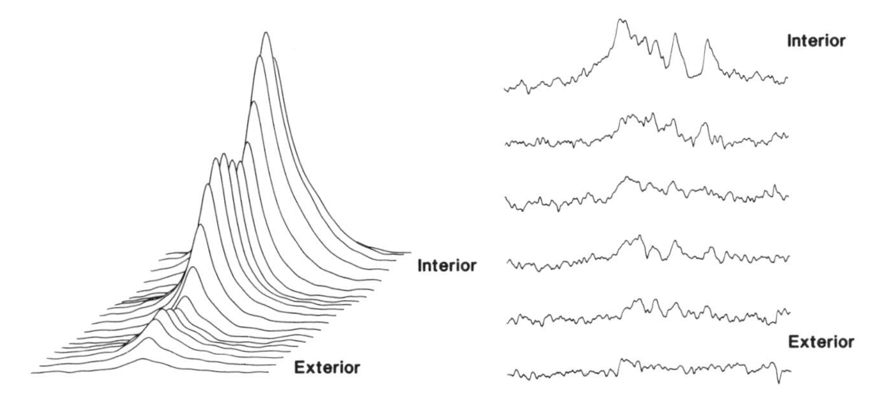

FIGURE 2. These two images are spectra from a normal dog kidney *in vivo* using an implanted coil. The left side of the figure is a ^{23}Na NMR Fourier series image indicating that [Na^+] increases deeper into kidney, consistent with the expected change from cortex to medulla. The right side of the figure shows six representative kidney spectra from a ^{31}P NMR Fourier series image. The inorganic phosphate and sugar phosphate resonances increase in intensity further from the coil.

recovered to 90% of control levels, but subsequently decreased to 60% of control values approximately 100 min after partial reperfusion occurred and remained low for the next hour.

FIGURE 2 presents Fourier series images of ^{23}Na and ^{31}P that demonstrate differences in the concentration of the sodium ion and phosphate metabolites in different regions of the kidney.

DISCUSSION

These preliminary results suggest that in two models of *in vivo* renal ischemia, when injury is limited to the kidney, ATP levels return to close to control levels within about 90 minutes after reperfusion, but may decrease slightly in the next hour. This could represent secondary damage due to injury occurring during the reperfusion period. Slight modifications to the implantable coil in our dog model should allow ATP to be followed for longer periods.

^{31}P-NMR in the Differential Diagnosis of Acute Renal Failure

LAURENCE CHAN AND JOSEPH I. SHAPIRO

Department of Medicine
School of Medicine
University of Colorado
Denver, Colorado 80262

High-energy phosphates play a key role in maintaining an appropriate metabolic milieu within cells and are thus important in the pathogenesis of acute renal failure (ARF). ^{31}P nuclear magnetic resonance (NMR) was therefore used to monitor *in vivo* changes in ATP and intracellular pH (pH_i) of rat kidney in various models of experimental ARF.[1,2] Further studies were performed to characterize the changes under different experimental conditions.

METHODS

All studies were performed in a 1.89 Tesla, 30 cm horizontal magnet using a Biospec spectrometer. The probe employed was a 1.5 cm diameter two-turn solenoid made of 2 mm diameter copper wire. The 90 and 180 degree pulse widths were determined on phantoms containing inorganic phosphate 100 mM by testing the effect of increasing pulse widths on the resultant ^{31}P NMR spectra using one scan for each pulse width studied. The accuracy of these pulse widths were later confirmed using kidney tissue *in situ*. Rats were anesthetized with pentobarbital given intraperitoneally and then subjected to a flank incision. The kidney was exposed and mobilized using blunt dissection. Kidney spectra were obtained by placing the solenoid over the externalized rat kidney. The probe was tuned to the frequency of ^{31}P (32.6 MHz) and the B_0 field was shimmed using the proton signal. Free induction decays were obtained using 60 degree pulses, 1 K array size, a 3,000 Hz sweep width, and 2 second delays. 256–512 scans were accumulated prior to exponential multiplication using 10 Hz line-broadening and Fourier transformation. Peak areas were calculated following baseline correction of the spectra. Comparison of spectra obtained with the above parameters with fully relaxed spectra revealed only minor differences in the relative peak areas obtained. pH_i was calculated from the chemical shift of inorganic phosphate.

RESULTS AND DISCUSSION

In pre-renal ARF, such as ischemia (renal artery clamp) or decreased perfusion due to hypotension (blood pressure less than 80 mm Hg), a low ATP content with intracellular acidosis (pH_i 6.5 to 6.9) was observed. Renal failure (increase serum BUN) developed in ischemic models when ischemia lasted 40 min or more. But a minimum pH of 6.5 and lowest ATP was observed after 5 min, returning to 70 to 90% of normal on restoring blood flow. In acute tubular necrosis (histological evidence) due to nephrotoxic agents (gentamicin 10 mg/kg or cyclosporine 30 mg/kg), no consistent

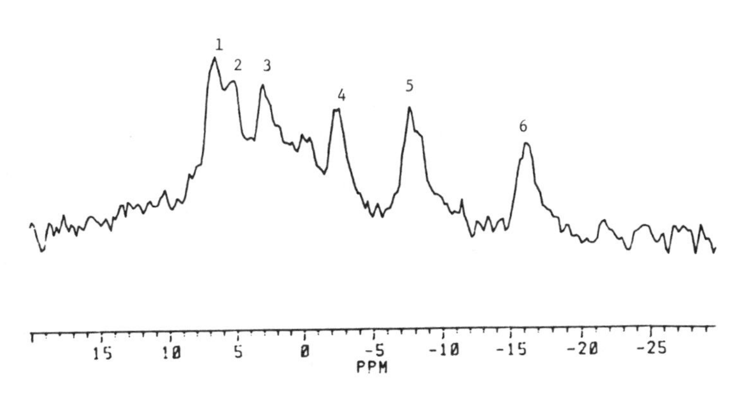

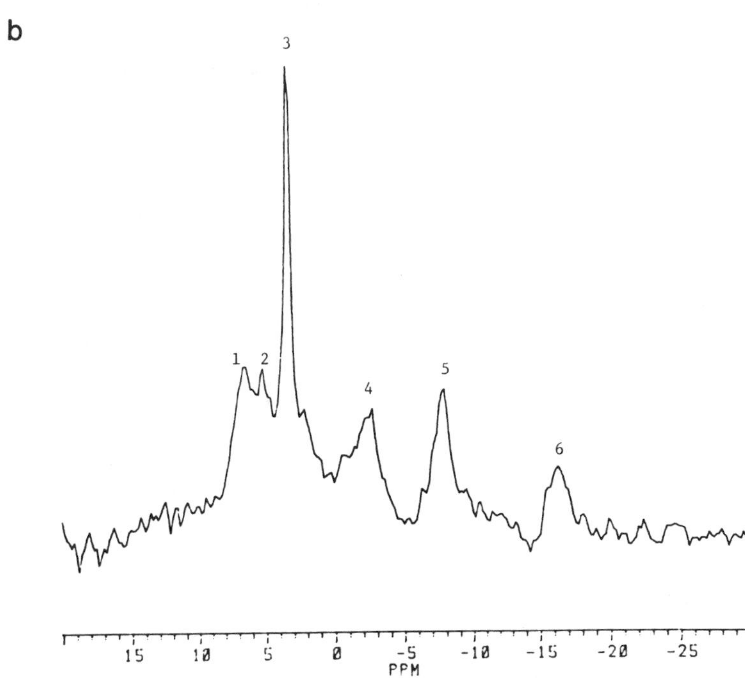

FIGURE 1. (a) Normal kidney spectrum. (b) Spectrum of kidney with urinary obstruction. Peak assignment: (1) sugar phosphate, AMP, and the 3 phosphate of 2,3-DPG; (2) inorganic phosphate and the 2 phosphate of 2,3-DPG; (3) phosphodiesters and urinary phosphate; (4) gamma phosphate of ATP; (5) alpha phosphate of ATP; and (6) beta phosphate of ATP.

changes in the steady-state ATP levels were detected. It is proposed that ATP concentration may be conserved by the kidney in spite of changes in transport work. Such conservation may be accomplished by a change in the energy turnover rate. Finally, a marked increase in a peak at the phosphodiesters region was observed in post-renal ARF due to urinary obstruction (FIG. 1). It was also demonstrated that these changes observed in the phosphodiester peak are mainly due to increases in the signal from urine P_i.

^{31}P NMR can therefore be used in the differential diagnosis and study of energy metabolism in ARF. Similar findings were also observed in kidney transplants. The different causes of graft dysfunction can be distinguished by ^{31}P NMR.[3] Furthermore, ischemia and rejection were both associated with decreases in ATP but were distinguishable from each other by differences in pH_i. These data suggest that ^{31}P NMR may have potential clinical application in differentiating among the causes of graft failure.

REFERENCES

1. CHAN, L., J. G. G. LEDINGHAM, J. A. DIXON, K. R. THULBORN, J. C. WATERTON, G. K. RADDA & B. D. ROSS. 1982. *In* Acute Renal Failure. Eliahou, Ed.: 35–41. John Libbey. London.
2. CHAN, L., J. C. WATERTON & G. K. RADDA. 1981. Trans. Biochem. Soc. **9:** 239–240.
3. SHAPIRO, J. I., C. E. HAUG, R. WEIL III & L. CHAN. 1987. Magnetic Resonance in Medicine (In press.)

Heart Transplantation to Groin:

A Model for NMR Study in the Rat

L. CHAN, J. I. SHAPIRO, C. E. HAUG,[a] AND R. WEIL III[a]

Departments of Medicine and [a]Surgery
University of Colorado School of Medicine
Denver, Colorado 80262

^{31}P NMR spectroscopy has emerged as a noninvasive technique to study the bioenergetics of living tissue. Most *in vivo* experiments can effectively be done by using a surface coil.[1] Internal organs, such as the heart, may present difficulties. To enable optimal application of NMR spectroscopy for *in vivo* study of myocardial metabolism, we have developed a new microsurgical model for NMR spectroscopy of a transplanted rat heart.

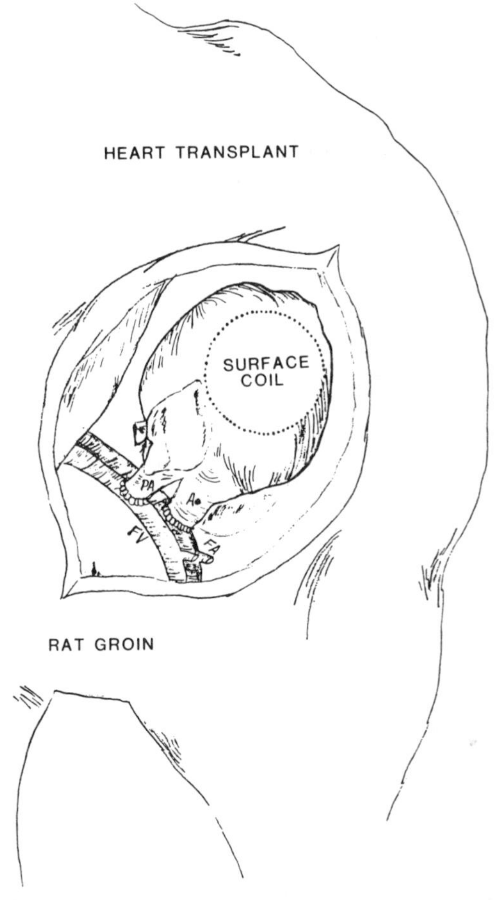

FIGURE 1. Technique of transplantation of rat heart to rat left groin using recipient femoral vessels for revascularization. The position for the placement of RF coil is shown.

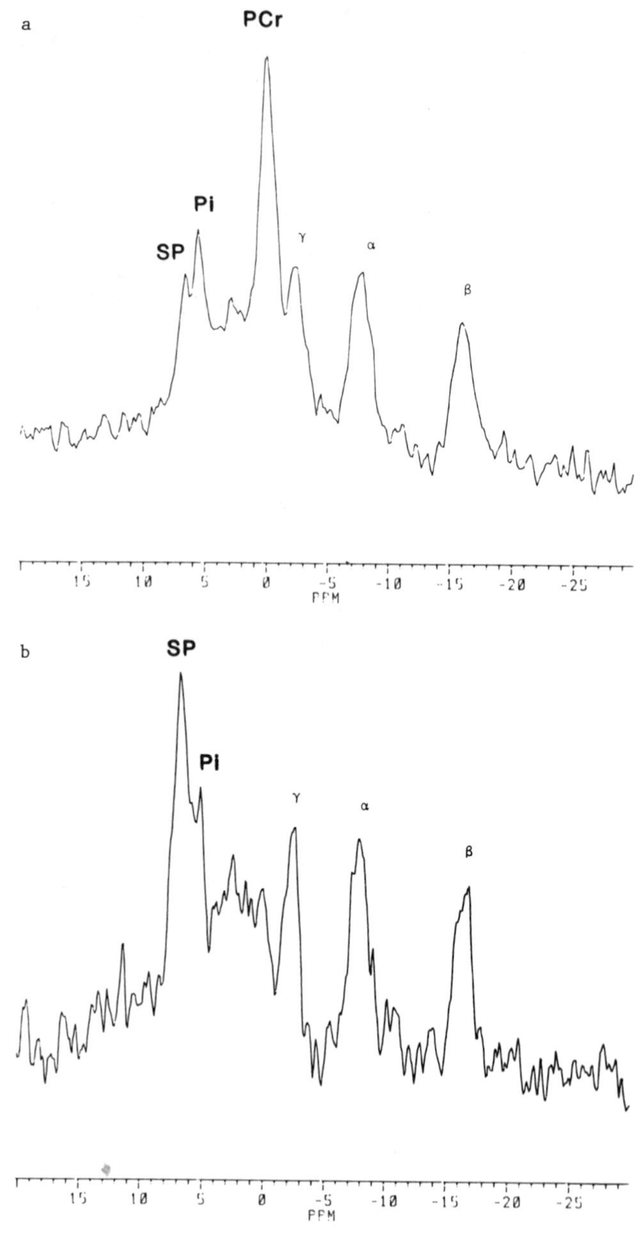

FIGURE 2. (a) ^{31}P NMR spectrum of control (nonrejected) cardiac allograft. 512 transients were taken with 60° pulses at 2 sec delays. Peak assignments: SP = sugar phosphates including AMP; PCr = phosphocreatine; $\gamma = \gamma$ phosphate of ATP; $\alpha = \alpha$ phosphate of ATP; and $\beta = \beta$ phosphate of ATP. (b) ^{31}P NMR spectrum of rejected cardiac allograft. Peak assignments and acquisition parameter as in FIG. 2a.

METHODS

Inbred rats were used for this study. Heterotopic cardiac allografts were transplanted to host femoral vessels using microvascular techniques.[2] Cyclosporin A (CsA, 15 mg/kg) was given subcutaneously for 7 days to prevent rejection. The allografts were studied in a 1.89 Tesla horizontal bore magnet and ^{31}P NMR spectroscopy was used to evaluate heterotopic cardiac allograft rejection. A 1.5 cm diameter two-turn solenoid (2 mm copper wire) tuned to the resonance frequency of phosphorus (32.6 MHz) was used. Ninety degree and 180 degree pulses were determined with phantoms that allowed for a significant ^{31}P signal on a single scan. This was confirmed with the *in vivo* preparation. The coil was positioned over the cardiac allograft (FIG. 1). The magnetic field homogeneity was adjusted using the H-1 signal from the water within the sample to be investigated.[1] The T_1 values of beating hearts for the ^{31}P spectra were determined with the inversion recovery method.

RESULTS AND DISCUSSION

Allografts treated with CsA did not show any histological evidence of rejection (NR) and had relatively high level of phosphocreatine (PCr), and low level of inorganic phosphate (P_i) in contrast to allograft with rejection (R) that had the opposite (FIG. 2). Study of five rats in each group revealed that the PCr/P_i as well as PCr/ATP ratios were signifcantly higher in NR than R allografts (5.20 ± .72 vs. 0.74 ± .16, and 2.48 ± .19 vs. 0.64 ± .21, respectively, both $p < .01$). Intracellular pH appeared to be higher in nonrejected hearts than rejected hearts. Intracellular pH was 7.25 ± 0.07 in nonrejected hearts compared with 7.09 ± 0.04 in rejected hearts. This difference approached but did not attain statistical significance.

Our data show that there are marked changes in the concentrations of intracellular high energy phosphates in rejected rat cardiac allografts. This study demonstrates the feasibility of investigating the transplanted rat heart with ^{31}P NMR. Furthermore, this particular transplant model can be used to investigate cardiac metabolism and graft rejection by NMR.[3] It is possible that other organs (such as kidney, liver, pancreas) could be transplanted to the groin to permit study of their spectroscopic characteristics under varying physiologic conditions.[4] If cellular biochemistry of transplanted organs can be non-invasively analyzed in animal models, it may then become possible to adapt these approaches for human application.

REFERENCES

1. ACKERMAN, J. J. H., T. H. GROVE, G. G. WONG, D. G. GADIAN & G. K. RADDA. 1980. Nature (London) **283:** 167–170.
2. RAO, V. K. & M. LISITZA. 1985. Transplantation **40:** 567–569.
3. HAUG, C. E., J. I. SHAPIRO, L. CHAN & R. WEIL III. 1987. Transplantation **44:**175–178.
4. SHAPIRO, J. I., C. E. HAUG, R. WEIL III & L. CHAN. 1987. Magnetic Resonance in Medicine. (In press.)

An Examination of the Specificity of *in Vivo* ^{31}P MRS Adriamycin Response Markers

J. L. EVELHOCH, N. A. KELLER, AND T. H. CORBETT

Division of Medical Oncology
Department of Internal Medicine
Wayne State University
Detroit, Michigan 48201

In vitro ^{31}P magnetic resonance spectroscopy (MRS) studies indicate that the ^{31}P MRS metabolic characteristics of MCF-7 human breast cancer cells differ markedly from those of a subline selected *in vitro* for adriamycin (ADR) resistance.[1] *In vivo* ^{31}P MRS studies have shown that the ^{31}P MRS metabolic characteristics of 16/C murine mammary adenocarcinomas (an ADR-sensitive tumor) change significantly in response to treatment with ADR prior to a change in tumor mass.[2] The results of these studies suggest that ^{31}P MRS may provide clinically applicable markers of ADR sensitivity. However, neither the presence in solid tumors of the differences observed *in vitro* between ADR-sensitive and ADR-resistant cells nor the absence in ADR-resistant tumors of treatment-induced changes observed in ADR-sensitive tumors (i.e., the specificity of the response markers) has been investigated. To address these

TABLE 1. Initial ^{31}P MRS Metabolic Characteristics of Murine Mammary Adenocarcinomas 17/A and 17/A/ADR[a]

	Mamm 17/A	Mamm 17/A/ADR
pH	7.18 ± 0.16	7.12 ± 0.19
	[6.88 to 7.40]	[6.76 to 7.31]
P_i:NTP	1.60 ± 1.04	1.65 ± 0.76
	[0.90 to 3.92]	[1.01 to 3.59]
PCr:NTP	0.49 ± 0.23	0.65 ± 0.55
	[0.27 to 1.20]	[0.31 to 2.39]
GPC:NTP[b]	1.74 ± 0.66	0.72 ± 0.24
	[0.79 to 3.06]	[0.24 to 1.19]
PC:NTP[b]	1.34 ± 0.52	0.69 ± 0.20
	[0.70 to 2.40]	[0.31 to 1.04]
PE:NTP[b]	1.52 ± 0.48	1.07 ± 0.25
	[0.69 to 2.24]	[0.62 to 1.25]
PC:PE[b]	0.88 ± 0.10	0.60 ± 0.12
	[0.69 to 1.21]	[0.49 to 0.88]

[a]Metabolic characteristics are pH determined by NMR and the peak height ratio of the metabolites indicated (P_i, inorganic phosphate; NTP, nucleoside triphosphates; PCr, phosphocreatine, GPC, glycerophosphocholine; PC, phosphocholine; PE, phosphoethanolamine). Values are reported as the mean ± std. dev. of the respective characteristics and the range of values observed are shown in brackets. For 17/A ($N = 16$), size = 505 mg ± 54 mg; for 17/A/ADR ($N = 15$), size = 481 mg ± 81 mg.

[b]Difference between initial characteristic in 17/A and 17/A/ADR is significant at the 99% confidence level ($p < 0.01$; Student's t test, two-tailed, independent samples).

TABLE 2. Early Changes Induced in the ^{31}P MRS Metabolic Characteristics of Murine Mammary Adenocarcinomas 17/A and 17/A/ADR by Treatment with 12 mg/kg Adriamycin[a]

	Mamm 17/A	Mamm 17/A/ADR
pH[b]	+0.09 ± 0.18	−0.07 ± 0.14
	[−0.16 to +0.38]	[−0.30 to +0.11]
P_i:NTP[c]	−0.50 ± 0.60[e]	+0.39 ± 0.80
	[−1.78 to −0.03]	[−0.85 to +1.68]
PCr:NTP	+0.01 ± 0.19	0.00 ± 0.19
	[−0.29 to +0.38]	[−0.30 to +0.31]
GPC:NTP[b]	−0.41 ± 0.55[d]	−0.03 ± 0.14
	[−1.89 to +0.18]	[−0.29 to +0.12]
PC:NTP[c]	−0.38 ± 0.38[e]	+0.12 ± 0.17
	[−1.12 to +0.02]	[−0.19 to +0.36]
PE:NTP	−0.06 ± 0.40	+0.20 ± 0.23[d]
	[−0.71 to +0.38]	[−0.14 to +0.65]
PC:PE[c]	−0.28 ± 0.14[e]	0.00 ± 0.16
	[−0.45 to −0.03]	[−0.32 to +0.18]

[a]Metabolic characteristics are the same as in TABLE 1. Values are reported as the mean ± std. dev. of the change in the respective characteristics (i.e., the pretreatment value minus the value observed 1 day (~24 hr) after treatment with 12 mg/kg ADR i.v.) with the range of values observed in brackets. For 17/A, $N = 10$, tumor growth delay (TGD) = 19 days; for 17/A/ADR, $N = 12$, TGD = 0 days.

[b]Difference between change in 17/A and 17/A/ADR is significant at the 95% confidence level ($p < 0.05$; Student's t test, two-tailed, independent samples).

[c]Difference between change in 17/A and 17/A/ADR is significant at the 99% confidence level ($p < 0.01$; Student's t test, two-tailed, independent samples).

[d]Change is significant at the 95% confidence level ($p < 0.05$; Student's t test, two-tailed, paired samples).

[e]Change is significant at the 99% confidence level ($p < 0.01$; Student's t test, two-tailed, paired samples).

questions, we have examined both the initial ^{31}P MRS metabolic characteristics [i.e., pH and the relative levels of phosphocreatine (PCr), inorganic phosphate (P_i), nucleoside triphosphates (NTP), glycerophosphocholine (GPC), phosphocholine (PC), and phosphoethanolamine (PE)] and the changes induced in those characteristics by treatment with 12 mg/kg ADR in an ADR-sensitive murine mammary adenocarcinoma (17/A) and a subline of that tumor in which ADR resistance has been induced *in vivo* (17/A/ADR[3]).

Before treatment, the following differences were present (TABLE 1): (1) GPC:NTP peak height ratio was higher in ADR-sensitive tumors; (2) peak height ratios of both PC:NTP and PE:NTP were higher in ADR-sensitive tumors [however, this difference was not observed for larger (untreated, control) tumors]; (3) PC:PE peak height ratio was higher in ADR-sensitive tumors. Several significant differences between ADR-sensitive and ADR-resistant tumors were observed for the treatment-induced changes in the ^{31}P MRS metabolic characteristics (pretreatment to 1 day after treatment; TABLE 2): (1) the P_i:NTP peak height ratio decreased in the ADR-sensitive tumors only. (2) The PC:NTP peak height ratio (and consequently the PC:PE peak height ratio) decreased in ADR-sensitive tumors only. (3) The GPC:NTP peak height ratio decreased in ADR-sensitive tumors only. However, it also decreased in control ADR-sensitive tumors. (4) Although the change in pH was not significant in either tumor line, the increase observed in the ADR-sensitive tumors was significantly different from the decrease observed in the ADR-resistant tumors. Other than the change noted for the GPC:NTP peak height ratio, none of these changes were observed in either ADR-sensitive or ADR-resistant tumors that were not treated with ADR ($N = 5$ and $N = 6$, respectively).

These results indicate that *in vivo* ^{31}P MRS provides *response*-specific predictive markers of ADR sensitivity. Markers observed in sensitive tumors include: (1) the PC:PE peak height ratio is higher and decreases in response to treatment; (2) the GPC:NTP peak height ratio is higher; (3) the P_i:NTP peak height ratio decreases to <1 in response to treatment; (4) the pH increases to >7.3 in response to treatment (or remains elevated).

REFERENCES

1. Cohen, J. S., R. C. Lyon, C. Chen, P. J. Faustino, G. Batist, M. Shoemaker, E. Rubalcaba & K. H. Cowan. 1986. Cancer Res. **46:** 4087–4090.
2. Evanochko, W. T., T. C. Ng, A. J. Lawson, T. H. Corbett, J. R. Durant & J. D. Glickson. 1983. Proc. Natl. Acad. Sci. USA **80:** 334–338.
3. Kessel, D. & T. H. Corbett. 1985. Cancer Lett. **28:** 187–193.

A Saturable Halothane Binding Site in Rat Brain Described by ^{19}F-NMR

ALEX S. EVERS AND D. ANDRÉ D'AVIGNON

Departments of Anesthesiology and Chemistry
Washington University
St. Louis, Missouri 63110

Fluorinated anesthetics can be observed in brain tissue using ^{19}F-NMR spectroscopy.[1] To further understand the chemical interaction of anesthetics with brain tissue, we have developed a method for quantitating brain halothane (CF_3-CHBrCl) concentration. Rats were anesthetized with various inspired concentrations of halothane for one hour and sacrificed. The cerebral hemispheres were placed in sealed capillaries and immersed in an external standard solution (methoxyflurane; $CHCl_2$—CF_2—OCH_3) in 10 mm NMR tubes. Brain halothane concentration was determined by obtaining ^{19}F spectra under quantitative conditions and comparing the integration of the halothane resonance to that of the external reference. The ratio obtained was corrected for the number of fluorine atoms in halothane and methoxyflurane, for the cross-sectional

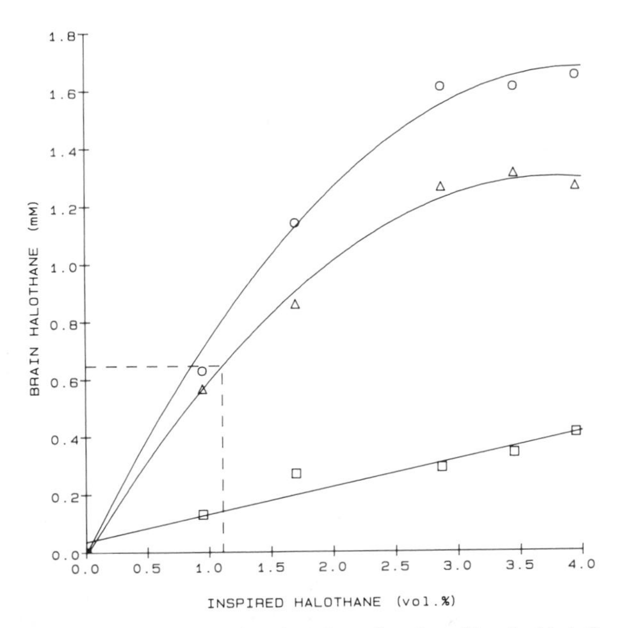

FIGURE 1. Brain halothane concentration plotted as a function of inspired halothane concentration. Circles represent total brain halothane concentration, triangles represent halothane concentration in the environment with $T_2 = 3.6$ msec, and squares represent halothane concentration in the environment with $T_2 = 43$ msec. Animals were anesthetized at the indicated inspired concentrations for 1 hour, and each point represents the mean of six determinations. In all cases standard deviations were $<10\%$ of the mean.

area of the sample and the external reference solution and for sample density, to obtain brain concentration. Using this method, steady-state brain halothane concentration was found to be a non-linear function of inspired concentration with apparent saturation of brain (1.6 mM) occurring at inspired concentrations above 2.5 vol % (FIG. 1). Saturation of brain occurred despite increasing plasma halothane concentration over the entire range (0–4 vol %) of inspired concentrations studied.

To further investigate the interaction of halothane with brain tissue, ^{19}F spin-spin relaxation times were measured. Spin-echo decay curves (using the Carr-Purcell-Meiboom-Gill method) revealed a non-linear relationship between the natural logarithm of the intensity of the halothane resonance and echo evolution time (FIG. 2).

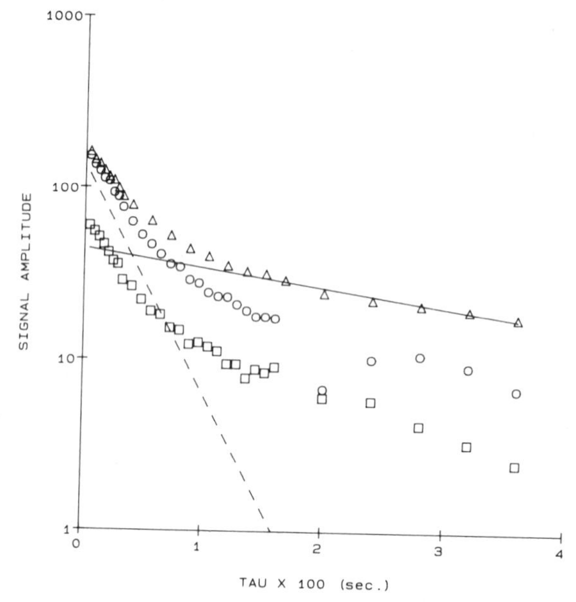

FIGURE 2. Representative spin-echo decay curves of the ^{19}F halothane resonance in rat brain. Animals were anesthetized for 1 hour at inspired halothane concentrations of 3.95% (△), 2.87% (○), or 0.95% (□), sacrificed and their brains prepared for spectroscopy. Spectra were obtained at 10°C using a 90°-τ-180° pulse sequence with 25 values of τ ranging from 0.0004–0.036 sec. The figure plots the intensity of the ^{19}F-halothane signal as a function of echo evolution time (τ).

These curves were well described by two exponentials indicating that halothane exists in two environments in brain, characterized by T_2 values of 3.6 and 43 msec. By quantitating the occupancy of these two environments as a function of inspired halothane concentration, it was found that the environment with T_2 = 3.6 msec saturated (1.2 mM) at an inspired concentration of ~2.5 vol % (FIG. 1). Half-maximal occupancy of this environment (0.6 mM) was achieved at 1.2 vol % inspired halothane, the known ED_{50} value for halothane in rats.[2] In contrast, occupancy of the environment with T_2 = 43 msec increased linearly from 0–4 vol % inspired halothane. Occupancy of the saturable environment for halothane occurred at a significantly faster rate than occupancy of the non-saturable environment. Following a 2-min exposure to 3% halothane, more than 90% of brain halothane was found in the saturable environment, and the rats were anesthetized. The combined results of these studies suggest that occupancy of a saturable halothane binding site in brain is responsible for the anesthetic effect of the drug.

REFERENCES

1. Wyrwicz, A. M., M. H. Pszenny, J. C. Schofield, P. C. Tillman, R. E. Gordon & P. A. Martin. 1983. Science **222:** 428–430.
2. Evers, A. S., W. J. Elliot, J. B. Lefkowith & P. Needleman. 1986. J. Clin. Invest. **77:** 1028–1033.

Cytosolic Inorganic Phosphate Does Not Appear To Regulate the Contractile Response in the Intact Rat Heart[a]

ARTHUR H. L. FROM, STEVAN D. ZIMMER,
AND MARC A. PETEIN

Cardiovascular Division
Department of Medicine
Minneapolis Veterans Administration Medical Center
University of Minnesota
Minneapolis, Minnesota 55417

STEPHEN P. MICHURSKI AND KAMIL UGURBIL

Department of Biochemistry
Gray Freshwater Biological Institute
University of Minnesota
Navarre, Minnesota 55392

It has been proposed that cytosolic inorganic phosphate (P_i) may be an important regulator of myocardial contractile tension development under ischemic or hypoxic conditions.[1] This suggestion is based on observations in skinned cardiac muscle that demonstrate that increasing P_i concentration depresses both maximal force generation and Ca^{2+} sensitivity[2] and in intact perfused hearts where, following hypoxia or ischemia, the elevation of P_i and the decline in the contractile function are correlated and precede decreases in cytosolic pH and ATP content.[1]

We therefore examined the relationship of P_i content (determined by ^{31}P NMR) to contractile function in Langendorff-perfused rat hearts by taking advantage of substrate-induced variations of cytosolic P_i.[3,4] Perfusate contained either 15 mM glucose (G; $N = 8$) or 10 mM glucose + 10 mM pyruvate (G + P; $N = 8$), another group of G + P hearts ($N = 5$) was studied following 18 minutes of warm ischemia. The ischemic period used was sufficient to cause large decreases in ATP levels but not long enough to induce post-ischemic mechanical dysfunction. Rate-pressure product (RPP), left ventricular systolic pressure (LV), and LV dp/dt were determined at several levels of cardiac activity achieved by varying heart rate and inotropic state (dobutamine infusion). FIGURE 1 shows RPP and P_i values for all groups of hearts. P_i values for G and G + P post-ischemia groups are comparable; however, both groups have P_i values significantly different ($p < .005$–$.001$) from the corresponding values of P_i in the G + P non-ischemic group. RPP products at each workload were comparable for all groups (P = NS). Similarly, when the relationship between RPP and MVO_2 is compared for all groups (FIG. 2) no differences are seen. There were also no differences between the three groups when LV and LV dp/dt were compared at comparable workloads. Thus, no matter how systolic performance is evaluated, contractile function

[a]This work was supported by the Veterans Administration Research Funds, the American Heart Association, Minnesota Affiliate, the United States Public Health Science Grants RO1-HL-33600, RCDA IKO4-HL-01241 (K.U.), and NRSA-F32-HL-07438 (S.D.Z.).

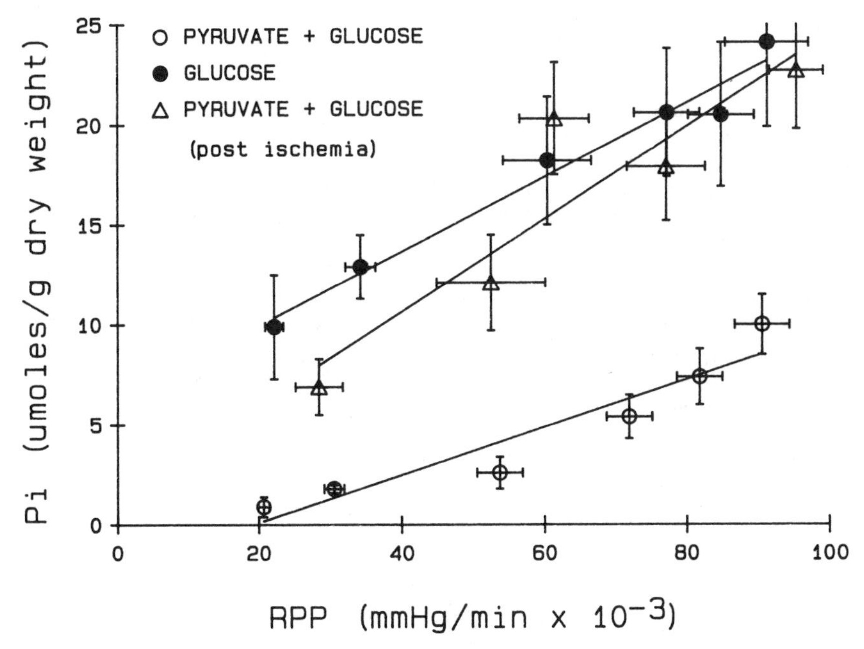

FIGURE 1. The relationship between P_i and myocardial oxygen consumption (MVO_2) is shown. Six workloads were achieved by varying heart rate (from 300 to 600 beats/min), ventricular filling pressure (from 8 to 17 mm Hg) by changing the volume of the intraventricular balloon from which pressure was recorded, and by the addition of dobutamine (40 to 80 ng/ml) to the perfusate. All P_i values for G and P + G post-ischemia groups were significantly different from corresponding P + G values ($p < .005$ to $.001$). The only P_i values that differed between the G and the P + G post-ischemia groups were those corresponding to their lowest common workload ($p < .025$).

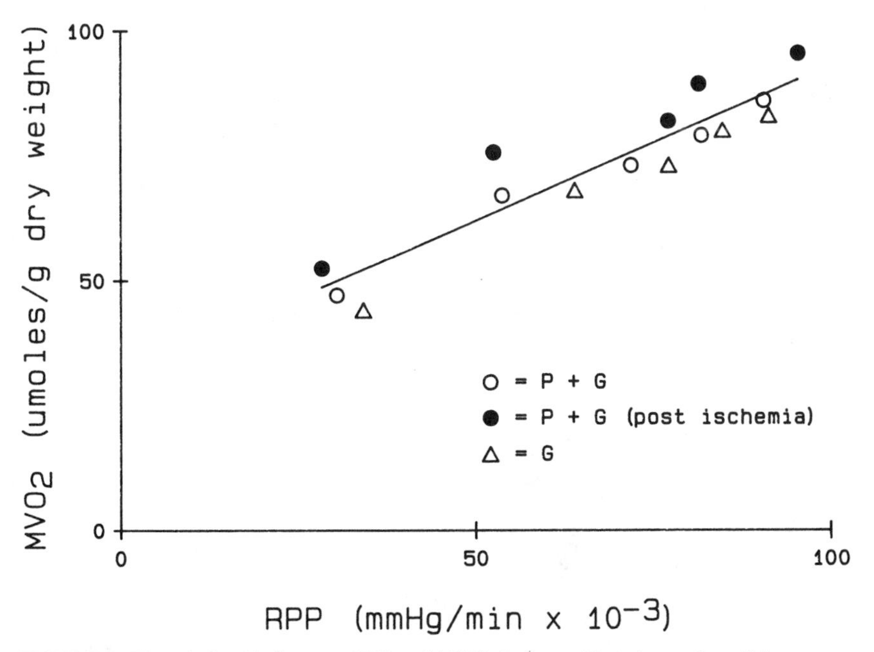

FIGURE 2. The relationship between RPP and MVO_2 is shown. Experimental conditions are as described in FIG. 1.

does not have a unique relationship to cytosolic P_i over the range of contents observed. Moreover, when P_i concentration is estimated from P_i content,[3,4] the P_i values fall within the range found to be regulatory in skinned preparations.[2] Opposing ADP effects cannot account for the lack of P_i-induced mechanical depression in the G group because ADP levels in the G + P post-ischemic group were two to five times lower (depending upon workload) and were comparable to those of the non-ischemic P + G group. Intracellular pH values were also comparable for all groups. Thus, pH variation cannot explain the lack of effect of P_i in the elevated P_i groups.

The current observations strongly suggest that in intact myocardium, moderate elevation of cytosolic P_i is not the cause of early contractile failure following the induction of ischemia or hypoxia, and that P_i may not significantly influence contractile function or the relationship between RPP and MVO_2 over a fairly wide range of concentrations. The reason(s) for the divergence of the present observations from the skinned muscle studies[2] is unclear. It is possible that the present observations can be explained by the apparently increased Ca^{2+} sensitivity of the intact cardiac muscle as compared to skinned[5] and/or a postulated decrease in the sensitivity of the intact preparation to P_i.

REFERENCES

1. Allen D. G., P. G. Morris & J. S. Pirolo. 1985. J. Physiol. (London) **361:** 100–107.
2. Kentish J. C. 1985. J. Physiol. (London) **370:** 585–604.
3. Ugurbil K., M. Petein, R. Maidan, S. Michurski & A. H. L. From. 1986. Biochemistry **25:** 100–107.
4. From A.H.L., M. A. Petein, S. P. Michursky, S. D. Zimmer & K. Ugurbil. 1986. FEBS Lett. **206:** 257–261.
5. Yue D. T., E. Marban & W. G. Wier. 1986. J. Gen. Physiol. **87:** 223–242.

Proton NMR *in Vivo* Imaging:

A Tool for Studying Embryo Development of Small Organisms

JOOST A. B. LOHMAN

Oxford Research Systems Ltd., Abingdon
Oxon, OX14, 1RY, England

GEORGE GASSNER

USDA/ARS/Metabolism and Radiation Research Laboratory
Fargo, North Dakota 58105

The study of developmental systems of plants and animals is crossing the threshold of a new era. In the past ninety years, a wide range of studies of development has included descriptive embryology, causal analysis, and biochemical analysis. All of these approaches require various degrees of invasive techniques. Now, using noninvasive NMR, the dynamics of pattern formation, compartmentalization, and biochemical and biophysical interactions within and between compartments can be examined from a new perspective. Our study using the locust (*Schistocerca gregaria* [Forskal]) shows that *in vivo* nuclear magnetic resonance imaging can be applied successfully to locust embryogenesis.

NMR experiments for the development of small organisms require a much higher spatial resolution than can normally be achieved on standard imaging equipment. First, stronger gradients are needed in order to obtain an adequate dispersion of NMR signals from the small volume elements. Second, since the small volume elements contain only a small amount of sample, an improvement to the sensitivity of the NMR experiment is necessary. Higher sensitivity may be obtained through an improved filling factor, which is achieved by limiting the dimensions of the radiofrequency coil to just accommodate the sample. Performing the NMR experiment at a higher static magnetic field strength will also contribute to improved sensitivity.

Cross-sectional NMR images of eggs of the locust were recorded at intervals of 12 hours over a period of 16 days from shortly after fertilization to a few days before they hatched. The experiments were performed in a 4.7 Tesla 54 mm bore vertical magnet, and a special 3-mm diameter solenoidal RF coil was constructed in order to produce an intense and homogeneous RF field. Gradients could be applied in three orthogonal directions, 0.1 T/m in the direction of the static magnetic field and 0.05 T/m in the other two directions. The imaging sequence used was based upon the STEAM technique, which allows for the multislice acquisition of separate water and lipid images.[1] The images in FIGURE 1 (page 436) illustrate, as an example, several developmental characteristics of the locust embryo at 77% development. Additional and more extensive details on procedures and developmental aspects of the locust embryo are under a separate title.[2]

REFERENCES

1. FRAHM, J., K. D. MARBOLDT, W. HANICKE & A. HAASE. 1985. Stimulated echo imaging. J. Mag. Reson. **64:** 81–93.
2. GASSNER, G. & J. A. B. LOHMAN. 1987. Combined proton NMR imaging and spectral analysis of locust embryonic development. Proc. Natl. Acad. Sci. USA **84:** 5297–5300.

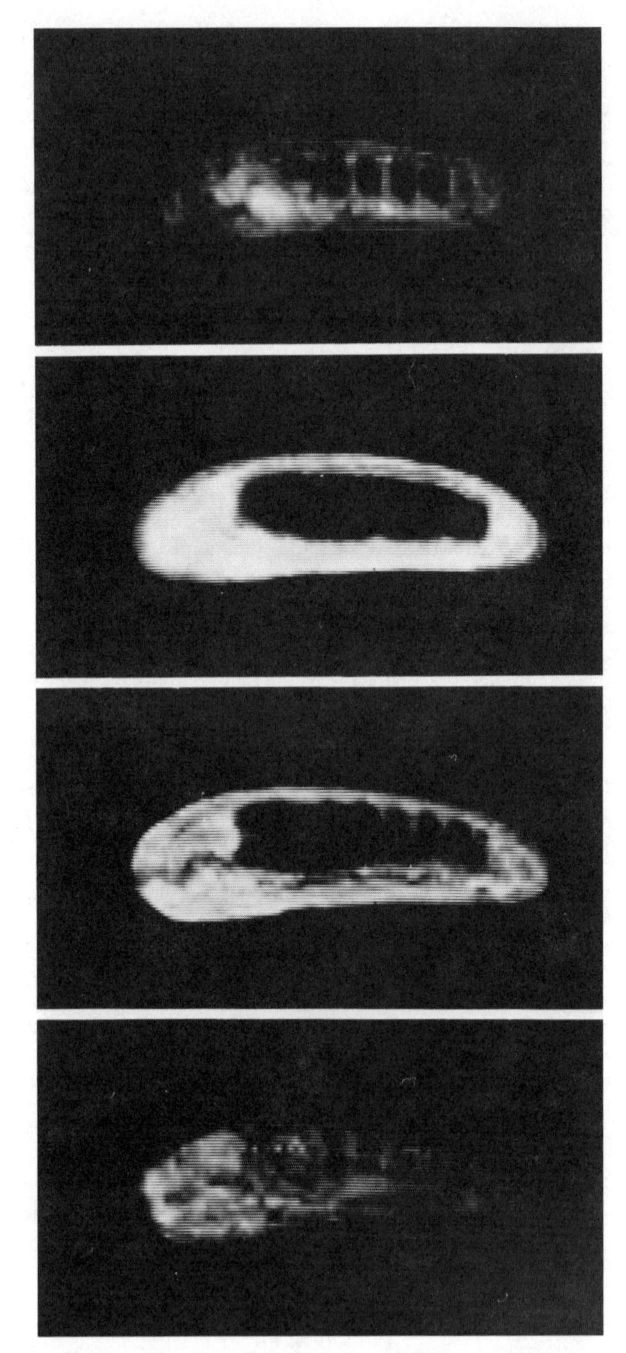

FIGURE 1. The four images show the water ^1H distribution in a locust embryo at 77% development. Four adjacent sagittal slices were obtained from a multislice, stimulated echo sequence capable of discriminating between water and lipid chemical shifts (data matrix: 128 × 128 × 4, 12 averages, 27 minutes acquisition time, 500 microns plane thickness, 100 microns in plane resolution). Details of segmentation and the development of limbs are clearly visible.

Comparison of Intra- and Extracellular Na^+ of Erythrocytes using ^{23}Na NMR in the Absence of Shift Reagent

L. S. OKERLUND AND R. J. GILLIES

Department of Biochemistry
Colorado State University
Fort Collins, Colorado 80523

Intracellular sodium levels (Na_{in}^+) are correlated with a number of cell functions, such as regulation of intracellular pH, proliferation, and volume. Measurement of intracellular Na^+ using ^{23}Na NMR is the most accurate technique available, since free and bound Na_{in}^+ are distinguishable. Na_{in}^+ is usually resolved from external Na^+ (Na_{ex}^+) through the use of shift reagents: impermeant, organically bound lanthinide rare earths such as $Dy(PPP)_2$[1] or $Dy(TTHA)_2$.[2] These compounds "contact" the extracellular Na^+ and thereby shift the resonance frequency away from the non-contacted intracellular peaks. Since the recent report[3] showing that the intra- and extracellular Na^+ resonances are equally visible to ^{23}Na NMR, these techniques are extremely useful in quantifying intracellular Na^+ levels. However, many systems are inappropriate for the use of shift reagents. It has been suggested by Fossel[4] to discriminate between intra- and extracellular Na^+ populations on the basis of T_2 differences.[3] This would preclude the use of shift reagents in the measurement of Na_{in}^+ and Na_{ex}^+. Such techniques will have applications to the study of *in vivo* systems such as neoplasms or tumors with possibly altered sodium content. It will also allow non-destructive measurement of changes in Na_{in}^+ levels over long time courses, which may occur by increased volume or hypertrophy.

In the present communication, we describe the discrimination of Na_{in}^+ from Na_{ex}^+ using T_2 differences in erythrocytes. Erythrocytes provide a relatively simple, anucleated system with (presumably) two relaxation components. Na_{ex}^+ are the more slowly relaxing, with T_2 values of >20 msec, whereas Na_{in}^+ T_2s are generally <10 msec.

T_2 determination of packed erythrocytes using the CPMG sequence gave rise to a multi-component plot of ln (magnetization) versus delay time. Due to instrument variability, there were a number of spurious points eliminated manually. As indicated in FIGURE 1, deconvolution of the multiexponential into two components yields a relaxation due to each of the components as determined from the slopes of a semi-log plot. Using the relation derived from the Bloch equations, the determined slope is $-1/T_2$ and the respective relaxation times and spin densities may thus be determined.

We have written a computer program to perform the T_2 calculations given the raw magnetization intensity data. This program calculates the natural log of the magnetization and performs a linear regression on the supplied data points by working with the first and last few points. The number of points used for the regression is variable and the best fitting equations, as determined on the basis of both correlation coefficients, are selected. By determining the slopes of the first and last portions of the plot, presumably corresponding to the two magnetization populations (the composite of the fast and slow relaxing, and slow relaxing, respectively), and subtracting the latter component from the earlier, it is possible to compute the slopes and corresponding intercepts of the separate fast relaxing (intracellular) and slow relaxing (extracellular) components. T_2s may be determined from the slopes and population ratios (Na_{ex}^+/Na_{in}^+) from the *y*-intercepts, since no relaxation has occurred at zero time, and, therefore, all nuclei are observable.

Our figure displays a two-component relaxation as illustrated by the co-plotted expected values from deconvolution. The T_2s determined from the slopes are sufficiently different to distinguish them and are similar to those obtained by other investigators. The sodium ratios determined from the intercepts of the deconvoluted relaxation data are, however, different than those determined from respective peak areas (FIG. 2). The reason for this is, as yet, unexplained. The approach described here provides a method by which to observe intra- and extracellular sodium ratios in systems that may be sensitive to the use of shift reagents. As it is presently described, discrimination of these components on the basis of their T_2 differences is not quantitative, when compared to data obtained using shift reagents. It is recommended,

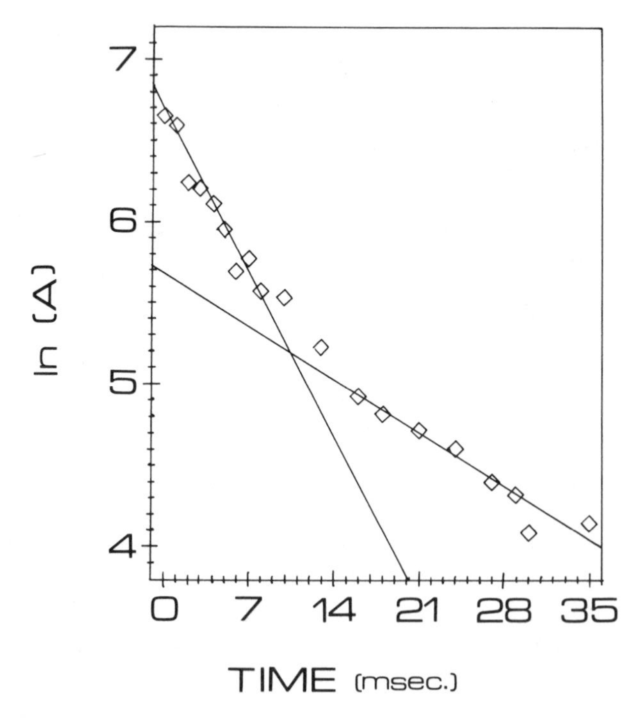

FIGURE 1. T_2 relaxation data from packed, fresh human erythrocytes in the absence of shift reagent. Magnetization of sample (A) was determined by a peak picking routine. The ln of these values is plotted against the delay time in CPMG (90-τ-180) pulse sequence for T_2 determination from linear slopes. The lines on the graph represent the best fit for the data using dual regression as indicated in the text.

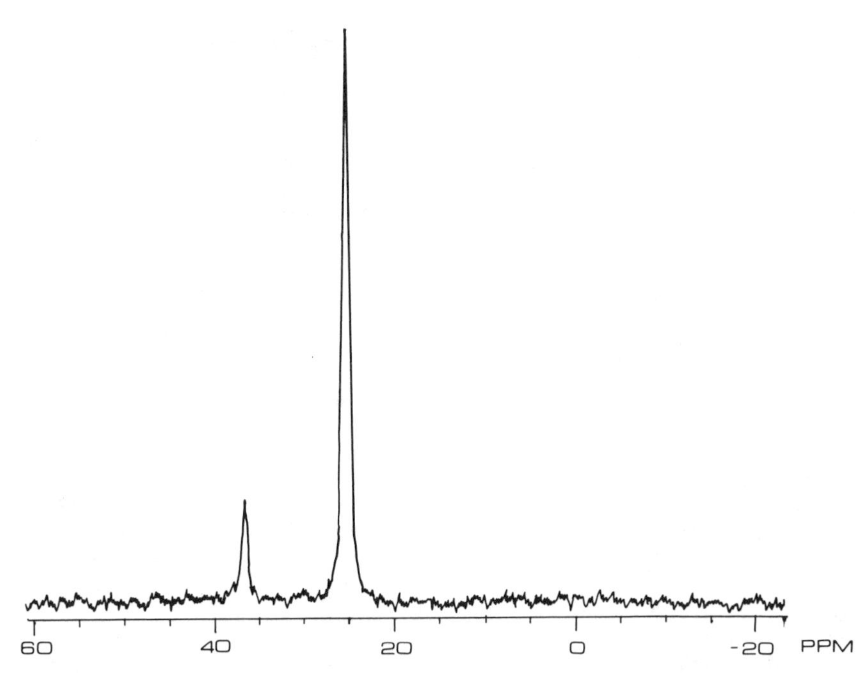

FIGURE 2. ^{23}Na NMR spectrum of packed, fresh human erythrocytes in the presence of 3 mM $Dy(PPP)_2$. Cells were harvested from fresh, whole blood, washed and resuspended in Hank's BSS buffered with 20 mM citrate at pH 7.2. The spectrum represents the transform of the FID arising from 48 90-degree pulses applied every 525 msec.

therefore, that use of this technique be confined to observing relative, and not absolute, changes in intracellular Na^+ activities.

REFERENCES

1. GUPTA, R. K. & R. GUPTA. 1982. J. Mag. Res. **47:** 344–350.
2. CHU, S. C., M. M. PIKE, E. T. FOSSEL, T. W. SMITH, J. A. BALSCHI & C. S. SPRINGER. 1984. J. Mag. Res. **56:** 33–47.
3. DUTTA, A., J. A. BALSCHI, J. S. INGWALL, T. S. SMITH & C. S. SPRINGER. 1985. Proc. Soc. Mag. Res. Med. **5** Suppl. 135–136.
4. FOSSEL, E. T. & H. HOEFELER. 1986. J. Mag. Reson. Med. **3:** 534–540.

^{31}P NMR Studies of Vascular Smooth Muscle from Rabbit Thoracic Aorta:

Oxidation of Glucose but not of Fatty Acid Is Required for Maintenance of Phosphagens

STEPHEN F. FLAIM,[a] MICHAEL BLUMENSTEIN, FRANKLIN C. CLAYTON, AND RUTH R. INNERS

Departments of Chemical Research and Biological Research
McNeil Pharmaceutical
Spring House, Pennsylvania 19477.

Whole rabbit thoracic aortas were studied using ^{31}P NMR. Baseline phosphagen concentrations determined in 11 different experiments were: phosphocreatine (PCR) 2.12 ± 0.22; adenosine triphosphate (ATP measured by γPO_4) 4.23 ± 0.24; monophosphate 6.23 ± 0.36 μmoles/g dry wt. Under control conditions, inorganic phosphate (P_i) was not measurable (FIG. 1A). These are similar to values obtained from vascular smooth muscle tissue extracts using either standard biochemical approaches[1] or ^{31}P NMR.[2,3] In those reports as well as in another using intact tissue,[4] qualitative but not quantitative comparisons were possible.

Tissue ATP was stable for greater than 20 hours and was unchanged by norepinephrine (10 μM), propranolol (1 μM), or glucose-free buffer. Glucose

TABLE 1. Effects of Metabolic Inhibitors over Time on Isometric Force Development in Isolated Rings of Rabbit Thoracic Aorta[a]

Condition	Time (hours) 0	(*N*)	20	(*N*)	p
Control	411 ± 68	(12)	324 ± 31	(11)	NS
Glucose-free	440 ± 44	(6)	196 ± 21	(6)	<0.01
TDGA	676 ± 56	(6)	293 ± 71	(6)	<0.01
Glucose-free + TDGA	545 ± 35	(6)	297 ± 51	(6)	<0.01
2-DOG	255 ± 10	(6)	239 ± 30	(6)	NS
4-PA	345 ± 44	(6)	155 ± 30	(5)	<0.01
TDGA + 4-PA	260 ± 35	(6)	174 ± 50	(4)	NS

[a]Data are mean ± standard error of the mean for maximum contractile response (g/g wet tissue weight) to 60 mM potassium chloride at 37°C. *N* = number of rings (1 ring/rabbit) tested; TDGA = 2-tetradecyloxirane-carboxylic acid (10 μM); 2-DOG = glucose replacement with 2-deoxyglucose (10 mM); 4-PA = 4-pentenoic acid (2 mM). p indicates level of statistical significance (Students nonpaired *t*-test) for comparison of 20-hour data to the 0-hour time control data.

NS = not statistically significant (significance defined as $p < 0.05$). Control = 2.5 mM HEPES buffer with 10 mM glucose.

[a]Current address: Department of Pharmacology, The Squibb Institute for Medical Research, P.O. Box 4000, Princeton, NJ 08543-4000.

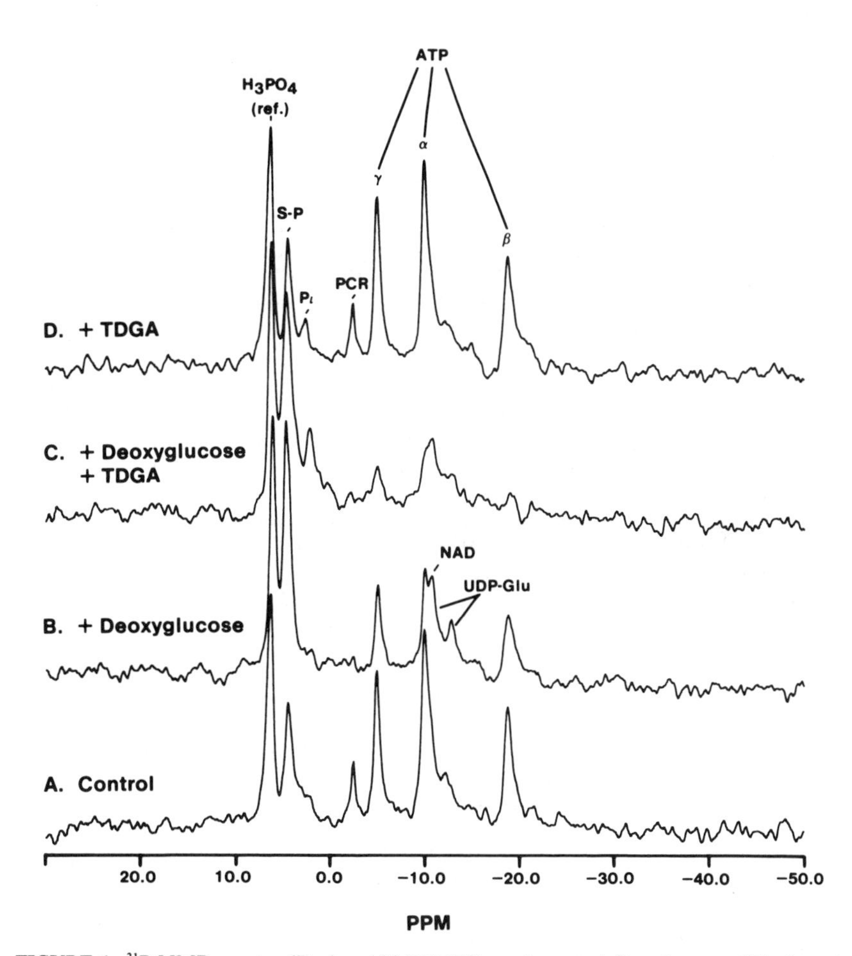

FIGURE 1. ^{31}P NMR spectra (Bruker AM-360 WB spectrometer) from intact rabbit thoracic aortas after ten hours of incubation under control conditions (A), under conditions of glucose replacement with 2-deoxyglucose to inhibit glucose oxidation (B), with 10 μM 2-tetradecyloxiranecarboxylic acid (TDGA) to inhibit fatty acid oxidation (D), and with both 2-deoxyglucose and TDGA (C). A single preparation for NMR study consisted of intact thoracic aortas from 6–9 New Zealand White Rabbits (body weight = 2.2–2.5 kg) with a combined tissue weight of 3.08 ± 0.17 g (wet) and 0.61 ± 0.02 g (dry). Aortas were studied under no-load conditions and constant superfusion with oxygenated (100% O_2) HEPES buffer (pH = 7.4, 22°C or 37°C) containing 0.5% fatty acid poor bovine serum albumin. Shown are the alpha (α), beta (β), and gamma (γ) phosphates of adenosine triphosphate (ATP), phosphocreatine (PCr), inorganic phosphate (P_i), and sugar phosphate (S-P). The reference standard (ref.) was phosphoric acid (H_3PO_4, 0.5 M) complexed to the down-field cation shift reagent, $PrCL_3$ (100 mM), contained in three sealed 5 μl capillary tubes spatially balanced within the experimental preparation. Peaks corresponding to nicotinamide adenine dinucleotide (NAD) and uridine diphosphate glucose (UDP-Glu) are also shown. Each spectrum is the result of 1,100 scans taken with 90° pulses and a repetition rate of 1.6 sec for a total time of 30 min.

(10 mM) replacement with 2-deoxyglucose (2-DOG) to inhibit glucose oxidation reduced ATP over 15 hours of incubation (control 4.2; 2-DOG 1.9 μmoles/g dry wt) (FIG. 1B). Addition of a long-chain fatty acid oxidation inhibitor (2-tetradecycloxirane-carboxylic acid, TDGA, 10 μM) had no effect on ATP without (FIG. 1D) or with glucose (control 4.8; TDGA 5.2) through ten hours. Addition of a mitochondrial beta-oxidation inhibitor (4-pentenoic acid, 2 mM) gave similar results. 2-DOG + TDGA reduced ATP more than 2-DOG alone at ten hours of incubation (control 3.0; 2-DOG + TDGA = 0.5) (FIG. 1C).

In separate experiments, isolated rings of rabbit thoracic aorta were exposed to conditions identical to those used in NMR studies for 20 hours, then placed on isometric force transducers, and tested for contractile capability. Contractile function was significantly reduced in rings exposed to glucose-free buffer, 10 μM TDGA, and 2 mM 4-PA but not 2-DOG (TABLE 1).

These data suggest that rabbit aortic smooth muscle is capable of maintaining normal ATP levels for up to 20 hours *in vitro* under resting conditions and that maintenance of phosphagen levels is not driven by either beta- or alpha-adrenergic receptor activity. The phosphocreatine/ATP ratio under resting conditions is less than 1.0 in rabbit thoracic aorta. Rabbit thoracic aorta can maintain 100% of control ATP levels during inhibition of fatty acid oxidation, but only 50% of control ATP during inhibition of glucose oxidation. Combined with function studies, these findings suggest that rabbit thoracic aorta relies primarily on fatty acid oxidation as an on-demand energy source for the support of contractile activity. Stored phosphagens arise primarily from glycolysis and do not contribute significantly to the support of contractile activity.

REFERENCES

1. BUTLER, T. M. & R. E. DAVIES. 1980. High-energy phosphates in smooth muscle. *In* Handbook of Physiology, Section 2: The Cardiovascular System, volume II, Vascular Smooth Muscle. D.F. Bohr, A.P. Somlyo & H.V. Sparks, Jr., Eds.: 237–252. American Physiology Society. Bethesda, MD.
2. HELLSTRAND, P. & H. J. VOGEL. 1985. Phosphagens and intracellular pH in intact rabbit smooth muscle studied by ^{31}P-NMR. Am. J. Physiol. **248:** C320–C329.
3. BARRON, J. T., T. GLONEK & J. V. MESSER. 1986. ^{31}P-Nuclear magnetic resonance analysis of extracts of vascular smooth muscle. Atherosclerosis **59:** 57–62.
4. DAWSON, M. J., N. C. SPURWAY & S. WRAY. 1985. A ^{31}P nuclear magnetic resonance (NMR) study of isolated rabbit arterial smooth muscle. J Physiol. **365:** 72p.

NMR Studies on a Renal Epithelial Cell Line: $LLC\text{-}PK_1/Cl_4$

A. W. H. JANS, E. KELLENBACH, B. GRIEWEL,
AND R. K. H. KINNE

Max-Planck-Institut für Systemphysiologie
4600 Dortmund 1, Federal Republic of Germany

$LLC\text{-}PK_1/Cl_4$ cells, which are derived from pig kidney, are widely used to study renal proximal tubular functions *in vitro*.[1-3] We have used different NMR techniques (^{31}P-, ^{23}Na-, and ^{13}C NMR) to investigate ion transport systems and metabolic pathways involved in sodium uptake and the regulation of intracellular pH (pH_i) in this cell line.

Changes in pH_i were measured by ^{31}P NMR. Intracellular sodium was determined by using $Dy(PPP_i)_2^{7-}$ as extracellular shift reagent. Substrate uptake was followed by using ^{13}C-D glucose. When suspended in bicarbonate-free Ringer's solution of pH 7.4, the intracellular pH was 7.24, the intracellular sodium concentration was 11.5 mM, and the rate of D-glucose utilization was 0.5×10^{-4} mM/min^{-1} mg protein. Addition

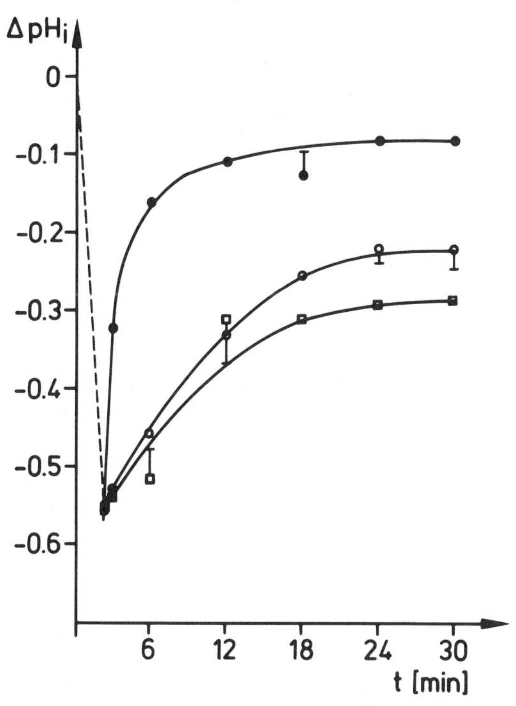

FIGURE 1. Time course of pH_i in $LLC\text{-}PK_1/Cl_4$ cells during exposure to 20% CO_2 at $t = 0$ min. (○; normal bicarbonate-free Ringer's solution with 1 mM amiloride, □; bicarbonate-free and sodium-free Ringer's solution).

of 20% CO_2 led to a rapid intracellular acidification, which was followed by a slower realkalinization. At 37°C, the initial rate of realkalinization was 0.4×10^{-4} mM/ min^{-1} mg protein, this rate was reduced in the absence of extracellular sodium and in the presence of 1 mM amiloride (amiloride-sensitive proton flux 0.3×10^{-4}) (FIG. 1). During the same time, intracellular sodium transiently increased to 13.9 mM. The initial rate of sodium entry during realkalinization at 25°C was 0.4×10^{-4} mM/min^{-1} mg protein, this rate was reduced to almost zero by 1 mM amiloride (FIG. 2). During realkalinization, D-glucose utilization was 0.52×10^{-4} mM/min^{-1} mg protein, 0.12×10^{-4} mM/min^{-1} mg protein was inhibitable by amiloride. These data demonstrate that in LLC/PK_1 cells the main route for sodium entry is a sodium proton exchanger, which is involved in recovery of the cell from intracellular acidification. Energy derived from D-glucose can be used to energize this process. Assuming that two sodium ions are

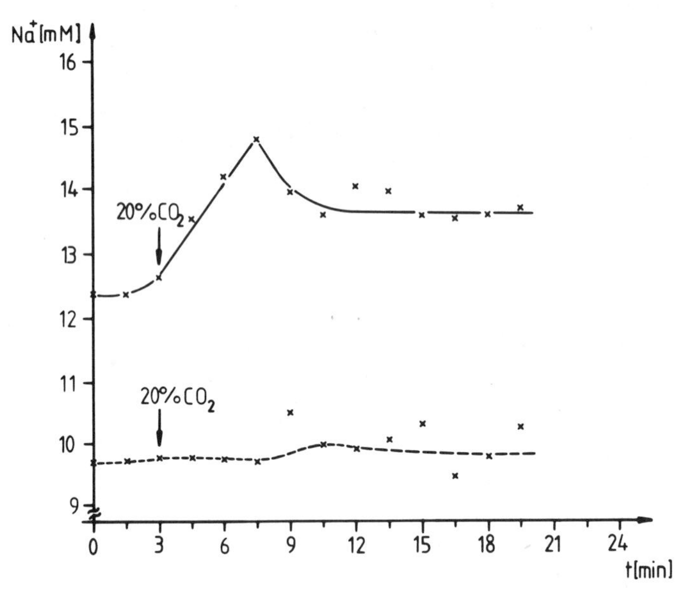

FIGURE 2. Time course of intracellular sodium in LLC-PK_1/Cl_4 cells during exposure to 20% CO_2 at $t = 3$ min. (x——x; normal bicarbonate-free Ringer's solution, x---x; normal bicarbonate-free Ringer's solution with 1 mM amiloride).

taken up into the cell per each glucose molecule utilized and taking into account the higher transport rate at 37°C,[4] the stoichiometry of the sodium proton exchanger in the intact cell is close to one.

REFERENCES

1. MULLIN, J. M., J. WEIBEL, J. DIAMOND & A. J. KLEINZELLER. 1980. J. Cell. Physiol. **104:** 375–389.
2. RABITO, C. A. 1986. Mineral. Electrolyte Metab. **12:** 32–41.
3. CHAILLET, J. R., K. AMSLER & W. F. BORON. 1986. Proc. Natl. Acad. Sci. USA **83:** 522–526.
4. DESMEDT, H. & R. KINNE. 1981. Biophys. Acta **648:** 247–253.

^{31}P NMR Evaluation of Rodent and Canine Myopathies[a]

Z. ARGOV,[b,e] U. GIGER,[c] Z. ROTH,[b] M. KORUDA,[d]
J. S. LEIGH,[b] AND B. CHANCE[b]

Departments of [b]Biochemistry and Biophysics, [d]Surgery, and [c]Veterinary School
University of Pennsylvania
Philadelpia, Pennsylvania 19104

A method currently employed in our laboratory for repeated *in vivo* studies of animal muscle bioenergetics[1] was used to evaluate chronic and acute myopathies in animal models of human diseases. The basic protocol includes monitoring ^{31}P NMR changes during rest, graded steady-state work, and during post exercise recovery. Muscle work is induced by percutaneous nerve stimulation at frequencies of 0.25–5 Hz (10 min periods at each rate to achieve a steady state.) No surgery is performed on the studied leg.

In normal rats, stimulation rate (four frequencies used: 0.25, 0.5, 1.0, and 2.0 Hz) versus P_i/PCr gave a linear plot with a slope of 2.2 ± 0.25 Hz/P_i/PCr (FIG. 1). This corresponded to a drop of phosphorylation potential (PP) from 80,700 ± 35,100 at rest to 6,700 ± 900 at 2 Hz, as estimated from fully relaxed spectra. The slope was reduced during substrate deficiency (6 days, of starvation in rats)[2] and during hormonal abnormality (6 weeks post ovariectomy)[3] to 1.25 ± 0.7 and 1.52 ± 0.1 Hz/P_i/PCr (±2 SD), respectively (FIG. 1). It did not change after induction of acute glycolytic block by iodoacetate (IA) (FIG. 1). Marked accumulation of sugar phosphates was noted during exercise in IA-injected rats, similar to that seen in an equivalent human disease-phosphofructokinase (PFK) deficiency.[4] Thus, the stimulation rate versus P_i/PCr is an effective physiological monitor of bioenergetic metabolism that can be used for ^{31}P NMR studies of various animal muscle diseases. The advantages of this method are: simplicity, it is suitable for repeated studies in the same animal, and it uses gradual activation of mitochondrial metabolism through several steady states. The deficiencies are: stimulation rate does not accurately describe work rate and thus, the linear slope is not according to the expected Michaelis-Menten kinetics (rectangular hyberbola). Wide range variations between animals calls for large samples in comparative experiments. We have tested the method in two available chronic models of human myopathies.

In dystrophic hamsters (Bio 14.6), stimulation rate versus P_i/PCr (0.5 ± 0.5 Hz/P_i/PCr), PCr/P_i at rest (5.3 ± 1.2) and post exercise recovery (0.5 ± 0.2 PCr/P_i units per min) were all abnormal compared with age matched (9 months) normal hamsters (1.05 ± 0.5 Hz/P_i/PCr; 6.55 ± 0.5; 1.42 ± 0.28 PCr/P_i per min, respectively). These findings are consistent with *in vivo* malfunction of muscle mitochondria.[5] This could be due to the primary muscle disease of this animal line or be secondary to their heart failure. The same technique was used to study larger animals;

[a]Supported by grants from the Muscular Dystrophy Association.

[e]Present address: Dept. of Neurology, Hadassah University Hospital, Jerusalem 91120, Israel.

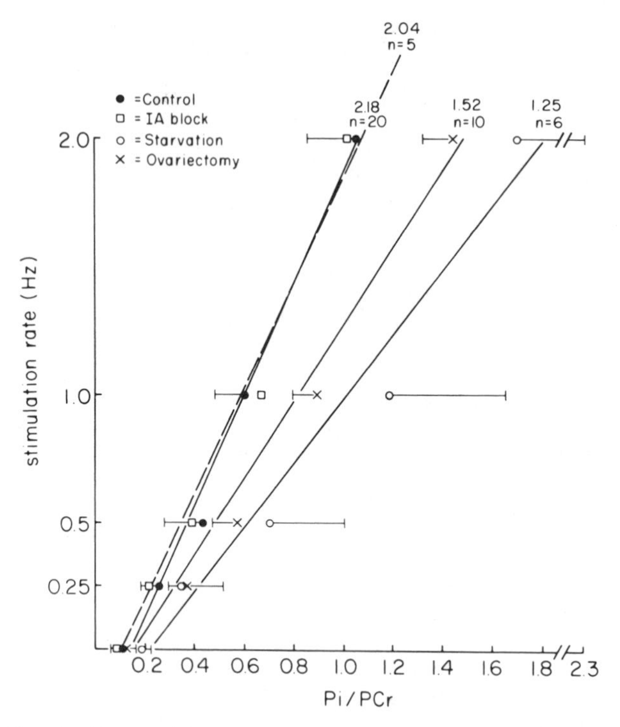

FIGURE 1. Exercise kinetics (stimulation rate vs. P_i/PCr) in normal rats and rats with three induced muscle disorders. Bars = 2 SD (were not plotted for iodoacetate-injected rats, IA, because they overlapped with controls). The average slope in Hz/P_i/PCr and the number of animals studied (N) are also indicated.

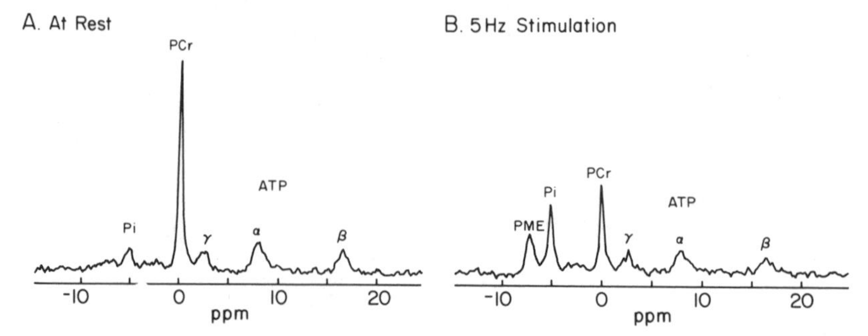

FIGURE 2. Fully relaxed ^{31}P NMR spectra (20 sec delay, 75 scans) in phosphofructokinase-deficient dogs at rest (A) and during 5 Hz stimulation (B). Note the presence of high PME peak and slight fall in ATP, neither occurred in normal dogs. Scale for both spectra taken from one animal is the same.

PFK deficient and normal dogs.[6] The exercise kinetics in the dogs showed a much higher slope (3.9 Hz/P_i/PCr in normal and 5.9 in PFK deficient) than that seen in rodents. This probably indicates higher oxidative capacity in dog muscle. Exercise in PFK-deficient dogs also induced rise in PME and a fall in ATP typical of distal glycolytic blocks (FIG. 2). These changes were associated with a drop of PP from 56,000 to 3,700. Thus, ^{31}P NMR studies offer a method of diagnosing and monitoring function in chronic animal myopathies.

REFERENCES

1. ARGOV, Z., J. MARIS, M. KORUDA, L. KING & B. CHANCE. 1985. Proc. Mag. Reson. Med. Soc. 4th mtg. (London) **1:** 432–433.
2. KORUDA, M., Z. ARGOV, J. MARIS, R. H. ROLANDI, R. G. SETTLE, D. O. JACOBS, B. CHANCE & J. I. ROMBEAU. 1985. Surg. Forum **36:** 61–63.
3. ROTH, Z., Z. ARGOV, J. MARIS, L. KING & B. CHANCE. 1985. Proc. Mag. Reson. Med. Soc. 4th mtg. (London) **1:** 534–535.
4. ARGOV, Z., D. NAGLE, J. MARIS, W. J. BANK & B. CHANCE. 1986. Neurology **36**(supp 1): 93.
5. ARGOV, Z., J. MARIS, J. LEIGH & B. CHANCE. 1986. Neurology **36**(supp 1): 171.
6. GIGER, U., Z. ARGOV, M. SCHNALL & B. CHANCE. 1986. Muscle & Nerve **19**(supp 5s): 241.

The Value of ^{31}P NMR in the Diagnosis and Monitoring the Course of Human Myopathies[a]

W. BANK,[b] Z. ARGOV,[b,c] J. S. LEIGH,[c] AND B. CHANCE[c]

Departments of [b]Neurology and [c]Biochemistry/Biophysics
University of Pennsylvania
Philadelphia, Pennsylvania 19104

^{31}P NMR proved to be a valuable research tool of human muscle bioenergetics in health[1] and disease.[2] Over the past five years, we have assessed its practical use in diagnosis and monitoring the course and therapy of myopathies. Abnormally low PCr/P_i at rest (usually due to high P_i) was first observed in mitochondrial myopathies (MM)[3-5] indicating a reduced phosphorylation potential.[6] Although 13 of 16 of MM patients showed this abnormality,[5] a low PCr/P_i at rest was also noted in several destructive myopathies (eg. muscle dystrophy and polymyositis), in advanced atrophic disorders, and following acute muscle injury.[7] Secondary mitochondrial damage may exist in all these conditions, reflecting a final common pathway in diseased muscle cells. Although this finding is not disease specific, it proved useful in the following situations: (1) Monitoring acquired muscle disease. In polymyositis PCr/P_i fell from

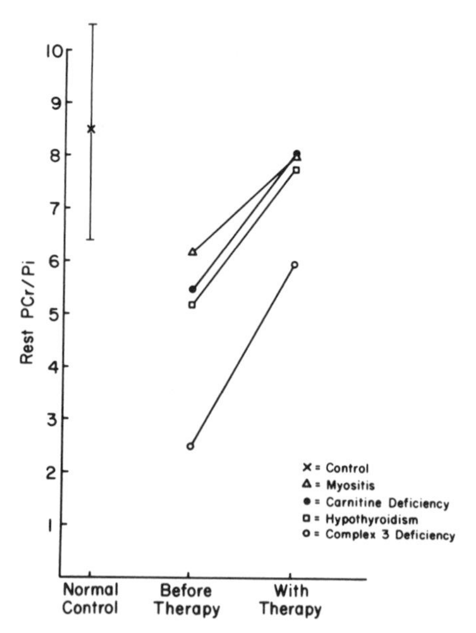

FIGURE 1. The change in PCr/P_i at rest following therapy. Bars = ±2 SD.

[a]Supported by the National Institutes of Health (NS 08075) and the Muscular Dystrophy Association.

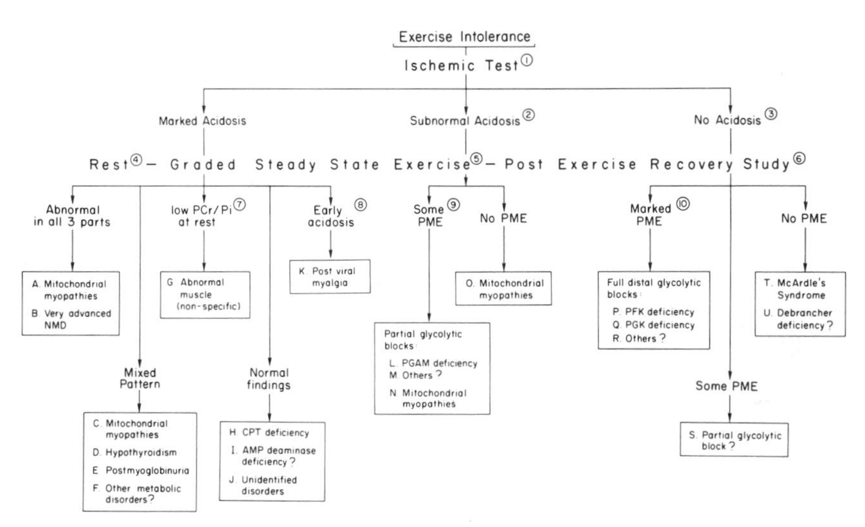

FIGURE 2. Diagnostic chart for the evaluation of exercise intolerance by muscle ^{31}P NMR. (1) Monitor intracellular pH during 3 min of maximal ischemic exercise confirmed by marked PCr depletion. Normally, pH falls to 6.65 or less. (2) Subnormal acidosis (pH 6.75–6.9) was found in MM during maximal but not ischemic exercise. (3) Diagnostic for complete glycolytic block. Associated with high PCr/ATP at rest (6.4–8.2 compared with 5.5 ± 0.7 in controls)[10] (4) 4.5–10 min (60–120 scans). (5) At least 2 levels of steady state are required to evaluate the work rate vs. P_i/PCr transfer function. (6) Post exercise recovery is studied from a common end exercise PCr/P_i in all cases (range: 0.5–0.9). (7) A nonspecific but significant finding indicating muscle disorder. (8) This phenomenon may also appear in persons with no complaint.[11] (9) Exercise dependent mild to moderate rise in PME (PME/ATP 1–2) with elimination $T_{1/2}$ of about 5 min. (10) PME/ATP >2 with elimination $T_{1/2}$ of 9–10 min. *Disease entities:* (?) Possible abnormalities that have not yet been confirmed. (B) Non specific in end stage neuromuscular diseases (NMO). (D) Phosphodiester peak may be elevated in this condition. (E) Seen in myoglobinuria of undetermined origin. (H) Five patients studied.[9] (K) This is a newly described but unclear entity[11] and the possibility of finding early acidosis in some normal persons casts doubt about the uniqueness of the finding. (N) Seen in mitochondrial ATPase deficiency. PME may be phosphoethanolamine or IMP, both may rise with exercise. (Q) Based on Ref. 12. (S). Based on our findings in dogs with partial PFK deficiency (enzyme activity about 6% of normal).[13]

4.5 to 3.3 during an acute clinical exacerbation associated with rise in serum CPK. (2) Following deterioration in progressive muscular dystrophy with age and disease state.[8] (3) Monitoring therapy in muscle diseases; PCr/P_i returned to normal as a result of specific therapy in several conditions. FIGURE 1 shows the change in PCr/P_i at rest as a result of medication in four conditions: mitochondrial complex 3 deficiency treated with vitamin K_3 and ascorbate (vitamin C),[3] carnitine deficiency,[9] and hypothyroidism responding to replacement therapy and polymyositis treated with high dose corticosteroids. Thus, the simple recording of PCr/P_i at rest may prove to be a valuable monitor of muscle recovery from various diseases and permit the objective evaluation of new therapies.

We have developed exercise protocols to diagnose and distinguish a variety of muscle disorders manifesting exercise intolerance. A flow chart (FIG. 2) illustrates ^{31}P NMR distinguishing features seen at rest, during exercise, and post exercise recovery. Exercise intolerance is currently the most suitable entity to be studied by physiological ^{31}P NMR spectroscopy. We expect the highest diagnostic yield of ^{31}P NMR in these conditions, and feel that the technique is now suitable for screening patients.

REFERENCES

1. CHANCE, B., B. J. CLARK, S. NIOKA, H. SUBRAMANIAN, J. M. MARIS, Z. ARGOV & H. BODE. 1985. Circulation **72**(supp 4): 103–110.
2. EDWARDS, R. H. T., D. R. WILKI, M. J. DAWSON, R. E. GORDON & D. SHAW. 1982. Lancet **1:** 725–730.
3. ELEFF, S., N. G. KENNAWAY, N. R. M. BUIST, V. M. DARLEY-USMAR, R. A. CAPALDI, W. J. BANK & B. CHANCE. 1984. Proc. Natl. Acad. Sci. USA **81:** 3529–3533.
4. ARNOLD, D. L., D. J. TAYLOR & G. K. RADDA. 1985. Ann. Neurol. **18:** 189–196.
5. ARGOV, Z., W. J. BANK, J. MARIS, P. PETERSON & B. CHANCE. 1986. Neurology (In press.)
6. CHANCE, B. 1984. Ann. N.Y. Acad. Sci. **428:** 318–332.
7. MCCULLY, K. K., B. BODEN, D. BROWN & B. CHANCE. 1986. Proc. Soc. Mag. Reson. Med. 5th Meeting (Montreal). **2:** 345–346.
8. YOUNKIN, D., P. BERMAN, J. SLADKY, C. CHEE, W. J. BANK & B. CHANCE. 1986. Neurology (in press.)
9. ARGOV, Z., J. MARIS, K. FISCBECK, W. BANK & B. CHANCE. 1985. Ann. Neurol. **18:** 119.
10. ARGOV, Z., J. MARIS, W. BANK & B. CHANCE. 1985. Ann. Neurol. **18:** 162.
11. ARNOLD, D. L., P. BORE, G. K. RADDA, P. STYLES & D. J. TAYLOR. 1984. Proc. Soc. Mag. Reson. Med. 3rd mtg. (New York) **1:** 12–13.
12. DUBOC, D., P. JEHENSON, M. FARDEAU, A. SYROTA & S. TRAN DINH. 1985. Proc. Soc. Mag. Reson. Med. 4th mtg. (London). **1:** 463–464.
13. GIGER, U., Z. ARGOV, M. SCHNALL & B. CHANCE. 1986. Muscle & Nerve **9**(suppl. 5S): 187.

Fluorodeoxyglucose Brain Metabolism Studied by NMR and PET

NICOLAS R. BOLO, KATHLEEN M. BRENNAN, REESE M. JONES, AND THOMAS F. BUDINGER

Donner Laboratory
Lawrence Berkeley Laboratory
University of California
Berkeley, California 94720

INTRODUCTION

2-^{18}F-2-deoxy-glucose (^{18}FDG) is used as a metabolic tracer for monitoring the local cerebral metabolic rate of glucose utilization measured by positron emission tomography (PET). The validity of these studies depends on the assumption that the tracer is trapped in the FDG-6-phosphate form and that FDG or FDG-6-P is not substantially metabolized along the other known metabolic pathways for glucose (FIG. 5). The possibility that fluorodeoxyglucose is metabolized past the fluorodeoxyglucose-phosphate state has been raised by a number of investigators who have noted metabolic products of 2-^{19}F-2-deoxy-glucose (^{19}FDG) using NMR spectroscopy studies of the animal heart and brain *in vivo*.[1-5] The major goals of our study are to discover the metabolic fate of FDG in the brain and to determine if the doses given for the NMR studies have an effect on brain metabolism. The kinetics of ^{18}FDG uptake and washout in the rabbit brain are monitored by ^{18}FDG positron emission tomography after injection of increasing doses of ^{19}FDG. ^{19}F-fluorinated metabolites in the brain are then quantitated by ^{19}F NMR. Fluorodeoxyglucose dose-dependent effects on brain metabolism are investigated.

MATERIALS AND METHODS

^{18}FDG Preparation

Radioactive ^{18}FDG is produced at the 76-inch Crocker cyclotron at the University of California at Davis. A deuteron beam of 23 MeV is used to irradiate a static neon gas target containing 0.1% F_2 to produce $^{18}F\text{-}F_2$. This is transferred from the cyclotron target vessel by neon gas flow into the reaction vessel, which contains tri-O-acetyl glucal in freon. The subsequent steps of silica gel column separation, acid hydrolysis, ion exchange resin, and alumina column purification are described elsewhere.[6] The ^{18}FDG preparation is analyzed by high-performance liquid chromatography and thin layer chromatography. The radiochemical purity of this ^{18}FDG is greater than 95%.

Isotope Injection and PET Kinetic Data Collection

Male New Zealand white rabbits (3–4 kg) were anesthetized with xylazine/ketamine/acepromazine i.m. Femoral catheters were surgically implanted in both femoral arteries of the rabbits in order to maintain separate lines for injection and

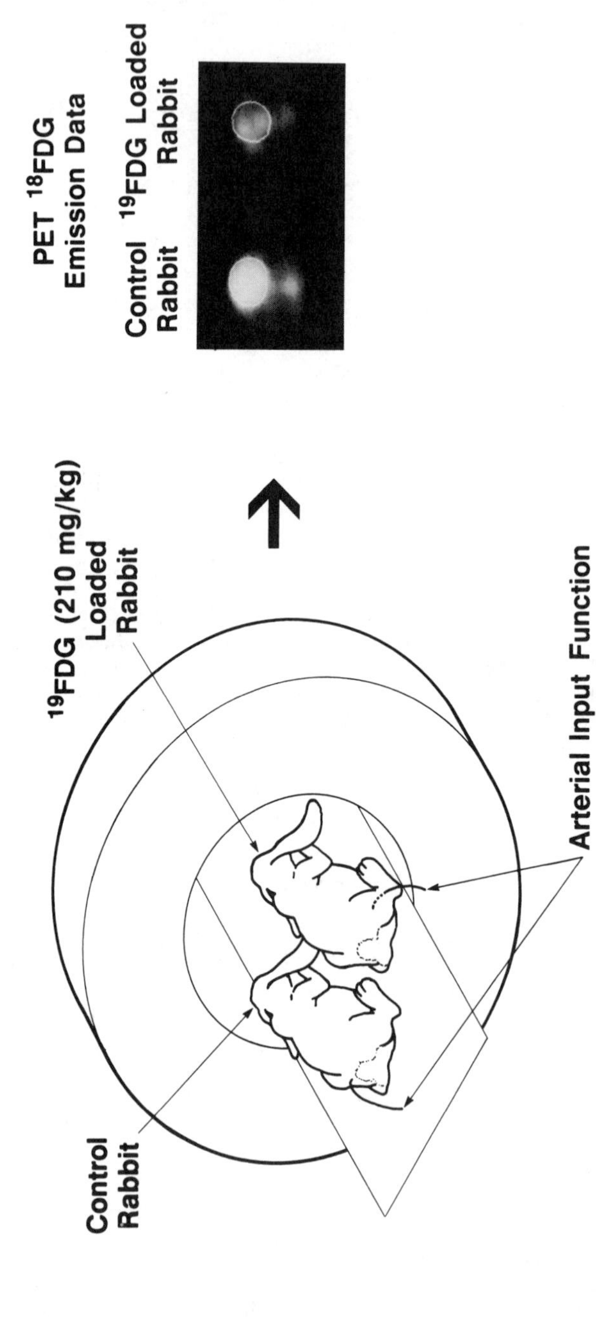

FIGURE 1. Experimental setup for the positron emission tomography study. Control and FDG-loaded rabbits are imaged simultaneously in the PET ring.

blood withdrawal. Two rabbits were given, respectively, 70 mg/kg and 210 mg/kg loading doses of cold ^{19}FDG 20 min prior to ^{18}FDG injections; the two control rabbits received no ^{19}FDG. For each dose, the loaded and control rabbits were positioned on their sternums side by side with both brains imaged simultaneously by PET using a 3 mm transverse section after identical doses (8.5 mCi) of ^{18}FDG were given in i.v. boluses (FIG. 1). Dynamic brain images were obtained continuously at 2.5–3.0-sec intervals for the first 4 min, then at 60-sec intervals until 10-min post injection followed by 5-min intervals until 40-min post injection. Data were also taken at the liver 40-min after injection. Immediately post ^{18}FDG injection, arterial input function samples (0.5 ml) were obtained from the opposite femoral line as rapidly as possible for the first 2 min then at progressively longer intervals out to 40 min post ^{18}FDG injection. These blood samples were then counted in a TM Tracor-gamma-well counter and the results were decay corrected back to the time of injection. The dynamic data were used along with arterial blood concentration data to derive the kinetic constants for the model of FIGURE 2. At the conclusion of the PET study, both rabbits were sacrificed by i.v. barbiturate overdose.

Tissue extraction and ^{19}F NMR Spectroscopy

The brains and livers of the 70 mg/kg and 210 mg/kg ^{19}FDG-loaded rabbits from the PET study described above were rapidly excised and frozen in liquid nitrogen immediately post euthanasia. An aqueous/organic extraction procedure developed for proton and carbon NMR spectroscopy of brain tissue was used.[9] Each frozen brain sample was homogenized in cold 1:2 chloroform:methanol (3 ml/g of brain tissue). The single-phase homogenate was vortexed with 1:1 chloroform:aqueous buffer (2 ml/g brain tissue). The added solvents force the mixture to separate into three phases after centrifuging: aqueous phase, organic phase, and protein pellet. Half of each brain was extracted with incorporation of 5 mM TFA in the extraction aqeous buffer (pH = 6.4 ± 0.1), the other brain half was extracted without added TFA. In order to determine whether TFA may be used as a valid internal reference, NMR spectra of extracts with and without added TFA were compared. All samples contained both white and gray matter. The aqueous fractions of the three-phase separation were lyophilized and redissolved each in 1 ml of D_2O. Phosphate-buffered D_2O solutions (pH = 6.4) of FDG, FDG + TFA, and TFA were observed for comparison to the extracts. All samples were placed in Wilmad 5 mm diameter NMR tubes. ^{19}F spectra were collected on the University of California, Berkeley, Chemistry Department AM-500 11.7 Tesla spectrometer at 470.5 MHz. Undecoupled spectra (64K data points) shown in FIGURE 3 were acquired using a one-pulse sequence with a sweep width of 100 kHz. Broadband WALTZ proton decoupled spectra were acquired using a bilevel decoupling sequence with the same sweep width and number of data points.

RESULTS

Effects of ^{19}FDG on ^{18}FDG Kinetics

In both studies, regions of interest were drawn over each rabbit's brain and the time activity curves compared for the ^{19}FDG loaded rabbit and the respective control rabbit. Results are shown in FIGURE 4 and TABLES 1 and 2.

In the study where a loading dose of 70 mg/kg ^{19}FDG was used, both rabbit brains

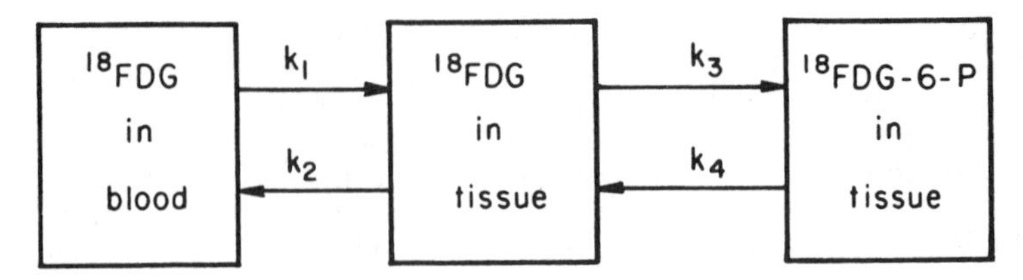

FIGURE 2. Three compartment kinetic model used for analysis of positron emission tomography data. Values of the kinetic constants k_{1-4} are found in TABLE 2.

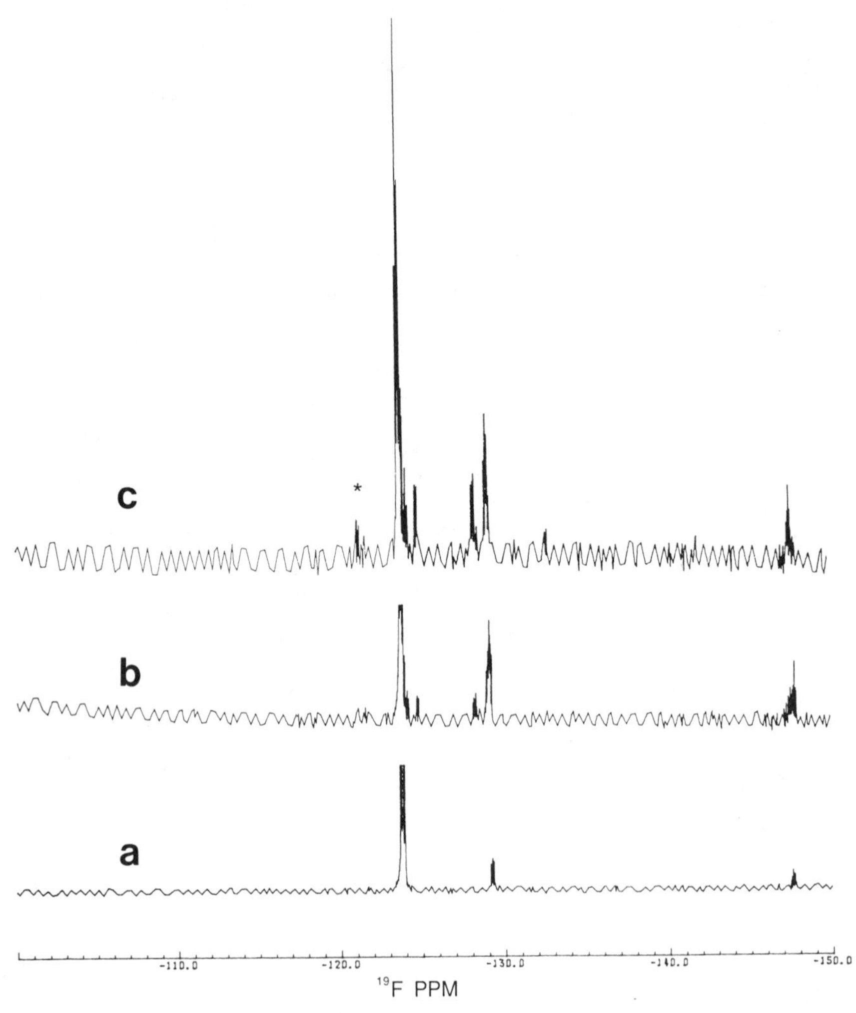

FIGURE 3. Undecoupled ^{19}FNMR spectra of (a) buffered standard of FDG, (b) 70 mg/kg loaded rabbit brain extract, (c) 210 mg/kg loaded rabbit brain extract. The asterisk indicates 6-phospho-2-F-gluconate at −121.11 ppm.

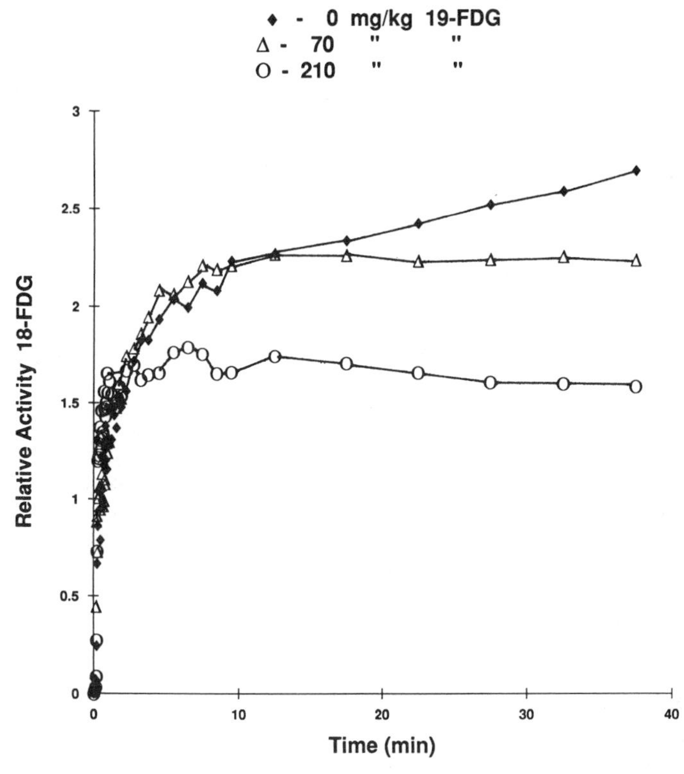

FIGURE 4. ^{18}FDG uptake curves comparing relative activity time course for control and ^{19}FDG loaded rabbits.

showed an identical ^{18}FDG uptake up to 15 min followed by a progressive decrease in activity in the ^{19}FDG rabbit brain up to 40 min. At 40 min, the ^{19}FDG brain had a decrease of 22% in ^{18}FDG activity compared to his respective control rabbit.

For the study where 210 mg/kg ^{19}FDG was used, uptakes were only similar for the first two minutes followed by an even more dramatic decrease in overall activity in the ^{19}FDG rabbit brain with the final activity being 43% less than the respective control brain.

In both studies, liver activity was found to be 20% greater in the ^{19}FDG-loaded rabbits compared to their respective controls (TABLE 1). Arterial input functions measuring ^{18}FDG were identical for both rabbits in each study.

TABLE 1. Comparison of Relative Organ Uptakes of ^{18}FDG in Control and ^{19}FDG-loaded Rabbits

Organ	Control	^{19}FDG (70 mg/kg)	^{19}FDG (210 mg/kg)
Brain	100%	78%	57%
Liver	100%	120%	120%

TABLE 2. Comparison of Kinetic Rate Constants[a] in Control and ^{19}FDG-loaded Rabbits

Parameter (min^{-1})	^{19}FDG (210 mg/kg) Rabbit	Control Rabbit	% Difference ^{19}FDG/Control
k_1	0.092	0.089	+3%
k_2	0.239	0.154	+55%
k_3	0.027	0.042	−36%
k_4	0.00015	0.0043	−97%
f_v	0.183	0.146	+25%

[a]See model in FIG. 2. The vascular fraction (percent of activity in a given organ due to blood) is represented by f_v.

Spectroscopy of Tissue Extracts

Spectral differences between the brain extracts were apparent as shown by FIGURE 3. No resonance shifts due to added TFA were observed in either the standards or the extracts. Thus, TFA was used as an internal chemical shift reference. Though an impurity at −93.83 ppm found in the TFA standard and in the extracts with added TFA calls for caution, it appears that the NMR extraction technique can be used quantitatively. Results are summarized in TABLE 3.

Buffered FDG standards show the β and α anomers of FDG whose resonances are centered respectively at −123.67 and −123.79 ppm. They appear as a multiplet in FIGURE 3a and are grouped together as a resonance at −123.7 ppm in TABLE 3. Two other multiplets centered at −129.14 and −147.53 ppm were found to be contaminants or chemical degradation products of the original product.

The ^{19}FDG-loaded rabbit brain extracts show the FDG peaks and also β and α anomers of FDG-6-P centered at −123.74 and −123.93 ppm. FDG-6-P is grouped with FDG in TABLE 3. The peaks at −129.14 and −147.53 ppm found in the standard

TABLE. 3. Comparison of Integrated Peak Areas from ^{19}F NMR Spectra in Standard ^{19}FDG-loaded Rabbit Brain Extracts[a]

Chemical Shift (ppm)	0	−93.83	Gluconate −121.11	FDG −123.70	−124.75	−128.18	−129.14	−147.53
Buffered TFA								
Peak Integrals	1000	15	0	0	0	0	0	0
Buffered FDG								
Peak integrals	0	0	0	30	0	0	2	0.8
Ratio to FDG integral	—	—	0	1	0	0	0.067	0.027
"70 mg/kg" extract								
Peak integrals	1000	15	0	31	1.6	3.4	40	16.5
Ratio to FDG integral	—	—	0	1	0.052	0.11	1.29	0.532
mg/g of tissue[b]	—	—	—	0.015	—	—	—	—
"210 mg/kg" extract								
Peak integrals	1000	15	1.2	16	1.3	2.3	5	2
Ratio to FDG integral	—	—	0.075	1	0.081	0.144	0.312	0.125
mg/g of tissue[b]	—	—	—	0.008	—	—	—	—

[a]FDG and FDG-6-P peaks were grouped for integration.

[b]Calculated considering FDG and FDG-6-P as FDG. Errors are evaluated at 5%.

are also found in both brain extracts, but at higher intensities relative to FDG than in the standard. They may be either impurities present in the original injection solution and have a different brain uptake than FDG, or they might also be metabolic products as has been reported.[1-5]

Both extracts show further resonances centered at −124.75 and −128.18 ppm, which remain unidentified. A peak centered at −121.11 ppm is present in the 210 mg/kg extract but not in the 70 mg/kg extract. This peak has been tentatively assigned

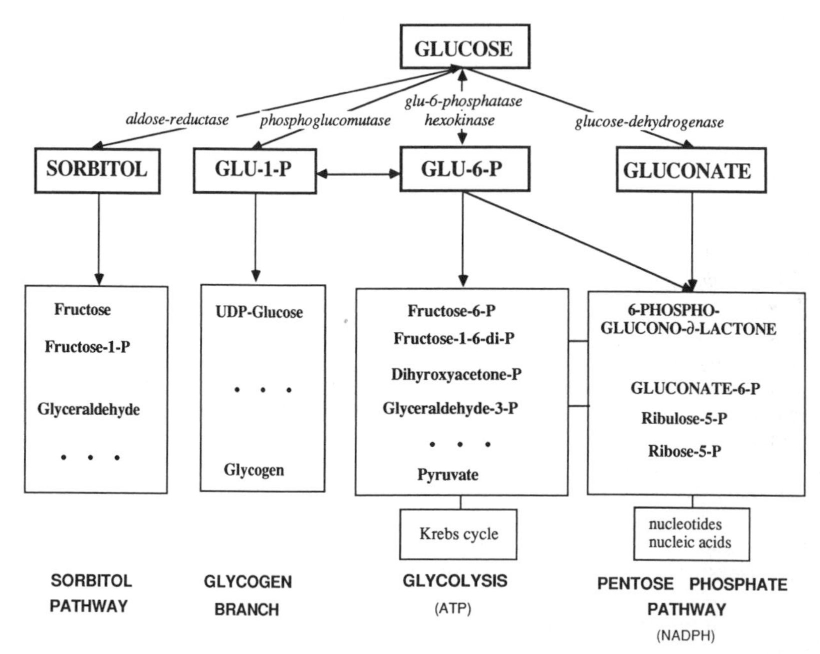

FIGURE 5. Metabolic pathways of D-glucose. In the brain, glucose is mainly catabolized along the glycolytic pathway for production of ATP to fulfill the cell's energetic needs. Generally in cell systems, if reductive power in the form of NADPH is needed e.g. for fatty acid biosynthesis, or ribose-5-P is needed in nucleic acid synthesis, metabolism may occur preferentially along the pentose phosphate pathway. Regulation between these two pathways occurs through intermediates at the fructose-6-P and glyceraldehyde-3-P levels. Storage of glucose is realized by formation of glycogen along the glycogen synthesis pathway (glycogen branch). Sorbitol is the product of glucose hydrogenation by aldose reductase.

to 2-F-6-phospho-gluconate, a metabolite that occurs if FDG enters the pentose phosphate pathway (also known as the pentose shunt). If the brain metabolic activity based on peak intensities relative to the FDG peak were the same in the 70 mg/kg loaded rabbit as in the 210 mg/kg loaded rabbit, one would expect a peak at −121.11 ppm with an intensity 7.5% that of FDG. This would have been detectable above baseline noise level but was not observed above noise in the −121.11 region of the 70 mg/kg sample.

DISCUSSION

This work is the basis for a study of fluorodeoxyglucose metabolism as a function of fluorodeoxyglucose tissue concentration currently in progress. Proton NMR spectra of these and other tissue extracts are presently being studied for effects on ^{1}H observable metabolites.

Studies of local cerebral glucose utilization using tracer amounts of radioactively labeled deoxyglucose (DG) are based on models in which deoxyglucose is trapped in the tissue in the DG-6-P form[7,8] after phosphorylation by hexokinase. However, fluorodeoxyglucose metabolism in the brain has recently been proposed to extend beyond the FDG-6-phosphate stage.[1–5] Some possible pathways are phosphatase activity, the pentose pathway, the aldose reductase sorbitol pathway, or the glycogen synthesis branch (FIG. 5). ^{19}F-NMR spectroscopy of brain tissue extracts was used in conjunction with ^{18}FDG PET to investigate these metabolic pathways.

The PET studies clearly show an effect of ^{19}FDG loading on ^{18}FDG kinetics. There was a decrease in brain accumulation of ^{18}FDG in animals who received a loading dose of ^{19}FDG and the rate of washout for ^{18}FDG is greater in the loaded than in the control animals. These animals were under anesthesia, thus the results might not be reflective of metabolic changes in the unanesthetized animal.

The NMR study enabled us to obtain biochemical information on the metabolism of FDG in the brain, although definite assignment of all resonances observed is necessary for a more complete understanding of the dose effects of FDG.

Our study supports evidence of metabolism beyond the FDG-6-P stage for the doses administered and suggests that there are dose-dependent effects on the brain biochemistry. In particular for the lower dose no significant peak at −121.11 ppm (tentatively assigned to 6-phospho-2-F-gluconate, a metabolite from the pentose phosphate shunt) was observed, suggesting that the pentose shunt becomes important only when FDG concentration exceeds some threshold, presumably greater than 70 mg/kg i.v. The highest carrier dose of ^{19}FDG during ^{18}F PET studies is 2 mg whereas doses of 200 to 500 mg/kg of FDG are administered in deoxy-glucose brain metabolism studies observing metabolites *in vivo* with ^{31}P, ^{19}F, or ^{1}H NMR spectroscopy.

Although the PET and NMR data presented here are consistent in showing that high doses of ^{19}FDG perturb brain metabolism and glucose transport kinetics, we recognize that a total of only four animals were used, two control and two ^{19}FDG-loaded animals. We are confident in the qualitative aspect of the data (presence or not of 6-phospho-2-F-gluconate in the brain extracts) for the animals studied but it is likely that statistical fluctuations in quantities will appear when a greater number of animals is studied.

REFERENCES

1. DEUEL, R., G. MUYUE, W. L. SHERMAN, D. J. SCHICKNER & J. J. H. ACKERMAN. 1985. Science **228:** 1329–1331.
2. WYRWICZ, A. M., R. MURPHY, I. PRAKASH, R. M. MORIARTY & T. DOUGHERTY. 1985. Abstr. 4th Annu. Mtg. Soc. Magnetic Resonance Med. **2:** 827.
3. BABCOCK, E. E. & R. L. NUNNALLY. 1985. Abstr. 4th Annu. Mtg. Magnetic Resonance Med. **2:** 751.
4. BERKOWITZ, B. A., S-G. KIM & J. J. H. ACKERMAN. 1986. Abstr. 5th Annu. Mtg. Soc. Magnetic Resonance Med. **1:** 257.
5. NAKADA, T. & I. L. KWEE. 1986. Abstr. 5th Annu. Mtg. Soc. of Magnetic Resonance Med. **1:** 259.

6. IDO T., C-N. WAN, V. CASELLA, J. S. FOWLER, A. P. WOLF, M. REIVICH & D. KHUL. 1978. J. Label. Comp. Radiopharm. **14:** 175–183.
7. SOKOLOFF L., M. REIVICH, C. KENNEDY, M. H. DES ROSIERS, C. S. PATLAK, K. D. PETTIGREW, O. SAKURADA & M. SHINOHARA. 1977. J. Neurochem. **28:**897–916.
8. NELSON T., G. LUCIGNANI, S. ATLAS, A. M. CRANE, G. A. DIENEL & L. SOKOLOFF. 1985. Science **229:** 60–62.
9. JONES R. M., T. RICHARDS & T. F. BUDINGER. 1984. Abstr. 3rd Annu. Mtg. Soc. Magnetic Resonance Med. 395.

Proton NMR Spectroscopic Studies of "Stunned" Myocardium

W. T. EVANOCHKO, R. C. REEVES, R. C. CANBY,
J. B. McMILLIN, AND G. M. POHOST

University of Alabama at Birmingham
Birmingham, Alabama 35294

The inability of the heart to rapidly recover function from a brief period of occlusion followed by reperfusion has been termed "stunned" or, more formally, myocardium with prolonged, postischemic ventricular dysfunction.[1,2] The mechanism for this transitory effect is still unclear. In view of the ease of measurement of myocardial mobile lipids (e.g., fatty acyl chains, free fatty acids, and glycerides) by proton NMR spectroscopy[3] and their involvement in ischemic processes,[4] high resolution ^{1}H NMR spectroscopy was performed on excised tissue following 15-min coronary occlusion followed by 3-hr reperfusion in order to gain insight into this important pathologic state. Alterations in lipids and other selected metabolites were determined, and these changes were correlated with myocardium function determined by crystal segment lengths studies and regional myocardial blood flow (RMBF) determined by radiolabeled microspheres.

Seven dogs were ventilated, a left thoractomy was performed, and the left anterior descending coronary (LAD) artery isolated. Segment length from the anterior region at risk and posterior control region was monitored along with left ventricular pressure. The LAD was occluded, after 15 min of ischemia the ligature was released, and reperfusion for 3 hr was continued. Hearts were arrested with KCl and removed. Samples from the endocardial and mid portion of the left ventricular wall from control, border, and previously ischemic regions were placed into 5-mm NMR tubes. NMR spectra were obtained using a Bruker AM360 employing a single frequency irradiation for water suppression and a Hahn spin-echo pulse sequence with a tau value of 60 msec. Samples that had occlusion RMBF less than 0.4 ml/g/min were considered ischemic.

Analysis of the average percent segment-length shortening demonstrates the anticipated systolic stretching during ischemia. Prompt partial recovery occurs with the onset of reperfusion, followed by slower recovery, which usually remains incomplete at 3 hr. Mobile lipid (ML) resonances are elevated in the sample from the ischemic-reperfused region. As anticipated, creatine showed no change between the post-ischemic (46 ± 43, arbitrary units, mean ± SEM) and control samples (47 ± 5, $p > .9$). Representative NMR spectra are shown in FIGURE 1. The ML/Cr ratio ($1.71 \pm .16$), as well as unnormalized ML (55 ± 7), was increased in the ischemic-reperfused samples as compared to the control samples (ML/Cr = $0.79 \pm .18$, $p = .012$; ML = 31 ± 8, $p = .003$). FIGURE 2 shows a noninfarcted proton spectrum from human myocardium obtained at autopsy, which demonstrates the potential to obtain similar NMR spectra in man.

The NMR-observed increase in ML content is consistent with the metabolic switch from fatty acid to glucose associated with an ischemic insult and the accumulation of nonoxidized fatty acid. Because NMR can image ML, these data suggest a noninvasive means to metabolically assess "stunned" myocardium.

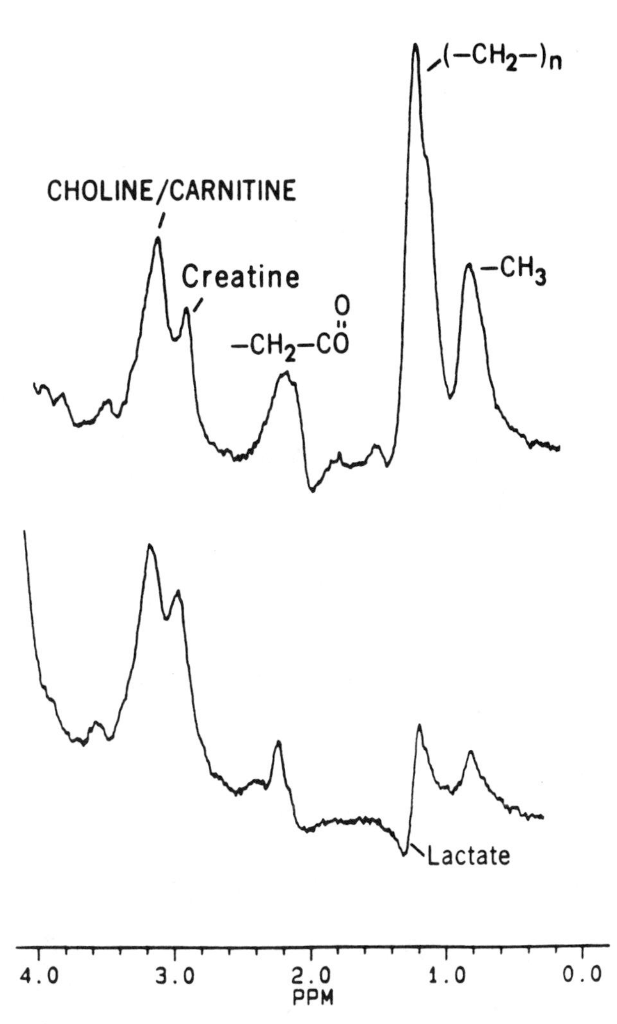

FIGURE 1. 360 MHz ^{1}H NMR spectra of excised canine myocardial samples demonstrating the response to "stunning." The lower spectrum is from control myocardium having an occlusion flow of 0.91 ml/g/min. The upper spectrum is from an occulsion zone following 15 min of ischemia and 3 hr of reperfusion having an occlusion flow of 0.4 ml/g/min and demonstrating the increase in the lipid resonances.

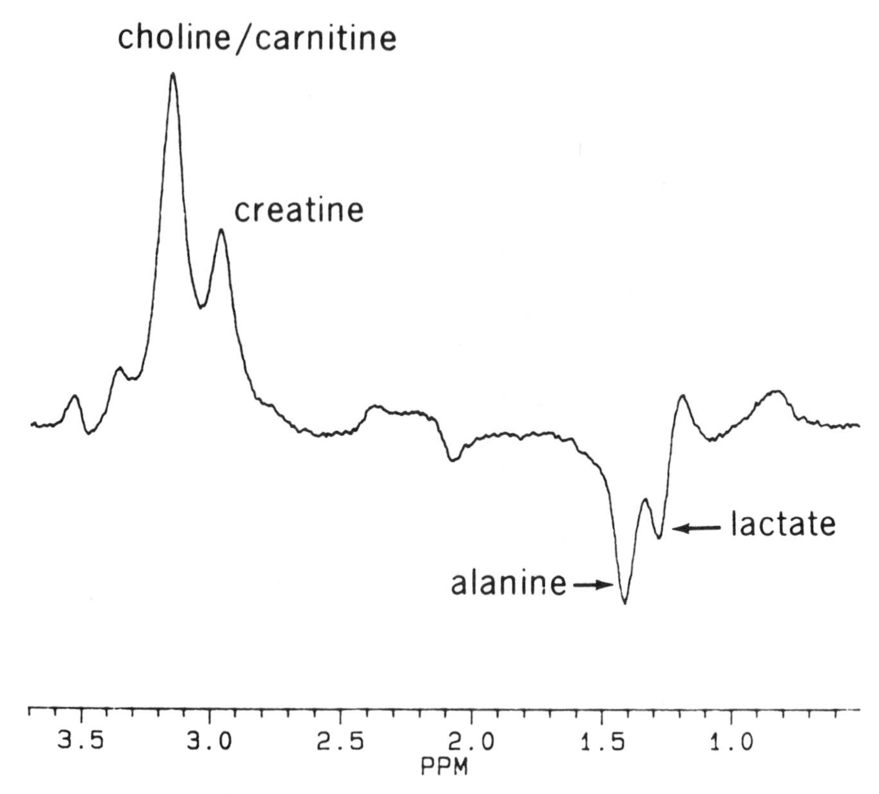

FIGURE 2. 360 MHz ^{1}H NMR spectrum of a sample of human myocardium obtained at autospy from a patient who died of cerebral hemorrhage demonstrating the potential to obtain similar NMR spectra in man.

REFERENCES

1. GENSINI, G. G., P. ESENTE, B. MARMOR, A. BLACK, M. BLACK, A. GIAMBARTOLOMEI, D. B. EFFLER, H. KREUZER, W. W. PARMLEY, P. RENTROP & H. W. HEISS, Eds. 1980. Gerhard Witzstrock Publishing. p. 593.
2. BRAUNWALD, E. & R. A. KLONER. 1982. Circulation **66:** 1146.
3. EVANOCHKO, W. T., R. C. REEVES & G. M. POHOST. Society of Magnetic Resonance in Medicine, Fourth Annual meeting. August 19–23, 1985. London. p. 650 (abstract).
4. KATZ, A. M. & F. C. MESSINEO. 1981. Circ. Res. **48:** 1.

Physiological Application of ^{17}O Promoted Proton T_2 Relaxation[a]

AMOS L. HOPKINS[b] AND RICHARD G. BARR[c,d]

[b]*Department of Biochemistry*
Case Western Reserve University
Cleveland, Ohio 44106

CLYDE B. BRATTON

Department of Physics
Cleveland State University
Cleveland, Ohio 44115

It has been demonstrated by Meiboom[1] that, near neutrality, the proton T_2 of water and buffer solutions is significantly shorter than T_1, due to small amount of naturally present $H_2{}^{17}O$ in $H_2{}^{16}O$. The change in T_2 resulting from $H_2{}^{17}O$ enrichment was indicative of the proton-^{17}O resident lifetime or exchange rate. In contrast to T_2, T_1 was not affected by enrichment or pH. We have shown that in both living muscle[2] and *in vivo* human ventricular CSF,[3] the T_2 is also less than T_1. Although it is improbable in the case of muscle or other tissue that the T_2 is significantly influenced by the naturally abundant ^{17}O, CSF may be an example of a biological fluid whose T_2 is significantly influenced by this isotope, suggesting the extension of Meiboom's methods to other biological systems. It is possible that in addition to following *in vivo* proton exchange phenomena, the distribution, diffusion, and macroscopic mobility of enriched water molecules can also be followed through the effect of ^{17}O on their protons. Since changes in T_2 can alter image intensity, localized variation in $H_2{}^{17}O$ concentrations could be directly visualized, or followed with surface coils as well as monitored in samples by the usual T_2 procedures. We have been unable to locate any previous reports of ^{17}O enrichment effects on proton T_2 in physiological environments or living systems.

The effect on the T_2 of biological solutions was determined with a Varian XL-200-H spectrometer using a Carr-Purcell-Meiboom-Gill six-echo pulse sequence. The effects of enrichment of 7% albumin solutions at pH 5.5 and 7.5 are shown in FIGURE 1. Predictably there is much less effect on the proton T_2 at pH 5.5 than there is at pH 7.5. In FIGURE 2, the difference in rate between the experimental values for a series of solutions (with added $H_2{}^{17}O$) and the rate for the same solutions with only the natural abundance is plotted against the concentration of $H_2{}^{17}O$ at a pH 7.2–7.4. Enrichment produces a marked effect on the relaxation rate in the presence of protein.

The injection of a 50% solution of $H_2{}^{17}O$ into the mealworm *Tenebrio molitor* produces a marked change in the width of the water proton peak but no effect on the lipid protons. When measured with a CPMG sequence, the T_2 was found to be shortened by 30%. There was no effect on survival rate of the experimental larvae as

[a]Supported in part by BRSG S07 RR-05410-24 awarded by the Biomedical Research Support Grant Program, Division of Research Resources, National Institutes of Health.

[d]Present address: The Cleveland Clinic Foundation, Department of Radiology, Cleveland, OH.

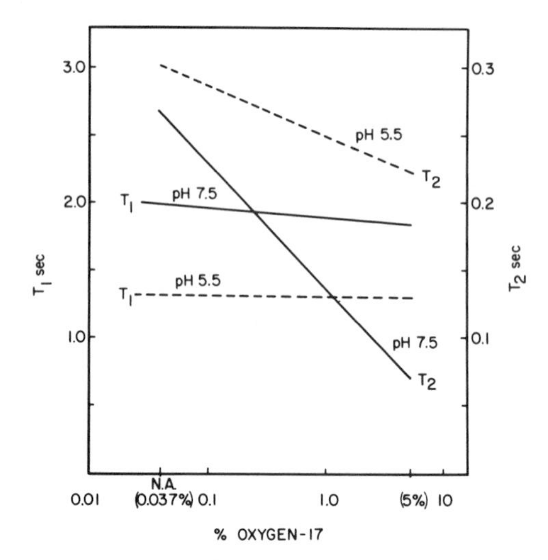

FIGURE 1. Comparison of the effects of ^{17}O on albumin solutions at pH 5.5 and 7.5 from 0.037% (natural abundance) up to 5% enrichment with $H_2{}^{17}O$. The difference in the magnitude of the T_2 change seen as a result of the difference in pH is expected from both the work of Meiboom *et al.* (J. Chem. Phys. 1966. **44:** 546–547), and is explainable on the basis of the pH-produced difference in proton exchange rate or resident life time (τ) of protons on ^{17}O. This similarity in ^{17}O-promoted proton relaxation mechanisms between dilute salt solutions and protein solutions suggests the possibility of using these methods in conjunction with proton imaging equipment to estimate *in vivo* proton exchange rates.

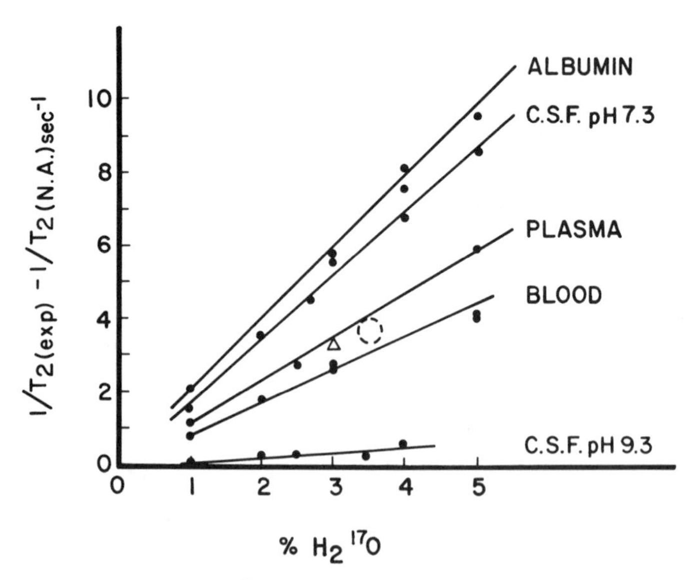

FIGURE 2. Effect of addition of $H_2{}^{17}O$ on the proton relaxation rate ($1/T_2$). In each case the rate for the solution having only the natural abundant $H_2{}^{17}O$ ($1/T_2$ N.A.) has been subtracted from the experimental value ($1/T_2$ exp). The CSF at pH 9.3 had been allowed to lose its CO_2 through equilibration with air. All other solutions were between pH 7.2 and 7.4. (Δ) denotes the value for albumin containing 3% ^{17}O with 0.05 M phosphate buffer added, pH unchanged. The rate is about one half that of the rate in the absence of phosphate, an effect predictable from Meiboom's results. The dashed circle encloses the data from enriched *Tenebrio molitor* larvae. The total body water had been enriched to 3.5%, the resulting T_2 indicated that the proton exchange rate was of the same order of magnitude as that seen in human plasma.

opposed to the controls. There are no reports in the literature indicating any toxic effects of $H_2{}^{17}O$.

The distribution of perfused $H_2{}^{17}O$ can be visualized in the living excised kidney with the use of clinical imaging equipment operating with a T_2 weighted pulse sequences. Under these conditions there is a 60% loss of image intensity in those areas of the tissue with the most $H_2{}^{17}O$.[4] The mean intensities of two echos can be used to estimate a mean T_2.[3] The T_2 values of the protons in the control and the ^{17}O-perfused kidney are found to be 70 ± 24 msec and 41 ± 24 msec, respectively.

Possibly the greatest merit of $H_2{}^{17}O$ will be in the area of measurement of capillary and membrane diffusion rates. However, in anticipation of other, more specific applications, we have incorporated ^{17}O into larger molecular species. For example, those compounds in which a high percentage of the oxygen is mobilized to water can be enriched with ^{17}O, possibly allowing visualization of the regions of highest metabolic activity. Experiments with enriched glucose, designed for application to the human cortex, are already underway.

The overall finding, that the results of enrichment with ^{17}O of protein solutions and tissues parallel those of Meiboom on buffer solutions, suggests that it is possible to approach problems of proton mobility and exchange in living tissue *in situ* through the use of proton-imaging equipment capable of determining T_2. At this time there is no other method capable of human application.

REFERENCES

1. MEIBOOM, S. 1961. J. Chem. Phys. **34:** 375–388.
2. BRATTON, C. B., A. L. HOPKINS & J. W. WEINBERG. 1965. Science. **147:** 738–739.
3. HOPKINS, A. L., H. N. YEUNG & C. B. BRATTON. 1986. Mag. Reson. Med. **3:** 301–311.
4. HOPKINS, A. L. & R. G. BARR. 1987. Mag. Reson. Med. **4:** 399–403.

A Comparison of Human Semen from Healthy, Subfertile, and Post-Vasectomy Donors by ^{31}P NMR Spectroscopy

ADAM S. LEVINE, NATALIE FOSTER,[a]
AND BARRY BEAN

Departments of Biology and [a]Chemistry
Lehigh University
Bethlehem, Pennyslvania 18015

Preliminary characterizations of the phosphorylated compounds present in human semen were obtained using a JEOL FX90-Q Multinuclear NMR with a variable temperature accessory. Semen samples of 2.0 ml containing 12.5% D_2O in a 10.0-mm sample tube were maintained at approximately 28.5°C. The spectra were recorded with a 250 msec pulse delay, 1 sec acquisition time, 0.95 sec acquisition delay, and over the range of -16 ppm to 11 ppm. Three distinguishable peaks were identified as glycerylphosphorylcholine (GPC) at 0.00 ppm (this point was used as an internal standard), inorganic phosphate (P_i) at -2.54 ppm to -2.00 ppm, and phosphorylcholine (PC) at -3.4 ppm, in accord with Arrata *et al.*[1] A temporal study (FIG. 1) showed that, in the hour immediately following ejaculation, the original ratio of PC to total phosphate (P) decreased by 60% while the original ratio of GPC to total P decreased by 30%. Following this hour and for the next 23 hours, the ratio of PC to total P underwent a gradual diminution, while the ratio of GPC to total P did not change by a significant amount. These data contrast with those of Arrata *et al.*[1] who reported that PC hydrolyzed to P_i within 30 minutes and that GPC was stable. Data reported below were collected after a minimum of 60 min post-ejaculation, following relative stabilization of GPC content. Semen specimens were evaluated either within 5 hours of collection (fresh) or quickly frozen without cryoprotectant and stored at -76°C (frozen). Ratios of GPC to total P differed with sample type. We observed the following ratios for GPC to total P: [0.17, 0.15, 0.16, 0.19, 0.099, 0.16, 0.15] $N = 7$ with a mean of 0.15 ± 0.028 for fresh healthy semen, [0.085, 0.070, 0.086] $N = 3$ with a mean of 0.080 ± 0.008 for frozen healthy semen, [0.074, 0.074] $N = 2$ with a mean of 0.074 for fresh subfertile semen (FIG. 2), [0.11, 0.040, 0.060, 0.00, 0.060, 0.00] $N =$ 6 with a mean of 0.045 ± 0.042 for frozen subfertile semen and a ratio of [0.00, 0.00, 0.00, 0.00] $N = 4$ with a mean of 0.00 for post-vasectomy specimens. While our data are generally consistent with those of Arrata *et al.*,[1] we detect a difference between fresh and frozen samples. This discrepancy may reflect partial biological hydrolysis of GPC and PC to P_i. The simple, reliable quantification of GPC is important for further investigation of its biological significance. We suggest that ^{31}P NMR analyses of human semen will play a valuable role in the understanding, detection, and treatment of male infertility.

ACKNOWLEDGMENT

The authors thank Dr. Douglas Brown for his expert help.

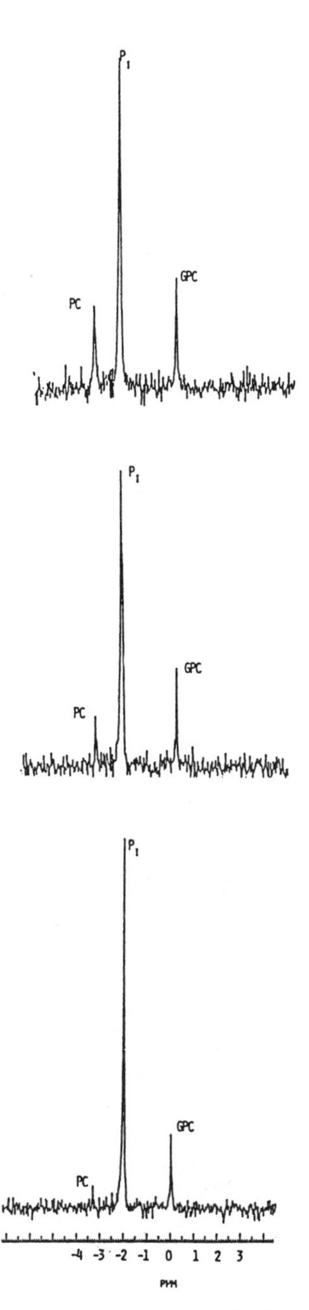

FIGURE 1. Typical spectrum of a healthy specimen. (A) Fresh, 20 min post-ejaculation. (B) Fresh, 60 min post-ejaculation. (C) 24 hr post-ejaculation.

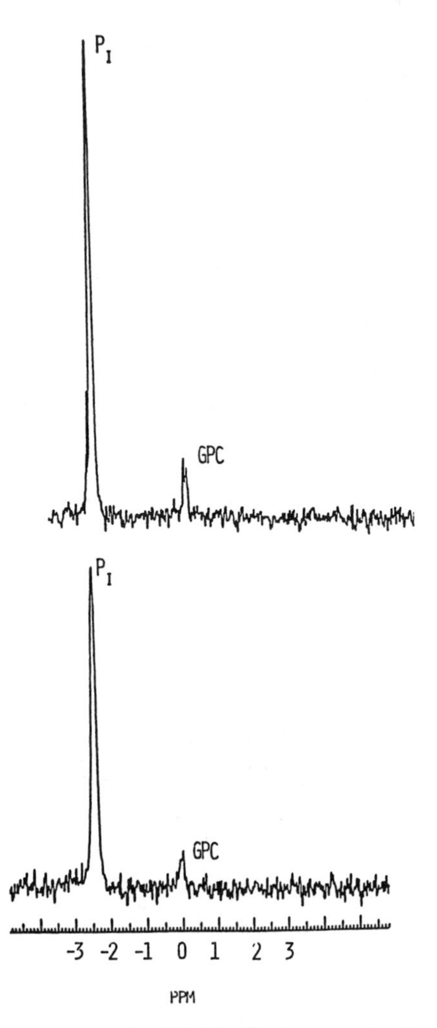

FIGURE 2. Ratios of GPC to total P. (Top) Subfertile, fresh sample. (Bottom) Subfertile, frozen sample.

REFERENCE

1. ARRATA, W. S., T. BURT & S. CORDER. 1978. The role of phosphate esters is male infertility. Fertil. Steril. **30:** 329–333.

Differentiation of Tumor from Non-malignant Human Tissue by Natural-Abundance ^{13}C NMR *in Vitro*

KAREN R. HALLIDAY,[a] CECILIA FENOGLIO-PREISER,[a]
AND LAUREL O. SILLERUD[b]

Life Sciences Division
Los Alamos National Laboratory
Los Alamos, New Mexico 87545

[a]*Department of Laboratory Medicine*
Veteran's Administration Medical Center
Albuquerque, New Mexico 87108

Department of Pathology
University of New Mexico Medical Center
Albuquerque, New Mexico 87108

In order to extend the range of applications of nuclear magnetic resonance (NMR) spectroscopy in the study of cancer, we have characterized the natural abundance ^{13}C spectra of several human tumors and determined that spectral differences exist between the tumors and adjacent, non-malignant control tissues. Tumors studied were prostatic adenocarcinoma, colonic adenocarcinoma, and colloid tumor, and a poorly differentiated lung carcinoma. This study was performed with a Bruker 400 MHz (9.4 Tesla) wide bore spectrometer using proton decoupling. Samples were obtained as residual pathology specimens stored at $-70°C$. For NMR analysis, tissue weighing an average of 2.4 g was placed in D_2O/phosphate-buffered saline at 310 K.

In the prostate, signals were present from triacylglycerols, sialic acid, P-choline, P-ethanolamine, alanine, lactate, and citrate. The spectrum from a poorly differentiated prostatic adenocarcinoma differed from that of the adjacent hyperplastic prostate in the following ways: the tumor spectrum showed 3.5 times as much fat; 1.5 times as much lactate; mucin peaks; and no substantial citrate signals. These assignments are consistent with the knowledge that normal and hyperplastic prostates maintain and secrete a significant quantity of citrate to semen and that citrate decreases in a differentiation-related manner in prostatic adenocarcinomas. In view of the increased lipid levels that accompany the decreased citrate, we have formulated the hypothesis that a transformation-associated enzyme alteration results in the metabolism of citrate to lipid in the tumors; a likely candidate for the altered enzyme is ATP-citrate lyase.

The spectra from five colonic adenocarcinomas and from matched normal colons were dominated by signals from triacylglycerols, although the tumor contained less neutral fat than did the normal control tissue. Signals for the head groups of phosphatidylcholine and phosphatidylethanolamine were increased in the tumors. The ratio of the lipid methylene peak height at 30.5 ppm to that of the choline signal at 55

[b]Address correspondence to: L.O.S. MS-M882 LS-7, Los Alamos National Laboratory, Los Alamos, NM 87545.

ppm was 3.9 for the tumor and 18.6 for the normal (TABLE 1). The mono- to poly-unsaturated lipid ratio as defined by the peak-height ratio of 130.4/128.8 ppm was 3.1 and 1.6 for the colon and adenocarcinoma respectively, reflecting a higher unsaturation figure for the tumor. With adjustments for scan number and tissue weight, lactate signals (21 ppm) were about 50% greater in the tumors than in the normal controls.

There is a colloid variant of colon adenocarcinoma, which makes abundant acidic mucin and has a worse prognosis than does the more ordinary adenocarcinoma just described. The NMR spectrum from a colonic tumor, which was half the standard histological pattern and half the colloid variant, was dominated by signals from the sugar and protein portions of mucins and was readily differentiated from normal colon, and from tumors composed entirely of the common pattern. Characteristic signals from the anomeric carbons of the sugars were found at 103 and 99.5 ppm and were assigned to the anomeric carbons from beta-linked galactose and galactosamine, and alpha-linked sialic acid moieties, respectively. A resonance from the glycosidic linkage mates was present at 81 ppm, from the methyl groups of N-acetylated sugars at 23.5 ppm and from a sugar ring amino carbon near 53 ppm. Other sugar ring carbon signals overlapped with resonances from amino acids.

TABLE 1. Comparison of Signal Intensity Ratios for Colonic Adenocarcinomas and Matched Normal Colon Tissue

Colon	$\frac{130^a}{128}$	$\frac{30.5}{55}$	$\frac{30.5}{40}$	$\frac{55}{40}$	$\frac{30.5}{21}$
Tumor	1.6 (0.5)[b]	3.9 (1.9)	3.9 (1.8)	1.0 (0.01)	9.3 (4.0)
Control	3.1 (0.2)	18.6 (6.7)	15.0 (4.0)	0.8 (0.15)	54.5 (14.1)

[a]Chemical shifts in ppm. [b]The mean (+/− S.D., $N = 5$) for the ratio of the peak heights for the signals whose chemical shifts are indicated.

An NMR spectrum of lung with moderate emphysematous changes was dominated by signals from protein amino acid side chains, including proline, glycine, aspartic acid, asparagine, glutamic acid, glutamine, leucine, isoleucine, valine, methionine, histidine, and phenylalanine. Signals from the sugar portions of glycoproteins were also noted. There were no apparent signals from triacylglycerols or from lactate. A poorly differentiated lung carcinoma demonstrated significant spectral differences from the non-malignant lung tissue. The tumor had prominent signals from triacylglycerols, with reduced intensities of signals from lung collagens and other proteins, including glycoproteins. As with the other tumors, lactate signals were noted.

In conclusion, this pilot study of human tumors has demonstrated that each of the evaluated tumors has a unique ^{13}C NMR spectrum that allows for the differentiation of these tumors from each other and from adjacent non-malignant tissue. These data indicate that natural abundance ^{13}C NMR spectroscopy can play a role in tumor diagnosis and in the study of tumor biology.

Insulin-Induced Effects on Glucose Metabolism by *Neurospora crassa* Studied Using ^{31}P Nuclear Magnetic Resonance

N. J. GREENFIELD,[a] M. McKENZIE,[a] S. FAWELL,[a]
F. ADEBODUN,[b] F. JORDAN,[b] AND J. LENARD[a]

[a]*Department of Physiology and Biophysics*
UMDNJ-Robert Wood Johnson Medical School
Piscataway, New Jersey 08854

[b]*Department of Chemistry*
Rutgers University
Newark, New Jersey 07102

Neurospora crassa and other "lower" eukaryotes have been found to produce molecules that cross-react immunologically with mammalian insulin and that possess insulin-like biological activity when tested against mammalian cells.[1] However, reports of effects of mammalian insulin on lower eukaryotic cells are almost nonexistent. It was recently found that a wall-less mutant of *Neurospora crassa* ("*N. crassa* slime") responded to mammalian insulin in several ways. Cells cultured in the presence of insulin were larger and more rounded, showed enhanced growth and viability, and possessed high-affinity, saturable insulin binding sites.[2]

In this study, ^{31}P nuclear magnetic resonance (NMR) was used to investigate both short and long term effects of bovine insulin on the metabolism of *N. crassa* slime. Spectra were taken directly from suspensions of intact, oxygenated, living cells (110 mg protein or 4×10^9 cells/ml) or from boiled lysates in a Bruker WP-200-SY spectrometer operating in the pulse Fourier transform mode at 81.0 MHz at 23°C.

FIGURE 1 shows spectra of the monophosphate region of lysates from cells grown in a rich defined medium[2] with or without added insulin (100 nM). Spectra were taken after resuspension in buffer and addition of glucose. The predominant peaks arise from intermediates in the hexose monophosphate shunt, in the glycolytic pathway, and in lipid metabolism. Measurement of the spectra of extracts as a function of pH, in the presence and absence of added standards, permitted unambiguous identification of the spectral peaks associated with a number of these metabolites, and thus provided evidence that both the glycolytic and the pentose phosphate pathways were operating at significant levels in these cells. In contrast with the situation in *Saccharomyces cerevisiae*,[3] levels of the glycolytic intermediate fructose-1,6-bisphosphate were very low in *N. crassa* slime (FIG. 1).

Quantitative analysis of spectra of both extracts and intact cells showed that cells grown in the presence of insulin contained 25–35% more total phosphomonoester metabolites than those grown in its absence (FIG. 1, $p < .01$). Some specific peaks also appeared to be quantitatively altered relative to the total, although the statistical significance of these observations could not be assessed.

If cells grown either with or without insulin were pelleted, resuspended in buffer, and glucose then added with or without insulin (100 nM), significant insulin-dependent changes in the average concentration of several metabolites were detected during the next 20–30 minutes. Metabolites showing the most pronounced differences included glucose-6-phosphate, fructose-6-phosphate, glyceraldehyde-3-phosphate, and

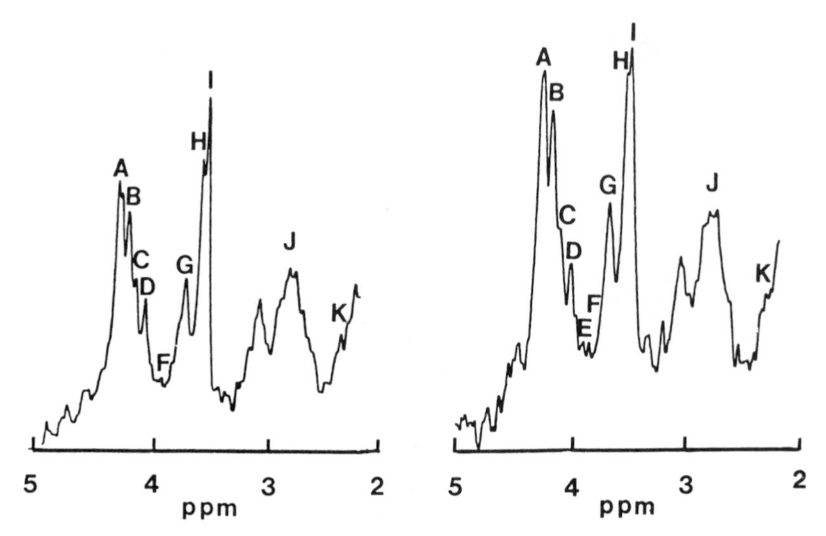

FIGURE 1. The phosphomonoester region of ^{31}P NMR spectra of aqueous boiled extracts of *N. crassa* cells prepared after addition of glucose to the cells. Cells were grown in supplemented defined media in the absence (left) or presence (right) of 100 nM insulin. They were resuspended in Vogels buffered salts containing reduced (0.075 mM) phosphate. D-glucose (100 mM) was added at time 0 and the cells were incubated with aeration for a total of 16 minutes. Key: A, Glucose-6-phosphate and 6-phosphogluconate; B, Ribulose and Xylulose-5-phosphate; C, α-Glycerol phosphate and Glyceraldehyde-3-phosphate; D, Phosphoethanolamine; E, Dihyroxyacetone phosphate (keto form); F, Fructose-1,6-diphosphate; G, Ribose-5-phosphate, Fructose-6-phosphate and β-Glycerol phosphate; H, 5′ AMP, 3-Phosphoglycerate and Fructose-1,6-diphosphate; I, Phosphocholine; J, Anomeric sugar phosphates; K, Glucose-1-phosphate.

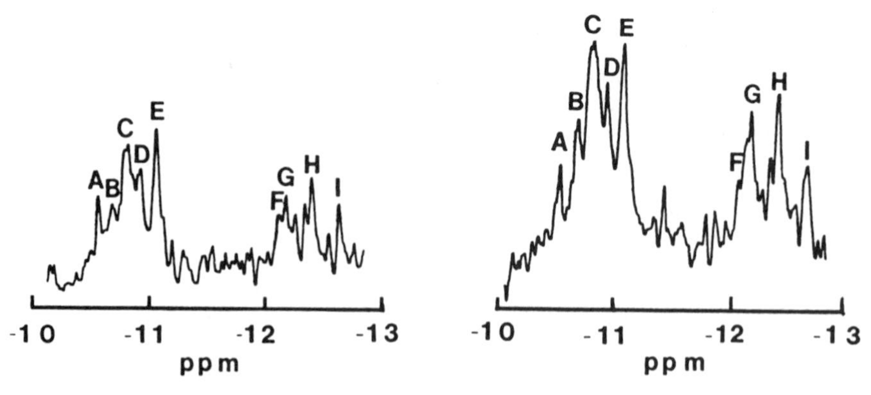

FIGURE 2. The diphosphate region of NMR spectra of aqueous extracts of *N. crassa* grown in the absence (left) and presence (right) of insulin, prepared after resuspension of the cells in Vogel's buffered salts containing 0.075 mM phosphate. Key: A, NTP; B, UDPG; C, NTP+UDPNAG+unknown; D, UDPG; E, UDPNAG+unknown; F, UDPGalactose; G, UDPG: H, UDPG+UDPGalatose+UDPNAG; I, UDPNAG.

3-phosphoglycerate. Insulin thus induced short-term as well as long-term changes in the intracellular concentrations of phosphomonoester compounds.

FIGURE 2 illustrates the diphosphate region of the spectra of extracts from cells cultured in the presence and absence of insulin. The spectrum of the glycogen precursor uridine diphosphoglucose (UDPG) is remarkably prominent in this region (Peaks B, D, G, and H). Cells grown with added insulin contained higher intracellular concentrations of UDPG (ca. 3.7 mM) than did controls (ca 2.4 mM, $p = .02$). The difference was abolished after addition of glucose to resuspended cells (UDPG = ca. 2.6 mM). If insulin was added along with glucose to resuspended cells grown in either condition, however, a lower UDPG level resulted, of ca. 2.1 mM ($p = 0.01$). Thus, both long-term and short-term effects of insulin on intracellular levels of this metabolite were seen, but were in opposite directions. Effects of insulin on glycogen metabolism in *N. crassa* slime were directly demonstrated by showing insulin-stimulated incorporation of radiolabeled glucose into glycogen, and insulin-stimulated conversion of the enzyme glycogen synthase from the relatively inactive glucose-6-phosphate–dependent (D) form to the more active indendent (I) form.[4] Insulin-dependent changes in the metabolism of *N. crassa* slime are thus strikingly similar to some of the effects of insulin in mammalian cells.

REFERENCES

1. LEROITH, D., J. SHILOACH, J. ROTH & M. A. LESNIAK. 1980. Proc. Natl. Acad. Sci. USA **77:** 6184–6188.
2. MCKENZIE, M. A., S. E. FAWELL, M. CHA & J. LENARD. 1988. Endocrinology. (In press.)
3. DEN HOLLANDER. J. A., K. UGURBIL & R. G. SCHULMAN. 1986. Biochemistry **25:** 212–219.
4. FAWELL, S., M. A. MCKENZIE, N. J. GREENFIELD, F. ADEBODUN, F. JORDAN & J. LENARD. 1988. Endocrinology. (In press.)

The Levels of Phosphoethanolamine and Phosphocholine in Intact Neuroblastoma Cells:

Effects of Cell Differentiation and of Exogenous Ethanolamine and Choline

P. GLYNN, T. M. LEE, AND S. OGAWA

AT&T Bell Laboratories
600 Mountain Avenue
Murray Hill, New Jersey 07974

High concentrations of phosphoethanolamine (PEt) have been reported in various brain tumor cells, such as mouse and human neuroblastoma,[1] human glioblastoma, and rat glioma cells. The PEt peak in *in vivo*[31] NMR is used to monitor responses to these tumor cells to chemotherapy and radiation therapy. Similarly, high PEt levels are observed in developing brains of neonates and near-term fetuses of various animals, ranging from rat to man. In contrast to tumor cells, the PEt content in these brain cells decreases along the course of the brain growth. Although PEt is a common metabolite present in many cells as an intermediate in phospholipid synthesis, the biological reason for its high level is not clear. In studying metabolic reactions in tumor or brain cells with *in vivo* NMR, there are difficulties in determining which contributions various cell types or organizations make to the observed development or metabolic processes, since tumors and brains are highly heterogeneous. Therefore, examining a homogeneous cell population can lead to a better understanding of these processes. *In vitro* study of cells in culture has, in addition, the advantage in controlling the cell environment quite easily. To this end, we have studied N1E-115 cells derived from murine neuroblastoma, a model cell line for sympathetic neurons.

These anchorage-dependent cells were maintained for periods of up to 3 days using a microcarrier-based system[2] while ^{31}P NMR profiles were obtained. During NMR experiments the cells were superfused with culture medium under physiological conditions. The present study involved undifferentiated cells and cells induced to differentiate with 2% DMSO and a reduced amount of fetal bovine serum in the culture medium prior to the NMR measurements. We found that undifferentiated N1E-115 cells had small PEt and PCh peaks in the phosphomonoester region, and that they became large when the cells were induced to differentiate. The ethanolamine kinase step, which is a step in the major pathway of phosphatidylethanolamine synthesis, was then perturbed by adding 200 μM ethanolamine to the medium to see the effect on the PEt levels in the cells. In differentiated cells, the PEt/ATP ratio increased with time and reached a steady state of .85 in several hours; a 2.4-fold increase from the initial ratio. This PEt accumulation rate is much higher than the PEt usage rate estimated for the cell growth in culture. Undifferentiated cells showed a much smaller rate of PEt increase and the steady-state level after 7 hours was only 35% of the final level observed in the differentiated cells. Similar events occurred during PCh accumulation.

We see an unusually high level of PEt accumulation *in vitro*, as shown in differentiated neuroblastoma cells exposed to an elevated level of ethanolamine. This ethanolamine level is slightly higher than that in fetal blood, which is an order of magnitude higher than in adult blood. The ethanolamine kinase activity also becomes high upon cell differentiation. The control of the size of the pools of PEt and PCh is obviously coupled to the cell differentiation process and the cell environment. These findings allow a better understanding of observations on the normal brain during development and on the tumor growth *in vivo*.

REFERENCES

1. Naruse, S., K. Hirakawa, Y. Harikawa, C. Tanaka, T. Higuchi, S. Ueda, H. Nishikawa & H. Watari. 1985. Measurements of *in vivo* ^{31}P nuclear magnetic resonance spectra in neuroectodermal tumors for the evaluation of the effects of chemotherapy. Cancer Res **45:** 2429–2433.
2. Ugurbil, K., D. L. Guernsey, T. R. Brown, P. Glynn, N. Tobkes & I. S. Edelman. 1981. ^{31}P NMR studies of intact anchorage-dependent mouse embryo fibroblasts. Proc. Natl. Acad. Sci. USA **76:** 1800–1804.

How Does Acute or Chronic Volume Loading Affect Myocardial Bioenergetics?

MARY OSBAKKEN,[a] LASZLO LIGETI, JOHN PIGOTT, MARIE YOUNG,[a] JANOS HAMAR, MEIR SHINNER, HARI SUBRAMANIAN, JACK LEIGH, AND BRITTON CHANCE

[a]*Anesthesia and Biochemistry/Biophysics Departments*
University of Pennsylvania
Philadelphia, Pennsylvania 19104

This study was designed to investigate the relationship between myocardial mechanical work and high energy phosphate energetics (with ^{31}P NMR) during acute and chronic volume loading.

METHODS

Animal Preparation

For the acute volume loading experiments, five dogs were sedated with innovar (Fentanyl, 0.08 mg/kg, and droperidol, 4 mg/kg) and anesthetized with Nembutal (10 mg/kg). Each dog was intubated and arterial and venous lines were placed for blood pressure and arterial blood gas monitoring and fluid and drug administration, respectively. An arterial-venous shunt was made by connecting the abdominal aorta to the vena cava with polyethylene tubing (5-mm orifice). The shunt could be opened or closed via a plastic regulator around the tubing. Dogs were heparinized (100 units/kg) to prevent the shunt from clotting. A mid-line thoracotomy was performed and a pericardial cradle was made for placement of a 2-turn, 2-cm diameter surface coil on the heart for NMR studies.

For chronic volume overload experiments, four dogs were prepared as follows. Cardiac windows were created by removing two ribs and accompanying skeletal muscle, placement of marlex mesh between the two exposed ribs, and closure of fascia and skin. Animals were allowed to recover three weeks before baseline NMR studies were performed. Volume overloads were created by surgical anastomosis of the abdominal aorta to vena cava. Dogs were followed for 2–6 months and studied periodically with ^{31}P NMR, using a 3-cm, 2-turn external surface coil. During these sequential NMR studies, dogs were anesthetized (as above), intubated, and arterial and venous catheters placed (percutaneously) for physiological monitoring and fluid and drug administration.

NMR Studies

For the NMR studies, acute and chronic animal preparations were placed in specially constructed plexiglass holders and introduced into a 2.1 Tesla, 25-cm bore magnet. After cardiac and respiratory gating were set up each animal-coil combination was tuned, shimmed, and pulse-width optimized. Each spectrum was an average of 100 scans, with a TR of approximately 4 sec.

Physiological Intervention

In the acute volume-loading condition, control spectra were obtained; the shunts were then opened and spectra were obtained for 30 minutes to 4.5 hours. In the chronically volume-loaded dogs, control spectra were obtained; then pressure loading (with norephinephrine, 1 μg/kg/min) was introduced for generation of work versus phosphate energetic curves. That is, for each ^{31}P NMR experiment the chronic dogs were in a stable state of volume overload; thus in order to evaluate the animal's myocardial reserve, a physiological stimulus was applied to the resident state. These data were used to determine the heart's metabolic stability.

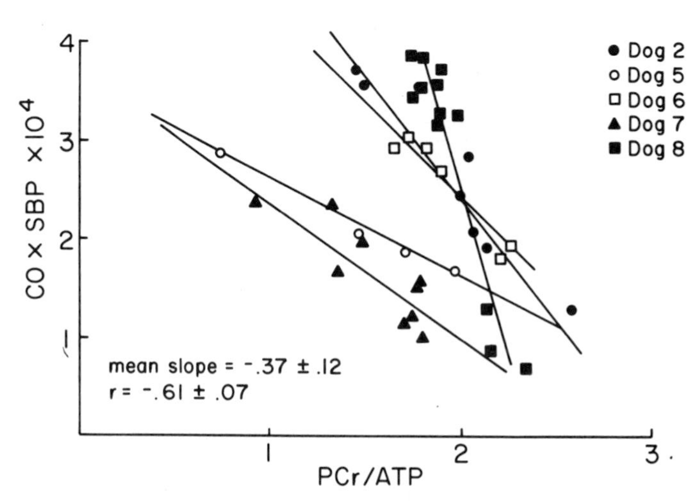

FIGURE 1. Transfer function (heart work vs. bioenergentics) of five dog hearts during acute volume loading. Note that as cardiac output × systolic pressure (CO × SBP) increases, PCr/ATP decreases (or ADP increases, based on the equation [ADP] = [ATP]/[PCr] × [Cr]/K_{CK} [H^+].) Generally CO × SBP increased early in acute volume loading, and decreased progressively with time after the shunt was opened.

Data Analysis

All spectra were phased and curve fitted with a least-squares fit algorithm. Peak areas of inorganic phosphate (P_i), phosphocreatine (PCr), and adenosine triphosphate (ATP) were obtained. If the creatine kinase reaction (PCr + ADP + H^+ $\rightleftarrows$ ATP + Cr) is assumed to be in equilibrium, adenosine diphosphate (ADP) can be obtained by the following calculation: [ADP] = [ATP]/[PCr] × [Cr]/K_{CK} [H^+] and phosphorylation potential can be calculated as: [ATP]/[ADP] [P_i] = [PCr]/[P_i] × K_{CK} [H^+]/[Cr]. Thus, PCr/ATP or P_i/PCr ratios can be calculated and used to estimate ADP or phosphorylation potential respectively.[1]

"Heart work" was estimated from cardiac output × systolic blood pressure (CO × SBP) (acute) or from heart rate × systolic blood pressure (HR × BP) (chronic). P_i/PCr or PCr/ATP ratios were then correlated with "heart work" to produce a "transfer function" that was used to estimate oxidative phosphorylation (V/V_{max}).

RESULTS

In acutely loaded dogs, opening of the shunts resulted in a maximal increase of cardiac output × systolic blood pressure of 77.6 ± 18% (Mean ± SD), associated with a maximal decrease in PCr/ATP ratio of 28.6 ± 7.9%. During the 30 minutes to 4.5 hours during which the shunts were kept open, both parameters fluctuated. However, the general trend was that as CO × SBP increased, PCr/ATP decreased with a slope of −.37 ± .12 (r = .61 ± .07) (FIG. 1).

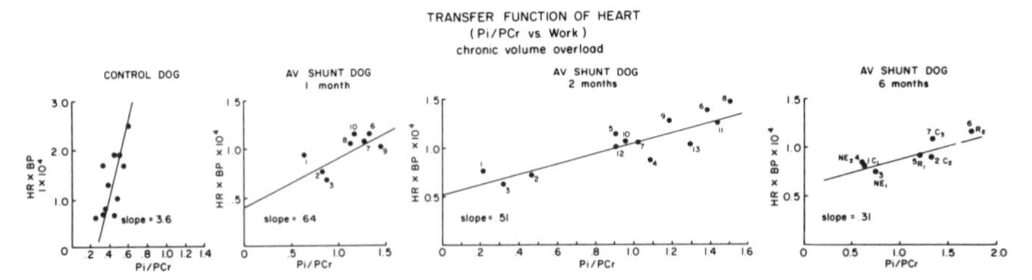

FIGURE 2. "Transfer function" of one dog heart during progressive volume overload (1–6 months). Heart rate × systolic blood pressure (HR × SBP) was used to estimate heart work and P_i/PCr was used to estimate phosphorylation potential based on the equation $[ATP]/[ADP][P_i] = [PCr]/[Pi] \times K_{CK} [H^+]/[Cr]$. In this animal, pressure loading (with norepinephrine, 1 μg/kg/min) was applied to the heart to create the transfer functions. Note that as the time of volume loading increases, the slope of the transfer function decreased, indicating less tight control of bioenergetics by mechanical function.

In the chronically loaded dogs, as workload was increased with pressure loading, P_i/PCr increased (PCr/ATP decreased). The slope of this relationship (the transfer function) became more horizontal as age of the shunt increased (FIG. 2).

CONCLUSION

These findings indicate that both acute and chronic volume loading effect a change in myocardial bioenergetics. In acute loading, where PCr/ATP was used to estimate [ADP], it appears that [ADP] may be an important regulator of energy metabolism during volume-loading conditions. In chronic volume loading, where the heart may have time to adapt to the increased load at rest, imposition of pressure loading to the volume loaded hearts is associated with a flattened response (i.e., the slope of the transfer functions were less steep), which indicates that there may be depressed metabolic recruitment (i.e, there may be less tight control of metabolism by work load).

REFERENCES

1. CHANCE, B., J. S. LEIGH, B. J. CLARK, J. MARIS, J. KENT, S. NIOKA & D. SMITH. 1985. Control of oxidative metabolism and oxygen delivery in human skeletal muscle: A steady-state analysis of the work-energy cost transfer function. Proc. Natl. Acad. Sci. USA **82:** 8384–8388.

^{7}Li-NMR Study of Human Erythrocytes[a]

J. F. M. POST, K. PANCHALINGAM,[b] G. WITHERS,[b]
D. E. WOESSNER,[c] AND J. W. PETTEGREW[b]

Department of Medicinal Chemistry
University of Pittsburgh
Pittsburgh, Pennsylvania 15261

[b]*Western Psychiatric Institute and Clinic*
University of Pittsburgh
Pittsburgh, Pennsylvania 15213

[c]*Mobil Research and Development Corporation*
Dallas, Texas 75244

Lithium (Li^+) is widely used as a psychotherapeutic drug. Little is known about the molecular basis for the biological action of Li^+ at present. It has been postulated that Li^+ competes for the binding sites of biological cations, such as Na^+, K^+, Mg^{2+}, and/or Ca^{2+}.[1] Another postulated action of Li^+ is that it interacts with membrane phospholipids.[2] To obtain more insight into the behavior of intracellular Li^+ at the molecular level, we undertook a NMR study of Li^+ inside red blood cells. For comparison, several other Li^+ solutions were studied.

Human erythrocytes were incubated with lithium chloride. The uptake of lithium ions was followed by ^{7}Li-NMR, using dysprosium shift reagent,[3] and followed single exponential kinetics with a time constant of 14.7 hours (FIG. 1). While for concentrated hemoglobin solutions the relaxation times T_1 and T_2 were of the same order of magnitude, it was found that $T_1 \simeq 5$ sec and $T_2 \simeq 0.15$ sec for intracellular lithium. T_1 is three times shorter than for free aqueous Li^+, which can be explained by intracellular viscosity. Relaxation rates R_1, observed for sucrose solutions of different concentrations, are linear with relative viscosity η as is to be expected for the fast motional region ($\tau_c \ll 2 \times 10^{-9}$ sec). From these experiments, we estimate the intracellular viscosity to be 4 cP. The very low T_2 value has to be caused by an additional low frequency interaction affecting the lithium ions. Our experiments show that lithium does not interact significantly with hemoglobin or phospholipid membranes. Exchange of lithium ions across the cell membrane appears to be very slow, with an exchange time of many seconds, which was concluded from a magnetization transfer experiment. In agar gels $T_2 \ll T_1$, and for the 20% gel we find similar values as for intracellular Li^+. The results of our experiment are summarized in TABLE 1.

Our work leads to the following five conclusions for intact normal human erythrocytes. (1) Intact erythrocyte membranes are permeable to Li^+ with an uptake following single exponential kinetics and a time constant of 14.7 hours. (2) Li^+ does not significantly bind to hemoglobin as demonstrated by experiments on concentrated hemoglobin solutions. (3) Preliminary ^{7}Li-NMR and differential scanning calorimetry results on dispersions of dipalmitoyl phosphatidyl choline or dipalmitoyl phosphatidyl glycerol do not indicate that the interactions of Li^+ with these phospholipids are

[a]Supported by IN-58W from the American Cancer Society, as well as by a grant from the United Cerebral Palsy Foundation, and by National Institute of Aging grants IR01 AG05657-01 and AG05133-01A1.

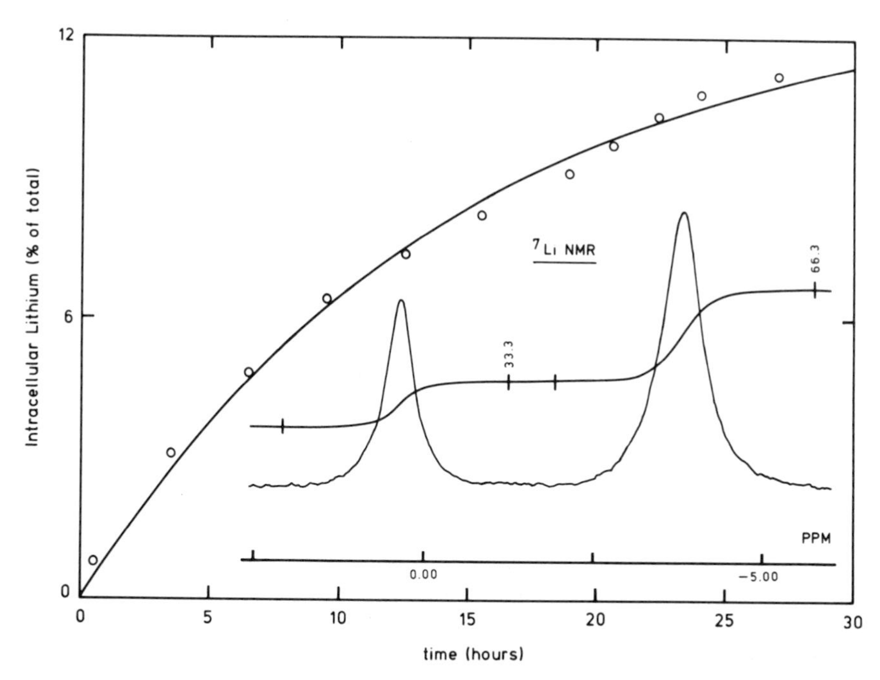

FIGURE 1. Intracellular Li^+ concentration as a percentage of the total Li^+ concentration, plotted as a function of time t (hours). The data were fitted with the function $A[1 - \exp(-t/T)]$, which yields a time constant $T = 14.7 \pm 1.2$ hours. (Insert) ^{7}Li-NMR spectrum of erythrocytes incubated for 24 hours in plasma containing 50 mM LiCl. Resonance frequency was 77.4 MHz. 5 mM dysprosium:tripolyphosphate shift reagent was added just before NMR experiments. Number of scans is 100.

TABLE 1. Summary of Results

Sample	T_1 (sec)	T_2 (sec)	T_1/T_2
LiCl (1 M)	16.65 ± 0.06	15.50 ± 0.06	1.07
LiCl (10 M)	6.80 ± 0.06	5.40 ± 0.05	1.26
LiCl (50 mM) in 26 wt% sucrose	6.89 ± 0.08	4.46 ± 0.14	1.54
LiCl (50 mM) in 38 wt% sucrose	4.03 ± 0.04	3.29 ± 0.08	1.22
LiCl (50 mM) in 46 wt% sucrose	2.29 ± 0.02	2.08 ± 0.03	1.10
1 M LiCl in 10% agar	6.63 ± 0.05	0.655 ± 0.02	10.12
1 M LiCl in 20% agar	4.09 ± 0.03	0.205 ± 0.01	19.95
5 mM LiCl in 32 wt% hemoglobin A	5.15 ± 0.06	2.29 ± 0.09	2.25
25 mM LiCl in 32 wt% hemoglobin A	4.86 ± 0.08	1.84 ± 0.04	2.64
25 mM LiCl in 32 wt% hemoglobin A, dilute 2× with Tris buffer	7.07 ± 0.11	3.69 ± 0.05	1.92
LiCl 50 mM in blood plasma ($N = 2$)[a]	4.89 ± 0.49	2.93 ± 0.09	1.67
Erythrocytes + 50 mM LiCl ($N = 5$)	5.10 ± 0.56	0.145 ± 0.02	35.17
Erythrocytes in isotonic sucrose/PIPES buffer	4.26 ± 0.07	0.089 ± 0.01	47.86

[a]Values given are the mean ± standard deviation. All values obtained at 25°C.
N = Number of human subjects.

significantly stronger than the interactions of Na^+ or K^+. Therefore, we do not expect such interactions to play a role in erythrocytes. (4) The presence of Li^+ does not affect the Na^+ NMR relaxation behavior in erythrocytes. ($T_1 = 25.0 \pm 1.8$ msec and 24.5 ± 1.2 msec, $T_2 = 11.2 \pm 0.9$ msec and 12.3 ± 0.9 msec, before and after incubation with LiCl, respectively). Therefore, competition for interaction sites between these ions does not seem to play an important role. (5) Intracellular viscosity, about 4 cP, can explain the reduction in the intracellular ^7Li T_1, but not in T_2. Slow exchange can be excluded as a relaxation mechanism. Because the T_1 and T_2 values of Li^+ in agar gels and erythrocytes are similar, we propose that heterogeneous diffusion through electrostatic field gradients imposed by the membrane-associated cytoskeleton may explain the relaxation behavior.[4] Association of the ions with this network cannot be ruled out.

In the future, more work will have to be done on the interaction between Li^+ and isolated cell constituents in order to explain the observed relaxation behavior. Because of the large difference between T_1 and T_2 for intracellular Li^+, ^7Li-NMR seems to be a promising method for further investigations of the intracellular environment. A complete report of the present work has been published elsewhere.[5]

REFERENCES

1. Frausto da Silva, J. J. R. & R. J. P. Williams. 1976. Nature **263:** 237–239.
2. Fossel, E. T., M. M. Sarasua & K. A. Koehler. 1985. J. Magn. Res. **64:** 536–540.
3. Gupta, R. K. & P. Gupta. 1982. J. Magn. Res. **47:** 344–349.
4. Berendsen, H. J. C. & H. T. Edzes. 1973. Ann. N.Y. Acad. Sci. **204:** 459–485.
5. Pettegrew, J. W., J. F. M. Post, K. Panchalingam, G. Withers & D. E. Woessner. 1987. J. Magn. Res. **71:** 504–519.

Crafted Pulses For Imaging and *in Vivo* NMR Spectroscopy[a]

F. LOAIZA, M. A. McCOY, AND W. S. WARREN

Department of Chemistry
Princeton University
Princeton, New Jersey 08544

M. S. SILVER AND H. EGLOFF

Siemens Medical Systems
Iselin, New Jersey

Biological applications of NMR, such as medical imaging or spectroscopy of complex molecules in aqueous solution, require a high level of specificity. In the case of multislice imaging, pulse sequences that excite only a well localized frequency range (slice selective pulses) do not generate distortions if signals at nearby frequencies (contiguous slices) are acquired with short delays; of course, the excitation within the slice should be nearly complete and uniform to maximize the signal-to-noise ratio and reduce artifacts. In the case of NMR spectroscopy of biomolecules, problems associated with low concentrations and large dynamic range requirements because of solvent signal are best handled by completely avoiding excitation at the solvent frequency. In some simple experiments, such as measuring an ordinary free induction decay, phase and amplitude distortions in the excited spectrum can be corrected by later data manipulation; however, in more complex sequences, such as those used in two-dimensional spectroscopy, such manipulations are not always possible. In general, then, creating a uniform and complete excitation of one spectral region with no excitation outside of this region (or generating the inverse of this distribution) is a common problem of importance in a wide variety of biological contexts.

We have demonstrated that radiofrequency pulse shaping provides a straightforward technique for exciting very narrow regions of a spectrum[1] or selectively suppressing specific regions.[2,3] These shapes are derived by a theoretical formalism, which has been presented in detail elsewhere.[4] In essence, it is a rapidly convergent perturbative expansion that describes the effects of an arbitrarily shaped pulse in terms of its Fourier transform and the Fourier transform of its autocorrelation function. In this work, we extend our previous studies, which presented applications of crafted pulses in laser spectroscopy or the NMR of small molecules, to the case of *in vivo* spectroscopy. We present spin-echo imaging results that start from our previously derived pulse shapes, and do a final optimization by computer to enhance multislice imaging.

The pulses, as calculated through the above-mentioned analytical formalism, excite a very narrow bandwidth about resonance and produce minimal excitation elsewhere. They are simple, symmetric, and only amplitude modulated, but they are optimized only for single pulse excitation. Since these pulses do not generate amplitude

[a]Supported by the National Institutes of Health under grant GM35253, by the New Jersey Commission on Science and Technology, and by Siemens Medical Systems. W.S.W. is an Alfred P. Sloan Fellow.

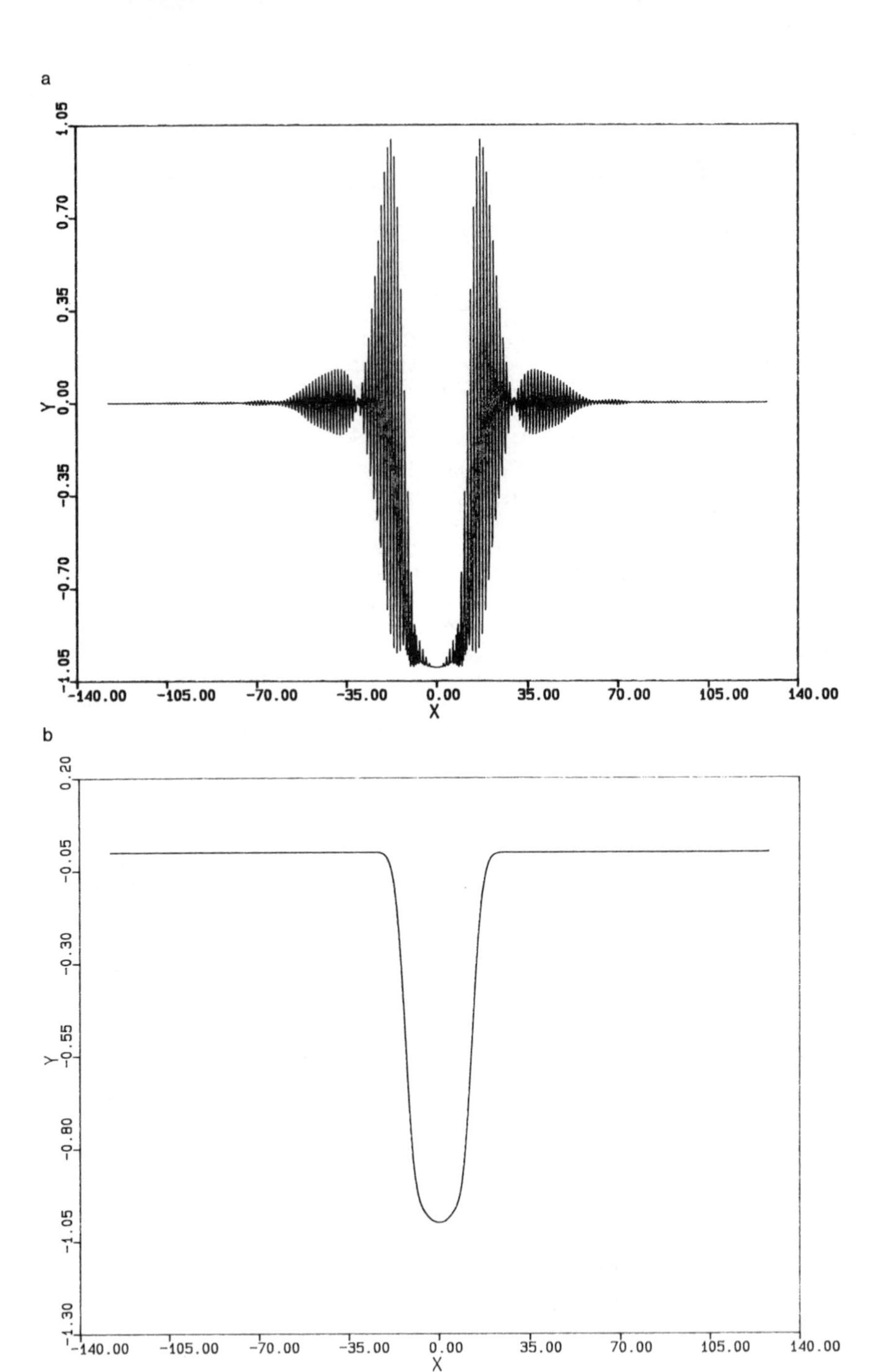

FIGURE 1. (a) Magnetization produced by the initial Hermite pulse sequence and (b) by the optimized pulses. As indicated in the text, there is a substantial difference between these and the actually observed echoes shown in (c) and (d). However, for multislice imaging it is precisely these perturbations, far from resonance, that are the ones responsible for deterioration of the image quality of contiguous slices. For each graph, *x* axis represents resonance offset and *y* axis represents $\mathbf{M}_y$.

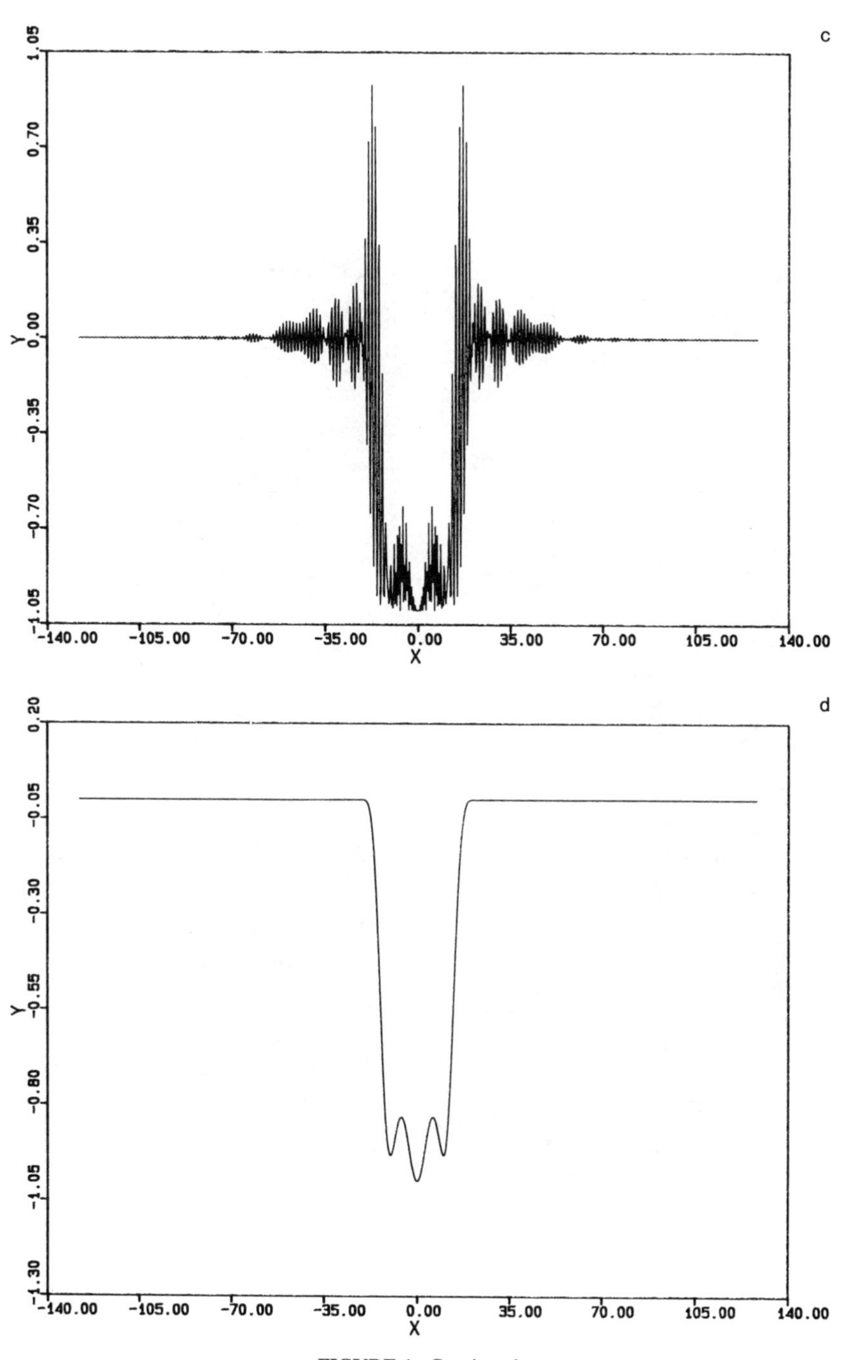

FIGURE 1. Continued

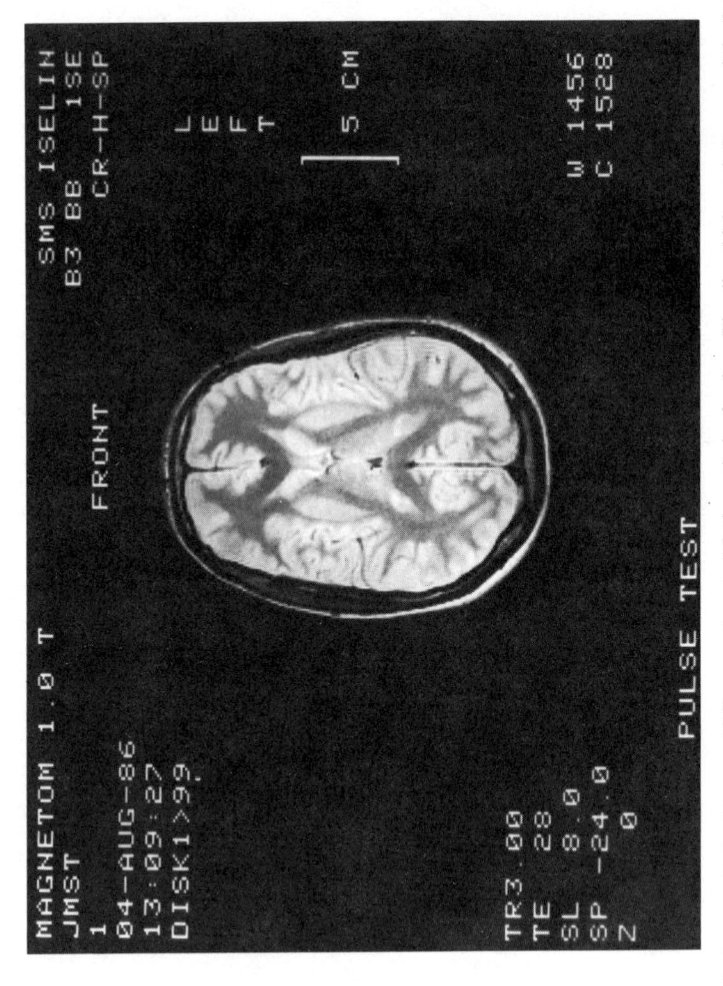

FIGURE 2. One of the images obtained in a multislice experiment with the optimized pulses. The slice thickness is 8 mm and the field strength is 1.5 Tesla. The total number of slices was 15 and two echoes were collected for each slice. The quality of all the other slices is comparable to the one shown.

and phase distortions they can be combined into multiple pulse sequences; however, they are not necessarily optimized for such sequences. In principle, a more detailed theoretical calculation could derive the small corrections needed to optimize (for example) a spin-echo sequence, but the elegance of the perturbative expansion is lost when higher terms are included in order to further increase their selectivity. Hence, the final refinement is better carried out using a simple method of steepest descent, whereby the spectral representation of the pulses is systematically varied and a gradient is determined—by calculating the response of the spin system and comparing it to a predetermined goal function—along which the pulse shape is altered. The algorithm proceeds to change the pulse shape along this path in frequency space until no further improvement is obtained and then starts a new calculation of the gradient. It loops through these steps until it matches the goal function, in which case the global minimum has been attained, or until a given number of spectral mutations has been carried out. The algorithm keeps track of the degree of approximation to the desired goal function and whenever the improvements are below a certain threshold it defines this new minimum as a local minimum and proceeds to make a big jump from there to a different pulse shape whence it continues its search.

The most subtle complication in this process is defining an appropriate target function. The optimization criteria for single slice imaging and multislice work are grossly different, as can be illustrated by a simple example (FIG. 1). The effects of any pulse sequence can be exactly calculated by solving Bloch equations or in a density matrix formalism. Such treatments can give the transverse magnetization components ($\mathbf{M}_x$ and $\mathbf{M}_y$) as a function of resonance offset $\Delta\omega$. FIGURE 1 (a and b) shows this theoretical frequency distribution for two different spin-echo sequences. After the sequence ends the macroscopic magnetization (for example, the formation and decay of an echo) is measured as a function of time. The Fourier transform of this temporal decay, which shows the observable frequency distribution, is shown for the same two sequences in FIGURE 1 (c and d, respectively). The top and bottom figures differ considerably, for reasons which are detailed elsewhere.[5] The bottom figures give the frequency selectivity of a single slice echo; the top figures show the normally unobservable distortions of equilibrium magnetization induced off resonance, but these distortions must be minimized for contiguous slice excitation. FIGURE 1b corresponds to a sequence designed to reduce the distortions from FIGURE 1a to a more tolerable level.

Using this computer algorithm we were able to optimize the initial Hermite pulse sequence. We implemented it in a commercial Siemens Magnetom imager to obtain contiguous multislice spin echo images with minimal distortion in the adjacent slices. One of these brain scans is shown in FIGURE 2. This zero-slice–gap image represents a substantial improvement over sequences derived to optimize single slice images.

REFERENCES

1. McCOY, M. A. & W. S. WARREN. 1985. J. Magn. Reson. **65:** 178.
2. GUTOW J. H., M. McCOY, F. SPANO & W. S. WARREN. 1985. Phys. Rev. Lett. **55:** 1090.
3. McCOY, M. A. & W. S. WARREN. 1980. Chem. Phy. Lett.
4. WARREN, W. S. 1984. J. Chem. Phys. **81:** 5437.
5. LOAIZA, F., M. McCOY, M. SILVER, M. LEVITT & W. S. WARREN. 1988. J. Mag. Res.

Studies of Intracellular pH Regulation in Cultured Mammalian Cells by ^{31}P NMR

BENJAMIN S. SZWERGOLD, JEROME J. FREED, AND TRUMAN R. BROWN

Fox Chase Cancer Center
Philadelphia, Pennsylvania

Intracellular pH is an inportant parameter that is closely regulated in essentially all cells studied.[1] Over the past several years there have been numerous reports correlating an increase in the intracellular pH with mitogenic stimulation of cultured mammalian cells brought to quiescence by serum deprivation. These results were obtained mostly by fluorescent techniques and the study of the distribution of radiolabeled weak acids. Application of NMR to the study of this phenomenon has been hampered by the difficulty of maintaining sufficient concentrations of viable cells in the spectrometer and by a lack of a suitable pH indicator for those cells, like the 3T3 mouse fibroblasts, that have very low levels of inorganic phosphate.

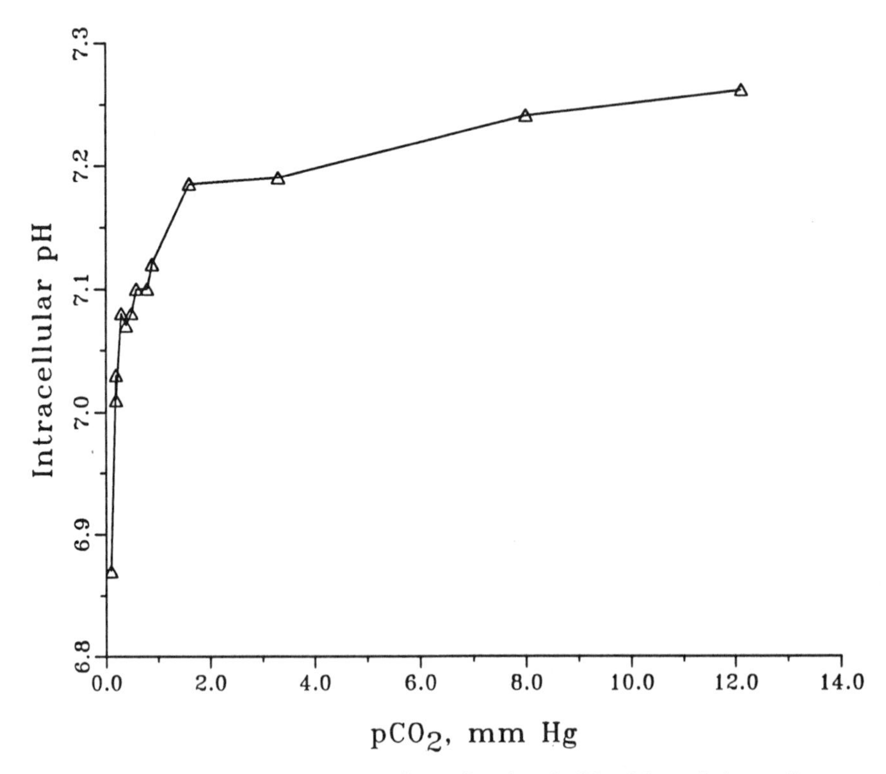

FIGURE 1. Intracellular pH in 3T3 cells as a function of pCO_2 of the perfusing medium.

TABLE 1. ^{31}P NMR pH Indicators

Compound	$\Delta\delta/\Delta pH$	pK_a	Uptake	Toxicity	Sensitivity to Temperature, Ionic Strength, and Divalent Cations
P_i	1.3	6.7	—	None	Relatively high
Deoxyglucose-6-P	1.9	6.2	Very good	Relatively high	Relatively high
Methyl phosphonate	−2.1	7.35	Very poor	Very low	Very low
2-Amino-4-phosphono butyric (APBA)	−1.7	6.9	Very good	Relatively high	Not done
2-Amino-5-phosphono valeric (APVA)	−1.9	7.35	Poor[a]	Low	Not done
2-Amino-6-phosphono hexanoic (APHA)	−2.0	7.61	Poor[a]	None	Not done

[a]Cells can be loaded with the indicator compound by incubation in a balanced saline medium.

In our laboratory we have adapted and refined a system of perfusing cells anchored to microcarrier beads[2] and have developed a series of novel ^{31}P NMR pH indicators (TABLE 1) that have allowed us to address this topic.

Using these compounds, most notably 2-amino-6-phosphono hexanoic acid (APHA), we have established that the intracellular pH (pH_i) in 3T3 cells is regulated at about 7.2 as the extracellular pH (pH_{ext}) of a bicarbonate-containing growth medium is varied between 7.0 and 7.5. This regulation is mediated not only by the well-studied Na^+/H^+ exchanger but also by a bicarbonate-dependent mechanism, since the pH_i regulation profile is critically dependent on the presence or absence of bicarbonate in the medium. The intracellular pH increases with increasing CO_2/HCO_3^- with an apparent K_m of 1 mm Hg CO_2. This corresponds to a total external carbonate concentration of 0.3 mM at pH 7.0 at 37°C (FIG. 1).

The reported intracellular alkalinization observed upon stimulation of quiescent cells with serum[3-5] occurs in these cells only at low levels of CO_2/HCO_3^- and is not observed when bicarbonate-buffered perfusion medium is used.

We have repeated these experiments five times at low CO_2/HCO_3^- levels and eight times in bicarbonate-buffered medium with results as summarized in TABLE 2. The values given in the table represent an average of FIDs acquired during the first 30–45 minutes following the addition of serum.

TABLE 2. ΔpH_i of 3T3 Cells following the Stimulation of Quiescent Cells with Serum

pCO_2 (mm Hg)	ΔpH_i
<1.0	0.17 ± 0.04 ($N = 5$)
<30.0	0.00 ± 0.03 ($N = 8$)

Our observations suggest that while pH_i is an important parameter regulated by at least two distinct mechanisms, the serum-induced intracellular alkalinization observed in the absence of bicarbonate is not a necessary part of the response of these cells to mitogenic stimulation.

REFERENCES

1. NUCCITELLI & DEAMER. 1982. Intracellular pH, its measurement, regulation and utilization in cellular functions. Kroc Foundation Series Vol. 5. Alan Liss. New York.
2. UGURBIL, K. *et al.* 1981. ^{31}P NMR studies of anchorage dependent mouse embryo fibroblasts. Proc. Natl. Acad. Sci. USA. **78:** 4843–4847.
3. POUYSSEGUR, JACQUES. 1985. The growth factor-activatable Na^+/H^+ exchange system: a genetic approach. Trends Biol. Sci. **10:** 453–455.
4. MOOLENAAR, W. H. *et al.* 1984. Phorbol ester and diacyl glycerol mimic growth factors in raising cytoplasmic pH. Nature **312:** 371–374.
5. MOOLENAAR, W. H. 1986. Effects of growth factors on intracellular pH regulation. Annu. Rev. Physiol. **48:** 363–376.

A Non-invasive Study of Drug Metabolism in Patients as Studied by ^{19}F NMR Spectroscopy of 5-Fluorouracil

WALTER WOLF,[a] MICHAEL S. SILVER,[b]
MICHAEL J. ALBRIGHT,[b] HORST WEBER,[c]
ULRICH REICHARDT,[d] AND ROLF SAUER[d]

[a]*Radiopharmacy Program*
University of Southern California
Los Angeles, California

[b]*Magnetic Resonance Research and Development*
Siemens Medical Systems, Inc.
Iselin, New Jersey

[c]*Cross Sectional Imaging Medical Engineering Group*
Siemens AG
Erlangen, Federal Republic of Germany

[d]*Department of Radiation Therapy*
University of Erlangen
Erlangen, Federal Republic of Germany

Fluorine-19 is a well known nucleus and has been used as a chemical probe for structure elucidation for many years.[1] The nucleus has a spin of ½ and has a 100% natural abundance. The gyromagnetic ratio of ^{19}F is only 6% lower than ^{1}H, making it accessible by retuning of standard MRI RF components. The ^{19}F sensitivity is nearly as high as ^{1}H, but unlike ^{1}H there is no large background signal *in vivo*. The extremely large chemical shift range of ^{19}F (over 200 ppm) facilitates spectroscopic differentiation of similar compounds, such as metabolites, even at the lower field strengths used for magnetic resonance imaging (MRI).

This report shows *in vivo* ^{19}F spectroscopic results of 1.5 T for humans undergoing chemotherapy with 5-fluorouracil (5-FU) for various forms of cancer. This study demonstrates that it is possible to observe fluorine-containing drugs in humans by ^{19}F NMR, as well as to directly estimate the time course of the drug's *in vivo* metabolism.

Three patients (one female and two males) scheduled to receive 5-FU as a chemotherapeutic agent were accessed to the present study. Spectra were taken on a 1.5T MAGNETOM MR imaging system equipped with a spectroscopic option (Siemens AG, Medical Engineering Group, Erlangen, FRG). A standard 15 cm spine coil was retuned to the ^{19}F frequency and was used in a transmit/receive mode. The spectrometer frequency for ^{19}F was 59.803000 MHz. The 90° pulse was optimized on a phantom for approximately 5 cm above the surface of the coil to assure selectivity from that depth out to one radius of the surface coil.

The sequence of spectra illustrated in FIGURE 1 is representative of those observed for each patient and were obtained from patient WK, male, 66 years old, with an unidentified carcinoma of the neck. The chemical shifts of the signals observed were

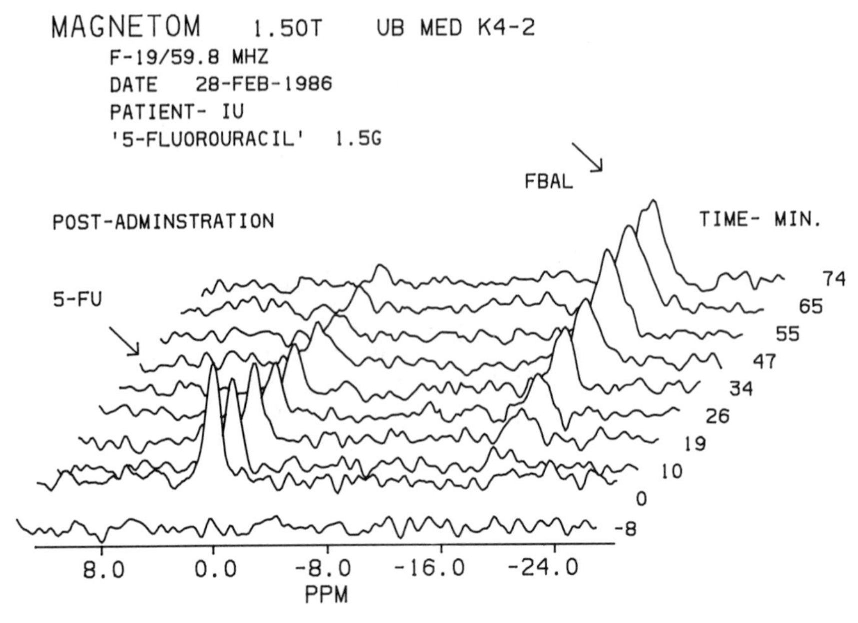

FIGURE 1. Serial ^{19}F spectra acquired with a surface coil over the liver of a patient IU (70-year-old female) in a 1.5 T MRI system, operating at 59.8 MHz for ^{19}F. Each spectrum is the result of 128 FIDs using a 250 μsec 90° pulse with 512 complex points collected over a period of 8.5 min. Actual spectral width was 24 ppm to -54 ppm. A shifted sine-bell squared apodization was used for signal-to-noise enhancement.

correlated with the known standards recorded under identical conditions. The chemical shifts are relative to 5-FU, which serves as a reference point for identification of the catabolites. FBAL, a major catabolite of 5-FU, appears at -18.84 ppm. The width at half-height for the FBAL is 1.94 ppm while the 5-FU half-height width is only 0.91 ppm. This broadening is due to the proton-fluorine spin-spin coupling in FBAL (FIG. 1), which is absent in 5-FU. No other signals were observed in the liver of the patients studied (total spectral range was 29 ppm to -54 ppm). The chemical shifts for 5-FU and its observed catabolites are shown in TABLE 1.

In conclusion, we have documented that *in vivo* ^{19}F NMR spectroscopy is feasible, in humans, and that it allows the determination of the time course of the metabolism of fluorinated drugs in individual patients. Such individual data may allow the estimation

TABLE 1. ^{19}F Chemical Shifts Observed

Patient	Field Strength	5-FU	FBAL	5-FUR
WK	1.5 T	0.0 ppm	-18.8 ppm	—
IU	1.5 T	0.0 ppm	-18.4 ppm	—
JB	1.5 T	0.0 ppm	-18.5 ppm	—
Phantom 1	1.5 T	0.0 ppm	-19.4 ppm	—
Phantom 2	1.0 T	0.0 ppm	-19.1 ppm	—

Abbreviations: 5-FU, 5-fluorouracil; FBAL, α-fluoro-β-alanine; and 5-FUL, 5-fluorouridine.

of drug availability in a specific patient, thereby optimizing and individualizing drug regimens.

REFERENCE

1. DUNGAN, C. H. & J. R. VAN WAZER. 1968. Compilation of Reported F19 NMR Chemical Shifts, 1951 to Mid-1967. Wiley-Interscience.

The Effect of Hypoglycemia on Brain Blood Flow and Brain Energy State During Neonatal Seizure

RICHARD S. K. YOUNG[a,b] BARRETT E. COWAN,[c]
OGNEN A. C. PETROFF,[b] RICHARD W. BRIGGS,[d]
AND EDWARD NOVOTNY[b]

Departments of [a]Pediatrics and [b]Neurology
Yale University
New Haven, Connecticut 06510

[c]Stanford University
Stanford, California 94305
and
Departments of [d]Radiology and Biological Chemistry
Pennsylvania State University
Hershey, Pennsylvania

The availability of glucose may become critical during neonatal seizure.[1] Brain glucose levels fall during neonatal seizure even when neonatal animals are normoglycemic. However, it is uncertain whether brain high energy phosphate reserves are depleted when hypoglycemia complicates seizure. The goal of this study was to test the hypothesis that hypoglycemia decreases cerebral blood flow (CBF) during neonatal seizure, thereby lowering levels of brain high energy phosphates.

Mongrel dogs (1–10 days old) were anesthetized, tracheostomized, mechanically ventilated, and subjected to bicuculline-induced seizure either with or without insulin-induced hypoglycemia. Bicuculline is an alkaloid that blocks the inhibitory neurotransmitter, γ-amino-butyric acid.

Electroencephalography showed a similar pattern of spike or spike and slow wave activity in all animals subject to seizure, regardless of plasma glucose level. CBF (measured autoradiographically with [^{14}C]iodoantipyrine in parallel groups of animals) was globally elevated in all animals injected with bicuculline.

In vivo ^{31}P NMR measurements (1.89 Tesla Oxford magnet) disclosed that the PCr/P_i and the ATP/P_i ratios were reduced to the same extent in all animals undergoing bicuculline-induced seizure, irrespective of brain glucose level (FIG. 1). *In vitro* biochemical determinations of *in situ* funnel-frozen brain showed brain glucose concentrations to be significantly reduced in animals subjected to seizure during hypoglycemia. Paradoxically, brain lactate concentrations were elevated in all bicuculline-injected animals (FIG. 2).

^{1}H NMR spectroscopic analysis (Brucker WM500) of funnel-frozen brain extracts confirmed the increase in lactate and decrease in phosphocreatine in all animals subjected to seizure. Alanine and β-hydroxybutyrate were substantially increased in the seizure-hypoglycemia group indicating possible utilization of non-carbohydrate substrates.

In summary (1) the compensatory increase in cerebral perfusion that occurs during neonatal seizure is not reduced by coexistent hypoglycemia; (2) consequently, the degree of metabolic perturbation during seizure-hypoglycemia is no greater than that

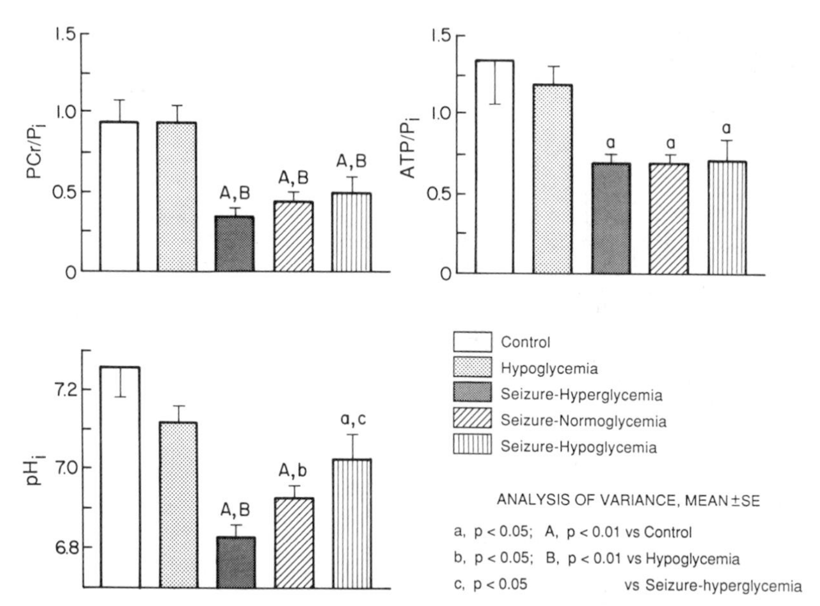

FIGURE 1. Brain high energy phosphate metabolism. Parallel changes in PCr/P_i and ATP/P_i ratios occur in the seizure, seizure-normoglycemia, and seizure-hypoglycemia groups.

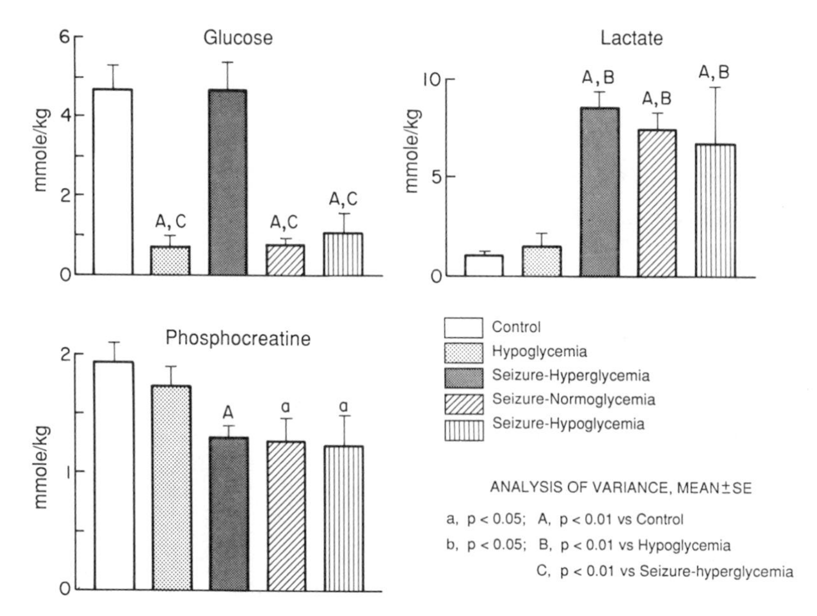

FIGURE 2. *In vitro* brain metabolites. There is marked reduction of brain glucose levels in hypoglycemia, seizure-normoglycemia, and seizure-hypoglycemia animals. Brain lactate is elevated in all animals undergoing seizure. Phosphocreatine is reduced in all seizure groups, irrespective of brain glucose level.

which occurs during seizure alone; (3) amino acids and ketone bodies may serve as metabolic fuel when hypoglycemia complicates neonatal seizures; (4) preserved levels of brain glucose do not ameliorate the reduction in high energy phosphates that accompany seizure.

REFERENCES

1. VOLPE, J. J. 1981. Neurology of the Newborn. Saunders. p. 301–320.
2. JONES, M. D. 1979. Energy metabolism in the developing brain. Sem. Perinat. **3:** 121–129.
3. YOUNG, R. S. K., M. D. OSBAKKEN, R. W. BRIGGS, S. K. YAGEL, D. W. RICE & S. GOLDBERG. 1985. ^{31}P NMR study of cerebral metabolism during prolonged seizures in the neonatal dog. Ann. Neurol. **18:** 14–20.
4. PETROFF, O. A. C., J. W. PRICHARD, T. OGINO, M. AVISON, J. R. ALGER & R. G. SHULMAN. 1986. Combined ^{1}H and ^{31}P nuclear magnetic resonance spectroscopic studies of bicuculline induced seizures in vivo. Ann. Neurol. **20:** 185–193.

Proton and Phosphorus NMR Studies of Traumatic Brain Injury in Rats

ROBERT VINK, TRACY K. McINTOSH,
SUSAN E. FERNYAK, MICHAEL W. WEINER,
AND ALAN I. FADEN

Center for Neural Injury
Magnetic Resonance Unit
San Francisco Veterans Administration Medical Center
University of California
San Francisco, California 94121

INTRODUCTION

Traumatic injury to the central nervous system (CNS) may result in functional deficits through both primary (mechanical) and secondary (delayed) mechanisms. Secondary damage to the CNS develops over a period of minutes to hours following trauma and may be related to alterations in cellular metabolism and associated tissue edema/ischemia. In the present study, we have utilized proton (^{1}H) and phosphorus (^{31}P) nuclear magnetic resonance (NMR) to determine whether alterations in high energy phosphates and/or lactate production are markers of irreversible tissue injury following experimental brain trauma.

METHODS

Male, Sprague-Dawley rats (350–400 g; $N = 10$) were anesthetized with pentobarbital, skin and temporal muscles retracted, and a plastic trauma screw inserted into a 2 mm craniotomy centered over the left hemisphere. A two-turn 5 × 9 mm NMR coil tuned to both ^{1}H and ^{31}P frequencies was positioned centrally around the trauma site, and sequential ^{1}H and ^{31}P spectra obtained prior to, and for 8 hr following moderate (1.5–2.5 atm) fluid percussion brain injury. ^{31}P NMR spectra were obtained in 10 min blocks using a 90° pulse and 0.7 sec repetition rate; lactate edited ^{1}H spectra were obtained using a binomial spin-echo sequence together with a selective (DANTE) excitation pulse and depth localization pulses.[1] All measurements were performed on a GE CSI 2T NMR spectrometer. Animals were monitored for mean arterial blood pressure (MABP) and for arterial blood gases throughout, and scored for posttraumatic neurological deficit at 24 and 48 hr.

RESULTS

Prior to injury, the PCr/P_i ratio was 4.86 ± 0.37 and intracellular brain pH was 7.10 ± 0.03. Following injury, PCr/P_i decreased to 2.86 ± 0.70 by 40 min, with a simultaneous decrease in pH to 6.86 ± 0.11 (FIG. 1). The changes in pH were

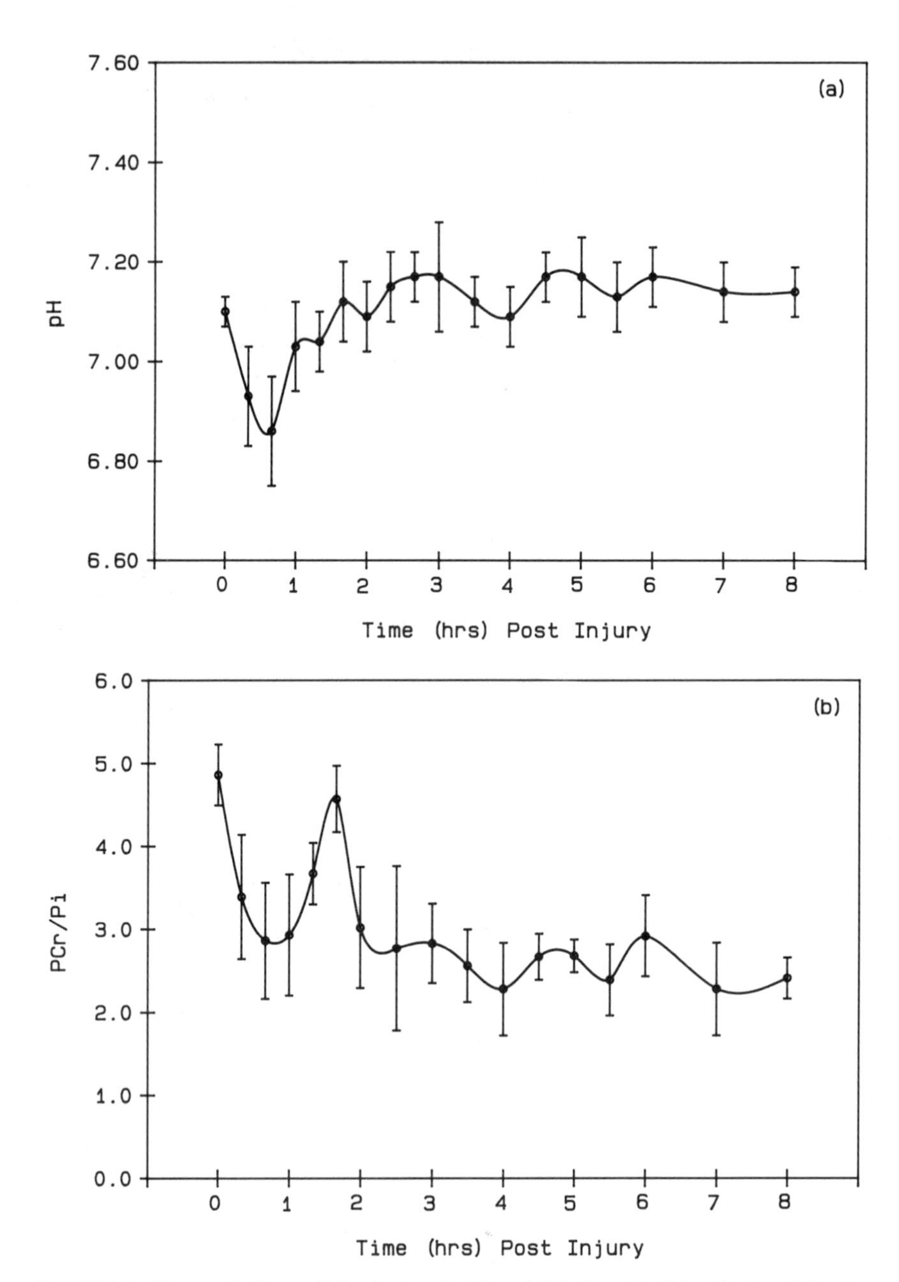

FIGURE 1. Changes in intracellular brain pH (a) and PCr/P_i ratio (b) with time following moderate fluid percussion brain injury in the rat. pH was determined from the chemical shift of P_i relative to PCr which was used as an internal reference. The relative concentrations of PCr and P_i were determined from their respective integrals following a computer "line-fitting" procedure of the original spectra. P_i: inorganic phosphate; PCr: phosphocreatine.

correlated with alterations in lactate concentration (FIG. 2). Both PCr/P_i and pH recovered by 100 min postinjury; PCr/P_i, but not pH, subsequently decreased to 2.41 ± 0.25 by 2 hr postinjury without recovery during the following 6 hr. No changes in ATP occurred at any time. Arterial blood gases and MABP before and after injury were not significantly different. Neurological deficit at 24 and 48 hr was moderate to severe in all animals.

DISCUSSION

A variety of mechanisms may contribute to irreversible tissue injury following CNS insult; these include ATP depletion, lactate accumulation, and tissue acidosis.[2] In

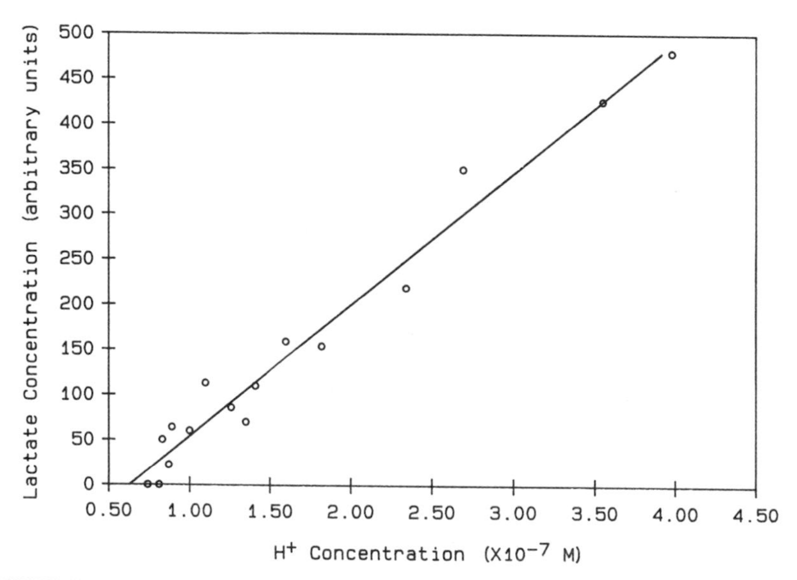

FIGURE 2. Relationship between intracellular brain pH and lactate concentration following fluid percussion head injury in the rat. Lactate concentration (expressed in arbitrary units) was determined from the integral of a computer "line-fitting" of the lactate edited proton spectra.

the present study, neurological dysfunction was observed after trauma without any alterations in ATP. Similarly, lactate accumulation and associated acidosis were transient in nature and of a magnitude not expected to result in tissue injury.[2] While increases in lactate and associated tissue acidosis may account for the transient PCr/P_i decrease in the immediate (0–2 hr) posttraumatic period, the apparent dissociation of PCr/P_i and pH in the 2–8 hr period precludes the possibility that the secondary reduction in PCr/P_i is a reflection of intracellular buffering by the creatine kinase reaction or of insufficient tissue oxygenation. Rather, we propose that the decreased PCr/P_i ratio in the absence of hypoxia or tissue acidosis is a reflection of mitochondrial dysfunction, and may be a marker of irreversible tissue injury.

REFERENCES

1. Hanstock, C. C., M. R. Bendall, D. P. Boisvert, H. P. Hetherington & P. S. Allen. 1986. Localization and spectral editing for *in-vivo* proton spectroscopy. Abstr. Soc. Magn. Reson. Med **5**(3): 977–978.
2. Siesjo, B. K. & T. Wieloch. 1985. Brain injury: Neurochemical aspects. *In* 1985 NIH Central Nervous System Trauma Status Report. D. P. Becker & J. T. Povlishock, Eds.: 513–532. William Byrd Press. Bethesda, MD.

Post-ischemic Mechanical Performance: Independence from ATP Levels[a]

STEVAN D. ZIMMER,[b] ARTHUR H. L. FROM,[b]
JOHN E. FOKER,[c] STEPHEN P. MICHURSKI[d]
AND KAMIL UGURBIL[d]

[b]*Cardiovascular Division*
Department of Medicine
Minneapolis Veterans Administration Medical Center
University of Minnesota
Minneapolis, Minnesota 55417

[c]*Department of Surgery*
University of Minnesota
Minneapolis, Minnesota 55455

[d]*Department of Biochemistry*
Gray Freshwater Biological Institute
University of Minnesota
Navarre, Minnesota 55392

Post-ischemic mechanical dysfunction is a well-established phenomenon, occurring after restoration of blood flow to otherwise viable myocardium.[1] While many potential mechanisms have been suggested, attention has been focused on post-ischemic reductions in ATP levels as a proximate cause of post-ischemic mechanical dysfunction.[2] Previous studies of this relationship have relied upon glucose ± insulin or fatty acids as substrates available to the myocardium pre-ischemia and upon reperfusion. However, other substrates (e.g. pyruvate) have been shown to be associated with little, if any, post-ischemic mechanical dysfunction after moderate periods of ischemia.[3,4] These previous studies have also relied on the spontaneous recovered mechanical activity (e.g. systolic pressure) as the measure of post-ischemic dysfunction. Also reductions in ATP levels that may not be limiting at relatively low workloads might be so at higher workloads. Therefore, it is important to assess the relationship between post-ischemic ATP levels and mechanical performance over a wide range of post-ischemic workloads and different substrate conditions.

To examine the relationship between substrate effect, ATP levels, and post-ischemic mechanical performance after a moderate degree of ischemia, we examined the effects of a normothermic, no-flow ischemic period on ATP levels (^{31}P NMR), LV systolic pressure (LV), myocardial oxygen consumption (MVO_2), and rate-pressure product (pacing rate vs. LV; RP) in isolated rat hearts (Langendorff) for two substrates [10 mM pyruvate + 10 mM glucose, $N = 5$; 10 mM glucose + 20 I.U./L

[a]This work was supported by the Veterans Administration Research Funds, the American Heart Association, Minnesota Affiliate, the United States Public Health Science Grants RO1-HL-33600, RCDA IK04-HL-01241 (K.U.), NRSA-F32-HL-07438 (S.D.Z.), and RO1-HL-26640 (J.E.F.).

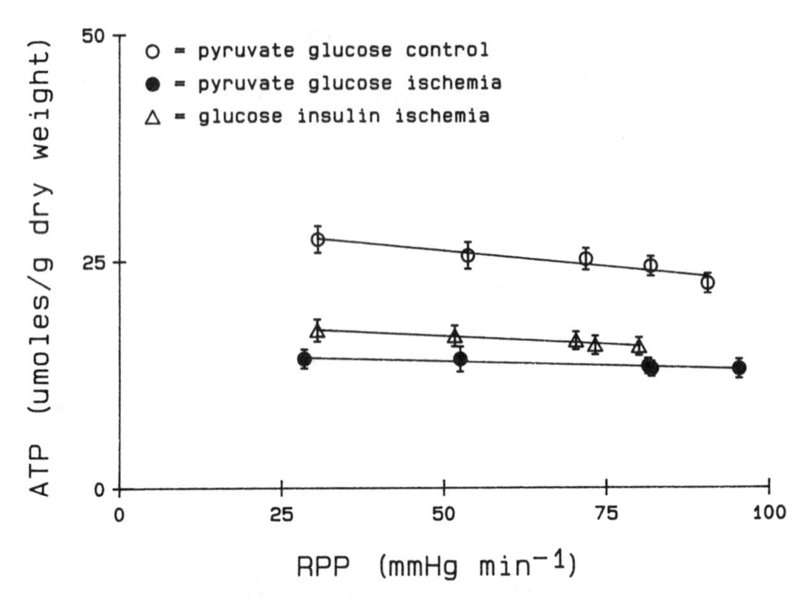

FIGURE 1. ATP content (^{31}P NMR) versus rate pressure product at each post-ischemic workload. The separation of the last three data points for post-ischemic glucose-insulin hearts from post-ischemic pyruvate-glucose hearts is indicative of significant mechanical dysfunction.

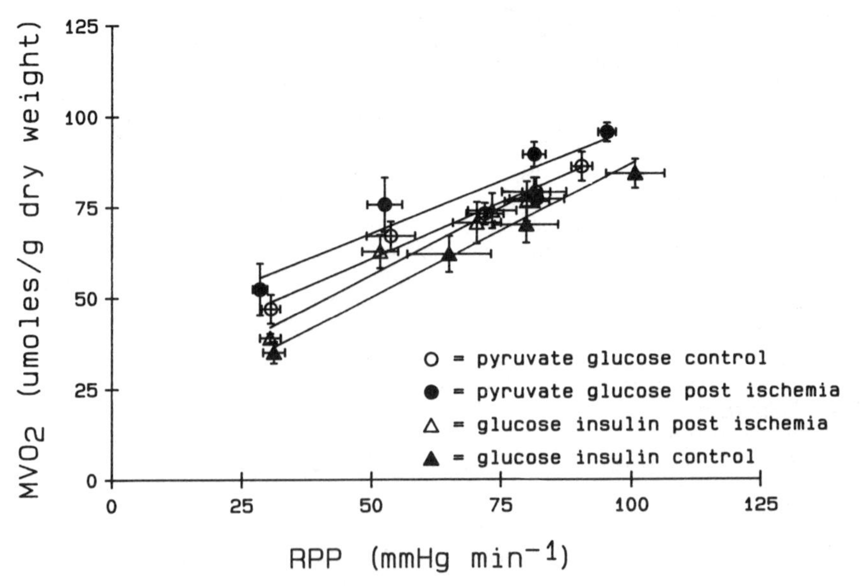

FIGURE 2. Myocardial oxygen consumption versus rate pressure product at each post-ischemic workload.

insulin, $N = 6$; 10 mM pyruvate + 10 mM glucose, $N = 6$ (non-ischemic controls)]. Data were obtained pre-ischemia (and post-ischemia after LV had stabilized) at one workload and at four progressively higher post-ischemic workloads identical to workloads for control hearts. Increases in workload were achieved in all groups by increases in pacing rate, ventricular filling pressure, and inotropic state (dobutamine, at four highest workloads). The ischemic duration for pyruvate-glucose hearts was 18 minutes. However, since glucose-insulin hearts showed the initial signs of ischemic contracture at 10.56 min (mean) they were reperfused after 8.8 min of ischemia to avoid the possibility of irreversible injury.

The relationship between post-ischemic ATP levels and RP for controls, post-ischemic pyruvate-glucose hearts and post-ischemic glucose-insulin hearts is shown in FIGURE 1. The relationship between MVO_2 and RP for the same group of hearts is shown in FIGURE 2. Pyruvate-glucose hearts showed no evidence of post-ischemic mechanical dysfunction over the range of post-ischemic workloads despite significantly lower levels of ATP than both controls and post-ischemic glucose-insulin hearts. In contrast, glucose-insulin hearts showed significant mechanical dysfunction despite less depletion of ATP (than post-ischemic pyruvate-glucose hearts) and a shorter period of ischemia. However, the MVO_2-RP relationship at any attainable workload in either ischemic group was not significantly different from their respective controls ($p =$ N.S.).

These results indicate that for the ischemic durations in this study substrate effects are significant in post-ischemic mechanical dysfunction and are independent of absolute levels of ATP, and the equivalence of the MVO_2-RP relationships indicates that mitochondrial function was unimpaired and hence did not contribute to the observed mechanical dysfunction in the glucose-insulin group. Thus, a moderate reduction of ATP center alone cannot explain post-ischemic mechanical dysfunction.

REFERENCES

1. BRAUNWALD, E. & R. A. KLONER. 1982. Circulation **66:** 1146–1149.
2. HUMPHREY, S. M., D. G. HOLLIS & R. N. SEELYE. 1985. Am. J. Physiol. **248:** H-644–651.
3. LIEDTKE, A. J., S. H. NELLIS, J. R. NEELY & H. C. HUGHES. 1976. Circ. Res. **39:** 378–387.
4. BUNGER, R., B. SWINDALL, D. BRODIE, D. ZDUNEK, H. STIEGLER & G. WALTER. 1986. J. Mol. Cell. Cardiol. **18:** 423–438.

^{1}H-NMR of Lactate by Homonuclear Transfer of Polarization

M. VON KIENLIN, J. P. ALBRAND, B. AUTHIER,
P. BLONDET, AND M. DÉCORPS

Département de Recherche Fondamentale
Service de Physique
Résonance Magnétique en Biologie et Médecine
Centre d'Etudes Nucléaires de Grenoble
38041 Grenoble Cedex, France

With the development of various schemes to suppress the large signal of water *in vivo*, ^{1}H-NMR spectroscopy should become a very powerful tool to study the metabolism of living tissues. The detection of lactate, which is the end product of anaerobic glycolysis, is of particular interest to follow the course of hypoxic or ischemic events and other impairments of the respiratory chain in the cells. Unfortunately, the detection of lactate *in vivo* can prove difficult, because the CH resonance in the flank of the water peak is usually suppressed together with it, while in many instances the CH_3 signal is hidden by a much larger lipid signal, which precludes direct observation.

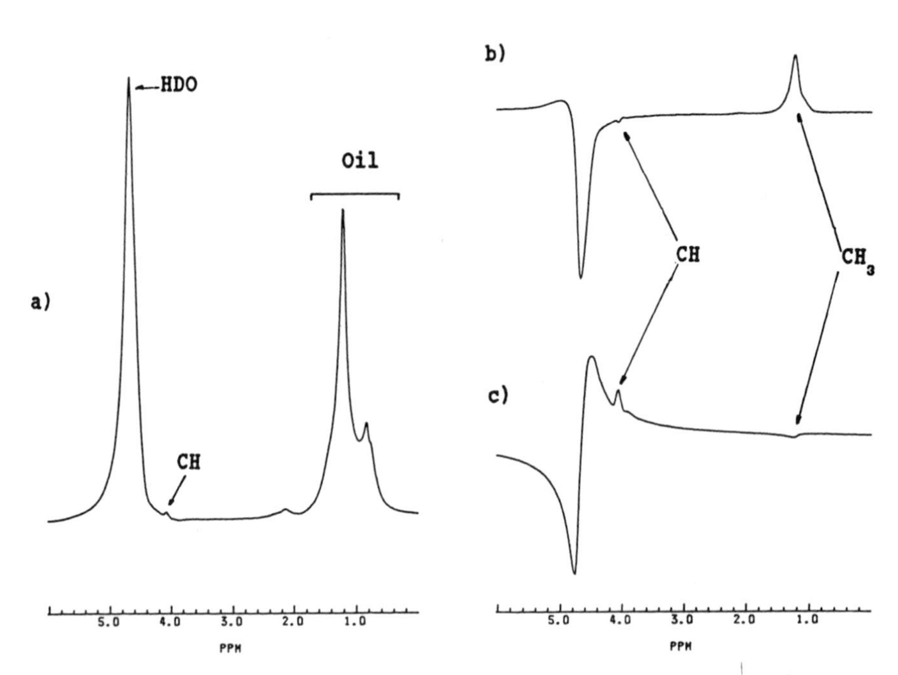

FIGURE 1. ^{1}H-NMR spectra obtained on the phantom (B_0 = 4.7 T, 128 scans). (a) Spin echo (272 msec), (b) homonuclear polarization transfer pulse sequence, and (c) control spectrum obtained with transfer pulse omitted.

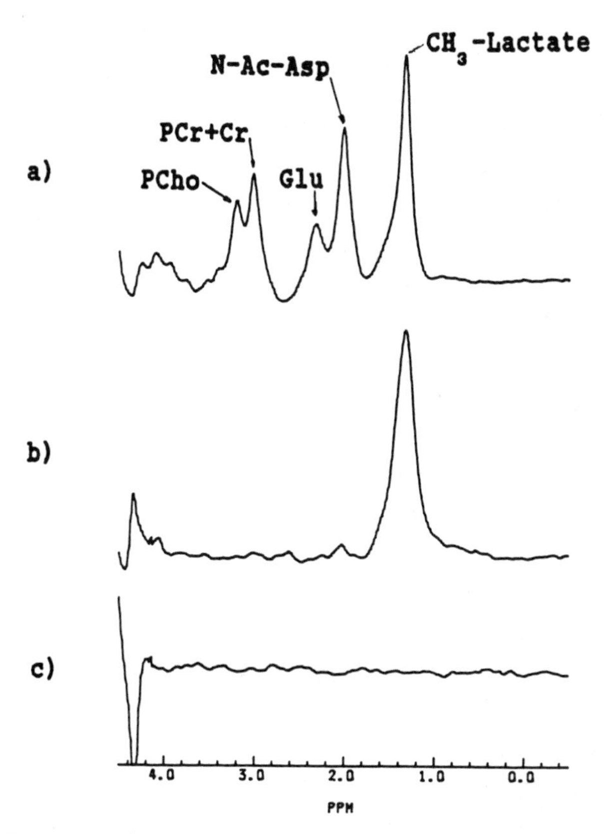

FIGURE 2. Post-mortem ^{1}H-NMR spectra of rat brain ($B_0 = 4.7$ Tesla, 128 scans). (a) Sequence $1\bar{1} - \tau - 4 - \tau - (2\tau = 272$ msec), (b) homonuclear polarization transfer pulse sequence, and (c) control spectrum obtained by omitting the selective excitation of the CH group.

Several methods,[1,2] based on spectral editing by difference spectroscopy, have been proposed to overcome this problem and subtract away the unwanted signal. To be effective, however, these methods require very stable conditions of measurements that can be difficult to meet with living samples, and with large residual lipid signals, the artefacts resulting from experimental drifts can still obscure the small signal to be detected.

To overcome these difficulties, we propose a new method to achieve the suppression of the lipid signal in a single acquisition using an homonuclear transfer of polarization[3] between the CH and the CH_3 protons of lactate. The pulse sequence used is the following: Saturation of lipids and CH_3—$90^\circ_{sel,x}$—t_1—180°_{Ex}—t_1—90°_y—t_2—180°_{Ex}—t_2—Acquisition. It begins with a selective saturation of the signals in the lipids region, including the CH_3 of lactate. Then a 90°_x selective pulse followed by an evolution period $2t_1 \simeq 1/5J$, under the influence of the homonuclear spin couplings, create antiphase magnetization of the CH spins, which is transferred to the CH_3 spins by the 90°_y pulse. A second free precession period, $2t_2 = 1/2J$, allows the CH_3 spins to rephase under the influence of the spin coupling with the CH spins. The acquisition can start at this point or can be preceded by a spin echo period with a 180° pulse selective on the CH_3 to achieve water suppression. The 180° pulses refocus chemical shifts and are phase-cycled following the EXORCYCLE scheme.

This method was first tested at 200 MHz on a phantom made of two concentric tubes. The outer 10 mm tube was filled with a 100 mM solution of calcium lactate in D_2O/H_2O and the inner 5 mm one with mineral oil. This concentric cell was positioned inside a 20 mm diameter saddle coil of a homemade probe.

FIGURE 1a shows the ^1H spectrum obtained with a standard Hahn spin echo sequence: even with a relaxation delay 2τ of 272 msec there is a large residual oil signal hiding the CH_3 signal of lactate. FIGURE 1b shows the spectrum resulting from the sequence of polarization transfer (excluding the water suppression): The only remaining signal at 1.3 ppm is due to the CH_3 of lactate. FIGURE 1c shows that this signal effectively disappears when the pulse of transfer is omitted.

The second example of this editing is illustrated in FIGURE 2. FIGURE 2a shows the post-mortem ^1H-NMR spectrum obtained on an intact rat with a chronically implanted surface coil,[4] using the spin echo sequence with water suppression: $\theta - t - \bar{\theta} - \tau - 4\theta - \tau -$ Acq. The delays are $t = 735$ μsec and $2\tau = 272$ msec.

FIGURE 2b shows the CH_3 lactate resonance edited with the sequence of transfer of polarization and a spin echo of 60 msec for water suppression. Finally, FIGURE 2c shows the spectrum obtained when omitting the selective excitation of the CH-group. As there is no CH-polarization, nothing can be transferred and hence there is no CH_3 signal.

REFERENCES

1. JUE, T., F. ARIAS-MENDOZA, N. C. GONNELLA, G. I. SHULMAN & R. G. SHULMAN. 1985. A ^1H NMR technique for observing metabolite signals in the spectrum of perfused liver. Proc. Natl. Acad. Sci. USA **82:** 5246–5249.
2. WILLIAMS, S. R., D. G. GADIAN & E. PROCTOR. 1986. A method for lactate detection in vivo by spectral editing without the need for double irradiation. J. Magn. Reson. **66:** 562–567.
3. DUMOULIN, C. L. & E. WILLIAMS. 1986. Suppression of uncoupled spins by single quantum homonuclear polarization transfer. J. Magn. Reson. **66:** 86–92.
4. DECORPS, M., J. F. LEBAS, J. L. LEVIEL, S. CONFORT, C. REMY & A. L. BENABID. 1984. Analysis of brain metabolism changes induced by acute potassium cyanide intoxication by ^{31}P NMR in vivo using chronically implanted surface coils. FEBS Lett. **168:** 1–6.

The Relationship between the Proton Electrochemical Potential and Phosphate Potential in Rat Liver Mitochondria:

Re-evaluation Using ^{31}P-NMR[a]

HAGAI ROTTENBERG

Pathology Department
Hahnemann University
Philadelphia, Pennsylvania 19102

Previous studies of the relationships between the proton electrochemical potential ($\Delta\tilde{\mu}_H$) (i.e. $\Delta\psi + \Delta pH$) and the phosphate potential (ΔG_p) (i.e. $\Delta G_p^o + RT\ln[ATP]/[ADP][P_i]$) in mitochondria led to conflicting results. Most investigators have observed that contrary to the predictions of the chemiosmotic theory, ΔG_p is not proportional to $\Delta\tilde{\mu}_H$.[1] More recently, it was claimed that earlier observations were erroneous because of various artifacts in the procedure of measuring ΔG_p and that $\Delta\tilde{\mu}_H$ is proportional to ΔG_p as expected.[2] It appears that the chemical techniques commonly used in these determinations may lead to various artifacts and large uncertainties. In this study, we used ^{31}P NMR to determine ΔG_p in mitochondrial suspension. At the intermediate range of values of ΔG_p ($4 \geq ATP/ADP \geq 0.25$) we evaluated ΔG_p directly from the chemical shift of P_i, ATP, and ADP. To estimate ADP concentration at high ATP/ADP ratios we added creatine and the enzyme creatine kinase. With excess enzyme the creatine kinase reaction is at equilibrium and the ratio $[ATP]/[ADP][P_i]$ is given by $[CrP]K_{crp}/[P_i][creatine]$.[3] Similarly, to obtain ATP concentration at low phosphate potential, we added excess myokinase and AMP. At steady state the ratio $[ATP]/[ADP][P_i]$ is given by $K_{myo}[ADP]/[AMP][P_i]$. FIGURE 1 shows the steady state (State 4) spectra obtained in dilute mitochondrial suspension (3 mg protein/ml) at different values of $\Delta\tilde{\mu}_H$. $\Delta\tilde{\mu}_H$ was manipulated by the use of valinomycin and potassium, which allow clamping of the mitochondrial membrane potential at any desirable value. At high $\Delta\tilde{\mu}_H$ (up to 3 mM K) all the nucleotides were converted to ATP and ΔG_p was calculated from the integrated signals of P_i and creatine phosphate as indicated above. At the intermediate value of $\Delta\tilde{\mu}_H$ (3 to 10 mM K), ΔG_p was calculated directly from the integrated signals of P_i, ADP, and ATP. At low $\Delta\tilde{\mu}_H$ (above 10 mM K), ΔG_p was calculated from the signals of AMP, P_i, and ADP. In parallel with the experiments shown in FIGURE 1, $\Delta\tilde{\mu}_H$ was determined from the distribution of ^3H-TPP^+ and ^{14}C-DMO.[4] FIGURE 2 shows the relationship between $\Delta\tilde{\mu}_H$ and the calculated ΔG_p from four different experiments (empty symbols). These results indicate that there are two distinct regions with different slopes: at low $\Delta\tilde{\mu}_H$ (up to about 120 mV) ΔG_p appears to be strictly proportional to $\Delta\tilde{\mu}_H$ with a slope of 4.0, as predicted by the chemiosmotic theory. However, from 120 mV to 200 mV (at the physiological range) the slope is close to 1, which could be fully accounted for by the effect of the electrogenic ADP/ATP translocator on the extramitochondrial ΔG_p.[5] To test this conclusion, we have studied the relationship between ΔG_p and $\Delta\tilde{\mu}_H$ as modulated by

[a]Supported by Public Health Service Grant GM-28173.

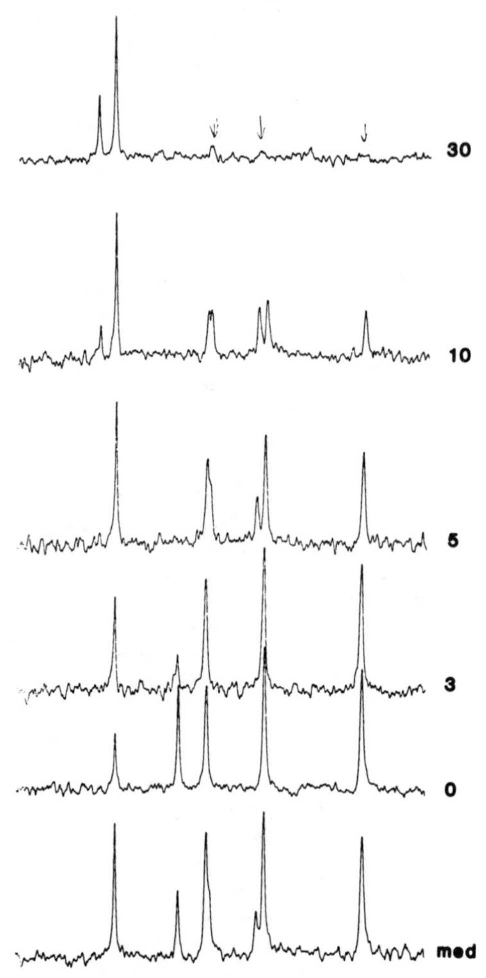

FIGURE 1. ^{31}P-NMR spectra of mitochondrial suspensions in steady state in the presence of valinomycin and potassium. Rat liver mitochondria (2.8 mg protein/ml) were incubated in a basic medium composed of 0.2 M sucrose, 15 mM $MgCl_2$, 10 mM sodium succinate, 10 mM HEPES (pH 7.4), 2×10^{-7} M valinomycin, and various concentrations of KCl (0,3,5,10,30 mM). At low potassium concentration (up to 3 mM) the medium also contained 0.5 mM ADP, 1.5 mM ATP, 1 mM creatine-phosphate, 4 mM creatine, and creatine kinase. At 5 mM potassium the medium contained, in addition to the basic medium, 1.5 mM ADP, 0.5 mM ATP, 5 mM P_i, and 5 mM creatine and creatine kinase. At 10 and 30 mM potassium, the medium contained, in addition to the basic medium, 2 mM AMP and 8 mM P_i. The medium was saturated with oxygen. Steady state was reached after 10–20 min of incubation (23°C) and lasted at least 15 min. Spectra were averaged at groups of 250 pulses (1.4 sec apart).

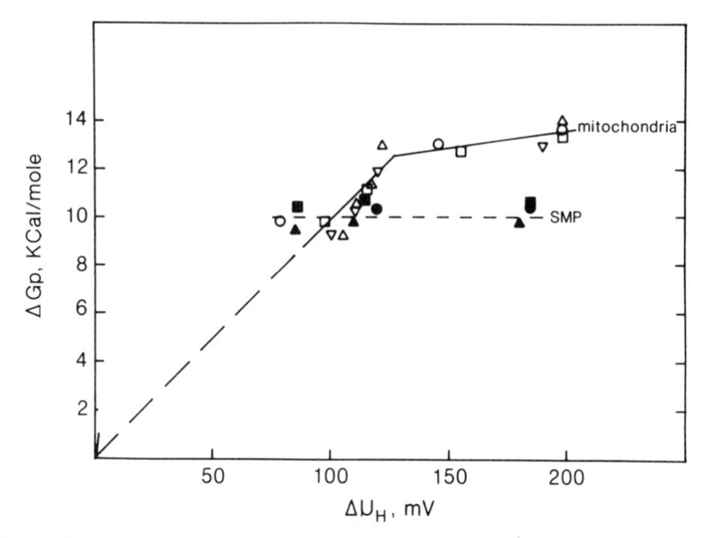

FIGURE 2. The dependence of the free energy of ATP hydrolysis on $\Delta\tilde{\mu}_H$ in suspension of mitochondria and submitochondrial particles. Results are taken from experiments similar to those shown in FIG. 1. $\Delta\tilde{\mu}_H$ in mitochondria was determined from the distribution of ^{3}H-TPP and ^{14}C-DMO. $\Delta\tilde{\mu}_H$ in submitochondrial particles was measured by flow-dialysis from the distribution of ^{14}C-SCN and ^{14}C-methylamine. Four separate experiments with mitochondria are shown (empty symbols) and three separate experiments with submitochondrial particles (full symbols). The medium for submitochondrial particles was composed of 0.2 M sucrose, 50 mM Kacetate, 5 mM $MgCl_2$, 5 mM succinate, 4 mM AMP, 5 mM P_i, 0.5 mM ADP, and myokinase. $\Delta\tilde{\mu}_H$ was modulated by the addition of nigericin (1×10^{-7}M) and gradual additions of valinomycin (1×10^{-10} to 1×10^{-7} M).

nigericin and valinomycin in inverted submitochondrial particles (FIG. 2, full symbols). It was indeed observed that over the tested range, $\Delta\tilde{\mu}_H$ has little effect on the value of ΔG_p. These preliminary results, if confirmed and validated, would indicate that at high $\Delta\tilde{\mu}_H$ (which is the physiological range) ATP synthesis by ATP synthase is not driven by the proton electrochemical potential.

REFERENCES

1. FERGUSON, J. S. 1985. Biochim. Biophys. Acta **811:** 47–95.
2. WOELDERS, H., W. J. VAN DER ZANDE, A. M. A. F. COLEN, R.-J. A. WANDERS & K. VAN DAM. 1985. FEBS Lett. **179:** 278–282.
3. GYULAI, L., Z. ROTH, J. S. LEIGH, JR. & B. CHANCE. 1985. J. Biol. Chem. **269:** 3947–3954.
4. ROTTENBERG, H. 1984. J. Membr. Biol. **81:** 127–138.
5. KLINGENBERG, M. & H. ROTTENBERG. 1977. Eur. J. Biochem. **73:** 125–130.

Diagnostic Markers of Experimental Pancreatitis: ^{31}P and ^{23}Na NMR Studies

O. KAPLAN,[a] T. KUSHNIR,[b] AND G. NAVON[b]

[a]*Department of Surgery*
Tel-Aviv Medical Center
Rokach Hospital
Tel-Aviv, Israel
[b]*School of Chemistry*
Tel-Aviv University
Tel-Aviv, Israel

Acute pancreatitis is characterized by progressive pathologic damage. There is no reliable quantitative means available at present to evaluate its severity and prognosis.[1] Such an assessment is essential for choosing the optimal therapeutic management and for studying the efficacy of treatment.

Previous studies in our laboratory[2] using ^{31}P NMR spectroscopy yielded information on the metabolic changes that took place in the nonperfused rat pancreas following induction of pancreatitis by intraparenchymal injections of 10% sodium taurocholate. These changes included a gradual depletion of high energy compounds ATP and creatine phosphate, which correlated with the extent of the pathologic lesion and provided a reliable indicator of the severity of the disease. A transient appearance of a signal at -0.16 ± 0.04 ppm was demonstrated. This signal (X in TABLE 1) was considered characteristic of the pathologic process.

Injection of 10% sodium taurocholate into the parenchyma of perfused rat pancreases resulted only in edematous pancreatitis. The peak at -0.16 ± 0.04 ppm was again detected. Studies are under way to further delineate this signal, which may elucidate the pathogenesis of pancreatitis. No other changes in the ^{31}P NMR spectra were noted in the perfused pancreas system, presumably due to the relatively high perfusion rates (6–8 ml/min) required to maintain a normal metabolic state. Similar attempts to induce experimental pancreatitis in perfused pancreases at low temperatures (10°C–30°C) and lower rates of perfusion yielded no ^{31}P NMR spectral changes whatsoever. We are currently studying other models of experimental pancreatitis using NMR spectroscopy, including intravascular triglycerides and oleic acid injections and intra bile duct sodium taurocholate injections in an attempt to detect and monitor the pathophysiologic processes.

Since interstitial edema is one of the most prominent pathologic features of acute pancreatitis,[3] ^{23}Na NMR measurements of experimental pancreatitis were carried out in nonperfused rat pancreases. The pancreases were placed in a chilled (0°C–2°C) isotonic, sodium free, mannitol solution (instead of saline) and both ^{31}P and ^{23}Na spectra of the whole organ were sequentially recorded. The sodium concentrations were compared to the metabolic status and the severity of the disease as indicated by the ^{31}P spectrum and macroscopic examination. TABLE 1 shows that ^{23}Na levels increased as the disease progressed concomitant with a decrease in the high energy compounds levels. The contribution of the hyperosmotic (372 mosmol) sodium of the Na-taurocholate to pancreatic sodium levels was studied by interparenchymal injections of 1 ml 372 mosmol NaCl. Sodium levels were high in those control animals but only in the first 45 min and never reached the levels of the diseased organs; sodium

TABLE 1. Diagnostic Markers of Pancreatitis

^{31}P NMR Spectra	Macroscopic Damage	Sodium Levels Per Dry Weight (mM/mg × 10^3)
c	+ + +	457
b (X)	+ +	203
a (PME, Pi, GPC, PCr, γATP, αATP, DPDE, βATP; δ, ppm: 10, 0, −10, −20)	0	60

[a]Normal pancreas. [b,c]Diseased pancreases.

levels returned to normal thereafter. In conclusion, ^{23}Na NMR may serve as an additional means in evaluating the severity and prognosis of acute experimental pancreatitis.

REFERENCES

1. SILEN, W. & W. L. STEER. 1984. Pancreas. *In* Principles of Surgery, 4th Edit. S. I. Schwartz, Ed.: 1350–1353. McGraw-Hill Co. New York.
2. KAPLAN, O., T. KUSHNIR, U. SANDBANK & G. NAVON. 1987. J. Sur. Res. (In press).
3. Pathologic Basis of Disease, 3rd Edit. 1984. S. L. Robbins, R. S. Cortan & V. Kumar, Eds.: 963-965. W. B. Saunders Co. Philadelphia.

The Fourier Series Window Method for Spatially Localized NMR Spectroscopy

Implementation with Composite Pulses and Multiple Coils

MICHAEL GARWOOD,[a,b] THOMAS SCHLEICH,[a,c] AND KAMIL UGURBIL[b]

[a]*Department of Chemistry*
University of California
Santa Cruz, California 95064

[b]*Gray Freshwater Biological Institute*
University of Minnesota
Navarre, Minnesota 55392

Localized NMR spectroscopy may be accomplished without the use of pulsed B_0 field gradients by employing rotating frame zeugmatography[1] or the depth pulse method.[2] The required gradient for spatial encoding is usually supplied by a surface coil antenna. The rotating frame method has been successfully demonstrated on phantoms and *ex vivo* tissue samples,[3–5] the *in vivo* rat kidney,[6] and the human thorax.[7] A $\sqrt{2}$ loss in signal-to-noise ratio relative to that attainable in the phase-modulated version occurs when the rotating-frame experiment is performed in an amplitude-modulated manner, as is typically the case.[1]

If a "window" or a set of several "windows" of nutational frequency are desired in place of the full metabolite map, the rotating-frame experiment may be reduced. This modification, known as the Fourier Series Window (FSW) method,[8–10] avoids Fourier transformation in the second dimension and permits the delineation of more than one spatial region by post-acquisition data reduction. No loss of sensitivity occurs at the center of the selected window when the nutation angle $n\theta$ equals 90°, 270°, 450°. . . . However, due to contaminating high flux signals in regions where θ = 270° and 450°, this particular FSW cannot be used to localize spectra to depths greater than one coil radius. High flux signals from regions where θ = 270° can be eliminated using incremented pulses of $n\theta$ = 60°, 120°, 240° . . . , while 270° and 450° signals may be eliminated with $n\theta$ = 45°, 90°, 135°, 225°, 270°. . . .

With these latter two FSW pulse increments, some longitudinal magnetization remains following the $n\theta$ pulse, and as a consequence a loss in sensitivity occurs, which can be obviated by converting the experiment to its phase-modulated equivalent.[1] This is accomplished by utilizing a 90° phase-shifted θ = 90° pulse subsequent to the incremented FSW pulse, which nutates the residual longitudinal magnetization at the window center into the transverse plane, thereby enabling detection of the full magnetization (FIG. 1). To avoid the detection of 270° signals with the $\theta[y]$ pulse, a phase alternated $\theta/3[\pm y]$ pulse is inserted in between the FSW pulse and the

[c]Address correspondence to: T. S., Department of Chemistry, University of California, Santa Cruz, CA 95064.

phase-shifted 90° pulse. With this pulse sequence, the magnetization in the region of interest is phase modulated as a function of nutation angle, $n\theta$. A chemical shift spectrum localized at the windowed region is processed by summing the free induction decays (FIDs) in a manner prescribed by the Fourier coefficients, i.e., the FIDs are multiplied by a set of calculated coefficients, or the number of transients collected at

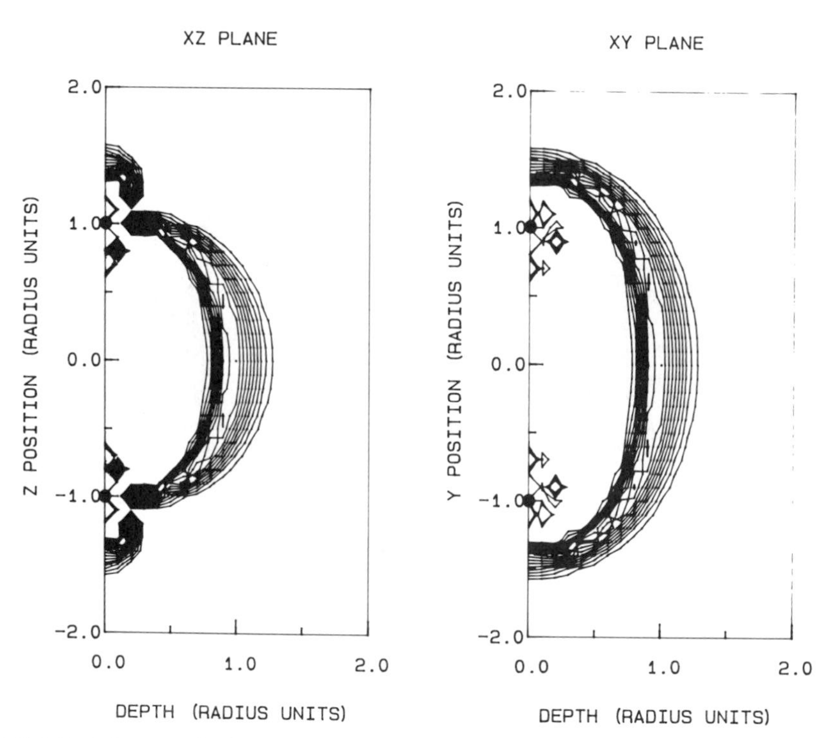

FIGURE 1. Phase-modulated FSW localized absorption signal amplitudes as a function of axial and radial position when a single surface coil is used for radiofrequency transmission and reception. The coil cross section is indicated by ● in the z and y' axes. Absorption and dispersion (not shown) correspond to the real and imaginary signals, respectively, which would be detected by quadrature phase detection. The pulse is: $\Sigma n(2\theta/3)[x]$; $\theta/3[\pm y]$; $\theta[y]$, where n ranges from 0 to 11. In data processing, the real and imaginary components are separated and multiplied by two different sets of coefficients, which define the odd and even window functions, respectively.

each pulse increment is made proportional to the corresponding coefficient, thus yielding maximal sensitivity.

Computer simulations have been used in this study to predict the degree of signal localization possible with single and multiple coils using FSW or combined FSW-depth pulse methods (FIG. 2). In addition, the simulations illustrate that the shape of the localized volume may be tailored by varying the coil configurations and FSW-depth pulse sequences.

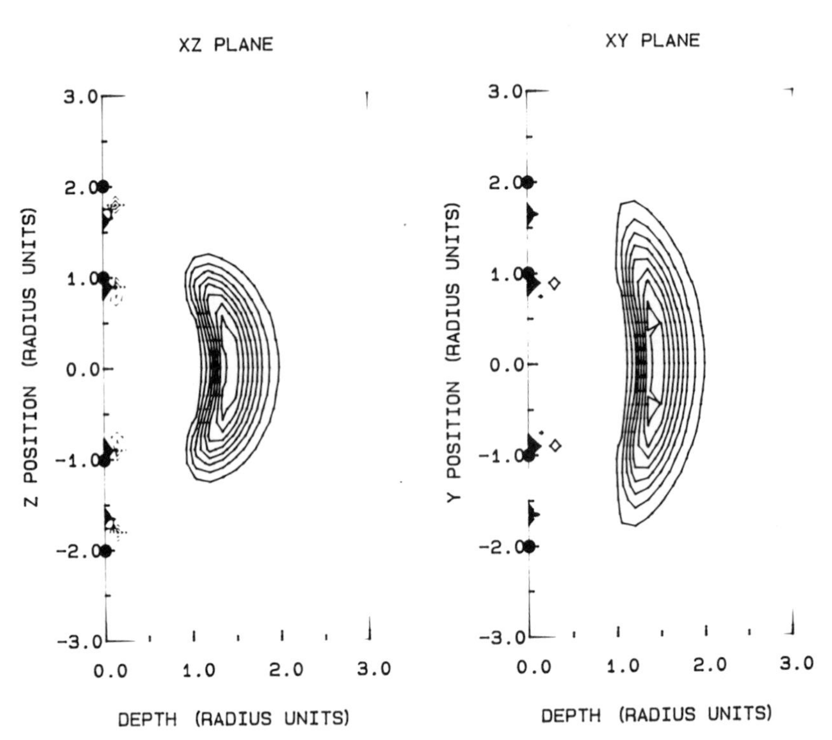

FIGURE 2. Simulation of a multiple surface coil transmit experiment using a FSW-depth pulse sequence: $\theta/3[\pm x]$; $2\theta[\pm x]$; $2\theta[\pm x,\bar{o}]$; $\Sigma n(\phi/2)[x]$; $\phi/3[\pm y]$; $\phi[y]$; acq. w/ϕ, where θ and ϕ are 90° pulses applied with the large coil (radius = 2) and small coil (radius = 1), respectively; n ranges from 0 to 7. Only the absorption (real) signal component is shown. Although this pulse sequence requires 256 transients to complete one cycle, the number of transients may be reduced (concomitant with reduced localization) by the omission of certain phase-cycled pulses (e.g., $\theta/3[\pm x]$ and/or using fewer terms in the Fourier series). In the simulations, contaminating signals outside the localized volume make up only 4.3% of the total root squared signal. Only absorption mode signals are shown, but dispersion signals in the $x'z$ and $x'y$ planes arise from nearly identically shaped volumes.

REFERENCES

1. HOULT, D. I. 1979. J. Magn. Reson. **33:** 183–197.
2. BENDALL, M. R. & R. E. GORDON. 1983. J. Magn. Reson. **53:** 365–385.
3. HAASE, A., C. MALLOY & G. K. Radda. 1983. J. Magn. Reson. **55:** 164–169.
4. GARWOOD, M., T. SCHLEICH, G. B. MATSON & G. ACOSTA. 1984. J. Magn. Reson. **60:** 268–279.
5. COX, S. J. & P. STYLES. 1980. J. Magn. Reson. **40:** 209–212.
6. BOGUSKY, R. T., M. GARWOOD, G. B. MATSON, G. ACOSTA, L. D. COWGILL & T. SCHLEICH. 1986. Magn. Reson. Med. **3:** 251–261.
7. STYLES, P., C. A. SCOTT & G. K. RADDA. 1985. Magn. Reson. Med. **2:** 402–409.
8. GARWOOD, M., T. SCHLEICH, B. D. ROSS, G. B. MATSON & W. D. WINTERS. 1985. J. Magn. Reson. **65:** 239–251.
9. GARWOOD, M., T. SCHLEICH, M. R. BENDALL & D. T. PEGG. 1985. J. Magn. Reson. **65:** 510–515.
10. METZ, K. R. & R. W. BRIGGS. 1985. J. Magn. Reson. **64:** 172–176.

NMR as a Noninvasive Tool in Meat Research[a]

PETER LUNDBERG AND HANS J. VOGEL

Division of Biochemistry
University of Calgary
Calgary, Canada T2N 1N4

INTRODUCTION

NMR has been used for some time for *in vivo* measurements of biological tissues and has established itself as a valuable and complementary method to be used in parallel with traditional extraction methods. To date most attention has been paid to problems such as central biochemical pathways, energy metabolism in simple organisms and organs, and of course to problems encountered in medicine. There has been relative little interest in applying NMR to problems in food technology, even if these questions can have a tremendous impact on everyday life. Following a suggestion by Gadian,[1] we show here that multinuclear metabolic NMR is a useful method for studying post-mortem events in carcasses of slaughtered animals.[2–7]

The treatment and storage of carcasses during the first hours after slaughter is of extreme importance for the final quality and tenderness of the meat. Mistreatment can cause large economical losses and waste of valuable food. For example, in order to reduce the risk of bacterial infections one would like to cool down a carcass as soon as possible. Nevertheless, if a muscle is cooled down below 15°C before the postmortem metabolism is completed it may shorten dramatically thus decreasing the tenderness of the meat.[8] Therefore it is important to measure the rates of postmortem metabolism and to study the efficiency of methods that are aimed at speeding up this process.

EXPERIMENTS

Samples

Muscles from bovine (M. longissimus dorsi, M. semitendinosus and M. biceps femoris), ovine (M. s.), porcine (M. s. and M. b. f.), mackerel (white swim muscle), and chicken (white breast muscle) carcasses were used for these experiments. Samples from the mammalian species were cut from the carcasses typically 15 min after slaughter. These samples were then transported to the NMR laboratory and placed in the magnet. The acquisition of data usually started within one hour from the time of slaughter. The fish and bird samples were from aged muscles. For bovine muscles, a comparison was made between isometrically and non-isometrically mounted samples. The isometrically mounted muscles were glued to two circular plastic plates with a polyacrylate glue.

[a]Supported by grants from the Alberta Heritage Foundation for Medical Research in the form of a studentship (P.L.) and a scholarship (H.J.V.).

NMR

Phosphorus-31 experiments were performed at 103.2 MHz on a homebuilt spectrometer equipped with an Oxford 6 T widebore magnet. The cycle time between acquisitions was 8 sec to allow for almost complete relaxation of the nuclei, which simplified the measurements of relative concentrations. Phosphocreatine in the samples was used as an internal shift reference, assigned to −2.35 ppm relative to 85% phosphoric acid. Two different homebuilt solenoidal probes of 10 and 20 mm were used.[2,3] Carbon-13 measurements were performed at 91 MHz using an NT 360 WB spectrometer. The cycle time was 1.4 sec, to maximize the S/N ratio for the resonances of interest. A 10 mm broadband Helmholtz probe or a 20 mm solenoidal probe (Nicolet) were used for these experiments. The WALTZ-16 decoupling sequence, software implemented, was used at low power levels in a bilevel scheme (0.2 W / 1 W).[4] Proton NMR was performed on the same spectrometer using a homebuilt 10 mm Helmholtz type probe at 361.8 MHz. The cycle time was 1.2 sec and a 90°-D1-90°-AQ ("jump and return") pulse sequence was used to decrease the water signal.[4] Peak areas were usually measured by cutting and weighing of the resonances but in a few cases, where applicable, only the peak heights were measured.

RESULTS AND DISCUSSION

Measurements of Metabolites

Using ^{31}P NMR, the levels of phosphocreatine (PCr), inorganic phosphate (P_i), ATP, and phosphomonoesters could be followed during the first 12 hr of the post-mortem process.[2,3] No unexpected phosphomonoesters or phosphodiesters were observed. As identified in perchloric acid extracts, the phosphomonoester peak consists initially mainly of sugar phosphates such as glucose-6-phosphate (G6P),[2] but as time progresses the composite resonance contains an increasing amount of IMP, which is the main degradation product from ATP (FIG. 1).[6] The initial levels of PCr in bovine muscles, extrapolated back from the first time point at 1 hr postmortem to the time of slaughter, was 15–20 μmols/g wet weight. ATP and the total amount of sugar phosphates (mainly G6P) were at the same time 5 ± 1 and 8 ± 2 μmoles/g wet weight, respectively. These numbers are in close agreement with those determined enzymatically in extracts.[2,6] The amount of free and mobile NMR-visible ADP can be measured as the difference in intensity between the γ and the β peaks of ATP (FIG. 1). The free ADP level was less than 10% of the ATP level indicating a concentration below 0.5 μmoles/g wet weight. The total ADP concentration determined in extracts is around 1.0 μmole/g wet weight.[6] Thus, almost one half of the total ADP is NMR-visible suggesting that the remainder is bound to proteins. NAD(H) was determined as the difference between the α and the γ ATP peaks. The obtained integral, was on average 10% of the ATP level (FIG. 1).

The ^{13}C NMR experiments allowed us to follow the degradation of glycogen to lactate. Also the ratio between PCr and creatine (Cr) was monitored qualitatively from the arginino resonances of PCr and Cr (158 ppm), which have slightly different chemical shifts. Proton NMR allowed the measurement of to the metabolites PCr, Cr, and lactate. Other prominent peaks in the spectra, such as total creatine, mobile fats, carnosine or carnitine, did not change in intensity to any large extent.[4] Amino acids, such as alanine, glutamine, glutamic acid, and valine, were all of low abundance (as measured in perchloric acid extracts of the water-soluble cellular components) and

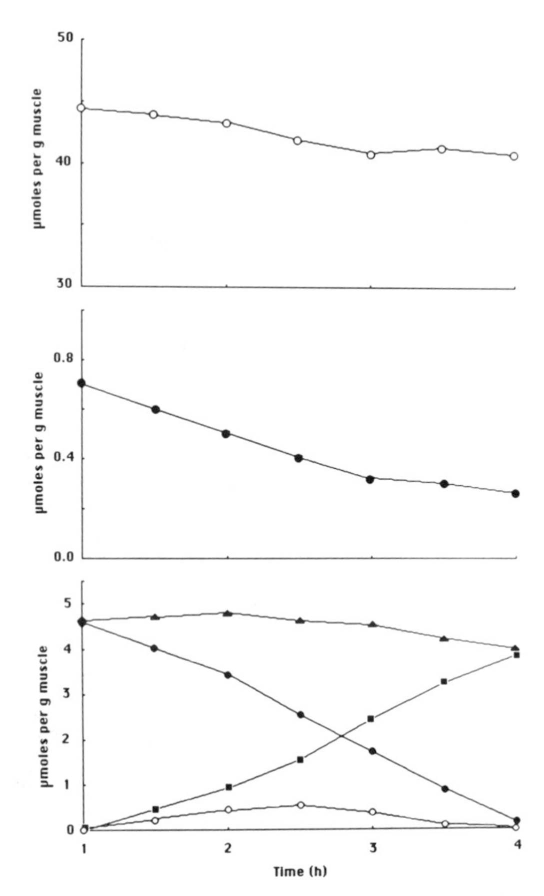

FIGURE 1. The concentrations in μmoles per g wet weight of (Bottom) ATP (●), ADP (○), AMP + IMP (■), and the total nucleotide pool (▲), (Middle) NAD (H) (●), (Top) the total phosphate pool (○) during postmortem metabolism of M. semitendinosus and M. biceps femoris (average value, $N = 4$) at 37°C. These data were obtained from NMR data as described in the text. IMP rather than AMP is the major product from ATP breakdown[6] and its level was determined by subtracting the initial value (at 1 hr) of the phosphomonoester resonance from all subsequent measurements. Note that the total nucleotide pool remains quite well preserved. The transient increase in ADP suggests that this might be an intermediate between ATP and IMP. The level of NAD(H) decreases to about half and also the total phosphorus intensity decreased by about 5%. Note that the curves run from 1 to 4 hr and that the zero timepoint is not included.

could only be followed in a qualitative but not in a quantitative manner in either ^{1}H or ^{13}C NMR.[4]

Postmortem Development in Bovine Muscles

The levels of PCr always decreased rapidly and the P_i showed a corresponding increase in intensity. The ATP level was usually constant until the PCr decreased below 2 μmoles/g wet weight (bovine muscles). The breakdown of ATP results in an almost quantitative build-up of IMP (FIG. 1). The metabolic rate varied considerably at different temperatures. At 37°C and 25°C, the metabolism of energy metabolites was finished after 4 and 10 hr, respectively. At 5°C and at 15°C, the metabolism was much slower and PCr was not depleted even after 10 hr and the level at that time was 1–2 μmoles/g wet weight.[5] No obvious difference between isometrically and nonisometrically mounted muscle samples were observed in these studies.[2,5] Samples, frozen within 15 min of slaughter, gave a different behavior. During thaw at room temperature, the metabolic rate was approximately a factor of ten higher than in nonfrozen samples.[3] This is presumably due to leakage from Ca^{2+}-containing compartments in

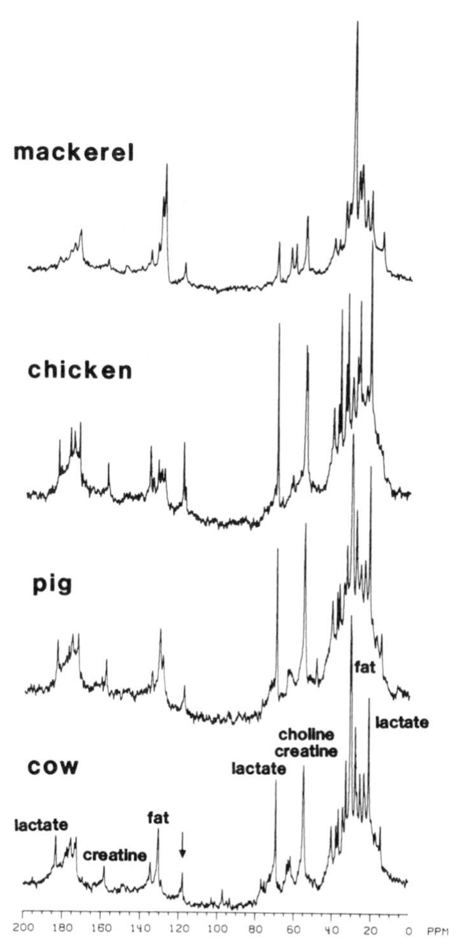

FIGURE 2. Proton decoupled ^{13}C NMR spectra of a series of meat samples from different species. The resonance marked with an arrow is from the dipeptide carnosine or anserine. The histidine moiety of this peptide provides for a convenient pH marker both in ^{13}C and proton NMR spectra.

the frozen state. During thaw the remaining energy metabolites induce an extreme Ca^{2+}-induced contraction of the muscle resulting in very tough meat. This observation indicates the necessity of depleting both ATP and PCr before the carcass can be frozen.

The total amount of (NMR visible) phosphorus was calculated for all samples to make sure that not a major part of the NMR-visible phosphate pool disappeared in the course of the measurements. Normally, the total phosphorus decreased slightly (less than 10%, see FIG. 1), whereas the total phosphorus in the thawing samples decreased by 40% from its maximum value. We believe that the latter observation represents a real immobilization of part of the phosphorus pool (presumably a complexing reaction with magnesium or calcium ions, which tends to broaden the phosphate resonance) as the tuning of the probe was controlled between each individual spectrum. We have found that in normal cases the total intensity of phosphorus is a useful indicator of the absolute concentration when the actual measured volume is not known. An average value of the total phosphorus of 43 μmoles per g wet weight was calculated for bovine muscle ($N = 8$).

From the ^{13}C NMR spectra it was determined that the glycogen level (as measured from the integral of the C2–C5 peaks) decreased about 30% during the first ten hours. Lactate built up in parallel with the glycogen decrease. Both ^{1}H and ^{13}C NMR showed the increase in Cr with a concomitant decrease of PCr as well as the dramatic increase in lactate.[4] Plots of lactate versus pH (^{13}C and ^{1}H NMR) showed a high linear correlation ($r = 0.99$, $N = 8$) between the two parameters in the measured range from pH 6.8 to pH 6.0. Thus, the buffering capacity is apparently not very large in this range.

pH Measurements

The intracellular pH was determined from the chemical shifts of P_i, G6P (^{31}P NMR),[2,3] and carnosine C_4-H (^{1}H NMR and ^{13}C NMR).[4] As a comparison, the pH was also determined electrochemically using the standard IAc/KCl method on homogenates.[2] We observed a broadening of the P_i peak only in the middle of the postmortem metabolism, suggesting that a pH range of about 0.3 units existed in the muscles. Considering the rather small chemical shift changes, the accuracy of the pH measurements remains somewhat of a problem. With this limitation in mind, we were happy to note that all the measurements showed a similar (within 0.2 pH units) appearance.[4] Moreover the results were in good agreement with those obtained with invasive methods.[2] As the pK_as for carnosine and P_i are 6.8 and that of G6P around 6.2, these three indicators together provide a good coverage in the pH range from 7.5 to 5.5, thus covering a physiological range. In the thawing bovine muscles the pH decreased below 6.0 after only 0.5 hr while it took 3.5 hr for the intracellular pH in the 37°C samples. At 25°C, pH 6.0 was reached after about 7.5 hr.

The effect of low voltage electrical stimulation,[6] a method used to increase the rate of postmortem metabolism, was very clear. The decrease of pH down to pH 6.0 (at 25°C) was 35% faster using this method as compared with control.[2] Thus, this seems to be an efficient way to deplete the pool of energy metabolites and allow for faster cooling of the carcass.

Species and Sex Variability

The initial ratio between PCr and P_i peaks intensities (at 1 hr post mortem) was very different for bovine, porcine, and ovine samples. In the bovine samples, the initial ratio was on average 3.2 while it was 0.3 in ovine and 0.01 in porcine muscles.[3] The ultimate pH values are however similar (5.5 to 5.7). Thus, the rate of postmortem metabolism in porcine and ovine samples was much faster than in bovine samples.[3] Furthermore, a significant difference in metabolic rate, as measured from the pH drop, was determined for different sexes of animals. The rate was very similar for cow and heifer but very different from that in bulls. The male bovine carcasses showed about 47% faster pH decline than the female bovine carcasses. The fastest metabolizing bull muscles reached pH 6.0 (at 25°C) after 6.5 hr while the average cow muscle was at pH 6.35 at this time.[4] Measurements on bovine muscles containing various amounts of "fast" and "slow-twitch" muscle fibers (red and white muscles) showed, within the experimental error, the same metabolic rate.[5] In contrast, in similar measurements on rabbit muscles the metabolic rate varied with the muscle type.[7]

ATP and pH were highly correlated in most bovine samples ($r = 0.93$).[5] pH 6.8 corresponded to an ATP concentration of 5.0–5.5 μmoles/g wet weight and pH 5.5 to totally depleted ATP. However in some bovine muscles, in spite of a similar initial level of ATP, the slope and intercept were very different with a large amount of ATP remaining at low pH (5.7).[5] This was not observed in porcine or ovine muscles, which also showed a clear linear correlation between the ATP level and the pH.[3] The reasons for this correlation are not clear at present. In FIGURE 2, the post rigor carbon-13 NMR spectra of some different types of muscles are shown. The main feature that differs is the varying amount of mobile fat (methylene carbons) within the muscles.

CONCLUSIONS

As the multinuclear NMR approach has allowed us to measure all the important parameters involved in muscle energy metabolism, we feel that it is a valuable complement to existing methods for analysis of postmortem metabolism. Each nucleus has certain advantages and disadvantages and we have found it very instructive to measure several different nuclei on the same type of samples. In particular the linear correlation between pH and ATP could be useful since a simple pH determination would be sufficient to determine the point of ATP depletion. In meat research there is also a need for physically measurable parameters that correlate with the tenderness and taste of meat. Therefore, we are now in the process[5] to attempt to correlate the NMR results with those obtained by standard methods of meat analysis, such as isometric tension and muscle shortening measurements, protein electrophoresis, and last but not least sensorical analysis by experienced taste panels. Finally, we want to point out that NMR is not only useful to study postmortem metabolism, but that it can also used for the noninvasive study of food additives such as polyphosphates,[9] for example.

ACKNOWLEDGMENTS

We are indebted to our colleagues Stefan Fabiansson, Eva Tornberg and Håkan Rudérus from the Swedish Meat Research Institute for introducing us to the various aspects of meat research and for fruitful collaborations and stimulating discussions.

REFERENCES

1. GADIAN, D. G. 1980. A physico-chemical approach to post-mortem changes in meat-Nuclear Magnetic Resonance. In Developments in Meat Science-1. R. Lawrie, Ed.: 89–113. Appl. Sci. Publ. Ltd. London.

2. VOGEL, H. J., P. LUNDBERG, S. FABIANSSON, H. RUDÉRUS & E. TORNBERG. 1985. Post-mortem energy metabolism in bovine muscles studied by non-invasive phosphorus-31 nuclear magnetic resonance. Meat Sci. **13:** 1–18.
3. LUNDBERG, P., H. J. VOGEL, S. FABIANSSON & H. RUDÉRUS. 1987. Postmortem metabolism in fresh porcine, ovine and frozen bovine muscle. Meat Sci. **19:** 1–14.
4. LUNDBERG, P., H. J. VOGEL & H. RUDÉRUS. 1986. Carbon-13 and proton NMR studies of postmortem metabolism in bovine muscles. Meat Sci. **18:** 133–160.
5. TORNBERG, E., G. LARSSON, P. LUNDBERG & H. J. VOGEL. 1987. The importance of the course of rigor on beef muscle tenderness. Meat Sci.
6. FABIANSSON, S. & S. LASER-REUTERSWÄRD. 1985. Low voltage electrical stimulation and post mortem metabolism in beef. Meat Sci. **12:** 205–215.
7. RENOU, J. P., P. CANIONI, P. GATELIER, C. VALIN & P. J. COZZONE. 1986. Phosphorus-31 NMR study of postmortem catabolism and intracellular pH in intact excised rabbit muscle. Biochimie **68:** 543–554.
8. HULTIN, H. O. 1984. Postmortem biochemistry of meat and fish. J Chem. Ed. **61:** 289–298.
9. DOUGLASS, M., M. P. MCDONALD, I. K. O'NEILL, R. C. OSNER & C. P. RICHARDS. 1979. Technical note: A study of the hydrolysis of polyphosphate additives in chicken flesh during frozen storage by ^{31}P-FTNMR spectroscopy. J. Fd. Techn. **14:** 193–197.

Uptake and Elimination of Isoflurane in Rabbit Tissue Studied with ^{19}F NMR Spectroscopy[a]

A. M. WYRWICZ AND C. B. CONBOY

Department of Chemistry
University of Illinois at Chicago
Chicago, Illinois 60680

P. H. EISELE

Animal Resources Service
School of Veterinary Medicine
University of California
Davis, California 95616

We have recently reported a prolonged residence of halothane and isoflurane in brain,[1,2] as compared to the predictions based on the accepted model[3] of inhalation anesthetic uptake and distribution. These measurements were carried out directly and noninvasively using *in vivo* ^{19}F NMR spectroscopy.[4] We compare here elimination of one of these anesthetics, isoflurane, from brain and paralumbar muscle, tissues that represent the two major pharmacokinetic compartments of the model, in order to evaluate their differences and/or similarities. We also report on the uptake of isoflurane by brain measured by a direct and an indirect technique.

METHODS

Animal Preparation

Elimination Studies

New Zealand white rabbits (2.5 to 3.5 kg) were intubated under ketamine (50 mg/kg) and xylazine (5 mg/kg) anesthesia. Catheters were placed in a femoral artery to collect blood samples. An additional catheter was placed in an ear vein for subsequent administration of maintenance anesthetic, pentobarbital. The animals were placed on a cradle and then positioned in the TMR magnet such that the surface coil was centered over the calvarium or paralumbar muscle. Isoflurane (1.5% in O_2) was administered for 90 min and then discontinued. The animals were continually ventilated with 100% oxygen. Body temperature was maintained with a heating pad. Ventilation was adjusted to maintain the arterial blood gases within normal range.

[a]Supported by National Institutes of Health Grants GM29520, GM33415, and RCDA K04GM00503 to A.M.W.

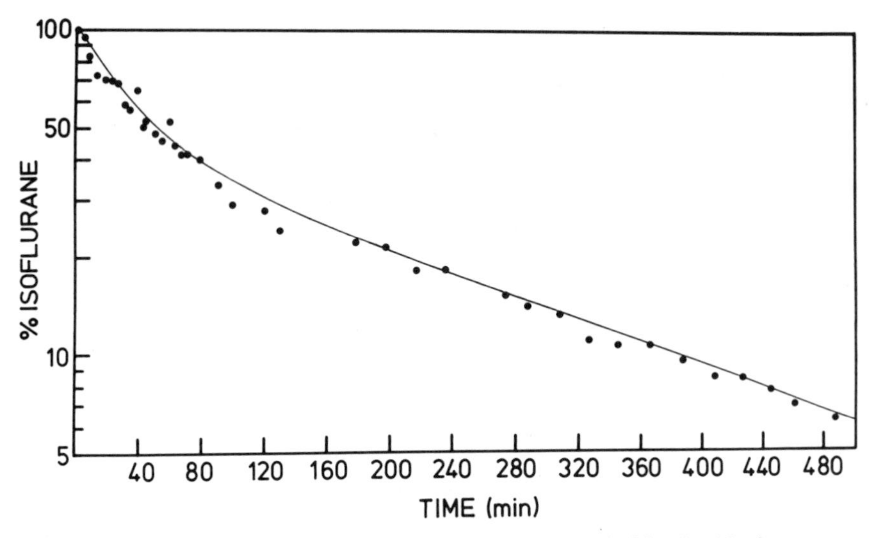

FIGURE 1. A time-course of isoflurane elimination from rabbit brain following 90 min exposure to the anesthetic. ^{19}F NMR signal areas are expressed as percent of isoflurane remaining in tissue. 100% represents the signal measured immediately following the termination of isoflurane anesthesia.

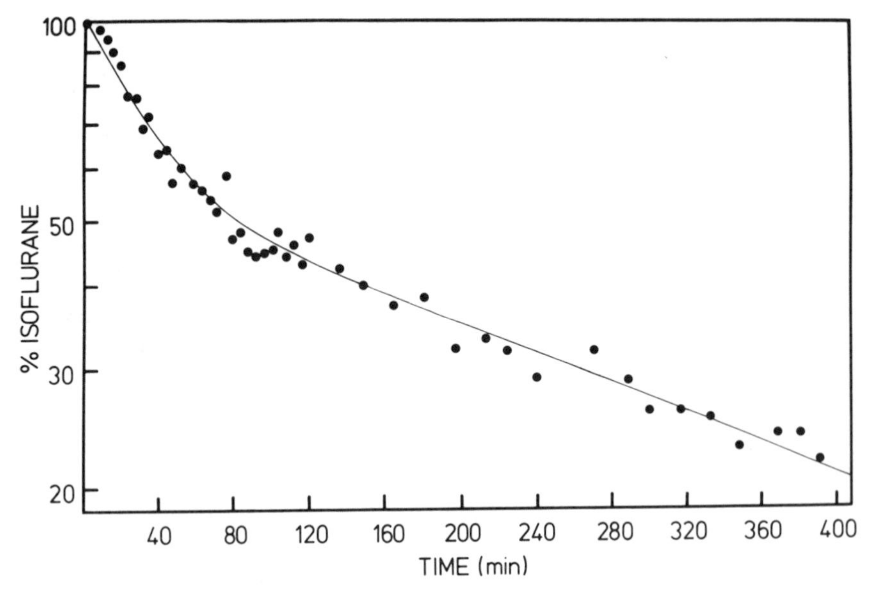

FIGURE 2. A time-course of isoflurane elimination from rabbit paralumbar muscle following 90 min of exposure to the anesthetic. The areas are expressed as percent ^{19}F NMR signal detected.

Uptake Studies

The animals were intubated under pentobarbital anesthesia (30 mg/kg) given i.v. The animals were then placed on a cradle and positioned in the magnet with the NMR surface coil centered over the calvarium. Isoflurane was administered at a constant delivered concentration (1.7% in oxygen) for the duration of the experiment. Inspired and expired gases were hand sampled from a catheter placed in the endotracheal tube to the level of the carina. They were analyzed with a Beckman LB2 infrared analyzer calibrated to room air and to 1% isoflurane standard. The body temperature was maintained and the blood gases were kept normal. The animals were mechanically ventilated.

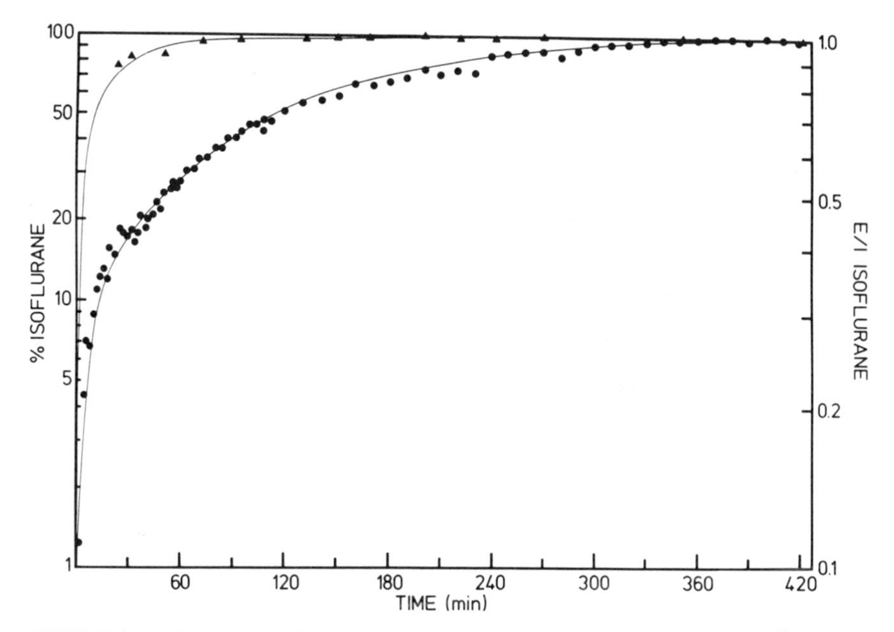

FIGURE 3. A time-course of isoflurane uptake measured for rabbit brain using ^{19}F NMR spectroscopy (●) and the expired/inspired rations of isoflurane measured using a gas analyzer (▲). The ^{19}F NMR signal obtained during the time course was compared to a standard peak area. The values from 330–420 min were averaged and the remainder of the values were normalized with respect to the average obtained.

NMR Measurements

Fluorine-19 NMR spectra were acquired on a TMR-32 spectrometer equipped with a 20-cm bore horizontal magnet and operating at 75.53 MHz. A two-turn 2.8-cm diameter surface coil was used. Magnetic field homogeneity was optimized by shimming on the water signal from brain tissue or muscle. Spectra were collected (without proton decoupling) continuously for the first 4 hr and at 10-min intervals from 4 to 6.5 hr following isoflurane administration. For the uptake studies, spectra were collected continuously for 7 hr. Acquisition parameters used were: 5,000 Hz sweep width, 20 μsec pulse width, 1.2 sec repetition time, and 200 transients. The number of scans was adjusted to ensure a sufficiently high signal/noise ratio. Peak areas were measured and chemical shifts were referenced to an external standard placed 1 cm from the center of the coil, above the animal. The standard consisted of 2.5% 1,2-dibromotetrafluoroethane in olive oil contained in a sealed 4 mm sphere. Peak areas were integrated and normalized with respect to the peak area of the standard within each individual spectrum.

RESULTS AND DISCUSSION

Representative time-courses of isoflurane elimination from the brain and paralumbar muscle of an intact animal are shown in FIGURES 1 and 2, respectively. A nonlinear decrease with time of the ^{19}F signal areas, proportional to the concentration of this agent, is observed in this semilogarithmic plot for both of these tissues. However, the relative amounts of isoflurane present in muscle following the same anesthesia regiment, were lower by a factor of two then those in brain.

The washout data can be best fit to a biexponential function, which gives two half-lives indicative of two major components. The half-lives for the fast-component are similar (26–28 min) for both brain and muscle. There is a substantial difference in the half-lives calculated for the slow-component: $t_{1/2}$ = 280 min for the muscle and $t_{1/2}$ = 174 min for the brain. The longer time constant for the muscle can be explained on the basis of greater solubility of isoflurane and lower blood perfusion as compared to the brain.[5] These observations suggest that kinetics of isoflurane elimination are biexponential, independent of the tissue type, and differ only in rates.

The time-course of isoflurane uptake is also non-linear on the semilogarithmic plot (FIG. 3). The ^{19}F NMR studies suggest a presence of fast and slow components as well. Concurrent measurements of inspired and expired concentrations indicate, by their ratio, a 99% equilibration much earlier than the NMR results. Since the expired/inspired ratio is indicative of the total body equilibration, it may not correctly reflect the pharmacokinetics of the brain alone.

REFERENCES

1. WYRWICZ, A. M., C. B. CONBOY, K. R. RYBACK, B. G. NICHOLS & P. EISELE. 1987. Biochim. Biophys. Acta **927:** 86–91.
2. WYRWICZ, A. M., C. B. CONBOY, K. R. RYBACK, B. G. NICHOLS & P. EISELE. 1987. Biochim. Biophys. Acta **929:** 271–277.
3. EGER, E. I., II. 1974. Anesthetic Uptake and Action. Williams and Wilkins. Baltimore, MD.
4. WYRWICZ, A. M., M. H. PSZENNY, J. C. SCHOFIELD, P. C. TILLMAN, R. E. GORDON & P. A. MARTIN. 1983. Science **222:** 428–30.
5. EGER, E. I., II. 1981. Isoflurane: A Compendium and reference. Airco, Inc. p. 16.

Use of Amplitude/Frequency Modulated Pulses To Conduct Phase-Modulated Rotating Frame Experiments

MICHAEL GARWOOD AND KÂMIL UĞURBIL

Department of Biochemistry
and
Gray Freshwater Biological Institute
University of Minnesota
Navarre, Minnesota 55392

A major advantage of the rotating-frame technique[1] is the ability to acquire data in a single experiment that can be processed into a set of chemical-shift spectra corresponding to localized volumes at successively greater depths. As implemented with surface coils,[2,3] reduced sensitivity occurs since the spatial information is encoded as the amplitude modulation frequency of the transverse magnetization as a function of the spatially dependent B_1 field. It has been pointed out[1] that a phase-modulated version of the experiment provides a $\sqrt{2}$ improvement in sensitivity. To conduct this latter experiment, a uniform 90° rotation is needed to take the $y'z'$-plane magnetization generated by the incremented inhomogeneous pulse and convert it to transverse magnetization (e.g. $(O, M_y, M_z) \rightarrow (M_z, M_y, O)$ or (M_y, M_z, O), corresponding to (x', y', z'), respectively). Because the B_1 field of a surface coil is highly inhomogeneous, uniform 90° rotations cannot be induced with conventional rectangular pulses; thus, the phase-modulated rotating-frame experiment has not been feasible in the past.

Here we describe frequency and amplitude pulses that can accomplish true 90° rotations in a manner that is highly insensitive to B_1 inhomogeneity. The first pulse, BIR-1 (B_1 independent rotation), is composed of two segments, each of which is based on the sin/cos modulated adiabatic half-passage.[4] The modulations of B_1 and the frequency, $\Delta\omega$, in the rotating frame of the pulse are:

$$\text{First Half: } B_1 = \hat{y}' \, Av\cos 2\pi\nu t, \quad \Delta\omega = A\sin 2\pi\nu t \qquad (0 < t < t_0/2)$$

$$\text{Second Half: } B_1 = \hat{x}' \, (-Av)\cos 2\pi\nu t, \quad \Delta\omega = -A\sin 2\pi\nu t \qquad (t_0/2 < t < t_0)$$

where $\hat{x}'$ and $\hat{y}'$ are unit vectors, B_1 and the parameter A are expressed in frequency units, v is a parameter that takes into account variations in B_1 strength through space, and $2\pi\nu t_0 = \pi$. This pulse transforms (M_x, M_y, M_z) to $(-M_y, -M_z, M_x)$, which is equivalent to a 90° rotation about y' plus a 90° phase shift. Pulse BIR-1 induces this transformation at all values of v where $d\alpha/dt \ll B_1^e$, in which B_1^e is the effective field and α is the polar angle subtending B_1^e and the transverse plane.

BIR-1 is extremely sensitive to chemical shift (off-resonance effects) as a result of $\Delta\omega$ being positive for half of the pulse and negative for the other half. In pulse BIR-2 given below, $\Delta\omega$ is always positive, which makes it relatively insensitive to chemical shift.

$$B_1 = \hat{y}' \, Av\cos 2\pi\nu t, \quad \Delta\omega = A\sin 2\pi\nu t \qquad (0 < t < t_0/2)$$

$$B_1 = \hat{x}' \, (-Av)\cos 2\pi\nu t, \quad \Delta\omega = A\sin 2\pi\nu t \qquad (t_0/2 < t < t_0)$$

$$B_1 = \hat{x}' \, Av\cos 2\pi\nu t, \quad \Delta\omega = -A\sin 2\pi\nu t \qquad (t_0 < t < 3t_0/2)$$

$$B_1 = \hat{x}'(-Av)\cos 2\pi\nu t, \quad \Delta\omega = -A\sin 2\pi\nu t \quad (3t_0/2 < t < 2t_0)$$

where $2\pi\nu t_0 = \pi$ as before. Pulse BIR-2 induces the transformation of (M_x, M_y, M_z) to $(M_y, -M_z, -M_x)$.

Finally, the range of v over which these pulses work can be improved substantially using second-order modulation functions of the type $\sin(\eta(t))/\cos(\eta(t))$, where $(\eta(t))$ is no longer a linear function of t, but has a more complex time dependence. One example

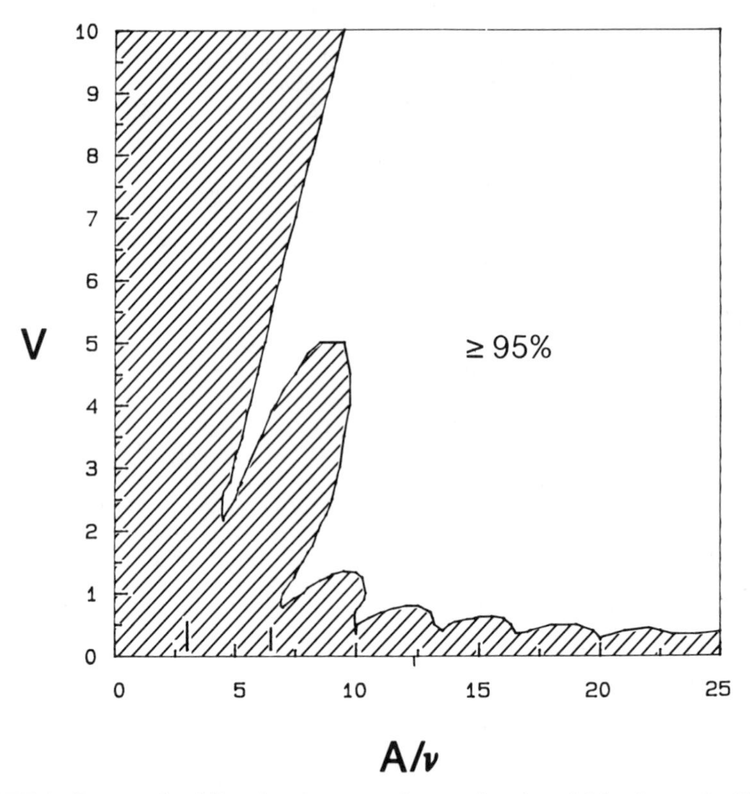

FIGURE 1. Contour plot delineating the range of v as a function of A/ν where pulse BIR-3 induces all three transformations with $\geq$ 95% efficiency. The calculations of a 42×32 grid of points for each rotation are based on numerical solutions of the Bloch equations.

is BIR-3, which is equivalent to BIR-2 with

$$\eta(t) = Q(\sin^{-1}[\tanh q(2t/t_0 - 1)]) + C$$

where C, q, and Q equal constants. FIGURE 1 shows a contour plot calculated using the Bloch equations for the three transformations induced by BIR-3 as a function of v and A/ν. FIGURE 2 illustrates the minimal effect that chemical shift produces on the 85% contour level at an A/ν value of 20 with $A = 5000$ Hz (which is the equivalent of a 50 μsec 90° broadband pulse).

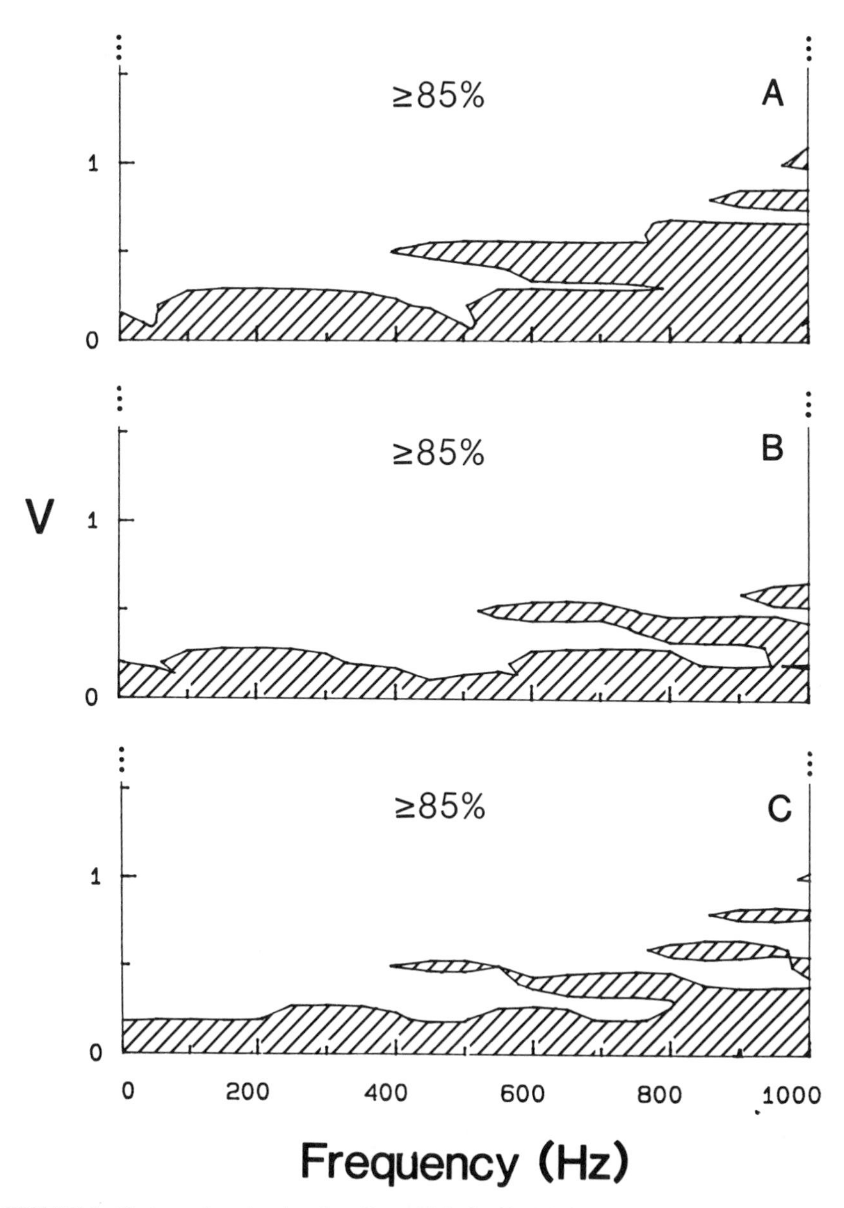

FIGURE 2. Contour plots showing the effect which chemical shift produces on the 85% contour level of pulse BIR-3 at $A/\nu = 20$ with $A = 5{,}000$ Hz. A, B, and C correspond to the $M_x \rightarrow M_z$, $M_y \rightarrow M_x$, and $M_z \rightarrow M_x$ transformations, respectively. Contours at larger values of v are not plotted because the transformations remained at > 85% level for all frequency-offset values up to 1,000 Hz for $v \leq 10$ (highest value calculated).

REFERENCES

1. HOULT, D. I. 1979. J. Magn. Reson. **33:** 183.
2. GARWOOD, M., T. SCHLEICH, G. B. MATSON, & G. ACOSTA. 1984. J. Magn. Reson. **60:** 268.
3. STYLES, P., C. A. SCOTT & G. K. RADDA. 1985. Magn. Reson. Med. **2:** 402.
4. BENDALL, M. R. & D. T. PEGG. 1986. J. Magn. Reson. **67:** 376.

NMR Studies of Intracellular Water

Use of the Amphibian Oocyte as a Model System[a]

G.A. MORRILL, A.B. KOSTELLOW, D. HOFFMAN,
AND R.K. GUPTA

Department of Physiology and Biophysics
Albert Einstein College of Medicine
Bronx, New York 10461

We have developed techniques to measure both fractional intracellular water content and true intracellular relaxation times (T_1, T_2) for water in isolated cells and tissues. Water relaxation measurements are generally complicated by the presence of varying amounts of extracellular water in the tissue, which exhibits relaxation times different from those of intracellular water. The method described here for measuring intracellular water content utilizes $^{35}Cl^-$ NMR as a measure of extracellular volume and ^2H NMR as a measure of the ratio of intra- and extracellular water volumes.[1] Knowledge of extracellular water volume also permits one to correct partially relaxed inversion-recovery FT NMR spectra of packed cell preparations for the presence of contaminating extracellular water to obtain the true T_1 and T_2 values for intracellular water. Fully grown oocytes (within their follicular envelopes) from the frog *Rana pipiens* are used here as a model system. Each *Rana* female contains 2,000–3,000 large (1.8 mm diameter) oocytes arrested in first meiotic prophase. They can be induced to undergo synchronous meiotic divisions by hormones, such as insulin and progesterone, as they are being superfused in the NMR tube.[2] This system is therefore ideal for investigating hormone-induced cell water and cell volume changes using NMR techniques. Because of the large size of the oocyte and slowness of the diffusion process, the exchange of intracellular water with extracellular water is slow on the NMR time scale.

Representative pairs of $^{35}Cl^-$ and ^2H spectra are illustrated in FIGURE 1. Intensities were measured by integrating the area under the resonance peaks. The method for measuring extracellular water volume is based on the observation of Rayson and Gupta[3] that the resonance of intracellular $^{35}Cl^-$ ions in most cells is so broad as to become NMR invisible. Therefore, the intensity of the $^{35}Cl^-$ NMR signal of a cell suspension is proportional to the volume of extracellular water within the sensitive volume of the NMR receiver coil. The intensity of the $^{35}Cl^-$ NMR signal of the cell-free medium in an identical sample-geometry is proportional to the total sensitive volume of the NMR coil. The ratio of the two intensities (f_{Cl}) represents the extracellular water volume as a fraction of the total suspension volume. The ratio of the intensities (f_D) of the water ^2H NMR signals of the cell suspension and the cell-free medium gives the water volume as a fraction of the total suspension volume. The intracellular water content (f_W), i.e. the volume of the intracellular water as a fraction of the total cell volume, is given by the equation: $f_W = (f_D - f_{Cl})/(1 - f_{Cl})$.[1] Using this method, the average water content of three samples of ovarian oocytes was found to be 42.6 ± 0.8% (vol/vol). In the intact follicle preparation used, the oocyte is closely

[a]Supported in part by research grant AM-32030, Biomedical Research Support Grant S07RR, and by National Cancer Institute Core grant P30-CA-13330. D. Hoffmann was a recipient of a fellowship from National Institutes of Health Training Grant HD 07053.

surrounded by epithelial envelopes composed of an outer thecal layer and an inner layer of follicle cells applied to the oocyte surface.[4] Previous gravimetric measurements[5] indicate that intracellular water content is 49.8 ± 0.4% (vol/weight) for a comparable follicle preparation. The gravimetric values may therefore include contributions from interstitial space and reflect difficulty in correcting for the density of the non-aqueous phase of the cells and tissue.

FIGURE 2 illustrates the semilog plots of inversion recovery T_1 and Carr-Purcell-Meiboom-Gill T_2 data for the medium (upper) showing single exponential recovery to equilibrium and the T_1 and T_2 of the follicular oocyte suspension (lower), showing recovery of the magnetization via the sum of two exponentials. The T_1 and T_2 values for the medium were 3.02 ± 0.03 and 2.17 ± 0.14 sec, respectively. The T_1 magnetization recovery and T_2 decay curves indicate the presence of slow and rapid components, the slow component yielding relaxation times close to those of the cell-free medium. When the contribution of the medium was subtracted, curves consistent with single exponential recovery of intracellular water were observed, corresponding to 460 ± 25 and 87 ± 13 msec for T_1 and T_2, respectively. These findings indicate: (1) a slow exchange between intra- and extracellular water on the NMR time-scale, (2) no evidence for

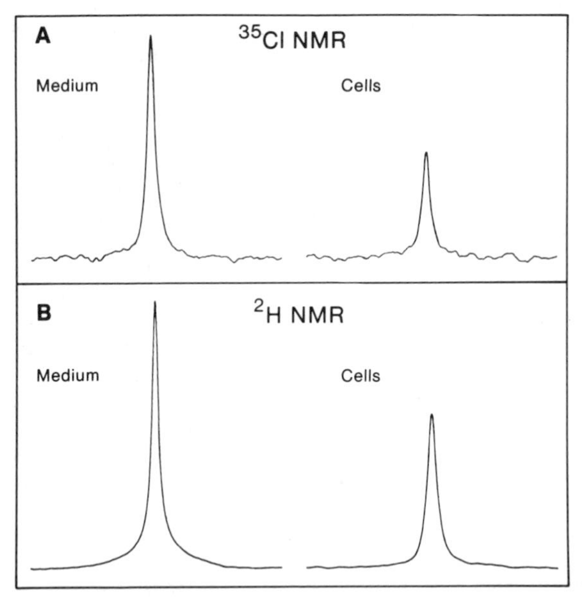

FIGURE 1. Representative pairs of $^{35}Cl^-$ (a) and ^2H (b) NMR spectra. The left traces correspond to the cell-free medium and the right traces to the oocyte suspension. *Rana pipiens* follicles were dissected from fresh ovaries with fine tipped forceps and approximately 50 oocytes with follicular envelopes were transferred to a 5 mm NMR tube with Ringer's solution.[2] Isolated follicular oocytes were preincubated in Ringer's solution containing 1% D_2O for 30 min. ^{35}Cl and ^2H measurements were carried out using a Varian XL-200 FT NMR instrument. The observing frequency was 19.6 MHz for the ^{35}Cl nucleus and 30.7 MHz for the ^2H nucleus. Measurements may be carried out without changing the NMR probe configuration and they do not require perturbation of the cell system by addition of membrane-impermeable reagents.[6,7] The method is equally applicable to situations of slow and fast water exchange.

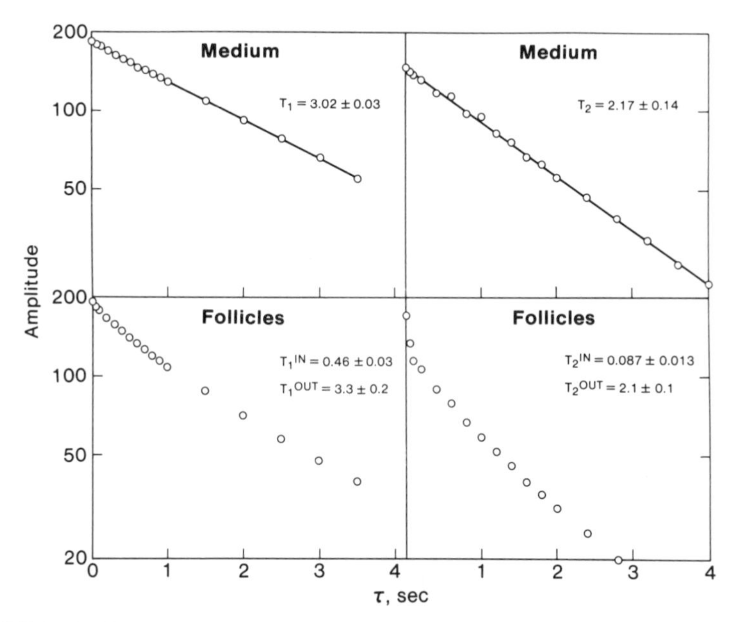

FIGURE 2. Semilog plots of inversion-recovery T_1 and Carr-Purcell-Meiboom-Gill T_2 data showing single exponential recovery for the medium (upper) and recovery via the sum of two exponentials for packed follicular oocytes (lower).

multiple intracellular water compartments, and (3) an increased rotational correlation time for water molecules in the oocyte cytoplasm compared to bulk water.

REFERENCES

1. HOFFMAN, D. & R. K. GUPTA. 1986. J. Magn. Res. **70:** 481–483.
2. MORRILL, G. A., A. B. KOSTELLOW, S. P. WEINSTEIN & R. K. GUPTA. 1983. Physiol. Chem. Physics Med. NMR **15:** 357–362.
3. RAYSON, B. M. & R. K. GUPTA. 1985. J. Biol. Chem. **260:** 7276–7281.
4. WEINSTEIN, S. P., G. A. MORRILL & A. B. KOSTELLOW. 1983. Develop. Growth Differ. **25:** 11–21.
5. MORRILL, G. A., D. H. ZIEGLER & V. S. ZABRENETZKY. 1977. J. Cell. Sci. **26:** 311–322.
6. COWAN, B. E., D. Y. SZE, M. T. MAI & O. JARDETZKY. 1985. FEBS Lett. **184:** 130–133.
7. SHINAR, H. & G. NAVON. 1985. FEBS Lett. **193:** 75–78.

Index of Contributors